Innovation in Food Engineering

New Techniques and Products

Edited by
Maria Laura Passos
Claudio P. Ribeiro

CRC Press
Taylor & Francis Group
Boca Raton London New York

CRC Press is an imprint of the
Taylor & Francis Group, an **informa** business

CRC Press
Taylor & Francis Group
6000 Broken Sound Parkway NW, Suite 300
Boca Raton, FL 33487-2742

Printed in the United States of America on acid-free paper
10 9 8 7 6 5 4 3 2 1

International Standard Book Number: 978-1-4200-8606-5 (Hardback)

Library of Congress Cataloging-in-Publication Data

Innovation in food engineering : new techniques and products / edited by Maria Laura Passos and Claudio P. Ribeiro.
p. cm. -- (Contemporary food engineering series)
Includes bibliographical references and index.
ISBN-13: 978-1-4200-8606-5
ISBN-10: 1-4200-8606-5
1. Food industry and trade. I. Passos, Laura. II. Ribeiro, Claudio P. III. Title. IV. Series.

TP370.I56 2010
664--dc22 2009015258

Visit the Taylor & Francis Web site at
http://www.taylorandfrancis.com

and the CRC Press Web site at
http://www.crcpress.com

Innovation in Food Engineering

New Techniques and Products

Contemporary Food Engineering

Series Editor

Professor Da-Wen Sun, Director

Food Refrigeration & Computerized Food Technology
National University of Ireland, Dublin
(University College Dublin)
Dublin, Ireland
http://www.ucd.ie/sun/

Innovation in Food Engineering: New Techniques and Products, *edited by Maria Laura Passos and Claudio P. Ribeiro* (2009)

Engineering Aspects of Milk and Dairy Products, *edited by Jane Selia dos Reis Coimbra and Jose A. Teixeira* (2009)

Processing Effects on Safety and Quality of Foods, *edited by Enrique Ortega-Rivas* (2009)

Engineering Aspects of Thermal Food Processing, *edited by Ricardo Simpson* (2009)

Ultraviolet Light in Food Technology: Principles and Applications, *Tatiana N. Koutchma, Larry J. Forney, and Carmen I. Moraru* (2009)

Advances in Deep-Fat Frying of Foods, *edited by Serpil Sahin and Servet Gülüm Sumnu* (2009)

Extracting Bioactive Compounds for Food Products: Theory and Applications, *edited by M. Angela A. Meireles* (2009)

Advances in Food Dehydration, *edited by Cristina Ratti* (2009)

Optimization in Food Engineering, *edited by Ferruh Erdoğdu* (2009)

Optical Monitoring of Fresh and Processed Agricultural Crops, *edited by Manuela Zude* (2009)

Food Engineering Aspects of Baking Sweet Goods, *edited by Servet Gülüm Sumnu and Serpil Sahin* (2008)

Computational Fluid Dynamics in Food Processing, *edited by Da-Wen Sun* (2007)

Dedicated
with respect and love
to my family and
to all my students (ex and new)

Maria Laura Passos

To my beloved parents,
Cláudio and Irenice,
for their constant support.

Cláudio P. Ribeiro, Jr.

Contents

PART II New Materials, Products, and Additives

Series Preface

Food engineering is the multidisciplinary field of applied physical sciences combined with the knowledge of product properties. Food engineers provide the technological knowledge transfer essential to the cost-effective production and commercialization of food products and services. In particular, food engineers develop and design processes and equipment in order to convert raw agricultural materials and ingredients into safe, convenient, and nutritious consumer food products. However, food engineering topics are continuously undergoing changes to meet diverse consumer demands, and the subject is being rapidly developed to reflect market needs.

In the development of food engineering, one of the many challenges is to employ modern tools and knowledge, such as computational materials science and nanotechnology, to develop new products and processes. Simultaneously, improving quality, safety, and security remain critical issues in the study of food engineering. New packaging materials and techniques are being developed to provide more protection to foods, and novel preservation technologies are emerging to enhance food security and defense. Additionally, process control and automation regularly appear among the top priorities identified in food engineering. Advanced monitoring and control systems are developed to facilitate automation and flexible food manufacturing. Furthermore, energy saving and minimization of environmental problems continue to be important issues in food engineering, and significant progress is being made in waste management, efficient utilization of energy, and reduction of effluents and emissions in food production.

The *Contemporary Food Engineering* book series, which consists of edited books, attempts to address some of the recent developments in food engineering. Advances in classical unit operations in engineering related to food manufacturing are covered as well as such topics as progress in the transport and storage of liquid and solid foods; heating, chilling, and freezing of foods; mass transfer in foods; chemical and biochemical aspects of food engineering and the use of kinetic analysis; dehydration, thermal processing, nonthermal processing, extrusion, liquid food concentration, membrane processes, and applications of membranes in food processing; shelf life, electronic indicators in inventory management, and sustainable technologies in food processing; and packaging, cleaning, and sanitation. These books are aimed at professional food scientists, academics researching food engineering problems, and graduate-level students.

The editors of these books are leading engineers and scientists from all parts of the world. All of them were asked to present their books in such a manner as to address the market needs and pinpoint the cutting-edge technologies in food engineering. Furthermore, all contributions are written by internationally renowned experts who have both academic and professional credentials. All authors have attempted to provide critical, comprehensive, and readily accessible information on the art and science of a relevant topic in each chapter, with reference lists for further information. Therefore, each book can serve as an essential reference source to students and researchers in universities and research institutions.

Da-Wen Sun
Series Editor

Series Editor

Born in southern China, Professor Da-Wen Sun is a world authority on food engineering research and education. His main research activities include cooling, drying, and refrigeration processes and systems; quality and safety of food products; bioprocess simulation and optimization; and computer vision technology. Especially, his innovative studies on vacuum cooling of cooked meats, pizza quality inspection by computer vision, and edible films for shelf-life extension of fruits and vegetables have been widely reported in national and international media. The results of his work have been published in over 200 peer-reviewed journal papers and more than 200 conference papers.

Sun received his BSc honors (first class), his MSc in mechanical engineering, and his PhD in chemical engineering in China before working in various universities in Europe. He became the first Chinese national to be permanently employed in an Irish university when he was appointed as college lecturer at the National University of Ireland, Dublin (University College Dublin), in 1995, and was then continously promoted in the shortest possible time to senior lecturer, associate professor, and full professor. He is currently the professor of food and biosystems engineering and the director of the Food Refrigeration and Computerized Food Technology Research Group at University College Dublin.

Sun has contributed significantly to the field of food engineering as a leading educator in this field. He has trained many PhD students who have made their own contributions to the industry and academia. He has also regularly given lectures on advances in food engineering in international academic institutions and delivered keynote speeches at international conferences. As a recognized authority in food engineering, he has been conferred adjunct/visiting/consulting professorships from 10 top universities in China including Zhejiang University, Shanghai Jiaotong University, Harbin Institute of Technology, China Agricultural University, South China University of Technology, and Jiangnan University. In recognition of his significant contribution to food engineering worldwide and for his outstanding leadership in this field, the International Commission of Agricultural Engineering (CIGR) awarded him the CIGR Merit Award in 2000 and again in 2006. The Institution of Mechanical Engineers (IMechE) based in the United Kingdom named him Food Engineer of the Year 2004. In 2008, he was awarded the CIGR Recognition Award in honor of his distinguished achievements in the top 1% of agricultural engineering scientists in the world.

Sun is a fellow of the Institution of Agricultural Engineers and a Fellow of Engineers Ireland. He has received numerous awards for teaching and research excellence, including the President's Research Fellowship and the President's Research Award of University College Dublin on two occasions. He is a member of the CIGR Executive Board and an honorary vice-president of CIGR; the editor in chief of *Food and Bioprocess Technology—An International Journal* (Springer); the former editor of the *Journal of Food Engineering* (Elsevier); and an editorial board member for the *Journal of Food Engineering* (Elsevier), the *Journal of Food Process Engineering* (Blackwell), *Sensing and Instrumentation for Food Quality and Safety* (Springer), and *Czech Journal of Food Sciences*. He is also a chartered engineer.

Preface

Food engineering has seen many developments since its establishment as an individual research area in the 1970s, but the food industry of the twenty-first century faces new challenges. Energy-related concerns are more important than ever, consumers have become much more quality conscious, nutritional aspects have acquired significant, if not dominant, importance, and the once-accepted compromise between convenience and quality for food products is no longer an option. In this book, we present an overview of the different routes that researchers are pursuing to provide solutions for these challenges. Without the presumption of being able to cover everything in a single volume but, at the same time, driven by the desire to contribute something new, we chose to address not only new or alternative techniques, but also new products, materials, and additives that have emerged as a response to the challenges faced by the food industry.

Part I deals with innovative or reformulated technologies applied to food processing. Considering the importance of thermal processing in the food industry, it seemed almost natural for us to start this first part with a chapter on nonthermal processes for food preservation. Thus, in Chapter 1, the theories, potential applications, and challenges of high-pressure processing, pulsed electric fields, and ultrasound are discussed. Chapter 2 focuses on innovation in one of the fastest-growing segments in the food industry—bakery products—in which low-temperature technology has been intensively investigated as a means to meet consumers' demands of convenience, health, and quality. Chapter 3 discusses biotechnology as an alternative to extraction for producing natural flavors, as well as analytical tools that can be used to distinguish between natural and synthetic products. Chapter 4 covers further applications of biotechnology in the food industry and focuses specifically on solid-state fermentation, which is an attractive option for efficient utilization and value addition of agro-industrial solid wastes. Chapter 5 addresses a similar philosophy of efficient utilization but applied to liquid residues and process streams. It provides the basic principles and potential applications of three membrane-based unit operations (pervaporation, nanofiltration, and electrodialysis) for the recovery, concentration, and purification of compounds with target bioactivity. Three other membrane-based unit operations (reverse osmosis, membrane distillation, and osmotic distillation), as well as some advanced thermal techniques are discussed in Chapter 6 as alternative techniques for fruit-juice concentration. Chapters 7 and 8 cover encapsulation technology. The former provides the basic theoretical aspects and a review of the current research on four types of delivery systems (gelled microparticles, spray-dried microparticles, emulsions, and liposomes), whereas the latter focuses on fluidized bed coating as an encapsulation method.

Water removal has long been adopted as a means of food preservation, and the importance of drying as a unit operation in the food industry is well recognized. Chapters 9 through 12 discuss drying-related innovations. Chapter 9 deals with recent developments in spray drying, the most traditional drying techniques, while

Chapter 10 provides an overview of superheated steam drying along with its advantages and potential applications to foods. Chapter 11 focuses on the use of spouted bed dryers to process tropical fruit pulps, while Chapter 12 presents some promising results concerning the use of microwave radiation to assist not only in the process of drying but also that of extraction. Chapter 13, in turn, is devoted to a different approach to water removal, that is, vacuum frying, which has emerged as a viable option for the production of snacks from fruits and vegetables.

Packaging is an important topic for the food industry and this book would not be complete without a discussion on some of the advances in this area. Chapter 14 reviews the state of the art in aseptic packaging, including the application of plasma decontamination. Chapter 15 discusses controlled- and modified-atmosphere packaging, with a special emphasis on applications in fresh-cut produce—a particularly difficult yet dynamic sector for these technologies due to several technical and food-safety challenges. Chapter 16 concludes Part I, and provides a critical review of the latest trends in the use of polymeric packages for food applications, including the use of nanocomposites, active packaging, and antimicrobial packaging.

Part II is dedicated to new materials, products, and additives. It starts with an in-depth analysis, in Chapter 17, of the development of biodegradable films for packaging materials that can be used as a substitute for petrochemical polymers. Chapter 18 illustrates the complex process involved in the establishment of a new product line through a case study related to the production of goat-milk powder in small agro-cooperatives. Functional foods have been successfully introduced into the market as a response to the consumers' demands for healthier products, and research in this area has been very active. This topic is covered in three chapters: Chapter 19 deals with meat products, Chapter 20 analyzes the inclusion of probiotic bacteria and prebiotic substrates into dairy foods, and, finally, Chapter 21 addresses the potential of cereals for the development of functional foods. Low-fat products, like functional foods, reflect the consumers' awareness of nutritional and health benefits, and Chapter 22 reviews the recent advances in this area. Last but not least, Chapter 23 discusses potential applications of biosurfactants in the food industry, as well as new approaches for their production and scale-up.

It has been a long journey from the conception to the realization of this book, and we could never have accomplished this without the dedicated work and commitment of all our contributors, to whom we are most grateful. The book, ultimately, is a result of their combined efforts. We would also like to thank Prof. Da-Wen Sun for inviting us to contribute this volume to the *Contemporary Food Engineering* series, an invitation that was the genesis of this entire project. In addition, we would like to thank our project coordinator at CRC Press, Patricia Roberson, who was always very efficient and helpful.

Maria Laura Passos
Cláudio P. Ribeiro, Jr.

Editors

Maria Laura Passos is currently a consultant in fluid-particle systems and drying technology. She acts as an associate researcher in the Chemical Engineering Drying Center of the Federal University of São Carlos and as a co-advisor at the Chemical Engineering Graduate Program of the Federal University of Rio Grande do Norte, as well as at the Technology Center of Minas Gerais (CETEC-MG). She was a full-time professor at the Engineering School of the Federal University of Minas Gerais (UFMG) in Belo Horizonte, Brazil. She received her BSc (with honors) in chemical engineering from UFMG, her MSc in chemical engineering from the Federal University of Rio de Janeiro, and her PhD from the chemical engineering department of McGill University, Montreal, Canada. She has participated, as a member, in official committees from governmental agencies to evaluate chemical engineering graduate and undergraduate education in Brazil. She has written over 180 scientific publications on fluid-particle systems and drying, and over 50 technical and educational reports on chemical engineering. She is a member of the editorial board of *Drying Technology*. She has participated as a reviewer in many international journals on chemical and food engineering, and has a large experience as an ad hoc consultant in several research agencies. She has worked as a member of organization and scientific committees of many symposiums, conferences, and congresses.

Cláudio P. Ribeiro, Jr. received his BSc (with honors) and MSc from the Federal University of Minas Gerais (UFMG), and his doctoral degree from the Federal University of Rio de Janeiro in 2005, all in chemical engineering. In the same year, he was awarded a fellowship from the Alexander von Humboldt Foundation to work as a visiting scholar at the University of Hannover, Germany. He returned to Brazil in 2006 and worked as a lecturer at the Federal University of Rio de Janeiro. In 2007, he moved to his current position as a postdoctoral researcher at the laboratory of Membrane Science and Technology from the Center of Energy and Environmental Resources at the University of Texas at Austin. He has published a total of 27 papers in peer-reviewed journals and 20 contributions to academic conferences. His research on an alternative route for fruit-juice concentration was chosen as the best Brazilian PhD thesis on engineering and exact sciences in 2005 by CAPES, the governmental agency responsible for evaluating graduate courses in Brazil. He has already participated as a reviewer in many international journals on chemical engineering.

Contributors

Odelsia L. S. Alsina
Department of Chemical Engineering
Federal University of Campina Grande
Campina Grande, Brazil

Lívia Vieira de Araújo
Department of Biochemistry
Federal University of Rio de Janeiro
Rio de Janeiro, Brazil

Cristiano P. Borges
Chemical Engineering Program
Federal University of Rio de Janeiro
Rio de Janeiro, Brazil

Elisabetta Brenna
Department of Chemistry, Materials and Chemical Engineering
Polytechnic University of Milan
Milan, Italy

Dimitris Charalampopoulos
Department of Food Biosciences
The University of Reading
Reading, United Kingdom

João G. Crespo
Department of Chemistry
Faculty of Science and Technology
New University of London

James R. Daniel
Department of Foods and Nutrition
Purdue University
West Lafayette, Indiana

Marcos Akira d'Ávila
School of Chemical Engineering
University of Campinas
Campinas, Brazil

Frédéric Depypere
Department of Food Safety and Food Quality
Ghent University
Ghent, Belgium

Koen Dewettinck
Department of Food Safety and Food Quality
Ghent University
Ghent, Belgium

Cynthia Ditchfield
Food Engineering Department
University of Sao Paulo
Pirassununga, Brazil

Liu Ping Fan
Department of Food Resource and Comprehensive Utilization Engineering
Jiangnan University
Wuxi, China

Juana Fernández López
Department of AgroFood Technology
Miguel Hernandez University
Alicante, Spain

Oliver Franken
Fraunhofer Institute for Laser Technology
Aachen, Germany

Denise Maria Guimarães Freire
Departament of Biochemistry
Federal University of Rio de Janeiro
Rio de Janeiro, Brazil

Giovanni Fronza
Institute of Chemistry of Molecular Recognition
National Research Council
Milan, Italy

Claudio Fuganti
Department of Chemistry, Materials and Chemical Engineering
Polytechnic University of Milan
Milan, Italy

Francesco G. Gatti
Department of Chemistry, Materials and Chemical Engineering
Polytechnic University of Milan
Milan, Italy

Clara González de los Reyes-Gavilán
Department of Microbiology and Biochemistry of Dairy Products
Dairy Products Institute of Asturias
Villaviciosa, Spain

Frederico de Araujo Kronemberger
Departament of Biochemistry
Federal University of Rio de Janeiro
Rio de Janeiro, Brazil

Jose M. Lagaron
Department of Food Quality and Safety
Institute of Agrochemistry and Food Technology
Spanish Council for Scientific Research
Burjassot, Spain

Paulo L. C. Lage
Chemical Engineering Program
Federal University of Rio de Janeiro (UFRJ)
Rio de Janeiro, Brazil

María A. Longo
Department of Chemical Engineering
University of Vigo
Vigo, Spain

Amparo López-Rubio
Department of Food Quality and Safety
Institute of Agrochemistry and Food Technology
Spanish Council for Scientific Research
Burjassot, Spain

Maria de Fátima D. Medeiros
Department of Chemical Engineering
Federal University of Rio Grande do Norte
Natal, Brazil

Uliana K. L. Medeiros
Department of Chemical Engineering
Federal University of Rio Grande do Norte
Natal, Brazil

Arun S. Mujumdar
Department of Mechanical Engineering
National University of Singapore
Singapore

Peter Muranyi
Fraunhofer Institute for Process Engineering and Packaging
Freising, Germany

Márcia Nitschke
Department of Physical Chemistry
University of São Paulo
São Carlos, Brazil

David O'Beirne
Department of Life Sciences
University of Limerick
Limerick, Ireland

Valérie Orsat
Department of Bioresource Engineering
Macdonald Campus of McGill University
Québec, Canada

Severino S. Pandiella
School of Chemical Engineering and Analytical Science
The University of Manchester
Manchester, United Kingdom

Maria Laura Passos
Department of Chemical Engineering
Federal University of Rio Grande do Norte
Natal, Brazil
and
Department of Chemical Engineering
Federal University of São Carlos
São Carlos, Brazil

José Angel Pérez Alvarez
Department of AgroFood Technology
Miguel Hernandez University
Alicante, Spain

Jan G. Pieters
Department of Biosystems Engineering
Ghent University
Ghent, Belgium

Somkiat Prachayawarakorn
Department of Chemical Engineering
King Mongkut's University of Technology Thonburi
Bangkok, Thailand

G. S. Vijaya Raghavan
Department of Bioresource Engineering
Macdonald Campus of McGill University
Québec, Canada

Navin K. Rastogi
Department of Food Engineering
Central Food Technological Research Institute
Mysore, India

Maria Inês Ré
Research Center of Albi on Particulate Solids, Energy and Environment
Albi School of Mines
Albi, France
and
Center of Processes and Products Technology
Institute of Technological Research of the São Paulo State
São Paulo, Brazil

Jorge Reinheimer
Faculty of Chemical Engineering
National University of Litoral
Santa Fe, Argentina

Cláudio P. Ribeiro, Jr.
Department of Chemical Engineering
University of Texas at Austin
Austin, Texas

Sandra C. S. Rocha
Department of Thermofluid Dynamics
University of Campinas
Campinas, Brazil

Cristina M. Rosell
Food Science Department
Institute of Agrochemistry and Food Technology
Burjasot, Spain

Mª Ángeles Sanromán
Department of Chemical Engineering
University of Vigo
Vigo, Spain

Maria Helena Andrade Santana
Department of Biotechnological Processes
University of Campinas
Campinas, Brazil

Stefano Serra
Institute of Chemistry of Molecular Recognition
National Research Council
Milan, Italy

Uma S. Shivhare
Department of Chemical Engineering and Technology
Panjab University
Chandigarh, India

Somchart Soponronnarit
School of Energy, Environment and Materials
King Mongkut's University of Technology Thonburi
Bangkok, Thailand

Ana Cristina de Souza
Department of Chemical Engineering
University of Sao Paulo
Sao Paulo, Brazil

Josilma S. Souza
Department of Chemical Engineering
Federal University of Rio Grande do Norte
Natal, Brazil

Carmen Cecilia Tadini
Department of Chemical Engineering
University of Sao Paulo
Sao Paulo, Brazil

Svetlozar Velizarov
Department of Chemistry
New University of Lisbon
Caparica, Portugal

Gabriel Vinderola
Faculty of Chemical Engineering
National University of Litoral
Santa Fe, Argentina

Colin Webb
School of Chemical Engineering and Analytical Science
The University of Manchester
Manchester, United Kingdom

Joachim Wunderlich
Fraunhofer Institute for Process Engineering and Packaging
Freising, Germany

Huang Li Xin
Research Institute of Chemical Industry of Forestry Products
Nanjing, China

Min Zhang
Department of Food Resource and Comprehensive Utilization Engineering
Jiangnan University
Wuxi, China

Nomenclature

a	activity (—)
a^*	redness (—)
A	area (L^2)
A^*	water permeability constant of a reverse osmosis membrane (tL^{-1})
A_i or a_i	constant (i = number)
b^*	yellowness (—)
B	geometric factor in the Laplace–Young equation (—)
B_i^*	permeability constant of solute i in a reserve osmosis membrane (Lt^{-1})
c_i	mass concentration of component i (ML^{-3})
c_p	specific isobaric heat capacity ($L^2t^{-2}T^{-1}$)
$c_{\mathrm{R\text{-}ab}}$	absorbed radiant heat capacity ($L^2t^{-2}T^{-1}$)
C_i	content (mass fraction) of component i (M/M)
C_i^{w}	content of component i per water content (w/w) (—)
C_{D}	drag coefficient (between fluid and particle) (—)
d	diameter (L)
d_{32}	mean particle Sauter diameter (L)
d_{eq}	equivalent (or hydraulic) diameter of a noncircular channel (L)
d_{p}	particle diameter (L)
d_{p}^*	pore diameter (L)
D_{m}	diffusion coefficient or mass diffusivity (L^2t^{-1})
E	electric field strength or electrical potential ($ML^2t^{-3}A^{-1}$)
fr	frequency (Cycles t^{-1})
F	force (MLt^{-2})
$\mathfrak{F}$	Faraday constant (At mol^{-1})
g	gravitational acceleration (Lt^{-2})
h	convective heat-transfer coefficient ($Mt^{-3}T^{-1}$)
H	height (L)
Id	dispersion index span (—)
j	complex operator (—)
J^{m}	mass flux ($ML^{-2}t^{-1}$)
J^{V}	volumetric flux (Lt^{-1})
J^{E}	electrical charge flux (AL^{-2})
k	convective mass-transfer coefficient (Lt^{-1})
K or K^{OV}	overall mass-transfer coefficient ($M^{-1}L^2t$)
kT	energy unit = Boltzman constant × absolute temperature (ML^2t^{-2})
l	thickness (L)
L	length (L)
L_{p}	permeability coefficient ($M^{-1}L^3t$)
L^*	lightness (—)
m	moisture content (expressed in d.b. or in w.b. when specified) (—)
M	mass (M)

$\dot{M}$ mass flow rate (Mt^{-1})
M_F momentum source per unit volume ($ML^{-2}t^{-2}$)
M_h energy source term per unit volume ($ML^{-1}t^{-3}$)
M_m mass source per unit volume ($ML^{-3}t^{-1}$)
MLCR mean logarithmic count reduction (—)
MW molecular weight (M/mol)
n number (—)
N_c constant drying rate (expressed in dry basis) (t^{-1})
N_k kinetic constant of a zeroth order reaction rate (t^{-1})
nc microbial count (—)
nt_{pk} number of test packages (—)
ns_{pk} number of sterile packages (—)
pd penetration depth (L)
p_i partial pressure of component i ($ML^{-1}t^{-2}$)
P pressure ($ML^{-1}t^{-2}$)
$\wp$ permeability ($M^{-1}L^3t$)
PW power dissipation (ML^2t^{-3})
q heat transfer rate (ML^2t^{-3})
Q volumetric flow rate (L^3t^{-1})
r radial distance (L)
R universal gas constant ($ML^2t^{-2}mol^{-1}T^{-1}$)
$\Re$ rejection coefficient (—)
R-sq coefficient of correlation (—)
RK risk of contamination from the packaging material (—)
s total number of permeated components (—)
S surface area (L^2)
SA_i area of the ith peak in a NMR spectrum (—)
SH_i height of the ith peak in a NMR spectrum (—)
t time (t)
T temperature (T)
u velocity (Lt^{-1})
U superficial velocity (Lt^{-1})
V volume (L^3)
wg weight (MLt^{-2})
x_i spatial coordinate (i = 1–3) (L)
y_i molar fraction of component i (—)
z incident radiant energy per surface area of the material (Mt^{-2})
Z_i valence of specie i (—)

GREEK LETTERS

α absorptance or absorptivity (—)
β liquid–air surface tension (Mt^{-2})
γ_i activity coefficient of component i (—)
Γ_{i-j} overall heat transfer coefficient from i to j phases or materials ($Mt^{-3}T^{-1}$)

δ_{ij} Kronecker delta (—)
Δt time interval (T)
Δtr decimal reduction time of the most resistant organism (T)
ΔH^{vap} latent heat (L^2t^{-2})
ΔP pressure drop ($ML^{-1}t^{-2}$)
ΔP_{cap} capillary penetration pressure of the liquid in the pores ($ML^{-1}t^{-2}$)
ΔP^{M} transmembrane pressure ($ML^{-1}t^{-2}$)
ΔP_{vac} gauge pressure for operation under vacuum ($ML^{-1}t^{-2}$)
$\Delta\pi$ osmotic pressure difference across the membrane ($ML^{-1}t^{-2}$)
ε voidage (= porosity) (—)
ϵ' dielectric constant (—)
ϵ'' dielectric loss factor (—)
ϵ^* complex permittivity (—)
ϵ_0 permittivity of free space ($A^2t^4M^{-1}L^{-3}$)
ζ solubility (—)
η efficiency (—)
θ angle of repose (—)
θ_c contact angle (—)
Θ total heat flux (Mt^{-3})
κ electrical conductivity ($A^2t^3M^{-1}L^{-3}$)
λ thermal conductivity ($MLt^{-3}T^{-1}$)
λ_L wavelength (L)
λ_{L0} free space wavelength (L)
μ dynamic viscosity ($ML^{-1}t^{-1}$)
$\bar{\mu}$ electrochemical potential (ML^2t^{-2} mol^{-1})
ν kinematic viscosity ($=\mu\ \rho^{-1}$) (L^2t^{-1})
ρ density (ML^{-3})
σ normal stress ($ML^{-1}t^{-2}$)
τ shear stress ($ML^{-1}t^{-2}$)
υ molar volume (=RT/P if ideal gas) (L^3mol^{-1})
ϕ sphericity (—)
φ packing density of the fibers in the module (—)
χ retention (—)
Ψ mean molecular free path (L)

DIMENSIONLESS GROUPS

Gz Graetz number ($=Re\ Pr\ d_{eq}\ L^{-1}$)
HR Hausner cohesion number ($=\rho_{ap\text{-}max}\ \rho_{ap}^{-1}$)
Kn Knudsen number ($=\Psi\ d_p^{*\ -1}$)
Nu Nusselt number ($=h\ d_p\ \lambda^{-1}$)
Pr Prandtl number ($=c_p\ \mu\ \lambda^{-1}$)
Re Reynolds number ($=u\ d_p \nu^{-1}$)
Sc Schmidt number ($=\nu\ D_m^{-1}$)
Sh Sherwood number ($=k\ d_p\ D_m^{-1}$)

SUBSCRIPTS

Ag	annulus gas
amb	ambient or environment
ap	apparent (bulk of solids)
ap-max	tapped (apparent maximum)
atom	atomizer
b	bulk of the liquid
bed	bed of particles
C	critical
cyc	cyclone outlet
D	disinfection
cond	condensed water
eq	equilibrium
F	final
fiber	fibers in a hollow-fiber module
fl	film
fry	frying
g	gas
glass	glass transition
H	heating
i	component i
in	inlet
inert	inert
l	liquid or suspension
m	liquid–membrane interface
max	maximum
melt	melting
milk	milk emulsion
mon	monolayer
ms	minimum spouting
op	operation
out	outlet
p	particle
p-bed	all particles in the bed
pd	powder
pd-ret	powder retained on inert surface
p-fl	wetted particles
p-dry	dry particles
wb	wet basis
pk	packaging material
pol	polymer
pp	pulp
pre-dry	predrying treatment
rd	radiator
ref	reference

rel	relative
s	solid
samp	sample
shell	shell of a hollow-fiber module
sor	water sorbed into solid
ss	dried solid (solid skeleton in the bed of particles without water)
ssp	stable spouting regime
st	steam
stor	storage
sur	surface
td	thermal decomposition
ur	urease
v	vapor
vac	vacuum
vap	vaporization
w	water
wall	equipment wall
0	initial

SUPERSCRIPTS

Eff	effective
F	feed
in	inside
M	membrane
out	outside
P	permeate
RF	receiving phase
sat	saturation condition

Abbreviations

A	ampere
ACE	angiotensin I-converting enzyme
ADA	American Dietetic Association
ADR	aromatic distribution ratio
AHA	American Heart Association
AITC	allyl isothiocyanate
ALA	alpha-linolenic acid
APET	amorphous polyethylene terephtalate
ASAE	American Society of Agricultural Engineers
ASTM	American Society for Testing of Materials
ATCC	American type culture collection
ATP	adenosine triphosphate
ATPase	an enzyme that hydrolyzes ATP
BCC	Business Communications Corporation
BOF	bleached oat fiber
CA	controlled atmosphere
CAP	cellulose acetate phthalate
CAM	crassulacean acid metabolism
CCM	calcium citrate malate
CDBD	cascaded dielectric barrier discharge
CDC	Centers for Disease Control and Prevention
CDRP	constant drying rate period
CFD	computational fluid dynamics
CFR	Code of Federal Regulations
CFU	Colony Forming Unit
CHD	coronary heart disease
CHDS	California Health Department Services
CMC	critical micelle concentration
CbMC	carboxymethyl cellulose
C/N	carbon to nitrogen ratio
COC	coconut oil cake
CRP	controlled release packaging
CSL	calcium stearoyl lactylate
CSTR	continuous stirred tank reactor
d.b.	dry basis
DA	Dalton or atomic mass unit is a mass equal to 1/12 the mass of carbon-12, i.e.:1.66×10^{-24} g
DATEM	di-acetyl tartaric acid esters of monoglyceride
DBD	dieletric barrier discharge
DBPC	double-blind placebo-controlled
DBS	sodium dodecylbenzene sulfonate

DCE	direct-contact evaporator
DDM	dialkyl dihexadecylmalonate
DF	dietary fiber
DHA	docosahexahenoic acid
DIN	Deutsches Institut für Normung (German National Standards Organization)
DLS	dynamic light scattering
DMF	dimethyl-formamide
DSC	differential scanning calorimetry
EA	elemental analysis
ED	electrodialysis
EPDM	ethylene-propylene-diene monomer
EPS	expanded polystyrene
EU	European Union
EVOH	ethylene-vinyl alcohol copolymer
FAO	Food and Agriculture Organization
FB	fluidized bed
FBC	fluidized bed coating
FDA	Food and Drug Administration
FDM	fat in dry matter
FDRP	falling drying rate period
FISH	fluorescent in situ hybridization
FMD	foam-mat drying
FOS	fructooligosaccharides
FSA	Food Standards Agency
GAP	good agricultural practice
GC	gas chromatography
GetStoffV	Gefahrstoffverordnung (German Ordinance on Hazardous Substances)
GI	gastrointestinal
GMP	good manufacturing practice
GOS	galactooligosaccharides
GRAS	generally recognized as safe
HACCP	hazard analysis and critical control point
HAOF	high absorption oat fiber
HDL	high density lipoprotein
HLB	hydrophilic–lipophilic balance
HPMC	hydroxypropyl methyl cellulose
HPP	high pressure processing
HPT	high pressure treatment
IBD	inflammatory bowel disease
IBGE	Instituto Brasileiro de Geografia e Estatistica (Brazilian Institute of Geography and Statistics)
IDF	International Dairy Federation
IFB	inner static fluid bed
ILT	Fraunhofer Institute for Laser Technology
IMO	isomaltooligosaccharides

IR	infrared radiation
IRMS	isotope ratio mass spectrometry
ISO	International Organization for Standardization
IVLV	Industrial Organization for Food Technology and Packaging
LAB	lactic acid bacteria
LCT	long chain triglycerides
LDL	low density lipoprotein
LUV	large unilamellar vesicle
MA	modified atmosphere
MAC	maximum allowable concentration
MAP	modified atmosphere packaging
MC	methyl cellulose
MCFA	medium chain fatty acids
MCT	medium chain triglycerides
MD	membrane distillation
MF	microfiltration
MFC	microfibrillated cellulose
ML	mean logarithmic
MLV	multilamellar vesicle
MMT	montmorillonite
MRS	Man, Rogosa, and Sharpe
MSD	multistage spray drying
MUFA	monounsaturated fatty acids
MUP	mupirocin (antibiotic)
MW	microwave
MWAE	microwave-assisted extraction
MWFD	microwave with freeze-drying
MWSBD	microwave with spouted-bed drying
MWVD	microwave vacuum drying
NF	nanofiltration
NFL	National Food Laboratory
NFPA	National Food Process Association
NMR	nuclear magnetic resonance
NRS	nonreducing sugars
NSP	nonstarch polysaccharide
OD	osmotic distillation
OF	oat fiber
OGMPA	Ontario Goat Milk Producers Association
OLV	oligolamellar vesicle
OMD	osmotic membrane distillation
OPP	oriented polypropylene
O/W	oil-in-water
ppb	parts per billion
ppm	parts per million
PA	polyamide
PCA	principal component analysis

PCL	polycaprolactone
PDMS	poly(dimethyl siloxane)
PE	polyethylene
PEA	polyesteramide
PEBA	polyether-polyamide block copolymers
PEF	pulsed electric field
PEN	polyethylene naphthalate
PET	polyethylene terephthalate
PHA	polyhydroxyalkanoates
PHBV	poly(3-hydroxybutyrate-*co*-3-hydroxyvalerate)
PID	propotional integral derivative
PIT	phase inversion temperature
PLA	polylactic acid
PLV	plurilamellar vesicle
POD	peroxidase
POE	polyoxyethylene
POMS	polyoctymethylsiloxame
PP	polypropylene
PPO	polyphenol oxidase
PS	polystyrene
PSI-cell	particle-source-in-cell
PTFE	polytetrafluorethylene
PVDF	poly(vinyliene difluoride)
PU	polyurethane
PUFA	polyunsaturated fatty acids
PV	pervaporation
PVC	polyvinyl chloride
PVdC	polyvinyledene chloride
PVOH	polyvinyl alcohol
RBO	rice bran oil
RDA	Recommended Dietary Allowance
RF	rice fiber
RJ	Rio de Janeiro (one of the Federal Brazilian States)
RN	Rio Grande do Norte (one of the Federal Brazilian States)
RNG	renormalization group theory
RO	reverse osmosis
RS	reducing sugars
RSM	Reynolds stress model
RSMt	response surface methodology
RS1	physically inaccessible starch
RS2	resistant starch granules
RS3	retrograded starch
RS4	chemically modified starch
RT-PCR	real-time polymerase chain reaction
RWE	refractance window evaporator
R1	monorhamnolipid

R2	dirhamnolipid
SAXS	small-angle x-ray scattering
SB	spouted bed
SCF	Scientific Committee for Food
SCFA	short chain fatty acids
SCFE	supercritical fluid extraction
SD	spray drying
SDB	sodium dodecylbenzene sulfonate
SEM	scanning electron microscopy
SFA	saturated fatty acids
SFME	solvent-free microwave extraction
SHIME	simulator of the human intestinal microbial ecosystem
SmF	submerged fermentation
SNIF	site-specific natural isotope fractionation
SOUR	specific oxygen uptake rate
SPE	sucrose polyester
SSD	superheated steam drying
SSF	solid-state fermentation
SUV	small unilamellar vesicle
TAA	total antioxidant activity
TAC	trialkoxycitrate
TAG	triacylglycerol
TASTE	thermally accelerated short time evaporator
TATCA	trialkoxytricarballylate
TBA	thiobarbituric acid
TGE	trialkoxyglyceryl ether
TMU	tetra-methyl-urea
TOP	thermodynamic operation point
TOS	transgalactosylated oligosaccharide
TPS	thermoplastic starch
TSS	total soluble solids
TTIs	time–temperature indicators
UHT	ultrahigh temperature
UF	ultrafiltration
USDA	United States Department of Agriculture
UV	ultraviolet
U/gds	units per gram of dry substrate
VDMA	Verband Deutscher Maschinen-und Anlagenbau (German Engineering Federation)
VF	vacuum frying
VFB	vibrated fluid bed
V-SMOW	Vienna standard ocean water
XOS	xylooligosaccharides
w.b.	wet basis
w/v	weight per volume
wt%	weight percent

WHO	World Health Organization
WVP	water vapor permeability
W/O	water-in-oil
W/O/W	water-in-oil-in-water
1D	one-dimensional
2D	two-dimensional
3D	three-dimensional

Part I

Innovative Techniques

1 Opportunities and Challenges in Nonthermal Processing of Foods

Navin K. Rastogi

CONTENTS

1.1 INTRODUCTION

Increasing consumer demand for high quality and convenient food products with natural flavor and taste, free from additives and preservatives, triggered the need for the development of nonthermal innovative approaches for processing foods with maximum safety and quality without the disadvantages of conventional processing. Quality and safety of food products are the two factors that most influence the choices made by today's increasingly demanding consumers. Consumer demand for minimally processed food products has presented particular challenges to food processors. New and alternative food processing methods, as well as novel combinations of existing methods, are continually being sought by industry in pursuit of producing better quality foods economically. Hence, new innovations, technologies, and concepts continue to emerge. However, regrettably, new methods tend to require higher investments.

Microorganisms and enzymes in foods are major factors in food spoilage and control of these factors can extend shelf life of foods. Most of the technologies used for food preservation involve the prevention or inhibition of microbial growth with the help of low temperature, reduction in water activity, acidification, or addition of preservatives. However, these technologies do not eliminate the existing microorganisms present in foods, so their use can introduce a level of uncertainty with respect to safety. The use of elevated temperatures may result in inactivation of microorganisms, leading to more stable and safer products, but it is detrimental to nutritional and sensorial properties of the foods. To meet the customer's demand of high quality and safe foods, it is necessary to implement new preservation technologies in the food industry. Emerging nonthermal processes, such as application of high pressures, pulsed electric fields (PEFs), and ultrasound, are such alternatives for processing foods with maximum safety and quality. Due to their important features such as killing of microorganisms and inactivation of enzymes at room temperature or even at lower temperatures, these technologies are regarded as potentially powerful tools in food processing. In recent years, there has been a significant increase in the number of scientific papers demonstrating novel and diversified uses of these technologies. They have many things to offer to the food industry, including efficiency enhancement of various operations and online detection of contaminants in foods.

This chapter discusses the effect of selected nonthermal processing technologies on quality and safety of foods. There is a tremendous innovative potential in using the benefits and advantages of these technologies to develop new processes and products or to improve existing ones, as demonstrated by the great number of applications of these nonthermal methods covered in this chapter. Current challenges that pose constraints for the industrial development of these technologies, as well as motivation for future research, are also presented.

1.2 HIGH-PRESSURE PROCESSING

It was discovered in 1899 that application of high pressure can inactivate microorganisms and preserve food. However, its commercial benefits in food processing were realized only in the late 1980s. The application of high-pressure processing (HPP) began with the first report published by Hite (1899), who demonstrated the application of high pressure for preservation of milk. Later, the applications were extended to preserve fruits and vegetables (Hite et al. 1914). Japanese and American food companies were highly encouraged by the ability of high pressure to inactivate microorganisms and spoilage catalyzing enzymes, while retaining other quality attributes, and launched high-pressure processed foods in the market (Mermelstein 1997, Hendrickx et al. 1998). The first high-pressure processed foods were introduced into the Japanese market in 1990 by Meidi-ya® (Meidi-Ya Co., Ltd., Tokyo, Kanto), who launched a line of jams, jellies, and sauces processed without the application of heat (Thakur and Nelson 1998). Other products included fruit preparations, fruit juices, rice cakes, and raw squid in Japan; fruit juices, especially apple and orange juice, in France and Portugal; and guacamole and oysters in the United States (Hugas et al. 2002).

The action of high pressure on microorganisms and proteins/enzymes was reported to be similar to that of high temperature. It enables transmittance of pressure rapidly and uniformly throughout the food, unlike spatial variations in preservation treatments associated with heat, microwave, or radiation. The effect of high pressure on protein/enzyme is reversible in the range 100–400 MPa, which is probably due to conformational changes and subunit dissociation and association process (Morild 1981). High-pressure application leads to the effective reduction of the activity of food-quality-related enzymes (oxidases), which ensures high-quality and shelf-stable products. Sometimes, food constituents offer piezoresistance to enzymes. At high pressures, microbial death is considered to be due to permeabilization of the cell membrane. The changes induced to cell morphology are reversible at low pressure and irreversible at high pressure. For instance, it was observed that in the case of *Saccharomyces cerevisiae*, at pressures of about 400 MPa, the structure and cytoplasmic organelles were grossly deformed and large quantities of intracellular material leaked out, whereas at 500 MPa, the nucleus could no longer be recognized and loss of intracellular material was almost complete (Farr 1990).

High pressure affects only noncovalent bonds (hydrogen, ionic, and hydrophobic bonds), causes unfolding of protein chains, and has little effect on chemical constituents associated with desirable food qualities such as flavor, color, or nutritional content. Small molecules such as amino acids, vitamins, and flavor compounds remain unaffected by high pressure, whereas the structure of large molecules such as proteins, enzymes, polysaccharides, and nucleic acid may be altered (Balci and Wilbey 1999, Rastogi et al. 2007).

The major benefits of HPP can be recapitulated in a nutshell as follows:

1. Uniform and instantaneous distribution of pressure irrespective of size and geometry
2. Effective at ambient temperature or even lower temperatures
3. Elimination of thermal damage and no use of chemical preservatives/additives
4. Operating cost of HPP is lower than that of thermal processing, typically being $0.1–0.5 per liter or kilogram

Over the past two decades, this technology has attracted significant research attention being considered for extension of shelf life, changing the physical and functional properties of food systems, exploiting the anomalous phase transitions of water under extreme pressures, and as a pretreatment to improve the efficiency of existing unit operations.

HPP can be applied to a range of foods, including juices and beverages, fruits and vegetables, meat-based products (cooked and dry ham, etc.), fish and precooked dishes, with meat and vegetables being the most popular applications (Norton and Sun 2008). Table 1.1 summarizes available commercial food product that are processed by high pressure. The product range is increasing and spreading from its origins in Japan, followed by the United States and now Europe (Hogan et al. 2005). The companies which can supply a wide variety of high-pressure equipment to the food industry are listed in Table 1.2.

1.2.1 Opportunities for High-Pressure Processing

Due to its energy and quality-related advantages, inclusion of high-pressure treatment (HPT) is gaining popularity as a complimentary processing step in the chain of integrated food processing. It can lead to novel products as well as new process development opportunities.

1.2.1.1 High-Pressure Blanching

High pressure at ambient temperatures can be used as a method of blanching similar to hot water or steam blanching, but without thermal degradation. This also minimizes the problems associated with the disposal of waste water after high-temperature blanching. Eshtiaghi and Knorr (1993) demonstrated that the application of pressure (400 MPa, 15 min, 20°C) in combination with 0.5 wt% citric acid solution resulted in complete inactivation of polyphenoloxidase of potato samples. Besides, it also resulted in a four-log cycle reduction in microbial count and higher retention of ascorbic acid. Buggenhout et al. (2006) indicated that application of

TABLE 1.1
High-Pressure-Processed Food Products Commercially Available Worldwide

Product	Name of the Company
Fruit and vegetable products	
Orange juice	M/s Ultifruit, Paris, France
Mandarin juice	M/s Wakayama Food Industries, Japan
Fruit juices	M/s Pampryl, La Courneuve, France
Fruit and vegetable juices	M/s Odwalla, Santa Cruz, CA
Apple juice	M/s Frubaca, Alcobaca, Portugal
Fruit juices and smoothies	M/s Orchard House, Northumberland, U.K.
Jams, fruit sauces, yoghurt, and jelly	M/s Meida-Ya, Tokyo, Japan
Fruit jams	M/s Solofruita, Milano, Italy
Tropical fruits	M/s Nishin Oil Mills, Tokyo, Japan
Guacamole, salsa dips, ready meals, and fruit juices	M/s Avomex, Fort Worth, TX
Animal products	
Beef	M/s Fuji Ciku Mutterham, Nagoya, Japan
Hummus	M/s Hannah International, Seabrook, NH
Ham	M/s Hormel Foods, Austin, MN
Poultry products	M/s Purdue Farms, Emporia, VA
Oysters	M/s Motivatit Seafoods, Houma, LA; Goose Point Oysters, Bay Center, WA; Joey Oysters, Amite, LA
Sliced ham and tapas	M/s Espuña, Olot, Spain

Source: Hogan, E. et al., High pressure processing of foods: An overview, in *Emerging Technologies for Food Processing*, Sun, D.W., Ed., Elsevier Ltd., London, 2005, 4–30. With permission.

high pressure at low temperatures (0.1–500 MPa at −26°C to 20°C) could inactivate microorganisms, but it failed to inactivate most food-quality-related enzymes. These results suggest that a blanching unit operation is required to prevent enzyme-related quality degradation during frozen storage. Kingsly et al. (2009a) indicated that HPP of peach fruits suspended in citric acid medium could be used as a potential alternative for hot-water blanching. Pressure treatment (>300 MPa) in combination with citric acid (1–1.2 wt%) has been found to be effective in inactivation of peach polyphenoloxidase enzymes. Castro et al. (2008) observed that HPP could be used as a pretreatment instead of blanching to produce frozen peppers with better nutritional (soluble protein and ascorbic acid) and texture (firmness) characteristics.

1.2.1.2 High-Pressure-Assisted Drying and Osmotic Dehydration

The application of high pressure damages the plant cell wall structure, leaving cells more permeable, which leads to significant changes in tissue architecture, resulting in increased mass transfer rates during dehydration as well as osmotic dehydration (Farr 1990, Dornenburg and Knorr 1993, Rastogi et al. 1994). High-pressure

TABLE 1.2
Main Suppliers of High-Pressure Processing Equipment and Services

Name of the Company	Specialization
M/s Resato International, Roden, Holland (http://www.resato.com)	Manufactures laboratory and industrial machines Pressure shift freezing systems Reciprocating intensifiers suitable for one or multiple autoclave systems (up to 1400 MPa)
M/s Avure Technologies, Inc., Kent, WA (http://www.avure.com)	Manufactures batch presses that pasteurize prepared ready-to-eat foods Unique pumping systems to enhance throughput (600 MPa)
M/s Elmhurst Research, Inc., Albany, NY (http://www.elmhurstresearch.com)	Designs and manufactures batch presses Patented vessel technology developed exclusively for food processing industry (689 MPa)
M/s Engineered Pressure Systems, Inc., Haverhill, MA (http://www.epsi-highpressure.com)	Manufactures laboratory and industrial equipment Manufacture hot, cold, and warm isostatic presses (100–900 MPa)
M/s Kobelco, Hyogo, Japan (http://www.kobelco.co.jp)	Manufactures laboratory and industrial equipment Manufacture hot and cold isostatic presses (98–686 MPa)
M/s Mitsubishi Heavy Industries, Hiroshima, Japan (http://www.mhi.co.jp)	Manufactures laboratory and industrial equipment Manufactures isostatic pressing system with large operating temperature range as option (686 MPa)
M/s NC Hyperbaric, Burgos, Spain (http://www.nchyperbaric.com)	Manufactures industrial equipment Designed a system to work with different volumes (600 MPa)
M/s Stansted Fluid Power Ltd., Essex, U.K. (http://www.sfp-4-hp.demon.co.uk)	Manufactures equipment for R&D and industrial scale equipment Single and multiple vessels with wide temperature (up to 1400 MPa)
M/s Uhde Hochdrucktechnik, Hagen, Germany (http://www.uhde-hpt.com)	Manufactures equipment for industry and research purposes Develops plant processes from initial testing to full-scale application (700 MPa)

Source: Norton, T. and Sun, D.W., *Food Bioprocess. Technol.*, 1, 2, 2008. With permission.
Note: Pressure capacity of standard machines is provided in the parenthesis.

pretreatment (600 MPa, 15 min at 70°C) to food drying reportedly resulted in a significant increase in drying rates of potato (Eshtiaghi et al. 1994). Similarly, Ade-Omowaye et al. (2001a) demonstrated that both the high pressure (400 MPa for 10 min at 25°C) and the chemical (NaOH and HCl) pretreatments were found to result in comparable increase in drying rates for paprika. This indicates that high-pressure pretreatments can be alternatives to chemical pretreatments thereby minimizing environmental pollution from chemicals.

Application of HPT (100–800 MPa) reportedly enhanced both water removal and solute gain during osmotic dehydration of pineapple (Rastogi and Niranjan 1998). Water and solute diffusivity values increased by a factor of four and two, respectively. The compression and decompression steps during the application and release of pressure, respectively, caused the removal of a significant amount of water, which was attributed to cell wall rupture. The microscopic examination showed that the extent of cell wall breakup increased with applied pressure (Rastogi and Niranjan 1998). The acceleration of mass transfer during ingredient infusion into foods due to application of high pressure (100–400 MPa) was also demonstrated in the case of potato by Sopanangkul et al. (2002). This was attributed to the fact that HPP was found to open up the tissue structure, thereby facilitating diffusion. However, pressures above 400 MPa induced starch gelatinization, which resulted in hindered diffusion. Rastogi et al. (2000a, b, 2003) demonstrated that HPP enhanced the rate of movement of the dehydration front in a potato samples subjected to subsequent osmotic dehydration. This was ascribed to the synergistic effect of cell permeabilization (due to high pressure) and osmotic stress (because of the concentration of the surrounding osmotic solution). Kingsly et al. (2009b) reported that in pineapple HPP reduced sample hardness, springiness, and chewiness, whereas it had no significant effect on cohesiveness. Moreover, the treatment reduced the drying time of pineapple slices. The effective moisture diffusivity was found to increase with an increase of pressure up to 500 MPa.

Villacis et al. (2008) demonstrated that in the case of high-pressure-assisted brining of turkey breast, water and sodium chloride diffused into the sample during high pressure come up time, an effect found to be maximum at 150 MPa. Holding the samples under high pressure resulted not only in the further infusion of sodium chloride into the sample, but also in the diffusion of moisture out of the sample. Within the range of experimental conditions studied, HPT at 150 MPa yielded meat samples with minimum hardness, gumminess, and chewiness.

1.2.1.3 High-Pressure-Assisted Rehydration

Loss of solids during rehydration is a major problem associated with use of dehydrated foods. The rehydration of high-pressure-treated, two-stages dried (osmotic dehydration and finish-drying at 25°C) pineapple indicated that the diffusion coefficients for water infusion as well as for solute diffusion during rehydration were significantly lower than those of the sample not subjected to HPT. The reduction of water infusion may be associated with the permeabilization of cell membranes due to application of high pressure (Rastogi and Niranjan 1998). The solid diffusion coefficient was also lower, and so was the release of the cellular components, which form a gel-network with divalent ions binding to de-esterified pectin (Eshtiaghi et al. 1994, Basak and Ramaswamy 1998, Rastogi et al. 2000c). The scheme of processing may be beneficially used to retain the solute (nutrients or color) within the food during rehydration.

1.2.1.4 High-Pressure-Assisted Frying

HPT was found to decrease oil uptake during frying of potato, which may be due to a reduction in moisture content caused by compression and decompression

(Rastogi and Niranjan 1998), as well as the prevalence of different oil mass transfer mechanisms (Rastogi et al. 2007).

1.2.1.5 High-Pressure-Assisted Solid–Liquid Extraction

HPT was shown to be an attractive alternative for solid–liquid extraction. For instance, in case of extraction of caffeine, combination of high pressures and moderate temperatures could become a viable alternative to the currently used process, eliminating the need for high extraction temperatures (Rastogi et al. 2007). HPT was also found to be a feasible method for extraction of trehalose from *S. cerevisiae*. The trehalase was inactivated at 700 MPa, whereas, trehalose was resistant to hydrolysis even at pressures of 1500 MPa (Kinefuchi et al. 1995). Perez et al. (2002) indicated that HPT could enhance the extraction of sugar and reduce the treatment time during the mashing stage in beer production. Working with tomato puree, Fernandez et al. (2001a) demonstrated that there was no change in the total concentration of β-carotene due to application of high pressure. However, after storage at 4°C for 21 days, stability of the antioxidant capacity of the water-soluble fraction increased compared to the untreated samples. Plaza et al. (2006) studied the stability of carotenoids and the antioxidative activity of Mediterranean vegetable soup (gazpacho) subjected to HPT (up to 300 MPa, 60°C) and stored at 4°C for 40 days. The results indicated that the treatment at 150 MPa led to better preservation of both properties than the treatment at 300 MPa did. Houska et al. (2006) demonstrated that high-pressure pasteurization is capable of preserving nutritional substances in juices, such as sulforaphane in broccoli juice, apart from inactivating microorganisms originally present in the raw juice.

Yutang et al. (2008) showed that the application of combined high pressure and microwave extraction of ginsenosides from Panax ginseng resulted in higher yields than other extraction methods, including soxhlet extraction, ultrasound-assisted extraction, and heat reflux extraction.

1.2.1.6 High-Pressure Shift Freezing and Pressure-Assisted Thawing

Slow freezing rates produce a few ice crystals of large sizes that alter the structure of the product and result in quality loss. Fast freezing rates, on the other hand, produce a large number of small ice crystals that cause less damage. However, rapid freezing using cryogens induces cracking because of the initial decrease of volume due to cooling and the subsequent increase in volume due to freezing (Kalichevsky et al. 1995). Application of high pressure during freezing can avoid these problems due to the instantaneous and homogenous formation of ice throughout the product, which eliminates internal stress. Reduction in freezing point under high pressure causes supercooling upon pressure release and promotes rapid ice nucleation and growth throughout the sample. Generally, thawing occurs more slowly than freezing, potentially causing further damage to the sample. High-pressure-induced thawing reduces the loss of water-holding capacity and improves both color and flavor retention in fruits.

Otero et al. (2000) have shown that entire volume of the sample (both surface and center) reached the initial freezing point at the same time in the case of high-pressure shift freezing, just before pressure release. The high level of supercooling during high-pressure shift freezing of peach and mango led to uniform and rapid ice nucleation throughout sample volume, which largely maintained the original tissue microstructure. Otero et al. (1998) demonstrated that high-pressure frozen eggplant samples had the highest firmness and the lowest rupture strain and drip loss compared to those of still-air frozen and air-blast frozen samples. Freezing of water under high pressure (100–700 MPa) resulted in a volume reduction, which led to the formation of several kinds of heavy ice polymorphs. Fuchigami et al. (1997a,b) verified that, under freezing conditions (−30°C), carrots pressurized at 200 MPa did not freeze, and when pressure was reduced to atmospheric pressure, carrots froze very quickly. These samples had better firmness, texture, and histological structure of frozen carrots than the ordinary frozen sample did. Similar results were obtained at 340 and 400 MPa, which was attributed to the shift of solid–liquid equilibrium line to lower temperatures, due to the presence of sugars in carrots. Similar results were later reported for Chinese cabbage (Fuchigami et al. 1998). Fernandez et al. (2006) showed that blanched and high-pressure frozen broccoli exhibited less cell damage, lower drip losses, and better texture than conventionally frozen samples.

Zhu et al. (2004a) indicated that high-pressure shift freezing of pork muscle resulted in small and regular crystals. Near the surface, there were many fine and regular intracellular ice crystals with well-preserved muscle tissue. From midway to the center, the ice crystals were larger in size and located extracellularly. Additionally, changes in color, a reduction in drip loss during thawing, considerable denaturation of myofibrillar proteins, and a reduction in muscle toughness were observed as a result of high-pressure shift freezing (Zhu et al. 2004b). In the case of salmon, Zhu et al. (2003) showed that high-pressure shift freezing produced a large amount of homogeneously distributed fine and regular intracellular ice crystals, which helped in the maintenance of muscle fibers in comparison to the frozen muscle structure. Fuchigami and Teramoto (1997) indicated that high-pressure freezing could be an effective means to improve the texture of frozen tofu. The rupture stress and strain of tofu frozen at 200 and 340 MPa were similar to untreated tofu.

Schubring et al. (2003) verified that the organoleptic characteristics of high-pressure-thawed fish fillets were superior to those of water thawed samples due to reduced drip loss, greater protein denaturation, improved microbial status, and textural parameters. Rouille et al. (2002) showed that high-pressure-thawed dogfish and scallops had better microbial quality and lower thawing time and drip volume than immersion thawed products. Chevalier et al. (1999) reported that high-pressure thawing of blue whiting was quicker and resulted in lower drip volume in comparison to conventional thawing.

1.2.1.7 High-Pressure-Assisted Thermal Processing

High-pressure-assisted thermal processing has recently emerged as a promising alternative technology for processing low-acid foods (Matser et al. 2004). The process, in general, involves simultaneous application of elevated pressures (500–700 MPa) and temperatures (90°C–120°C) to a preheated food (Matser et al. 2004, Rajan et al. 2006a, Nguyen et al. 2007, Rastogi et al. 2008a). The compression heating during pressurization and rapid cooling on depressurization help reduce the severity of thermal effects encountered with conventional thermal processing

(Ting et al. 2002, Rasanayagam et al. 2003). Microbial efficacy of pressure-assisted thermal processing technology on inactivating pathogenic and spoilage bacterial spores has been widely reported (Gola et al. 1996, Rovere et al. 1998, Heinz and Knorr 2001, Reddy et al. 2003, 2006, Margosch et al. 2004, Patazca et al. 2006, Rajan et al. 2006a,b). The technology reduces process time and preserves food quality, especially texture, color, and flavor as compared to those of retorted products (Hoogland et al. 2001, Krebbers et al. 2002, 2003, Juliano et al. 2006).

Nguyen et al. (2007) reported less quality degradation (texture, color, and carotene content) for pressure-assisted thermal processed carrots, whose hardness was later improved by a combined pretreatment involving calcium infusion, heating, and pressurization (Rastogi et al. 2008a,b). Leadley et al. (2008) compared the high-pressure sterilization with the equivalent thermal processing and concluded that the high-pressure-sterilized green bean samples were darker and greener in appearance, and also twice as firm as the thermally processed samples were. Wilson et al. (2008) reviewed the current status of HPT for inactivation of bacterial spores, and particularly examined the requirement for a combination of high-pressure and high-temperature processing to achieve sterilization of foods.

1.2.1.8 Specific Application of High Pressure in Fruit and Vegetable Products

Various researchers have reported extended shelf life of high-pressure-processed orange juice under refrigeration, with increased flavor retention depending upon the processing and storage conditions (Donsi et al. 1996, Parish 1998, Takahashi et al. 1998, Strolham et al. 2000). Fernandez et al. (2001b) showed that there was no significant difference in antioxidative capacity and sugar and carotene contents between high-pressure and thermally pasteurized orange juice. The rate of degradation of ascorbic acid was lower for orange juice subjected to HPT than for the thermally pasteurized juice (Polydera et al. 2005). HPT was also found to result in cloud stabilization (Goodner et al. 1999), increased extraction of flavanones, and higher retention of potential health-promoting attributes in orange juice (Sanchez et al. 2005). In addition, HPT resulted in no change in color, browning index, viscosity, concentration and titratable acidity, levels of alcohol insoluble acids, ascorbic acid, and β-carotene (Bull et al. 2004).

McInerney et al. (2007) indicated that antioxidant capacity and total carotenoid content differed among vegetables but were unaffected by HPT. Gow and Hsin (1999) demonstrated that high-pressure-processed guava puree could be stored up to 40 days at 4°C without any change in color, cloudiness, ascorbic acid content, flavor distribution, and viscosity. Novotna et al. (1999) showed that sensory quality (aroma) of apple juice subjected to HPT was better than that of the pasteurized juice in terms of aroma. Lambert et al. (1999) indicated that HPP resulted in the inactivation of enzymes responsible for the degradation of food quality such as polyphenoloxidase (PPO) and peroxidase (POD) in case of strawberry puree. No major changes in strawberry aroma profiles were observed up to 500 MPa, but higher pressures induced significant changes in the aroma profiles due to the synthesis of new compounds.

Moio et al. (1994) demonstrated that white grape must was sterilized at 500 MPa pressure for 3 min, with little changes in physicochemical properties. However, red grape must was not sterilized at this pressure due to higher stability of the natural microflora present in this fruit. Phunchaisri and Apichartsrangkoon (2005) demonstrated that HPT (600 MPa at 60°C for 20 min) resulted in less loss of visual quality in both fresh and syrup-processed lychee compared to thermal processing. HPT led to extensive inactivation of POD and PPO in fresh lychee, even though these effects were less significant when the sample was processed in syrup. Prestamo and Arroyo (2000) indicated that HPP of melon did not cause any browning, whereas peaches and pears underwent browning. This browning could be prevented by the addition of ascorbic acid. After processing, PPO and POD enzymes could not be inhibited, but textures of all fruits were acceptable.

Watanabe et al. (1991) explained the method for production of high-pressure-processed jam. Powdered sugar, pectin, citric acid, and freeze concentrated strawberry juice were mixed, degassed, and then pressurized (400 MPa at room temperature for 5 min). The texture of jam was similar to the one of conventional jam, and the product had a bright red color with the original flavor. Kimura et al. (1994) have shown that pressure-treated jam had better quality than heat-treated jam. The pressure-treated jam could be stored at refrigeration temperature up to 3 months.

Arroyo et al. (1997, 1999) demonstrated that HPT (300–400 MPa) of selected vegetables resulted in a decrease in viable aerobic mesophiles, fungi, and yeasts, also affecting organoleptic properties. In case of tomato, loosening and peeling away of skin were observed, with no change in color and flavor. However, lettuce and cauliflower remained firm but underwent browning due to displacement of POD. Wennberg and Nyman (2004) showed that HPP had a marked effect on the distribution of soluble and insoluble fiber in case of white cabbage.

Basak and Ramaswamy (1998) showed that high pressure had a dual effect on texture of fruits and vegetables. The instantaneous loss of texture due to application of high pressure was followed by a gradual recovery of the instantaneous initial loss of texture after holding the sample up to 200 MPa. The extent of this initial loss was more prominent at higher pressures, while partial recovery of texture was more prominent at lower pressures. Sila et al. (2004) indicated that HPT of carrots combined with $CaCl_2$ infusion improved texture during thermal processing. In a later work (Sila et al. 2005), carrot subjected to HPT showed less loss of texture when further processed at high temperatures (100°C–125°C). The textural properties were significantly improved when calcium infusion was combined with low-temperature blanching.

Butz et al. (2002) demonstrated that HPP did not affect chlorophyll *a* and *b* in broccoli, lycopene and β-carotene in tomatoes, and antioxidative activities of carrot and tomato homogenates. Antimutagenicity of carrot, leek, spinach, kohlrabi, and cauliflower juices were also unaffected by HPP (Butz et al. 1997). Plaza et al. (2003) explored HPP along with citric acid and sodium chloride for the manufacture of minimally processed tomato products with optimal sensory and microbiological characteristics. Qiu et al. (2006) indicated that the highest stability of lycopene in tomato puree was obtained by HPP at 500 MPa and (4 ± 1)°C storage. Sanchez et al. (2006) verified that high-pressure-processed tomato puree had higher redness, carotenoids, and vitamin C than a thermally pasteurized sample.

Kadlec et al. (2006) pointed out that HPT (500 MPa, 10 min) of germinated chickpea seeds resulted in reduction in the total number of microorganisms without any quality and sensory changes during 21 days of storage. Penas et al. (2008) optimized the combination of treatment time, pressure, and temperature applied to mung bean and alfalfa seeds in reduction in the native microbial load in sprouts without affecting their germination capacity. The optimal treatment conditions were 40°C, 100 and 250 MPa for alfalfa and mung bean seeds, respectively.

1.2.1.9 Specific Application of High Pressure in Dairy Products

Research on application of high pressure on milk was initiated with a view to developing an alternative process for pasteurization. Huppertz et al. (2002) reviewed the effect of high pressure on properties and contents of milk. A number of researchers have studied inactivation of microorganisms (such as *Listeria monocytogenes*, *Staphylococcus aureus*, or *Listeria innocua*) either naturally present or introduced in milk (Styles et al. 1991, Erkman and Karatas 1997, Gervilla et al. 1997). Hite et al. (1914) pointed out that the HPP resulted in a significant reduction in microorganisms and the combination of high pressure with temperature resulted in increased shelf life. Vachon et al. (2002) demonstrated that periodic oscillation of high pressure was very effective for the destruction of pathogens such as *L. monocytogenes*, *Escherichia coli*, and *Salmonella enteritidis*. Pandey et al. (2003) showed that higher pressures, longer holding time, and lower temperature resulted in greater destruction of microorganisms in raw milk. Mussa and Ramaswamy (1997) studied the kinetics of microorganism destruction, alkaline phosphatase degradation, and changes in the color and viscosity to establish a pasteurization condition for fresh raw milk.

Capellas et al. (1996) reported that samples of goat milk cheese inoculated with 10^8 CFU/g and subjected to HPP (400–500 MPa) showed no surviving *E. coli* even after 15, 30, or 60 days of storage at 2°C–4°C. O'Reilly et al. (2000) indicated high sensitivity of *E. coli* in cheddar cheese at pressures above 200 MPa, possibly due to acid injury during cheese fermentation. Carminati et al. (2004) showed that HPT (400–700 MPa) was effective in the reducing *L. monocytogenes* in gorgonzola cheese rinds without significantly changing its sensory properties. Viazis et al. (2008) evaluated the efficacy of HPP to inactivate pathogens in human milk without loss of any important nutritional biomolecules. Koseki et al. (2008) indicated that a mild heat treatment (37°C for 240 min or 50°C for 10 min) inhibited the recovery of *L. monocytogenes* in high-pressure-processed milk, and the product was safely stored for 70 days at 25°C.

Evrendilek et al. (2008) indicated that, in case of Turkish white cheese, the maximum reduction in *L. monocytogenes* counts, of about 4.9 log CFU/g, was achieved at 600 MPa. HPP resulted in total reduction in molds, yeasts, and *Enterobacteriaceae* counts for the cheese samples produced from raw and pasteurized milk. This suggests that HPP can be effectively used to reduce the microbial load in Turkish white cheese.

Morgan et al. (2000) indicated that the combination of HPP with a bacteriocin (lacticin) resulted in a synergistic effect in controlling microbial flora of milk without significantly influencing its cheese-making properties. Black et al. (2005) also demonstrated that combining HPP and an antibiotic (nisin) resulted in a greater inactivation of gram-positive bacteria. The gram-negative bacteria, in this case, were found to be more sensitive to high pressure, either alone or in combination with nisin, than gram-positive bacteria. This HPP–antibiotic combination may allow lower pressures and shorter processing times to be used without compromising product safety. Sierra et al. (2000) showed that HPP of milk is a gentler process than conventional procedures for extending shelf life, because no significant variation in the content of B_1 and B_6 vitamins was observed.

HPP up to 300 MPa was found to have little effect on the β-lactoglobulin in whey, whereas further increase in pressure above 600 MPa resulted in a decrease in the levels of β-lactoglobulin, which indicated the pressure denaturation of this protein (Brooker et al. 1998, Pandey and Ramaswamy 1998). Short-time exposure to high pressure reportedly enhanced activity of lipoprotein lipase and glutamyl transferase of milk. However, long-time (100 min) pressure exposure did not bring about any inactivation of lipase, while glutamyl transferase followed first-order inactivation kinetics (Pandey and Ramaswamy 2004).

Borda et al. (2004a,b) pointed out that combined high pressure and thermal inactivation of plasmin and plasminogen from milk was found to follow first-order kinetics. A synergistic effect of temperature and high pressure was observed in the range 300–600 MPa. However, an antagonistic effect of temperature and pressure was observed at pressures greater than 600 MPa, because the enzymes were stabilized by the disruption of disulfide bonds.

HPT of milk affects its coagulation process and cheese-making properties indirectly through a number of effects on milk proteins, including reduction in the size of casein micelles, probably followed by interaction with micellar κ-casein. Pressures lower than 150 MPa did not have any influence on the rennet coagulation time, but this parameter decreased at higher pressures (Derobry et al. 1994). Kolakowski et al. (2000), Needs et al. (2000), and Zobrist et al. (2005) also reported a decrease in the rennet coagulation time in the case of milk is due to the high-pressure-induced association of whey proteins with casein micelles. Trujillo et al. (1999a) and Needs et al. (2000) pointed out that the rennet coagulation time of pressure-treated milk was higher than the one of thermally pasteurized milk.

HPP was found to accelerate the rate of curd formation and curd firming of rennet milk (Ohmiya et al. 1987). The rate of curd firming increased up to 200 MPa, but further increases in pressure resulted in its decrease (Lopez et al. 1996). The rate of curd formation was also found to have a maximum at 200 MPa (Needs et al. 2000).

O'Reilly et al. (2001) investigated the use of HPP in cheese making. High-pressure-induced disruption of casein micelles and denaturation of whey proteins. Furthermore, HPT increased pH of milk, reduced rennet coagulation time, and increased cheese yield, thereby indicating the potential of HPP in this application. Drake et al. (1997) demonstrated that HPT of cheese milk resulted in increased cheese yield due to denaturation of whey proteins as well as enhancement of water-holding capacity. Arias et al. (2000) and Huppertz et al. (2004) also reported an increase in cheese yield due to HPT. Molina et al. (2000) reported increased yield of pressurization of pasteurized milk due to improvement in the coagulation properties of proteins. Huppertz et al. (2005) showed that yield of cheese curd from high-pressure-treated and subsequently heated milk was greater than that from unheated and unpressurized milk.

Yokohama et al. (1991) showed that HPT resulted in accelerated ripening of cheese due to milk protein proteolysis, which resulted in an increase in free amino acid content and taste of pressure-treated cheese was described as excellent. O'Reilly et al. (2000) also showed that HPP led to an increase in ripening of cheese due to the degradation of casein. Trujillo et al. (1999b) showed that small peptides and free amino acids content indicated a higher extent of proteolysis in cheese made from high-pressure-treated milk.

Buffa et al. (2001) showed that cheese prepared from raw and pressure-treated goat milk was firmer, less fracturable, and less cohesive than pasteurized milk. HPT resulted in a more elastic, regular, and compact protein matrix with smaller and uniform fat globules, which resembled the structure of cheese prepared from raw milk. The level of free amino acids was also shown to increase (Saldo et al. 2002).

1.2.1.10 Specific Application of High Pressure in Animal Products

Microorganisms in meat can be inactivated by HPT, an effect whose extent depends on several parameters such as type of microorganism, pressure level, process temperature and time, pH and composition of food, or dispersion medium. In general, gram-negative bacteria are more sensitive to pressure than gram-positive bacteria, but large differences in pressure resistance are apparent among various strains of the same species. Bacterial spores are highly resistant to pressure (unless pressurization is carried out at temperatures close to 100°C).

Carlez et al. (1993, 1994), for instance, demonstrated that HPP of beef resulted in inactivation of microorganisms, which increased shelf-life by 2–6 days upon subsequent storage of meat at low temperatures (3°C). Morales et al. (2008) observed that a multiple-cycle HPT (400 MPa, 1 min, 4 cycles) was more effective than the single-cycle HPT (400 MPa, 20 min, 1 cycle) to inactivate *E. coli* in ground beef.

Jung et al. (2003) also reported the similar results, along with an improvement in meat color by increased redness. Hayman et al. (2004) showed that in the case of high-pressure-processed ready-to-eat meats, counts of aerobic and anaerobic mesophiles, lactic acid bacteria, *Listeria* spp., *Staphylococci*, *Brochothrix thermosphacta*, *Coliforms*, and fungi were undetectable when stored at 4°C for 98 days. Consumer acceptability and sensory quality of the product was found to be very high. Garriga et al. (2004) demonstrated that, in the case of ham, HPP prevented the growth of *Enterobacteriaceae*, yeasts, and lactic acid bacteria, resulting in increased shelf life. In addition, food safety risks associated with *L. monocytogenes* and *Salmonella* were also reduced. Koseki et al. (2007) observed that HPT (550 MPa, 10 min) of sliced cooked ham inoculated with *L. monocytogenes* reduced the microbial count below detectable limits. The bacterial count gradually increased during storage, and exceeded the initial inoculum level at the end of a 70 day period.

Jofre et al. (2008) showed that antimicrobial packaging, HPP, and refrigerated storage could be a very effective combination to obtain value added, ready-to-eat products from ham with a safe long-term storage up to 3 months at 6°C. Marcos et al. (2008) confirmed the suitability of combining HPP and antimicrobial biodegradable packaging technologies to control *L. monocytogenes* growth to increase shelf life of cooked ham.

Gomez et al. (2007) indicated that combination of HPP and edible films yielded the best results in terms of preventing oxidation and inhibiting microbial growth in cold-smoked sardine, thereby increasing its shelf life. Coating the muscle with films enriched with oregano or rosemary extracts increased phenol content and antioxidant power when used in combination with HPP.

Jung et al. (2000) showed that HPT in case of beef induced a significant increase in the activity of lysosomal enzymes, but did not improve beef tenderness or reduce the aging period. In fact, pressurization increased toughness, due to modification in myofibrillar components.

Suzuki et al. (1993) reported pressure-induced tenderization of bovine liver cells, which was caused due to improvement in actomyosin toughness. HPT of ovine and bovine muscles resulted in firmness and contraction, but, after cooking, the meat was tender and had higher moisture content. Suzuki et al. (1992) indicated that the HPP resulted in meat tenderization without heating. Pressures up to 300 MPa caused increased myofibril fragmentation and marked modification in its ultrastructure. Ayo et al. (2005) observed that textural properties of meat batters with walnuts were not affected by HPP. However, hardness, cohesiveness, springiness, and chewiness of the cooked products were reduced by addition of walnut. Bai et al. (2004) showed that sensory properties such as color, extension, and flavor varied considerably after HPT of beef and mutton. Furthermore, shrinkage of sarcomere and reduction in shear force was observed. Han and Ledward (2004) found that, in the case of beef muscle, hardness increased with increasing pressure (200–800 MPa) at a constant temperature (up to 40°C), whereas it decreased significantly with application of pressure (200 MPa) at higher temperatures (60°C or 70°C). Accelerated proteolysis may be the major contributing factor to loss in hardness of beef.

Calpain and cathepsins enzymes have an influence on tenderization of meat. Qin et al. (2001) showed that total calpain activity decreased due to HPP, but acid phosphatase and alkaline phosphatase activities were not significantly reduced up to 300 MPa. Homma et al. (1995) found that total activity of calpains in pressurized muscle increased due to a reduction in the levels of calpastatin because of its pressure sensitivity, and this resulted in meat tenderization. Carballo et al. (1996) indicated that HPT of finely comminuted bovine meat resulted in the formation of gels with smooth cohesive texture and high water retention.

HPT in combination with low storage temperature (300–900 MPa, 14°C–28°C) was shown to be a potential technology for the extension of shelf life of fresh raw ground chicken. The expected shelf life of chicken in sealed polyfilm pouches processed at 888 MPa was reported to be more than 98 days (O'Brien and Marshall 1996). Jimenez-Castro et al. (1998) showed that HPP resulted in increased water and fat-binding properties of chicken and pork batters. The samples were found to be softer, cohesive, springy, or chewy than nonpressurized samples. Orlien et al. (2000) verified that chicken breast muscle subjected to HPT (up to 500 MPa) showed no rancidity during chilled storage and was found to be similar to untreated meat. Pressures beyond 500 MPa resulted in increased lipid oxidation, which was related to cell membrane damage. Bragagnolo et al. (2006) established the formation of free radicals during HPP of chicken breast and thigh. The formation of free radical was found to increase with increasing pressure and processing time. Radical formation was more significant in thigh meat than it was in breast meat, and salt addition further promoted radical formation, especially in chicken thigh. El Moueffak et al. (2001) indicated that combined effect of high pressure and temperature could be used to give a product of similar microbiological quality to that obtained by pasteurization.

HPT (700 MPa) of salmon spread extended shelf life at low temperature (3°C–8°C) without significant chemical, microbiological, or sensory changes. HPP completely inactivated pathogens present in the inoculated sample (Carpi et al. 1995). High-pressure-induced gels (200–420 MPa) of blue whiting were found to have lower adhesiveness, higher water-holding capacity, and less yellowness than heat-induced gels did (Perez and Montero 2000). Vacuum treatment and HPT extended the shelf life of prawn samples, although it did affect muscle color very slightly, giving it a whiter appearance. The viable shelf life of 1 week for air-stored samples was extended to 21, 28, and 35 days for vacuum-packaged samples, samples treated at 200 and 400 MPa, respectively (Lopez-Caballero et al. 2000). Angsupanich and Ledward (1998) verified that pressure-treated fish was harder, chewier, and gummier than both the raw and cooked products were. Lakshmanan and Dalgaard (2004) showed that HPP up to 250 MPa could not inactivate *L. monocytogenes* in smoked salmon, but it had a marked effect on both color and texture of the product. Yagiz et al. (2007) indicated that a pressure of 300 MPa effectively reduced the initial microbial population in rainbow trout and mahi-mahi up to 6- and 4-log reduction, respectively. The redness of rainbow trout was lower as compared to mahi-mahi. The lipid oxidation for rainbow trout increased with increasing pressure, whereas, in the case of mahi-mahi, the maximum oxidation was found at 300 MPa and then it declined with increase in pressure. The optimum high pressure for influencing lipid oxidation, microbial load, and color changes were found to be 300 MPa for rainbow trout and 450 MPa for mahi-mahi.

Mor and Yuste (2003) showed that high-pressure-processed sausages were less firm, more cohesive, had lower weight loss, and higher preference scores than heat-treated samples, without any effects on the color attributes. Carpi et al. (1999) demonstrated that shear resistance of pressure-treated and vacuum-packaged raw ham was lower than that of the control sample, without changes in the sensory attributes and shelf life. Tanzi et al. (2004) demonstrated that HPP is a useful technique for control of *L. monocytogenes* in sliced Parma ham. The treated sample had less red color and intense salty taste.

1.2.2 Challenges in High-Pressure Processing

During pressurization, temperature of food material increases as a result of physical compression. This has been ignored in most of the studies available in the literature. This temperature increase may have an effect on gelling of food components, stability of proteins, migration of fat, etc., and its magnitude depends, in part, upon the initial temperature, material compressibility, specific heat, and target pressure. All compressible substances change temperature during physical compression and this is an unavoidable thermodynamic effect (Ting et al. 2002). Water has the lowest compression heating values, whereas fats and oils have the highest; this indicates that there is a difference in the rates of heating food components under pressure.

Similarly, the temperature of the pressure-transmitting fluid also changes after compression depending on its own thermophysical properties, such as thermal conductivity, viscosity, and specific heat, which influences the sample temperature. This phenomenon can introduce additional temperature gradients in the product (Denys et al. 2000a). Thus, the change in the temperature of the pressure transmitting fluid as a result of compression heating and subsequent heat transfer may be an important parameter for microbial inactivation.

The main difficulty in monitoring or modeling heat transfer in HPP is to properly account for the variation of thermophysical properties of food material under pressure. The determination of food properties under high pressure is a complex task. Otero et al. (2002) compiled a list of the thermophysical properties of liquid water and ice over a considerable range of pressures and temperatures. Denys et al. (2000b) estimated the density of apple sauce and tomato paste under high pressure by the displacement method. Denys et al. (2000a,b) evaluated the thermal expansion coefficient of various products such as apple sauce, tomato paste, and agar gel subjected to pressure up to 400 MPa. Denys and Hendrickx (1999) used the line heat source probe for determining the thermal conductivity of foods up to 400 MPa. Kubašek et al. (2006) estimated the thermal diffusivity of olive oil using the numerical analysis method.

Minerich and Labuza (2003) demonstrated that, during HPP, the sample at the geometric center of a large food product received a pressure lower than the pressure delivered by the processing system, which challenged the assumption that all foods follow the isostatic rule. This finding may have greater implications when determining the microbial lethality for large food items pasteurized or sterilized using high pressures.

Although the temperature of a homogeneous food increases uniformly due to compression, a variation in temperature gradient within the food can be developed during the holding period because of heat transfer between the food and the pressurizing fluid, as well as across the walls of the pressure vessel (Farkas and Hoover 2001). After pressurization, temperature of the product is normally greater than temperature of the pressure vessel. As a result, heat must be lost to the pressure-vessel wall, and the product regions close to these walls may not achieve the final temperature reached at the center of the food product (De Heij et al. 2001, Ting et al. 2002). It is important to note that the nonuniformity of the thermal-related process has a major impact primarily on pressure sterilization when a combination of elevated pressures and modest temperatures are used for the sterilization of low-acid foods.

The effectiveness of HPP is greatly influenced by the physical and mechanical properties of the packaging material. The packaging material must be able to withstand the operating pressures, have good sealing properties, and the ability to prevent quality deterioration during the application of pressure. The headspace must also be minimized while sealing the package to ensure efficient utilization of package as well as space within the pressure vessel. This also minimizes the time taken to reach the target pressure and avoids bursting of package during pressurization.

The film-barrier properties and structural characteristics of a polymer-based packaging material were unaffected when exposed to pressures up to 400 MPa (Nachamansion 1995). Dobias et al. (2004) demonstrated that HPP affected the sealability of single-layered films and the overall migration, whereas multilayered packaging was found to be more suitable in terms of mechanical properties, transparency, water vapor permeability, and migration characteristics. Schauwecker et al. (2002) demonstrated that there was no detectable migration of 1,2-propanediol through polyester/nylon/aluminum/polypropylene meal ready-to-eat type pouches. Caner et al. (2000) concluded that the water vapor transmission, as well as gas transmission, in the case of metalized polyethylene teraphthalate (PET), was most adversely affected by high pressure. Caner et al. (2004) reviewed the effects of HPP on barrier and mechanical properties of packaging films and suitable packaging materials for HPP. Lambert et al. (2000) observed that the package prepared by cast coextrusion was susceptible to delamination during HPP, whereas the packages prepared by tubular extrusion process were more robust in terms of barrier properties, migration, and overall integrity. Le Bail et al. (2006) studied the influence of HPP on mechanical and barrier properties of various packaging materials such as polyethylene, low-density polyethylene, polyamide/surlyn, and PET/biaxially oriented polyamide/polyethylene. The properties examined included maximum and rupture stress, strain at rupture, and water vapor permeability. HPP was found to have minimal effect on mechanical strength and water vapor barrier properties of the different packaging materials, although slight improvements in water vapor barrier properties were reported for low-density polyethylene.

It is often difficult to compare results of experiments produced in different laboratories due to variations in features and configurations of the equipment, as pointed out by Hugas et al. (2002). It is necessary to provide an adequate description of the equipment such as vessel size, chamber dimensions, material of construction, wall thickness, pressure-transmitting fluid, heating and cooling system, power specification, data acquisition system, and any other relevant information necessary to reproduce the results. It is important to document thermal conditions and temperature distribution within the processed sample volume (Balasubramaniam et al. 2004). Variation in parameters such as come-up, processing, and decompression times leads to different results. For instance, slow pressurization rates might result in induction of stress response, which may render the process less effective, whereas faster pressurization rates might result in higher inactivation rates. The coldest sample region is located near vessel walls or vessel closures, whereas temperature sensors are often located at the axial center of the pressure vessel. Thus, the location of the thermocouple sensor needs to be specified (Balasubramanian and Balasubramaniam 2003). The temperature of the sample, pressure vessel, and pressure-transmitting fluid can also affect the results and hence these parameters need to be reported in the study.

HPP can cause changes in the activity of enzymes, structure of proteins, hydrogen bonds, and hydrophobic and intermolecular interactions. Information relating to the effects of high pressures on generation of toxins, allergens, and nutrients is scarce. In the case of thermally processed foods, protein denaturation reduces the allergenicity of many foods, whereas in case of high-pressure-processed foods, it has to be established regarding putative allergenicity, because it is a key concern in the safety assessment of novel foods.

The development of methods and techniques to validate high-pressure pasteurization and sterilization may be challenging due to nonavailability of information on minimum temperature and time requirements for processing foods. It is important to establish microbiological criteria for safe production of foods by HPP.

1.3 PULSED ELECTRIC FIELD

PEF involves application of a short burst of high voltage to a food placed between two electrodes. Electric current flows only for microseconds through the food. In the traditional method, electrical energy is converted into thermal energy within the food, which causes microbial inactivation. When electrical energy is applied in the form of short pulses, bacterial cell membranes are destroyed by mechanical effects with no significant heating of the food. A high-voltage generator produces a high-voltage charge, which supercharges the process capacitor. The capacitor is then discharged through the food material by a switch releasing a pulse of duration in the microsecond to millisecond range in a treatment chamber between parallel electrodes (Rastogi 2003).

PEF technology has potential for economic and efficient energy use, as well as to provide consumers with microbiologically safe, minimally processed, nutritious, and fresh-like foods. Its potential applications include cold sterilization of liquid foods such as juices, cream soups, milk, and egg products.

PEF processing is a nonthermal technique which has been shown to inactivate microorganisms with minimum loss of flavor and food quality, potentially making it the answer to current consumer demands for products with high organoleptic and nutritional qualities. It offers almost fresh, minimally processed foods with a little loss of color, flavor, and nutrients. The low processing temperatures used in PEF render the process energy efficient, which translates into lower cost and fewer environmental impacts. Studies on energy requirements have concluded that PEF is an energy-efficient processing technique compared to thermal pasteurization particularly when a continuous system is used (Qin et al. 1995a,b). PEF has been mainly applied to preserve quality of foods, for example, to improve the shelf life of bread, milk, orange juice, apple juice, and liquid eggs as well as to improve the fermentation properties of brewer's yeast.

Exposure of microbial cells to an electric field for a few microseconds leads to electrical breakdown and change in the structure of the cell membrane, which results in a drastic increase in permeability. This nonthermal inactivation of microorganisms by PEF can prove to be beneficial for the development of preservation processes in the food industry that retain high food quality. Many researchers have demonstrated that PEF processing is a nonthermal way to maintain food safety by inactivating spoilage and pathogenic microorganisms (Sale and Hamilton 1967a,b, Mizuno and Hori 1988, Jayaram et al. 1992, Qin et al. 1994, Vega-Mercado et al. 1997). PEF can be used for pasteurization and possibly also sterilization, with the integration of other processing parameters such as pH, ionic strength, temperature, and HPP (Jeyamkondan et al. 1999). Moreover, it is conducted at ambient, subambient, or slightly above ambient temperature for less than a second, and the energy loss due to heating of food is minimal.

PEF is advantageous because the change in product color, flavor, and nutritive value is minimal (Dunn and Pearlman 1987, Jin and Zhang 1999). It is considered to be superior to the traditional heat treatment of foods because it avoids or greatly reduces changes in the sensory and physical properties of the food (Barbosa-Canovas et al. 2001). Available commercial food products processed by PEF are summarized in Table 1.3. The companies which can supply a wide variety of PEF equipment to the food industry are listed in Table 1.4.

1.3.1 Opportunities in Pulsed Electric Field Processing

PEF technology has experienced considerable success and holds further promise in a variety of applications. It is an exciting emerging technology that offers not only

TABLE 1.3
PEF-Processed Food Products Commercially Available Worldwide

Product	Name of the Company
Fruits juices (preservation)	M/s Genesis Juice Corp., Eugene, OR, United States
Ginger honey lemonade	
Strawberry honey lemonade	
Apple strawberry juice	
Fruit juices (cell disintegration)	M/s Beckers Bester, Eilsleben, Germany
Apple juice	

Sources: Clark, J.P., *Food Technol.*, 60(1), 66, 2006; Company Web sites; Günther, U. and Kern, M., Personal communication Prof. Stefan Toepfl, September 12, 2008.

TABLE 1.4
Main Suppliers of PEF Processing Equipment

Name of the Company	Affiliation	Specialization
M/s eL-Crack, Inc., Quakenbruck, Germany (http://www.elcrack.de)	German Institute of Food Technology, Quakenbruck, Germany	Manufactures equipment up to a capacity of 1500 L/h of feed Manufactures laboratory and industrial machines Peak voltage 30 kV 5 and 30 kW system Batch or continuous chamber
M/s Diversified Technologies, Inc., Ridgeland, MS (http://www.divtecs.com)	Ohio State University Research Foundation, Columbus, OH	Manufactures commercial units as well as a smaller R&D units Capacity between 1,000 and 20,000 L/h Peak voltage 65 kV Up to 75 kW system

Sources: Clark, J.P., *Food Technol.*, 60(1), 66, 2006; Company Web sites.

enhanced potential for preservation of food but can also be utilized for enhancing the rate of unit operations such as osmotic dehydration and conventional dehydration. Mostly, it has been successfully applied in pasteurization of selected fluid foods. Recent applications of PEF have shown that even nonfluid foods can be processed. It may be used to modify existing processes or to develop new, energy-efficient, environment-friendly technology options for the food and drink industry, as well as for pharmaceutical and biotechnological applications. The potential for continuous application and the short processing time makes it an attractive and novel nonthermal unit operation.

1.3.1.1 Pulsed Electric Field-Assisted Osmotic Dehydration

PEF treatment has been reported to increase the permeability of plant cells (Geulen et al. 1994, Knorr et al. 1994, Knorr and Angersbach 1998), which resulted in improved mass transfer during osmotic dehydration (Angersbach et al. 1997, Rastogi and Niranjan 1998). The effective diffusion coefficients of water and solute were found to increase exponentially with electric field strength. PEF-induced cell damage also resulted in tissue softening due to loss of turgor pressure, leading to a reduction in compressive strength. Taiwo et al. (2003) demonstrated that the cell membrane permeabilization increased with increasing field strength and higher pulse number, thus facilitating water loss during osmotic dehydration. Taiwo et al. (2001) also studied the effect of PEF treatment on the osmotic dehydration of apple slices. PEF treatment resulted in increased water loss, which was attributed to increased permeability of the cell membrane, whereas the effect on solid gain was minimal. Further drying of PEF-treated and osmotically dehydrated samples yielded good-quality products having firmer texture, brighter color, and better retention of vitamin C than samples that were either blanched or frozen. Tedjo et al. (2002) reported that PEF pretreatment resulted in higher moisture loss and solid gain in mangoes during subsequent osmotic dehydration.

Ade-Omowaye et al. (2003a,b) studied the influence of varying number of pulses during PEF on subsequent osmotic dehydration of fresh red bell peppers (*Capsicum annuum* L.). Significant differences in water loss of the samples subjected to 1 and 5 pulses were reported during osmotic dehydration, without a statistically significant difference beyond 5 pulses. Similarly, a steady increase in solid gain was also observed up to 5 pulses, but there was no difference observed in solid gain of the samples subjected to 10–50 pulses. Ade-Omowaye et al. (2002) indicated that the PEF pretreatment of bell peppers at varying field strengths was comparable to the pretreatment at elevated temperatures for osmotic dehydration, with the advantage of avoiding the excessive tissue softening or enzymatic browning associated with the latter. Additionally, the retention of ascorbic acid and carotenoids in the case of osmotic dehydration of PEF-pretreated pepper was higher. Amami et al. (2007a,b) concluded that PEF (0.60 kV/cm, time 0.05 s, 500 rectangular monopolar pulses each of 100 μs) enhanced the water loss and solid gain during osmotic dehydration (under stirring or centrifugation) of carrot tissue, as well as its rehydration capacity. However, the firmness of rehydrated product decreased. El-Belghitia et al. (2007) showed that exposure of carrot gratings to PEF (0.67 kV/cm, 300 pulses of 100 μs) followed by centrifugal separation resulted in an increased extraction of carrot solids.

Such results have created a lot of research interest toward exploring the potential of PEF as a pretreatment during osmotic dehydration of plant foods, and this technique can become a useful method to enhance osmotic water loss.

1.3.1.2 Pulsed Electric Field-Assisted Hot Air Drying

As PEF results in permeabilization of plant cells, it can improve the efficiency of the drying process of fruits or vegetables. This would not only increase the production capacity of an existing industrial plant without further investment but also improve the retention of nutritional components in the final dried product. There are many reports in the literature regarding the increased drying rates during air drying due to the prior application of PEF. Angersbach et al. (1997) demonstrated that a PEF treatment could reduce the drying time of potato cubes by approximately one-third. Ade-Omowaye et al. (2001a) verified that use of PEF reduced drying time (or improved drying rate) for coconut dehydration compared to that of untreated samples. The PEF-pretreated sample was reported to have improved mass and heat transfer coefficients compared to those concerning the blanched or chemical pretreated samples. A reduction of approximately 25% in the drying time for PEF-pretreated paprika was reported. It was suggested that PEF could be an alternative to the conventional chemical or thermal pretreatment, thus minimizing environmental pollution from applied chemicals and reducing leaching and thermal destruction of nutrients. Ade-Omowaye et al. (2001a) and Lebovka et al. (2007) showed that pretreatment of potato samples at 70°C did not have any beneficial effect on drying rate. On the other hand, pretreatment at 50°C increased the effective moisture diffusion coefficient, which was comparable to the one associated with the PEF-pretreated samples.

1.3.1.3 Pulsed Electric Field-Assisted Rehydration

The rehydration capacity is defined as maximum amount of water that the product is able to absorb upon immersion in water. PEF-pretreated samples subjected to air drying present a lower rehydration capacity, probably because of the greater sample shrinkage caused by the faster water loss during air drying, due to increasing membrane permeabilization (Tedjo et al. 2002). However, as reported by Taiwo et al. (2002), PEF-pretreated samples subjected to osmotic dehydration and finally air drying present a higher rehydration capacity. Coupling osmotic dehydration with PEF as a pretreatment, has the potential of improving the rehydration behavior of many air-dried foods. The enhanced rehydration capacity, in this case, has been attributed to the less compact structures due to absorption of sugar during osmotic dehydration.

1.3.1.4 Pulsed Electric Field-Assisted Preservation

It has been demonstrated that permeabilization of plant cells can be achieved at lower field strengths (up to 5 kV/cm), whereas a higher field strength is required to inactivate bacterial cells (~20 kV/cm) and enzymes (~50 kV/cm). This indicates that inactivation of microorganisms, as well as enzymes, is much more difficult than the permeabilization of plant cells. Consequently, PEF can be used as an alternative technology for traditional thermal pasteurization, because it can only inactivate the vegetative cells and spores appear to be resistant to it. High acid foods or foods with low water activity may be the best candidates for PEF processing. Generally, this technique does not seem to influence the flavor and taste of any product. The changes observed after PEF may usually be attributed to storage conditions or growth of microorganisms.

Apple juice treated with PEF (50kV/cm, 10 pulses of 2 μs, 45°C) was found to have a shelf life of 28 days as compared to 21 days for fresh-squeezed apple juice. There were no changes in ascorbic acid and sugars due to PEF treatment. A sensory panel found no significant differences between the fresh and PEF-treated juices (Ade-Omowaye et al. 2001c). PEF did not significantly affect the color (Evrendilek et al. 2000) or flavor (Harrison et al. 2001) of apple juice. Schilling et al. (2007) demonstrated that a PEF treatment (1, 3, 5kV/cm, 30 pulses) of apple mash increased juice yield. Overall composition as well as the nutritive value with respect to polyphenol contents and antioxidant capacities of the PEF-treated apple juice did not significantly differ from fresh apple juice.

Sitzmann (1995) reported that, in the case of orange juice, the natural microflora was reduced by 3-log cycles with an applied electric field of 15kV/cm without significant quality changes. Zhang et al. (1997) indicated that the square waveform was the most effective pulse for the processing of reconstituted orange juice. The aerobic counts were reduced by 3- to 4-log cycles at 32kV/cm and shelf life of juice stored at 4°C was reported to be about 5 months. The retention of vitamin C and color of PEF-treated juice was found to be better as compared to heat-treated juice. Moreover, PEF-treated orange juice had much better taste than heat-treated orange juice. Zhang et al. (1997) showed that PEF caused less change in the color of orange juice shortly after treatment and during the initial storage period (at 4°C), even though darkening of the color was observed during prolonged storage, which was attributed to conversion of ascorbic acid to furfural. The thermally pasteurized juice also showed the same effect. Yeom et al. (2000) indicated that PEF-treated orange juice showed less browning than the thermally pasteurized one stored at 4°C for up to 112 days. Esteve et al. (2001) reported that there was no adverse effect due to PEF on carotene content in orange–carrot juice. Cortes et al. (2008) observed less nonenzymatic browning in PEF-treated orange juice than in the thermally pasteurized one. In addition, there was no significant increase in browning index and hydroxymethylfurfural content of the juices pasteurized by PEF. Elez and Martin (2007) demonstrated that PEF-treated orange juice and gazpacho showed 87.5%–98.2% and 84.3%–97.1% retention of vitamin C, respectively, which is higher than that of the thermally pasteurized products. There was no difference in antioxidant capacity between PEF-treated and untreated products, whereas heat-treated foods showed lower values of antioxidant capacity. Cserhalmi et al. (2006) pointed out that the PEF (50 pulses at 28kV/cm) did not have any influence on pH, brix, electric conductivity, viscosity, nonenzymatic browning index, hydroxymethylfurfurol, color, organic acid content, and volatile flavor compounds of freshly squeezed citrus juices (grapefruit, lemon, orange, tangerine).

Raw skimmed milk treated with PEF (40kV/cm, 40 exponential pulses, 2 μs) had a shelf life of 14 days at 4°C, and the processing temperature did not exceed 28°C (Barbosa-Canovas et al. 2001). Qin et al. (1995b) reported that milk subjected to two stages of PEF processing (7 pulses and 6 pulses of 40kV/cm) achieved a shelf life of 2 weeks under refrigerated conditions. There was no apparent change in the physical and chemical properties, with no significant difference between thermally pasteurized and PEF-treated milk. Yogurt and yogurt-based products did not change color after the PEF treatment (Yeom et al. 2001).

The preservation of foods by PEF does not affect proteins under conditions required for destruction of vegetative bacteria (Barsotti et al. 2002), so it can be advantageously used for the preservation of eggs and liquid egg products. Qin et al. (1995b) indicated that the PEF treatment of whole liquid egg resulted in a reduction of viscosity and enhancement of color compared to fresh eggs. A sensory panel did not find any significant difference between scrambled eggs prepared from fresh and PEF-treated eggs. Barbosa-Canovas et al. (1999) also reported that color of beaten eggs became more orange. In another report (Gongora-Nieto et al. 2001) no influence of PEF on the color of whole liquid egg was found, which may probably be due to varied processing conditions.

Vega-Mercado et al. (1996) demonstrated that the application of PEF (two steps of 16 pulses each at 35 kV/cm) to pea soup enhanced its shelf life by 4 weeks, without any adverse effect on flavor, during storage at refrigeration temperature.

Min and Zhang (2003) concluded that PEF processed (40 kV/cm for 57 μs) tomato juice retained more flavor compounds than the thermally processed product. PEF-processed juice also had significantly lower nonenzymatic browning and higher redness. Moreover, sensory analysis indicated that the flavor of the PEF-processed juice was preferred. Odriozola et al. (2007) showed that the application of PEF (35 kV/cm, pulse width 1 μs for 1000 μs, 250 Hz in bipolar mode) may be appropriate to achieve nutritious tomato juice, because it resulted in maximum retention of lycopene (131.8%), vitamin C contents (90.2%), and antioxidant capacity (89.4%).

1.3.1.5 Pulsed Electric Field-Assisted Extraction

Even though solid–liquid extraction (pressing) of fruits and vegetables is an economical method of juice extraction that provides fresh-like juices, it is not sufficient to ensure rupture of all cells to obtain a high extraction yield. A large number of cells still remain intact during solid–liquid expression and the juice contained in these cells cannot be extracted. The application of PEF as a pretreatment before pressing allows significant increase in the juice yield and leads to products with higher quality. PEF can also be useful for the recovery of desired substances from plant cells without the use of chemical or thermal treatments. This procedure ensures not only an increase in the yields in terms of recovered solids but also a reduction in damage to several nutrients (vitamins, antioxidants) compared to traditional processing methods. PEF-assisted extraction from plant foods can be a real alternative to thermal extraction because it offers higher quality products without any thermal degradation. Moreover, electrical damage of cellular membranes of the majority of fruit and vegetable plants can be achieved at moderate electric fields (0.5–1.0 kV/cm). Thus, industrial implementation of PEF devices seems to be quite probable in the near future.

Dornenburg and Knorr (1993) demonstrated the usefulness of PEF (1.6 kV and 10 pulses) in the complete release of red pigment (amaranthin) from *Cheopodium rubrum* cells. The extent of pigment release was more sensitive to an increase in field strength than the pulse number. In addition, PEF treatment affected the yield of anthraquinone release from a *Morinda citrifolia* suspension.

Angersbach and Knorr (1997) showed that application of PEF led to the release of cell content due to electropermeabilization of cells. Flaumenbaum (1986) and McLellan et al. (1991) also reported an increase in juice yield (10%–12%) by subjecting apple mash to electroplasmolysis, and the product was lighter in color and less oxidized than in the case of the enzyme or heat-treated samples. Bazhal and Vorobiev (2000) demonstrated that the application of a PEF treatment to mechanically precompressed apple slices resulted in an increase in juice yield. Schilling et al. (2008) indicated the use of PEF as an alternative to enzymatic treatment for apple mash. PEF led to an enhanced release of nutritionally valuable phenols into the juice and the quality of pectin was retained, which allowed sustainable pomace utilization.

PEF treatment was found to increase the yield of carrot juice due to permeabilization of carrot cells (Knorr and Angersbach 1998). In the case of finely ground carrot (particle size <1.5 mm), the juice yield increased from 51.3% to 76.1% as compared to the control, whereas an increase from 30.0% to 70.3% was found with coarsely ground (<3.0 mm) carrots. Bouzrara and Vorobiev (2000, 2001) corroborated that the juice yield during solid–liquid expression of carrot slices after the first pressing (25.6%) could be increased up to 72.4% with the use of PEF (360 V/cm, 100 μs). Praporscic et al. (2007) demonstrated that a PEF treatment (250–400 V/cm) for juice expression from soft plant tissues resulted in enhanced juice yields, together with a noticeable decrease in cloudiness and increase of dry matter content for both apple and carrot juices. Chalermchat and Dejmek (2005) showed that PEF is an effective method for increasing pressing efficiency during extraction from potato. The treatment (0.68 kV/cm, duration of 1 ms) was found to lead to a fivefold reduction of the compressive force.

Bouzrara (2001) reported that, in the case of spinach, an intermediate PEF treatment (875 V/cm, 100 μs, 100 Hz for 5 s) during juice extraction at a pressure of 10 bar enhanced the juice yield from 30% to 60.6%. Pui et al. (2008) verified that the combination of PEF (580 V/cm) with vacuum infusion of trehalose in spinach leaves drastically improved their freezing tolerance. Yin et al. (2007) showed that a PEF treatment could increase the stability of chlorophyll by adding zinc ion and stabilizers. Spinach puree treated by PEF with acetate zinc below 75 ppm could be stored for a long time at room temperature.

Eshtiaghi and Knorr (1999) demonstrated that PEF could be applied as a processing step in the extraction of sugar beet juice. The PEF treatment enhanced the dry solids content of the pulp from 15.25% to 24.91%, which indicated its enormous potential in the sugar industry. Bouzrara and Vorobiev (2001) utilized PEF (427 V/cm, 500 pulses, 100 μs) as an intermediate treatment after mechanical precompression of sliced particles in the case of sugar beet and observed as increased juice yield up to 79%. Jemai and Vorobiev (2006) studied the feasibility of using PEF as an intermediate treatment to assist cold pressing of sugar beet. An increase in yield and purity of the sugar juice, as well as reduced losses, was verified as a result of PEF.

Gachovska et al. (2006) showed that PEF treatment increased extraction of juice from alfalfa mash by 38% as compared to untreated samples. Moreover, dry matter, protein and mineral contents of PEF-treated samples were significantly higher.

PEF can also be used as a powerful tool in the wet processing of coconut. Ade-Omowaye et al. (2001b) reported that PEF pretreatment (2.5 kV/cm, 20 pulses, 575 μs) enhanced the yield of coconut milk by 20%, along with 50% and 58% increase in protein and fat contents, respectively. The yield obtained was comparable to the one yield from freeze–thaw process, which is an energy-intensive process. The PEF treatment resulted in maximum permeabilization at low energy input and a shorter treatment time. Ade-Omowaye et al. (2001c) indicated that PEF treatment (1.7 kV/cm, 30 pulses) of red pepper resulted in higher yield and quality of juice as compared to those obtained by the pectolytic enzyme treatment. The redness and extracted carotene were higher for PEF-treated paprika.

1.3.2 Challenges in Pulsed Electric Field Processing

Before an industrial adoption of this emerging technique for food preservation, it is necessary to show that the process is an economically interesting alternative in comparison to existing technologies in terms of operational cost, investment, quality of product, and consumer acceptance. Some of the current limitations of PEF include limited availability of vendors who can supply industrial scale equipment and high initial equipment cost. Presently, PEF configurations are primarily optimized for fluid foods and further research may be needed to evaluate various PEF chamber configurations that can provide optimal solid product handling for subsequent dehydration or extraction experiments. Attention must be paid to potential safety problems due to the presence of entrapped air bubbles in the food matrix (that can cause dielectric breakdown during the treatment). Addressing these problems will make the PEF technology industrially competitive to the conventional thermal technology and will lead to its wider acceptance by the food industry. There is a tremendous innovative potential in utilizing the benefits and advantages of emerging technologies to develop new processes and products or to improve existing ones.

As detailed in Section 1.3.1, the potential of PEF as an alternative to conventional thermal processing of foods has been widely demonstrated. However, most of the reported studies have been conducted at the laboratory level, and differences between results from various researchers still exist. Hence, before this technique can be applied for industrial purposes, further research at the commercial level is necessary. Still, it is required to conduct substantial research and development activities to understand, optimize, and apply this complex technology to its full potential. Therefore, the PEF technique needs to be rigorously tested and proven to be safe while being commercially viable.

The existing juice industries may not be in a comfortable situation to adopt this new technology because they already have process lines with large capacities, which can be the one of the barriers. However, these may be overcome by the additional benefits offered by the technology in terms of quality and nutrition. If the objective of PEF is the permeabilization of plant cells to improve the efficiency of an existing unit operation, such as dehydration or extraction of fruits or vegetables, then the investments in this direction may be more justifiable. As the PEF treatment itself is fairly simple and requires little space, it may be attractive economically. Because of the relatively mild conditions (field strength between 200 and 500 V/mm) for plant cell permeabilization, the life time of electronic components used in the technology would most likely be very long, and the additional process costs are likely to be compensated by the reduction in dehydration time or cost.

In recent years few researchers have discussed the problems due to electrochemical reactions in the vicinity of electrode and medium interfaces, which triggers the need for seeking alternative electrode materials to replace stainless steel so as to control the electrochemical reactions.

It is very important to note that the systems and equipment used by the individual laboratories reporting PEF studies are often quite dissimilar and these differences are in most instances both important and fundamental. They vary over a broad range in chamber design features, and pulser and pulse characteristics. It is important to keep these differences in equipment and conditions in mind when comparing PEF results from individual laboratories; often they are so wide ranging as to make direct comparisons difficult. Researchers need to work together to harmonize processing and experimental setup. Furthermore, the chemical aspects of PEF should also be considered to demonstrate that the process is not harmful to food and food products. It is easy to convince the industry and authorities regarding the appropriateness of a new technology when the proposed process provides safety and stability from both microbiological and chemical points of view.

1.4 ULTRASOUND

When ultrasound travels through a medium, like any sound wave, it results in a series of compression and rarefaction. At sufficiently high power, the rarefaction may exceed the attractive forces between molecules in a liquid phase and, subsequently, result in the formation of cavitation bubbles. Each bubble affects the localized field experienced by neighboring bubbles. Under such situations, the irregular field causes the cavitation bubble to become unstable and collapse, thereby releasing energy for chemical and mechanical effects. For example, in aqueous systems, at an ultrasonic frequency of 20 kHz, the collapse of each cavitation bubble acts as a localized "hot spot," generating enough energy to increase the temperature to about 4000 K and the pressure to values higher than 1000 atm. This bubble collapse, distributed through the medium, has a variety of effects within the system depending upon the type of material involved.

The cavitation collapse can occur in the bulk liquid immediately surrounding the bubble where the rapid bubble collapse generates shear forces that can produce mechanical effects. It can also occur in the bubble itself where any species introduced during its formation will be subjected to extreme conditions of temperature and pressure on collapse, also leading to chemical effects.

Unlike cavitation bubble collapse in the bulk liquid, collapse of a cavitation bubble on or near a surface is asymmetrical because the surface provides resistance to liquid flow from that side, which result in an inrush of liquid predominantly from the side of the bubble remotes from the surface, leading to the formation of a powerful liquid jet targeted at the surface. The effect is equivalent to a high-pressure jetting and is the reason why ultrasound is used for cleaning. This effect can also increase mass and heat transfer to the surface by disruption of the interfacial boundary layers.

Cavitation bubble collapse in the liquid phase near a particle can force it into rapid motion. Under these circumstances the general dispersive effect is accompanied by interparticle collisions that can lead to erosion, surface cleaning, and wetting of the particles, as well as particle size reduction.

Ultrasound produces very rapid localized changes in pressure and temperature that cause shear disruption, cavitation, thinning of cell membranes, localized heating, and free radical production, which have a lethal effect on microorganisms. Details of cavitation are given by Leighton (1998) and the physics of ultrasound are described by Suslick (1988). A review of the effects of ultrasound on microorganisms is given by Sala et al. (1995). McClements (1995), Mason (2000), and Mason et al. (2005) outline the increasing number of industrial processes that use power ultrasound as a processing aid including the mixing of materials, foam formation or destruction, agglomeration and precipitation of airborne powders, the improvement in efficiency of filtration, drying and extraction techniques in solid materials, and the enhanced extraction of valuable compounds from vegetables and food products.

Different authors have investigated the inactivation effect of combining ultrasound and heat treatments. It has been observed that the thermoresistance of several bacterial spores such as *Bacillus cereus*, *B. licheniformis*, or *B. stearothermophilus* and thermoduric *Streptococci* decreased when heat treatment followed ultrasound treatment at 20 kHz (Burgos et al. 1972, Ordonez et al. 1984, Sanz et al. 1985). This effect was more pronounced when heat and ultrasound were applied simultaneously (Ordonez et al. 1987, Garcıa et al. 1989, Guerrero et al. 2001). Simultaneous application of ultrasound under pressure alone (manosonication) or in combination with pressure and thermal treatment (manothermosonication) led to increased microbial inactivation. A synergistic and lethal effect of ultrasound treatment in combination with pressure and thermal treatment was shown in the case of *Enterococcus faecium* and *B. subtilis* spores (Raso et al. 1998a,b, Pagan et al. 1999a). This synergistic effect in case of *B. subtilis* spores was attributed to the permeabilization of the outer membranes of bacterial spores by the combined treatment of pressure and ultrasound (Raso et al. 1998b).

The shearing and compression effects of ultrasound cause denaturation of proteins that result in reduced enzyme activity. However, short bursts of ultrasound may increase enzyme activity, possibly by breaking down large molecular structures and making the enzymes more accessible for reactions with substrates. The effects of ultrasound on meat proteins produced tenderization in meat tissues after prolonged exposure, and the release of myofibrillar proteins in meat products results in improved water-binding capacity, tenderness, and cohesiveness (McClements 1995). Since the resistance of most microorganisms and enzymes to ultrasound is too high, it requires higher intensity of ultrasound treatment that may produce adverse changes to the texture and other physical properties of the food, thereby affecting the sensory attributes. This may render ultrasound not very useful in case of food preservation. However, this technique may find its use in many other food-processing applications.

The applications of ultrasound in the food industry can be divided into two distinct categories, depending on the adopted ultrasound intensity. Low-intensity ultrasound uses power levels less than 1 W/cm^2, which are so small that the ultrasonic waves cause no physical or chemical alterations in the properties of the material through which it passes. It is referred to as nondestructive use of ultrasound. The most common application of low-intensity, nondestructive ultrasound is as an analytical technique for providing information about the physicochemical properties of foods, such as composition, structure, and physical state. In addition, operating conditions such as flow rate also can be monitored. In contrast, the power levels used in high intensity applications are so large (in the range from 10 to 1000 W/cm^2) that they cause physical disruption of the material to which they are applied, or promote certain chemical reactions (e.g., oxidation). High-intensity ultrasound can be used for the inactivation of microorganisms and enzymes, generation of emulsions, tenderization of meat, and enhancement of drying, extraction, and filtration operations.

1.4.1 Opportunities in Ultrasound Processing

1.4.1.1 Ultrasound-Assisted Inactivation of Microorganisms and Enzymes

Ultrasound-assisted inactivation of microorganisms and enzymes results in the extension of shelf life of raw materials or prepared foods. The effect of ultrasound on different microbial species is known to depend on the shape and size of the microorganisms (bigger cells being more sensitive than smaller ones, and coccal forms are more resistant than rod-shaped bacteria), type of cells (gram positive being more resistant than gram negative, aerobic being more resistant than anaerobic), and physiological state (younger cells being more sensitive than older ones, spores being much more resistant than vegetative cells) (Piyasena et al. 2003).

Ordonez et al. (1987) indicated a reduction in the effectiveness of ultrasound on inactivation of bacteria at elevated temperatures (50°C–60°C). Garcia et al. (1989) suggested that this efficiency loss could be due to the elevation of vapor pressure in the sonicated medium that impairs or reduces the intensity of cavitational collapse, which was avoided by increasing the applied pressure of the sonicated medium (Burgos et al. 1992) and termed as manothermosonication. The potential of ultrasound to inactivate emerging pathogenic microorganisms such as *L. monocytogenes*, a number of strains of *Salmonella* spp., *E. coli*, or *S. aureus*, which are increasingly found in outbreaks of food poisoning, has been verified (Sala et al., 1995, Pagan et al. 1999a,b). Ferrante et al. (2007) demonstrated that the control of *L. monocytogenes* in orange juice could be achieved by combining high intensity ultrasound with mild heat treatment and natural antimicrobials.

Inactivation of *Saccharomyces* with the combination of heat and ultrasound has been found to be almost independent of pH (Guerrero et al. 2001). Guerrero et al. (2005) indicated inactivation of *S. cerevisiae* could be enhanced by incubating with low molecular weight chitosan prior to ultrasound.

A continuous ultrasound system in combination with steam injection has shown up to fourfold higher inactivation rates of *E. coli* and *Lactobacillus acidophilus* in several liquid foods such as milk and fruit juices (Zenker et al. 2003). Ugarte et al. (2006) explored the use of acoustic energy to secure apple juice safety. Sonication was found to increase *E. coli* K12 cell destruction by 5.3-log, 5.0-log, and 0.1-log cycles at 40°C, 50°C, and 60°C, respectively. D'Amico et al. (2006) pointed out that ultrasound treatments with or without mild heating were effective in reducing *L. monocytogenes* in raw milk and *E. coli* O157:H7 in apple cider. Continuous flow ultrasound treatment combined with mild heat (57°C) for 18 min resulted in a 5-log reduction of *L. monocytogenes* in milk, a 5-log reduction in total aerobic bacteria in raw milk and a 6-log reduction in *E. coli* O157:H7 in pasteurized apple cider. Seymour et al. (2002) reported microbial decontamination on the surface of minimally processed fruits and vegetables (lettuce, cucumber, carrots, parsley, and others) by ultrasound. The microorganisms were subsequently killed by the use of chemical sanitizers such as chlorine. Scouten and Beuchat (2002) indicated the decontamination of alfalfa seeds inoculated with *Salmonella* or *E. coli* O157 by combined treatments of ultrasound and $Ca(OH)_2$, which could be an alternative to chlorine treatments to avoid contamination. The combination of ultrasound with chlorine treatment was found to result in 4-log reduction of *Salmonella* on poultry surfaces (Lillard 1994), whereas Sams and Feria (1991) found that there was no decrease in total aerobic count from broiler drumsticks sonicated with and without heat.

The use of ultrasound at ambient pressure has also been successful to inactivate food-quality-related enzymes. POD was inactivated by combinations of heat and ultrasound at neutral (Gennaro et al. 1999) or low pH (Yoon et al. 2000) and lipoxygenase was inactivated at low sonication intensities (Thakur and Nelson 1997). Villamiel and Jong (2000) reported no effect of ultrasound on endogenous milk enzymes such as alkaline phosphatase, g-glutamyltranspeptidase and lactoperoxidase at room temperature, but synergistic inactivation was obtained at higher temperatures (60°C–75°C). Lopez et al. (1994) demonstrated that the monothermosonication treatment was much more efficient than heat treatment alone for inactivating enzymes such as lipoxygenase, POD, and PPO in buffer systems. Proteases and lipases are the contributing factors for the shelf life of ultraheat-treated milk. These enzymes were inactivated up to 10 times faster by monothermosonication treatment (Vercet et al. 1997). Pectin methylesterase from orange, a thermostable enzyme, was inactivated almost 500 times faster by monothermosonication treatment than by heat treatment at the same temperature (Vercet et al. 1999). Pectin methylesterase and polygalacturonase in tomato juice were also inactivated by monothermosonication treatment with higher efficiency (Vercet et al. 2002).

The combination of ultrasound and pressure and/or heat shows considerable promise for the inactivation of microorganisms and enzymes. Therefore, techniques like thermosonication, manosonication, and manothermosonication may be of more relevance in the future as an energy-efficient processing alternative for the food industry.

1.4.1.2 Ultrasound-Assisted Drying

The use of ultrasound in drying is a very lucrative option because it can act without affecting the main characteristics and quality of the product. Heat-sensitive foods can be dehydrated more rapidly and at a lower temperature when ultrasound is used in combination with hot air drying.

Gallego-Juarez et al. (2007) indicated that high intensity ultrasound in combination with hot air systems resulted in adequate drying rates for vegetable drying even at lower temperatures. Fernandes and Rodrigues (2007) and Rodrigues and Fernandes (2007) showed that ultrasonic treatment prior to air drying of banana and melon (*Curcumis melo*) led to an increase in water effective diffusivity (of about 11%)

during air drying, which translated into a reduction in the drying time (of about 25%). During the ultrasonic pretreatment, banana and melon were found to lose sugar, it was suggested that the pretreatment could be used to produce dried fruits with lower sugar contents. Garcia-Perez et al. (2007) demonstrated that ultrasound not only enhanced the mass transfer during drying of carrot, persimmon, and lemon peel, but also improved product quality, since it did not significantly heat the material. The rate of drying was found to depend on air velocity and raw material characteristics. Carcel et al. (2007a) also reported that the use of high-intensity ultrasound increased the drying rate of persimmon. Jambrak et al. (2007) indicated that the use of ultrasound as a treatment method prior to drying of mushrooms, Brussels sprouts, and cauliflower was helpful in reducing the drying time. The rehydration studies showed that the percentage rehydration for ultrasound-treated samples was higher than that of untreated samples. The potential benefits due to the application of ultrasound regarding enhancement of rate of mass transfer in case of dehydration and osmotic dehydration have been summarized by Mason et al. (2005).

1.4.1.3 Ultrasound-Assisted Osmotic Dehydration

Osmotic dehydration is widely used for partial removal of water from food materials by immersion in a hypertonic solution. It is generally a slow operation, and additional ways are needed to increase mass transfer without adversely affecting product quality. A classical way to increase mass transfer rates is the use of mechanical stirring. Another possibility could be the use of power ultrasound to enhance rate of mass transfer due to acoustic streaming, which is the net movement of fluids in the presence of high-intensity ultrasonic fields. Simal et al. (1998) have observed that the application of ultrasound to osmotic dehydration of porous fruit, such as apple cubes, accelerates mass transfer rates. Ultrasonic osmotic dehydration technology can be carried out at lower solution temperature to obtain higher rate of water loss and solute gain, while preserving the natural flavor, color, and heat-sensitive nutritive components.

Carcel et al. (2007a) showed that, in the case of apple, the ultrasonic treatments increased water as well as solute diffusivity up to 117% and 137%, respectively. Stojanovic and Silva (2007) indicated that application of high-frequency ultrasound (850 kHz) on osmotic dehydration of rabbiteye blueberries increased the water diffusion rate, but it resulted in loss of anthocyanins and phenolics.

Fernandes et al. (2008) evaluated the effect of ultrasound pretreatment applied at atmospheric pressure during osmotic dehydration of melon. Osmotic dehydration for less than 30 min resulted in a decrease in water diffusivity due to incorporation of sugar, whereas water diffusivity increased when the osmotic treatment was carried out for more than 1 h, due to the breakdown of cells, which lowered the resistance to water diffusion. The ultrasound treatment increased water diffusivity owing to the formation of microscopic channels, which lowered the resistance to water diffusion. Yun and Yanyun (2008) investigated the influence of pulsed-vacuum and ultrasound on the osmodehydration kinetics and microstructure of apples (Fuji). Pulsed-vacuum treatment resulted in less loss of firmness and higher amount of solid gain, whereas ultrasound pretreatment resulted in the highest water and firmness loss. Analysis of microstructure using scanning electron microscopy confirmed the severe cell deformation and structure collapse in ultrasound-treated samples in comparison to pulsed-vacuum-treated samples.

Gabaldon-Leyva et al. (2007) indicated that application of ultrasound resulted in increased loss of water and uptake of soluble solids during brining of bell pepper. The increase was attributed to the increased cell wall permeability, which facilitated the transport of water and solute. Sanchez et al. (1999) demonstrated that rate of water removal and sodium chloride gain increased when ultrasound was applied during brining of cheese. Carcel et al. (2007b) demonstrated that ultrasound was found to influence the moisture and sodium chloride contents of samples of pork loin (longissimus dorsi) during brining. The final moisture content was significantly higher than initial one, whereas the sodium chloride content was proportional to applied ultrasonic intensity.

1.4.1.4 Ultrasound-Assisted Extraction

Ultrasound is a promising method to assist in the extraction of valuable compounds from food products. It is attributed to the propagation of ultrasound pressure waves, which result in cavitation phenomena. It is particularly useful in combination with conventional solvent extraction. The beneficial effects of ultrasound are derived from its mechanical effects on the process by increasing penetration of solvent into the product and enhancing the mass transfer process to and from interfaces. Such benefits are supposed to be related to enhancement of cellular contents diffusion through disruption of the cell walls produced by acoustical cavitation (Chendke and Fogler 1975). Furthermore, the ultrasound-assisted extraction is achieved at lower temperatures, which are more favorable for thermally unstable compounds (Wu et al. 2001).

Bing et al. (2006) have shown that the application of ultrasound (0.15 W/cm^2) increased extraction yield of Geniposide up to 16.5% from Gardenia fruit. Albu et al. (2004) demonstrated that application of ultrasound improved the performance of ethanol on the extraction of carnosic acid from rosemary. Ultrasound-assisted extraction of ginsenosides (tri-terpene saponins) from ginseng roots was found to be approximately three times faster than the traditional extraction method (Tang and Eisenbrand 1992). Similarly, extraction of carvone and limonene from caraway seeds was doubled (Chemat et al. 2004). Hemwimol et al. (2006) verified that ultrasound-assisted extraction of anthraquinones from roots of *M. citrifolia* in an ethanol–water system with a 75% reduction in extraction time and yield comparable to the control sample.

The combination of ultrasound and supercritical carbon dioxide extraction was a feasible option for an improvement in extraction rate or yield of amaranth oil from seeds (Bruni et al. 2002), almond oil (Riera et al. 2004), tea seed oil (Rajaei et al. 2005), and gingerols from ginger (Balachandran et al. 2006).

This potential of ultrasound to increase yield during extraction of edible oil was also demonstrated in the case of soybean (Haizhou et al. 2004, Babaei et al. 2006). Similarly, higher yields of carvone and limonene from caraway seeds were also reported by Chemat et al. (2004).

Vilkhu et al. (2008) demonstrated that ultrasound extraction of grape marc (a solid waste of the winemaking industry) resulted in up to 35% increase in phenolic compounds. However, extraction of these compounds yielded much higher recovery from their respective seeds, which provides the characteristic color and flavor to wine.

Jing et al. (2008) optimized the extraction of health-promoting phenolic compounds from wheat bran. An ethanol concentration of 64% (w/v), an extraction temperature of 60°C, and an extraction time equal to 25 min resulted in a release of 3.12 mg gallic acid equivalents per gram of phenolic content. Rodrigues and Pinto (2007) indicated that high amounts of phenolic compounds could be obtained from coconut shell by ultrasound-assisted technology. This technique also resulted in enhanced content of tea polyphenols, amino acids, and caffeine in tea, as well as improved sensory quality of tea in comparison with those obtained by conventional extraction (Xia et al. 2006).

Ultrasound-assisted extraction of crushed grapes increased the total anthocyanins content by 15%–18% in grape juice (Vilkhu et al. 2008). A combination of microwave and ultrasound treatments was also found to increase the extraction of pigments from strawberries (Cai et al. 2003). Xiaohua et al. (2006) demonstrated that ultrasound-assisted solvent extraction was more effective for extraction of lutein from egg yolk compared to traditional solvent extraction.

Cocito et al. (1995) verified that ultrasound could be used to extract aroma compounds which impart flavor to wines. Ultrasound-assisted extraction of ground soybeans was found to increase the extraction of isoflavones by 15% (Rostagno et al. 2003). Furuki et al. (2003) showed a rapid and selective extraction of phycocyanin from *Spirulina platensis* using ultrasound. Phycocyanin with higher purity was extracted at 28 kHz in its crude extract, due to the selective extraction of active component at this frequency.

The use of ultrasound was reported to accelerate the efficiency of extraction and preserve structural and molecular properties of hemicellulose from buckwheat hulls (Hromadkova and Ebringerova 2003), cellulose from sugarcane bagasse (Sun et al. 2004), and xyloglucan from apple pomace (Caili et al. 2006). Extraction of hemicelluloses from sugarcane bagasse was improved by destruction of cell walls and cleavage of links between lignin and hemicelluloses (Jing et al. 2004). Ebringerova et al. (1998) indicated that water-soluble xylans extracted from corn cob waste using ultrasound had a similar primary structure to classically extracted xylans, but had slightly higher immunostimulatory activities. Xu et al. (2000) demonstrated that optimum extraction of flavonoids from bamboo leaves could be performed at lower temperature in combination with ultrasound, rather than using hot water bath extraction at 80°C. Jacques et al. (2007) pointed out that the ultrasonic treatment resulted in improved extraction of caffeine and palmitic acid from leaves of *Ilex paraguariensis*.

Entezari et al. (2004) successfully optimized ultrasonic operating conditions in laboratory trials of date syrup extract, which led to a higher extraction in a shorter time with reduced microbial count. Application of ultrasound treatment to corn in the conventional wet-milling process enhanced starch separation, providing an increase in final starch yield from 6.35% to 7.02%. In addition, ultrasound-treated starches exhibited higher paste viscosities and whiteness (Zhang et al. 2005).

1.4.1.5 Ultrasound-Assisted Detection of Foreign Bodies

It is naturally desirable by the food industry that all foreign bodies such as any pieces of solid matter, including metal, glass, stone, and plastic are detected and removed before the product reaches the consumer. Using ultrasound to detect foreign bodies in a bottle poses a challenging task for signal processing. When glass fragments settle down on the bottom or are stuck to the wall of a bottle, ultrasound signals returned from the fragments will superimpose onto the echoes from the inner surface of the bottle. This superposition makes it impossible to distinguish the signals by spectrum analysis.

Zhao et al. (2004) developed a foreign body detector based on the ultrasound pulse/echo method in which a foreign body is detected by examining the amplitude ratios between the echoes from the container's outer and inner surfaces. Zhao et al. (2006) developed a method to detect glass fragments behind glass walls by combining radial basis function neural networks with short-time Fourier transform for ultrasound signal classification. Zhao et al. (2007a) demonstrated that ultrasound backscattered signals for object detection could be too weak to be perceived when superimposed to strong reflection signals, and could also be complicated either in time or frequency domain. These peculiarities raise a challenge for signal-processing methods. The root mean squares method in combination with other methods was found to be the best option for successful object detection. Zhao et al. (2007b) further developed a method involving maximum component integration based on the short-time Fourier transform algorithm to detect small objects in containers. The proposed scheme was able to make selective and full use of multiecho information and demonstrated improved detection ability that it could detect small glass fragments contained inside glass containers.

Haeggstrom and Luukkala (2001) used an ultrasound reflection technique with an echo classifier to detect and identify foreign bodies such as wooden cubes, stone glass, and spheres of wood, plastic, bone, and steel in commercial food samples (processed cheese, margarine, and cherry marmalades). It was concluded that detection of selected foreign bodies was possible in homogeneous food products at 20–75 mm probing depth, but nonhomogeneous food products restricted the probing depth to 50 mm due to poor signal-to-noise ratio.

Cho and Irudayaraj (2003a,b) developed a noncontact type ultrasound imaging technique for detection of foreign objects in cheese and poultry. The results indicated that the technique provided a rapid, nondestructive detection of impurities, foreign objects, and defects in the selected food materials.

Lagrain et al. (2006) analyzed two different bread types, a fine-grain bread with small cells and a coarse-grain bread with larger gas cells, using both image analysis and ultrasonic techniques. The ultrasonic technique used noncontact air coupled transducers at ultrasonic frequencies. The phase velocity and attenuation of ultrasonic waves were measured in transmission and reflection experiments. These measurements confirmed the structural differences between the two different bread types and the usefulness of the Biot–Allard model for description of wave propagation in bread crumbs for a wide frequency range. In addition, the use of noncontact ultrasound allowed estimating flow resistivity, open porosity, a measure for the size of intersections in the crumb cell walls, and tortuosity, which could be considered as a structural form factor.

1.4.1.6 Ultrasound-Assisted Filtration

The removal of suspended or dissolved solids is generally accomplished using conventional type of filters consisting of filter medium or semipermeable membrane.

These conventional and membrane filters are susceptible to clogging. As a result, flux rate goes down, and then these filter mediums must be either replaced, stopping the operation, or cleaned on periodic basis. Ultrasound-assisted filtration has been demonstrated to operate more efficiently and for much longer periods without maintenance. This occurs because sonication results in agglomeration of fine particles and also supplies sufficient vibrational energy to keep particles suspended, hence leaving more free channels for liquid flow.

Fairbanks and Chen (1971) verified that ultrasound-assisted filtration increased the filtration rate of motor oil through a sandstone filter by 18 times. There have been a number of developments in acoustic filtration and separation processes (Tarleton and Wakeman 1997). One example is the application of an electrical potential across the slurry mixture while acoustic filtration is performed (Senapati 1991). The filter itself is made the cathode, while the anode on the top of the slurry, functions as a source of attraction for the predominantly negatively charged particulate material. Such technique is applied, for instance, in the dewatering of coal slurry (50 wt% moisture content). Conventional filtration reduces the moisture to 40%wt, whereas using ultrasound this was improved to 25%wt, a value that was further reduced to 15%wt by electroacoustic filtration. The potential for this technique is clearly enormous when applied to a continuous belt drying process in the deliquoring of such extremes as sewage sludge or fruit pulps.

1.4.1.7 Ultrasound-Assisted Freezing

Cavitation results in the occurrence of microstreaming, which is able to enhance heat and mass transfer accompanying the freezing operation. Cavitation also helps to favor nucleation and to increase its rate since the gas bubbles produced can act as nucleating agents. Crystal fragmentation is another significant acoustic phenomenon, which can lead to crystal size reduction, which is one of the most important aspects that many freezing processes target. These effects of ultrasound have been proven useful in assisting food freezing. Ultrasound can be utilized to induce nucleation and to control crystal size distribution in frozen products during solidification of fluid food. Not only can it increase the freezing rate, but it also improves the quality of the freezing product. Application of ultrasound can also benefit prevention of incrustation on the freezing surface. The ability of power ultrasound in performing these functions is affected by a number of parameters, such as the duration, intensity, or frequency of ultrasonic waves. Zheng and Sun (2005, 2006) have reviewed the application of ultrasonic waves in freezing including initiation of ice nucleation, control of crystal size distribution, prevention of incrustation on freezing surface, improvement in product quality and still exiting problems.

1.4.1.8 Miscellaneous Opportunities in Ultrasound Processing

The ultrasonic analysis of foods can be done by establishing relationships between their physicochemical properties (e.g., composition, structure, and physical state) and the measurable ultrasonic properties (velocity, attenuation coefficient, and impedance). Ultrasound has been used to measure a wide variety of different properties of foods such as sugar concentration of aqueous solutions, salt concentration of brine, triacylglycerols in oils, droplet concentration of emulsions, alcohol content of beverages, air bubbles in aerated foods, composition of milk, ratio of fat to lean in meals, and biopolymer concentration in gels (McClements 1995).

Thickness of materials can also be measured using ultrasonic devices. This is particularly interesting in the case for which thickness is difficult to measure using conventional techniques, as occurs, for instance, with pipes, chocolate layers on confectionery, fat or lean tissue in meat, and liquids in can and eggshells.

A number of ultrasonic devices have been developed that can be used to measure the speed at which a food material flows through a pipe. Ultrasonic flowmeters are suitable for determining flow rates of up to a few meters per second in systems with dimensions ranging from a few millimeters (e.g., the flow of blood in a vein) to more than a kilometer (e.g., the flow of water in rivers or oceans).

Ultrasound can be used to determine particle sizes in emulsions or suspensions in a manner that is analogous to light scattering. An ultrasonic wave incident upon an ensemble of particles is scattered by an amount that depends on the size and concentration. The ultrasonic velocity and attenuation coefficient both depend on the degree of scattering, and can therefore be used to provide information about particle size.

The ultrasonic properties of a material change significantly when it melts or crystallizes, and hence ultrasound can be used to monitor these phase transitions. One of the most commonly used methods is to measure changes in the ultrasonic velocity with time or temperature. The ultrasonic velocity in solids is significantly greater than that in liquids; thus, the ultrasonic velocity in a sample increases when a component crystallizes, and decreases when it melts. Ultrasonic velocity measurements have been used to monitor crystallization and melting behavior in a variety of bulk and emulsified food fats, including margarine, butter, meat, and shortening.

Canselier et al. (2002) have demonstrated that ultrasound can be used as a means of mixing materials in many industrial processes. Products such as tomato ketchup or soups can be subjected to a liquid vessel in which mixture is passed through a narrow opening where the ultrasonic vibrations are generated due to liquid flow. The treatment of oil-in-water emulsions by ultrasound produces much smaller drop sizes than mechanical agitation does under the same conditions, which leads to more stable emulsions. In addition, for a given drop size, less surfactant is required (Canselier et al. 2002).

Mongenot et al. (2000) studied the effect of ultrasound emulsification on encapsulation of liquid cheese aroma in different carbohydrate matrices by spray drying. The use of ultrasound resulted in a lower microcapsule size and higher aroma retention than those obtained by the mechanical mixing. Mahdi et al. (2007) indicated that emulsions with droplet size below 0.5 μm could be produced by ultrasound in combination with microfluidization technique. Otherwise, it was possible to obtain 1.0 μm droplet size by microfluidizer alone. For ultrasound emulsification, increasing the energy input through improving sonication time helped to reduce droplet size with minimum recoalescence of new droplets.

Lin and Chen (2007) have demonstrated that the use of ultrasound prevents fouling of milk in a concentric pipe heat exchanger over 2 h of heating. For similar heating conditions and in the absence of ultrasound, serious fouling developed within 1 h only. Ultrasound reduces the solid–liquid interface temperature for the same heating duty, leading to considerable reduction in fouling. The ultrasonically induced movement of the depositing species in near the wall region is also thought to be responsible for fouling reduction, as it does not allow the molecules to stay at the wall longer than the time needed for them to form a firm deposit. Lu et al. (2005) have indicated that application of ultrasound in the case of evaporation of sugar juice using a tubular evaporator markedly reduced scaling and increases cleaning speed by softening the scale layer. Laboratory tests have shown that the inhibition of scale formation was due to effect on crystal size and morphology.

1.4.2 Challenges in Ultrasound Processing

The potential of ultrasound as a tool in food processing has been proven at the laboratory scale, and there are a number of examples of scale up. Most of the studies are based on the frequency of ultrasound available commercially (20 or 40 kHz). Attention should be focused for the use of different frequencies with varying parameters such as temperature, treatment time, and acoustic power.

Ultrasound is certainly not applicable for characterizing all food materials. However, there are many foods for which the technique has considerable advantages over existing technologies. It is a rapid, precise, nondestructive, and noninvasive technique that can be applied to a concentrated or optically opaque system. In addition, it can easily be adapted for online measurements, which would prove useful for monitoring food processing operations. The application of ultrasound techniques has been proven to be more fruitful for the monitoring the concentration of aqueous solutions and suspensions, determining droplet size and concentration in emulsions, monitoring crystallization in fats, and monitoring creaming profiles in emulsions and suspensions; in particular, for online determination of these properties during processing.

The presence of small gas bubbles in a sample can attenuate ultrasound so much that sometimes an ultrasound wave cannot propagate through the sample. This problem can be overcome by taking reflection rather than transmission measurements, even though the signal from the bubbles may interfere with other components. A lot of information about the thermophysical properties such as densities, compressibilities, heat capacities, and thermal conductivities of a material is needed to make theoretical predictions of its ultrasonic properties. Theoretical analyses of the data from systems containing many components with unknown properties are therefore scanty.

Low-intensity ultrasound is somewhat inexpensive and finds its use in the food industry. The usefulness of high-intensity ultrasound for modifying certain physical and chemical properties of foods has been realized for many years. Nevertheless, it was only very recently that manufacturers have begun to adapt laboratory-scale equipment for large-scale processing operations. The increasing use of high-intensity ultrasound depends largely on the availability of low-cost instrumentation that is proven to have significant advantages over alternative technologies.

Many foods such as plant tissues, aerated foods, and some semicrystalline fats (chocolates) have a very high level of attenuation, which can make measurement extremely difficult. The use of a shorter path length may not be feasible in a real process due to cleaning, fouling, and other practical restrictions. The use of lower frequency reduces the spatial resolution. In some cases, there are number of sample variables changing simultaneously, and that will affect the ultrasonic properties. In this situation, the simple sensor may not be enough, resulting in broad and difficult to resolve peaks. If it is difficult to get very precise and uniform temperature control throughout the sample, additional errors in further measuring the property based on temperature may be introduced. The presence of air in the sample results in huge impedance mismatch between gas bubbles and other food materials, which causes reflection by air bubbles and a very strong scattering. Ultrasound can thus be used as a technique for detecting included air, which is not otherwise readily visible.

1.5 CONCLUDING REMARKS

Destruction of microorganisms and inactivation of enzymes at low or moderate temperatures without changing organoleptic and nutritional properties show that the nonthermal technologies discussed here have a high potential to be used in the development of new generation value-added foods. Even if these technologies are not likely to completely replace traditional processing methods, they can certainly complement or be integrated with the existing ones. Nevertheless, their improved physicochemical and sensory properties imparted to foods offer emerging and exciting opportunities for the food industry. These technologies can be integrated to other unit operations such as blanching, drying, osmotic dehydration, rehydration, frying, extraction, freezing, and thawing, for improving food processing.

New nonthermal technologies are under investigation because of their excessive potential to inactivate microorganisms with little harm to the product. New methods regrettably tend to require investments. High capital expenditure may limit their application initially, but this will be offset by higher product quality as well as by selection of appropriate economies of scale. The progress of each technology and its commercialization will further result in the reduction of equipment cost in the near future, and safe and nutritious products will be available to all consumers at an affordable cost. It is important to note that to increase the changes that consumers will buy the improved product and industry will consider investing in new technologies, issues such as product safety, high retention of nutrition, regulatory acceptance, occupational safety, investment costs, and energy costs should be addressed for each technology.

The results discussed in this chapter should provide an encouragement to future scientists to begin more comprehensive research and development leading to the introduction of newer, safer, energy efficient, and economic techniques that can result in the development of industrially significant food processing unit operations.

ACKNOWLEDGMENTS

The author is grateful to Dr. V. Prakash, director of CFTRI for constant encouragement. Thanks are also due to Dr. K. S. M. S. Raghavarao, head of the food engineering department for support. The author acknowledges Prof. Dietrich Knorr, TU Berlin, Germany; Prof. Gauri S. Mittal, University of Guelph, Canada; Prof. V. M. Balasubramaniam, Ohio State University, United States; and Dr. K. Vilkhu, Food

Science Australia, Australia, for their help in providing useful information. Prof. Stefan Toepfl deserves special thanks for helping the author bring out the commercialization aspect of pulsed electric field by contacting user industries on behalf of the author.

REFERENCES

Ade-Omowaye, B.I.O., Rastogi, N.K., Angersbach, A., and Knorr, D. 2001a. Effect of high pressure or high electrical field pulse pretreatment on dehydration characteristics of paprika. *Innovat. Food Sci. Emerg. Technol.* 2: 1–7.

Ade-Omowaye, B.I.O., Angersbach, A., Eshtiaghi, M.N., and Knorr, D. 2001b. Impact of high electrical field pulses on cell permeabilization and as pre-processing step in coconut processing. *Innovat. Food Sci. Emerg. Technol.* 1: 203–209.

Ade-Omowaye, B.I.O., Angersbach, A., Taiwo, K.A., and Knorr, D. 2001c. The use of pulsed electric fields in producing juice from paprika (*Capsicum annuum* L.). *J. Food Process. Preserv.* 25: 353–365.

Ade-Omowaye, B.I.O., Rastogi, N.K., Angersbach, A., and Knorr, D. 2002. Osmotic dehydration of bell peppers: Influence of high intensity electric field pulses and elevated temperature treatment. *J. Food Eng.* 54: 35–43.

Ade-Omowaye, B.I.O., Taiwo, K.A., Eshtiaghi, M.N., Angersbach, A., and Knorr, D. 2003a. Comparative evaluation of the effects of pulsed electric field and freezing on cell membrane permeabilization and mass transfer during dehydration of red bell peppers. *Innovat. Food Sci. Emerg. Technol.* 4: 177–188.

Ade-Omowaye, B.I.O., Talens, P., Angersbach, A., and Knorr, D. 2003b. Kinetics of osmotic dehydration of red bell peppers as influenced by pulsed electric field pretreatment. *Food Res. Int.* 36: 475–483.

Albu, S., Joyce, E., Paniwnyk, L., Lorimer, P., and Mason, J. 2004. Potential for the use of ultrasound in the extraction of antioxidants from *Rosmarinus officinalis* for the food and pharmaceutical industry. *Ultrason. Sonochem.* 11: 261–265.

Amami, E., Fersi, A., Vorobiev, E., and Kechaou, N. 2007a. Osmotic dehydration of carrot tissue enhanced by pulsed electric field, salt and centrifugal force. *J. Food Eng.* 83: 605–613.

Amami, E., Fersi, A., Khezami, L., Vorobiev, E., and Kechaou, N. 2007b. Centrifugal osmotic dehydration and rehydration of carrot tissue pre treated by pulsed electric field. *Lebensm. Wiss. Technol.* 40: 1156–1166.

Angersbach, A. and Knorr, D. 1997. High electric field pulses as pre-treatment for affecting dehydration characteristics and rehydration properties of potato cubes. *Nahrung-Food* 41: 194–200.

Angersbach, A., Heinz, V., and Knorr, D. 1997. Elektrische Leitfähigkeit als Maß des Zellaufschlußgrades von Zellulären Materialien durch Verarbeitungsprozesse. (Equipment for the measurement of electrical conductivity during processing) *Lebensm. Verpackungstechnik (LVT)*. 42: 195–200.

Angsupanich, K. and Ledward, D.A. 1998. High pressure treatment effects on cod (*Gadus morhua*) muscle. *Food Chem.* 63: 39–50.

Arias, M., Lopez, F.R., and Olano, A. 2000. Influence of pH on the effects of high pressure on milk proteins. *Milchwissenschaft* 55: 191–194.

Arroyo, G., Sanz, P.D., and Prestamo, G. 1997. Effect of high pressure on the reduction of microbial populations in vegetables. *J. Appl. Microbiol.* 82: 735–742.

Arroyo, G., Sanz, P.D., and Prestamo, G. 1999. Response to high pressure, low temperature treatment in vegetables: Determination of survival rates of microbial populations using flow cytometry and detection of peroxidase activity using confocal microscopy. *J. Appl. Microbiol.* 86: 544–556.

Ayo, J., Carballo, J., Solas, M.T., and Jiménez, C.F. 2005. High pressure processing of meat batters with added walnuts. *Int. J. Food Sci. Technol.* 40: 47–54.

Babaei, R., Jabbari, A., and Yamini, Y. 2006. Solid–liquid extraction of fatty acids of some variety of Iranian rice in closed vessel in the absence and presence of ultrasonic waves. *Asian J. Chem.* 18: 57–64.

Bai, Y., Zhao, D.D., and Yang, G. 2004. Changes of microscopic structure and shear force value of bovine and mutton skeletal muscle under hydrostatic high-pressure treatment. *Food Sci. China* 25: 27–31.

Balachandran, S., Kentish, E., Mawson, R., and Ashokkumar, M. 2006. Ultrasonic enhancement of the supercritical extraction from ginger. *Ultrason. Sonochem.* 13: 471–479.

Balasubramanian, S. and Balasubramaniam, V.M. 2003. Compression heating influence of pressure transmitting fluids on bacteria inactivation during high pressure processing. *Food Res. Int.* 36: 661–668.

Balasubramaniam, V.M., Ting, E.Y., Stewart, C.M., and Robbins, J.A. 2004. Recommended laboratory practices for conducting high-pressure microbial inactivation experiments. *Innovat. Food Sci. Emerg. Technol.* 5: 299–306.

Balci, A.T. and Wilbey, R.A. 1999. High pressure processing of milk—The first 100 years in the development of new technology. *Int. J. Dairy Technol.* 52: 149–155.

Barbosa-Canovas, G.V., Gongora-Nieto, M.M., Pothakamury, U.R., and Swanson, B.G. 1999. *Preservation of Foods with Pulsed Electric Fields*. Academic Press, London, U.K., pp. 1–9, 76–107, 108–155.

Barbosa-Canovas, G.V., Gongora-Nieto, M.M., Pothakamury, U.R., and Swanson, B.G. 2001. Pulsed electric fields. *J. Food Sci.* Special supplement, 65–79.

Barsotti, L., Dumay, E., Mu, T.H., Fernandez Dias, M.D., and Cheftel, J.C. 2002. Effects of high voltage pulses on protein-based food constituents and structures. *Trends Food Sci. Technol.* 12: 136–144.

Basak, S. and Ramaswamy, H.S. 1998. Effect of high pressure processing on the texture of selected fruits and vegetables. *J. Texture Stud.* 29: 587–601.

Bazhal, M. and Vorobiev, E. 2000. Electrical treatment of apple cossettes for intensifying juice pressing. *J. Sci. Food Agric.* 80: 1668–1674.

Bing, J.J., Xiang, H.L., Mei, Q.C., and Zhi, C.X. 2006. Improvement of leaching process of Geniposide with ultrasound. *Ultrason. Sonochem.* 13: 455–462.

Black, E.P., Kelly, A.L., and Fitzgerald, G.F. 2005. The combined effect of high pressure and nisin on inactivation of microorganisms in milk. *Innovat. Food Sci. Emerg. Technol.* 6: 286–292.

Borda, D., Indrawati, Smout C., Van Loey, A.M., and Hendrickx, M. 2004a. High-pressure thermal inactivation kinetics of a plasmin system. *J. Dairy Sci.* 87: 2351–2358.

Borda, D., Van Loey, A.M., Smout C., and Hendrickx, M. 2004b. Mathematical models for combined high pressure and thermal plasmin inactivation kinetics in two model systems. *J. Dairy Sci.* 87: 4042–4049.

Bouzrara, H. 2001. Amélioration du pressage de produits végétaux par Champ Electrique Pulse. Cas de la betterave à sucre. PhD thesis, Université de Technologie de Compiègne, France.

Bouzrara, H. and Vorobiev, E. 2000. Beet juice extraction by pressing and pulsed electric fields. *Int. Sugar J.* 1216: 194–200.

Bouzrara, H. and Vorobiev, E. 2001. Non-thermal pressing and washing of fresh sugarbeet cossettes combined with a pulsed electrical field. *Zucker* 126: 463–466.

Bragagnolo, N., Danielsen, B., and Skibsted, L.H. 2006. Combined effect of salt addition and high-pressure processing on formation of free radicals in chicken thigh and breast muscle. *Eur. Food Res. Technol.* 223: 669–673.

Brooker, B., Ferragut, V., Gill, A., and Needs, E. 1998. Properties of rennet gel formed from high pressure treated milk. In *Proceedings of the VTT Symposium, Fresh Novel Foods by High Pressure*, ed., K. Autio, Helsinki, Finland, pp. 55–61.

Bruni, R., Guerrini, A., Scalia, S., Romagnoli, C., and Sacchetti, G. 2002. Rapid techniques for the extraction of vitamin E isomers from *Amaranthus caudatus* seeds. *Ultrason. Supercrit. Fluid Ext.* 13: 257–261.

Buffa, M.N., Trujillo, A.J., Pavia, M., and Guamis, B. 2001. Changes in textural, microstructural, and colour characteristics during ripening of cheeses made from raw, pasteurized or high pressure treated goats' milk. *Int. Dairy J.* 11: 927–934.

Buggenhout, S., Messagie, I., Plancken, I., and Hendrickx, M. 2006. Influence of high pressure low temperature treatments on fruit and vegetable quality related enzymes. *Euro. Food Res. Technol.* 223: 475–485.

Bull, M. K., Zerdin, K., Howe, E. et al. 2004. The effect of high pressure processing on the microbial, physical and chemical properties of Valencia and Navel orange juice. *Innovat. Food Sci. Emerg. Technol.* 5: 135–149.

Burgos, J., Ordonez, J.A., and Sala, F.J. 1972. Effect of ultrasonic waves on the heat resistance of *Bacillus cereus* and *Bacillus licheniformis* spores. *Appl. Microbiol.* 24: 497–498.

Burgos, J., Condon, S., Lopez, P., Ordonez, J.A., Raso, J., and Sala, F.J. 1992. Method for the destruction of microorganisms and enzymes: MTS process (manothermo-sonication). World Patent WO 93/19619, filed March 31, 1992 and issued March 31, 1993.

Butz, P., Edenharder, R., Fister, H., and Tauscher, B. 1997. The influence of high pressure processing on antimutagenic activities of fruit and vegetable juices. *Food Res. Int.* 30: 287–291.

Butz, P., Edenharder, R., Fernandez Garcia, A., Fister, H., Merkel, C., and Tauscher, B. 2002. Changes in functional properties of vegetables induced by high pressure treatment. *Food Res. Int.* 35: 295–300.

Cai, J., Liu, X., Li, Z., and An, C. 2003. Study on extraction technology of strawberry pigments and its physicochemical properties. *Food Ferment. Ind.* 29: 69–73.

Caili, F., Haijun, T., Quanhong, L., Tongyi, C., and Wenjuan, D. 2006. Ultrasound-assisted extraction of xyloglucan from apple pomace. *Ultrason. Sonochem.* 13: 511–516.

Caner, C., Hernandez, R.J., and Pascall, M.A. 2000. Effect of high pressure processing on the permeance of selected high barrier laminated films. *Packaging Technol. Sci.* 13: 183–195.

Caner, C., Hernandez, R.J., and Harte, B.R. 2004. High-pressure processing effects on the mechanical, barrier and mass transfer properties of food packaging flexible structures: A critical review. *Packaging Technol. Sci.* 17: 23–29.

Canselier J.R., Delmas, H., Wilhelm, A.M., and Abismail, B. 2002. Ultrasound emulsification—An overview. *J. Dispersion Sci. Technol.* 23: 333–349.

Capellas, M., Mor-Mur, M., Sendra, E., Pla, R., and Guamis, B. 1996. Population of aerobic mesophiles and inoculated *E. coli* during storage of fresh goat's milk cheese treated with high pressure. *J. Food Protect.* 59: 82–87.

Carballo, J., Fernandez, P., and Colmenero, F.J. 1996. Texture of uncooked and cooked low and high fat meat batters as affected by high hydrostatic pressure. *J. Agric. Food Chem.* 44: 1624.

Carcel, J.A., Garcia Perez, J.V., Riera, E., and Mulet, A. 2007a. Influence of high-intensity ultrasound on drying kinetics of persimmon. *Drying Technol.* 25: 185–193.

Carcel, J.A., Benedito, J., Bon, J., and Mulet, A. 2007b. High intensity ultrasound effects on meat brining. *Meat Sci.* 76: 611–619.

Carlez, A., Rosec, J.P., Richard, N., and Cheftel, J.C. 1993. High pressure inactivation of *Citrobacter freundii, Pseudomonas fluorescens* and *Listeria innocua* in inoculated minced beef muscle. *Lebensm. Wiss. Technol.* 26: 357–363.

Carlez, A., Rosec, J.P., Richard, N., and Cheftel, J.C. 1994. Bacterial growth during chilled storage of pressure treated minced meat. *Lebensm. Wiss. Technol.* 27: 48–54.

Carminati, D., Gatti, M., Bonvini, B., Neviani, E., and Mucchetti, G. 2004. High-pressure processing of Gorgonzola cheese: Influence on *Listeria monocytogenes* inactivation and on sensory characteristics. *J. Food Protect.* 67: 1671–1675.

Carpi, G., Gola, S., Maggi, A., Rovere, P., and Buzzoni, M. 1995. Microbial and chemical shelf life of high pressure treated salmon cream at refrigeration temperatures. *Industria Conserve* 70: 386–397.

Carpi, G., Squarcina, N., Gola, S., Rovere, P., Pedrielli, R., and Bergamaschi, M. 1999. Application of high pressure treatment to extend the refrigerated shelf life of sliced cooked ham. *Industria Conserve* 74: 327–339.

Castro, S.M., Saraiva, J.A., Lopes Da Silva, J.A. et al. 2008. Effect of thermal blanching and of high pressure treatments on sweet green and red bell pepper fruits (*Capsicum annuum* L.). *Food Chem.* 107: 1436–1449.

Chalermchat, Y. and Dejmek, P. 2005. Effect of pulsed electric field pretreatment on solid liquid expression from potato tissue. *J. Food Eng.* 71: 164–169.

Chemat, S., Lagha, A., Ait Amar, H., Bartels, V., and Chemat, F. 2004. Comparison of conventional and ultrasound-assisted extraction of carvone and limonene from caraway seeds. *Flavour Fragrance J.* 19: 188–195.

Chendke, P.K. and Fogler, H.S. 1975. Macrosonics in industry. 4. Chemical processing. *Ultrasonics* 13: 31–37.

Chevalier, D., Bail, A.L., Chourot, J.M., and Chantreau, P. 1999. High pressure thawing of fish (whiting): Influence of the process parameters on drip losses. *Lebens. Wiss. Technol.* 32: 25–31.

Cho, B.K. and Irudayaraj, J.M.K. 2003a. A noncontact ultrasound approach for mechanical property determination of cheeses. *J. Food Sci.* 68: 2243–2247.

Cho, B.K. and Irudayaraj, J.M.K. 2003b. Foreign object and internal disorder detection in food materials using noncontact ultrasound imaging. *J. Food Sci.* 68: 967–974.

Clark, J.P., 2006. Pulsed electric field processing. *Food Technol.* 60(1): 66–67.

Cocito, C., Gaetano, G., and Delfini, C. 1995. Rapid extraction of aroma compounds in must and wine by means of ultrasound. *Food Chem.* 2: 311–320.

Cortes, C., Esteve, M.J., and Frigola, A. 2008. Color of orange juice treated by high intensity pulsed electric fields during refrigerated storage and comparison with pasteurized juice. *Food Control* 19: 151–158.

Cserhalmi, Z., Sass, K.A., Toth, M.M., and Lechner, N. 2006. Study of pulsed electric field treated citrus juices. *Innovat. Food Sci. Emerg. Technol.* 7: 49–54.

D'Amico, D.J., Silk, T.M., Junru, W., and Mingruo, G. 2006. Inactivation of microorganisms in milk and apple cider treated with ultrasound. *J. Food Protect.* 69: 556–563.

De Heij, W.B.C., Van Schepdael, L.J.M.M., and van der Berg, R.W. 2001. Increasing preservation efficiency and product quality through homogeneous temperature profile in high pressure application. Paper presented at the XXXIX European High Pressure Research Group Meeting, Advances on High Pressure Research, Santander (Spain), September 16–19, 2001.

Denys, S. and Hendrickx, M.E. 1999. Measurement of the thermal conductivity of foods at high pressure. *J. Food Sci.* 64: 709–713.

Denys, S., Ludikhuyze, L., Van Loey, A.M., and Hendrickx, M.E. 2000a. Modelling conductive heat transfer and process uniformity during batch high-pressure processing of foods. *Biotechnol. Prog.* 16: 92–101.

Denys, S., Van Loey, A.M., and Hendrickx, M.E. 2000b. A modeling approach for evaluating process uniformity during batch high hydrostatic pressure processing: Combination of a numerical heat transfer model and enzyme inactivation kinetics. *Innovat. Food Sci. Emerg. Technol.* 1: 5–19.

Derobry, B.S., Richard, F., and Hardy, J. 1994. Study of acid and rennet coagulation of high pressurized milk. *J. Dairy Sci.* 77: 3267–3274.

Dobias, J., Voldrich, M., Marek, M., and Chudackova, K. 2004. Changes of properties of polymer packaging films during high pressure treatment. *J. Food Eng.* 61: 545–549.

Donsi, G., Ferrari, G., and Matteo, M.D. 1996. High pressure stabilization of orange juice: Evaluation of the effects of process conditions. *Ital. Food Beverage Technol.* 8: 10–14.

Dornenburg, H. and Knorr, D. 1993. Cellular permeabilization of cultured plant tissue by high electric field pulse and ultra high pressure for recovery of secondary metabolites. *Food Biotechnol.* 7: 35–48.

Drake, M.A., Harrison, S.L., Asplund, M., Barbosa-Canovas, G., and Swanson, B.G. 1997. High pressure treatment of milk and effects on microbiological and sensory quality of cheddar cheese. *J. Food Sci.* 62: 843–845.

Dunn, J.E. and Pearlman, J.S. 1987. Methods and apparatus of extending the shelf-life of fluid food products. U.S. Patent 4,695,472, filed March 31, 1985 and issued September 22, 1987.

Ebringerova, A., Hromadkova, Z., Alfodi, J., and Hribalova, V. 1998. The immunologically active xylan from ultrasound-treated corn cobs: Extractability, structure and properties. *Carbohydr. Polym.* 37: 231–239.

El-Belghitia, K., Rabhia, Z., and Vorobiev, E. 2007. Effect of process parameters on solute centrifugal extraction from electropermeabilized carrot gratings. *Food Bioprod. Process.* 85: 24–28.

El Moueffak, A., Cruz, C., Antoine, M. et al. 2001. Stabilization of duck fatty liver by high pressure treatment. Inactivation of *Enterococcus faecalis*. *Food Sci.* 21: 71–76.

Elez, M.P. and Martin B.O. 2007. Effects of high intensity pulsed electric field processing conditions on vitamin C and antioxidant capacity of orange juice and gazpacho, a cold vegetable soup. *Food Chem.* 102: 201–209.

Entezari, H., Nazary, H., and Khodaparast, H. 2004. The direct effect of ultrasound on the extraction of date syrup and its micro-organisms. *Ultrason. Sonochem.* 11: 379–384.

Erkman, O. and Karatas, S. 1997. Effect of high hydrostatic pressure on *Staphylococcus aureus* in milk. *J. Food Eng.* 33: 257–262.

Eshtiaghi, M.N. and Knorr, D. 1993. Potato cube response to water belching and high hydrostatic pressure. *J. Food Sci.* 58: 1371–1374.

Eshtiaghi, M.N. and Knorr, D. 1999. Method for treating sugar beet, World Patent WO 99/64634, filed June 09, 1999 and issued November 02, 1999.

Eshtiaghi, M.N., Stute, R., and Knorr, D. 1994. High pressure and freezing pretreatment effects on drying, rehydration, texture and color of green beans, carrots and potatoes. *J. Food Sci.* 59: 1168–1170.

Esteve, M.J., Frígola, A., Rodrigo, D., Rodrigo, M., and Torregrosa, F. 2001. Pulsed electric fields inactivation kinetics of carotenes in mixed orange–carrot juice. Paper presented at the Eurocaft2001 Conference, Berlin.

Evrendilek, G.A., Jin, Z.T., Ruhlman, K.T., Qiu, X., Zhang, Q.H., and Richter, E.R., 2000 Microbial safety and shelf-life of apple juice and cider processed by bench and pilot scale PEF systems. *Innovat. Food Sci. Emerg. Technol.* 1: 77–86.

Evrendilek, G.A., Koca, N., Harper, J.W., and Balasubramanian, V.M. 2008. High-pressure processing of Turkish white cheese for microbial inactivation. *J. Food Prot.* 71: 102–108.

Fairbanks, H.V. and Chen, W.I. 1971. Ultrasonic acceleration of liquid flow through porous media. *Chem. Eng. Symp. Ser.* 67: 108.

Farkas, D.F. and Hoover, D.G. 2001. High pressure processing. *J. Food Sci.*, Special supplement on kinetics of microbial inactivation for alternative food processing technologies. 47–64.

Farr, D. 1990. High pressure technology in food industry. *Trends Food Sci. Technol.* 1: 14–16.

Fernandes, F.A.N. and Rodrigues, S. 2007. Ultrasound as pre-treatment for drying of fruits: Dehydration of banana. *J. Food Eng.* 82: 261–267.

Fernandez, G.A., Butz, P., and Tauscher, B. 2001a. Effects of high-pressure processing on carotenoid extractability, antioxidant activity, glucose diffusion, and water binding of tomato puree (*Lycopersicon esculentum* Mill.). *J. Food Sci.* 66: 1033–1038.

Fernandez, G.A., Butz, P., Bognar, A., and Tauscher, B. 2001b. Antioxidative capacity, nutrient content and sensory quality of orange juice and an orange-lemon-carrot juice product after high pressure treatment and storage in different packaging. *Eur. Food Res. Technol.* 213: 290–296.

Fernandez, P.P., Prestamo, G., Otero, L., and Sanz, P.D. 2006. Assessment of cell damage in high pressure shift frozen broccoli: Comparison with market samples. *Z. Lebensm. Unter. Forsch. A.* 224: 101–107.

Fernandes, F.A.N., Gallao, M.I., and Rodrigues, S. 2008. Effect of osmotic dehydration and ultrasound pre-treatment on cell structure: Melon dehydration. *Lebensm. Wiss. Technol.* 41: 604–610.

Ferrante, S., Guerrero, S., and Alzamora, S.M. 2007. Combined use of ultrasound and natural antimicrobials to inactivate *Listeria monocytogenes* in orange juice. *J. Food Prot.* 70: 1850–1856.

Flaumenbaum, B.L. 1986. Anwendung der Electoplasmolyse bei der Hertstallung von Fructschaften. (Application of electroplasmolysis for the extraction of fruit juice) *Flussiges Obst.* 35: 19–20.

Fuchigami, M. and Teramoto, A. 1997. Structural and textural changes in kinu tofu due to high pressure freezing. *J. Food Sci.* 62: 828–832, 837.

Fuchigami, M., Miyazaki, K., Kato, N., and Teramoto, A. 1997a. Histological changes in high pressure frozen carrots. *J. Food Sci.* 62: 809–812.

Fuchigami, M., Kato, N., and Teramoto, A. 1997b. High pressure freezing effects on textural quality of carrots. *J. Food Sci.* 62: 804–808.

Fuchigami, M., Kato, N., and Teramoto, A. 1998. High pressure freezing effects on textural quality of Chinese cabbage. *J. Food Sci.* 63: 122–125.

Furuki, T., Maeda, S., Imajo, S., Hiroi, T., Amaya, T., and Hirokawa, T. 2003. Rapid and selective extraction of phycocyanin from *Spirulina platensis* with ultrasonic cell disruption. *J. Appl. Phycol.* 15: 319–324.

Gabaldon-Leyva, C.A., Quintero-Ramos, A., Barnard, J., Balandran-Quintana, R.R., Talamas-Abbud, R., and Jimenez-Castro, J. 2007. Effect of ultrasound on the mass transfer and physical changes in brine bell pepper at different temperatures. *J. Food Eng.* 81: 374–379.

Gachovska, T.K., Ngadi, M.O., and Raghavan, G.S.V. 2006. Pulsed electric field assisted juice extraction from alfalfa. *Can. Biosys. Eng.* 48: 3.33–3.37.

Gallego-Juarez, J.A., Riera, E., Fuente Blanco, S.D.L., Rodriguez, C.G., Acosta, A.V.M., and Blanco, A. 2007. Application of high-power ultrasound for dehydration of vegetables: Processes and devices. *Drying Technol.* 25: 1893–1901.

Garcia, M.L., Burgos, J., Sanz, B., and Ordoñez, J.A. 1989. Effect of heat and ultrasonic waves on the survival of two strains of *Bacillus subtilis. J. Appl. Bacteriol.* 67: 619–628.

Garcia-Perez, J.V., Carcel, J.A., Benedito, J., and Mulet, A. 2007. Power ultrasound mass transfer enhancement in food drying. *Food Bioprod. Process.* 85: 247–254.

Garriga, M., Grebol, N., Aymerich, M.T., Monfort, J.M., and Hugas, M. 2004. Microbial inactivation after high-pressure processing at 600 MPa in commercial meat products over its shelf life. *Innovat. Food Sci. Emerg. Technol.* 5: 451–457.

Gennaro, L.D., Cavella, S., Romano, R., and Masi, P. 1999. The use of ultrasound in food technology. I. Inactivation of peroxidase by thermosonication. *J. Food Eng.* 39: 401–407.

Gervilla, R., Capellas, M., Ferragut, V., and Guamis, B. 1997. Effect of high hydrostatic pressure on *Listeria innocua* 910 CECT inoculated into ewe's milk. *J. Food Prot.* 60: 33–37.

Geulen, M., Teichgräber, P., and Knorr, D. 1994. High electric field pulses for cell permeabilization. *Z. Lebensmittelwirtschaft.* 45: 24–27.

Gola, S., Foman, C., Carpi, G., Maggi, A., Cassara, A., and Rovere, P. 1996. Inactivation of bacterial spores in phosphate buffer and in vegetable cream treated with high pressure. In *High Pressure Bioscience and Biotechnology*, eds. R. Hayashi and C. Balny, pp. 253–269, Amsterdam, The Netherlands: Elsevier Science.

Gomez, E.J., Montero, P., Gimenez, B., and Gomez-Guillen, M.C. 2007. Effect of functional edible films and high pressure processing on microbial and oxidative spoilage in cold-smoked sardine (*Sardina pilchardus*). *Food Chem.* 105: 511–520.

Gongora-Nieto, M.M., Seignour, L., Riquet, P., Davidson, P.M., Barbosa-Canovas, G.V., and Swanson, B.G. 2001. Nonthermal inactivation of *Pseudomonas* fluorescens in liquid whole egg. In *Pulsed Electric Fields in Food Processing*, eds. G.V. Barbosa-Canovas and Q.H. Zhang, pp. 193–211. Lancaster: Technomic Publishing Company.

Goodner, J.K., Braddock, R.J., Parish, M.E., and Sims, C.A. 1999. Cloud stabilization of orange juice by high pressure processing. *J. Food Sci.* 64: 699–700.

Gow, C.Y. and Hsin, T.L. 1999. Changes in volatile flavor components of guava juice with high-pressure treatment and heat processing and during storage. *J. Agric. Food Chem.* 47: 2082–2087.

Guerrero, S., Lopez Malo, A., and Alzamora, S.M. 2001. Effect of ultrasound on the survival of *Saccharomyces cerevisiae*: Influence of temperature, pH and amplitude. *Innovat. Food Sci. Emerg. Technol.* 2: 31–39.

Guerrero, S., Tognon, M., and Alzamora, S.M. 2005. Response of *Saccharomyces cerevisiae* to the combined action of ultrasound and low weight chitosan. *Food Control* 16: 131–139.

Günther, U. and Kern, M., Personal communication, Prof. Stefan Toepfl, September 12, 2008.

Haeggstrom, E. and Luukkala, M. 2001. Ultrasound detection and identification of foreign bodies in food products. *Food Control* 12: 37–45.

Haizhou, L., Pordesimo, L., and Weiss, J. 2004. High intensity ultrasound-assisted extraction of oil from soybeans. *Food Res. Int.* 37: 731–738.

Han J.M. and Ledward, D.A. 2004. High pressure/thermal treatment effects on the texture of beef muscle. *Meat Sci.* 68: 347–355.

Harrison, S.L., Chang, F.J., Boylston, T., Barbosa-Canovas, G.V., and Swanson, B.G. 2001. Shelf stability, sensory analysis, and volatile flavor profile of raw apple juice after pulsed electric field, hydrostatic pressure, or heat exchanger processing. In *Pulsed Electric Fields in Food Processing*, eds. G.V. Barbosa-Canovas and Q.H. Zhang, pp. 241–257. Lancaster: Technomic Publishing Company.

Hayman, M.M., Baxter, I., O'Riordan, P.J., and Stewart, C.M. 2004. Effects of high-pressure processing on the safety, quality, and shelf life of ready-to-eat meats. *J. Food Prot.* 67: 1709–1718.

Heinz, V. and Knorr, D. 2001. Effect of high pressure on spores. In *Ultrahigh Pressure Treatment of Foods*, eds. M.E.C. Hendrickx and D. Knorr, pp. 77–116. New York: Kluwer Academic/Plenum Publishers.

Hemwimol, S., Pavasant, P., and Shotipruk, A. 2006. Ultrasonic-assisted extraction of anthraquinones from roots of *Morinda citrifolia*. *Ultrason. Sonochem.* 13: 543–548.

Hendrickx, M., Ludikhuyze, L., Broeck Van den I., and Weemaes, C. 1998. Effect of high pressure on enzymes related to food quality. *Trends Food Sci. Technol.* 9: 97–203.

Hite, B.H. 1899. The effect of pressure in the preservation of milk. *W. Va. Univ., Agric. Exp. Stat. Bull.* 58: 15–35.

Hite, B.H., Giddings, N.J., and Weakly, C.E. 1914. The effects of pressure on certain micro-organisms encountered in the preservation of fruits and vegetables. *W. Va. Univ., Agric. Exp. Stat. Bull.* 146: 1–67.

Hogan, E., Kelly, A.L., and Sun, D.W. 2005. High pressure processing of foods: An overview, In *Emerging Technologies for Food Processing*, ed. D.W. Sun, pp. 4–30. London, U.K.: Elsevier Ltd.

Homma, N., Ikeuchi, Y., and Suzuki, A. 1995. Effects of high pressure treatment on the proteolytic enzymes in meat. *Meat Sci.* 38: 219–228.

Hoogland, H., de Heij, W., and van Schepdael, L. 2001. High pressure sterilization: Novel technology, new products, new opportunities. *New Food.* 4: 21–26.

Houska, M., Strohalm, J., Kocurova, K. et al. 2006. High pressure and foods-fruit/vegetable juices. *J. Food Eng.* 77: 386–398.

Hromadkova, Z. and Ebringerova, A. 2003. Ultrasonic extraction of plant materials—Investigation of hemicellulose release from buckwheat hulls. *Ultrason. Sonochem.* 10: 127–133.

Hugas, M., Garriga, M., and Monfort, J.M. 2002. New mild technologies in meat processing: High pressure as a model technology. *Meat Sci.* 62: 359–371.

Huppertz, T., Kelly, A.L., and Fox, P.F. 2002. Effects of high pressure on constituents and properties of milk. *Int. Dairy J.* 12: 561–572.

Huppertz, T., Fox, P.F., and Kelly, A.L. 2004. Effects of high pressure treatment on the yield of cheese curd from bovine milk. *Innovat. Food Sci. Emerg. Technol.* 5: 1–8.

Huppertz, T., Hinz, K., Zobrist, M.R., Uniacke, T., Kelly, A.L., and Fox, P.F. 2005. Effects of high pressure treatment on the rennet coagulation and cheese-making properties of heated milk. *Innovat. Food Sci. Emerg. Technol.* 6: 279–285.

Jacques, R.A., Freitas, L.S., Perez, V.F. et al. 2007. The use of ultrasound in the extraction of *Ilex paraguariensis* leaves: A comparison with maceration. *Ultrason. Sonochem.* 14: 6–12.

Jambrak, A.R., Mason, T.J., Lelas, V., Herceg, Z., and Herceg, I.L. 2007. Effect of ultrasound treatment on solubility and foaming properties of whey protein suspensions. *J. Food Eng.* 86: 281–287.

Jayaram, S., Castle, G.S.P., and Margaritis, A. 1992. Kinetics of sterilization of *Lactobacillus brevis* cells by the application of high voltage pulses. *Biotech. Bioeng.* 40: 1412–1420.

Jemai, AB. and Vorobiev, E. 2006. Pulsed electric field assisted pressing of sugar beet slices: Towards a novel process of cold juice extraction. *Biosyst. Eng.* 93: 57–68.

Jeyamkondan, S., Jayas, D.S., and Holley, R.A. 1999. Pulsed electric field processing of foods: A review. *J. Food Prot.* 62: 1088–1096.

Jimenez-Castro, F., Fernandez, P., Carballo, J., and Fernandez, M.F. 1998. High pressure cooked low fat pork and chicken batters as affected by salt levels and cooking temperature. *J. Food Sci.* 63: 656–659.

Jin, Z.T. and Zhang, Q.H. 1999. Pulsed electric field treatment inactivates microorganisms and preserves quality of cranberry juice. *J. Food Process. Preserv.* 23: 481–499.

Jing, S., Run, C.S., Xiao, S., and Yin, Q.S. 2004. Fractional and physicochemical characterization of hemicelluloses from ultrasonic irradiated sugarcane bagasse. *Carbohydr. Res.* 339: 291–300.

Jing, W., Baoguo, S., Yanping, C., Yuan, T., and Xuehong, L. 2008. Optimization of ultrasound-assisted extraction of phenolic compounds from wheat bran. *Food Chem.* 106: 804–810.

Jofre, A., Aymerich, T., and Garriga, M. 2008. Assessment of the effectiveness of antimicrobial packaging combined with high pressure to control *Salmonella* sp. in cooked ham. *Food Control* 19: 634–638.

Juliano, P., Toldrág, M., Koutchma, T., Balasubramaniam, V.M., Clark, S., Mathews, J.W., Dunne, C.P., Sadlerand, G., and Barbosa-Canovas G.V. 2006. Texture and water retention improvement in high-pressure thermally treated scrambled egg patties. *J. Food Sci.* 71: E52–E61.

Jung, S., Ghoul, M., and de Lamballerie-Anton, M. 2000. Changes in lysosomal enzyme activities and shear values of high pressure treated meat during ageing. *Meat Sci.* 56: 239–246.

Jung, S., Ghoul, M., and de Lamballerie-Anton, M. 2003. Influence of high pressure on the color and microbial quality of beef meat. *Lebensm. Wiss. Technol.* 36: 625–631.

Kadlec, P., Dostalova, J., Houska, M., Strohalm, J., Culkova, J., Hinkova, A. and Starhova, H. 2006. High pressure treatment of germinated chickpea *Cicer arietinum* L. seeds. *J. Food Eng.* 77: 445–448.

Kalichevsky, M.T., Knorr, D., and Lillford, P.J. 1995. Potential food applications of high-pressure effects on ice-water transitions. *Trends Food Sci. Technol.* 6: 253–259.

Kimura, K., Ida, M., Yosida, Y., Okhi, K., Fukumoto, T., and Sakui, N. 1994. Comparison of keeping quality between pressure processed jam and heat processed jam: Changes in flavor components, hue and nutrients during storage. *Biosci. Biotechnol. Biochem.* 58: 1386–1391.

Kinefuchi, M., Yamazaki, A., and Yamamoto, K., 1995. Extraction of trehalose from *Saccharomyces cerevisiae* by high-pressure treatment. *J. Appl. Glycosci.* 42: 237–242.

Kingsly, A.R.P., Balasubramaniam, V.M., and Rastogi, N.K. 2009a. Influence of high pressure blanching on polyphenoloxidase activity of peach fruits and its drying behavior, *Int. J. Food Properties.* 12: 671–680.

Kingsly, A.R.P., Balasubramaniam, V.M., and Rastogi, N.K. 2009b. Effect of high pressure processing on texture and drying behavior of pineapple, *J. Food Process Eng.* 32: 369–381.

Knorr, D. and Angersbach, A. 1998. Impact of high electric field pulses on plant membrane permeabilization. *Trends Food Sci. Technol.* 9: 185–191.

Knorr, D., Geulen, W., Grahl, T., and Sitzmann, W. 1994. Food application of high electric field pulses. *Trends Food Sci. Technol.* 5: 71–75.

Kolakowski, P., Reps, A., and Fetlinski, A. 2000. Microbial quality and some physicochemical properties of high pressure processed cow milk. *Pol. J. Food Nutr. Sci.* 9: 19–26.

Koseki, S., Mizuno, Y., and Yamamoto, K. 2007. Predictive modeling of the recovery of *Listeria monocytogenes* on sliced cooked ham after high pressure processing. *Int. J. Food Microbiol.* 119: 300–307.

Koseki, S., Mizuno, Y., and Yamamoto, K. 2008. Use of mild heat treatment following high pressure processing to prevent recovery of pressure injured *Listeria monocytogenes* in milk. *Food Microbiol.* 25: 288–293.

Krebbers, B., Matser, A.M., Koets, M., and van den Berg, R.W. 2002. Quality and storage stability of high-pressure preserved green beans. *J. Food Eng.* 54: 27–33.

Krebbers, B., Matser, A.M., Hoogerwerf, S.W., Moezelaar, R., Tomassen, M.M.M., and van den Berg, R.W. 2003. Combined high-pressure and thermal treatments for processing of tomato puree: Evaluation of microbial inactivation and quality parameters. *Innovat. Food Sci. Emerg. Technol.* 4: 377–385.

Kubasek, M., Houska, M., Landfeld, A., Strohalm, J., Kamarad, J., and Zitny, R. 2006. Thermal diffusivity estimation of the olive oil during its high-pressure treatment. *J. Food Eng.* 74: 286–291.

Lagrain, B., Boeckx, L., Wilderjans, E., Delcour, J.A., and Lauriks, W. 2006. Non-contact ultrasound characterization of bread crumb: Application of the Biot–Allard model. *Food Res. Int.* 39: 1067–1075.

Lakshmanan, R. and Dalgaard, P. 2004. Effects of high-pressure processing on *Listeria monocytogenes*, spoilage microflora and multiple compound quality indices in chilled cold-smoked salmon. *J. Appl. Microbiol.* 96: 398–408.

Lambert, Y., Demazeau, G., Largeteau, A., and Bouvier, J.M. 1999. Changes in aromatic volatile composition of strawberry after high pressure treatment. *Food Chem.* 67: 7–16.

Lambert, Y., Demazeau, G., Largeteau, A., Bouvier, J.M., Laborde Croubit, S., and Cabannes, M. 2000. Packaging for high pressure treatments in the food industry. *Packaging Technol. Sci.* 13: 63–71.

Le Bail, A., Hamadami, N., and Bahuad, S. 2006. Effect of high-pressure processing on the mechanical and barrier properties of selected packaging. *Packaging Technol. Sci.* 19: 237–243.

Leadley, C., Tucker, G., and Fryer, P.A. 2008. Comparative study of high pressure sterilization and conventional thermal sterilization: Quality effects in green beans. *Innovat. Food Sci. Emerg. Technol.* 9: 70–79.

Lebovka, N.I., Shynkaryk, N.V., and Vorobiev, E. 2007. Pulsed electric field enhanced drying of potato tissue. *J. Food Eng.* 78: 606–613.

Leighton, T.G. 1998. The principles of cavitation. In *Ultrasound in Food Processing*, eds. M.J.W. Powey and T.J. Mason, pp. 151–182. London, U.K.: Blackie Academic and Professional.

Lillard, H.S. 1994. Decontamination of poultry skin by sonication. *Food Technol.* 44: 73–74.

Lin, S.X.Q. and Chen, X.D. 2007. A laboratory investigation of milk fouling under the influence of ultrasound. *Food Bioprod. Process.* 85: 57–62.

Lopez, P., Sala, F.J., Fuente, J.L., Condon, S., Raso, J., and Burgos, J. 1994. Inactivation of peroxidase, lipoxygenase, and polyphenoloxidase by manothermosonication. *J. Agric. Food Chem.* 42: 252–256.

Lopez, F.R., Carrascosa, A.V., and Olano, A. 1996. The effect of high pressure on whey protein denaturation and cheese making properties of raw milk. *J. Dairy Sci.* 79: 929–936.

Lopez-Caballero, M.E., Perez Mateos, M., Borderias, J.A., and Montero, P. 2000. Extension of the shelf life of prawns (*Penaeus japonicus*) by vacuum packaging and high pressure treatment. *J. Food Prot.* 63: 1381–1388.

Lu, H.Q., Xie, C.F., Yang, R.F., and Qiu, T.Q. 2005. Preventing fouling in evaporators through ultrasound. *Int. Sugar J.* 107: 456–461.

Mahdi, J.S., Yinghe, H., and Bhandari, B. 2007. Production of sub-micron emulsions by ultrasound and microfluidization techniques. *J. Food Eng.* 82: 478–488.

Marcos, B., Aymerich, T., Monfort, J.M., and Garriga, M. 2008. High pressure processing and antimicrobial biodegradable packaging to control *Listeria monocytogenes* during storage of cooked ham. *Food Microbiol.* 25: 177–182.

Margosch, D., Ehrmann, M.A., Gaenzle, M.G., and Vogel, R.F. 2004. Comparison of pressure and heat resistance of *Clostridium botulinum* and other endospores in mashed carrots. *J. Food Prot.* 67: 2530–2537.

Mason, T.J. 2000. Large scale sonochemical processing: Aspiration and actuality. *Ultrason. Sonochem.* 7: 145–149.

Mason, T.J., Riera, E., Vercet, A., and Buesa, P.L. 2005. Application of ultrasound, In *Emerging Technologies for Food Processing*, ed. D.W. Sun, pp. 323–332, London, U.K.: Elsevier Ltd.

Matser, A.M., Krebbers, B., Berg, R.W., and Bartels, P.V. 2004. Advantages of high pressure sterilization on quality of food products. *Trends Food Sci. Technol.* 15: 79–85.

McClements, D.J. 1995. Advances in the application of ultrasound in food analysis and processing. *Trends Food Sci. Technol.* 6: 293–299.

McInerney, J.K., Seccafien, C.A., Stewart, C.M., and Bird, A.R. 2007. Effects of high pressure processing on antioxidant activity, and total carotenoid content and availability, in vegetables. *Innovat. Food Sci. Emerg. Technol.* 8: 543–548.

McLellan, M.R., Kime, R.L., and Lind, L.R. 1991. Electroplasmolysis and other treatments to improve apple juice yield. *J. Sci. Food Agric.* 57: 303–306.

Mermelstein, N.H. 1997. High pressure processing reaches the U.S. market. *Food Technol.* 51: 95–96.

Min, Z. and Zhang, Q.H. 2003. Effects of commercial scale pulsed electric field processing on flavor and color of tomato juice. *J. Food Sci.* 68: 1600–1606.

Minerich, P.L. and Labuza, T.P. 2003. Development of a pressure indicator for high hydrostatic pressure processing of foods. *Innovat. Food Sci. Emerg. Technol.* 4: 235–243.

Mizuno, A. and Hori, Y. 1988. Destruction of living cells by pulsed high voltage application. *IEEE Trans. Ind. Appl.* 24: 387–394.

Moio, L., Masi, P., Pietra, L.l., Cacace, D., Palmieri, L., Martino, E.d., Carpi, G., and Dall'Aglio, G. 1994. Stabilization of mustes by high pressure treatment. *Industrie delle Bevande.* 23: 436–441.

Molina, E., Alvarez, M.D., Ramos, M., Olano, A., and Lopez, F.R. 2000. Use of high pressure treated milk for the production of reduced fat cheese. *Int. Dairy J.* 10: 467–475.

Mongenot, N., Charrier, S., and Chalier, P. 2000. Effect of ultrasound emulsification on cheese aroma encapsulation by carbohydrates. *J. Agric. Food Chem.* 48: 861–867.

Mor, M.M. and Yuste, J. 2003. High pressure processing applied to cooked sausage manufacture: Physical properties and sensory analysis. *Meat Sci.* 65: 1187–1191.

Morales, P., Calzada, J., Avila, M., and Nunez, M., 2008. Inactivation of *Escherichia coli* O157:H7 in ground beef by single-cycle and multiple-cycle high-pressure treatments. *J. Food Prot.* 71: 811–815.

Morgan, S.M., Ross, R.P., Beresford, T., and Hill, C. 2000. Combination of high hydrostatic pressure and lacticin 3147 causes increased killing of *Staphylococcus* and *Listeria. J. Appl. Microbiol.* 88: 414–420.

Morild, E. 1981. The theory of pressure effects on proteins. *Adv. Prot. Chem.* 35: 93–166.

Mussa, D.M. and Ramaswamy, H.S. 1997. Ultra high pressure pasteurization of milk: Kinetics of microbial destruction and changes in physicochemical characteristics. *Lebensm. Wiss. Technol.* 30: 551–557.

Nachamansion, J. 1995. Packaging solutions for high quality foods processed by high hydrostatic pressure. In *Proceedings of Europack*, Vol. 7, Poland pp. 390–401.

Needs, E.C., Stenning, R.A., Gill, A.L., Ferragut, V., and Rich, G.T. 2000. High pressure treatment of milk: Effects on casein micelle structure and on enzyme coagulation. *J. Dairy Res.* 67: 31–42.

Nguyen, L.T., Rastogi, N. K., and Balasubramaniam, V.M. 2007. Evaluation of instrumental quality of pressure-assisted thermally processed carrots. *J. Food Sci.* 72: E264–E270.

Norton, T. and Sun, D.W. 2008. Recent advances in the use of high pressure as an effective processing technique in the food industry. *Food Bioprocess Technol.* 1: 2–34.

Novotna, P., Valentova, H., Strohalm, J., Kyhos, K., Landfeld, A., and Houska, M. 1999. Sensory evaluation of high pressure treated apple juice during its storage, *Czech J. Food Sci.* 17: 196–198.

O'Brien, J.K. and Marshall, R.T. 1996. Microbiological quality of raw ground chicken processed at high hydrostatic pressure. *J. Food Prot.* 59: 146–150.

O'Reilly, C.E., O'Connor, P.M., Kelly, A.L., Beresford, T.P., and Murphy, P.M. 2000. Use of high hydrostatic pressure for inactivation of microbial contaminants in cheese. *Appl. Environ. Microbiol.* 66: 4890–4896.

O'Reilly, C.E., Kelly, A.L., Murphy, P.M., and Beresford, T.P. 2001. High-pressure treatment: Applications in cheese manufacture and ripening. *Trends Food Sci. Technol.* 12: 51–59.

Odriozola, S.I., Aguilo A.I., Soliva, F.R., Gimeno, A.V., and Martin, B.O. 2007. Lycopene, vitamin C and antioxidant capacity of tomato juice as affected by high intensity pulsed electric fields critical parameters. *J. Agric. Food Chem.* 55: 9036–9042.

Ohmiya, K., Fukami, K., Shimizu, S., and Gekko, K. 1987. Milk curdling by rennet under high pressure. *J. Food Sci.* 52: 84–87.

Ordonez, J.A., Sanz, B., Hernandez, P.E., and Lopez-Lorenzo, P. 1984. A note on the effect of combined ultrasonic and heat treatments on the survival of thermoduric streptococci. *J. Appl. Bacteriol.* 56: 175–177.

Ordonez, J.A., Aguilera, M.A., Garcia, M.L., and Sanz, B. 1987. Effect of combined ultrasonic and heat treatment (thermosonication) on the survival of a strain of *Staphylococcus aureus. J. Dairy Res.* 54: 61–67.

Orlien, V., Hansen, E., and Skibsted, L.H. 2000. Lipid oxidation in high pressure processed chicken breast muscle during chill storage: Critical working pressure in relation to oxidation mechanism. *Eur. Food Res. Technol.* 211: 99–104.

Otero, L., Solas, M.T., Sanz, P.D., Elvira, C. de, and Carrasco, J.A. 1998. Contrasting effects of high pressure assisted freezing and conventional air freezing on eggplant tissue microstructure. *Z. Lebensm. Untersuch. Forsch. A.* 206: 338–342.

Otero, L., Martino, M., Zaritzky, N., Solas, M., and Sanz, P.D. 2000. Preservation of microstructure in peach and mango during high pressure shift freezing. *J. Food Sci.* 65: 466–470.

Otero, L., Molina, A.D., and Sanz, P.D. 2002. Some interrelated thermo physical properties of liquid water and ice I: A user-friendly modeling review for food high-pressure processing. *Crit. Rev. Food Sci. Nutr.* 42: 339–352.

Pagan, R., Mañas, P., Alvarez, I., and Condon, S. 1999a. Resistance of *Listeria monocytogenes* to ultrasonic waves under pressure at sublethal (manosonication) and lethal (manothermosonication) temperatures. *Food Microbiol.* 16: 139–148.

Pagan, R., Manas, P., Raso, J., and Condon, S., 1999b. Bacterial resistance to ultrasonic waves under pressure (manosonication) and lethal (manothermosonication) temperatures. *Appl. Environ. Microbiol.* 65: 297–300.

Pandey, P.K. and Ramaswamy, H.S. 1998. Effect of high pressure of milk on textural properties, moisture content and yield of Cheddar cheese. Paper presented at *The IFT Annual Meeting*, Atlanta, GA.

Pandey, P.K. and Ramaswamy, H.S. 2004. Effect of high-pressure treatment of milk on lipase and gamma-glutamyl transferase activity. *J. Food Biochem.* 28: 449–462.

Pandey, P.K., Ramaswamy, H.S., and Idziak, E. 2003. High pressure destruction kinetics of indigenous microflora and Escherichia coli in raw milk at two temperatures. *J. Food Process Eng.* 26: 265–283.

Parish, M.E. 1998. High pressure inactivation of *Saccharomyces cerevisiae*, endogenous microflora and pectinmethylesterase in orange juice. *J. Food Safety*. 18: 57–65.

Patazca, E., Koutchma, T., and Ramaswamy, H.S. 2006. Inactivation kinetics of *Geobacillus stearothermophilus* spores in water using high pressure processing at elevated temperatures. *J. Food Sci.* 71: M110–M116.

Penas, E., Gomez, R., Frias, J., and Vidal, V.C. 2008. Application of high-pressure treatment on alfalfa (*Medicago sativa*) and mung bean (*Vigna radiata*) seeds to enhance the microbiological safety of their sprouts. *Food Control* 19: 698–705.

Perez, M.M. and Montero, P. 2000. Response surface methodology multivariate analysis of properties of high pressure induced fish mince gel. *Euro. Food Res. Technol.* 211: 79–85.

Perez, L.C., Ledward, D.A., Reed, R.J.R., and Simal, G.J. 2002. Application of high-pressure treatment in the mashing of white malt in the elaboration process of beer. *J. Sci. Food Agric.* 82: 258–262.

Phunchaisri, C. and Apichartsrangkoon, A. 2005. Effects of ultra-high pressure on biochemical and physical modification of lychee (*Litchi chinensis* Sonn.). *Food Chem.* 93: 57–64.

Piyasena, P., Mohareb, E., and McKellar, R.C. 2003. Inactivation of microbes using ultrasound: A review. *Int. J. Food Microbiol.* 87: 207–216.

Plaza, L., Munoz, M., de Ancos, B., and Cano, M.P. 2003. Effect of combined treatments of high pressure, citric acid and sodium chloride on quality parameters of tomato puree. *Eur. Food Res. Technol.* 216: 514–519.

Plaza, L., Sanchez-Moreno, C., De Ancos, B., and Cano, M.P. 2006. Carotenoid content and antioxidant capacity of Mediterranean vegetable soup (gazpacho) treated by high-pressure/temperature during refrigerated storage. *Eur. Food Res. Technol.* 223: 210–215.

Polydera, A.C., Stoforos, N.G., and Taoukis, P.S. 2005. Effect of high hydrostatic pressure treatment on post processing antioxidant activity of fresh Navel orange juice. *Food Chem.* 91: 495–503.

Praporscic, I., Shynkaryk, M.V., Lebovka, N.I., and Vorobiev, E. 2007. Analysis of juice colour and dry matter content during pulsed electric field enhanced expression of soft plant tissues. *J. Food Eng.* 79: 662–670.

Prestamo, G. and Arroyo, G. 2000. Preparation of preserves with fruits treated by high pressure. *Alimentaria* 318: 25–30.

Pui, Y.P., Galindo, F.G., Vicente, A., and Dejmek, P. 2008. Pulsed electric field in combination with vacuum impregnation with trehalose improves the freezing tolerance of spinach leaves. *J. Food Eng.* 88: 144–148.

Qin, B.L., Zhang, Q., Barbosa-Canovas, G.V., Swanson, B.G., and Pedrow, P.D. 1994. Inactivation of microorganisms by pulsed electric fields of different voltage waveforms. *IEEE Trans. Dielectr. Electr. Insul.* 1: 1047–1057.

Qin, B.L., Chang, F., Barbosa-Canovas, G.V., and Swanson, B.G. 1995a. Nonthermal inactivation of *S. cerevisiae* in apple juice using high electrical field pulses. *Lebensm. Wiss. Technol.* 49: 55–60.

Qin, B., Pothakamury, U.R., Vega, H., Martin, O., Barbosa-Canovas, G.V., and Swanson, B.G. 1995b. Food pasteurization using high intensity pulsed electric fields. *J. Food Technol.* 49: 55–60.

Qin, H., Nan, Q.X., and Che, R.Z. 2001. Effects of high pressure on the activity of major enzymes in beef. *Meat Res.* 3: 13–16.

Qiu, W.F., Jiang, H.H., Wang, H.F., and Gao, Y.L. 2006. Effect of high hydrostatic pressure on lycopene stability. *Food Chem.* 97: 516–523.

Rajaei, A., Barzegar, M., and Yamini, Y. 2005. Supercritical fluid extraction of tea seed oil and its comparison with solvent extraction. *Euro. Food Res. Technol.* 220: 401–405.

Rajan, S., Ahn, J., Balasubramaniam, V.M., and Yousef, A.E. 2006a. Combined pressure-thermal inactivation kinetics of *Bacillus amyloliquefaciens* spores in egg patty mince. *J. Food Prot.* 69: 853–860.

Rajan, S., Pandrangi, S., Balasubramaniam, V.M., and Yousef, A.E. 2006b. Inactivation of *Bacillus stearothermophilus* spores in egg patties by pressure-assisted thermal processing. *Lebensm. Wiss. Technol.* 39: 844–851.

Rasanayagam, V., Balasubramaniam, V.M., Ting, E., Sizer, C.E., Bush, C., and Anderson, C. 2003. Compression heating of selected fatty food materials during high-pressure processing. *J. Food Sci.* 68: 254–259.

Raso, J., Calderon, M.L., Gongora, M., Barbosa-Canovas, G.V., and Swanson, B.G. 1998a. Inactivation of *Zygosaccharomyces bailii* in fruit juices by heat, high hydrostatic pressure and pulsed electric fields. *J. Food Sci.* 63: 1042–1044.

Raso, J., Pagan, R., Condon, S., and Sala, F.J. 1998b. Influence of temperature and pressure on the lethality of ultrasound. *Appl. Environ. Microbiol.* 64: 465–471.

Rastogi, N.K. 2003. Application of high intensity pulses in food processing, *Food Rev. Int.* 19: 229–251.

Rastogi, N.K. and Niranjan, K. 1998. Enhanced mass transfer during osmotic dehydration of high pressure treated pineapples. *J. Food Sci.* 63: 508–511.

Rastogi, N.K., Subramanian, R., and Raghavarao, K.S.M.S. 1994. Application of high pressure technology in food processing. *Indian Food Indus.* 13: 30–34.

Rastogi, N.K., Angersbach, A., and Knorr, D. 2000a. Combined effect of high hydrostatic pressure pretreatment and osmotic stress on mass transfer during osmotic dehydration, Paper presented at the *8th International Congress of Food and Engineering (ICEF'8)*, Mexico. April 9–13.

Rastogi, N.K., Angersbach, A., and Knorr, D. 2000b. Synergistic effect of high hydrostatic pressure pretreatment and osmotic stress on mass transfer during osmotic dehydration. *J. Food Eng.* 45: 25–31.

Rastogi, N.K., Angersbach, A., Niranjan, K., and Knorr, D. 2000c. Rehydration kinetics of high pressure treated and osmotically dehydrated pineapple. *J. Food Sci.* 65: 838–841.

Rastogi, N.K., Angersbach, A., and Knorr, D. 2003. Combined effect of high hydrostatic pressure pretreatment and osmotic stress on mass transfer during osmotic dehydration. In *Transport Phenomena in Food Processing*, eds. J. Welti-Chanes, J.F.V. Ruis, and G.V. Barbosa-Canovas, pp. 109–121. Boca Raton, FL: CRC Press.

Rastogi, N.K., Raghavarao, K.S.M.S., Balasubramaniam, V.M., Niranjan, K., and Knorr, D. 2007. Opportunities and challenges in high pressure processing of foods. *Crit. Rev. Food Sci. Nutr.* 47: 69–112.

Rastogi, N.K., Nguyen, L.T., and Balasubramaniam, V.M. 2008a. Effect of pretreatments on carrot texture after thermal and pressure-assisted thermal processing. *J. Food Eng.* 88: 541–547.

Rastogi, N.K., Nguyen, L.T., and Balasubramaniam, V.M. 2008b. Improvement in texture of pressure-assisted thermally processed carrots using response surface methodology. *Food Bioprocess. Technol.* (in press).

Reddy, N.R., Solomon, H.M., Tetzloff, R.C., and Rhodehamel, E.J. 2003. Inactivation of *Clostridium botulinum* type A spores by high-pressure processing at elevated temperatures. *J. Food Prot.* 66: 1402–1407.

Reddy, N.R., Tetzloff, R.C., Solomon, H.M., and Larkin, J.W. 2006. Inactivation of *Clostridium botulinum* nonproteolytic type B spores by high pressure processing at moderate to elevated high temperatures. *Innovat. Food Sci. Emerg. Technol.* 7: 169–175.

Riera, E., Golás, Y., Blanco, A., Gallego, A., Blasco, M., and Mulet, A. 2004. Mass transfer enhancement in supercritical fluids extraction by means of power ultrasound. *Ultrason. Sonochem.* 11: 241–244.

Rodrigues, S. and Fernandes, F.A.N. 2007. Use of ultrasound as pretreatment for dehydration of melons. *Drying Technol.* 25: 1791–1796.

Rodrigues, S. and Pinto, G.A.S. 2007. Ultrasound extraction of phenolic compounds from coconut (*Cocos nucifera*) shell powder. *J. Food Eng.* 80: 869–872.

Rostagno, M.A., Palma, M., and Barroso, C.G. 2003. Ultrasound-assisted extraction of soy isoflavones. *J. Chromatogr. A* 1012: 119–128.

Rouille, J., Lebail, A., Ramaswamy, H.S., and Leclerc, L. 2002. High pressure thawing of fish and shellfish. *J. Food Eng.* 53: 83–88.

Rovere, P., Gola, S., Maggi, A., Scaramuzza, N., and Miglioli, L. 1998. Studies on bacterial spores by combined pressure-heat treatments: Possibility to sterilize low-acid foods. In *High Pressure Food Science, Bioscience and Chemistry*, ed. N.S. Isaacs, pp. 354–363. Cambridge, U.K.: The Royal Society of Chemistry.

Sala, F.J., Burgos, J., Condón, S., López, P., and Raso, J. 1995. Effect of heat and ultrasound on microorganisms and enzymes. In *New Methods of Food Preservation*, ed. W. Gould, pp. 176–204. London, U.K.: Blackie Academic and Professional.

Saldo, J., McSweeney, P.L.H., Sendra, E., Kelly, A.L., and Guamis, B. 2002. Proteolysis in caprine milk cheese treated by high pressure to accelerate cheese ripening. *Int. Dairy J.* 12: 35–44.

Sale, A.J.H. and Hamilton, W.A. 1967a. Effects of high electric fields on microorganisms. I. Killing of bacteria and yeasts. *Biochim. Biophys. Acta* 148: 781–788.

Sale, A.J.H. and Hamilton, W.A. 1967b. Effects of high electric fields on microorganisms. II. Mechanism of action of the lethal effect. *Biochim. Biophys. Acta* 148: 789–800.

Sams, A.R. and Feria, R. 1991. Microbial effects of ultrasonication of broiler drumstick skin. *J. Food Sci.* 56: 247–248.

Sanchez, E.S., Simal, S., Femenia, A., Benedito, J., and Rossello, C. 1999. Influence of ultrasound on mass transport during cheese brining. *Eur. Food Res. Technol.* 209: 215–219.

Sanchez, M.C., Plaza, L., Elez, M.P., de Ancos, B., Martin, B.O., and Cano, M.P. 2005. Impact of high-pressure and pulsed electric field on bioactive compounds and antioxidant activity of orange juice in comparison with traditional thermal processing. *J. Agric. Food Chem.* 53: 4403–4409.

Sanchez, M.C., Plaza, L., de Ancos, B., and Cano, M.P. 2006. Impact of high-pressure and traditional thermal processing of tomato puree on carotenoids, vitamin C and antioxidant activity. *J. Sci. Food Agric.* 86: 171–179.

Sanz, B., Palacios, P., Lopez, P., and Ordonez, J.A. 1985. Effect of ultrasonic waves on the heat resistance of *Bacillus stearothermophilus* spores. In *Fundamental and Applied Aspects of Bacterial Spores*, eds. G.J. Dring, D.J. Ellar, and G.W. Gould, pp. 251–259. New York: Academic Press.

Schauwecker, A., Balasubramaniam, V.M., Sadler, G., Pascall, M.A., and Adhikari, C. 2002. Influence of high pressure processing on selected polymeric materials and on the migration of a pressure transmitting fluid. *Pack. Technol. Sci.* 15: 255–262.

Schilling, S., Alber, T., Toepfl, S., Neidhart, S., Knorr, D., Schieber, A., and Carle, R. 2007. Effects of pulsed electric field treatment of apple mash on juice yield and quality attributes of apple juices. *Innovat. Food Sci. Emerg. Technol.* 8: 127–134.

Schilling, S., Toepfl, S., Ludwig, M., Dietrich, H., Knorr, D., Neidhart, S., Schieber, A., and Carle, R. 2008. Comparative study of juice production by pulsed electric field treatment and enzymatic maceration of apple mash. *Eur. Food Res. Technol.* 226: 1389–1398.

Schubring, R., Meyer, C., Schlueter, O., Boguslawski, S., and Knorr, D. 2003. Impact of high pressure assisted thawing on the quality of fillets from various fish species. *Innovat. Food Sci. Emerg. Technol.* 4: 257–267.

Scouten, A.J. and Beuchat, L.R. 2002. Combined effects of chemical, heat and ultrasound treatments to kill *Salmonella* and *Escherichia coli* O157:H7 on alfalfa seeds. *J. Appl. Microbiol.* 92: 668–674.

Senapati, N. 1991. Ultrasound in chemical processing. In *Advances in Sonochemistry*, ed. T.J. Mason, pp. 187–210. London, U.K.: JAI Press.

Seymour, I.J., Burfoot, D., Smith, R.L., Cox, L.A., and Lockwood, A. 2002. Ultrasound decontamination of minimally processed fruits and vegetables. *Int. J. Food Sci. Technol.* 37: 547–557.

Sierra, I., Vidal, V.C., and Lopez, F.R. 2000. Effect of high pressure on the vitamin B1 and B6 content of milk. *Milchwissenschaft* 55: 365–367.

Sila, D.N., Smout, C., Vu, T.S., and Hendrickx, M.E. 2004. Effects of high-pressure pretreatment and calcium soaking on the texture degradation kinetics of carrots during thermal processing. *J. Food Sci.* 69: E205–E211.

Sila, D.N., Smout, C., Vu, S.T., Van Loey, A.M., and Hendrickx, M. 2005. Influence of pretreatment conditions on the texture and cell wall components of carrots during thermal processing. *J. Food Sci.* 70: E85–E91.

Simal, S., Benedito, J., Sanchez, E.S., and Rossello, C. 1998. Use of ultrasound to increase mass transport rates during osmotic dehydration. *J. Food Eng.* 36: 323–336.

Sitzmann, W. 1995. High voltage pulse techniques for food preservation. In *New Methods for Food Preservation*, ed. G.W. Gould, pp. 236–252. London, U.K.: Blackie Academic and Professional.

Sopanangkul, A., Ledward, D.A., and Niranjan, K. 2002. Mass transfer during sucrose infusion into potatoes under high pressure. *J. Food Sci.* 67: 2217–2220.

Stojanovic, J. and Silva, J.L. 2007. Influence of osmotic concentration, continuous high frequency ultrasound and dehydration on antioxidants, color and chemical properties of rabbit eye blueberries. *Food Chem.* 101: 898–906.

Strolham, J., Valentova, H., Houska, M., Novotna, P., Landfeld, A., Kyhos, K., and Gree, R. 2000. Changes in quality of natural orange juice pasteurized by high pressure during storage. *Czech J. Food Sci.* 18: 187–193.

Styles, M.F., Hoover, D.G., and Farkas, D.F. 1991. Response of *Listeria monocytogenes* and *Vibrio Parahaemolyticus* to high hydrostatic pressure. *J. Food Sci.* 56: 1404–1407.

Sun, X., Sun, F., Zhao, H., and Sun, C. 2004. Isolation and characterization of cellulose from sugarcane bagasse. *Poly Degrad. Stability* 84: 331–339.

Suslick, K.S. Homogeneous sonochemistry. 1988. In *Ultrasound. It's Chemical, Physical, and Biological Effects*, ed. K.S. Suslick, pp. 123–163. New York: VCH Publishers, Inc.

Suzuki, A., Kim, K., Homma, N., Ikeuchi, Y., and Saito, M. 1992. Acceleration of meat conditioning by high pressure treatment. In *High Pressure and Biotechnology*, eds. C. Balny, R. Hayashi, K. Heremans, and P. Mason, pp. 224, 219–227. Paris: Coll. Inserm.

Suzuki, A., Watanabe, M., Ikeuchi, Y., Saito, M., and Takahashi, K. 1993. Effects of high pressure treatment on the ultrastructure and thermal behavior of beef intramuscular collagen. *Meat Sci.* 35: 17–25.

Taiwo, K.A., Angersbach, A., Ade-Omowaye, B.I.O., and Knorr, D. 2001. Effect of pretreatment on the diffusion kinetics and some quality parameters of osmotically dehydrated apple slices. *J. Agric. Food Chem.* 49: 2804–2811.

Taiwo, K.A., Angersbach, A., and Knorr, D. 2002. Influence of high electric field pulses and osmotic dehydration characteristics of apple slices at different temperatures. *J. Food Eng.* 52: 185–192.

Taiwo, K.A., Angersbach, A., and Knorr, D. 2003. Effects of pulsed electric field on quality factors and mass transfer during osmotic dehydration of apples. *J. Food Process Eng.* 26: 31–48.

Takahashi, F., Pehrsson, P.E., Rovere, P., and Squarcina, N. 1998. High-pressure processing of fresh orange juice. *Industria Conserve* 73: 363–368.

Tang, W. and Eisenbrand, G. 1992. *Chinese Drugs from Plant Origin.* Springer, Berlin, Germany, pp. 710–737.

Tanzi, E., Saccani, G., Barbuti, S., Grisenti, M.S., Lori, D., Bolzoni, S., and Parolari, G. 2004. High pressure treatment of raw ham. Sanitation and impact on quality. *Industria Conserve* 79: 37–50.

Tarleton, E.S. and Wakeman, R.J. 1997. Ultrasonically assisted separation processes. In *Ultrasound in Food Processing*, eds. M.J.W. Povey and T.J. Mason, pp. 193–218. London, U.K.: Thomson Science.

Tedjo, W., Taiwo, K.A., Eshtiaghi, M.N., and Knorr, D. 2002. Comparison of pretreatment methods on water and solid diffusion kinetics of osmotically dehydrated mangoes. *J. Food Eng.* 53: 133–142.

Thakur, B.R. and Nelson, P.E. 1997. Inactivation of lipoxygenase in whole soy flour suspension by ultrasonic cavitation. *Die Nahrung.* 41: 299–301.

Thakur, B.R. and Nelson, P.E. 1998. High pressure processing and preservation of foods. *Food Rev. Int.* 14: 427–447.

Ting, E., Balasubramaniam, V.M., and Raghubeer, E. 2002. Determining thermal effects in high-pressure processing. *Food Technol.* 56: 31–35.

Trujillo, A.J., Royo, B., Guamis, B., and Ferragut, V. 1999a. Influence of pressurization on goat milk and cheese composition and yield. *Milchwissenschaft* 54: 197–199.

Trujillo, A.J., Royo, C., Ferragut, V., and Guamis, B. 1999b. Ripening profiles of goat cheese produced from milk treated with high pressure. *J. Food Sci.* 64: 833–837.

Ugarte-Romero, E., Hao, F., Martin, S.E., Cadwallader, K.R., and Robinson, S.J. 2006. Inactivation of *Escherichia coli* with power ultrasound in apple cider. *J. Food Sci.* 71: E102–E108.

Vachon, J.F., Kheadr, E.E., Giasson, J., Paquin, P., and Fliss, I. 2002. Inactivation of foodborne pathogens in milk using dynamic high pressure. *J. Food Prot.* 65: 345–352.

Vega-Mercado, H., Martin Belloso, O., Chang, F.J., Barbosa-Canovas, G.V., and Swanson, B.G. 1996. Inactivation of *E. coli* and *Bacillus subtilis* suspended in pea soup using pulsed electric field. *J. Food Process. Preserv.* 20: 117–121.

Vega-Mercado, H., Martin-Belloso, O., Qin, B.L., Barbosa-Canovas, G.V., and Swanson, B.G. 1997. Non thermal food preservation: Pulsed electric fields. *Trends Food Sci. Technol.* 8: 151–157.

Vercet, A., Lopez, P., and Burgos, J. 1997. Inactivation of heat-resistant lipase and protease from *Pseudomonas fluorescens* by manothermosonication. *J. Dairy Sci.* 80: 29–36.

Vercet, A., Lopez, P., and Burgos, J. 1999. Inactivation of heat-resistant pectinmethylesterase from orange by manothermosonication. *J. Agric. Food Chem.* 47: 432–437.

Vercet, A., Sanchez, C., Burgos, J., Montañes, L., and Lopez Buesa, P. 2002. The effects of manothermosonication on tomato pectic enzymes and tomato paste rheological properties. *J. Food Eng.* 53: 273–278.

Viazis, S., Farkas, B.E., and Jaykus, L.A. 2008. Inactivation of bacterial pathogens in human milk by high-pressure processing. *J. Food Prot.* 71: 109–118.

Vilkhu, K.S., Mawson, R., Simons, L., and Bates, D. 2008. Applications and opportunities for ultrasound assisted extraction in the food industry: A review. *Innovat. Food Sci. Emerg. Technol.* 9: 161–169.

Villacis, M.F., Rastogi, N.K., and Balasubramaniam, V.M. 2008. Effect of high pressure on moisture and NaCl diffusion into turkey breast. *Lebensm. Wiss. Technol.* 41: 836–844.

Villamiel, M. and Jong, P.D. 2000. Influence of high-intensity ultrasound and heat treatment in continuous flow on fat, proteins, and native enzymes of milk. *J. Agric. Food Chem.* 48: 472–478.

Watanabe, M., Arai, E., Kumeno, K., and Honma, K. 1991. A new method for producing a non heated jam sample: The use of freeze concentration and high pressure sterilization. *Agric. Biol. Chem.* 55: 2175–2176.

Wennberg, M. and Nyman, M. 2004. On the possibility of using high pressure treatment to modify physico-chemical properties of dietary fibre in white cabbage (*Brassica oleracea* var. capitata). *Innovat. Food Sci. Emerg. Technol.* 5: 171–177.

Wilson, D.R., Dabrowski, L., Stringer, S., Moezelaar, R., and Brocklehurst, T.F. 2008. High pressure in combination with elevated temperature as a method for the sterilization of food. *Trends Food Sci. Technol.* 19: 289–299.

Wu, H., Hulbert, G.J., and Mount, J.R. 2001. Effects of ultrasound on milk homogenization and fermentation with yoghurt starter. *Innovat. Food Sci. Emerg. Technol.* 1: 211–218.

Xia, T., Shi, S., and Wan, X. 2006. Impact of ultrasonic-assisted extraction on the chemical and sensory quality of tea infusion. *J. Food Eng.* 74: 557–560.

Xiaohua, Y., Zhimin, X., Witoon, P., and Joan, K. 2006. Improving extraction of lutein from egg yolk using an ultrasound-assisted solvent method. *J. Food Sci.* 71: C239–C241.

Xu, G., Zhang, H., and Hu, J. 2000. Leaching method of flavone from bamboo leaves. *Chin. J. Anal. Chem.* 28: 857–859.

Yagiz, Y., Kristinsson, H.G., Balaban, M.O., and Marshall, M.R. 2007. Effect of high pressure treatment on the quality of rainbow trout (*Oncorhynchus mykiss*) and mahi mahi (*Coryphaena hippurus*). *J. Food Sci.* 72: C509–C515.

Yeom, H.W., Streaker, C.B., Zhang, Q.H., and Min, D.B. 2000. Effects of pulsed electric fields on the quality of orange juice and comparison with heat pasteurization. *J. Agric. Food Chem.* 48: 4597–4605.

Yeom, H.W., Evrendilek, G.A., Jin, Z.T., and Zhang, Q.H. 2001. Processing of yogurt based product with pulsed electric fields. Paper presented in *IFT Annual Meeting*, New Orleans, LA.

Yin, Y., Han, Y., and Liu, J. 2007. A novel protecting method for visual green color in spinach puree treated by high intensity pulsed electric fields. *J. Food Eng.* 79: 1256–1260.

Yokohama, H., Swamura, N., and Motobyashi, N. 1991. Method for accelerating cheese ripening. European Patent EP 0 469 857 A1, filed July 30, 1991 and issued February 5, 1992.

Yoon, K.J., Park, S.O., and Bong, S.N. 2000. Inactivation of peroxidase by hurdle technology. *Food Sci. Biotechnol.* 9: 124–129.

Yun, D. and Yanyun, Z. 2008. Effects of pulsed-vacuum and ultrasound on the osmodehydration kinetics and microstructure of apples (Fuji). *J. Food Eng.* 85: 84–93.

Yutang, W., Jingyan, Y., Yong, Y., Chenling, Q., Huarong, Z., Lan, D., Hanqi, Z., and Xuwen, L. 2008. Analysis of ginsenosides in Panax ginseng in high pressure microwave-assisted extraction. *Food Chem.* 110: 161–167.

Zenker, M., Heinz, V., and Knorr, D. 2003. Application of ultrasound-assisted thermal processing for preservation and quality retention of liquid foods. *J. Food Prot.* 66: 1642–1649.

Zhang, Q.H., Qiu, X., and Sharma, S.K. 1997. Recent developments in pulsed electric field processing. In *New Technologies Year Book*, pp. 31–42. Washington, DC: National Food Processors Association.

Zhang, T., Niu, X., Eckhoff, R., and Feng, H. 2005. Sonication enhanced cornstarch separation. *Starch-Starke* 57: 240–245.

Zhao, B., Jiang, Y., Basir, O.A., and Mittal, G.S. 2004. Foreign body detection in foods using the ultrasound pulse/echo method. *J. Food Qual.* 27: 274–288.

Zhao, B., Yang, P., Basir, O.A., and Mittal, G.S. 2006. Ultrasound based glass fragments detection in glass containers filled with beverages using neural networks and short time Fourier transform. *Food Res. Int.* 39: 686–695.

Zhao, B., Basir, O.A., and Mittal, G.S. 2007a. Maximum components integration for image processing: An application of ultrasound for detection of small objects in containers. *J. Food Pro. Eng.* 30: 393–405.

Zhao, B., Basir, O.A., and Mittal, G.S. 2007b. Hybrid of multi-signal processing methods for detection of small objects in containers filled with beverages using ultrasound. *Lebensm. Wiss. Technol.* 40: 655–660.

Zheng, L. and Sun, D.W. 2005. Ultrasonic assistance of food freezing. In *Emerging Technologies for Food Processing*, ed. D.W. Sun, pp. 603–627. London, U.K.: Elsevier Ltd.

Zheng, L. and Sun, D.W. 2006. Innovative applications of power ultrasound during food freezing processes—A review. *Trends Food Sci. Technol.* 17: 16–23.

Zhu, S., Le Bail, A., Ramaswamy, H.S., and Chapleau, N. 2004a. Characterization of ice crystals in pork muscle formed by pressure-shift freezing as compared with classical freezing methods. *J. Food Sci.* 69: E190–E197.

Zhu, S.M., Le Bail, A., Chapleau, N., Ramaswamy, H.S., and de Lamballerie-Anton, M. 2004b. Pressure shift freezing of pork muscle: Effect on color, drip loss, texture, and protein stability. *Biotechnol. Prog.* 20: 939–945.

Zhu, S., Ramaswamy, H.S., and Simpson, B.K. 2004c. Effect of high-pressure versus conventional thawing on color, drip loss and texture of Atlantic salmon frozen by different methods. *Lebensm. Wiss. Technol.* 37: 291–299.

Zhu, S.M., Bail, A.L., and Ramaswamy, H.S. 2003. Ice crystal formation in pressure shift freezing of Atlantic salmon (Salmo salar) as compared to classical freezing methods. *J. Food Process. Preserv.* 27: 427–444.

Zobrist, M.R., Huppertz, T., Uniacke, T., Fox, P.F., and Kelly, A.L. 2005. High pressure induced changes in the rennet coagulation properties of bovine milk. *Int. Dairy J.* 15: 655–662.

2 Trends in Breadmaking: Low and Subzero Temperatures

Cristina M. Rosell

CONTENTS

2.1 INTRODUCTION

Some of the fastest growing segments in the food processing industry are bakery products. Commercial bakeries have very soon understood the changes in consumer lifestyles and have shifted their production processes, products, and even distribution channels to meet the new requirements. Consumers' demands and needs become more important and bakeries are facing new challenges for satisfying them. Fresh bread all day long and a wide variety of baked products with new flavors, shapes, and sizes are some of the consumers' demands. The progress in food science and technology has also played a fundamental role in this business success. The main players in this picture are the key consumers demanding product quality and innovation, the manufacturers offering a range of new products, and the retailers focused on product differentiation and distribution, and they constitute the market drivers identifying the emerging opportunities of the global bakery market. A first-glance overview of bakery evolution shows that the already defined three mega-trends (Lord 2002), convenience, health, and indulgence or pleasure, are the driving forces of the global bakery market. Those mega-trends are embodied within eight particular categories: functional and fortified, low and light, natural and organic, which are related to health; snacks, easy to prepare and the use of bake off technology (BOT) based on convenience; and finally ethnic and exotic, indulgent and premium, and novelty and fun that covered pleasure or indulgence (Lord 2002). Manufacturers and commercial bakeries cannot ignore these trends when planning new product development strategies.

Convenience bakery products solve the problem of food preparation and shopping imposed by accelerated consumer lifestyles, increasing number of women in the workforce, fragmentation of the traditional family, and growing single-occupancy households.

Nevertheless, the demand for convenience products is not limited to consumers in developed countries, as reflected by the rapid growth of the fast food sectors in developing countries. Food retailers and manufacturers in developing country markets are adjusting themselves to satisfy consumer demand for different food products. In fact, retail food markets in Latin America and developing countries in Asia are facing changes similar to those observed in Eastern Europe. Some of the Asian countries like China and Vietnam are at the initial stages of this transformation with smaller penetration of multinational companies (Regmi and Gehlhar 2005).

Traditional baking processes are sometimes too limited and inflexible to fully satisfy manufacturer's requirements and consumer's demands. One of the most significant technological developments in the bakery sector in the last few decades is the BOT applied to frozen or low temperature stored products allowing a continuous growth of easily prepared bakery items at home and in institutional markets. Freezing dough, either fermented or not, par-baked or fully baked, becomes in many cases necessary to face the present demands. Low-temperature technology has been initially applied to bakery products to solve the economical losses associated with the bread staling problem that decreases consumer acceptance (Hebeda et al. 1990). These low-temperature goods allow fresh products to be available at any time of day. Additionally, since BOT does not require highly trained people, it saves costs and ensures a product of uniform quality at any time.

The use of freezing technology in this market is not new, but in the last years it has been converted into a mainstream business motivated by the diversification of bakery products and the centralization of bread production with numerous distribution points. Low-temperature technology can be applied to all the bakery products compiled within the three mega-trends. The application of low temperatures in breadmaking easily provides different types of bread and forms (fresh, refrigerated, and frozen) that increase bread attractiveness, and, as a result, it contributes to the growth of the market. Bakeries and retailers are looking to frozen technology as the way to grow their business in the future since this technology responds to the increased demand of convenience, health, and quality.

2.2 ROLE OF LOW TEMPERATURES IN THE BREADMAKING PROCESS

2.2.1 Conventional Breadmaking Process

Although there is extensive literature about the process of breadmaking, a brief description will be given here to clarify the role of low temperatures in this process. Figure 2.1 shows the different stages of breadmaking. These basic stages include

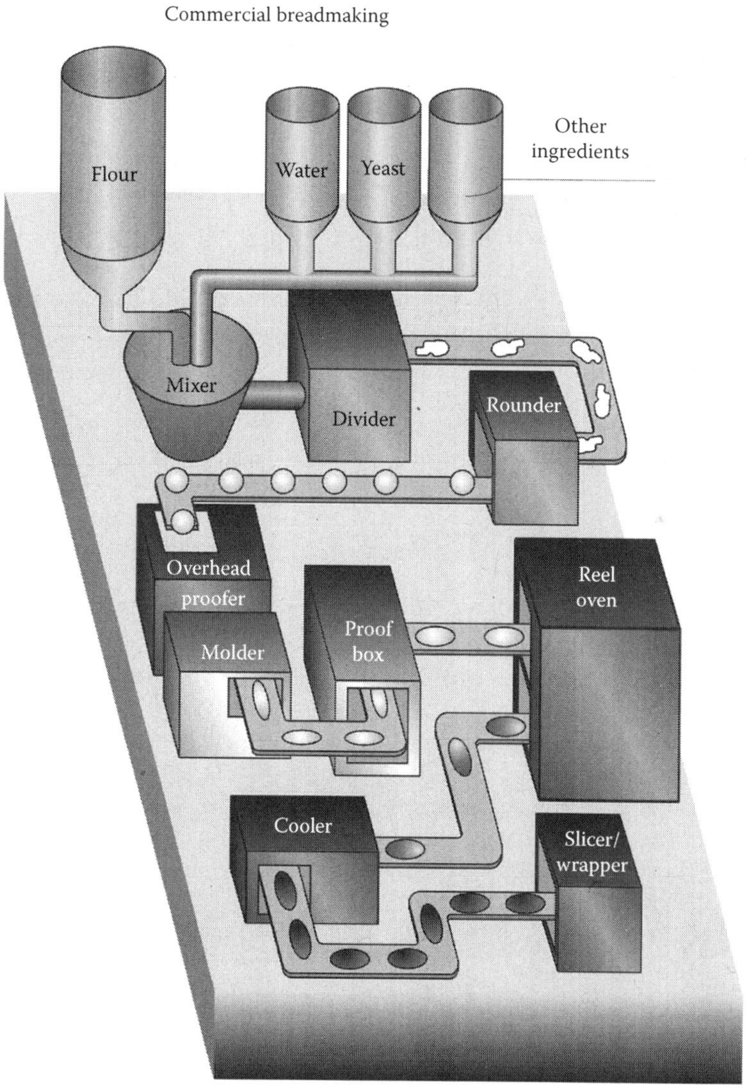

FIGURE 2.1 Commercial breadmaking at different stages.

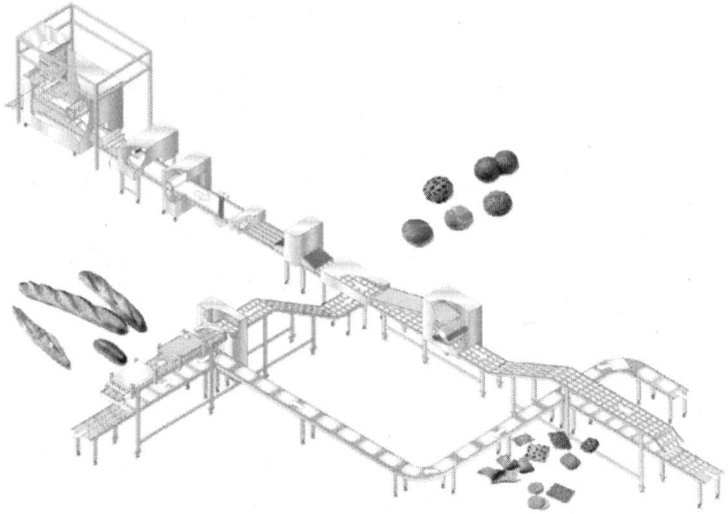

FIGURE 2.2 Scheme of breadmaking production for different types of bakery products. (From Rondo Doge, Switzerland, http://www.sermont.es/rondo/Rondo_ind_panificacion.htm)

mixing the ingredients, dough resting, dividing and shaping, proofing, and baking, with a great diversification in the intermediate stage depending on the type of product, as schematized in Figure 2.2. Globally, wheat nourishes more people and is the base of more foods than any other grain. In addition, wheat suits the preparation of leavened bread (Rosell 2007). Thus, hereafter, the breadmaking process should refer to wheat bread products.

Returning to Figure 2.1, one should note that breadmaking is a dynamic process with continuous physicochemical changes in its components, the most important of which occur along mixing, proofing, and baking. During initial mixing, wheat dough is exposed to large uni- and biaxial deformations (Bollaín and Collar 2004). Moreover, the material distribution, the disruption of the initially spherical protein particles, and the flour compound hydration occur simultaneously and together with the stretching and alignment of the proteins, leading to the formation of a three-dimensional viscoelastic structure with gas-retaining properties. The properties of this network are governed by the quaternary structures resulting from disulfide-linked polymer proteins and hydrogen bonding aggregates (Aussenac et al. 2001). Bread dough is a viscoelastic material that exhibits an intermediate rheological behavior between a viscous liquid and an elastic solid, which enables it to retain the gas produced during the fermentation stage, yielding an aerated crumb bread structure. The viscoelastic network plays a predominant role on both dough machinability and textural characteristics of the finished bread (Collar and Armero 1996, Uthayakumaran et al. 2000, Rosell and Foegeding 2007).

During proofing or fermentation, the yeast metabolism results in carbon dioxide release and growth of air bubbles previously incorporated during mixing, yielding the expansion of the dough, which inflates to larger volumes and thinner cell walls before collapsing. Dough tensile and biaxial large deformations (associated with polymer chain branching and entanglement interactions between high-molecular-weight polymers) have significant relationships to bread quality parameters such as volume and form ratio (Bollaín et al. 2006). During proofing and baking, the growth of gas bubbles determines the characteristics of the bread's aerated structure and thus the ultimate volume and texture of the baked product. During the baking process, flour compounds are subjected to mechanical work and heat treatment that promote changes in their physicochemical properties (Collar 2003, Rosell et al. 2007, Rosell and Foegeding 2007).

2.2.2 Low Temperature Breadmaking Processes

Low and subzero temperatures can be applied to the breadmaking performance at different points of the process as shown in Figure 2.3. Chilling or refrigeration can be used as a short storage practice in bakeries, when products are baked on the following day. This technique can be applied to dough, partially baked bread, or full-baked products, although the starch retrogradation, and thus staling, is accelerated at temperatures between the glass transition (T_{glass}) and the melting temperature (T_{melt}) where nucleation and propagation phenomena are balanced (Slade and Levine 1987).

The most extended practice of small artisan bakers is the use of controlled proofing or fermentation. This technique requires a special cabinet with temperature and humidity control that can retard or accelerate the fermentation operation (Figure 2.4), allowing almost a complete stop of proofing during the night and yeast activation during the early morning for having the loaves ready for consumers.

However, the main representative bakery products that use frozen technology are frozen dough, par-baked bread, and frozen breads. Frozen dough is the major alternative to conventional or traditional breadmaking. Frozen dough allows large-scale centralized production, storage, and distribution in the frozen state, and proofing and baking in retailers or bakeries. It requires additional cost for freezing, transportation, and frozen storage, as well as some training and experience for finishing the product in the in-store bakery. The use of frozen dough is very attractive because of the low volume of the unfermented dough, which is very convenient when storage is involved. During the first 2 months of frozen storage, samples undergo the main changes, reaching an asymptotic behavior after 2–3 months and remaining stable thenceforth, having satisfying quality characteristics even after 9 months of storage (Giannou and Tzia 2007). Frozen dough should have 16 weeks shelf life if the dough is not subjected to temperature oscillations during transportation and storage. Temperature fluctuations during frozen storage can cause dramatic changes in dough and bread quality. Phimolsiripol et al. (2008) have indicated that large temperature fluctuations (from −18°C up to −8°C) during frozen storage should result in significantly more rapid loss of dough and bread quality than storage at constant and controlled temperatures (−18°C ± 0.1°C). These temperature fluctuations are responsible for an increase in breadcrumb firmness, a low CO_2 production, and loss of yeast viability and bread specific volume. A variation of frozen dough is the preproofed frozen dough that has been developed to shorten the time that it takes to prepare

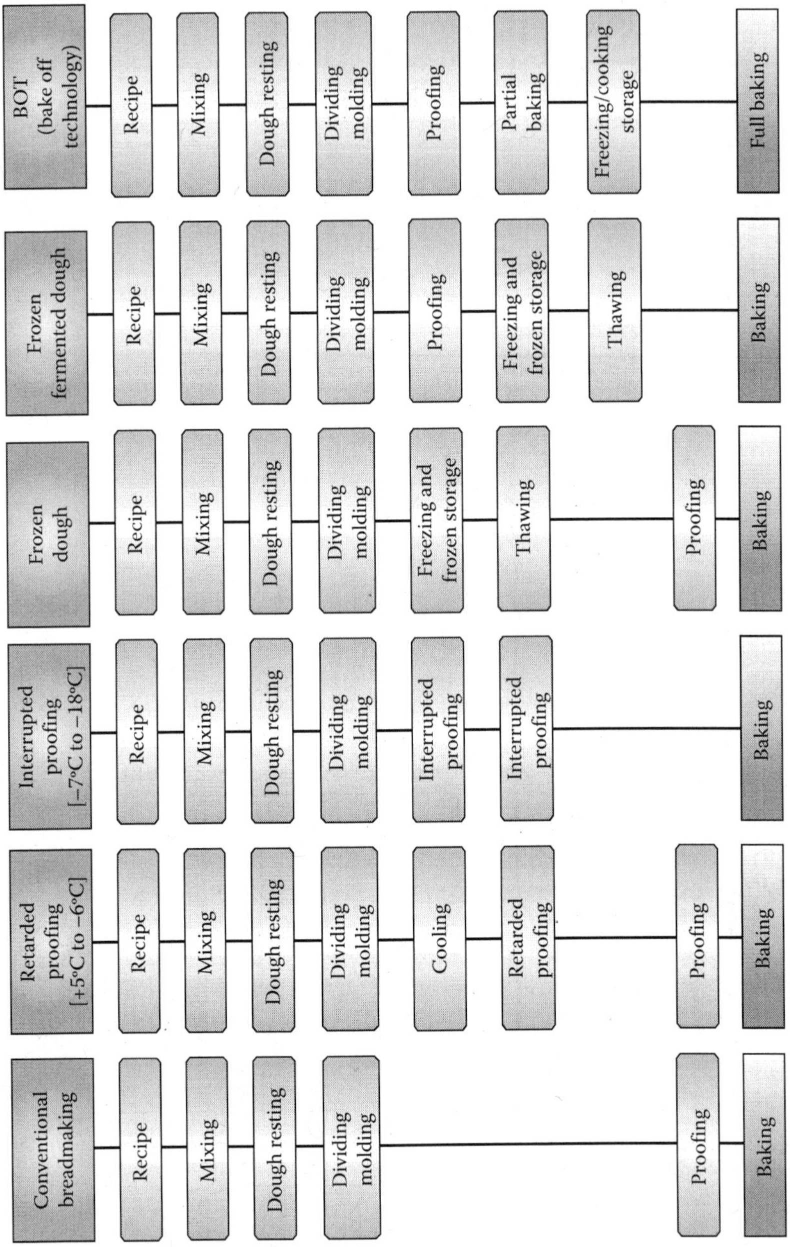

FIGURE 2.3 Methods of breadmaking with the application of subzero or low temperatures.

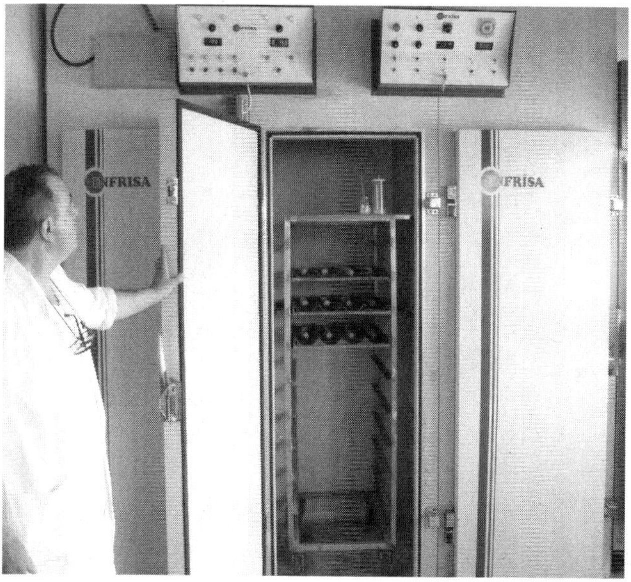

FIGURE 2.4 Fermentation cabinet with temperature and humidity controller. (Photo from C. Rosell.)

fresh bread from frozen dough. These products are fermented before freezing and they can be placed in a programmable oven while frozen. The main disadvantages are the higher frozen storage cost due to the larger volume of fermented products, the low volume of the final bread, and the cost of the special ingredients and improvers for optimizing the product quality (Lallemand Baking Update 2008).

A different alternative is the partially baked bread (part-baked, par-baked bread, or prebaked bread), produced in large plant bakeries following the conventional process, with the exception of baking. This technology is also called BOT and it has proved itself as a good method for preventing the staling process and obtaining a product at any time of the day for which the quality is close to that of fresh bread. The partial baking or interrupted baking method consists in baking the bread dough until the structure is fixed, giving the product a structured crumb but without a crunchy crust (Figure 2.5). In the retail bakery, the partially baked bread only requires a very short baking time to generate the crust and to release an attractive flavor. The market of partially baked bread is rapidly growing because the product is already sized, shaped, and partially baked; thus no skilled personnel are needed at the retails for finishing the product. Nevertheless, several studies have been conducted for assessing the effect of proofing (Le Bail et al. 2005), partial baking (Fik and Surowka 2002, Bárcenas et al. 2003a,b), and chilling and freezing conditions (Bárcenas and Rosell 2006a,b) on bread quality. A major problem is the critical time and temperature control required for the two steps of baking. Frozen bread with 71% fraction of baking time shows high stability of sensory features and of rheological parameters during a storage period of 11 weeks and after thawing. Full baking has a superior quality in comparison with its frozen and thawed full-baked counterparts (Fik and

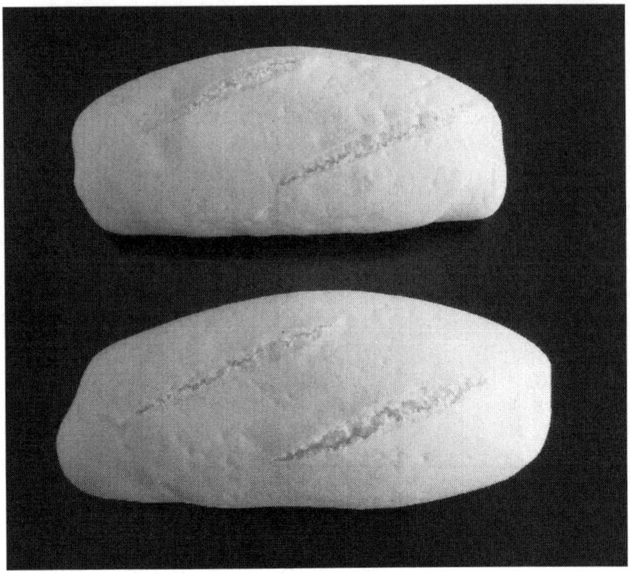

FIGURE 2.5 Par-baked breads. (Photo from M. Barcenas.)

Surowka 2002). The optimal time for the initial partial baking lies within the range from 74% to 86% of the time needed for the full baking in conventional breadmaking (Fik and Surowka 2002). Par-baked bread can be stored under frozen or low temperature conditions. Par-baked frozen bread is suited to produce crusty bread; the major disadvantage is the high storage cost for the large product volume. Par-baked loaves stored at low temperatures show progressive crumb hardening and rapid crystallization of the amylopectin chains, but the heat applied during the full-baking process can reverse those processes, and the extent of that improvement is directly related to the duration of par-baked bread storage (Bárcenas and Rosell 2007). When par-baked loaves are stored under subzero temperatures, no retrogradation of amylopectin is detected during storage, but some structural changes are produced in the starch as indicated by the increase in amylopectin enthalpy observed during aging of the full-baked breads (Bárcenas et al. 2003a,b). Another alternative is the Milton-Keynes process, designed to stabilize the par-baked breads without freezing or refrigeration (Lallemand Baking Update 2008). It consists in using a vacuum cooler to stabilize the crusty par-baked structure without producing wrinkling or shriveling of the loaf during storage at ambient temperatures.

In order to extend the freshness of bread, a household practice is to keep the bakery products frozen and use a microwave oven for thawing them. Sometimes bakeries freeze the products to facilitate production planning, in which case bread should be frozen immediately after baking and should be frozen and thawed as quickly as possible. The resulting reheated products have been frequently associated with a low quality, derived from their lack of flavor, nonuniform heating, tough and rubbery crust, and rapid staling due to the recrystallization of the starch (Sumnu 2001). However, chilling affects bread to different extents depending on which stage of breadmaking it is applied in (Mandala et al. 2007).

2.2.3 Physicochemical Changes during Breadmaking Associated with Low Temperatures

Freezing affects the baking performance of frozen bread dough. The volume of the dough during proofing is related to two main factors, namely, yeast and the gluten network. Concerning yeast, a slow freezing rate is usually recommended in the literature so as to preserve its activity (Rosell and Gómez 2007). The gluten network is damaged during freezing, affecting its ability to retain CO_2, which is reflected in lowered specific volumes. Freezing and thawing operations exert some stress on the dough that cause a deterioration in the baked product quality, mainly affecting the wheat flour protein fraction and the baker yeast shelf life. Consequently, extended proof times are needed and reduced loaf volumes are obtained from frozen dough (Phimolsiripol et al. 2008).

Yeast activity is strongly related to its location inside the dough, being higher at the surface section whatever the freezing rate and the frozen storage duration may be (Havet et al. 2000). The freezing rate and the frozen storage conditions have strong influence on yeast activity. Le Bail et al. (1998) have reported that low freezing rate conditions (air velocity of 1 m/s: −20°C) result in the highest yeast activity, although it decreases after a 4-week storage. In contrast, high freezing rate conditions (air velocity of 4 m/s: −40°C) result in the lowest gassing power.

During breadmaking, the survival yeast counts are constant in several steps of the process before baking in nonfrozen and frozen dough. Phimolsiripol et al. (2008) noticed a 9% decrease in the yeast viability due to the freezing process alone. Freezing and frozen storage of dough made with fresh yeast has more negative effects on baking quality than the addition of frozen yeast to dough (Ribotta et al. 2003a). When compressed yeast is frozen and stored at −18°C, the CO_2 production decreases, whereas the amount of dead cells, total protein, and reducing substances leached from the yeast increases. The yeast leakages have an adverse effect on loaf volume (Ribotta et al. 2003a).

Concerning the dough structure deterioration, the presence of ice crystals formed during freezing can produce the rupture of the gluten structure leading to baked breads with a flat surface, harder crumbs with a coarse texture, and uneven distribution of great air cells. Because of this, the rate of freezing and the frozen temperature have an important effect on the frozen dough quality (Sharadanant and Khan 2003a,b).

Frozen doughs undergo rheological and microstructural changes from the first day of frozen storage. Inoue et al. (1994) observed a steady decrease in the maximum resistance from the first day of frozen storage, whereas dough extensibility increased significantly only after 70 days of frozen storage and it was related to the decrease in gassing power. In addition, the adhesiveness of thawed dough increased with the length of the storage period (Takasaki and Karasawa 1992). Rheological strength of the frozen dough (extensigraph resistance and storage modulus) declined throughout frozen dough storage (Lu and Grant 1999).

Fundamental rheological studies have shown that during the freezing step, ice crystals compress the gluten and a significant phase separation is observed between gluten and ice. Moreover, frozen storage induces a decrease in stress and strain at breaking, but an increase in modulus (stress/strain) under uniaxial deformation

(Nicolas et al. 2003). The gluten network loosens and decreases its water retention ability; the lipid fraction is removed from the gluten protein due to the decrease in water in the continuous protein phase promoting the fusion of lipid droplets and an increase in their size (Aibara et al. 2005).

Some fermentation should occur during dough thawing as indicated by the decrease in reducing sugar content (Inoue et al. 1994). After prolonged frozen storage, the proportion of alcohol-soluble protein increases (Inoue et al. 1994), with a concomitant decrease in the electrophoretic pattern of high molecular glutenin subunits, which suggests depolymerization of the protein matrix during frozen storage (Ribotta et al. 2001).

Doughs, examined after mixing, frozen, and thawing, show changes in the main flour components, proteins and starch. After mixing and 24 h frozen storage, starch granules appeared firmly embedded in the gluten network, which was mostly intact, as observed in fresh dough (Rojas et al. 2000). After prolonged frozen storage (6 months) and freeze–thaw cycles, some starch granules exhibited internal damage and were more separated from the gluten matrix. In addition, less water appeared to be associated with the gluten matrix and starch (Berglund et al. 1991). Nuclear magnetic resonance studies revealed an increase in the water mobility, due to its release from the gluten matrix, and, as a result, the dehydration of the gluten component could be observed in infrared spectroscopy studies (Esselink et al. 2003).

Water transport from the gluten to the starch paste occurs during frozen storage (Bot 2003). When gluten was stored at −15°C, the water content in the gluten phase decreased by approximately 1% over the first 3 weeks of storage, after which an apparent steady state was attained and the amount of ice did not increase further. In contrast, these changes were not observed at −25°C (Bot 2003). These structural changes were accompanied by a significant increase in the gelatinization enthalpy after 5 months of storage (Ribotta et al. 2003b), indicating water migration and ice crystallization (Lu and Grant 1999). The amount of freezable water in frozen doughs increased with frozen storage.

Moreover, aging in differential scanning calorimeter of dough from different frozen storage periods showed increasing retrogradation temperature range and retrogradation enthalpy of starch (Ribotta et al. 2003b). Conversely, during frozen storage of par-baked dough, no retrogradation of amylopectin was detected, but aging studies of full-baked bread showed progressive increase in the retrogradation temperature range of amylopectin, and large increase in retrogradation enthalpy with frozen storage period. These results indicate the existence of structural changes of amylopectin produced during frozen storage (Bárcenas et al. 2003a,b).

The thermal and microstructural properties of frozen hydrated gluten have revealed that gluten forms a continuous, homogeneous and nonfibrous network (Kontogiorgos and Goff 2006). There is a wide temperature range over which the gluten matrix vitrifies, making it impossible to assign a T_{glass} value for hydrated frozen gluten. Therefore, the construction of state diagrams is not feasible at subzero temperatures. Two different mechanisms have been proposed for explaining the gluten deterioration during freezing, namely, the growth of ice that is confined in dough matrix capillaries and the growth of bulk ice, both of them associated with ice recrystallization (Kontogiorgos and Goff 2006). At the micro-structural level, it is observed that ice crystals are preferably formed in gas cells (Esselink et al. 2003). Specifically, confocal laser scanning microscopy has shown that ice formation starts at the gas pore interface, where large ice crystals are initially located. Gas pore interfaces in dough are preferential sites for ice nucleation, favoring the growth of ice crystals in these regions and the redistribution of water in dough (Baier-Schenk et al. 2005a).

The amount, morphology, and distribution of ice in prefermented frozen bread dough change during frozen storage. Under constant storage conditions, the ice fraction of frozen dough stored at −22°C accounted for 53% of the total water and remained constant even over a period of 56 days, although changing the morphology and distribution of ice. Ice crystals were observed in the gas pores of the dough and were already present at 1 h after freezing. Crystal growth was observed after 1 day of frozen storage and led to a redistribution of water in the dough, which affected the properties of polymeric compounds in dough and reduced the baking performance of prefermented frozen doughs (Baier-Schenk et al. 2005b). Conversely, in wheat doughs obtained from different cultivars, an increase in the amount of freezable water was observed during frozen storage, probably due to differences in protein content and gluten quality (Lu and Grant 1999).

All these changes in dough level led to damaged crumb grain with broken gluten fibrils forming the skeletal framework of coarse pore walls, and many knots were generated on gluten fibrils derived from freeze damage. As the number of freeze–thaw cycles increased, both the coarseness of the gluten fibrils and the size of the knots augmented (Naito et al. 2004).

The reduction in the dough resistance induced by freezing and thawing operations has been partially related to certain compounds released from dead yeast cells after freezing and thawing.

During freezing and frozen storage, the number of viable yeast cells decreases and, as a consequence, a reducing compound (glutathion) is released, which can break down the disulphide bonds among proteins leading to a weakening effect on the gluten (Inoue et al. 1994, Ribotta et al. 2001). Besides, hydrophobic interactions become weak when temperature decreases, which can partially explain the steady deterioration of the gluten network during frozen storage.

Crust properties have been considered as important factors for the quality characterization of bread. One of the quality problems of the fresh bread obtained from partially baked bread is the crust flaking resulting from the detachment of some part of the crust. It can be related to excessive drying of the surface at the end of the postbaking chilling and freezing operations (Lucas et al. 2005, Hamdami et al. 2007). This phenomenon has been ascribed to two different processes: the concentration of water as ice under the crust due to the presence of the freezing front, and the interfacial differences between crust and crumb associated with the tensile forces and stresses induced by the thermomechanical shock (Lucas et al. 2005). Le Bail et al. (2005) have reported that chilling conditions after partial baking are the most determinant parameter on the crust flaking, followed by the proofing conditions. In general, high air humidity during those processes tends to reduce crust flaking.

2.3 TECHNOLOGICAL IMPLEMENTATION OF LOW TEMPERATURES IN BREADMAKING

2.3.1 Matching the Formulation to New Requirements

Refrigerated or frozen dough products are less perishable than bakery fresh products; thus the timing of product distribution is not crucial. Nevertheless, it should be taken into account that prolonged storage will kill the yeasts, thereby impairing the raising of the dough after thawing, besides the dough structure deterioration due to ice recrystallization.

In addition, the application of low or subzero temperatures to the breadmaking process must conform to new requirements regarding raw materials, machinery, package, and transport (Rosell and Gómez 2007).

Gluten weakening leads to an increase in the proofing time, a reduction in both the oven spring and the dough resistance to stress conditions (Gelinas et al. 1996). Therefore, the production of frozen dough requires flour with superior quality compared to the one used in conventional breadmaking processes. Specific wheat varieties with strong gluten or appropriated wheat blends are needed for obtaining the required strength (Gelinas et al. 1996, Boehm et al. 2004).

Dough strength dramatically decreases during freezing and thawing, and it undergoes gradual changes during frozen storage. The same trend is observed for the resulting loaf volume. Loss of dough strength on freezing, thawing, and during frozen storage is the main reason for the decline in bread loaf volume (Inoue and Bushuk 1992). Accordingly, very strong flours are recommended for improving bread quality from frozen doughs and in the flour selection; protein content appears to be less important than protein quality (Inoue and Bushuk 1992). A more detailed definition of the flour characteristics indicates that flours with high starch swelling characteristics, along with moderately high gluten strength, are the most appropriate for producing optimum quality frozen doughs, with good shelf life and baking properties (Bhattacharya et al. 2003).

Yeasts have different sensitivities to freezing, and with the growth of the freezing technology, a number of freeze-tolerant yeasts have been selected (Rosell and Gómez 2007). *Torulaspora delbrueckii* is a freeze-tolerant yeast that showed constant survival yeast counts for 10 weeks under frozen conditions, but there was a decrease in the fermentative ability per yeast cell during longer storage (Takasaki and Karasawa 1992). Different attempts have been proposed for improving the yeast resistance to low temperatures, like the use of cold-resistant yeast treated by high-osmotic-pressure solution for making bread (Huang and Lu 2006a), or using yeast treated with glycerol for conferring antifreezing properties (Huang and Lu 2006b). The first method consists in preparing a high-osmotic-pressure solution from yeast cream, acetone, and sugar cane, dispersing active dry yeast in it, stirring and filtering, whereas in the second method yeast is prepared by dispersing the active dry yeast in water, adding glycerin, stirring, storage, centrifugation, and filtering. In both cases, the resulting sediment is mixed with the rest of the breadmaking ingredients. In addition, the presence of gluten and trehalose (10 wt%) in the bread recipe when doughs are subjected to mechanical freezing and frozen storage has a beneficial effect on the cell survival rate for both yeasts (Salas-Mellado and Chang 2003).

An attractive alternative is the yeast immobilization based on thermogelling immobilization for creating a physical barrier between the substrate and the yeast in order to reduce the production of CO_2 at subzero temperatures (Gugerli et al. 2004). Global dough immobilization with gelatin allows controlling and improving dough fermentation. This method permits to preserve gluten network from freezing damage by slowing down the fermentation at low temperatures (<10°C), while enabling it at ambient temperature, at which the fermentation activity was recovered by up to 83% and the leavened dough stability was improved. The gelatin catalyzes maltose uptake, and thus acts both as a stopping agent and as a proofing improver (Gugerli et al. 2004). Even some practices during breadmaking can modify the yeast behavior. For instance, prefermentation before freezing, punching and remolding, or resheeting and molding treatments increased loaf volume by 10%–110% for the baked breads using freeze-tolerant yeast, whereas these treatments decreased loaf volume by 70% using freeze-sensitive yeast (Takano et al. 2002).

The addition of hydrophilic gums to the frozen dough recipe has been reported for controlling ice crystallization and recrystallization to improve the shelf-life stability of frozen dough, partially baked bread and also frozen bread. Indeed, hydrophilic gums (carboxymethyl cellulose, arabic gum, kappa carrageenan, and locust bean gum) lowered the enthalpy of the freezable water endothermic transition, indicating a decrease in freezable water (Sharadanant and Khan 2003a). Scanning electron micrographs of frozen dough have clearly shown that the inclusion of locust bean gum and arabic gum confers better retention of the gluten network within dough frozen storage. The locust bean gum decreases the amount of sodium dodecyl sulphate soluble proteins but keeps the values of residue proteins amount high, thus protein breakdown is not completely hindered by this hydrocolloid (Sharadanant and Khan 2006). Recipes with locust bean gum, arabic gum, and carboxymethyl cellulose improve the dough characteristics to varying degrees and also the specific loaf volume (Sharadanant and Khan 2003a,b). Bread loaves containing locust bean gum have brighter crust color, lighter crumb color, and better moisture retention (Sharadanant and Khan 2003b). K-carrageenan is the only gum that shows a detrimental effect on frozen dough (Sharadanant and Khan 2003a,b), and also on partially baked bread kept under frozen conditions (Bárcenas et al. 2004).

The presence of bread improvers (alpha-amylase, sourdough, hydroxypropylmethylcellulose [HPMC]) minimizes the negative effect (increase in the retrogradation temperature range) during frozen storage observed in partially baked bread (Bárcenas et al. 2003a,b). Concerning the aging of the full-baked loaf after freezing and rebaking, these three bread improvers decrease the retrogradation enthalpy of amylopectin, retarding the staling. Full-baked loaves from partially baked frozen bread containing HPMC have great specific volume and moisture retention, and low hardness of breadcrumb and water activity; thus the hydrocolloid inhibits the effect of the frozen storage on bread staling (Bárcenas et al. 2004, Bárcenas and Rosell 2006a,b).

HPMC also decreased the hardening rate of the breadcrumb and retarded the amylopectin retrogradation, acting through the possible interaction between HPMC and the main bread constituents, gluten and starch (Bárcenas and Rosell 2005, 2006a,b, Rosell and Foegeding 2007). Xanthan addition (0.16 g/100 g flour) also improves bread properties after frozen storage and subsequent thawing by microwave heating, probably due to inhibition of the rapid water loss induced by the microwave oven (Mandala 2005).

Calcium stearoyl lactylate (CSL) and diacetyl tartaric acid esters of monoglyceride (DATEM) produce significant improvement in baking properties, increasing the loaf volume for the nonfrozen and one-day frozen dough treatments (Inoue et al. 1995). Frozen doughs supplemented with DATEM, gluten, and guar gum also produce breads of greater volume and more open crumb structure than those prepared without additives (Ribotta et al. 2001). Although the combination of DATEM and gum guar has improved the volume and texture of bread obtained from frozen dough, it could not avoid the effect of frozen storage on the dynamic rheological parameters and microstructure damage (Ribotta et al. 2004). Some shortenings also improve the quality of the loaves from frozen dough (Aibara et al. 2005). Possibly, shortenings are only effective when their lipid fraction interacts with the gluten proteins and thereby protects the gluten matrix from frozen damage (Aibara et al. 2005).

An addition of 2% vital wheat gluten improves bread loaf volume and grain structure, conferring dough stability throughout several freeze–thaw cycles. Therefore, lower protein flour supplemented with vital wheat gluten can replace higher protein flour in the production of frozen doughs (Wang and Ponte 1994). Additionally, wheat gluten hydrolysate can be effective in increasing loaf volume for both frozen and nonfrozen dough, and the addition of peptides is effective for improving the baking quality of frozen dough (Koh et al. 2005). By adding protein and polysaccharides to conventional dough, it is possible to minimize the negative effects on bread quality caused by freezing (Eun et al. 2005).

Recently, the use of technological aids with strengthening action has been recommended for reducing dough sensitivity to freezing and frozen storage. In fact, doughs containing transglutaminase (enzyme with transferase activity) keep the gluten network with the starch granules embedded in the matrix after 5 weeks of frozen storage (Huang et al. 2008). This is attributed to the ability of this enzyme to polymerize proteins maintaining the integrity of the protein network (Bonet et al. 2005, 2006).

2.3.2 Necessary Equipment for Breadmaking Lines with Cold Temperature Control

Frozen dough performance requires a reduction in the amount of water (although longer fermentation time after thawing must be applied), which improves both the fermentation stability and form ratio of breads obtained from frozen doughs (Rasanen et al. 1997). During the breadmaking process, dough consistency, and thus the amount of water, is a limiting factor for protein hydration and later for starch gelatinization and gelling, which has a direct effect on the mixing, fermenting, cooking and cooling performance of the wheat bread dough (Rosell and Collar 2009).

During mixing, time and temperature also play a significant role on bread volume (Rouille et al. 2000, Zounis et al. 2002). The final mixing temperature was found to influence the structure stability of frozen bread. Loaves baked from stored frozen dough with a final mixing temperature of 31°C were poorer in both gassing power and overall loaf quality than those baked from doughs mixed at 16°C (Zounis et al. 2002).

In general, too warm doughs at the end of mixing tend to be sticky, whereas too cold doughs can have molding problems. It is possible to regulate dough consistency or doughfeel by controlling dough temperature.

There are different systems for controlling the temperature during mixing, which include bowl jacket cooling, bowl end cooling, adding ice or water, bar breaker cooling and agitator bar cooling (Rigik 2007). Other alternatives comprise the use of chilling ingredients. The use of chilled water is the cheapest and most common way to cool down the dough, although chipped ice has a greater cooling capacity than water. Cubed ice should be avoided because it can lead to ice pockets due to the presence of incompletely melted points, which can yield wet, soggy spots and create large holes or blisters. Another possibility for chilling dough is the use of a jacketed-bowl, which consists of a double-layered mixing bowl with food-grade glycol running between the two bowl layers. A complement to bowl cooling is the bowl end cooling method that chills the ends of the mixing bowl and provides additional cooling. Breaker bars and agitator bars are relatively new; the former consist of solid bars that run across the top of the bowl, whereas the latter are placed inside the bowl mixer. Flour can be also chilled to counteract the heat resulted from flour friction due to its movement through dispensing tubes. Flour cooling before mixing can be reached by injecting liquid carbon dioxide into the flour lines or blowing it directly into the mixer as cryogenic snow; carbon dioxide in solid form (dry ice) or liquid nitrogen can be also effective methods for cooling the dough. The use of liquid carbon dioxide does not impair the dough because it is one of the natural gases released during fermentation and it is not cold enough to damage yeast (Rigik 2007).

Even intermediate stages of the breadmaking process have a significant effect on the bread quality obtained from frozen dough. In fact, the volume of bread prepared from ball-shaped frozen doughs (not sheet-molded or bench-rested) was significantly lower compared to cylindrical or flat sheet doughs ($P<0.05$), even though respective dough proof times and bread scores were not significantly different (Gélinas et al. 1995). In order to maintain the dough temperature during postmixing operations, sometimes cooling systems are adapted to rollers to ensure the proper dough temperature during dividing and shaping. Late innovations in dough cooling include the addition of temperature probes in the mixer for monitoring the dough temperature during mixing, which allows online actions for obtaining the correct temperature.

Dough freezing can be accomplished by a variety of commercial freezers that can be included in the production line of breadmaking. For reduced spaces, like in-store bakeries, a mechanical blast freezer is the ideal solution. Conversely, cryogenic freezers in spirals or tunnels are the best alternative when considering large bakery plants. Cryogenic systems freeze products with nitrogen or carbon dioxide, depending on the location and the available supply. Cryogenic tunnels are used for many applications; they use high velocity cryogenic liquid to attain very low temperatures (−196°C), whereas spiral freezers reach temperatures of −150°C with increased capacities, saving as much as 50% more space than a mechanical freezer (Seiz 2006).

Different devices have been proposed for baking frozen dough. Recently, Canicas and Chateau (2006) have described a machine with one zone for storing frozen dough portions and another one for baking by infrared heaters operating at a wavelength of 1–1.4 μm, which reach a temperature of up to 800°C in a few seconds while blowers create an air flow around the products. In summary, the dough portions are conducted through zones that allowed thawing and rising. Another alternative for refrigerated dough is a controller that regulates the operation of heating and cooling unit in a cooking chamber based on the output of humidity, temperature, and height sensors to adjust the temperature and humidity inside the cooking chamber (Yamakawa and Tsukamoto 2003).

After frozen storage, dough must thaw under appropriate humidity and temperature control. Fermentation of thawed doughs takes place in proofing cabinets similar to the ones used in the conventional breadmaking process. Some equipment has been developed for controlling the thawing and proofing conditions concerning relative humidity and temperature that allow desired combinations of time, temperature, and humidity, which may be programmed by the operator (Parker 1995).

The production of par-baked breads requires some modification of the breadmaking lines, particularly in baking, cooling, and freezing (Hillebrand 2005). The main difference between the conventional and par-baking processes relies on how products are baked. Par-baked products remain in the oven until the crumb structure has been formed, so it is necessary to reduce the oven temperature or the baking time. Any type of oven can be used for making par-baked bread because time and temperature changes are only minor modifications. Usually tunnel ovens are used for making par-baked breads, although vertical ovens are also recommended because they use less floor space and their compact structure minimizes the amount of steam necessary for obtaining a thin and soft crust. The amount of steam is decisive, since with insufficient steam the product will dry out completely, and the crust will detach during the second baking (Hillebrand 2005). Slight modifications are also necessary during the cooling and freezing processes of these products. Par-baked breads should be cooled to room temperature after baking and before entering the freezer. Too warm a core temperature generates a dry shell that will flake off later, and also supplementary energy and more time will be necessary to freeze the breads. Humidity of the ambient air is also a key factor because when too much humid air is blown to cool the bread, the mold spores that are circulated can develop and grow on the surface of par-baked products.

Par-baked breads should be about 90% frozen when they exit the freezer because too much time could produce freezer burns, and it is also true that the product may collapse if not frozen enough (Hillebrand 2005). Another consideration is that par-baked bread should remain frozen until its full baking, because if the frozen par-baked bread partially thaws and then refreezes, its quality suffers.

2.4 OTHER INNOVATIVE TECHNOLOGICAL ALTERNATIVES IN BREADMAKING

Breadmaking lines are continuously introducing some changes to improve or optimize the process regarding energy consumption and product quality. An alternative for optimizing the freezing of par-baked breads focused on the improvement of product quality is the use of two-stage freezing—changing air temperature and velocity (Hamdami et al. 2007). It has been reported that the use of rapid freezing

(5 m/s and 233 K) of par-baked bread loaves tends to minimize their weight loss and also the freezing time. Conversely, slow freezing (0.5 m/s and 253 K) applied at the beginning of the freezing operation has a tendency to decrease the ice content at the crust–crumb interface. Therefore the combination of slow freezing at the beginning with a rapid freezing in the second stage improves the quality of the full-baked product (Hamdami et al. 2007).

The application of frozen technology has been extended to other ethnic-baked goods like parotta, after appropriate research for optimizing the process. The effect of storage of frozen parotta dough and ready-to-bake frozen parotta dough at –20°C for 3 months on rheological and baking characteristics has shown that optimum thawing conditions are 1 min in the microwave oven or 16 h in the refrigerator. The overall rating of parotta has been good even after 3 months of storage as indicated by the sensory evaluation of fresh parottas (Indrani et al. 2002).

New product launches within the bakery (baked or par-baked products) category are experiencing an exponential growth. Europe accounted for 40% of launches, North America for almost 25%, and Asia Pacific and Latin America for more than 16% each (Launois 2008). Healthy and natural products are becoming more popular worldwide as consumers become aware of the potential dangers or beneficial aspects of some ingredients on health. The wholegrain category accounted for most launches over the past 2 years; in fact, from a total of 861 bread products launched last year in Europe, around 15% of them contained wholegrains. The other tendency in product innovation is the replacement of additives or preservatives for healthy alternatives, which accounts for the increasing number of bakery products claiming "no additives/preservatives," that was around 14% of the total number of launched products. Those product categories were followed by vegetarian bread products, low allergen, and organic bakery products (Launois 2008). Focused on breadmaking aids, different enzymes produced by bacteria living at temperatures close to 0°C have been isolated and characterized, which open up new perspectives for using fully active enzymes at low temperatures.

2.5 CONCLUDING REMARKS

Innovation in the food industry combines technological innovation with social and cultural innovation. It occurs throughout the entire food chain, including production, harvesting, primary and secondary processing, manufacturing, packaging, and distribution. Innovations can be focused on any point of the chain, but they promote changes in other parts of the food system, even in consumer eating patterns and in general social and cultural areas.

The market for subzero and low-temperature bakery products has shown a steady increase in recent years due to the popularity of frozen technology and also because nowadays this technology allows product diversification. This technology helps bakeries bring new products to the market quickly and successfully.

There is intense research in product formulation that goes from the use of different flours passing for additives and technological aids. Research tends to introduce novel flour and traditional grains, like amaranth, quinoa, sorghum, or spelt, which are particularly adequate for consumers with gluten intolerance.

REFERENCES

Aibara, S., Ogawa, N., and Hirose, M. 2005. Microstructures of bread dough and the effects of shortening on frozen dough. *Biosci. Biotechnol. Biochem.* 69:397–402.

Aussenac, T., Carceller, J.L., and Kleiber, D. 2001. Changes in SDS solubility of glutenin polymers during dough mixing and resting. *Cereal Chem.* 78:39–45.

Baier-Schenk, A., Handschin, S., von Schonau, M., Bittermann, A.G., Bachi, T., and Conde-Petit, B. 2005a. In situ observation of the freezing process in wheat dough by confocal laser scanning microscopy (CLSM): Formation of ice and changes in the gluten network. *J. Cereal Sci.* 42:255–260.

Baier-Schenk, A., Handschin, S., and Conde-Petit, B. 2005b. Ice in prefermented frozen bread dough—An investigation based on calorimetry and microscopy. *Cereal Chem.* 82:251–255.

Bárcenas, M.E. and Rosell, C.M. 2005. Effect of HPMC addition on the microstructure, quality and aging of wheat bread. *Food Hydrocoll.* 19:1037–1043.

Bárcenas, M.E. and Rosell, C.M. 2006a. Different approaches for improving the quality and extending the shelf life of the partially baked bread: Low temperatures and HPMC addition. *J. Food Eng.* 72:92–99.

Bárcenas, M.E. and Rosell, C.M. 2006b. Effect of frozen storage time on the quality and aging of par-baked bread. *Food Chem.* 95:438–445.

Bárcenas, M.E. and Rosell, C.M. 2007. Different approaches for increasing the shelf life of partially baked bread: Low temperatures and hydrocolloid addition. *Food Chem.* 100:1594–1601.

Bárcenas, M.E., Haros, M., Benedito, C., and Rosell, C.M. 2003a. Effect of freezing and frozen storage on the staling of part-baked bread. *Food Res. Int.* 36:863–869.

Bárcenas, M.E., Haros, M., and Rosell, C.M. 2003b. An approach to study the effect of different bread improvers on the staling of prebaked frozen bread. *Eur. Food Res. Technol.* 218:56–61.

Bárcenas, M.E., Benedito, C., and Rosell, C.M. 2004. Use of hydrocolloids as bread improvers in interrupted baking process with frozen storage. *Food Hydrocoll.* 18:769–774.

Berglund, P.T., Shelton, D.R., and Freeman T.P. 1991. Frozen bread dough ultrastructure as affected by duration of frozen storage and freeze-thaw cycles. *Cereal Chem.* 68:105–107.

Bhattacharya, M., Langstaff, T.M., and Berzonsky, W.A. 2003. Effect of frozen storage and freeze-thaw cycles on the rheological and baking properties of frozen doughs. *Food Res. Int.* 36:365–372.

Boehm, D.J., Berzonsky, W.A., and Bhattacharya, M. 2004. Influence of nitrogen fertilizer treatments on spring wheat (*Triticum aestivum* L.) flour characteristics and effect on fresh and frozen dough quality. *Cereal Chem.* 81:51–54.

Bollaín, C. and Collar, C. 2004. Dough viscoelastic response of hydrocolloid/enzyme/surfactant blends assessed by uni- and bi-axial extension measurements. *Food Hydrocoll.* 18:499–507.

Bollaín, C., Angioloni, A., and Collar, C. 2006. Relationships between dough and bread viscoelastic properties in enzyme supplemented wheat samples. *J. Food Eng.* 77:665–671.

Bonet, A., Caballero, P., Rosell, C.M., and Gómez, M. 2005. Microbial transglutaminase as a tool to restore the functionality of gluten from insect damaged wheat. *Cereal Chem.* 82:425–430.

Bonet, A., Blaszczak, W., and Rosell, C.M. 2006. Formation of homopolymers and heteropolymers between wheat flour and several protein sources by transglutaminase catalyzed crosslinking. *Cereal Chem.* 83:655–662.

Bot, A. 2003. Differential scanning calorimetric study on the effects of frozen storage on gluten and dough. *Cereal Chem.* 80:366–370.

Canicas, E. and Chateau, D. 2006. Automatic machine for dispensing dough-based products, especially freshly-baked bread, from frozen and preserved dough portions. International Patent 2006,087,478-A2, filed February 17, 2006, and issued August 24, 2006.

Collar, C. 2003. Significance of viscosity profile of pasted and gelled formulated wheat doughs on bread staling. *Eur. Food Res. Technol.* 216:505–513.

Collar, C. and Armero, E. 1996. Physico-chemical mechanisms of bread staling during storage: Formulated doughs as a technological issue for improvement of bread functionality and keeping quality. *Recent Res. Dev. Nutr.* 1:115–143.

Esselink, E.F.J., Van Aalst, H., Maliepaard, M., and Van Duynhoven, J.P.M. 2003. Long-term storage effect in frozen dough spectroscopy and microscopy. *Cereal Chem.* 80:396–403.

Eun, J.B., Jung, J.I., and Yoon, Y. 2005. Production of frozen dough for bakery products, containing protein and polysaccharides. KR Patent 2004076547-A, filed February 25, 2003, and issued September 1, 2004.

Fik, M. and Surowka, K. 2002. Effect of prebaking and frozen storage on the sensory quality and instrumental texture of bread. *J. Sci. Food Agric.* 82:1268–1275.

Gélinas, P., Deaudelin, I., and Grenier, M. 1995. Frozen dough: Effects of dough shape, water content, and sheeting-molding conditions. *Cereal Foods World* 40:124–126.

Gélinas, P., McKinnon, C.M., Lukow, O.M., and Townley-Smith, F. 1996. Rapid evaluation of frozen and fresh dough involving stress conditions. *Cereal Chem.* 73:767–769.

Giannou, V. and Tzia, C. 2007. Frozen dough bread: Quality and textural behavior during prolonged storage—Prediction of final product characteristics. *J. Food Eng.* 79:929–934.

Gugerli, R., Breguet, V., von Stockar, U., and Marison, I.W. 2004. Immobilization as a tool to control fermentation in yeast-leavened refrigerated dough. *Food Hydrocoll.* 18:703–715.

Hamdami, N., Tuan Pham, Q., Le-Bail, A., and Monteau, J.Y. 2007. Two stage freezing of part baked breads : Application and optimization. *J. Food Eng.* 82:418–426.

Havet, M., Mankai, M., and Le Bail, A. 2000. Influence of the freezing condition on the baking performances of French frozen dough. *J. Food Eng.* 45:139–145.

Hebeda, R.E., Bowles, L.K., and Teague, W.M. 1990. Developments in enzymes for retarding staling of baked goods. *Cereal Foods World* 35:453–457.

Hillebrand, M. 2005. Altering equipment to produce par-baked breads. *Baking Management*, http://baking-management.com/equipment/bm_imp_7877/ (accessed April 24, 2008).

Huang, W. and Lu, Y. 2006a. Production of refrigerated bread dough using yeast treated by high osmotic pressure solution. CN Patent 1692748-A, filed June 6, 2005, and issued November 9, 2005.

Huang, W. and Lu, Y. 2006b. Fermentation production of refrigerated bread dough by using yeast treated by glycerol for anti-freezing. CN Patent 1692749-A. filed June 6, 2005, and issued November 9, 2005.

Huang, W.N., Yuan, Y.L., Kim, Y.S., and Chung, O.K. 2008. Effects of transglutaminase on rheology, microstructure, and baking properties of frozen dough. *Cereal Chem.* 85:301–306.

Indrani, D., Prabhasankar, P., Rajiv, J., and Rao, G.V. 2002. Effect of storage on the rheological and parotta-making characteristics of frozen parotta dough. *Eur. Food Res. Technol.* 215:484–488.

Inoue, Y. and Bushuk, W. 1992. Studies on frozen doughs. II. Flour quality requirements for bread production from frozen dough. *Cereal Chem.* 69:423–428.

Inoue, Y., Sapirstein, H.D., Takayanagi, S., and Bushuk, W. 1994. Studies on frozen doughs. III. Some factors involved in dough weakening during frozen storage and thaw-freeze cycles. *Cereal Chem.* 71:118–121

Inoue, Y., Sapirstein, H.D., and Bushuk, W. 1995. Studies on frozen doughs. IV. Effect of shortening systems on baking and rheological properties. *Cereal Chem.* 72:221–226.

Koh, B.K., Lee, G.C., and Lim, S.T. 2005. Effect of amino acids and peptides on mixing and frozen dough properties of wheat flour. *J. Food Sci.* 70:S359–S364.

Kontogiorgos, V. and Goff, H.D. 2006. Calorimetric and microstructural investigation of frozen hydrated gluten. *Food Biophys.* 1:202–215.

Lallemand Baking Update. 2008. Alternatives to scratch baking. Volume 2, number 14. http://www.lallemand.com/Home/eng/index.shtm (accessed April 24, 2008).

Launois, A. 2008. New bread products focus on health. http://www.bakeryandsnacks.com/news/ng.asp?n=84586-intel-gndp (accessed April 30, 2008).

Le Bail, A., Havet, M., and Pasco, M. 1998. Influence of the freezing rate and of storage duration on the gassing power of frozen bread dough. In *Proceedings of Conference on Hygiene, Quality and Safety in the Cold Chain and Air-Conditioning*, Publisher IIF, pp. 181–188. Nantes (France): Hygiene, Quality & Safety in the Cold Chain and Air Conditioning ISSN 0151-1637.

Le Bail, A., Monteau, J.Y., Margerie, F., Lucas, T., Chargelegue, A., and Reverdy, Y. 2005. Impact of selected process parameters on crust flaking of frozen part baked bread. *J. Food Eng.* 69:503–509.

Lord, D. 2002. Future innovations in bakery. MBA Group Limited: Business Insights Ltd.

Lu, W. and Grant, L.A. 1999. Effects of prolonged storage at freezing temperatures on starch and baking quality of frozen doughs. *Cereal Chem.* 76:656–662.

Lucas, T., Quellec, S., Le Bail, A., and Davenel, A. 2005. Chilling and freezing of part-baked breads. II. Experimental assessment of water phase changes and of structure collapse. *J. Food Eng.* 70:151–164.

Mandala, I.G. 2005. Physical properties of fresh and frozen stored, microwave-reheated breads, containing hydrocolloids. *J. Food Eng.* 66:291–300.

Mandala, I., Karabela, D., and Kostaropoulos, A. 2007. Physical properties of breads containing hydrocolloids stored at low temperature. I. Effect of chilling. *Food Hydrocoll.* 21:1397–1406.

Naito, S., Fukami, S., Mizokami, Y., Ishida, N., Takano, H., Koizumi, M., and Kano, H. 2004. Effect of freeze-thaw cycles on the gluten fibrils and crumb grain structures of breads made from frozen doughs. *Cereal Chem.* 81:80–86.

Nicolas, Y., Smit, R.J.M., van Aalst, H., Esselink, F.J., Weegels, P.L., and Agterof, W.G.M. 2003. Effect of storage time and temperature on rheological and microstructural properties of gluten. *Cereal Chem.* 80:371–377.

Parker, T.W. 1995. Quick thawing dehumidification proof box. U.S. Patent 5442994, Assignee M. Raubvogel Co., Inc, filed March 1, 1993, and issued August 22, 1995.

Phimolsiripol, Y., Siripatrawan, U., Tulyathan, V., and Cleland, D.J. 2008. Effects of freezing and temperature fluctuations during frozen storage on frozen dough and bread quality. *J. Food Eng.* 84:48–56.

Rasanen, J., Laurikainen, T., and Autio, K. 1997. Fermentation stability and pore size distribution of frozen prefermented lean wheat doughs. *Cereal Chem.* 74:56–62.

Regmi, A. and Gehlhar, M. 2005. New directions in global food markets. USDA. Electronic report, http://www.ers.usda.gov/publications/aib7947aib794.pdf (accessed March 28, 2008).

Ribotta, P.D., Leon, A.E., and Añon, M.C. 2001. Effect of freezing and frozen storage of doughs on bread quality. *J. Agric. Food Chem.* 49:913–918.

Ribotta, P.D., Leon, A.E., and Añon, M.C. 2003a. Effect of yeast freezing in frozen dough. *Cereal Chem.* 80:454–458.

Ribotta, P.D., Leon, A.E., and Añon, M.C. 2003b. Effect of freezing and frozen storage on the gelatinization and retrogradation of amylopectin in dough baked in a differential scanning calorimeter. *Food Res. Int.* 36:357–363.

Ribotta, P.D., Perez, G.T., Leon, A.E., and Anon, M.C. 2004. Effect of emulsifier and guar gum on microstructural, rheological and baking performance of frozen bread dough. *Food Hydrocoll.* 18:305–313.

Rigik, E. 2007. Dough cooling 101: The cold truth on mixing matters. *Baking Management*, http://baking-management.com/equipment/bm_imp_17144/ (accessed April 24, 2008).

Rojas, J.A., Rosell, C.M., Benedito, C., Pérez-Munuera, I., and Lluch, M.A. 2000. The baking process of wheat rolls followed by cryo scanning electron microscopy. *Eur. Food Res. Technol.* 212:57–63.

Rosell, C.M. 2007. Cereals and health worldwide: Adapting cereals to the social requirements. In *ICC Conference Proceedings, Cereals-The Future Challenges*, Rosario, Argentina.

Rosell, C.M. and Collar, C. 2009. Effect of temperature and consistency on wheat dough performance. *Int. J. Food Sci. Technol.* 44:493–502.

Rosell, C.M. and Foegeding, A. 2007. Interaction of hydroxypropylmethylcellulose with gluten proteins: Small deformation properties during thermal treatment. *Food Hydrocoll.* 21:1092–1100.

Rosell, C.M. and Gómez, M. 2007. Freezing in breadmaking performance: Frozen dough and part-baked bread. *Food Rev. Int.* 23:303–319.

Rosell, C.M., Collar, C., and Haros, M. 2007. Assessment of hydrocolloid effects on the thermo-mechanical properties of wheat using the Mixolab. *Food Hydrocoll.* 21:452–462.

Rouille, J., Le Bail, A., and Courcoux, P. 2000. Influence of formulation and mixing conditions on breadmaking qualities of French frozen dough. *J. Food Eng.* 43:197–203.

Salas-Mellado, M.M. and Chang, Y.K. 2003. Effect of formulation on the quality of frozen bread dough. *Braz. Arch. Biol. Technol.* 46:461–468.

Seiz, K. 2006. New opportunities, new freezer considerations. *Baking Management*, http://baking-management.com/equipment/bm_imp_17144/ (accessed April 24, 2008).

Sharadanant, R. and Khan, K. 2003a. Effect of hydrophilic gums on the quality of frozen dough. I. Dough quality. *Cereal Chem.* 80:764–772.

Sharadanant, R. and Khan, K. 2003b. Effect of hydrophilic gums on the quality of frozen dough. II. Bread characteristics. *Cereal Chem.* 80:773–780.

Sharadanant, R. and Khan, K. 2006. Effect of hydrophilic gums on the quality of frozen dough: Electron microscopy, protein solubility, and electrophoresis studies. *Cereal Chem.* 83:411–417.

Slade, L. and Levine, H. 1987. Recent advances in starch retrogradation. In *Industrial Polysaccharides. The Impact of Biotechnology and Advanced Methodologies*, Eds. S.S. Stivala, V. Crescenzi, and I.C. Dea, pp. 387–430. New York: Gordon and Breach.

Sumnu, G. 2001. A review on microwave baking of foods. *Int. J. Food Sci. Technol.* 36:117–127.

Takano, H., Naito, S., Ishida, N., Koizumi, M., and Kano, H. 2002. Fermentation process and grain structure of baked breads from frozen dough using freeze-tolerant yeasts. *J. Food Sci.* 67:2725–2733.

Takasaki, S. and Karasawa, K. 1992. Effect of storage period on survival of freeze tolerant yeast and on the rheological properties of frozen dough. *J. Jpn. Soc. Food Sci. Technol.* 39:813–820.

Uthayakumaran, S., Newberry, M., Keentok, M., Stoddard, F.L., and Bekes, F. 2000. Basic rheology of bread dough with modified protein content and glutenin-to gliadin ratios. *Cereal Chem.* 77:744–749.

Wang, Z.J. and Ponte, J.G. 1994. Improving frozen dough qualities with the addition of vital wheat gluten. *Cereal Foods World* 39:500–503.

Yamakawa, Y. and Tsukamoto, H. 2003. Bread baking processing device adjusts temperature inside cooking chamber so that bread dough refrigerated after completion of primary fermentation process is cooked. JP Patent 2003000441-A, filed June 20, 2001, and issued January 7, 2003.

Zounis, S., Quail, K.J., Wootton, M., and Dickson, M.R. 2002. Effect of final dough temperature on the microstructure of frozen bread dough. *J. Cereal Sci.* 36:135–146.

3 Biotechnological Tools to Produce Natural Flavors and Methods to Authenticate Their Origin

Elisabetta Brenna, Giovanni Fronza, Claudio Fuganti, Francesco G. Gatti, and Stefano Serra

CONTENTS

3.1 INTRODUCTION

For several years the only sources of natural flavors have been essential oils, fruit juices, vegetable extracts, and some products of animal origin, such as musk, zibet, and amber. Extraction from nature often consists in low yield, expensive, and troublesome processes. The supply of plant materials is subject to several factors: seasonal variation, variability of weather, risks of plant diseases, political stability of the producing countries, and trade restrictions. Increasing concern for animal protection has forbidden the use of animal-derived odorous compounds. Consumers show an increasing preference for what is defined "natural," especially when dealing with food and beverages. Commercial, industrial products gain higher added value when the flavoring substances they contain are termed as "natural." Within the European Union, flavorings used to fall under directive 88/388, but the regulatory situation was changed at the end of 2008. On July 28, 2006, the European Commission presented a proposal (European Commission 2006) to update the legislation governing food flavorings in order to reflect technological and scientific developments in this field. The current definitions of flavorings (e.g., flavoring substance, flavoring preparation, thermal process flavoring) have been refined, and some new definitions have been introduced for "food ingredients with flavoring properties," "source material," "flavor precursor," and "other flavoring." The proposal introduces stricter conditions for the use of the term "natural" when describing flavorings, and it removes the reference to "nature-identical," which is considered to be misleading for the consumer and which applies to natural flavors prepared by synthetic routes. Only two categories of flavorings for labeling purposes would remain, "natural" and "artificial." In the proposal, it is specified that "natural flavoring substance" shall mean a flavoring substance obtained by appropriate physical, enzymatic, or microbiological processes from the material of a vegetable, animal, or microbiological origin, either in the raw state or after processing for human consumption by one or more traditional food preparation operations.

This new classification will boost the optimization of new biotechnological processes for the preparation of natural flavorings, which have been developed so far (Krings and Berger 1998, Lomascolo et al. 1999a, Schrader et al. 2004, Serra et al. 2005) according to the following approaches: plant cell cultures; microbial processes; enzyme-catalyzed reactions.

- Plant tissue culture has been used to produce a wide range of flavors typical of the plant from which the tissues derive. Every cell contains all the genetic information necessary to produce the several chemical components of the aroma obtained from that plant. Feeding intermediates of the biosynthetic pathway can enhance the production of flavor metabolites by precursor biotransformation. The best results have been achieved when the aroma of the fruit or plant is mainly or entirely due to a single compound or only to a few compounds with similar structures and properties. This is the case of vanilla, whose major flavor component is vanillin, of raspberries, whose key aroma compound is raspberry ketone, and of onions and garlic, for which the characteristic flavor consists of alliin derivatives. The best results in this field have been recently reviewed (Ramachandra Rao and Ravishankar 2002, Hrazdina 2006).
- Microbial processes (Feron et al. 1996, Vandamme and Soetaert 2002, Aguedo et al. 2004, Xu et al. 2007) consist either in the *de novo synthesis* of natural flavors by normal microbial metabolism, or in the bioconversion

of appropriate precursor compounds. It is possible to promote *in situ* flavor generation, as an integral part of food or beverage production processes, or to develop microbial cultures to produce specific aroma compounds, which can be isolated and employed as additives in food manufacture.

- Enzyme processes (Longo and Sanromán 2006, Waché et al. 2006) can take advantage of a large number of isolated enzymes (lipases, proteases, and glucosidases) that are able to catalyze the production of aroma compounds from suitable precursors. These enzymic reactions are usually characterized by high substrate specificity, high enantio- and regio-selectivity, and they can be performed in mild conditions.

Only biocatalytic processes based on the microbial conversion of suitable precursor molecules are considered in this chapter, according to the following classification: aroma compounds from phenyl propanoid precursors (vanillin and raspberry ketone); aroma compounds from amino acids (benzaldehyde and 2-phenylethanol); and aroma compounds from fatty acids (green notes and γ-decanolide).

The optimization of cost and efficiency of processes for the biosynthesis of natural flavors has to be supported by the concomitant development of analytical tools for a quick and reliable distinction between natural and synthetic products. Most of the frauds in food industry are related to the adulteration of expensive natural flavors with readily available natural-identical products of petrochemical origin. This widespread illegal procedure produces great economical losses to both producers and consumers. Therefore, the topic of the authentication of the synthetic origin of flavors is addressed in the second part of this chapter, taking into consideration the application in this field of analytical techniques for determining stable isotope ratios ($^2H/^1H$, $^{13}C/^{12}C$, and $^{18}O/^{16}O$) at natural abundance level.

3.2 MICROBIAL PRODUCTION OF NATURAL FLAVORS

3.2.1 From Phenyl Propanoid Precursors

3.2.1.1 Vanillin

Vanillin is one of the most widely employed flavoring agents. Less than 1% of the world's production, which is estimated to be around 12,000 tons (Walton et al. 2003), is natural vanillin obtained from the cured pods mainly of *Vanilla planifolia*. The green seedpods contain vanillin as the β-D-glucoside, and they do not show the characteristic vanilla flavor. Vanillin is released through a long curing process, during which enzyme hydrolysis takes place and the pods become shriveled and brown. As a consequence of the limited supply and high price of natural vanilla, many efforts have been devoted to the optimization of biotechnological process routes to natural vanillin (Ramachandra Rao and Ravishankar 2000, Walton et al. 2000, Priefert et al. 2001), in order to find a convenient alternative to the chemically synthesized cheaper material.

Vanillin was found to be the intermediate of the microbial degradation of several substrates, such as eugenol, isoeugenol, ferulic acid, phenolic stilbenes, and lignin. These metabolic pathways have been exploited to develop biotechnological processes for the production of the flavor.

3.2.1.1.1 From Eugenol or Isoeugenol

Eugenol is the principal component of clove oil. Its low market price makes it a potential feedstock for vanillin production, but conversion yields are still very poor. Isoeugenol could provide better production yields, but natural isoeugenol is not available in large amounts. Research has been focused on the direct use of ferulic acid, which is decidedly more abundant (Figure 3.1).

The conversion of isoeugenol and eugenol into vanillin was developed by Haarmann and Reimer using *Serratia marcescens* DSM 30126 (Rabenhorst and Hopp 1991). When isoeugenol was employed as a substrate, the yields were optimized around 20% (3.8 g/L), while eugenol was transformed affording only 0.018 g/L. The biotransformation of eugenol into vanillin was improved to 280 mg/L with a strain of *Pseudomonas* spp. TK2102 (Washisu et al. 1993).

In 2000, a *Bacillus subtilis* strain was found to be able to transform isoeugenol into vanillin (Shimoni et al. 2000). The bioconversion capabilities of this strain were tested in growing cultures and cell free extracts, producing 0.61 g/L (molar yield of 12.4%) and 0.9 g/L (molar yield of 14%) of vanillin, respectively.

Pseudomonas putida I58 (Furukawa et al. 2003), *P. chlororaphis* (Kasana et al. 2007), *B. subtilis* HS8 (Zhang et al. 2006), and *B. pumilus* S-1 (Hua et al. 2007a) have been reported to transform isoeugenol into vanillin. Biphasic biotransformations of isoeugenol with different organic solvents have been described using strains of *B. fusiformis* SW-B9 (Zhao et al. 2005), *B. fusiformis* CGMCC1347 (Zhao et al. 2006) and *P. putida* IE27 (Yamada et al. 2007). The highest vanillin yield from isoeugenol by microbial biotransformation was achieved by using 60% (v/v) isoeugenol as substrate and solvent at pH 4.0, and the strain *B. fusiformis* SW-B9: vanillin was produced at 32.5 g/L over 72 h.

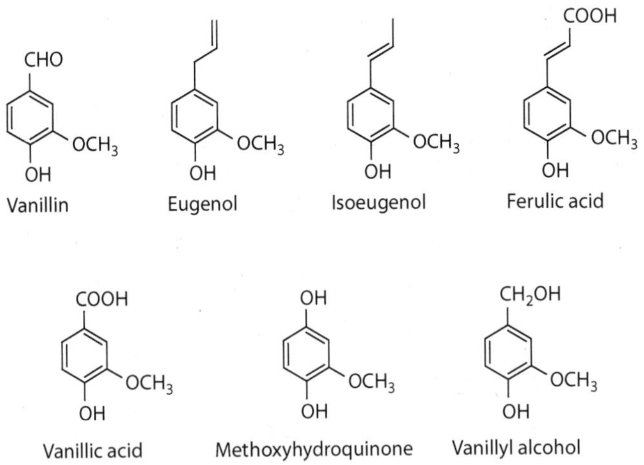

FIGURE 3.1 Vanillin and related compounds.

3.2.1.1.2 From Ferulic Acid

Ferulic acid is very abundant in nature, being a component of the cell-wall material, and a product of the microbial oxidation of lignin. A process for converting ferulic acid into vanillin using *Pycnoporus cinnabarinus* was patented in 1993 (Gross et al. 1993). Only 64 mg/L of vanillin could be produced after 7 days of culture, with a molar yield of 27.5%. Yields could be improved by developing a new two-step procedure employing two filamentous fungi with complementary bioconversion abilities (Lesage-Meessen et al. 1996a,b). *Aspergillus niger* first transformed ferulic acid into vanillic acid (Figure 3.1), and then *P. cinnabarinus* or *Phanerochaete chrysosporium* reduced vanillic acid to vanillin. Vanillic acid production by *A. niger* from ferulic acid reached 920 mg/L with a molar yield of 88% and vanillin was best obtained with *P. cinnabarinus*, attaining 237 mg/L with a molar yield of 22%. However, the vanillic acid oxidative system produced methoxyhydroquinone as a by-product, especially in *P. cinnabarinus* cultures. When cellobiose was used as the carbon source and/or added to the culture medium of *P. cinnabarinus* strains (Lesage-Meessen et al. 1997) just before the addition of vanillic acid, it directed the vanillic acid metabolism via the reductive route leading to vanillin. Adding 3.5 g/L cellobiose to 3-day-old maltose cultures of *P. cinnabarinus* MUCL39532, and 2.5 g/L cellobiose to 3-day-old cellobiose cultures of *P. cinnabarinus* MUCL38467 yielded 510 and 560 mg/L vanillin with a molar yield of 50.2% and 51.7%, respectively. Cellobiose might either have acted as an easily metabolizable carbon source, required for the reductive pathway to occur, or as an inducer of cellobiosequinone oxidoreductase, which is known to inhibit vanillic acid decarboxylation. The use of high-density cultures of *P. cinnabarinus* in glucose–phospholipid medium produced 760 mg/L of vanillin from ferulic acid in 15 days (Oddou et al. 1999).

The biotechnological process of vanillin production from vanillic acid by *P. cinnabarinus* was scaled-up (Stentelaire et al. 2000) at a laboratory level in two different kinds of bioreactors, a mechanically stirred and an airlifted one. In the mechanically stirred bioreactor, vanillin was produced in greater quantities. It was also observed that vanillin was highly toxic to the growth of *P. cinnabarinus* on agar medium over a concentration of 1000 mg/L. The application of selective XAD-2 resin led to a reduction of vanillin concentration in the medium, thus limiting its toxicity toward the fungal biomass, and preventing the formation of unwanted by-products. The concentration of vanillin produced reached 1575 mg/L.

Using *P. chrysosporium* in the two-step process of vanillin production, vanillic acid was preferentially converted into vanillyl alcohol via vanillin. The *in situ* adsorption of vanillin on XAD2 polystyrene resin was found to be a suitable method to recover vanillin before its transformation into vanillyl alcohol (Stentelaire et al. 1998), obtaining 500 mg/L of vanillin, with a molar yield of 47%.

Recently, high vanillin productivity was achieved (Hua et al. 2007b) in the batch biotransformation of ferulic acid by *Streptomyces* sp. strain V-1. Due to the toxicity of vanillin and the product inhibition, fed-batch biotransformation with high concentration of ferulic acid was found to be unsuccessful. To overcome this problem, a biotransformation strategy using adsorbent resin was developed. Resin DM11 was selected because it adsorbed vanillin the most and ferulic acid the least. When 8%

resin DM11 (wet w/v) was added to the biotransformation system, 45 g/L ferulic acid could be added continually and 19.2 g/L vanillin was obtained within 55 h, with a molar yield of nearly 50%.

The problem now is to optimize the process to produce ferulic acid enzymatically from maize bran or sugar beet pulp, in order to sustain the supply of natural ferulic acid.

3.2.1.2 Raspberry Ketone

Raspberry ketone is the main flavor of raspberries and it is characterized by a very low odor threshold: 1–10 ppb. Its occurrence in nature is rather scarce (3.7 mg in 1 kg of berries, Vandamme and Soetaert 2002), thus biocatalytic processes have been investigated for the preparation of natural raspberry ketone in order to find suitable alternatives to extraction (Figure 3.2).

Raspberry ketone could be obtained from a culture of a *Nidularia* species (~1 mg/L) in a peptone nutrient medium (Tiefel and Berger 1993). The addition of *p*-coumaric acid to these cultures increased the production up to 10 mg/L. The hydrolysis of betuloside, the *O*-glucoside of betuligenol found in rhododendron, birch and maple, could be catalyzed by commercial *A. niger* β-glucosidase and afforded betuligenol (Dumont et al. 1994). This was later oxidized to raspberry ketone by the secondary-alcohol dehydrogenase activity of the yeast *Candida boidinii*. The molar yield of enzymatic oxidation of betuligenol to raspberry ketone reached 20%–45%. The fungus *Beauveria bassiana* was shown to transform 4-(4-hydroxyphenyl)-but-3-en-2-one into raspberry ketone, which was further metabolized to tyrosol (Fuganti et al. 1996).

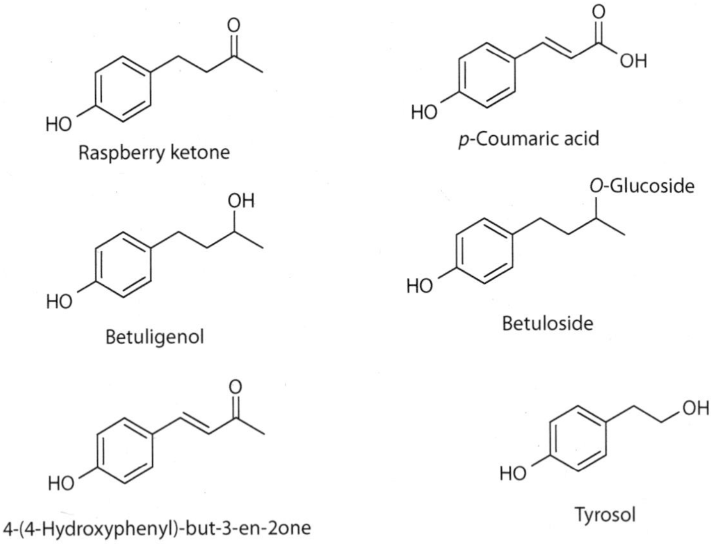

FIGURE 3.2 Raspberry ketone and related compounds.

3.2.2 From Amino Acids

3.2.2.1 Benzaldehyde

Benzaldehyde is the second most important flavor after vanillin for annual production, with 7000 tons and 100 tons/year, respectively, of the synthetic and natural compound (Lomascolo et al. 1999a). Most of the natural benzaldehyde (ca. 80 ton/year) is obtained by retroaldol reaction of natural cinnamaldehyde, obtained mainly from cassia oil (Buck et al. 1987) (Figure 3.3). However, some have protested (Feron et al. 1996) against the "natural" status of this benzaldehyde, because of the use of an acid or base catalyst is involved to promote retroaldol reaction. The remaining 20 tons are produced from the kernels of apricots, peaches, prunes and bitter almonds, by enzymatic hydrolysis of amygdalin and concurrent formation of toxic HCN as a side product. Thus, the bioconversion of a suitable natural precursor, i.e., phenylalanine, has been investigated to find an alternative route for natural benzaldehyde.

Natural L-phenylalanine has become available cheaply and in large quantities as an intermediate of the synthesis of the high-intensity sweetener aspartame, and it can be converted into benzaldehyde by biotechnological processes using different microorganisms, such as *Ischnoderma benzoinum* (Fabre et al. 1996), *Bjerkandera adusta* (Lapadatescu et al. 1997), and *Polyporus tuberaster* (Kawabe and Morita 1994). The main problem is that benzaldehyde is toxic toward microbial metabolism, and its accumulation in the culture medium may strongly inhibit cell growth (Lomascolo et al. 1999b). The conversion of phenylalanine into benzaldehyde occurs through different metabolic paths and leads to the formation of different coproducts: 2-phenylethanol in the case of *I. benzoinum* ATCC 26314 (Fabre et al. 1996), benzyl alcohol in the case of *I. benzoinum* INRA 33 (Lapadatescu et al. 1997) and of *P. tuberaster* (Kawabe and Morita 1994), and 3-phenylpropanol for *I. benzoinum* CBS 311.29 (Krings et al. 1996). With these fungi benzaldehyde was produced in quantity ranging from 71 to 836 mg/L, according to the strain used. When *B. adusta* CBS 595.78 was immobilized onto solid carriers, the production yield increased from 71 to 587 mg/L compared with nonimmobilized cells (Lapadatescu et al. 1997).

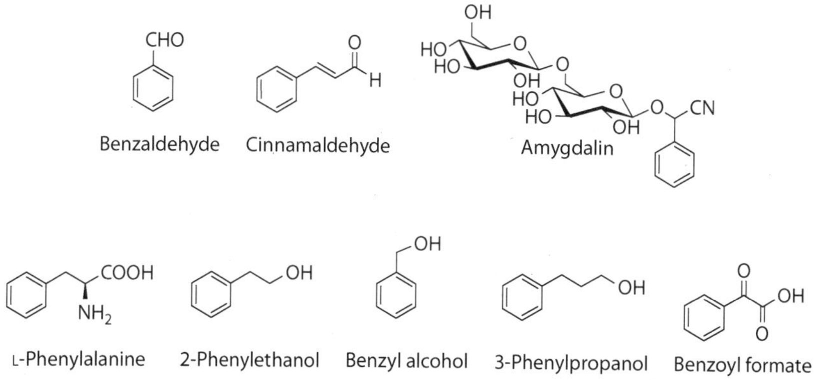

FIGURE 3.3 Benzaldehyde and related compounds.

Immobilization of benzaldehyde onto resins in bioreactors (Krings and Berger 1995) was also experimented to limit its toxicity. A culture medium of *Poria cocos* was continuously ultrafiltered and the biomass reinjected into the bioreactor, where the extraction of benzaldehyde was performed with resin in an adsorption column. The adsorption yield of benzaldehyde from a 3-day-old culture of *P. cocos* reached 62.3% of the total benzaldehyde. The technique of pervaporation, i.e., the separation of the components of a liquid medium through a selective membrane (see Chapter 5 for further details), was employed in a culture medium of *B. adusta* with a polydimethylsiloxane membrane, concentrating benzaldehyde up to 90 times (Lamer et al. 1996). The improvement obtained by continuous extraction with resins makes basidiomycetes a valuable means for the industrial production of benzaldehyde from L-phenylalanine.

Park and Jung (2002) described that calcium alginate-encapsulated whole-cell enzymes from *P. putida* could convert benzoyl formate into benzaldehyde, which accumulated in the capsule core, thus avoiding further reduction to benzyl alcohol by the action of alcohol dehydrogenase, and providing continuous production of benzaldehyde until reactant exhaustion.

3.2.2.2 2-Phenylethanol

2-Phenylethanol is a flavor and fragrance compound with a typical rose-like odor, widely employed in perfumes, cosmetics, and food. Although essential rose oil can contain up to 60% of this alcohol, it is too valuable to be used as a source of natural 2-phenylethanol for food flavorings. Many microorganisms, especially yeasts, are capable of producing 2-phenylethanol by normal metabolism (*de novo* synthesis), but the final concentration of the flavor in the culture broth generally remains very low (Etschmann et al. 2002). Higher yields can be obtained by bioconversion of phenylalanine. The pathways of phenylalanine metabolism in microorganisms are described by Etschmann et al. (2002), the Erlich pathway being the most important (Figure 3.4). According to this mechanism, 2-phenylethanol can be obtained from L-phenylalanine by transamination to phenylpyruvate, followed by decarboxylation to phenylacetaldehyde and reduction to alcohol.

It was described that in 4-week-old cultures of *I. benzoinum* ATCC 26314 2-phenylethanol could be obtained, as an intermediate in the conversion of L-phenylalanine into benzaldehyde, in concentration of 450 mg/L (Fabre et al. 1996). Fabre et al. (1997) screened 21 yeast strains for 2-phenylethanol production and found that *Kluyveromyces marxianus* was particularly interesting in spite of the fact that growth inhibition of the yeast was observed when the concentration of 2-phenylethyl alcohol reached a critical value (near 1.4 g/L) (Fabre et al. 1998). Two-phase

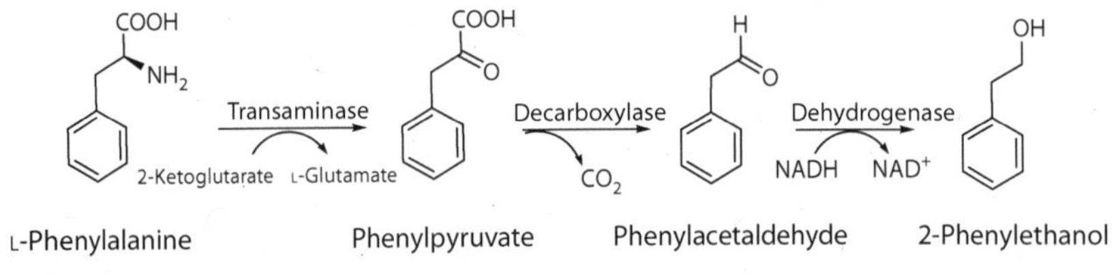

FIGURE 3.4 Ehrlich pathway for 2-phenylethanol synthesis.

fermentation of *K. marxianus* CBS 600 with oleyl alcohol for the *in situ* product removal yielded 3 g/L of 2-phenylethanol (Etschmann et al. 2003). An increase to 5.6 g/L was achieved by medium optimization using a genetic algorithm (Etschmann et al. 2004).

It was observed that the toxicity of the product limited the bioconversion of L-phenylalanine into 2-phenylethanol by the yeast *Saccharomyces cerevisiae* (Stark et al. 2002). Phenylethanol extraction by a separate organic phase in the fermentor was developed as an *in situ* product recovery technique to enhance productivity. Oleic acid was chosen as organic phase for two-phase fed-batch cultures, and 2-phenylethanol was produced in a concentration of 12.6 g/L.

An alternative *in situ* product recovery procedure was developed by Serp et al. (2003) by entrapping an organic solvent (dibutyl sebacate) within a polyethylene matrix, to form a highly absorbent, chemically and mechanically stable composite resin. The use of this technique allowed a twofold increase in the volumetric productivity of 2-phenylethanol.

3.2.3 From Fatty Acids

3.2.3.1 Green Notes

The green notes employed for flavor and fragrances are aldehydes and alcohols, mainly of six carbon atoms, produced in nature from lipids according to the lipoxygenase pathway (Figure 3.5). The first step is the deoxygenation of linoleic and linolenic acids promoted by lipoxygenase, to give 13-hydroperoxylinoleic acid (C13-HPOD) and 13-hydroperoxylinolenic acid (C13-HPOT). These latter intermediates are then cleaved by hydroperoxide lyase (HPO lyase), to produce the corresponding aldehydes. The aldehydes can be further enzymatically isomerized and reduced by alcohol dehydrogenase, to afford the corresponding alcohols. These flavor compounds are

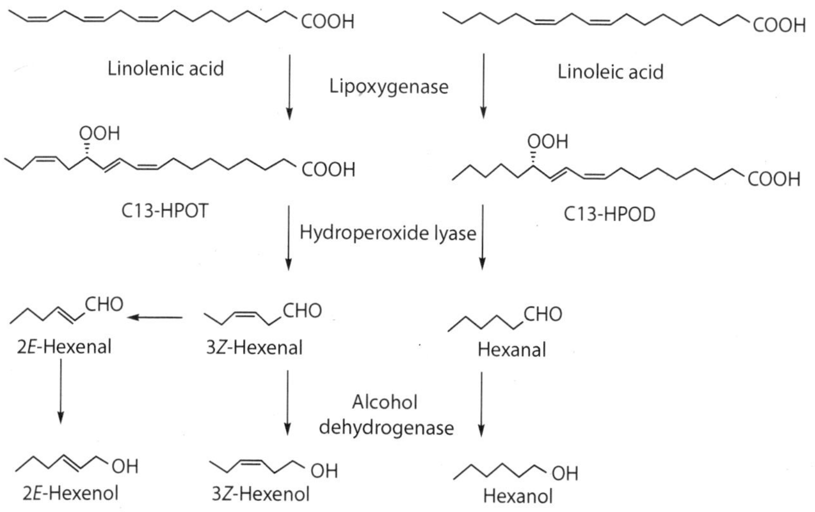

FIGURE 3.5 Lipoxygenase pathway to green notes.

widely used by the food industry to reconstitute the fresh green odor of fruits and vegetables, which gets lost during processing. Distillation of plant oils is expensive and it cannot satisfy the market demand for natural flavors. Biotechnological procedures have been developed (Brunerie 1991, Muller et al. 1995, Brunerie and Koziet 1997) based on the reaction of linoleic and linolenic acids with soy flour containing lipoxygenase to obtain C13-HPOD or C13HPOT. Subsequently, the incubation mixture is treated with a plant homogenate as a source of lyase for the conversion into *n*-hexanal or 3Z-hexen-1-al. These products could be reduced or isomerized by *S. cerevisiae*. The lyase reaction was found to be rate-limiting, and a novel production strategy, based on genetically modified *S. cerevisiae* coexpressing the HPO lyase gene from banana and the lipoxygenase gene, was reported, thus unifying all enzyme activities needed in one host (Muheim et al. 1997, Häusler et al. 2001).

HPO lyase from green bell pepper was expressed in *Yarrowia lipolytica* and C6 aldehydes have been produced in the culture medium (Bourel et al. 2004), affording a total of 350 mg of aldehydes per liter culture medium when the HPO substrate was introduced directly during the growth phase of the yeast.

3.2.3.2 Decanolides

γ- and δ-Lactones are important flavors for food applications. They give relevant contributions to taste and flavor nuances such as fruity, coconut-like, buttery, creamy, sweet, or nutty.

γ-Decalactone is the most important lactone for flavor application with a market volume of several hundred tons per year. It is characterized by an oily peachy aroma, with a very tenacious and powerful odor. Okui et al. (1963) first investigated the formation of γ-decalactone in the catabolism of ricinoleic acid by yeasts of the genus *Candida*. Four successive cycles of β-oxidation are responsible for the degradation of ricinoleic acid into 4-hydroxydecanoic acid, which lactonizes at lower pH to γ-decalactone (Gatfield et al. 1993) (Figure 3.6). The microbial γ-decalactone has

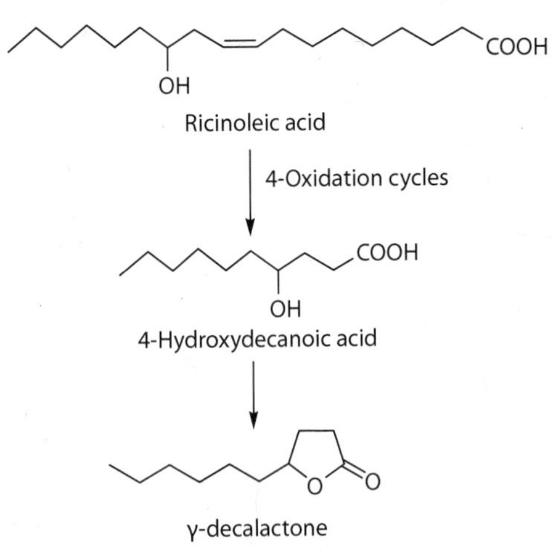

FIGURE 3.6 Synthesis of γ-decalactone from ricinoleic acid.

the same absolute configuration of the one found in peach and other fruits. The highest product concentrations have been achieved with *Y. lipolytica* strains. Nicaud et al. (1995) obtained γ-decalactone from ricinoleic acid methyl ester in high yields using a genetically engineered multiple auxotrophic mutant. In 2000, Haarmann and Reimer GmbH established a process affording up to 11 g/L γ-decalactone in 55 h without a genetically modified production strain and with raw castor oil as substrate (Rabenhorst and Gatfield 2000). The production of γ-lactone was also investigated with a number of different *Sporidiobolus* spp., such as *Sporidiobolus salmonicolor*, *Sporidiobolus ruinenii*, *Sporidiobolus johnsonii*, and *Sporidiobolus pararoseus* (Dufossé et al. 1998). Castor oil was employed as a substrate in a process using *A. niger*, *Pichia etchellsii*, or *Cladosporium suaveolens* to produce γ-decalactone (Cardillo et al. 1989).

3.3 METHODS FOR THE ASSESSMENT OF THE NATURAL ORIGIN OF FLAVORS

Natural and natural-identical flavors are chemically identical molecules, but they can be distinguished by the different isotopic enrichment of the atoms building up the molecular skeleton. The values of stable isotope ratios, such as $^{2}H/^{1}H$, $^{13}C/^{12}C$, and $^{18}O/^{16}O$, depend upon the way the molecule has been generated. They can discriminate between an enzymatic and a nonenzymatic synthesis. In addition, they can also be influenced by a particular biosynthetic route and by the environmental (including climatic) conditions under which the biosynthesis occurs. Isotopic abundances and patterns are now the most powerful means of assessing the authenticity, the botanical and also the geographic origin of a wide range of natural products (Schmidt et al. 1998). For many compounds the approach has been purely empirical, based on the comparison of unknown samples with authentic references. In a few cases the knowledge of the biosynthetic routes and of the reaction mechanisms allow the establishment of a set of systematic rules for the interpretation, on a scientific basis, of the experimental isotopic data (Schmidt et al. 1995, 2001, 2003, Schimdt and Ersenreich 2001, Schmidt 2003).

Chemical elemental analysis procedures coupled with isotope ratio mass spectrometry (IRMS) have been used to determine the isotopic ratios of the elements, mainly carbon ($^{13}C/^{12}C$), oxygen ($^{18}O/^{16}O$), and hydrogen ($^{2}H/^{1}H$). An average isotopic ratio of the whole molecule is obtained,* thereby losing site-specific information. Nevertheless, experiments have been carried out in which positional

---

* The ratio between the heavy and light isotopes of an element (H, C, or O) in a sample is measured and compared to that measured in a reference compound. The isotopic abundance of a sample is expressed by the equation: $\delta = \left(\frac{R_{samp}}{R_{ref}} - 1\right) \times 1000$, where δ is the isotope ratio of the sample expressed in delta units (‰, per mil) relative to the reference material. R_{samp} and R_{ref} are the absolute isotope ratios of the sample and of the reference material. Example for carbon: $\delta^{13}C = \left(\frac{(^{13}C/^{12}C)samp}{(^{13}C/^{12}C)ref} - 1\right) \times 1000$ and R_{ref} is the absolute value of the $^{13}C/^{12}C$ ratio of Vienna Pee Dee Belemnite ($R_{VPDB} = 0.0112372$).

isotopic information has been collected from chemical treatment of the molecule under study.

Site-specific isotope ratios for hydrogen and carbon can be obtained by using nuclear magnetic resonance,* according to a well established technique (SNIF–NMR®, Site-Specific Natural Isotope Fractionation by Nuclear Magnetic Resonance, Eurofins, Nantes, France) (Martin and Martin 1990). The European Community official method (Commission Regulation N.2348/91) for the detection of illegal chaptalization of wine is based on this technique (Martin and Martin 1981). The ^{2}H and ^{13}C enrichment of each hydrogen and carbon atom of the molecular skeleton can be measured to gain a great deal of information on the synthetic history of the molecule.

The use of these two different approaches for the authentication of the main natural flavors is herein discussed.

3.3.1 Vanillin

Natural-identical vanillin is mainly prepared according to two synthetic routes (Figure 3.7): (1) from lignin-containing waste liquors obtained from acid sulfite pulping of wood (Hocking, 1997); (2) from guaiacol of petrochemical origin, by reaction with glyoxylic acid, followed by oxidation and decarboxylation (Bauer et al. 1990).

The first attempt to characterize the origin of vanillin samples by ^{2}H-NMR was performed by Toulemonde and coworkers in 1983 (Toulemonde et al. 1983), and a few years later Martin and coworkers proposed the use of the SNIF-NMR technique for the same aim (Maubert et al. 1988). The considerable progress achieved in the field of magnetic resonance technology allowed Martin's group to obtain, in 1997, (Remaud et al 1997a) highly reproducible values of site-specific *D*/*H* ratios. These data (see atom numbering in Figure 3.7) show a very clear-cut discrimination of vanillin samples obtained from *Vanilla planifolia* beans, from lignin and from guaiacol: (1) ex-beans: $(D/H)_1$ = 130.8 ppm; $(D/H)_3$ = 157.3 ppm; $(D/H)_4$ = 196.4 ppm; $(D/H)_5$ = 126.6 ppm; (2) ex-lignin: $(D/H)_1$ = 119.9 ppm; $(D/H)_3$ = 132.1 ppm; $(D/H)_4$ = 168.8 ppm; $(D/H)_5$ = 105.9 ppm; (3) ex-guaiacol: $(D/H)_1$ = 315.2 ppm; $(D/H)_3$ = 138.8 ppm; $(D/H)_4$ = 143.8 ppm; $(D/H)_5$ = 139.1 ppm. It was also shown that

* The absolute values of the site-specific (*D*/*H*) ratios are calculated according to the following formula:

$$(D/H)_i = \frac{n_{\mathrm{ref}}\,\mathrm{wg}_{\mathrm{ref}}\,(\mathrm{MW})_{\mathrm{samp}}\,\mathrm{SA}_i\,(D/H)_{\mathrm{ref}}}{n_i\,\mathrm{wg}_{\mathrm{samp}}\,(\mathrm{MW})_{\mathrm{ref}}\,\mathrm{SA}_{\mathrm{ref}}}$$

where the reference standard is usually tetramethylurea (TMU), with a known isotope ratio $(D/H)_{\mathrm{ref}}$; n_{ref} and n_i are the number of equivalent deuterium atoms of TMU and of the *i*th peak of the sample; $\mathrm{wg}_{\mathrm{ref}}$ and $\mathrm{wg}_{\mathrm{samp}}$ are the weights of the standard and the sample; $\mathrm{MW}_{\mathrm{samp}}$ and $\mathrm{MW}_{\mathrm{ref}}$ are the corresponding molecular weights; SA_i and $\mathrm{SA}_{\mathrm{ref}}$ are the areas of the *i*th NMR peak and of the standard; $(D/H)_{\mathrm{ref}}$ is the reference standard isotope ratio as determined by isotope ratio mass spectrometry on the Vienna Standard Ocean Water (V-SMOW) scale ($R_{\mathrm{V\text{-}SMOW}} = {}^{2}H/{}^{1}H = 1.5576 \times 10^{-4}$). $(D/H)_i$ ratios are expressed in ppm. For the treatment of ^{13}C NMR spectral data see Tenailleau et al. (2004a).

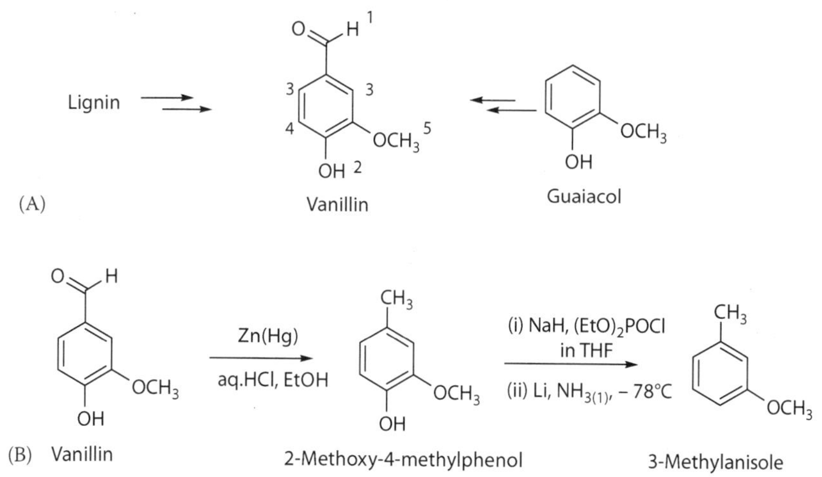

FIGURE 3.7 (A) Chemical routes to vanillin; (B) degradation of vanillin samples to obtain positional values of $\delta^{18}O$. (Adapted from Fronza, G. et al., *Helv. Chim. Acta*, 84, 351, 2001.)

the method could detect 5% of vanillin ex-guaiacol in vanillin ex-bean and about 10% of vanillin ex-lignin in vanillin ex-bean. In 2004, $(D/H)_i$ values were employed (John and Jamin 2004) to discriminate vanillin obtained from vanilla plantations in India, which had been found to show significant deviations from the aromatic profile of common natural vanillin.

A rapid continuous-flow technique for the quantitative determination of hydrogen isotope ratios in organic materials at natural abundance levels was developed by Kelly et al. (1998) and applied to the analysis of vanillin. This technique is based on the pyrolytic conversion of organic materials directly to hydrogen and carbon monoxide or carbon using an elemental analyzer coupled to a commercially available, high-resolution isotope ratio mass spectrometer. Samples were pyrolyzed in a helium stream at 1080°C over an inert form of carbon. Hydrogen was separated from the other pyrolysis gases by gas chromatography and entered the ion source of the isotope ratio mass spectrometer. The following δ^2H‰ values were obtained: (1) natural vanillin: −100.7‰; (2) vanillin from turmeric: −114.7‰; (3) synthetic vanillin: −58.9‰; vanillin (Sigma): 50.1‰.

$\delta^{18}O$ and $\delta^{13}C$ values of vanillin samples were determined simultaneously by a pyrolytic technique, generating carbon monoxide under continuous flow conditions, to be analyzed by IRMS (Py-CF-IRMS) (Dennis et al. 1998). Carbon monoxide gave information on the oxygen isotopes present in vanillin (average value for CHO, OH, and OCH_3 oxygen atoms). Moreover, according to the authors, the carbon–oxygen bond of the parent vanillin molecule was retained during pyrolysis, so that $\delta^{13}C$ should represent an average value for the carbonyl, methoxy and C-4 benzene ring positions. The results are the following: (1) Vanillin A (unknown origin): $\delta^{18}O$ = +12.7‰, $\delta^{13}C$ = −28.4‰; (2) Vanillin B (unknown origin): $\delta^{18}O$ = −2.2‰, $\delta^{13}C$ = −31.3‰; (3) Vanillin C (unknown origin): $\delta^{18}O$ = −2.7‰, $\delta^{13}C$ = −13.0‰; (4) Vanillin from vanilla beans: $\delta^{18}O$ = +14.7‰; (5) Synthetic vanillin (petroleum source): $\delta^{18}O$ = −0.2‰; (6) Vanillin from turmeric: $\delta^{18}O$ = +15.0‰.

Vanillin C resulted to be enriched in ^{13}C and it was clearly differentiated by this procedure. The work highlighted that synthetic vanillin derived from petroleum source was significantly depleted in ^{18}O compared with vanillin prepared from turmeric or from vanilla beans. Thus, the authors predicted that the oxygen in Vanillin A arose from a natural source, whereas the oxygen in Vanillin B and C arose from petroleum source.

In 2001, the positional values of $\delta^{18}O$ for vanillin were measured (Fronza et al. 2001) on the following samples: one synthetic vanillin of undefined origin, one synthetic sample from guaiacol, a synthetic sample from lignin, two extractive samples from *Vanilla* beans. The positional values could be obtained from the global $\delta^{18}O$ of vanillin samples by selective and progressive removal of the O functionalities. According to the scheme reported in Figure 3.7, each sample of vanillin was converted into 2-methoxy-4-methylphenol upon Clemmensen reduction, and subsequent removal of the phenolic O-atom gave 3-methylanisole. Samples derived from guaiacol could be distinguished from those possessing the aromatic moiety of natural origin because they showed a different ^{18}O enrichment at OCH_3 and OH. The values for the phenolic oxygen atom varied from +8.9‰ and +12‰ for the synthetic materials (one of undefined origin, and the other one from guaiacol), to +6.5‰, +5.3‰, and +6.3‰ for the sample from lignin and the two samples from *Vanilla* pods. The O-atoms of the methoxy groups showed the following values: −2.9‰, −3.2‰ for the synthetic samples, +3.5‰, +3.1‰, and +2.3‰ for the sample from lignin and for those obtained from *Vanilla* pods. Extractive vanillin differed from the product from lignin on the basis of the $\delta^{18}O$ values of the carbonylic oxygen atom, ranging from +25.5 and +26.2 in the natural material to + 19.7‰ in the lignin-based sample. The global ^{18}O values of the analyzed samples were found to be very similar (11.6‰, 11.9‰ for the synthetic vanillin molecules, 9.9‰ for that obtained from lignin, and 11.3‰, 11.6‰ for the extractive samples), thus confirming the clear diagnostic information that the positional $\delta^{18}O$ values can provide on the origin of different vanillin samples.

$\delta^{13}C$ isotope ratios were determined by EA–IRMS (Elemental Analysis-Isotope Ratio Mass Spectrometry) (Lamprecht et al. 1994) on vanillin samples extracted and purified by semi-preparative high-performance liquid chromatography. These $\delta^{13}C$ values were found to fall in the range from −21.5‰ to −20.2‰ for vanillin from vanilla extracts, and in the range from −29.1‰ to −26.89‰ for synthetic vanillin samples from lignin and from guaiacol.

In 1997, Kaunzinger et al. (1997) employed a gas chromatograph (GC) coupled online via combustion interface with IRMS to measure $\delta^{13}C$ ratios of vanillin and of minor components of natural vanilla extracts, such as 4-hydroxybenzaldehyde, 4-hydroxybenzylalcohol, vanillic acid, anisic alcohol, and 4-hydroxybenzoic acid. According to this technique, the substances eluting from the GC column are converted into carbon dioxide in a combustion oven and then directly analyzed in the isotope mass spectrometer for the simultaneous determination of masses 44, 45, and 46. By correlating isotopic data with concentrations of the compounds considered in the work, a characteristic three-dimensional authenticity profile was obtained for

V. planifolia and *V. tahitensis*. In this work the $\delta^{13}C$ values of vanillin from different *Vanilla* varieties were found to be rather different. The $\delta^{13}C$ value was −15.5‰ for *V. tahitensis* and −19.0‰ for *V. planifolia*, whereas the $\delta^{13}C$ values for 4-hydroxybenzaldehyde were quite similar.

Samples of biotechnological origin were considered in a work published by Bensaid et al. (2002). The formyl group of vanillin, whose oxygen atom easily undergoes exchange with water during any preparation and purification steps, was eliminated by conversion into guaiacol. The degradation reaction (Pd/C, 210°C, 2 h) was shown to proceed without significant isotopic fractionation at the sites of interest. $\delta^{13}C$ and $\delta^{18}O$ values of the corresponding guaiacol samples were evaluated by IRMS, and the site-specific $\delta^{13}C$ ratios of the formyl group were evaluated by comparison with the $\delta^{13}C$ values of the starting vanillin samples. The analysis of these results has shown that the formyl group of vanillin is more enriched in ^{13}C in synthetic products compared to those of natural origin. Particularly interesting are the data obtained for vanillin samples prepared through biotechnological routes from ferulic acid. ^{13}C depletion of the formyl group is higher for vanillin derived from ferulic acid of natural origin ($\delta^{13}C$ = −38.2‰) than for vanillin obtained from ferulic acid from fossil precursor ($\delta^{13}C$ = −26.4‰).

The site-specific isotopic $^{13}C/^{12}C$ ratios were determined by ^{13}C NMR spectroscopy (Tenailleau et al. 2004b) on 21 vanillin samples with the following known origin: 2 were extracted from beans; 3 were prepared by synthesis from lignin, 10 from different batches of guaiacol; 4 were obtained from natural ferulic acid by biotechnological route, and two were prepared from natural ferulic acid by synthetic procedures.

Total reduced molar fractions (y_i/Y_i) and specific isotopic deviations δ_i (‰) were calculated for the 21 samples from ^{13}C NMR spectra. Principal component analysis (PCA) was performed on the data set of the 84 analyses (21 vanillin samples and 4 spectra per vanillin sample) in the space of eight total reduced molar fractions. A variance analysis showed that major contributions came only from CHO and OCH_3. Similarly, PCA was used to analyze the 84-data-point set in the space of the eight specific isotopic deviations δ_1–δ_8, calculated from the ^{13}C NMR spectra: a variance analysis revealed that CHO and OCH_3 were again the most discriminating.

Very distinct groupings have been obtained in both cases in the PCA plane. Guaiacol-derived samples are well separated from those from other sources. Ex-lignin samples are efficiently distinguished from natural vanillin (ex-beans). As for the samples obtained from natural ferulic acid, the groupings of biotechnology-derived and synthetic-derived samples are distinct, although relatively closed to one another. One of the strengths of the method is that blended samples will show deviations from these established patterns, which could be easily detected. The repeatability for the determination of site-specific $^{13}C/^{12}C$ ratios at natural abundance by quantitative ^{13}C NMR spectroscopy has been recently tested (Caytan et al. 2007) and the technique has been shown to allow the discrimination of very small variations in $^{13}C/^{12}C$ ratios between carbon positions.

$^{13}C/^{12}C$ ratios are considered to be the most reliable data to establish an isotopic fingerprint for vanillin. As a matter of fact, *V. planifolia* exploits the crassulacean acid metabolism (CAM) pathway of photosynthesis to fix CO_2 (Walton et al. 2003).

CAM produces compounds with $\delta^{13}C$ values intermediate to those for C3 or C4 plants: C3 plants typically produce phenylpropanoids with $\delta^{13}C$ in the range −26‰ to −30‰; C4 plants generate phenylpropanoids with $\delta^{13}C$ from −12‰ to −15‰, vanillin shows $\delta^{13}C$ of about −21‰ (Tenailleau et al. 2004b).

The values of natural vanillin are quite different from those determined for samples of vanillin obtained from lignin, where CO_2 is not fixed by CAM, or from fossil fuel sources (between ca. −25‰ and −37‰). Vanillin samples produced by microbial fermentation will display $\delta^{13}C$ values depending on the metabolic pathway involved and on the origin of the feedstock: for example, ferulic acid can be obtained by C3 photosynthesis in wheat bran or sugar beet, or by C4 photosynthesis in maize.

3.3.2 Raspberry Ketone

Stable isotope characterization was performed by Fronza et al. (1999a) on the following samples of raspberry ketone: (1) two synthetic commercial samples; (2) two samples obtained from one of the synthetic sample upon aqueous basic and acid treatment, respectively; (3) five natural samples obtained upon microbial reduction of the natural unsaturated ketone 4-(4-hydroxyphenyl)-but-3-en-2-one (Figure 3.8); (4) one sample obtained upon catalytic hydrogenation of the natural unsaturated ketone; (5) one sample obtained upon catalytic hydrogenation of the synthetic modification of the unsaturated ketone; (6) one sample obtained upon baker's yeast reduction of the synthetic unsaturated ketone. In the discussion, the data collected in another laboratory for a commercial sample of raspberry ketone sold as natural have been also considered.

The values of site-specific *D*/*H* isotope ratio were evaluated by ^2H NMR, and the collected data allowed the distinction of the analyzed samples into two sets. The former, possessing $(D/H)_3/(D/H)_2 > 1$ (see atom numbering in Figure 3.8) included all of the samples derived from the unsaturated precursor (obtained by condensing extractive 4-hydroxybenzaldehyde with acetone from sugar fermentation), and the sample of unknown origin. The second set included the commercial synthetic materials and three samples sold as natural. A further distinction among the products of the first set, derived from the natural precursor, was then possible. The sample prepared by catalytic hydrogenation shows a $(D/H)_5/(D/H)_4$ value much higher that those obtained by bioreduction. Both $(D/H)_4$ and $(D/H)_5$ were higher for raspberry ketone obtained from the natural precursor by baker's yeast reduction than for the samples produced with other microbial systems. The following conclusions have also been drawn: (1) the mean value of $(D/H)_3$ for natural samples is much higher than that for synthetic

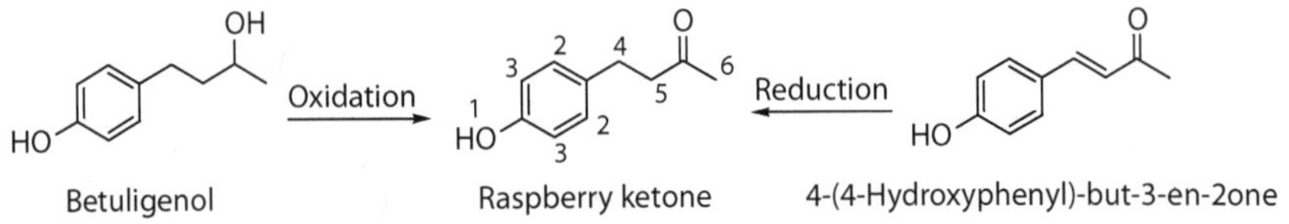

FIGURE 3.8 Synthetic approaches to the raspberry ketone samples investigated by Fronza et al. (1999a). (From Fronza, G. et al., *J. Agric. Food Chem.*, 46, 248,1999a.)

samples; (2) the mean $(D/H)_2$ values do not allow a discrimination between natural and synthetic raspberry ketones; (3) the $(D/H)_4$ mean value for natural samples is smaller than for synthetic samples. A development of this work (Fronza et al. 1999b) was the evaluation of the *D/H* values of a sample of raspberry ketone extracted from Himalayan *Taxus baccata*, and of two samples obtained by oxidation with *Candida boidinii* and CrO_3, respectively, of the alcohol betuligenol isolated from the same *T. baccata* (Figure 3.8). When the ratio $(D/H)_3/(D/H)_2$ was plotted vs. $(D/H)_5/(D/H)_4$ for these three samples and for those of the previous investigation, three well defined regions could be identified: (1) extractive from *Taxus* and biogenerated by oxidation of betuligenol; (2) biogenerated upon enzymic saturation of unsaturated ketone; and (3) synthetic. In the latter group, falls the three commercial samples sold as natural were found in the third region.

3.3.3 Benzaldehyde

Deuterium NMR spectra were recorded for benzaldehyde samples from petrochemical and botanical sources (Hagedorn 1992). The petrochemical materials were obtained either from the oxidation of toluene or from the chlorination of toluene to benzal chloride, followed by hydrolysis (Figure 3.9). The botanical samples were prepared either by retroaldol reaction of cinnamaldehyde from cassia bark (termed cassia benzaldehyde) or by emulsin-mediated reaction of amygdalin from fruit kernels (termed bitter almond oil). The molar fractions* of deuterium for the formyl hydrogen and for the ortho, and meta + para hydrogen atoms were determined from the deuterium NMR spectra integrals. Differences in the distribution of deuterium among the aromatic sites were found to be best emphasized by using the ADR value (aromatic distribution ratio = $y_{[ortho]}/y_{[meta+para]}$).

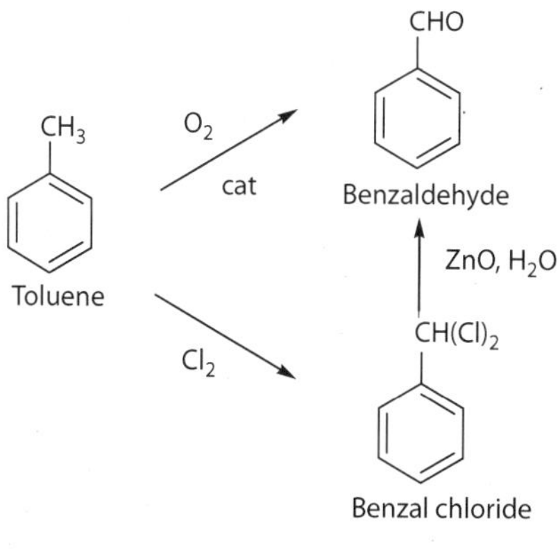

FIGURE 3.9 Chemical routes to benzaldehyde.

* Molar fractions y_i were calculated according to the following formula: $y_i = \frac{SA_i}{\Sigma_i SA_i}$ where SA_i is the area of the *i*th NMR peak.

Benzaldehyde samples prepared by oxidation of toluene showed very high deuterium enrichment at the formyl position ($y_{[formyl]}$ = 0.5548), probably due to a large primary kinetic isotope effect expected for the reaction. In benzaldehyde samples obtained from benzal chloride, the deuterium level of the CHO group was near statistical ($y_{[formyl]}$ = 0.1695; statistical = 1/6 or 0.1667). For the petrochemical samples the aromatic distribution was very similar (ADR from 0.6681 to 0.6900), quite close to the statistical value (statistical ADR = 2/3 = 0.6667), which is the value found for toluene itself. Toluene is obtained by catalytic reforming of paraffins, and this process is expected to provide random deuterium distribution in the aromatic ring.

The deuterium distribution for cassia benzaldehyde samples should be unchanged from that of original cinnamaldehyde. The biosynthesis of cinnamaldehyde, which involves several different reactions before complete aromatization, is expected to produce nonrandom deuterium distribution at the aromatic hydrogen atoms: $y_{[ortho]}$ = 0.3630 and ADR = 0.7518 were found for cassia benzaldehyde. Bitter almond oil was prepared by steam distillation of benzaldehyde released from amygdalin without any change of the isotopic distribution of the parent compound. As amygdalin is formed from phenylalanine via mandelonitrile, a nonstatistical aromatic distribution of deuterium would be expected.

The technique could detect adulteration of bitter almond oil known mixtures. An ADR value greater than 0.6084 for an unknown sample could indicate 10% adulteration of bitter almond oil with the other commercial sources of benzaldehyde.

Increase in reference data for benzaldehyde samples was achieved in 1997 by Remaud et al. (1997b). Site-specific $(D/H)_i$ ratios were measured from ^{2}H NMR experiments on benzaldehyde samples from toluene, benzal chloride, cinnamaldehyde and bitter almond oil. Three types of parameters could be obtained in this study: (1) site-specific deuterium ratios $(D/H)_i$ (in ppm); (2) molar fractions, y_i; (3) the relative $R_{i/j}$ ratios ($R_{i/j} = F_j$ SH_i/SH_j, where SH is the signal height), which represent the real number of deuterium atoms in site i with respect to site j, to which its stoichiometric number of hydrogen atoms F_j is given. Parameters $R_{2/1}$, $R_{3/1}$, $R_{4/2}$, and $R_{4/3}$ are therefore the deuterium probability factors of sites 2 (ortho), 3 (para), and 4 (meta) in a situation characterized by stoichiometric number $F_j = 1$ for reference site $j = 1$ (CHO group) and $j = 3$ (para H) and by stoichiometric number $F_j = 2$ when the reference site is $j = 2$ (ortho). For a random distribution of deuterium, the four relative ratios $R_{i/j}$ would be $R_{3/1} = 1$ and $R_{2/1} = R_{4/2} = R_{4/3} = 2$. $R_{2/1}$ and $R_{3/1}$ reflect the ratio between the deuterium content of aromatic and formyl sites; $R_{4/2}$ and $R_{4/3}$ provide information on the relative deuterium distribution of the meta and ortho positions, and the meta and para positions, respectively.

Principal component analyses were performed with the isotope ratios only, and then with molar fractions y_i and relative ratios $R_{i/j}$. In both case groupings of the analytical data an agreement with the sample origins was observed, without any overlapping. A random deuterium distribution on the hydrogen atoms in the aromatic ring was observed for the synthetic samples: i.e., the ratios $R_{4/2}$ and $R_{4/3}$ were close to 2. The deuterium ratio between the formyl group and the benzene ring was found to be very different when compared to the synthetic products on the one hand and the benzaldehyde from cassia and from bitter almond oil on the other hand. Random distribution of deuterium was observed for the ex-bitter almond oil and ex-cassia products with $R_{3/1}$ close to 1, while the aldehyde function showed a higher deuterium content in the synthetic samples, with $R_{3/1} = 0.125$ for benzaldehyde from toluene, and $R_{3/1} = 0.673$ for samples from benzal chloride. These data are in agreement with the occurrence of a large primary kinetic isotope effect in the oxidation reaction of toluene: the elimination of the light hydrogen atom from the methyl is favored, and the remaining hydrogen in the aldehydic group is strongly enriched. The main difference between the benzaldehyde ex-cassia and the benzaldehyde ex-bitter almond oil is found to be the lower deuterium content of the meta position and the higher deuterium content of the para position for the former.

Ruff et al. (2000) determined the δ^2H values of benzaldehyde samples of different origin by using a mass spectrometer online coupled to a GC via a pyrolysis interface (HRGC—P-IRMS). The compounds, after separation and elution from the GC column, yielded H_2 in the pyrolysis interface. Simultaneous recording of masses 2 ($^1H\,^1H$) and 3 ($^1H\,^2H$) was performed, and the isotope ratios were expressed in ‰. The following ranges of δ^2H values were determined on benzaldehyde samples of different origin: (1) synthetic from benzal chloride: δ^2H from −85‰ to −78‰; (2) synthetic from toluene: δ^2H from +420‰ to + 668‰; (3) "natural," including "ex-cassia:" δ^2H from −144‰ to −83‰; (4) bitter almond oils: δ^2H from −148‰ to −113‰; (5) from fruits: δ^2H from −146‰ to −111 ‰; (6) from kernels: δ^2H from −188‰ to −115‰; (7) from leaves: δ^2H from −189‰ to −165‰. The same authors applied this technique (Ruff et al. 2001) for a screening of the origin of benzaldehyde in commercial food, such as processed cherries, processed milk product with cherry flavor, alcohol-free beverages, and aromatized teas.

The deuterium isotopic pattern of the aromatic ring in benzenoid and phenylpropanoid flavors has been deeply investigated and related, respectively, to shikimate metabolic pathway for natural samples and to the chemical reactions involved in the manufacturing process for the synthetic materials (Martin et al. 2006a,b, Schmidt et al. 2006).

3.3.4 2-Phenylethanol

A natural abundance ^2H nuclear magnetic resonance study of the origin of 2-phenylethanol was carried out by Fronza et al. (1995a) using the following samples: (1) synthetic commercial samples from different suppliers; (2) the product obtained from synthetic phenylacetaldehyde upon bakers' yeast reduction; (3) the product obtained from the same aldehyde upon $NaBH_4$ reduction; (4) two samples of commercial natural 2-phenylethanol; (5) a sample prepared from natural L-phenylalanine according to the synthetic scheme shown in Figure 3.10.

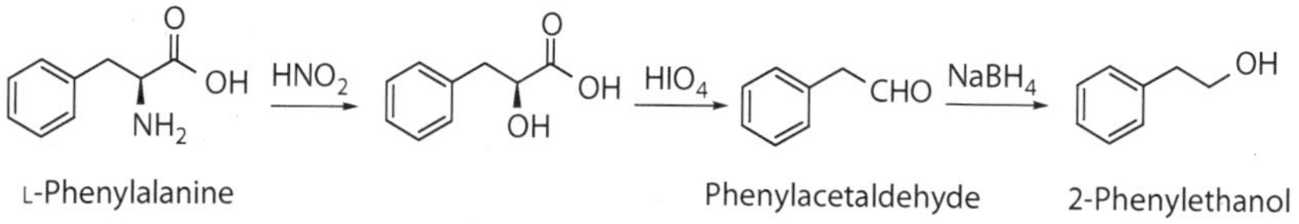

FIGURE 3.10 Synthesis of phenylethanol from L-phenylalanine. (Adapted from Fronza G. et al., *J. Agric. Food Chem.*, 43, 439, 1995a.)

The molar fraction of the aromatic hydrogen atoms did not show any significant variation, while molar fractions y_2 (for CH_2O) and y_3 (for $PhCH_2$) showed relevant changes according to the origin of the samples. The natural samples displayed the lowest values of the molar fraction y_2 (ca. 0.13) and the highest values of y_3 (ca. 0.22). The synthetic samples were characterized by a high deuterium content at the CH_2O group (y_2 = ca. 0.20). The sample obtained from the reduction of synthetic phenylacetaldehyde by bakers' yeast showed a deuterium distribution very similar to that of synthetic samples, while phenylethanol prepared from phenylacetaldehyde by $NaBH_4$ reduction showed a significantly lower deuterium content at the CH_2O moiety ($y_2 = 0.18$). The sample obtained from natural phenylalanine through nonenzymic controlled reaction showed a rather high value of y_3 (0.2051), very similar to that measured for the natural samples, but in contrast with them it displayed a higher value of the mole fraction y_2 (0.17). This y_2 value is quite similar to that of the sample obtained from synthetic phenylacetaldehyde upon $NaBH_4$ reduction.

Thus, the deuterium content at the CH_2O group resulted to be particularly diagnostic for the authentication of the naturalness of phenylethanol, which was influenced by the way phenylacetaldehyde has been reduced.

3.3.5 Green Notes

A SNIF NMR investigation was performed by Fronza et al. (1996) on *n*-hexanol, promoted by the observation that the fungus *Colletotrichum gloeosporioides* CBS 193.82 was able to aerobically reduce a set of carboxylic acids, including hexanoic acid, to the corresponding alcohols (Fronza et al., 1995b). When the biotransformation was applied to natural hexanoic acid, the yield of natural hexanol reached 2 g/L after 24 h of incubation. For experimental simplification, the ^2H NMR spectra were recorded on the acetate ester obtained from the alcohol samples by treatment with acetic anhydride in pyridine. Site-specific D/H ratios and molar fractions y_i were evaluated for the following samples of acetate: (1) synthetic commercial samples from different suppliers and lots; (2) natural commercial samples; (3) the compound produced in *C. gloeosporioides* from natural hexanoic acid; and (4) the compound produced from the same natural acid upon $LiAlH_4$ reduction.

The values of molar fraction y_1 (for the CH_2O position) and y_4 (for the three CH_2 in position 3, 4, and 5) were found to be the most discriminating. In particular, the sample obtained by microbial reduction of natural hexanoic acid showed the highest value of y_1 (0.260), much higher than that of the sample (0.045) prepared from the same precursor upon metal hydride reduction. The values y_1 for the synthetic samples and for the natural commercial ones were not very different, ranging from 0.127 to 0.145 for the former group and from 0.151 to 0.154 for the latter. However, a differentiation between the commercial synthetic and natural materials could be made by the y_4 values. These values ranged from 0.407 to 0.374 for the first set and from 0.312 to 0.324 for the second one. Thus, the natural (extractive) samples showed higher y_1 and lower y_4 than the synthetic materials.

A short communication by Muller and Gautier in 1994 (Muller and Gautier 1994) reported the D/H values for three samples of 3Z-hexenol, one of synthetic origin, one extracted from mint, and one prepared from linolenic acid by a biotechnological route.

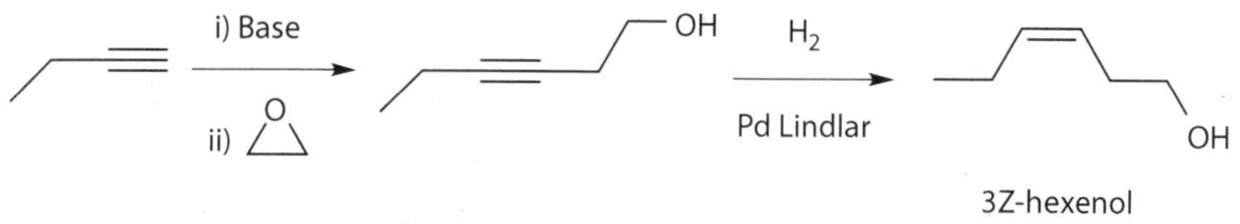

FIGURE 3.11 Chemical route to 3Z-hexenol.

A more extensive work was performed on 3Z-hexenol in 1997 (Barbeni et al. 1997), using the following samples: (1) five synthetic commercial samples; (2) two natural commercial samples; (3) four natural samples obtained by biodegradation of linolenic acid according to a procedure with baker's yeast-mediated reduction of the intermediate 3Z-hexenal (Brunerie 1991).

The y_1 (molar fraction of the two vinylic hydrogen atoms), and y_3 (molar fraction of CH_2O) allowed a distinct differentiation between synthetic and biosynthetic samples. The y_1 values fell in the range 0.047–0.057 and 0.141–0.152, respectively, for these two sample groups, and y_3 values were between 0.208 and 0.252 in the former set, and between 0.209 and 0.099 for the second one. These data were explained on the basis of the two different origins of these hydrogen atoms. The vinylic hydrogen atoms are inserted by *syn* catalytic partial hydrogenation of the triple bond of the acetylenic intermediate in the synthetic sample (Figure 3.11), while they belong to the starting linolenic acid in the biosynthetic compounds. The y_3 (CH_2O) values of the synthetic materials are due to the deuterium content of the ethylene oxide employed in the synthesis. In the bio-generated products, CH_2O is created at the end of the sequence by enzymatic reduction of the intermediate 3Z-hexenal. The origin of natural commercial samples could be tentatively defined by considering $(D/H)_1$ and $(D/H)_3$ values: one of them seemed to be of extractive origin (from mint) by comparison with the data reported by Muller and Gautier (1994); the other one appeared to derive from a biotechnological route.

3.4 CONCLUDING REMARKS

Biotechnological production of natural flavors shows interesting topics to be addressed and developed. The increasing preference of the flavor industry for natural products requires the optimization of efficient and cheap procedures for their production. This research effort has to be supported by the development of analytical techniques to authenticate the natural origin of flavors, in order to fight fraud and counterfeiting. The analysis of stable isotope ratios has been found to be a very reliable technique and to offer a lot of information on the synthetic procedures of natural products.

REFERENCES

Aguedo, M., Huong Ly, M., Belo, I., et al. 2004. The use of enzymes and microorganisms for the production of aroma compounds from lipids. *Food Technol. Biotechnol.* 42: 327–336.

Barbeni, M., Cisero, M., and Fuganti, C. 1997. Natural abundance ^2H nuclear magnetic resonance study of the origin of (Z)-3-hexenol. *J. Agric. Food Chem.* 45: 237–241.

Bauer, K., Garbe, D., and Surburg, H. 1990. Single fragrance and flavor compounds. In *Common Fragrance and Flavor Materials*, Weinheim: VCH.

Bensaid, F.F., Wietzerbin, K., and Martin, G.J. 2002. Authentication of natural vanilla flavorings: Isotopic characterization using degradation of vanillin into guaiacol. *J. Agric. Food Chem.* 50: 6271–6275.

Bourel, G., Nicaud, J.M., Nthangeni, B., et al. 2004. Fatty acid hydroperoxide lyase of green bell pepper: Cloning in *Yarrowia lipolytica* and biogenesis of volatile aldehydes. *Enzyme Microb. Technol.* 35: 293–299.

Brunerie, P. 1991. Procedure for the synthesis of *cis*-3-hexen-1-ol. French Patent 2,652,587, filed October 3, 1989, and issued April 5, 1991 (in French).

Brunerie, P. and Koziet, Y. 1997. Process for producing *cis*-3-hexenol from unsaturated fatty acids. U.S. Patent 5,620,879, filed June 30, 1995, and issued April 15, 1997.

Buck, K.T., Boeing, A.J., and Dolfini, J.E. 1987. Method of producing benzaldehyde. U.S. Patent 4,673,766, filed April 25, 1986, and issued June 16, 1987.

Cardillo, R., Fuganti, C., Sacerdote, G., et al. 1989. Process for the microbiological production of gammadecalactone (*R*) and gamma-octalactone (*R*). European Patent 356,291, filed August 3, 1989, and issued February 28, 1990.

Caytan, E., Botosoa, E.P., Silvestre, V., et al. 2007. Accurate quantitative ^{13}C NMR spectroscopy: Repeatability over time of site-specific ^{13}C isotope ratio determination. *Anal. Chem.* 79: 8266–8269.

Dennis, M.J., Wilson, P., Kelly, S., et al. 1998. The use of pyrolytic techniques to estimate site specific isotope data of vanillin. *J. Anal. Appl. Pyrol.* 47: 95–103.

Dufossé, L., Feron, G., Mauvais, G., et al. 1998. Production of γ-decalactone and 4-hydroxydecanoic acid in the genus *Sporidiobolus*. *J. Ferment. Bioeng.* 86: 169–173.

Dumont, B., Hugueny, P., and Belin, J.M. 1994. Preparation of raspberry-like ketones by bioconversion. French Patent 2,724,666, filed September 19, 1994, and issued March 22, 1996.

European Commission. 2006. Package of proposals for new legislation on food additives, flavourings and enzymes. http://ec.europa.eu/food/food/ chemicalsafety/additives/prop_leg_en.htm (accessed September 26, 2008).

Etschmann, M.M.W., Bluemke, W., Sell, D., et al. 2002. Biotechnological production of 2-phenylethanol. *Appl. Microbiol. Biotechnol.* 59: 1–8.

Etschmann, M.M.W., Sell, D., and Schrader, J. 2003. Screening of yeasts for the production of the aroma compound 2-phenylethanol in a molasses-based medium. *Biotechnol. Lett.* 25: 531–536.

Etschmann, M.M.W., Sell, D., and Schrader, J. 2004. Medium optimization for the production of the aroma compound 2-phenylethanol using a genetic algorithm. 2004. *J. Mol. Cat. B.: Enzyme* 29: 187–193.

Fabre, C., Blanc, P., and Goma, G. 1996. Production of benzaldehyde by several strains of *Ischnoderma benzoinum*. *Sci. Aliments* 16: 61–68.

Fabre, C.E., Blanc, P.J., and Goma, G. 1997. Screening of yeasts producing 2-phenylethanol. *Biotechnol. Tech.* 11: 523–525.

Fabre, C.E., Blanc, P.J., and Goma, G. 1998. Production of 2-phenylethyl alcohol by *Kluyveromyces marxianus*. *Biotechnol. Prog.* 14: 270–274.

Feron, G., Bonnarme, P., and Durand, A. 1996. Prospects for the microbial production of food flavours. *Trends Food Sci. Technol.* 71: 285–293.

Fronza, G., Fuganti, C., Grasselli, P., et al. 1995a. Natural abundance ^{2}H nuclear magnetic resonance study of the origin of 2-phenylethanol and 2-phenylethyl acetate. *J. Agric. Food Chem.* 43: 439–443.

Fronza, G., Fuganti, C., Grasselli, P., et al. 1995b. Reduction of carboxylates to alkanols catalyzed by *Colletotrichum gloeosporoides*. *J. Chem. Soc., Chem. Commun.* 439–440.

Fronza, G., Fuganti, C., Zucchi, G., et al. 1996. Natural abundance ^2H nuclear magnetic resonance study of the origin of *n*-hexanol. *J. Agric. Food Chem.* 44: 887–891.

Fronza, G., Fuganti, C., Guillou, C., et al. 1999a. Natural abundance ^2H nuclear magnetic resonance study of the origin of raspberry ketone. *J. Agric. Food Chem.* 46: 248–254.

Fronza, G., Fuganti, C., Pedrocchi-Fantoni, G., et al. 1999b. Stable isotope characterization of raspberry ketone extracted from *Taxus baccata* and obtained by oxidation of the accompanying alcohol (betuligenol). *J. Agric. Food Chem.* 47: 1150–1155.

Fronza, G., Fuganti, C., Serra, S., et al. 2001. The positional $\delta^{18}O$ values of extracted and synthetic vanillin. *Helv. Chim. Acta* 84: 351–359.

Fuganti, C., Mendozza, M., Joulain, D., et al. 1996. Biogeneration and biodegradation of raspberry ketone in the fungus *Beauveria bassiana. J. Agric. Food Chem.* 44: 3616–3619.

Furukawa, H., Morita, H., Yoshida, T., et al. 2003. Conversion of isoeugenol into vanillic acid by *Pseudomonas putida* I58 cells exhibiting high isoeugenol-degrading activity. *J. Biosci. Bioeng.* 96: 401–403.

Gatfield, I.L., Güntert, M., Sommer, H., et al. 1993. Some aspects of microbiological manufacture of flavor-active lactones with particular reference to γ-decalactone. *Chem. Mikrobiol. Technol. Lebensm.* 15: 165–170.

Gross, B., Ashter, M., Corrieu, G., et al. 1993. Production of vanillin by bioconversion of benzenoid precursors by *Pycnoporus*. U.S. patent 5,262,315, filed May 1, 1992, and issued November 16, 1993.

Hagedorn, M.L. 1992. Differentiation of natural and synthetic benzaldehydes by ^2H nuclear magnetic resonance. *J. Agric. Food Chem.* 40: 634–637.

Häusler, A., Lerch, K., Muheim, A., et al. 2001. Hydroperoxide lyases. U.S. patent 6,238,898, filed October 14, 1999, and issued May 29, 2001.

Hocking, M. 1997. Vanillin: Synthetic flavoring from spent sulfite liquor. *J. Chem. Educ.* 74: 1055–1059.

Hrazdina, G.J. 2006. Aroma production by tissue cultures. *J. Agric. Food Chem.* 54: 1116–1123.

Hua, D., Ma, C., Lin, S., et al. 2007a. Biotransformation of isoeugenol to vanillin in a newly isolated *Bacillus pumilus* strain: Identification of major metabolites. *J. Biotechnol.* 130: 463–470.

Hua, D., Ma, C., Song, L., et al. 2007b. Enhanced vanillin production from ferulic acid using adsorbent resin. *Appl. Microbiol. Biotechnol.* 74: 783–790.

John, T.V. and Jamin, E. 2004. Chemical investigation and authenticity of Indian vanilla beans. *J. Agric. Food Chem.* 52: 7644–7650.

Kasana, R.C., Sharma, U.K., Sharma, N., et al. 2007. Isolation and identification of a novel strain of *Pseudomonas chlororaphis* capable of transforming isoeugenol to vanillin. *Curr. Microbiol.* 54: 457–461.

Kaunzinger, A., Juchelka, D., and Mosandl, A. 1997. Progress in the authenticity assessment of vanilla. 1. Initiation of authenticity profiles. *J. Agric. Food Chem.* 45: 1752–1757.

Kawabe, T. and Morita, H. 1994. Production of benzaldehyde and benzyl alcohol by the mushroom *Polyporus tuberaster* K2606. *J. Agric. Food Chem.* 42: 2556–2560.

Kelly, S.D., Parker, I.G., Sharman, M., et al. 1998. On-line quantitative determination of $^2H/^1H$ isotope ratios in organic and water samples using an elemental analyzer coupled to an isotope ratio mass spectrometer. *J. Mass Spectrom.* 33: 735–738.

Krings, U. and Berger, R.G. 1995. Porous polymers for fixed bed adsorption of aroma compounds in fermentation processes. *Biotechnol. Tech.* 9: 19–24.

Krings, U. and Berger, R.G. 1998. Biotechnological production of flavours and fragrances. *Appl. Microbiol. Biotechnol.* 49: 1–8.

Krings, U., Hinz, M., and Berger, R.G. 1996. Degradation of [^2H] phenylalanine by the basidiomycete *Ischnoderma benzoinum. J. Biotechnol.* 51: 123–129.

Lamer, T., Spinnler, H.E., Souchon, L., et al. 1996. Extraction of benzaldehyde from fermentation broth by pervaporation. *Process Biochem.* 31: 533–542.

Lamprecht, G., Pichlmayer, F., and Schmidt, E.R. 1994. Determination of the authenticity of vanilla extracts by stable isotope ratio analysis and component analysis by HPLC. *J. Agric. Food Chem.* 42: 1722–1727.

Lapadatescu, C., Feron, G., Vergoignan, C., et al. 1997. Influence of cell immobilization on the production of benzaldehyde and benzyl alcohol by the white-rot fungi *Bjerkandera adusta, Ischnoderma benzoinum* and *Dichomitus squalens. Appl. Microbiol. Biotechnol.* 47: 708–714.

Lesage-Meessen, L., Delattre, M., Haon, M., et al. 1996a. Method for obtaining vanillic acid and vanillin by bioconversion by an association of filamentous microorganisms. WO 96/08576, filed September 13, 1994, and issued March 121, 1996.

Lesage-Meessen, L., Delattre, M., Haona, M., et al. 1996b. A two-step bioconversion process for vanillin production from ferulic acid combining *Aspergillus niger* and *Pycnoporus cinnabarinus. J. Biotechnol.* 50: 107–113.

Lesage-Meessen, L., Haon M., Delattre, M., et al. 1997. An attempt to channel the transformation of vanillic acid into vanillin by controlling methoxyhydroquinone formation in *Pycnoporus cinnabarinus* with cellobiose. *Appl. Microbiol. Biotechnol.* 47: 39–397.

Lomascolo, A., Stentelaire C., Asther, M., et al. 1999a. Basidiomycetes as new biotechnological tools to generate natural aromatic flavours for the food industry. *Trends Biotechnol.* 17: 282–289.

Lomascolo, A., Lesage-Meessen, L., Labat, M., et al. 1999b. Enhanced benzaldehyde formation by a monokaryotic strain of *Pycnoporus cinnabarinus* using a selective solid adsorbent in the culture medium. *Can. J. Microbiol.* 45: 653–657.

Longo, M.A. and Sanromán, M.A. 2006. Production of food aroma compounds: Microbial and enzymatic methodologies. *Food Technol. Biotechnol.* 44: 335–353.

Martin, G.J. and Martin, M.L. 1981. Deuterium labeling at the natural abundance level as studied by high field quantitative ^{2}H NMR. *Tetrahedron Lett.* 22: 3525–3528.

Martin, G.J. and Martin, M.L. 1990. In *NMR Basic Principles and Progress*, Ed. H. Günther. Heidelberg: Springer-Verlag.

Martin, G.J., Heck, G., Djamaris-Zainal, R., et al. 2006a. Isotopic criteria in the characterization of aromatic molecules. 1. Hydrogen affiliation in natural benzenoid/phenylpropanoid molecules. *J. Agric. Food Chem.* 54: 10112–10119.

Martin, G.J., Heck, G., Djamaris-Zainal, R., et al. 2006b. Isotopic criteria in the characterization of aromatic molecules. 2. Influence of the chemical elaboration process. *J. Agric. Food Chem.* 54: 10120–10128.

Maubert, C., Guérin, C., Mabon, F., et al. 1988. Détermination de l'origine de la vanilline par analyse multidimensionnelle du fractionnement isotopique naturel spécifique de l'hydrogène. (Determination of the origin of vanillin by multivariate analysis of the site-specific natural isotope fractionation factors of hydrogen. Original language: French). *Analusis* 16: 434–439.

Muheim, A., Häusler, A., Schilling, B., et al. 1997. The impact of recombinant DNA-technology on the flavour and fragrance industry. In *Flavours and Fragrances: Proceeding of a Conference*, pp. 11–20. Cambridge, U.K.: Royal Society of Chemistry.

Muller, B.L. and Gautier, A.E. 1994. Green note production: A challenge for biotechnology. In *Trends in Flavour Research*, Eds. H. Maarse and D.G. van der Heij, pp. 475–479. Amsterdam, the Netherlands: Elsevier.

Muller, B., Gautier, A., Dean, C., et al. 1995. Process for the enzymatic preparation of aliphatic alcohols and aldehydes from linoleic acid, linolenic acid, or a natural precursor. U.S. patent 5,464,761, filed December 9, 1993, and issued January 21, 1994.

Nicaud, J.M., Belin, J.M., Pagot, Y., et al. 1995. Bioconversion of substrate with microbe auxotrophic for specific compound in medium deficient in this compound. French patent 2,734,843, filed June 6, 1995, and issued December 12, 1996.

Oddou, J., Stentelaire, C., Lesage-Meessen, L., et al. 1999. Improvement of ferulic acid bioconversion into vanillin by use of high-density cultures of *Pycnoporus cinnabarinus*. *Appl. Microbiol. Biotechnol.* 53: 1–6.

Okui, S., Uchiyama, M., and Mizugaki, M. 1963. Metabolism of hydroxyl fatty acids. II. Intermediates of the oxidative breakdown of ricinoleic acid by genus. *Candida. J. Biochem.* 54: 536–540.

Park, J.K. and Jung, J.Y. 2002. Production of benzaldehyde by encapsulated whole-cell benzoylformate decarboxylase. *Enzyme Microb. Technol.* 30: 726–733.

Priefert, H., Rabenhorst, J., and Steinbüchel, A. 2001. Biotechnological production of vanillin. *Appl. Microbiol. Biotechnol.* 56: 296–314.

Rabenhorst, J. and Hopp, R. 1991. Procedure for the production of natural vanillin. European Patent 0,405,197 A1, filed June 6, 1989, and issued January 2, 1991 (in German).

Rabenhorst, J. and Gatfield, I. 2000. Method of producing γ-decalactone. World Patent 0024920, filed October 24, 1998, and issued May 4, 2000.

Ramachandra Rao, S. and Ravishankar, G.A. 2000. Vanilla flavour: Production by conventional and biotechnological routes. *J. Sci. Food Agric.* 80: 289–304.

Ramachandra Rao, S. and Ravishankar G.A. 2002. Plant cell cultures: Chemical factories of secondary metabolites. *Biotechnol. Adv.* 20: 101–153.

Remaud, G., Martin, Y.-L., Martin, G.G., et al. 1997a. Detection of sophisticated adulterations of natural vanilla flavors and extracts: Application of the SNIF-NMR method to vanillin and p-hydroxybenzaldehyde. *J. Agric. Food Chem.* 45: 859–866.

Remaud, G., Debon, A.A., Martin, Y., et al. 1997b. Authentication of bitter almond oil and cinnamon oil: Application of the SNIF-NMR method to benzaldehyde. *J. Agric. Food Chem.* 45: 4042–4048.

Ruff, C., Hör, K., Weckerle, B., et al. 2000. $^{2}H/^{1}H$ Ratio analysis of flavor compounds by online gas chromatography pyrolysis isotope ratio mass spectrometry (HRGC-P-IRMS): Benzaldehyde. *J. High Resolut. Chromatogr.* 23: 357–359.

Ruff, C., Hör, K., Weckerle, B., et al. 2001. Authenticity control of aromatic substances. Gas chromatographic isotope ratio-mass spectrometry for the determination of $^{2}H/^{1}H$-ratios of benzaldehyde in food. *Deutsche Lebensmittel-Rundschau* 96: 243–247.

Schmidt, H.-L., 2003. Fundamentals and systematics of the non-statistical distributions of isotopes in natural compounds. *Naturwissenschaften* 90: 537–552 (erratum: 2004, 91, 148).

Schmidt, H.-L. and Eisenreich, W., 2001. Systematic and regularities in the origin of ^{2}H patterns in natural compounds. *Isot. Environ. Health Stud.* 37: 253–254.

Schmidt, H.-L., Kexel, H., Butzenlechner, M., et al. 1995. Non-statistical isotope distribution in natural compounds: Mirror of their biosynthesis and key for their origin assignment. In *Stable Isotopes in the Biosphere*, Eds. E. Wada, T. Yoneyama, M. Minagawa, T. Ando, and B.D. Fry, pp. 17–35. Kyoto, Japan: Kyoto University Press.

Schmidt, H.-L., Roßmann, A., and Werner, R.A. 1998. Stable isotope ratio analysis in quality control of flavourings. In *Flavourings, Production, Composition, Applications, Regulations*, Eds. E. Ziegler and H. Ziegler, pp. 539–594, Weinheim: Wiley-VCH.

Schmidt, H.-L., Werner, R.A., and Roßmann, A. 2001. ^{18}O Pattern and biosynthesis of natural plant products. *Phytochemistry* 58: 9–32.

Schmidt, H.-L., Werner, R.A., and Eisenreich, W. 2003. Systematics of ^{2}H patterns in natural compounds and its importance for the elucidation of biosynthetic pathways. *Phytochem. Rev.* 2: 61–85.

Schmidt, H.-L., Werner, R.A., Eisenreich, W., et al. 2006. The prediction of isotopic patterns in phenylpropanoids from their precursors and the mechanism of the NIH-shift: Basis of the isotopic characteristics of natural aromatic compounds. *Phytochemistry* 67: 1094–1103.

Schrader, J., Etschmann, M.M.W., Sell, D., et al. 2004. Applied biocatalysis for the synthesis of natural flavour compounds—Current industrial processes and future prospects. *Biotechnol. Lett.* 26: 463–472.

Serp, D., von Stockar, U., and Marison, I.W. 2003. Enhancement of 2-phenylethanol productivity by *Saccharomyces cerevisiae* in two-phase fed-batch fermentations using solvent immobilization. *Biotechnol. Bioeng*. 82: 103–110.

Serra, S., Fuganti, C., and Brenna, E. 2005. Biocatalytic preparation of natural flavours and fragrances. *Trends Biotechnol*. 23: 193–198.

Shimoni, E., Ravid, U., and Shoham, Y. 2000. Isolation of a *Bacillus* sp. capable of transforming isoeugenol to vanillin. *J. Biotechnol*. 78: 1–9.

Stark, D., Münch, T., Sonnleitner, B., et al. 2002. Extractive bioconversion of 2-phenylethanol from L-phenylalanine by *Saccharomyces cerevisiae. Biotechnol. Prog*. 18: 514–523.

Stentelaire, C., Lesage-Meessen, L., Delattre, M., et al. 1998. By-passing of unwanted vanillyl alcohol formation using selective adsorbents to improve vanillin production with *Phanerochaete chrysosporium. World J. Microbiol. Biotechnol*. 14: 285–287.

Stentelaire, C., Lesage-Meessen, L., Oddou, J., et al. 2000. Design of a fungal bioprocess for vanillin production from vanillic acid at scalable level by *Pycnoporus cinnabarinus. J. Biosci. Bioeng*. 89: 223–230.

Tenailleau, E., Lancelin, P., Robins, R.J., et al. 2004a. NMR approach to the quantification of nonstatistical 13C distribution in natural products: Vanillin. *Anal. Chem*. 76: 3818–3825.

Tenailleau, E.J., Lancelin, P., Robins, R.J., et al. 2004b. Authentication of the origin of vanillin using quantitative natural abundance ^{13}C NMR. *J. Agric. Food Chem*. 52: 7782–7787.

Tiefel, P. and Berger, R.G. 1993. Volatiles in precursor fed cultures of basidiomycetes. *Progress in Flavour Precursor Studies*, Eds. P. Schreier and P. Winterhalter, p. 439, Berlin: Allured Carol Stream.

Toulemonde, B., Horman, I., Egli, H., et al. 1983. Applications of high resolution NMR. Part II. Differentiation between natural and synthetic vanillin samples using ^{2}H-NMR. *Helv. Chim. Acta* 66: 2342–2345.

Vandamme, E.J. and Soetaert, W. 2002. Bioflavours and fragrances via fermentation and biocatalysis. *J. Chem. Technol. Biotechnol*. 77: 1323–1332.

Waché, Y., Husson, F., Feron, G., et al. 2006. Yeast as an efficient biocatalyst for the production of lipid-derived flavours and fragrances. *Antonie van Leeuwenhoek* 89: 405–416.

Walton, N.J., Narbad, A., Faulds, C.B., et al. 2000. Novel approaches to the biosynthesis of vanillin. *Curr. Opin. Biotechnol*. 11: 490–496.

Walton, N.J., Mayer, M.J., and Narbad, A. 2003. Molecules of interest: Vanillin. *Phytochemistry* 63: 505–515.

Washisu, Y., Tetsushi, A., Hashimoto, N., et al. 1993. Manufacture of vanillin and related compounds with *Pseudomonas*. Japanese Patent 5,227,980, filed February 21, 1992, and issued September 7, 1993.

Xu, P., Hua, D., and Ma, C. 2007. Microbial transformation of propenylbenzenes for natural flavour production. *Trends Biotechnol*. 25: 571–576.

Yamada, M., Okada, Y., Yoshida, T., et al. 2007. Biotransformation of isoeugenol to vanillin by *Pseudomonas putida* IE27 cells. *Appl. Microbiol. Biotechnol*. 73: 1025–1030.

Zhang, Y., Xu, P., Han, S., et al. 2006. Metabolism of isoeugenol via isoeugenol-diol by a newly isolated strain of *Bacillus subtilis* HS8. *Appl. Microbiol. Biotechnol*. 73: 771–779.

Zhao, L.-Q., Sun, Z.-H., Zheng, P., et al. 2005. Biotransformation of isoeugenol to vanillin by a novel strain of *Bacillus fusiformis. Biotechnol. Lett*. 27: 1505–1509.

Zhao, L.-Q., Sun, Z.-H., Zheng, P., et al. 2006. Biotransformation of isoeugenol to vanillin by *Bacillus fusiformis* CGMCC1347 with the addition of resin HD-8. 2006. *Process Biochem*. 41: 1673–1676.

4 Application of Solid-State Fermentation to Food Industry

María A. Longo and Mª Ángeles Sanromán

CONTENTS

4.1 INTRODUCTION

Nowadays, there is great interest in the development and use of natural food and additives derived from microorganisms, since they are more desirable than the synthetic ones produced by chemical processes. Solid-state fermentation (SSF) is defined as the cultivation of microorganisms on solid, moist substrates in the absence of a free aqueous phase; that is, at average water activities (defined as the relative humidity of the gaseous phase in equilibrium with the moist solid) significantly below 1 (Hölker and Lenz 2005). The low moisture content means that fermentation can only be carried out by a limited number of microorganisms, mainly yeasts and fungi, although some bacteria have also been used (Pandey et al. 2000a).

SSF resembles natural microbiological processes like composting and ensiling. A wide range of solid materials can be employed as supports in SSF and they are usually classified into two great categories: noninert and inert materials (Krishna 2005). In the former, a divided and humidified solid (e.g., cereal grain, flour, bran, sawdust) functions both as support and nutrient source, while in the latter a nutritionally inert solid (e.g., synthetic foam), acting exclusively as a support, is soaked in a nutrient solution. It is noteworthy the growing interest in utilizing agro-industrial wastes as noninert SSF supports, making the whole process much more economical (Pandey et al. 2000a,b). In industrial applications, this natural process can be utilized in a controlled way to produce a desired product. SSF is considered as an advantageous technology in a number of processes related to different industrial sectors.

As it can be seen in Figure 4.1, this process has been known since ancient times and different fungi have been cultivated in SSF for the production of food. Traditional food fermentation such as Japanese "koji," Indonesian "tempeh," Indian "ragi," and French "blue cheese" can be cited as some of the oldest documented applications of SSF. Also, in China, SSF has been used extensively to produce brewed foods (such as Chinese wine, soy sauce, and vinegar) since ancient times (Chen 1992). In addition, SSF is commercially used in Japan to produce industrial enzymes (Suryanarayan 2003). Since 1986, a series of research projects for the value-addition of tropical agricultural products and subproducts by SSF has been developed in Brazil due to the high amounts of agricultural residues generated by this country (Soccol and Vandenberghe 2003).

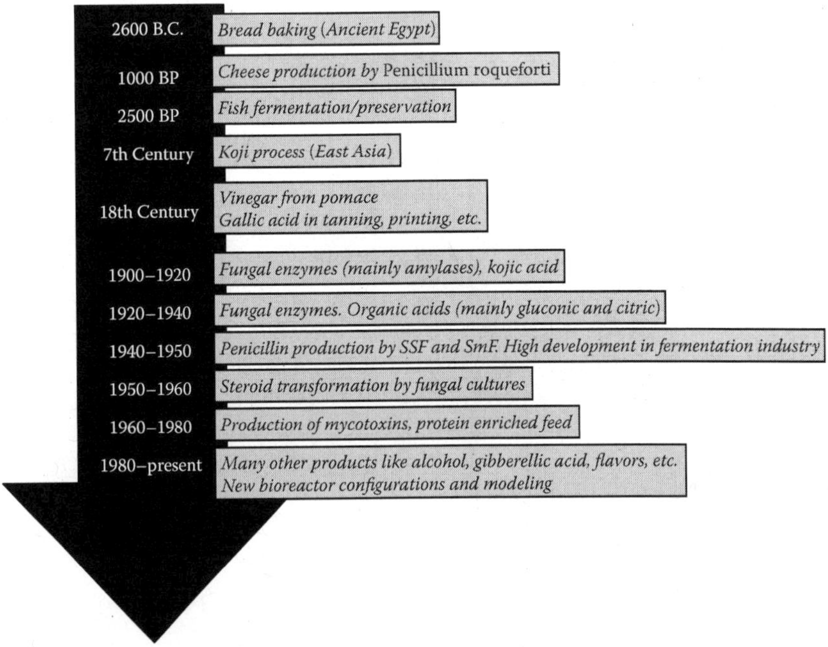

FIGURE 4.1 Milestones in the development of SSF. (Based upon data from Pandey, A., *Proc. Biochem.*, 27, 110, 1992. With permission.)

In recent years, there has been an increasing trend toward efficient utilization and value-addition of agro-industrial wastes in solid-state cultivation. Not only does the utilization of these wastes in bioprocesses provide alternative substrates, but it also helps in solving pollution problems. The selection of an adequate support for performing solid-state cultivation is essential, since the success of the process depends on it.

SSF constitutes an interesting alternative since the metabolites so produced are concentrated and purification procedures are less costly. SSF is preferred to submerged fermentation (SmF) because of its simplicity, low capital investment, lower levels of catabolite repression and end product inhibition, low wastewater output, better product recovery, and high quality production (Gangadharan et al. 2006). In addition, it presents several advantages over the traditionally employed SmF (Table 4.1). A detailed economic analysis of the production of *Penicillium restrictum* lipase in both SmF and SSF has been performed by Castilho et al. (2000). Their results show that the great advantage of SSF processes is the extremely cheap raw material used as main substrate. Moreover in several processes the composition of the raw material induces the production of bulk chemicals and enzymes (Lorenzo et al. 2002). However, the main disadvantage is the fact that there are few designs available in the literature for bioreactors operating in solid-state conditions. This is principally due to several problems encountered in the control of parameters such as pH, temperature, aeration and oxygen transfer, and moisture. Nevertheless, in recent years a great effort has been made to solve these limitations.

The ability of microorganisms to grow on a solid substrate is a function of their requirements of water activity, their capacity of adherence and penetration into the

TABLE 4.1
Advantages and Disadvantages of SSF over SmF

Advantages	Disadvantages
Higher volumetric productivities	Problems in scale-up
Low water consumption of SSF processes, thereby reducing liquid effluent treatment costs	Difficult control of process parameters (pH, heat, nutrient conditions, etc.)
Reduced contamination by bacteria and yeasts	The scientists's inadequate knowledge of the SSF process
Reduced metabolite degradation during the fermentation process	Higher impurity product, increasing recovery product costs
High concentration of metabolites in the final solid substrate renders further processing needless	
Certain metabolites are expressed in SSF only	
SSF fermentation products (enzymes, spores, and metabolites) generally have a longer shelf life	
Fermentation time is usually shorter	
SSF is better adapted to mixed-microbial cultures	
SSF can use agricultural wastes as substrates	

substrate, and their ability to assimilate mixtures of different polysaccharides due to the nature, often complex, of the substrates used (Perez-Guerra et al. 2003).

Filamentous fungi are the best-adapted microorganisms for SSF. The hyphal mode of fungal growth and their good tolerance to low water activity and high osmotic pressure conditions (high nutrient concentration) give fungi major advantages over unicellular microorganisms in the colonization of the solid substrate and the utilization of the available nutrients (Krishna 2005).

Although few bacteria and yeasts can grow in environments with low water activity, several bacteria have been used in SSF for enzymes production, composting, ensiling and some food processes (e.g., sausages, Japanese natto, fermented soybean paste, Chinese vinegar), while yeasts have been mainly used for ethanol production and protein enrichment of agricultural residues (Perez-Guerra et al. 2003).

The aim of this chapter is to review the potential application of SSF for the production of several metabolites of great interest to the food industry, including enzymes (α-amylase, fructosyl transferase, lipase, pectinase), organic acids (lactic, citric, and gluconic acids), and xanthan gum. Special attention is paid to the choice of raw materials and the main variables involved in process optimization. In addition, different types of bioreactors for SSF processes as well as new bioreactor configurations and the main variables involved in process optimization are described.

4.2 UPGRADING NUTRITIONAL VALUE OF CHEAP RAW MATERIALS BY SSF

SSF is now being used for upgrading the feed values of waste cellulosic materials, for enzyme production, and for enhancing the quality of existing foods, especially oriental foods (Pandey et al. 2000a). Iyayi and Aderolu (2004) reported the improvement of the nutritive value of some agro-industrial by-products through fungal fermentation, and their subsequent utilization for feeding layers. In their study the nutrient compositions of some agro-by-products such as brewer's dried grains, rice bran, palm kernel meal and corn bran, after biodegradation with *Trichoderma viride,* were investigated. Hen diets in which the biodegraded by-products replaced maize resulted in lower cost of egg production than the standard commercial diet. Furthermore, the nutritional status of the diet was thereby enhanced. Using such by-products to feed layers spared half of the maize in the diet and produced better laying performance.

There are several low-value agro-industrial wastes, whose nutritional value can be increased by the application of SSF. With a suitable pretreatment, these wastes can be used as animal feed. Villas- Bôas et al. (2003) studied the sequential fermentation of apple pomace by *Candida utilis* and *Pleurotus ostreatus*, and found a significant enrichment of crude protein and mineral level, which permitted its use as protein supplement in cattle feeding. The limited utilization of palm kernel cake in fish feeds, due to its high fiber and low protein content, could be overcome by processing this material under solid-state culture conditions with some fungal strains (i.e., *T. longiobrachiatum*). This treatment led to a significant increase in protein level, as well as to a decrease in cellulose and hemicellulose contents (Iluyemi et al. 2006).

Monogastric animals, such as pigs and poultry, need a certain amount of phosphorous for their skeletal growth, but they lack the enzyme needed to efficiently digest phytate in their feed. Consequently, they excrete large amounts of phosphorous into the environment, thus causing an important pollution problem. Supplementation of phytase to their feed appears as an interesting strategy for upgrading the nutritional quality of phytate-rich feed, reducing inorganic phosphorous supplementation and consequently fecal phosphorous excretion (Pandey et al. 2001). Phytases also improve the bioavailability of phytate phosphorous in plant foods for humans (Martínez et al. 1996). The influence of microbial phytases on digestibility and other related properties of sorghum, wheat, maize, and barley-based feeds for broiler chickens has been evaluated, individually or in combination with glycanases (Wu et al. 2003, 2004a,b).

Orozco et al. (2008) analyzed the ability of three *Streptomyces* strains to upgrade the nutritional value of coffee pulp residues from Nicaragua in SSF conditions. The presence, in these residues, of compounds such as polyphenols, tannins, chlorogenic acids, and caffeine prevents their utilization as domestic fodder. The characteristic pyrolysis products derived from polyphenols and polysaccharides were identified both in control and treated coffee pulp, the decrease achieved in the total polyphenol-derived compounds after the growth of the strains being remarkable. The analysis of these compounds demonstrated that both monomethoxy- and dimethoxy-phenols were degraded. Thus, changes evidenced in coffee pulp treated by *Streptomyces* through the application of analytical pyrolysis reveal the biotechnological interest of these bacteria to upgrade a useless and pollutant residue to be used for feeding purposes.

Grape pomace produced by viticulture is generally disposed of in open areas. It can be used as animal feed, especially in the dry season when pastures are scarce. Its use is restricted to only up to 30% of the feed for ruminants due to its very low nutritional value and its antinutritional factors such as phenolic components that inhibit the ruminal symbionts. High costs of handling and transportation usually limit the direct benefit of feeding animals with many agricultural by-products (Sanchez et al. 2002). However, the decomposition of these residues is possible by the action of fungi, thus releasing available nutrients. Sanchez et al. (2002) determined that the recycling of viticulture residues through SSF by *Pleurotus* has great potential to produce an available high-fiber feed for use in ruminants. Moreover, *Pleurotus ostreatus-complex*, which is the third most important edible mushroom cultivated worldwide, was produced with a good quality for human consumption.

Edible mushrooms have been cultivated for many years, and it is expected that their production will further increase in the future due to market demand. Mushrooms have been eaten and appreciated for their flavor, economic and ecological values, and medicinal properties for many years. They have a chemical composition, which is attractive from the nutritional point of view, with 90% water and 10% dry matter, which is constituted by 27%–48% protein, less than 60% carbohydrates, and 2%–8% lipids. The most cultivated mushroom worldwide is *Agaricus bisporus* (button mushroom), followed by *Lentinus edodes* (shiitake), *Pleurotus* spp. (oyster mushrooms), *Auricula auricula* (wood ear mushroom), *Flammulina velutipes* (winter mushroom), and *Volvariella volvacea* (straw mushroom) (Sanchez 2004).

Mushrooms can be produced from natural materials, such as agricultural waste, woodland, animal husbandry, and from manufacturing industries. Still, it has been necessary to develop more efficient, faster, and reliable production systems. Thus, highest biological efficiency has been obtained when sawdust and alternative substrates such as sugarcane and coffee residues were used (Morales et al. 1991, Morais et al. 2000, Pire et al. 2001, Royse 1985, 1996, Salmones et al. 1999). The bioconversion of vineyard pruning, barley straw and wheat straw by four strains of *Lentinula edodes* using SSF was evaluated by Gaitán-Hernández et al. (2006). The results showed that vineyard pruning has great potential for edible mushroom production due to its low cost, short production cycles, and high biological efficiency.

Currently, the great interest evinced by the people on adequate nutrition habits has caused a considerable increase in the investigation of the functional properties of foods. The manufacture of prebiotic and probiotic products has skyrocketed, and the supposed ability of some compounds to prevent certain diseases has strongly promoted their consumption. For the propagation of probiotics, the process of SmF is contemporarily dominant in both research and industry, although SSF has commonly been used in many other sectors of food fermentation to ameliorate the nutritional quality of cereals (Patel et al. 2004). An additional advantage over SmF is the fact that SSF and semi-SSF processes establish an environment comparable to the human large intestine, the source of most human-derived probiotics. Patel et al. (2004) applied SSF to introduce probiotic functionalities to breakfast cereals and similar food products, by cultivating a *Lactobacillus plantarum* strain on oat bran and spent oats after lipid extraction and limited hydrolysis of the raw materials by *Aspergillus awamori* and *A. oryzae*. These topics are discussed in Chapters 20 and 21, in which the feasibility of incorporating dried cell-free fraction of fermented milks into food matrixes and the possibilities of using cereals for developing new functional foods are analyzed.

4.3 ENZYME PRODUCTION

Nowadays, large-scale production of most of the usually demanded enzymes is carried out by SmF, using genetically modified strains. The high production costs hinder the use of many enzymes in industrial processes. As an attractive alternative method, SSF holds tremendous potential for the production of enzymes (Pandey et al. 2000a). In this section, a brief review of the more recent developments in the production of several enzymes by SSF is reported.

4.3.1 α-Amylases

Since the 1950s, fungal amylases have been used to manufacture sugar syrups containing specific mixtures of sugars that could not be produced by conventional acid hydrolysis of starch. Amylases are extensively employed in processed-food industry such as baking, brewing, preparation of digestive aids, production of cakes, fruit juices, starch syrups, etc. α-Amylases (endo-1,4-α-D-glucan glucanohydrolase EC 3.2.1.1) are extracellular endoenzymes that randomly cleave the 1,4-α linkages between adjacent glucose units in the linear amylose chain, and ultimately generate glucose, maltose, and maltotriose units. Microbial α-amylases are mainly secreted as primary metabolites, and their production is reported to be growth associated (Francis et al. 2003). The production of α-amylases has traditionally been carried out by SmF; however, SSF systems appear as a promising technology. Selvakumar et al. (1996) reviewed the microbial synthesis of starch-saccharifying enzyme in SSF.

The biotechnological potential of SSF for the production of this enzyme has been reported using several wastes and microorganisms. Krishna and Chandrasekaran (1996) used banana fruit stalk as a substrate in SSF with *Bacillus subtilis*. Different factors such as initial moisture content, particle size, thermal treatment time and temperature, pH, incubation temperature, additional nutrients, inoculum size, and incubation period on the production of α-amylase were characterized. Results obtained for the optimization of process parameters clearly showed the impact of such parameters on the gross yield of enzymes, as well as their independent nature in influencing the organism's ability to synthesize the enzyme.

Bogar et al. (2002) evaluated the SSF conditions to 10 *A. oryzae* strains grown on spent brewing grain and on corn fiber, verifying that the former substrate led to higher enzyme productions in all cases. Moreover, the whole solid substrate fermentation material (crude enzyme, *in situ* enzyme) may be considered a cheap biocatalytic material for animal feed rations and for alcohol production from starchy materials. In another study, Francis et al. (2003) used spent brewing grains in SSF for the production of α-amylase by *A. oryzae*, and determined that the supplement of fermentation media with Tween-80 or calcium ions enhanced α-amylase activity. Ramachandran et al. (2004), in turn, reported the use of coconut oil cake (COC) as a substrate for the production of α-amylase by *A. oryzae* under SSF conditions. Raw COC supported the growth of the culture, resulting in the production of 1372 U/gds (units per gram of dry substrate) α-amylase in 24 h. Supplementation of 0.5% starch and 1% peptone to the substrate positively enhanced the enzyme synthesis producing 3388 U/gds, proving COC a promising substrate for α-amylase production.

Gangadharan et al. (2006) screened 14 different agro-residues for α-amylase production using *B. amyloliquefaciens* ATCC 23842, and found excellent results with a mixture of wheat bran and groundnut oil cake, supplemented with KH_2PO_4 and 1% soluble starch. A maximum enzyme titer of 62,470 U/gds was obtained after 72 h of fermentation at 37°C by using the above solid substrate mixture (5 g) with an initial moisture of 85%. Balkan and Ertan (2007) evaluated the ability of *P. chrysogenum* to produce α-amylase by SSF using corncob leaf, rye straw, wheat straw, and wheat bran as substrates. The effects of moisture level, particle size, and inoculum concentration on enzyme synthesis were investigated. Under optimum conditions, wheat bran showed the highest enzyme production with values around 160 U/mL.

Currently, gelatinization is coupled with liquefaction, which is possible by the action of thermostable amylases, which have been reported in both SmF (Stamford et al. 2001) and SSF (Babu and Satyanarayana 1995). Sodhi et al. (2005) determined that the productivity of thermostable amylases from *Bacillus* sp. was affected by the nature of the solid substrate (wheat bran, rice bran, corn bran, and combination of two brans), the nature of the moistening agent, the level of moisture content, the incubation temperature, and the presence or absence of surfactant, carbon, nitrogen, mineral, amino acid, and vitamin supplements. Maximum enzyme production of 464,000 U/g dry bacterial bran was obtained on wheat bran supplemented with glycerol (1.0%, w/w), soybean meal (1.0%, w/w), L-proline (0.1%, w/w), vitamin B-complex (0.01%), and moistened with tap water containing 1% Tween-40.

Similarly, Anto et al. (2006) found that the highest amylase activity in *B. cereus* cultures was obtained when wheat bran was supplemented with glucose, although the addition of different nitrogen sources led to a decline in amylase production. Balkan and Ertan (2007) produced α-amylase by *B. amyloliquefaciens*, under SSF in shaken-culture. The results showed that α-amylase production in a medium with corn gluten meal was five times higher than that in the medium containing starch and other components. Moreover, the temperature of fermentation was found to be a crucial factor in α-amylase production.

Shukla and Kar (2006) carried out a comparative study of the ability of *B. licheniformis* and *B. subtilis* to produce this enzyme on potato peel and wheat bran under SSF, and reported the superiority of the former substrate.

Recently, Xu et al. (2008) studied the optimization of nutrient levels for the production of α-amylase by *A. oryzae* As 3951 in SSF with spent brewing grains, an inexpensive substrate and solid support, using response surface methodology based on Plackett–Burman and Box–Behnken designs. It was determined that corn steep liquor, $CaCl_2$, and $MgSO_4$ were the most compatible supplements, and their optimized concentrations were 1.8%, 0.22%, and 0.2%, respectively. Under these optimized conditions, around 17.5% increase in enzyme yield was observed.

4.3.2 Pectinases

Pectinases constitute a heterogeneous group of enzymes that hydrolyze pectic substances, which occur as structural polysaccharides in the middle lamella and primary cell walls of higher plants and are responsible for maintaining the plant tissue integrity (Alkorta et al. 1998). These enzymes include polygalacturonase, pectin esterase, pectin lyase, and pectate lyase on the basis of their mode of action. Pectinases are widely used in the food industry to clarify fruit juices and wine, to improve oil extraction, to remove the peel from citrus fruit, to increase the firmness of several fruits, and to degum fibers (Lozano et al. 1988). They have also been utilized in other processes in which the elimination of pectin is essential, such as wine production, coffee and tea processing plants, and vegetable tissue maceration (Soares et al. 2001).

The use of SSF for pectinase production has been proposed using different solid agricultural and agro-industrial residues as substrates such as wheat bran (Castilho et al. 1999, Singh et al. 1999), soy bran (Castilho et al. 2000), cranberry and strawberry pomace (Zheng and Shetty 2000), coffee pulp and coffee husk (Antier et al. 1993a), husk (Antier et al. 1993b), cocoa (Schwan et al. 1997), lemon and orange peel (Garzón and Hours 1991, Ismail 1996, Maldonado et al. 1986), orange bagasse, sugarcane bagasse, and wheat bran (Martins et al. 2002), sugarcane bagasse (Acuña-Arguelles et al. 1994), and apple pomace (Hours et al. 1988a,b). Also, Bai et al. (2004) produced pectinase from *A. niger* by SSF using sugar beet pulp as a carbon source and wastewater from monosodium glutamate production as nitrogen and water source. This allowed not only to reduce production costs but also to decrease the pollution source.

Thermostable enzymes are highly specific and thus have considerable potential for many industrial applications, since their stability and activities at high temperatures have become desirable properties for industrial processing. The use of such enzymes in maximizing reactions accomplished in the food and paper industry, detergents, drugs, toxic wastes removal, and drilling for oil is being studied extensively. The enzymes can be produced by thermophilic strains through either optimized fermentation of the microorganisms or cloning in fast-growing mesophiles by recombinant DNA technology. Martins et al. (2002) demonstrated that it is feasible to use agro-industrial residues for pectin lyase and polygalacturonase production by a newly isolated thermophile, *Thermoascus aurantiacus*. This fungus is able to produce high levels of extracellular enzymes during the SSF of mixtures of sugarcane and orange bagasses. Pectinases produced by *T. aurantiacus* have a high optimum temperature (65°C) with good thermostability, especially pectin lyase, which was stable for 5h at 60°C.

Several pectin-rich agro-wastes, viz., lemon peel, sorghum stem, and sunflower head have been studied as substrates for the production of pectinase by *A. niger* DMF 27 and *A. niger* DMF 45 in SmF and SSF systems, respectively (Patil and Dayanand 2006). It was determined that in SSF when the agro-wastes were supplemented with additional carbon and nitrogen sources (the addition of sucrose being more effective than glucose) an increased level in the production of pectinases was detected.

A very high level of alkalophilic and thermostable pectinases has been produced from newly isolated strains of *B. subtilis* under SSF, using combinations of cheap agricultural residues such as, wheat bran and cotton seed cake, wheat bran and citrus waste, wheat bran and alpha-alpha leaves, cotton seed cake and citrus waste, cotton seed cake and alpha-alpha leaves, citrus waste and alpha-alpha leaves, and also mixture of the four substrates (Ahlawat et al. 2007). Among these substrates, the highest yield of pectinase production was observed by using a combination of wheat bran and citrus waste (6592U/g of dry substrate) supplemented with 4% yeast extract, when incubated at 37°C for 72h using deionized water at pH 7.0 as moistening agent.

It has been found that, in the case of pectinases, SSF is more productive than SmF and, in addition, the enzymes produced by SSF showed more interesting properties: they had a higher stability to pH and temperature, and they were also less affected by catabolic repression than pectinases produced by SmF (Acuña-Arguelles et al. 1995, Díaz-Godínez et al. 2001, Joshi et al. 2006, Kapoor and Kuhad 2002).

4.3.3 Lipases

Lipases (triacylglycerol acylhydrolases, EC 3.1.1.3) hydrolyze triglycerides to fatty acid and glycerol, and under certain conditions, catalyze the reverse reaction forming glycerides from glycerol and fatty acids. Some lipases are also able to catalyze both transesterification and enantioselective hydrolysis reactions. Lipases are nowadays widely used at industrial scale with applications as additives in detergents, in the elaboration of dietetic foods for use in the food industry, to obtain bioactive molecules in the pharmaceutical industry and pure optical compounds in chemical synthesis processes, and also to modify fats and lipids by hydrolysis and esterification reactions (Hernaiz and Sinisterra 1999, Kazlauskas 1994).

Most studies on lipolytic enzyme production by bacteria, fungi, and yeasts have been performed in submerged cultures; however, there are a few reports on lipase synthesis in solid state cultures. Fungi are broadly recognized as one of the best lipase sources, and they are widely used in the food industry. Several reports deal with extracellular lipase production by fungi such as *Rhizopus* sp., *Aspergillus* sp., and *Penicillium* sp. on different solid substrates such as gingelly oil cake, babassu oil cake, olive cake, soy cake, COC, wheat bran, rice bran, almond meal, and sugarcane bagasse (Adinarayana et al. 2004, Cordova et al. 1998, Di Luccio et al. 2004, Gombert et al. 1999, Kamini et al. 1998, Mahadik et al. 2002, Miranda et al. 1999, Ortiz-Vazquez et al. 1993, Rao et al. 1993a, Ul-Haq et al. 2002). In order to improve the productivity of the process, the use of mixed solid substrates, such as COC and wheat bran (1:1) (Benjamin and Pandey 1998) or wheat bran and olive oil (9:1) (Nagy et al. 2006), and the supplementation of the medium with lipase inducers, such as olive oil (Di Luccio et al. 2004, Mahadik et al. 2002, Palma et al. 2000), have been proposed.

Comparative studies between SmF and SSF systems for lipase production by fungi (Christen et al. 1995, Ohnishi et al. 1994, Rivera-Muñoz et al. 1991) showed that enzyme yields were higher and more stable in SSF. In addition, Mateos Diaz et al. (2006) studied the lipase from the thermotolerant *Rhizopus homothallicus* obtained in SmF and SSF cultures. They have found that the two enzymes produced are monomers with similar molecular mass and an identical protein structure. However, some of their properties are different, namely the specific activity on trioctanoin, the temperature at which maximum activity occurs, and the thermal stability. Their results indicated that the enzyme from SSF was more thermostable than that obtained in SmF. In accordance with these results, Falony et al. (2006) determined the good stability with respect to temperature and pH of the lipase produced by *A. niger* J-1 grown on wheat bran.

Deoiled *Jatropha* seed cake was assessed for its suitability as substrate for enzyme production by SSF by a solvent tolerant *Pseudomonas aeruginosa* PseA strain. Maximum lipase production was observed at 50% substrate moisture, a growth period of 120 h, and a substrate pH of 7.0. Enrichment with maltose as carbon source increased lipase production by 1.6-fold. Nitrogen supplementation with peptone for lipase production also enhanced the enzyme yield reaching 1084 U lipase activity per gram of *Jatropha* seed cake (Mahanta et al. 2008).

In recent years, several authors reported the extracellular lipase production by yeasts or bacteria on different solid substrates. For instance, Domínguez et al. (2003) determined the great potential of food-agro-industrial wastes (groundnut and barley bran) as support-substrates for lipase production in solid-state cultures of the yeast *Yarrowia lipolytica*, since they led to much higher activities than those found using an inert support. Babu and Rao (2007) performed a comparative study on the extracellular lipase production in SSF using *Y. lipolytica* NCIM 3589 with various mixed substrates. Different parameters such as moisture content, carbon level, and nitrogen level of the medium were optimized. A maximum lipase activity of 9.3 units per gram of dry fermented substrate was observed with a mixed substrate of sugarcane bagasse and wheat bran in 7 days of fermentation.

Previously, Bhusan et al. (1994) reported lipase production in SSF system by an alkalophilic yeast strain belonging to *Candida* sp., using several solid substrates such as rice and wheat bran, oiled with different concentrations of rice bran oil, the former being the best substrate, with higher lipase yields. For *C. rugosa*, Rao et al. (1993b) determined that the carbon to nitrogen, C/N, ratio of the medium is an important parameter for lipase production. Afterward, comparative studies for lipase production were carried out by Benjamin and Pandey (1996a,b, 1997a,b), who cultivated *C. rugosa* on COC and demonstrated the higher enzyme production in the SSF system compared to SmF.

4.3.4 Tannases

Tannases (tannin acyl hydrolases, EC 3.1.1.20) catalyze the hydrolysis of tannic acid by breaking its ester and depside bonds, releasing glucose and gallic acid. The fundamental application of tannases in nature is litter degradation and as a defensive mechanism against the attack of phytopathogens. These enzymes are widely used in the food and pharmaceutical industry, in the processing of tea, and also for production of pharmaceutically important compounds like gallic acid. This tannin product is the substrate for the chemical synthesis of propyl gallate and trimethoprim, which are important in the food and pharmaceutical industries. Tannases are also used for the stabilization of malt polyphenols, clarification of beer and fruit juices, for the prevention of phenol-induced madeirization in wine and fruit juices, and for the reduction of antinutritional effects of tannins in animal feed (Aguilar et al. 2001a, Lekha and Lonsane 1997, Sabu et al. 2005).

These enzymes are synthesized by various filamentous fungi such as *Aspergillus* and *Penicillium*, and also by some bacteria and yeasts, but in all cases in the presence of tannic acid (Aoki et al. 1976, Kumar et al. 1999, Mondal and Pati 2000). Although these enzymes have been mainly produced by SmF, the advantages of SSF over traditional methods have been discussed in recent reports. By SSF, higher extracellular tannase activities can be reached with lower cost (Lekha and Lonsane 1994, 1997).

Tannase-producing fungal strains have been isolated from different locations including garbage, forests, orchards, etc. The strain giving maximum enzyme yield was identified to be *A. ruber*. The tannase production under solid-state fermentation by *A. ruber* was studied by Kumar et al. (2007) using different tannin-rich substrates like ber leaves (*Zizyphus mauritiana*), jamun leaves (*Syzygium cumini*), amla leaves (*Phyllanthus emblica*), and jawar leaves (*Sorghum vulgaris*). Jamun leaves were found to be the best substrate for enzyme production under SSF. Addition of carbon and nitrogen sources to the medium did not increase tannase production.

Several reports (Díaz-Godínez et al. 2001, Romero-Gómez et al. 2000) demonstrated the ability of *A. niger* for tannase production under SSF employing polyurethane foam as inert support. The enzyme secreted in this kind of system was purified to homogeneity and characterized, and it was shown to have β-glucosidase activity in the absence of tannic acid (Ramírez Coronel et al. 2003). In addition, some support-substrates have been tested: sugarcane pith bagasse (Lekha and Lonsane 1994), pearl barley (Seth and Chand 2000), forest residues (Kar and Banerjee 2000), and wheat bran (Sabu et al. 2005). Sabu et al. (2006) isolated a tannase-yielding bacterial strain from sheep excreta identified as *Lactobacillus* sp. ASR S1, which was cultivated using tamarind seed powder, wheat bran, palm kernel cake, and coffee husk as substrates. Maximum tannase production (0.85 U/gds) was obtained when SSF was carried out using coffee husk, supplemented with 0.6% tannic acid and 50% (w/v) moisture, inoculated with 1 mL cell suspension and incubated at 33°C for 72 h.

Comparative studies of the tannase production by *R. oryzae* in SmF, SSF, and SSF-modified systems, using the tannin-rich forest residue *Caesalpinia digyna* as substrate (Kar and Banerjee 2000), demonstrated that the SSF-modified system gave the best results. The bioreactor used in SSF-modified system is a cylindrical vessel. A float is provided inside the vessel, which consists of a base of glass wool cloth. Fermentation of the solid substrate is carried out on the float by the induced inoculum. One advantage of this kind of float is that the microorganism added to the substrate converts the substrate into the desired product, which leaches into the liquid medium, thus improving the heat transfer in the system.

4.3.5 Phytases

Phytases [myo-inositol(1,2,3,4,5,6)hexakisphosphate phosphohydrolases] have been identified in plants, microorganisms, and in some animal tissues. They represent a subgroup of phosphatases, which are capable of initiating the stepwise dephosphorylation of phytate [myo-inositol(1, 2,3,4,5, 6)hexakisphosphate], the most abundant inositol phosphate in nature. There is also a great potential for the use of phytases in processing and manufacturing of food for human consumption. As described by Greiner and Konietzny (2006), technical improvements by adding phytases during food processing have been reported for breadmaking, production of plant protein isolates, corn wet milling, and the fractionation of cereal bran.

Phytases can be synthesized by a number of bacteria, yeasts, and fungi. Pandey et al. (2001) published an interesting review on the production, purification, and properties of microbial phytases, in which both SmF and SSF systems were considered.

Papagianni et al. (2000) compared phytase production by *A. niger* in SmF and SSF, with especial emphasis on the influence of medium composition and fungal morphology. Addition of wheat bran, a slow releasing organic phosphate source, enhanced *A. niger* growth and enzyme synthesis. Agitation during inoculum production for SSF was found to be relevant for enzyme biosynthesis (Papagianni et al. 2001). Krishna and Nokes (2001) cultivated *A. niger* under SSF on wheat bran and full-fat soybean flour, and indicated the significant influence of liquid inoculum age and cultivation time on enzyme levels.

Thermostability, mycelial morphology, inoculum age and quality, media composition, and fermentation duration have been cited as important and useful criteria in the industrial application of phytases. Thus, *A. fumigatus* was reported to produce a potentially interesting heat-stable enzyme, and its expression in *Pichia pastoris* was investigated (Pasamontes et al. 1997a, Rodríguez et al. 2000). The phytase from *B. amyloliquefaciens* DS11 showed a half-life of 42 min at 80°C (Kim et al. 1999). Pasamontes et al. (1997b), in turn, investigated a heat-stable phytase able to withstand temperatures up to 100°C over a period of 20 min. Moreover, Mandviwala and Khire (2000) studied the extracellular phytase production by a thermotolerant strain *A. niger* NCIM 563 that was cultivated on agricultural residues such as wheat bran, mustard cake, cowpea meal, groundnut cake, coconut cake, cotton cake, and black bean flour. In these systems, the addition of KH_2PO_4 and Triton X-100 increased the enzyme yield. In the case of *A. niger* A-98 cultivated in rapeseed meal, El-Batal and Abdel Karem (2001) analyzed the influence of different factors on phytase production, including moisture content, glucose, phosphate, surfactants, gamma irradiation, and the addition of some surfactants (Tween 80 or Triton X-100). The results showed that the system with 0.3% (v/w) Tween 80 produced the largest amount of phytase (5.0 U/g). The increase in phytase production may be explained by a possible effect of Tween 80 on the cell permeability, which resulted in a higher release of the enzyme.

Bogar et al. (2003) evaluated three *Mucor* and eight *Rhizopus* strains, and several natural supports (canola meal, COC, and wheat bran) for the production of phytase by SSF. The highest yield was obtained with *Mucor racemosus* NRRL 1994 on COC, supplemented with glucose, casein, and ammonium sulfate. This support, together with palm kernel cake, was also chosen as the most adequate by Ramachandran et al. (2005), after comparing SSF phytase production by three *Rhizopus* strains on six types of oil cakes (coconut, sesame, palm kernel, groundnut, cottonseed, and olive). A similar work (Roopesh et al. 2006) considered the mixture of oil cakes and wheat bran as a solid SSF support for phytase production by *M. racemosus*.

4.3.6 Xylanases

Xylanases hydrolyze β-1,4-linkages in xylan, which has great significance in the conversion of hemicellulose to pentose sugars. Xylan is a major component of plant hemicellulose, and, after cellulose, it is the most abundant renewable polysaccharide in nature. Xylanases break down the nonstarch polysaccharides in high-fiber rye- and barley-based feeds, thus reducing viscosity and increasing adsorption (Bedford and Classen 1992). The use of these enzymes has been proposed for clarifying juices and wines, for extracting coffee, plant oils, and starch, for improving the nutritional properties of agriculture silage and grain feed, for macerating cell walls, for producing food thickeners, and for providing different textures to bakery products (Bajpai 1997).

Although xylanases have been produced by fungi and bacteria, filamentous fungi are preferred for commercial production as the levels of enzyme biosynthesis by fungal cultures are higher than those obtained from other organisms. Milagres et al. (2004) optimized xylanase production by *T. aurantiacus* ATCC 204492 under SSF on sugarcane bagasse, using a glass-column reactor with forced aeration. They concluded that airflow rates during fermentation had a significant effect on enzyme activity, whereas the influence of initial mass of bagasse was negligible.

On the other hand, thermostability is an important property for industrial application of xylanases. Strains of the thermophilic fungi *Thermomyces lanuginosus* have been reported to be among the best xylanase producers in nature. Xylanase activity has been found in organisms in association with cellulases, β-glucosidases, or other enzymes (Krishna 2005).

Recently, Sanghi et al. (2008) produced high levels of xylanase by alkalophilic *B. subtilis* ASH using easily available inexpensive agricultural waste residues such as wheat bran, wheat straw, rice husk, sawdust, gram bran, groundnut, and maize bran in SSF. Among these, wheat bran was found to be the best substrate. The enzyme production was stimulated by the addition of nutrients such as yeast extract, peptone, and beef extract. In contrast, addition of glucose and xylose repressed the production of xylanase. The extent of repression by glucose (10% w/v) was 81% and it was concentration dependent. Supplementation of the medium with 4% xylose caused 59% repression. Under optimized conditions, xylanase production in SSF (8964 U of xylanase/g dry wheat bran) was about twofold greater than in SmF. Thus, *B. subtilis* produced a very high level of xylanase in SSF using inexpensive agro-residues, a level which is much higher than that reported by any other bacterial isolate.

4.4 PRODUCTION OF ORGANIC ACIDS

Organic acids have been utilized for a long time by the food industry as food additives and preservatives for preventing deterioration and extending the shelf life of perishable food ingredients. Here, three common organic acids widely used in the food industry have been considered.

4.4.1 Lactic Acid

Lactic acid fermentation has received extensive attention. It has wide applications in food, pharmaceutical, leather, and textile industries, and also as a chemical feed stock. Nowadays, its use as a precursor for the biodegradable plastic polylactic acid, used in medical, industrial, and consumer products (Gross and Kalra 2002, John et al. 2006, Senthuran et al. 1997, Vickroy 1985, Wee et al. 2006), has received much consideration. Biologically active L(+) form, as well as D(−) or DL lactic acid, can be produced by fermentation, depending on the organism used. Generally, L(+)-lactic acid is utilized by human metabolism because of the presence of L-lactate dehydrogenase, and it is hence preferred for food applications.

Lactic acid production through SSF has been carried out using fungal and bacterial strains, mainly by strains of *Rhizopus* and *Lactobacillus* sp. Soccol et al. (1994) obtained a slightly higher productivity than in submerged cultivation when L(+)-lactic acid was produced by *Rhizopus oryzae* in solid-state conditions operating with sugarcane bagasse as a support. Lactic acid bacteria like *L. delbrueckii* or *L. casei* can also be used in SSF for lactic acid production. *L. delbrueckii* and *L. casei* were able to grow on a solid support such as sugarcane bagasse or cassava hydrolysate prepared from the bagasse (John et al. 2006, Rojan et al. 2005).

Naveena et al. (2005a,b) found that *L. amylophilus* GV6 are able to produce L(+)-lactic acid under SSF conditions using wheat bran as both support and substrate. In a previous paper (Naveena et al. 2003), wheat bran and different brans of pigeon pea, green gram, black gram, and corn fiber were tested. Wheat bran was found to be the best among all solid substrates tested. A similar study was reported for SSF of *L. amylovorus* NRRL B-4542 (Nagarjun et al. 2005). In this work the optimization of physical parameters and medium composition was performed using an orthogonal experimental design, and a software for automatic design and analysis of the experiments, both based on the Taguchi protocol. It was determined that among the physical parameters, temperature had the highest influence, and among media components, yeast extract, $MgSO_4{\cdot}7H_2O$, and Tween 80 played important roles in the conversion of starch to lactic acid.

Recently, John et al. (2007) used polyurethane foam impregnated with cassava bagasse starch hydrolysate as major carbon source for the production of L-lactic acid using *L. casei* in solid-state condition. The key parameters, such as reducing sugar, inoculum size, and nutrient mixture, were optimized by a statistical approach using response surface methodology. More than 95% conversion of sugars to lactic acid was attained after 72 h. While considering the lactate yield based on the solid support used, a very high yield of 3.88 g lactic acid per gram polyurethane foam was achieved.

4.4.2 Citric Acid

Citric acid is an important commercial product with a global production in 2004 around 1.4 million tons per year, as estimated by Business Communications Co. (BCC). Citric acid is one of the most commonly used organic acids in food and pharmaceutical industries. The food industry is the largest consumer of citric acid, using almost 70% of the total production, followed by about 12% for the pharmaceutical industry and 18% for other applications (Shah et al. 1993). Its pleasant taste, high solubility, and flavor-enhancing properties have ensured its dominant position in the market. Although citric acid can be obtained by chemical synthesis, the cost is much higher than when obtained using fermentation. As a result, almost all citric acid is commercially produced by fermentation, mainly through SmF of starch- or sucrose-based media, using the fungus *A. niger*. Recently, in order to increase the efficiency of citric acid production using *A. niger*, SSF has been studied as a potential alternative to SmF (Soccol et al. 2006).

The production of citric acid depends strongly on an appropriate strain and on operational conditions. Oxygen level is an important parameter for citric acid fermentation. Several researchers (Pintado et al. 1998, Prado et al. 2004) have studied the influence of forced aeration on citric acid production and the metabolic activity of *A. niger* in SSF by respirometric analysis. They showed that citric acid production was favored by a limited biomass production, which occurred with low aeration rates. Both works showed the feasibility of using the strain *A. niger* for citric acid production by SSF. Recently, Khosravi-Darani and Zoghi (2008) carried out SSF to compare efficiency of acid, alkaline, and urea pretreatment of sugarcane bagasse for production of citric acid using *A. niger* ATCC 9142. It was concluded that among these three pretreatments, which had an important role in increasing citric acid productivity from sugarcane bagasse, urea pretreatment gave the most influential support for acid production.

Several agro-industrial residues, such as apple pomace, coffee husk, wheat straw, pineapple waste, mixed fruit, maosmi waste, cassava bagasse, banana, sugar beet cosset, and kiwi fruit peel, have been investigated for their potential to be used as substrates (Hang and Woodams 1985, Khare et al. 1995, Kumar et al. 2003a,b, Shojaosadati and Babaripour 2002). In addition, SSF gave high citric acid yield without inhibition related to the presence of certain metal ions such as Fe^{2+}, Mn^{2+}, and

Zn^{2+} (Gutierrez-Rozas et al. 1995). In fact, Shankaranand and Lonsane (1994) reported that addition of these minerals into the production media to a certain level enhanced citric acid production by 1.4–1.9 fold with respect to SmF. Therefore SSF is a good way of using nutrient-rich solid wastes as a substrate. In addition, Imandi et al. (2008) used statistical experimental designs to optimize medium constituents of the fermentation medium for the production of citric acid from pineapple waste by SSF using *Y. lipolytica* NCIM 3589. The following four variables: yeast extract, moisture content of substrate, KH_2PO_4, and Na_2HPO_4 were identified, by Plackett–Burman design, as significant for citric acid production. All four variables tested for the correlation between their concentrations and the production of citric acid showed significant influence on the production. The methodology, as a whole, proved to be adequate for the design and optimization of the bioprocess in order to obtain a commercially valuable product like citric acid from a low-grade agro-waste like pineapple waste.

4.4.3 Gluconic Acid

Gluconic acid is a multifunctional compound that has potential use as a bulk chemical mainly for the food, feed, pharmaceutical, beverage, and textile industries. Due to the enormous demand of about 50,000–60,000 tons per year, the microbial fermentation processes are exclusively used for commercial gluconic acid production using glucose as the major carbohydrate source (Roehr et al. 1996, Roukas 2000, Singh et al. 2003, Singh and Singh 2006).

Singh et al. (2003) analyzed the production of gluconic acid with respect to varying substrate concentrations in submerged, semisolid-state, surface and SSF, using chopped sugarcane bagasse as substrate. They demonstrated that an overproduction of gluconic acid was obtained under the solid-state process conditions. More recently, Singh and Singh (2006) achieved semicontinuous production of gluconic acid with pseudoimmobilized mycelia of *A. niger* ORS-4·410, with a promising yield (95.8%) under SSF conditions. Moreover, Roukas (2000) has reported that fig is an attractive substrate, and the addition of methanol at concentrations up to 6% (w/v), resulted in a marked increase in the production of citric and gluconic acid by *A. niger*. On the other hand, Sharma et al. (2008) have evaluated the production of gluconic acid by SSF of a metal resistant *A. niger* (ARNU-4) strain using tea waste as solid support and a molasses-based fermentation medium. Various crucial parameters such as moisture content, temperature, aeration, and inoculum size were investigated. A nonclarified molasses-based fermentation media was utilized by strain ARNU-4, and maximum gluconic acid production was observed following 8–12 days of fermentation. Different concentrations of additives (viz. oil cake, soya oil, jaggery, yeast extract, cheese whey, and mustard oil) were supplemented for further enhancement of the production ability of microorganism. Among them, the addition of yeast extract enhanced gluconic acid production.

4.5 SSF BIOREACTORS: TYPES AND MODELING

In biotechnology production processes, the bioreactor provides the environment for growth and activity for the microorganisms, which cause the biological reaction.

There are several parameters (aeration, pH, humidity, agitation, temperature) relevant for the selection of the suitable bioreactor for each particular fermentation process. Bioreactor design is an integral part of biotechnology. With the development of the biotechnology industry, there has been a large improvement in bioreactor design, with sophisticated automatism controlled by computer in SmF process. This is not true, however, in the case of SSF, for which only limited types of bioreactors and control systems have emerged (Rodríguez Couto and Sanromán 2005).

There are different types of bioreactors for fermentation processes and each design tries to make conditions more favorable for production of different biologicals. The design of an efficient industrial-level reactor for SSF is of significance because this process is more environmental friendly than SmF. However, heat and mass transfer in SSF reactors are not as efficient as in the case of SmF. In order to reduce this problem, sometimes agitation and rotation systems are introduced in solid-state fermentors, but they disrupt the fungal mycelia, and porosity of the medium is adversely affected. Therefore, the scarcity of bioreactor types, together with the different disadvantages detected in the already existing bioreactor types to perform SSF processes, have promoted the necessity of developing new bioreactor configurations.

Different bioreactor types have been used in SSF processes and each in its own design tries to make conditions more favorable for fermentation under solid-state conditions. The bioreactors commonly used can be distinguished by the type of aeration or the mixing system employed, including packed beds, rotating drums, gas–solid fluidized beds, and other stirred bioreactors (Rodríguez Couto and Sanromán 2005).

Packed-bed: A glass column is filled with the solid substrate, retained on a perforated base. Through the bed of substrate, humidified air is continuously forced (Durand et al. 1993, Kang et al. 2005, Rodríguez Couto et al. 2000). Usually the temperature control during fermentation is performed by a jacket with water circulation. However, several difficulties are detected in this design, such as nonuniform growth, poor heat removal, poor product recovery, and scale-up problems.

Tray: In this type of bioreactor, the substrate is spread out flatly in a container especially intended for this purpose, and it is incubated in an especially air-conditioned room. The system can be used for the production of large amounts of the product. Moreover, reactor and method are very space and labor intensive. The main disadvantage of this configuration is that the fermented substrate has to be moved manually within the containers. Numerous trays and large volumes are required, making it an unattractive design for large-scale production (Bhanja et al. 2007, Vandenberghe et al. 2004).

Rotary horizontal drum: This type of bioreactor consists of a cylindrical container, which is allocated horizontally and pivoted (Figure 4.2). The container is filled up to no more than one-third of its volume with a granular cultivation substrate, where the microorganism grows. The heat generated by the growth of the microorganism can be dissipated to a large extent by the partially cooled shell of the container. This design allows adequate aeration and mixing of the substrate, while limiting the damage to the inoculum or product. Mixing is performed by rotating the entire vessel or by various agitation devices such as paddles and baffles (Domínguez et al. 2001,

FIGURE 4.2 Photograph of a rotary horizontal drum bioreactor for SSF.

Nagel et al. 2001a,b, Prado et al. 2004, Stuart et al. 1999). Its main disadvantage is that the drum can be filled to only one-third of its volume, otherwise mixing is inefficient.

Air solid fluidized bed: The substrate is constantly kept under a fluidization condition by a continuous airflow, thereby avoiding adhesion and aggregation of substrate particles. However, a relatively large reactor volume is necessary. The air required for keeping the fluidized bed up is recirculated, and its humidity must be precisely controlled. This procedure requires a lot of energy for keeping the fluidized bed up. Although the mass and heat transfer rates, as well as aeration and mixing of the substrate are all enhanced, damage to inoculum and heat build-up through sheer forces may affect the final product yield (Mitchell et al. 2003, Tengerdy and Szakács 2003).

Navarrete-Bolaños et al. (2004) have employed a modular rotating drum bioreactor (equipped with inlet air injection, variable speed pumps, humidifier, and gas analyzer) for xanthophylls extraction from marigold flowers. Marigold extracts have been commercialized internationally and are used as additives for poultry feed, as they provide bright colors in egg yolks, skin, and fatty tissues. As mentioned in Section 4.3.6, Milagres et al. (2004) showed that *T. aurantiacus* was able to produce a high level of thermostable xylanase when sugarcane bagasse was used as a substrate in a glass-column reactor with forced aeration. They determined that the airflow rate had a significant effect on enzyme activity, whereas initial mass of bagasse had none.

The different disadvantages detected in the above mentioned types of bioreactors have promoted the necessity for developing new bioreactor configurations or modifying the already existing ones. These new bioreactor configurations should be able to operate in continuous mode with high productivity for prolonged periods of time without operational problems, as well as offer easy scale-up for attending any industrial process. Our research group has been working in this field, resulting in the design of a new bioreactor, called immersion bioreactor. This bioreactor consists of a jacketed cylindrical glass vessel (22.5 cm height and 23.2 cm in inner diameter) with a round bottom (working volume of 2.5 L), inside which several wire mesh baskets (3 cm height and 6.5 cm in diameter; mesh size 1.5 mm) filled with support colonized by the fungus are placed. These baskets move upward and downward by means of a pneumatic system, remaining outside the medium for more time than inside. (Rivela et al. 2000) (Figure 4.3). It is noteworthy that this bioreactor configuration has been also able to run in continuous mode without operational problems, attaining high ligninolytic enzyme activities (Rodríguez Couto et al. 2002).

In order to avoid the cooling problem in SSF with *A. oryzae* and the negative effect of water content that is a limiting factor for fungal growth in SSF, a modified SSF (mSSF) process was proposed by Bhanja et al. (2007). mSSF was carried out in a GROWTEK bioreactor developed at IIT Kharagpur. In this bioreactor system, solid substrate was kept on a mesh that was in direct contact with the liquid nutrient medium, located below the mesh. The advantage of this configuration was the fact that this mSSF process could simulate the SSF one, in which wheat bran was used as the insoluble substrate. Comparative studies between SSF and mSSF permit to conclude that *A. oryzae* IFO-30103 produced very high levels of α-amylase (15,833 U/gds) by mSSF compared to SSF (12,899 U/gds) carried out in enamel-coated metallic trays utilizing wheat bran as substrate.

Recently, Liu et al. (2007) have developed a novel SSF bioreactor with air pressure pulsation. By developing a measurement and control system under the Virtual Instrument concept, the performance of the SSF bioreactor with pressure pulsation was studied by cultivating *T. koningii* in a solid medium made of wheat bran and corncob. The cooling effects of pressure pulsation on solid porous beds were

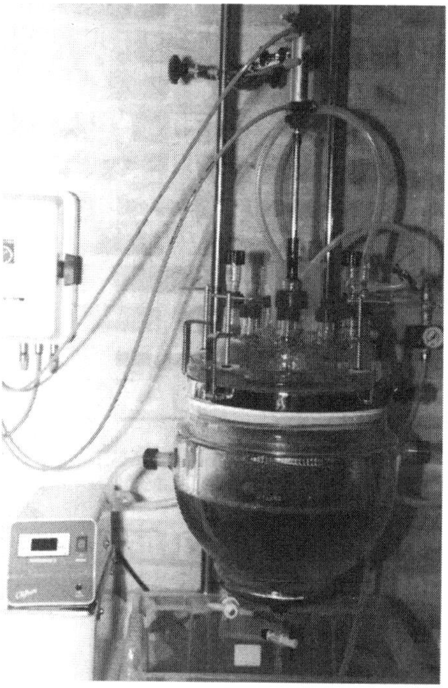

FIGURE 4.3 Photograph of an immersion bioreactor for SSF.

discussed. Experimental results showed that pressure pulsation enhances medium moisture evaporation, and hence, heat dissipation. Furthermore, through changing the pressure pulsation directions, it was possible to mitigate the temperature gradients in the bioreactor. To sum up, pressure pulsation can provide the microbes with a growing environment at optimal temperature and medium water content.

Bhanja et al. (2008) tried to overcome three major problems of SSF (oxygen supply, heat accumulation, and water activity), and designed a new bioreactor to improve the nutraceutical properties of the fermented foods through phenolic enrichment. In this study, the performance of this new bioreactor was compared with that of the traditional SSF one by *A. oryzae* using brown rice as the substrate for enrichment of phenolics and antioxidant potential. Interestingly, rice fermented in this new bioreactor resulted in a higher yield of phenolics and DPPH (1-1 diphenyl-2-picrylhydrazyl) scavenging activity in each of the extraction media as compared to SSF and control.

Accurate models describing the biological, mass-, and heat-transfer phenomena are necessary to optimize SSF bioreactor operations (Gelmi et al. 2002). However, such models are difficult to construct given the complexity of bioreactor behavior and the scarcity of process measurement. In fact, only a few mathematical models have been proposed in the area of SSF to date, which can be divided into two categories: macroscale and microscale models. Durand (2003) reviewed the types of bioreactor for SSF and highlighted the main configurations that have emerged over the last few years and the potential for scaling-up in each category of reactor. Banarjee and Bhattacharyya (2003) have described evolutionary operation as an important tool for the optimization of complex processes like SSF. In addition, Mitchell et al. (2004) have reviewed the development in modeling of microbial growth kinetics and intraparticular phenomena in SSF, discussing the insights achieved through the modeling work and the improvements to models that will be necessary in the future. Recently, Fernández-Fernández and Pérez-Correa (2007) have proposed a combined, realistic phenomenological-noise model fitted with experimental data for SSF bioreactors. Incorporating the realistic noise model makes it possible to attain a usefully close representation of the complex behavior observed in real measurements from a pilot scale SSF bioreactor.

4.6 CONCLUDING REMARKS

In this chapter, a critical analysis of the literature on the main, current applications of SSF, in relation to the production of relevant compounds for the food processing industry has been realized. Several fungi, bacteria, and yeasts are reported as very efficient producers in SSF process of biotechnologically interesting enzymes and food products, which can be produced by using a support-substrate of null cost, thus reducing the global production cost. The importance of the adequate selection of the main variables involved in the process, such as raw materials, physical parameters, medium additions, and bioreactor configuration is outstanding. For this reason, in all cases it is necessary that there should be a preliminary stage in which the effect of these variables should be studied, and their optimal value determined. Due to the limited types of bioreactors and control systems that have emerged in this moment, more studies to scale-up the SSF process are underway in several research laboratories.

REFERENCES

Acuña-Arguelles, M.E., Gutierrez-Rojas, M., Viniegra-González, G., and Favela-Torres, E. 1994. Effect of water activity on exo-pectinase production by *Aspergillus niger* CH4 on solid state fermentation. *Biotechnol. Lett.* 16:23–28.

Acuña-Arguelles, M.E., Gutierrez-Rojas, M., Viniegra-González, G., and Favela-Torres, E. 1995. Production and properties of three pectinolytic activities produced by *Aspergillus niger* in submerged and solid-state fermentation. *Appl. Microbiol. Biotechnol.* 43:808–814.

Adinarayana, K., Raju, K.V.V., Zargar, M.I., Devi, R.B., Lakshmi, P.J., and Eillaiah, P. 2004. Optimization of process parameters for production of lipase in solid-state fermentation by newly isolated *Aspergillus* species. *Indian J. Biotechnol.* 3:65–69.

Aguilar, C.N. and Gutiérrez-Sánchez, G. 2001a. Review: Sources, properties, applications and potential uses of tannin acyl hydrolase. *Food Sci. Technol. Int.* 7:373–382.

Ahlawat, S., Battan, B., Dhiman, S.S., Sharma, J., and Mandhan, R.P. 2007. Production of thermostable pectinase and xylanase for their potential application in bleaching of kraft pulp. *J. Ind. Microbiol. Biotechnol.* 34:763–770.

Alkorta, I., Garbisu, G., Llama, M.J., and Serra, J.L. 1998. Industrial applications of pectic enzymes: A review. *Process Biochem.* 33:21–28.

Antier, P., Minjares, A., Roussos, S., and Viniegra-Gonzalez, G. 1993a. New approach for selecting pectinase producing mutants of *Aspergillus niger* well adapted to solid state fermentation. *Biotechnol. Adv.* 11:429–440.

Antier, P., Minjares, A., Roussos, S., Raimbault, M., and Viniegra-González, G. 1993b. Pectinase-hyperproducing mutants of *Aspergillus niger* C28B25 for solid-state fermentation of coffee pulp. *Enzyme Microb. Technol.* 15:254–260.

Anto, H., Trivedi, U., and Patel, K. 2006. Alpha amylase production by *Bacillus cereus* MTCC 1305 using solid-state fermentation. *Food Technol. Biotechnol.* 44:241–245.

Aoki, K., Shinke, R., and Nishira, H., 1976. Chemical composition and molecular weight of yeast tannase. *Agric. Biol. Chem.* 40:297–302.

Babu, I.S. and Rao, G.H. 2007. Lipase production by *Yarrowia lipolytica* NCIM 3589 in solid state fermentation using mixed substrate. *Res. J. Microbiol.* 2:469–474.

Babu, K.R. and Satyanarayana, T. 1995. α-Amylase production by thermophilic *Bacillus coagulans* in solid state fermentation. *Process Biochem.* 30:305–309.

Bai, Z.H., Zhang, H.X., Qi, H.Y., Peng, X.W., and Li, B.J. 2004. Pectinase production by *Aspergillus niger* using wastewater in solid state fermentation for eliciting plant disease resistance. *Bioresource Technol.* 95:49–52.

Bajpai, P. 1997. Microbial xylanolytic enzyme system: Properties and applications. *Adv. Appl. Microbiol.* 43:141–194.

Balkan, B. and Ertan, F. 2007. Production of α-amylase from *Penicillium chrysogenum* under solid-state fermentation by using some agricultural by-products. *Food Technol. Biotechnol.* 45:439–442.

Banarjee, R. and Bhattacharyya, B.C. 2003. Evolutionary operation as a tool optimization for solid state fermentation. *Biochem. Eng. J.* 13:149–155.

Bedford, M.R. and Classen, H.L. 1992. Reduction of intestinal viscosity through manipulation of dietary rye and pentosanases concentration is effected through changes in the carbohydrate composition of the intestinal aqueous phase and results in improved growth rate and food conversion efficiency of broiler chicks. *J. Nutr.* 26:560–569.

Benjamin, S. and Pandey, A. 1996a. Lipase production by *Candida rugosa* on copra waste extract. *Indian J. Microbiol.* 36:201–204.

Benjamin, S. and Pandey, A. 1996b. Optimization of liquid media for lipase production by *Candida rugosa. Bioresource Technol.* 55:167–170.

Benjamin, S. and Pandey, A. 1997a. Coconut cake—A potent substrate for the production of lipase by *Candida rugosa* in solid state fermentation. *Acta Biotechnol.* 17:241–251.

Benjamin, S. and Pandey, A. 1997b. Enhancement of lipase production during repeated batch cultivation using immobilised *Candida rugosa*. *Process Biochem*. 32:437–440.

Benjamin, S. and Pandey, A. 1998. Mixed-solid substrate fermentation. A novel process for enhanced lipase production by *Candida rugosa*. *Acta Biotechnol*. 18:315–324.

Bhanja, T., Rout, S., Banerjee, R., and Bhattacharyya, B.C. 2007. Comparative profiles of α-amylase production in conventional tray reactor and GROWTEK bioreactor. *Bioprocess Biosyst. Eng*. 30:369–376.

Bhanja, T., Rout, S., Banerjee, R., and Bhattacharyya, B.C. 2008. Studies on the performance of a new bioreactor for improving antioxidant potential of rice. *LWT—Food Sci. Technol*. 41:1459–1465.

Bhusan, B., Dosanjih, N.S., Kumar, K., and Hoondal, G.S. 1994. Lipase production from an alkalophilic yeast sp. by solid state fermentation. *Biotechnol. Lett*. 16:841–842.

Bogar, B., Szakacs, G., Tengerdy, R.P., Linden, J.C., and Pandey, A. 2002. Production of α-amylase with *Aspergillus oryzae* on spent brewing grain by solid substrate fermentation. *Appl. Biochem. Biotechnol*. 102–103:453–461.

Bogar, B., Szakacs, G., Pandey, A., Abdulhameed, S., Linden, J.C., and Tengerdy, R.P. 2003. Production of phytase by *Mucor racemosus* in solid-state fermentation. *Biotechnol. Prog*. 19:312–319.

Castilho, L.R., Alves, T.L.M., and Medronho, R.A. 1999. Recovery of pectinolytic enzymes produced by solid state culture of *Aspergillus niger*. *Process Biochem*. 34:181–186.

Castilho, L.R., Alves, T.L.M., and Medronho, R.A. 2000 Production and extraction of pectinases obtained by solid state fermentation of agro-industrial residues with *Aspergillus niger*. *Bioresource Technol*. 71:45–50.

Chen, H.Z. 1992. Advances in solid-state fermentation. *Res. Appl. Microbiol. (China)* 3:7–10.

Christen, P., Angeles, N., Corzo, G., Farres, A., and Revah, S. 1995. Microbial lipase production on a polymeric resin. *Biotechnol. Tech*. 9:597–600.

Cordova, J., Nemmaoui, M., Ismaili-Alaoui, M., et al. 1998. Lipase production by solid state fermentation of olive cake and sugar cane bagasse. *J. Mol. Catalysis B: Enzymatic* 5:75–78.

Di Luccio, M., Capra, F., Ribeiro, N.P., Vargas, G.D.L.P., Freire, D.M.G., and De Oliveira, D. 2004. Effect of temperature, moisture, and carbon supplementation on lipase production by solid-state fermentation of soy cake by *Penicillium simplicissimum*. *Appl. Biochem. Biotechnol*. 113:173–180.

Díaz-Godínez, G., Soriano-Santos, J., Augur, C., and Viniegra-González, G. 2001. Exopectinases produced by *Aspergillus niger* in solid-state and submerged fermentation: A comparative study. *J. Ind. Microbiol. Biotechnol*. 26:271–275.

Domínguez, A., Rivela, I., Rodríguez Couto, S., and Sanromán, M.A. 2001. Design of a new rotating drum bioreactor for ligninolytic enzyme production by *Phanerochaete chrysosporium* grown on an inert support. *Process Biochem*. 37:549–554.

Domínguez, A., Costas, M., Longo, M.A., and Sanroman, A. 2003. A novel application of solid state culture: Production of lipases by *Yarrowia lipolytica*. *Biotechnol. Lett*. 25:1225–1229.

Durand, A. 2003. Bioreactor designs for solid state fermentation. *Biochem. Eng. J*. 13:113–125.

Durand, A., Renaud, R., Almanza, S., Maratray, J., Diez, M., and Desgranges, C. 1993. Solid-state fermentation reactors: From lab scale to pilot plant. *Biotechnol. Adv*. 11:591–597.

El-Batal, A.I. and Abdel Karem, H. 2001. Phytase production and phytic acid reduction in rapeseed meal by *Aspergillus niger* during solid state fermentation. *Food Res. Int*. 34:715–720.

Falony, G., Coca Armas, J., Dustet Mendoza, J.C., and Martínez Hernández, J.L. 2006. Production of extracellular lipase from *Aspergillus níger* by solid-state fermentation. *Food Technol. Biotechnol*. 44:235–240.

Fernández-Fernández, M. and Pérez-Correa, J.R. 2007. Realistic model of a solid substrate fermentation packed-bed pilot bioreactor. *Proc. Biochem.* 42:224–234.

Francis, F., Sabu, A., Nampoothiri, K.M., et al. 2003. Use of response surface methodology for optimizing process parameters for the production of α-amylase by *Aspergillus oryzae. Biochem. Eng. J.* 15:107–115.

Gaitán-Hernández, R., Esqueda, M., Gutiérrez, A., Sánchez, A., Beltrán-García, M., and Mata, G. 2006. Bioconversion of agrowastes by *Lentinula edodes*: The high potential of viticulture residues. *Appl. Microbiol. Biotechnol.* 71:432–439.

Gangadharan, D., Sivaramakrishnan, S., Nampoothiri, K.M., and Pandey, A. 2006. Solid culturing of *Bacillus amyloliquefaciens* for alpha amylase production. *Food Technol. Biotechnol.* 44:269–274.

Garzón, C.G. and Hours, R.A. 1991. Citrus waste: An alternative substrate for pectinase production in solid-state culture. *Bioresource Technol.* 39:93–95.

Gelmi, C., Perez-Correa, R., and Agosin, E. 2002. Modelling *Gibberella fujikuroi* growth and GA3 production in solid-state fermentation. *Process. Biochem.* 37:1033–1040.

Gombert, A.K., Pinto, A.L., Castilho, L.R., and Freire, D.M.G. 1999. Lipase production by *Penicillium restrictum* in solid-state fermentation using babassu oil cake as substrate. *Process Biochem.* 35:85–90.

Greiner, R. and Konietzny, U. 2006. Phytase for food application. *Food Technol. Biotechnol.* 44:125–140.

Gross, R.A. and Kalra, B. 2002. Biodegradable polymers for the environment. *Science* 297:803–807.

Gutiérrez-Rozas, M., Cordova, J., Auria, R., Revah, S., and Favela-Torres, E. 1995. Citric acid and polyols production by *Aspergillus niger* at high glucose concentration in solid state fermentation on inert support. *Biotechnol. Lett.* 17:219–224.

Hang, Y.D. and Woodams, E.E. 1985. Grape pomace a novel substrate for microbial production of citric acid. *Biotechnol. Lett.* 7:253–254.

Hernaiz, M.J. and Sinisterra, J.M. 1999. Modification of purified lipase from *Candida rugosa* with polyethylene glycol: A systematic study. *Enzyme Microb. Technol.* 24:181–190.

Hölker, U. and Lenz, J. 2005. Solid state fermentation—Are there any biotechnological advantages? *Curr. Opin. Microbiol.* 8:301–306.

Hours, R.A., Voget, C.E., and Ertola, R.J. 1988a. Some factors affecting pectinase production from apple pomace in solid-state cultures. *Biol. Wastes* 24:147–157.

Hours, R.A., Voget, C.E., and Ertola, R.J. 1988b. Apple pomace as raw material for pectinases production in solid state culture. *Biol. Wastes* 23:221–228.

Iluyemi, F.B., Hanafi, M.M., Radziah, O., and Kamarudin, M.S. 2006. Fungal solid state culture of palm kernel cake. *Bioresource Technol.* 97:477–482.

Imandi, S.B., Bandaru, V.V.R., Somalanka, S.R., Bandaru, S.R., and Garapati, H.R. 2008. Application of statistical experimental designs for the optimization of medium constituents for the production of citric acid from pineapple waste. *Bioresource Technol.* 99 (10):4445–4450.

Ismail, A.S. 1996. Utilization of orange peels for the production of multi-enzyme complexes by some fungal strains. *Process Biochem.* 1:645–650.

Iyayi, E.A. and Aderolu, Z.A. 2004. Enhancement of the feeding value of some agro-industrial by-products for laying hens after their solid state fermentation with *Trichoderma viride. Afr. J. Biotechnol.* 3:182–185.

John, R.P., Nampoothiri, K.M., and Pandey, A. 2006. Solid-state fermentation for L-lactic acid production from agro wastes using *Lactobacillus delbrueckii. Process Biochem.* 41:759–763.

John, R.P., Nampoothiri, K.M., and Pandey, A. 2007. Polyurethane foam as an inert carrier for the production of L(+)-lactic acid by *Lactobacillus casei* under solid-state fermentation. *Lett. Appl. Microbiol.* 44:582–587.

Joshi, V.K., Parmar, M., and Ran, N.S. 2006. Pectin esterase production from apple pomace in solid-state and submerged fermentations. *Food Technol. Biotechnol.* 44:253–256.

Kamini, N.R., Mala, J.G.S., and Puvanakrishnan, R. 1998. Lipase production from *Aspergillus niger* by solid-state fermentation using gingelly oil cake. *Process Biochem.* 33:505–511.

Kang, S.W., Lee, S.H., Yoon, C.S., and Kim, S.W. 2005. Conidia production by *Beauveria bassiana* (for the biocontrol of a diamondback moth) during solid-state fermentation in a packed-bed bioreactor. *Biotechnol. Lett.* 27:135–139.

Kapoor, M. and Kuhad, R.C. 2002. Improved polygalacturonase production from *Bacillus* sp. MG-cp-2 under submerged (SmF) and solid state (SSF) fermentation. *Lett. Appl. Microbiol.* 34:317–322.

Kar, B. and Banerjee, R. 2000. Biosynthesis of tannin acyl hydrolase from tannin-rich forest residue under different fermentation conditions. *J. Ind. Microbiol. Biotechnol.* 25:29–38.

Kazlauskas, R. 1994. Elucidating structure mechanism relationship in lipases, prospects for predicting and engineering catalytic properties. *Trends Biotechnol.* 12:464–472.

Khare, S.K., Krishana, J., and Gandhi, A.P. 1995. Citric acid production from Okara (soy residue) by solid state fermentation. *Bioresource Technol.* 54:323–325.

Khosravi-Darani, K. and Zoghi, A. 2008. Comparison of pretreatment strategies of sugarcane baggase: Experimental design for citric acid production. *Bioresource Technol.* 99:6986–6993.

Kim, D.H., Oh, B.C., Choi, W.C., Lee, J.K., and Oh, T.K. 1999. Enzymatic evaluation of *Bacillus amyloliquefaciensphytase* as a feed additive. *Biotechnol. Lett.* 21:925–927.

Krishna, C. 2005. Solid-state fermentation systems—An overview. *Crit. Rev. Biotechnol.* 25:1–30.

Krishna, C. and Chandrasekaran, M. 1996. Banana waste as substrate for α-amylase production by *Bacillus subtilis* (CBTK 106) under solid-state fermentation. *Appl. Microbiol. Biotechnol.* 46:106–111.

Krishna, C. and Nokes, S.E. 2001. Predicting vegetative inoculum performance to maximize phytase production in solid-state fermentation using response surface methodology. *J. Ind. Microbiol. Biotechnol.* 26:161–170.

Kumar, D., Jain, V.K., Shanker, G., and Srivastava, A. 2003a. Utilisation of fruits waste for citric acid production by solid state fermentation. *Process Biochem.* 38:1725–1729.

Kumar, D., Jain, V.K., Shanker, G., and Srivastava, A. 2003b. Citric acid production by solid state fermentation using sugarcane bagasse. *Process Biochem.* 38:1731–1738.

Kumar, R., Sharma, J., and Singh, R. 2007. Production of tannase from *Aspergillus ruber* under solid-state fermentation using jamun (*Syzygium cumini*) leaves. *Microbiol. Res.* 162:384–390.

Kumar, R.A., Gunasekaran, P., and Lakshmanan, M. 1999. Biodegradation of tannic acid by *Citrobacter freundii* isolated from a tannery effluent. *J. Basic Microbiol.* 39:161–168.

Lekha, P.K. and Lonsane, B.K. 1994. Comparative titres, location and properties of tannin acyl hydrolase produced by *Aspergillus niger* PKL 104 in solid-state, liquid surface and submerged fermentations. *Process Biochem.* 29:497–503.

Lekha, P.K. and Lonsane, B.K. 1997. Production and application of tannin acyl hydrolase: State of the art. *Adv. Appl. Microbiol.* 44:215–260.

Liu, J., Li, D.B., and Yang, J.C. 2007. Operating characteristics of solid-state fermentation bioreactor with air pressure pulsation. *Appl. Biochem. Microbiol.* 43:211–216.

Lorenzo, M., Moldes, D., Rodríguez Couto, S., and Sanromán, A. 2002. Improving laccase production by employing different lignocellulosic wastes in submerged cultures of *Trametes versicolor. Bioresource Technol.* 82:109–113.

Lozano, P., Manjón, F., Romojaro, F., and Iborra, J. 1988. Properties of pectolytic enzymes covalently bound to nylon for apricot juice clarification. *Process Biochem.* 23:75–78.

Mahadik, N.D., Puntambekar, U.S., Bastawde, K.B., Khire, J.M., and Gokhale, D.V. 2002. Production of acidic lipase by *Aspergillus niger* in solid state fermentation. *Process Biochem.* 38:715–721.

Mahanta, N., Gupta, A., and Khare, S.K. 2008. Production of protease and lipase by solvent tolerant *Pseudomonas aeruginosa* PseA in solid-state fermentation using *Jatropha curcas* seed cake as substrate. *Bioresource Technol.* 99:1729–1735.

Maldonado, M.C., Navarro, A., and Callieri, D.A.S. 1986. Production of pectinases by *Aspergillus* sp. using differently pretreated lemon peel as the carbon source. *Biotechnol. Lett.* 8:501–504.

Mandviwala, T.N. and Khire, J.M. 2000. Production of high activity thermostable phytase from thermotolerant *Aspergillus niger* in solid state fermentation. *J. Ind. Microbiol. Biotechnol.* 24:237–243.

Martínez, C., Ros, G., Periago, M.J., Lopez, G., and Ortuño y Rincón, J. 1996. Phytic acid in human nutrition. *Food Sci. Technol. Int.* 2:201–209.

Martins, E.S., Silva, D., Da Silva, R., and Gomes, E. 2002. Solid state production of thermostable pectinases from thermophilic *Thermoascus aurantiacus. Process Biochem.* 37:949–954.

Mateos Diaz, J.C., Rodríguez, J.A., Roussos, S., et al. 2006. Lipase from the thermotolerant fungus *Rhizopus homothallicus* is more thermostable when produced using solid state fermentation than liquid fermentation procedures. *Enzyme Microb. Technol.* 39:1042–1050.

Milagres, A.M.F., Santos, E., Piovan, T., and Roberto, I.C. 2004. Production of xylanase by *Thermoascus aurantiacus* from sugar cane bagasse in an aerated growth fermentor. *Process Biochem.* 39:1387–1391.

Miranda, O.A., Salgueiro, A.A., Pimentel, M.C.B., Lima Filho, J.L., Melo, E.H.M., and Duran, N. 1999. Lipase production by a Brazilian strain of *Penicillium citrinum* using an industrial residue. *Bioresource Technol.* 69:145–147.

Mitchell, D.A., Von Meien, O.F., and Krieger, N. 2003. Recent developments in modeling of solid-state fermentation: Heat and mass transfer in bioreactors. *Biochem. Eng. J.* 13:137–147.

Mitchell, D.A., von Meien, O.F., Krieger, N., and Dalsenter, F.D.H. 2004. A review of recent developments in modeling of microbial growth kinetics and intraparticle phenomena in solid-state fermentation. *Biochem. Eng. J.* 17:15–26.

Mondal, K.C. and Pati, B.R. 2000. Studies on the extracellular tannase from newly isolated *Bacillus licheniformis* KBR 6. *J. Basic Microbiol.* 40:223–232.

Morais, M.H., Ramos, A.C., Matos, N., and Santos-Olivera, E.J. 2000. Production of shiitake mushroom (*Lentinula edodes*) on lignocellulosic residues. *Food Sci. Technol. Int.* 6:123–128.

Morales, P., Martínez-Carrera, D., and Martínez-Sánchez, W. 1991. Cultivation of shiitake on several substrates in Mexico. *Micologia Neotropical Aplicada* 4:75–81 (In Spanish).

Nagarjun, P.A., Rao, R.S., Rajesham, S., and Rao, L.V. 2005. Optimization of lactic acid production in SSF by *Lactobacillus amylovorus* NRRL B-4542 using Taguchi methodology. *J. Microbiol.* 43:38–43.

Nagel, F.J., Tramper, J., Bakker, M.S., and Rinzema, A. 2001a. Temperature control in a continuously mixed bioreactor for solid-state fermentation. *Biotechnol. Bioeng.* 72:219–230.

Nagel, F.J., Tramper, J., Bakker, M.S., and Rinzema, A. 2001b. Model for on-line moisture-content control during solid-state fermentation. *Biotechnol. Bioeng.* 72:231–243.

Nagy, V., Toke, E.R., Keong, L.C., et al. 2006. Kinetic resolutions with novel, highly enantioselective fungal lipases produced by solid state fermentation. *J. Mol. Catalysis B: Enzymatic* 39:141–148.

Navarrete-Bolaños, J.L., Jiménez-Islas, H., Botello-Alvarez, E., Rico-Martínez, R., and Paredes-López, O. 2004. An optimization study of solid-state fermentation: Xanthophylls extraction from marigold flowers. *Appl. Microbiol. Biotechnol.* 65:383–390.

Naveena, B.J., Vishnu, C., Altaf, M.D., and Reddy, G. 2003. Wheat bran an inexpensive substrate for production of lactic acid in solid-state fermentation by *Lactobacillus amylophilus* GV6: Optimization of fermentation conditions. *J. Sci. Ind. Res.* 62:453–456.

Naveena, B.J., Altaf, M.D., Bhadrayya, K., Madhavendra, S.S., and Reddy, G. 2005a. Direct fermentation of starch to L(+) lactic acid in SSF by *Lactobacillus amylophilus* GV6 using wheat bran as support and substrate: Medium optimization using RSM. *Process Biochem.* 40:681–690.

Naveena, B.J., Altaf, M.D., Bhadriah, K., and Reddy, G. 2005b. Selection of medium components by Plackett-Burman design for production of L(+)lactic acid by *Lactobacillus amylophilus* GV6 in SSF using wheat bran. *Bioresource Technol.* 96:485–490.

Ohnishi, K., Yoshida, Y., and Sekiguchi, J. 1994. Lipase production of *Aspergillus oryzae. J. Fermentation Bioeng.* 77:490–495.

Orozco, A.L., Pérez, M.I., Guevara, O., et al. 2008. Biotechnological enhancement of coffee pulp residues by solid-state fermentation with *Streptomyces* Py-GC/MS analysis. *J. Anal. Appl. Pyrol.* 81(2):247–252.

Ortiz-Vazquez, E., Granados-Baeza, M., and Rivera-Munoz, G. 1993. Effect of culture conditions on lipolytic enzyme production by *Penicillium candidum* in a solid state fermentation. *Biotechnol. Adv.* 11:409–416.

Palma, M.B., Pinto, A.L., Gombert, A.K., et al. 2000. Lipase production by *Penicillium restrictum* using solid waste of industrial babassu oil production as substrate. *Appl. Biochem. Biotechnol. A: Enzyme Eng. Biotechnol.* 84–86:1137–1145.

Pandey, A. 1992. Recent process developments in solid-state fermentation. *Process. Biochem.* 27:109–117.

Pandey, A., Soccol, C.R., and Mitchell, D. 2000a. New developments in solid state fermentation. I. Bioprocesses and products. *Process Biochem.* 35:1153–1169.

Pandey, A., Soccol, C.R., Nigam, P., Brand, D., Mohan, R., and Roussos, S. 2000b. Biotechnological potential of coffee pulp and coffee husk for bioprocesses. *Biochem. Eng. J.* 6:153–162.

Pandey, A., Szakacs, G., Zoclo, C.R., Rodríguez-León, J.A., and Zoclo, V.T. 2001. Production, purification and properties of microbial phytases. *Bioresource Technol.* 77:203–214.

Papagianni, M., Nokes, S.E., and Filer, K. 2000. Production of phytase by *Aspergillus niger* in submerged and solid-state fermentation. *Process Biochem.* 35:397–402.

Papagianni, M., Noke, S.E., and Filer, K. 2001. Submerged and solid-state phytase fermentation by *Aspergillus niger*: Effects of agitation and medium viscosity on phytase production, fungal morphology and inoculum performance. *Food Technol. Biotechnol.* 39:319–326.

Pasamontes, L., Haiker, M., Wyss, M., Tessier, M., and vanLoon, A.P.G.M. 1997a. Gene cloning, purification, and characterization of a heat-stable phytase from the fungus *Aspergillus fumigatus. Appl. Environ. Microbiol.* 63:1696–1700.

Pasamontes, L., Haiker, M., Henríquez Huecas, M., Mitchell, D.B., and vanLoon, A.P.G.M. 1997b. Cloning of the phytases from *Emericella nidulans* and the thermophilic fungus *Talaromyces thermophilus. Biochim. Biophys. Acta—Gene Struct. Expr.* 1353:217–223.

Patel, H.M., Wang, R., Chandrashekar, O., Pandiella, S.S., and Webb, C. 2004. Proliferation of *Lactobacillus plantarum* in solid-state fermentation of oats. *Biotechnol. Prog.* 20:110–116.

Patil, S.R. and Dayanand, A. 2006. Exploration of regional agrowastes for the production of pectinase by *Aspergillus niger. Food Technol. Biotechnol.* 44:289–292.

Pérez-Guerra, N., Torrado-Agrasar, A., López-Macias, C., and Pastrana, L. 2003. Main characteristics and applications of solid substrate fermentation. *Electron. J. Environ. Agric. Food Chem.* 2:343–350.

Pintado, J., Lonsane, B.K., Gaime-Perraud, I., and Roussos, S. 1998. On-line monitoring of citric acid production in solid-state culture by respirometry. *Process Biochem.* 33:513–518.

Pire, D.G., Wright, J.E., and Albertó, E. 2001. Cultivation of shiitake using sawdust from widely available local woods in Argentine. *Micol. Aplicada Int.* 13:87–91.

Prado, F.C., Vandenberghe, L.P.S., Lisboa, C., Paca, J., Pandey, A., and Soccol, C.R. 2004. Relation between citric acid production and respiration rate of *Aspergillus niger* in solid-state fermentation. *Eng. Life Sci.* 4:179–186.

Ramachandran, S., Patel, A.K., Nampoothiri, K.M. et al. 2004. Coconut oil cake—A potential raw material for the production of α-amylase. *Bioresource Technol.* 93:169–174.

Ramachandran, S., Roopesh, K., Nampoothiri, K.M., Szakacs, G., and Pandey, A. 2005. Mixed substrate fermentation for the production of phytase by *Rhizopus* spp. using oilcakes as substrates. *Process Biochem.* 40:1749–1754.

Ramírez-Coronel, M.A., Viniegra-González, G., Darvill, A., and Augur, C. 2003. A novel tannase from *Aspergillus niger* with β-glucosidase activity. *Microbiology* 149:2941–2946.

Rao, P.V., Jayaraman, K., and Lakshmanan, C.M. 1993a. Production of lipase by *Candida rugosa* in solid-state fermentation. 1. Determination of significant process variables. *Process Biochem.* 28:385–389.

Rao, P.V., Jayaraman, K., and Lakshmanan, C.M. 1993b. Production of lipase by *Candida rugosa* in solid state fermentation. 2. Medium optimization and effect of aeration. *Process Biochem.* 28:391–395.

Rivela, I., Rodríguez Couto, S., and Sanromán, A. 2000. Extracellular ligninolytic enzyme production by *Phanerochaete chrysosporium* in a new solid-state bioreactor. *Biotechnol. Lett.* 22:1443–1447.

Rivera-Muñoz, G., Tinoco-Valencia, J.R., Sanchez, S., and Farres, A. 1991. Production of microbial lipases in a solid-state fermentation system. *Biotechnol. Lett.* 13:277–280.

Rodríguez Couto, S. and Sanromán, M.A. 2005. Application of solid-state fermentation to ligninolytic enzyme production. *Biochem. Eng. J.* 22:211–219.

Rodríguez, E., Mullaney, E.J., and Lei, X.G. 2000. Expression of the *Aspergillus fumigatus* phytase gene in *Pichia pastoris* and characterization of the recombinant enzyme. *Biochem. Biophy. Res. Commun.* 268:373–378.

Rodríguez Couto, S., Rivela, I., Muñoz, M.R., and Sanromán A. 2000. Ligninolytic enzyme production and the ability of decolourisation of Poly R-478 in packed-bed bioreactors by *Phanerochaete chrysosporium. Bioprocess Eng.* 23:287–293.

Rodríguez Couto, S. Barreiro, M., Rivela, I., Longo, M.A., and Sanromán, A. 2002. Performance of a solid-state immersion bioreactor for ligninolytic enzyme production: Evaluation of different operational variables. *Process Biochem.* 38:219–227.

Roehr, M., Kubicek, C.P., and Kominek, J. 1996. Gluconic acid. In *Biotechnology, Product of Primary Metabolism*, vol. 6, H. J.Rehm, and G. Reed, eds., Weinhein: Verlag-Chemie Press, pp. 347–362.

Rojan, P.J., Nampoothiri, K.M., Nair, A.S., and Pandey, A. 2005. L(+)-Lactic acid production using *Lactobacillus casei* in solid-state fermentation. *Biotechnol. Lett.* 27:1685–1688.

Romero-Gómez, S., Augur, C., and Viniegra-González, G. 2000. Invertase production by *Aspergillus niger* in submerged and solid state fermentation. *Biotechnol. Lett.* 22:1255–1258.

Roopesh, K., Ramachandran, S., Nampoothiri, K.M., Szakacs, G., and Pandey, A. 2006. Comparison of phytase production on wheat bran and oilcakes in solid-state fermentation by *Mucor racemosus. Bioresource Technol.* 97:506–511.

Roukas, T. 2000. Citric and gluconic acid production from fig by *Aspergillus niger* using solid-state fermentation. *J. Ind. Microbiol. Biotechnol.* 25:298–304.

Royse, D.J. 1985. Effect of spawn run time and substrate nutrition on yield and size of shiitake mushroom. *Mycologia* 77:756–762.

Royse, D.J. 1996. Yield stimulation of shiitake by millet supplementation of wood chip substrate. In *Mushroom Biology and Mushroom Products*, D. J. Royse, ed., University Park, PA: Penn State University Press, pp. 277–283.

Sabu, A., Kiran, G.S., and Pandey, A. 2005. Purification and characterization of tannin acyl hydrolase from *Aspergillus niger* ATCC 16620. *Food Technol. Biotechnol.* 43:133–138.

Sabu, A., Augur, C., Swati, C., and Pandey, A. 2006. Tannase production by *Lactobacillus* sp. ASR-S1 under solid-state fermentation. *Process Biochem.* 41:575–580.

Salmones, D., Mata, G., Ramos, L.M., and Waliszewski, K.N. 1999. Cultivation of shiitake mushroom, *Lentinula edodes*, in several lignocellulosic materials originating from the subtropics. *Agronomie* 19:13–19.

Sanchez, A., Ysunza, F., Beltrán-García, M.J., and Esqueda, M. 2002. Biodegradation of viticulture wastes by *Pleurotus*: A source of microbial and human food and its potential use in animal feeding. *J. Agric. Food Chem.* 50:2537–2542.

Sanchez, C. 2004. Modern aspects of mushroom culture technology. *Appl. Microbiol. Biotechnol.* 64:756–762.

Sanghi, A., Garg, N., Sharma, J., Kuhar, K., Kuhad, R.C., and Gupta, V.K. 2008. Optimization of xylanase production using inexpensive agro-residues by alkalophilic *Bacillus subtilis* ASH in solid-state fermentation. *World J. Microbiol. Biotechnol.* 24:633–640.

Schwan, R.F., Cooper, R.M., and Wheals, A.E. 1997. Endopolygalacturonase secretion by *Kluyveromyces marxianus* and other cocoa pulp-degrading yeasts. *Enzyme Microb. Technol.* 21:234–244.

Selvakumar, P., Ashakumary, L., and Pandey, A. 1996. Microbial synthesis of starch saccharifying enzyme in solid state fermentation. *J. Sci. Ind. Res.* 55:443–449.

Senthuran, A., Senthuran, V., Mattiasson, B., and Kaul, R. 1997. Lactic acid fermentation in a recycle batch reactor using immobilized *Lactobacillus casei. Biotechnol. Bioeng.* 55:843–853.

Seth, M. and Chand, S. 2000. Biosynthesis of tannase and hydrolysis of tannins to gallic acid by *Aspergillus awamori*—Optimisation of process parameters. *Process Biochem.* 36:39–44.

Shah, N.D., Chattoo, B.B., Baroda, R.M., and Patiala, V.M. 1993. Starch hydrolysate, an optimal and economical source of carbon for the secretion of citric acid by *Yarrowia lipolytica. Starch* 45:104–109.

Shankaranand, V.S. and Lonsane, B.K. 1994. Ability of *Aspergillus niger* to tolerate metal ions and minerals in solid state fermentation system for production of citric acid. *Process Biochem.* 29:29–37.

Sharma, A., Vivekanand, V., and Singh, R.P. 2008. Solid-state fermentation for gluconic acid production from sugarcane molasses by *Aspergillus niger* ARNU-4 employing tea waste as the novel solid support. *Bioresource Technol.* 99:3444–3450.

Shojaosadati, S.A. and Babaripour, V. 2002. Citric acid production from apple pomace in multi layer packed bed solid state bioreactor. *Process Biochem.* 37:909–914.

Shukla, J. and Kar, R. 2006. Potato peel as a solid state substrate for thermostable α-amylase production by thermophilic *Bacillus* isolates. *World J. Microbiol. Biotechnol.* 22:417–422.

Singh, O.V. and Singh, R.P. 2006. Bioconversion of grape must into modulated gluconic acid production by *Aspergillus niger* ORS-4·410. *J. Appl. Microbiol.* 100:1114–1122.

Singh, O.V., Jain, R.K., and Singh, R.P. 2003. Gluconic acid production under varying fermentation conditions by *Aspergillus niger. J. Chem. Technol. Biotechnol.* 78:208–212.

Singh, S.A., Plattner, H., and Diekmann, H. 1999. Exopolygalacturonate lyase from a thermophilic *Bacillus* sp. *Enzyme Microb. Technol.* 25:420–425.

Soares, M.M.C.N., Da Silva, R., Carmona, E.C., and Gomes, E. 2001. Pectinolytic enzyme production by *Bacillus* species and their potential application on juice extraction. *World J. Microbiol. Biotechnol.* 17:79–82.

Soccol, R.S. and Vandenberghe, L.P.S. 2003. Overview of applied solid-state fermentation in Brazil. *Biochem. Eng. J.* 13:205–218.

Soccol, C.R., Marin, B., Rimbault, M., and Labeault, J.M. 1994. Potential of solid state fermentation for production of L(+) lactic acid by *Rhizopus oryzae. Appl. Microbiol. Biotechnol.* 41:286–290.

Soccol, C.R., Vandenberghe, L.P.S., Rodrigues, C., and Pandey, A. 2006. New perspectives for citric acid production and application. *Food Technol. Biotechnol.* 44:141–149.

Sodhi, H.K., Sharma, K., Gupta, J.K., and Soni, S.K. 2005. Production of a thermostable α-amylase from *Bacillus* sp. PS-7 by solid state fermentation and its synergistic use in the hydrolysis of malt starch for alcohol production. *Process Biochem.* 40:525–534.

Stamford, T.L., Stamford, N.P., Coelho, L.C., and Araujo, J.M. 2001. Production and characterization of a thermostable alpha-amylase from *Nocardiopsis* sp. endophyte of yam bean. *Bioresource Technol.* 76:137–141.

Stuart, D.M., Mitchell, D.A., Johns, M.R., and Litster, J.D. 1999. Solid-state fermentation in rotating drum bioreactors: Operating variables affect performance through their effects on transport phenomena. *Biotechnol. Bioeng.* 63:383–391.

Suryanarayan, S. 2003. Current industrial practice in solid state fermentations for secondary metabolite production: The Biocon India experience. *Biochem. Eng. J.* 13:189–195.

Tengerdy, R.P. and Szakacs, G. 2003. Bioconversion of lignocellulose in solid substrate fermentation. *Biochem. Eng. J.* 13:169–179.

Ul-Haq, I., Idrees, S., and Rajoka, M.I. 2002. Production of lipases by *Rhizopus oligosporous* by solid-state fermentation. *Process Biochem.* 37:637–641.

Vandenberghe, L.P.S., Soccol, C.R., Prado, F.C., and Pandey, A. 2004. Comparison of citric acid production by solid-state fermentation in flask, column, tray, and drum bioreactors. *Appl. Biochem. Biotechnol. A: Enzyme Eng. Biotechnol.* 118:293–303.

Vickroy, T.B. 1985. Lactic acid. In *Comprehensive Biotechnology: The Principles, Applications and Regulation of Biotechnology in Industry, Agriculture and Medicine,* vol. 3, H.W. Blanch, S. Drew, and D.I.C. Wang, eds., Toronto ON, Canada: Dic Pergamon Press, pp. 761–776.

Villas-Bôas, S.G., Esposito, E., and Matos de Mendonça, M. 2003. Bioconversion of apple pomace into a nutritionally enriched substrate by *Candida utilis* and *Pleurotus ostreatus. World J. Microbiol. Biotechnol.* 19:461–467.

Wee, Y.J., Kim, J.N., and Ryu, H.W. 2006. Biotechnological production of lactic acid and its recent applications. *Food Technol. Biotechnol.* 44:163–172.

Wu, Y.B., Ravindran, V., and Hendriks, W.H. 2003. Effects of microbial phytase, produced by solid-state fermentation, on the performance and nutrient utilisation of broilers fed maize- and wheat-based diets. *Br. Poult. Sci.* 44:710–718.

Wu, Y.B., Ravindran, V., and Hendriks, W.H. 2004a. Influence of exogenous enzyme supplementation on energy utilisation and nutrient digestibility of cereals for broilers. *J. Sci. Food Agric.* 84:1817–1822.

Wu, Y.B., Ravindran, V., Thomas, D.G., Birtles, M.J., and Hendriks, W.H. 2004b. Influence of phytase and xylanase, individually or in combination, on performance, apparent metabolisable energy, digestive tract measurements and gut morphology in broilers fed wheat-based diets containing adequate level of phosphorus. *Br. Poult. Sci.* 45:76–84.

Xu, H., Sun, L., Zhao, D., Zhang, B., Shi, Y., and Wu, Y. 2008. Production of α-amylase by *Aspergillus oryzae* As 3951 in solid state fermentation using spent brewing grains as substrate. *J. Sci. Food Agric.* 88:529–535.

Zheng, Z. and Shetty, K. 2000. Solid state production of polygalacturonase by *Lentinus edodes* using fruit processing wastes. *Process Biochem.* 35:825–830.

5 Membrane Processing for the Recovery of Bioactive Compounds in Agro-Industries

Svetlozar Velizarov and João G. Crespo

CONTENTS

5.1 INTRODUCTION

This chapter discusses the use of membrane processes for the recovery, concentration, and purification of compounds with target bioactivity from complex agro-industrial process streams and residues. Considering the large diversity of membrane processes used nowadays at industrial scale, the authors decided to focus this chapter on membrane processes particularly suitable to deal with the recovery of small bioactive molecules (molecular weight below 1 kDa). This option derives from the fact that the literature about recovery of large bioactive molecules, such as proteins, by membrane and other downstream processes is rather abundant. Small molecules started recently to attract a growing interest, which is certainly linked to the recognition of their diverse and powerful bioactive properties, including antioxidant and anti-carcinogenic activities. Small bioactive molecules may be recovered from agro-industrial process streams by different membrane operations, namely pervaporation, nanofiltration, and electrodialysis, usually integrated in multistep schemes. This chapter describes firstly the basic principles, features, and limitations of these membrane operations and, secondly, illustrates how these processes may be used for the recovery of valuable bioactive compounds.

5.2 NEW OPPORTUNITIES FOR THE RECOVERY OF BIOACTIVE COMPOUNDS BY MEMBRANE PROCESSING

Significant advances have been made on the understanding of the role of bioactive compounds in reducing the risk of major chronic diseases and the underlying biological mechanisms that account for these effects. Numerous bioactive compounds appear to have beneficial health effects (Kris-Etherton et al. 2002), although the concept of "bioactive" compound is not unanimously accepted and used. For most authors, it should comprise constituents which are "extra-nutritional" and typically naturally occurring in small quantities in plant products and lipid-rich food stocks (Kitts 1994). However, in a broader sense, many compounds which do possess nutritional value (e.g., amino acids, peptides) are also frequently referred to as bioactive (Wang and de Mejia 2005).

The reason for the increasing demand of bioactive compounds stems from the growing consumers concern with their life quality. Additionally, with the increasing population aging over the past decades, the concern with a careful and dedicated diet has augmented noticeably. This explains why the functional foods market, as well as the nutraceutical and cosmeceutical, have grown so rapidly.

The different physicochemical nature of bioactive compounds makes their selective and efficient separation and recovery one of the most important steps in their production. The challenge here is the development of suitable downstream processing techniques, allowing for the recovery of these compounds from complex streams without affecting their structure and function, which ultimately translates

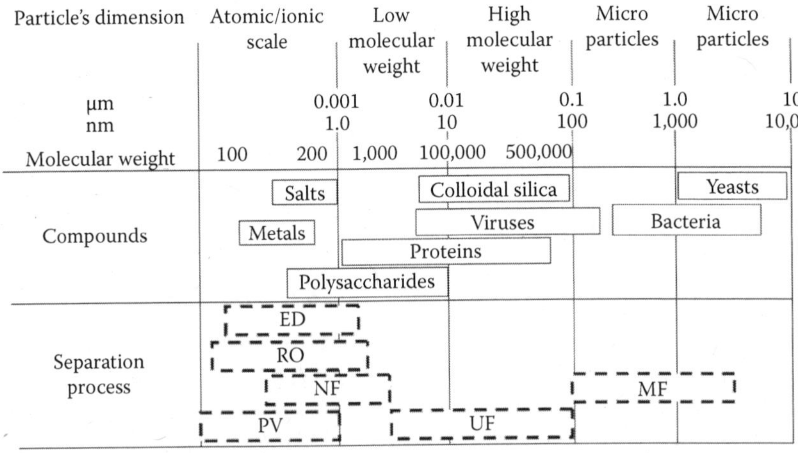

FIGURE 5.1 Diagram illustrating the common range of application of some membrane separation processes depending on the size of target particles and/or compounds.

into their bioactivity. Membrane processes offer potential solutions for this problem due to their intrinsic mild character, without the need of using extracting solvents, operating at ambient temperatures and without using high mechanical stress conditions. Additionally, the large diversity of membrane materials and processes make possible a judicious selection of the best technique and corresponding operating conditions for a given type of target solute in a defined matrix. Figure 5.1 shows how different membrane processes may be selected to deal with specific target compounds.

5.3 FUNDAMENTAL CONCEPTS IN MEMBRANE PROCESSING

Membrane-based unit operations such as reverse osmosis (RO), nanofiltration (NF), ultrafiltration (UF), microfiltration (MF), electrodialysis (ED), and pervaporation (PV), if properly applied, offer the advantage of performing efficient and selective separations of a number of valuable products. In many cases, one membrane operation can be followed by another to obtain a product stream of even higher quality or with predefined composition and properties. Thus, the successful introduction of a membrane-based operation into the production line of a bioactive compound ultimately relies on understanding its principles, application constraints and best combination with other separation techniques in view of the overall bioactive compound production process.

Membrane separations are particularly suitable for the recovery of bioactive compounds from complex media on account of the following aspects:

1. Operation generally occurs at ambient temperatures and under smooth conditions, thus maintaining the desired properties of the target bioactive compound.
2. The use of extracting agents, such as organic solvents, is not required, which avoids secondary contamination and the necessity for additional purification.
3. The separation unit is usually rather suitable for scale-up due to its most common physical arrangement as one or more separate membrane modules.

In these unit operations, the membrane can be viewed as a semipermeable barrier between a feed and a permeate stream. The two bulk phases can be deliberately introduced to the two sides of the membrane (as in ED), but the permeate stream, in most membrane separations (RO, NF, UF, MF, etc.), is generated by transport from the feed phase. This physical phase separation of the two streams often allows for operation with no or minimal chemical pretreatment, which otherwise may form deleterious by-products. Particles and dissolved components may be (partially) retained based on properties such as size, shape, and charge. Therefore, membrane-based operations can be used to separate and purify a wide range of components, ranging from suspended solids (MF) to small organic compounds and ions (RO and NF). However, selecting the most appropriate membrane separation technique is very important, as it determines not only the product quality, but also the treatment costs. Therefore, a good understanding of the mechanisms governing a given membrane operation, its specific advantages and/or constraints for performing a target separation, the effect of relevant operational parameters, and the best strategies for process monitoring and control are essential for a proper process selection and successful implementation.

The thermodynamics of irreversible processes, through which a system of equations for the fluxes of the various components can be set up, has been so far the most widely used theoretical formalism for describing membrane transport phenomena (Katchalsky and Curran 1965, Mulder 1996, Strathmann 2000, Wesselingh and Krishna 2000). The flux equations are conceptually based on the action of a "driving force" that "drives" a component across the membrane barrier. This "driving force" (per mole of component i) is directly proportional to the electrochemical potential gradient ($\nabla\bar{\mu}$), which can be split into contributions from composition, pressure, and electrical potential gradients, as shown in Equation 5.1 for the unidirectional flux of a general component i across a flat membrane perpendicular to the x-axis:

$$-\frac{\mathrm{d}\bar{\mu}_i}{\mathrm{d}x} = -RT\frac{\mathrm{d}}{\mathrm{d}x}(\gamma_i y_i) - \frac{\mathrm{d}}{\mathrm{d}x}(\upsilon_i P) - \Im Z_i \frac{\mathrm{d}E}{\mathrm{d}x} \tag{5.1}$$

where

- γ_i is the activity coefficient of component i
- y_i and υ_i are its mole fraction and molar volume, respectively
- $\Im$ is the Faraday constant
- Z_i is the valence of specie i
- E is the electrical potential

Under the action of the driving force, the membrane causes different components to move in the system with different average velocities, resulting in average (viscous) friction between the components. While in some cases all three terms in Equation 5.1 are relevant, most often one of them dominates. For example, in PV, this is the compositional term of the chemical potential gradient; in ED the electrical potential gradient term, and in UF the pressure gradient term, as schematically illustrated in Figure 5.2. The dominant contribution of one major driving force often allows for using well known simplified equations relating a target component flux with the corresponding driving force through the use of phenomenological proportionality coefficients, as in the case of a mass flux of a component i, J_i^{m}—Fick's law (Equation 5.2); a volume flux, J^{V}—Darcy's law (Equation 5.3); and an electrical charge flux, J^{E}—Ohm's law (Equation 5.4):

$$J_i^{\mathrm{m}} = -D_{\mathrm{m},i}\frac{\mathrm{d}c_i}{\mathrm{d}x} \tag{5.2}$$

$$J^{\mathrm{V}} = -L_{\mathrm{p}}\frac{\mathrm{d}P}{\mathrm{d}x} \tag{5.3}$$

$$J^{\mathrm{E}} = \kappa\frac{\mathrm{d}E}{\mathrm{d}x} \tag{5.4}$$

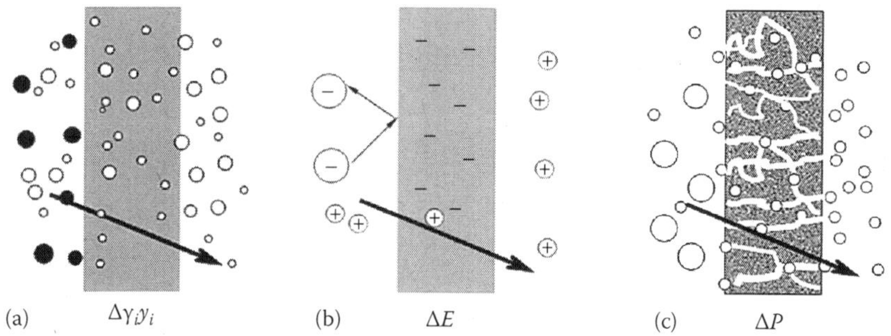

FIGURE 5.2 Schematic representation of the operation principle of some membrane separation processes, with membranes separating a feed (left) from a permeate phase (right). Major driving forces acting upon solutes are indicated by arrows illustrating gradients of activity, $\gamma_i y_i$, electrical potential, E, or pressure, P. In pervaporation, (a) closed circles represent nonvolatile compounds and the open circles volatile compounds. In electrodialysis, (b), only a negatively charged cation-exchange membrane, thus excluding the passage of anions, is presented. In ultrafiltration (c) bigger circles denote compounds with a molecular weight higher than the average molecular cut-off of the membrane.

where

$D_{m,i}$ and c_i are, respectively, the diffusion coefficient and the mass concentration (kg/m^3) of component i

L_p is the permeability coefficient

κ is the electrical conductivity

While the mathematical representation of the transmembrane fluxes is usually straightforward, the most frequently referred practical problem in all membrane separation processes, which still limits their wider application and often frightens newcomers to the field, is the frequently encountered flux decline due to concentration polarization effects and/or fouling by feed solution compounds retained at the membrane surface. Fouling may occur due to concentration polarization or as a consequence of adsorption of feed solution compounds at the membrane surface and, especially for porous membranes, within the membrane structure. Therefore, the control of concentration polarization and membrane fouling is a major task in the design of all membrane separation processes (Mulder 1996, Strathmann 2000, Baker 2004).

In spite of their different nature, the presentation of each membrane separation process can be generally organized around a four-step scheme:

1. Identifying the target compound, its physicochemical characteristics (molecular weight, solute geometry, dipole moment, diffusivity, hydrophilicity/hydrophobicity, dissociation constant, etc.), and the desired degree of its separation (purification)
2. Analyzing the feed matrix regarding its composition in terms of other organic and/or inorganic compounds and physicochemical properties such as pH, ionic strength, etc.

3. Selecting the proper membrane (morphology, surface charge, hydrophilicity/hydrophobicity, surface properties, etc.), which is compatible with the separation goal
4. Selecting and optimizing the relevant operation conditions, aiming at obtaining a desired product with defined specifications.

In the following sections, it is demonstrated how this goal could be effectively accomplished.

5.3.1 Pervaporation

PV is a unit operation in which a dense membrane separates a liquid feed solution from a vapor permeate (Néel 1991). This operation, schematically represented in Figure 5.3, is therefore suitable for separating low molecular weight target bioactive compounds, which are volatile, e.g., aroma compounds. The transport across the membrane is most frequently explained and modeled in the light of the so-called solution–diffusion mechanism (Mulder 1996, Strathmann 2000, Baker 2004), in which solute–membrane interactions play an important role. The driving force for the occurrence of this separation is the gradient in the chemical potential of the compounds between the feed and the receiving (permeate) compartment. A compound first sorbs (partitions) at the liquid or membrane interface (according to the corresponding thermodynamic equilibrium) and then diffuses across the membrane material down its concentration gradient. Either vacuum (Figure 5.3a) or a carrier (sweep) gas stream (Figure 5.3b) desorbs the molecules reaching the permeate side, maintaining the driving force, and thus the flux of the target compound across the membrane. If the degree of sorption (dissolution) of the target compound into the membrane and the velocity of its transport (expressed in terms of diffusivity) through the membrane are different from those of other compound(s) present in the liquid feed, separation will occur.

PV is a mild process operation and therefore particularly indicated for separating mixtures, which cannot withstand the harsh conditions of distillation. Compared to classical evaporation, PV has the additional advantage of being a more selective technique due to the use of a semipermeable membrane barrier between the feed liquid and the (permeate) vapor. Although this selectivity is usually obtained at the price of a lower flux (the membrane is a transport resistance), this can be compensated for by using a larger membrane area. The costs of commercial PV membranes are relatively low, and hollow-fiber membrane configurations, in particular, allow for high packing densities (high membrane area to volume ratio) and, therefore, compact membrane modules.

The membrane material of choice for organophilic PV, if recovery of organic compounds from aqueous streams is aimed, is usually polydimethylsiloxane (PDMS), including chemically modified derivates through introduction of bulky side groups designed to reduce the partial water flux (e.g., polyoctylmethylsiloxane [POMS]). Additionally, other elastomeric materials such as polyether-polyamide block copolymers (PEBA), ethylene-propylene-diene monomer (EPDM) elastomers, and filler-type membranes can be used (Böddeker 1994). In order to decrease the contribution of diffusion and minimize the water flux, most organophilic PV membranes in

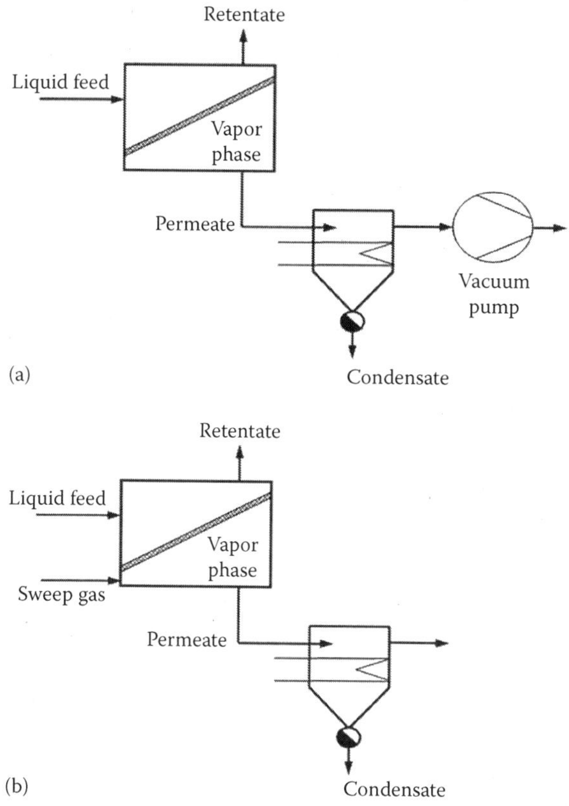

FIGURE 5.3 Schematic drawing of a pervaporation process with an applied downstream vacuum (a) or a sweep gas (b).

practice are composites consisting of a thin and highly selective membrane layer on a macro-porous support for an enhanced mechanical stability.

Concentration polarization can occur on both sides of the membrane, especially in organophilic PV. This phenomenon is schematically illustrated in Figure 5.4. Feed-side concentration polarization may become particularly relevant for a compound with a high sorption affinity toward the membrane, which may lead to its depletion near the membrane interface if the external mass transfer conditions are not sufficiently good to guarantee its fast transport from the bulk feed to the interface. As a consequence of this depletion near the interface, the driving force for transport, and the resulting partial (component) flux, becomes lower. At the permeate side, the laminar boundary layer is at least as thick as the porous supporting substructure, where, sometimes, capillary condensation may also occur (Strathmann 2000).

Better module design and new approaches for improved mass transfer conditions, without significantly increasing the energy expenses, are needed in this case. The use of Dean vortices (Moulin et al. 2001), application of computational fluid dynamics (CFD) (Liu et al. 2004, Ghidossi et al. 2006), and the assessment of full-scale vibrating PV units (Vane and Alvarez 2002) are examples of such works.

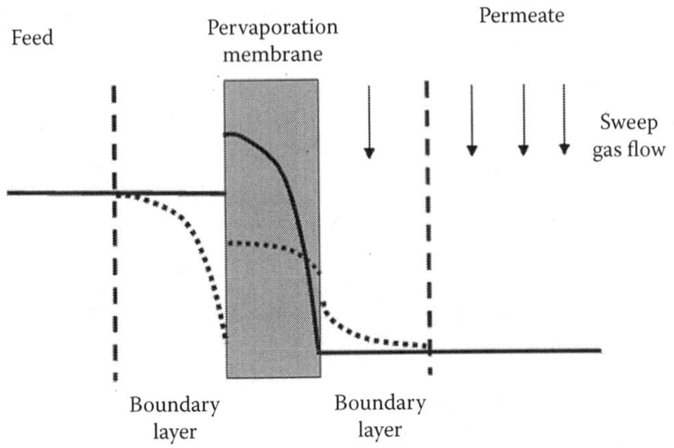

FIGURE 5.4 Concentration gradients (concentration polarization) in the liquid boundary layers adjacent to the membrane surfaces formed as a result of permeation of a compound through a selective pervaporation membrane. The full straight lines in the boundary layer regions indicate an ideal process performance not affected by concentration polarization. Permeate-side concentration polarization may occur at an insufficient sweep gas flow. Hypothetical concentration gradients of a permeating compound through a pervaporation membrane without and with concentration polarization in the adjacent boundary layers are presented by a full and a punctuated dash line, respectively.

5.3.2 Nanofiltration

Pressure-driven membrane operations use pressure difference between the feed (retentate) and a permeate side as the driving force to transport the solvent (usually water) alone and/or the solvent together with a target compound(s) across the membrane (Mulder 1996, Strathmann 2000, Baker 2004). These membrane-based operations include RO, NF, UF, and MF. The operating transmembrane pressure ranges significantly but it is usually 20–100 bar for RO, 5–20 bar for NF, 2–5 bar for UF, and 0.1–2 bar for MF. Another possible classification can be based on the molecular weight of the compound to be separated by a given membrane operation and it is usually up to 100 Da (RO), 100–1,000 Da (NF), and 1,000–100,000 Da (UF) (see Figure 5.1).

Obviously, UF and MF are not suitable for the direct membrane processing of low molecular weight bioactive compounds from a solution. However, they can be implemented in a hybrid process operation, which, for example, would first produce larger aggregates that might be subsequently filtered.

RO is already a well-established technology used for many years in water desalination (Fritzmann et al. 2007). RO membranes are generally asymmetric, i.e., consist of a thin separating polymer (nowadays, mostly polyamide) layer combined with a porous support to guarantee the membrane mechanical stability. RO membranes discriminate on the basis of molecular size and charge, and very high (often close to 100%) retention of low molecular weight compounds and ions (total desalination) can be achieved due to the dense properties of the separating layer. Moreover, RO units can be easily automated and controlled.

In contrast to RO membranes, NF membranes can provide more selective separation, which is not exclusively based on the differences in the molecular size of the compounds to be separated (sieving effect), e.g., for the partial separation of monovalent from multivalent ions. NF membranes are sometimes designated as "loose" RO membranes (Ho and Sirkar 1992), since they provide higher solvent (most commonly water) fluxes at lower transmembrane pressures. However, this more open membrane polymeric structure makes not only the size but also the molecular shape (geometry and orientation toward the membrane) of the permeating compound much more important parameters in NF separations (Santos et al. 2006).

NF membranes are usually asymmetric and negatively charged at neutral and alkaline pH. Therefore, separation of negatively charged compounds is based not only on different rates of their diffusion through the membrane, but also on the electrostatic repulsion (Donnan exclusion) between anions in solution and negatively charged membrane surface groups, which is obviously stronger for multivalent anions (Levenstein et al. 1996). Therefore, NF is much more sensitive than RO to the ionic strength and pH of the feed solution. Recently, a more general theoretical examination of the effects of transmembrane pressure and Donnan exclusion on flux in NF has been reported (Gilron et al. 2006). The advantage of introducing this additional mechanism of charged compounds exclusion (in addition to the size-based exclusion) is that higher separation degrees (rejections), similar to those in RO, can be achieved but at higher solvent fluxes through the membrane.

The membrane surface charge is mainly due to the adsorption of ions from the feed solution rather than due to the presence of fixed charged groups (as for an ion exchange membrane). Therefore, it depends strongly on the total electrolyte concentration in the treated solution (Hagmeyer and Gimbel 1998). Furthermore, it changes from negative to zero net charge at the membrane isoelectric point, and then to positive at lower pH values (usually <4) due to adsorption of cations from the solution. This pH dependence can strongly affect the separation of a target bioactive compound. Therefore, the selection of adequate operating conditions is critical for NF applications.

NF membranes are also rather sensitive to concentration polarization phenomena near the membrane surface contacting the feed (retentate) side, which is illustrated in Figure 5.5. Due to the pressure-driven nature of the NF separation process, accumulation of compounds (contrary to their depletion as in PV) is encountered within the liquid laminar boundary layer on the retentate side. Therefore, the solubility products of some salts, if present in the retentate, can be exceeded, thus forming precipitates (mineral fouling), which adds to possible organic and biological fouling due to organic matter and/or the microorganisms present. The presence of inorganic components, such as hardness (Ca^{2+}, Mg^{2+}) and SO_4^{2-} anions, can also interfere with the separation of the target compound due to problems that these components may cause with the change in the osmotic pressure of the solution (Ritchie and Bhattacharyya 2002). A gel or cake layer at the surface of a NF membrane not only reduces the transmembrane flux, but, even more importantly, it may severely alter the separation properties of the membrane, since such a layer may act as an additional semipermeable membrane, which retains low molecular weight compounds that are not supposed to be rejected by the membrane.

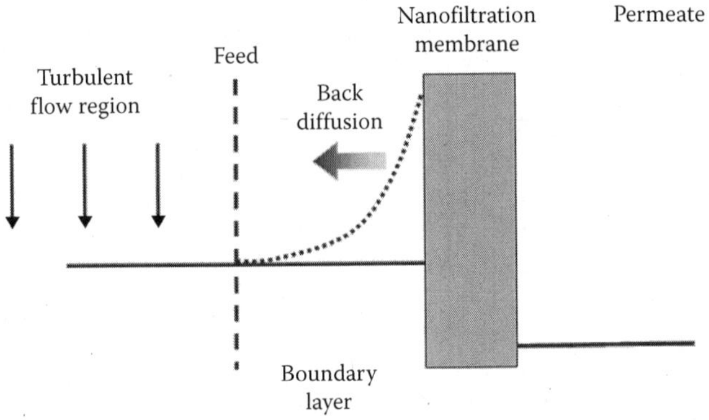

FIGURE 5.5 Concentration polarization in nanofiltration illustrated by a punctuated dash line in the boundary layer region. The full straight line indicates an ideal process performance not affected by concentration polarization. Accumulation of a compound near the membrane surface causes its back diffusion into the bulk feed phase.

Strathmann (2000) defines four main strategies that can be applied to avoid or control fouling:

1. Pretreatment of the feed solution
2. Hydrodynamic optimization of the membrane module
3. Membrane surface modifications
4. Membrane cleaning with proper chemical agents

Pretreatment of the feed solution may involve chemical precipitation, prefiltration (other membrane operations like MF and/or UF are often used) before entering the NF module(s), pH-adjustment, chlorination, etc.

A proper module design, an increase in the shear stress at the membrane surface and/or the use of appropriate spacers are often effective ways for minimizing membrane fouling problems. Modeling strategies, especially those using CFD, can be rather useful for identifying the best spacer geometry and arrangement (Santos et al. 2007a). If necessary, membrane surface modifications can also be performed. Such modifications usually introduce hydrophilic moieties or charged groups on the membrane surface. A recent, and rather interesting area of research, deals with the development of membranes that may reverse the hydrophobic/hydrophilic character of their surface when exposed to an external stimulus. Such stimulus-responsive membranes may, for example, change their character reversibly when exposed to temperature shifts, electromagnetic fields or alternate UV/visible light radiation conditions. This approach opens new possibilities for reversing fouling without the use of any external mass agent (Lim et al. 2006, Nayak et al. 2006).

Despite the above mentioned challenges, NF is rapidly becoming more and more an attractive alternative for the separation (purification) of bioactive compounds in agro-industries. To a great extent, this is due to the introduction of highly efficient NF membranes and module configurations, which allow for lower capital investment and operating or maintenance costs.

5.3.3 Electrodialysis

Dialysis in general can be defined as an operation that separates compounds due to their different diffusional rates across a semipermeable membrane (Mulder 1996, Strathmann 2000, Baker 2004). Therefore, in practice, dialysis can be used for the separation of compounds which differ significantly in size to guarantee a sufficiently large difference in their diffusional rates; the classical example is hemodialysis (artificial kidney) for purifying human blood by removing small solutes such as urea, phosphates, chloride, etc., while retaining large proteins and other components in the blood. Therefore, the ability of dialysis to discriminate between bioactive compounds of similar molecular weight is limited. Moreover, as the mass transport rate of a given compound depends on its concentration gradient across the membrane (as in PV), dialysis is characterized by considerably lower maximum possible fluxes in comparison to pressure-driven membrane operations, such as NF. Therefore, despite its relevance in small-scale pharmaceutical applications, dialysis is rarely a technique of choice for large-scale separations.

In ED, the transport of charged organic compounds and/or ions present in the feed is accelerated due to an electric potential difference applied externally by means of electrodes (anode and cathode), between which polymeric cation-exchange membranes bearing fixed negative electric charges and anion-exchange membranes bearing fixed positive charges (Xu 2005) are placed in an alternating series. The ED operational principle is schematically presented in Figure 5.6. When an electric potential difference between the electrodes is established, the cations migrate

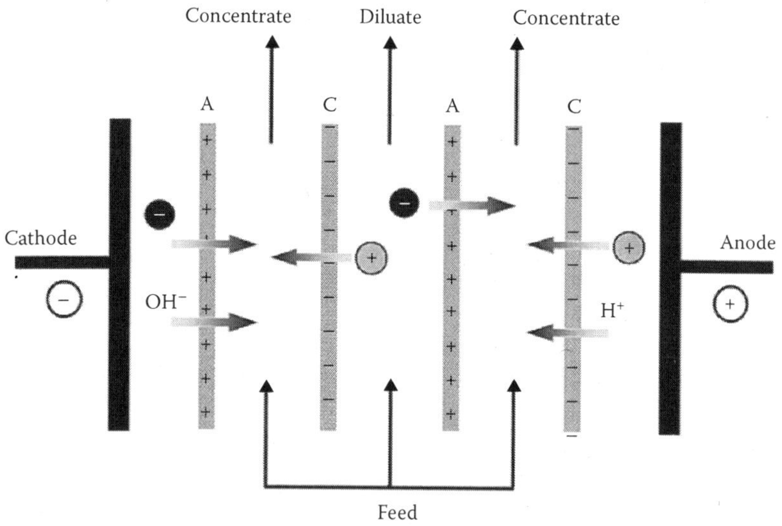

FIGURE 5.6 Schematic representation of an electrodialysis process. Alternating anion (A) and cation (C) exchange membranes are arranged in a plate-and-frame module. Periodical change of electrode polarity (electrodialysis reversal), thus temporarily transforming the diluate stream into a concentrate stream and vice versa, is often applied for reducing concentration polarization and/or membrane scaling problems.

toward the cathode while the anions migrate towards the anode. Cations permeate a cation-exchange membrane but are retained by an anion-exchange membrane, and *vice versa* (anions permeate an anion-exchange membrane but are retained by a cation-exchange membrane). Thus, ions are accumulated in alternating cells, forming a concentrate solution, while the other cells are depleted of ions, thereby forming the so-called diluate solution (Mulder 1996, Strathmann 2000, Baker 2004).

Traditionally, it has been widely accepted (Helfferich 1962) that homogenous (prepared from coherent gels) ion exchange membranes are superior in terms of their conductivity in comparison with heterogeneous ones (consisting of ion exchange particles, embedded in an inert binder like polystyrene, polyethylene, etc.). Nevertheless, recent attempts based on creating advanced polymeric matrixes containing highly ordered ion-exchange particles have demonstrated interesting results, which may allow for the development of novel heterogeneous ion exchange membranes with highly improved performance (Oren et al. 2004).

Bipolar ion exchange membranes, composed of a cation and anion exchange layer and an aqueous interphase between them, have also gained considerable attention as a method for producing acids and bases from their corresponding salts by electrically enforced water dissociation (Nagasubramanian et al. 1977, Mulder 1996, Strathmann 2000, Xu 2005). When an electrical potential difference is established between the electrodes, all charged components are removed from the aqueous interphase between the two ion-exchange layers. If only water is left in the solution between the layers, transport of electrical charge can only be accomplished by protons and hydroxyl ions, which are available in very low concentrations of approximately 10^{-7} mol/L in completely deionized water. However, protons and hydroxyl ions removed from the interphase between the cation- and anion-exchange membranes are continuously replenished in this process because of the water dissociation equilibrium.

In ED, concentration polarization has two main consequences. First, it leads to a decrease in the ion concentration at the membrane surface facing the diluate compartments. Second, it increases the ion concentration at the membrane surface facing the concentrate compartments of the electrodialysis stack. The latter increase may result in salts precipitation. On the other hand, low ionic concentrations in the diluate compartments are undesirable, since they decrease the limiting current density. Concentration polarization effects can be minimized by optimizing the cell design and using adequate liquid flow velocities, which reduces the thickness of the laminar boundary layers at the membrane surfaces.

Process performance deterioration due to membrane scaling is a frequently observed problem. As a result, ED systems are often operated in the so-called electrodialysis reversal mode, in which the polarity of the electrodes is reversed (usually several times per hour) to change the direction of ion movement (Mulder 1996, Strathmann 2000, Baker 2004).

The external electrical potential driving force allows for obtaining much higher fluxes than those in simple dialysis, but a different degree of demineralization (inorganic ions are also removed from the solution) is also obtained depending on the voltage and the type of membranes used. When the purpose is the purification of a bioactive compound from accompanying salts, reduction in water hardness can be a desired effect, but if the ions must be retained, ED may cause too deep a "softening" (as in a RO process). Therefore, the suitability of ED depends strongly on the desired ionic composition of the product stream containing the target bioactive compound(s).

Due to the presence of fixed positive charges, anion-exchange membranes are generally considered more sensitive to fouling since most naturally occurring organic compounds usually bear a net negative charge. Moreover, the permeability of commercial anion-exchange membranes is usually limited to low molecular weight (below 350 Da) compounds (Strathmann 2000). Therefore, fouling of anion-exchange membranes often occurs when an anion is still small enough to penetrate into the membrane matrix, but its mobility is so low that the membrane becomes blocked. Another possible cause for fouling is adsorption of humic acids, polysaccharides, proteins, and surfactants from the feed solution. To overcome this problem, antifouling anion-exchange membranes have been developed. For example, a membrane coated with a thin layer of cation-exchange groups, thus causing electrostatic repulsion of negatively charged compounds with higher molecular weight, has been manufactured under the name Neosepta® (Astom Corporation, Tokyo, Japan) by Astom, now a part of the Tokuyama Corporation. It should be noted, however, that fouling of a cation-exchange membrane under alkaline conditions by formation of calcium carbonate and calcium hydroxide has also been documented, using both optical and electron scanning microscopy images (Ayala-Bribiesca et al. 2006).

Reduction of the price of currently available ion-exchange membranes, especially those having monovalent ion perm-selectivity, which are more expensive compared to the membranes used in pressure-driven processes, would contribute for the competitiveness of ED processes.

In summary, ED can provide an efficient separation of low molecular bioactive compounds from agro-industrial process streams. For monovalent compounds, the use of monovalent perm-selective (anion and/or cation exchange) membranes seems to be especially attractive. Situations, in which ED appears to be less applicable, include the treatment of feed solutions of low salinity (conductivity of less than 0.5 mS/cm), due to correspondingly low possible limiting current densities. For such situations, a pressure-driven membrane technique (e.g., NF) might be preferable. Finally, when the bioactive compounds are either noncharged or have high molecular weight, an ED treatment can still be applied for removing undesired ionic compounds (usually inorganic ions) from the feed solution, thus contributing to purify the product stream.

5.4 SELECTED CASE STUDIES

Since it is rather difficult to identify which membrane applications might be referred to as specifically selective and indicated for bioactive compounds (other medium constituents are often also retained or transported across the membrane), the perspective presented here might be subjective. Moreover, the discussion neither pretends to be exhaustive nor does it cover the whole range of different possibilities for using membrane processes in the recovery of bioactive compounds. However, it does show by selected examples the main directions of possible successful membrane applications in this field.

In most cases, only recent examples are referred to since the works selected usually familiarize the interested reader with the general history of a specific problem. Where it is appropriate, the discussion first starts with a given membrane separation used as a single operation, and then moves on to its integration in processing schemes. An attempt is made to identify the advantages, limitations and future research needs in each case. Some of the applications discussed could still be considered far from practical implementation, but nevertheless, they might provide useful information on future developments.

5.4.1 Pervaporation

PV is a unique technique for the selective recovery of volatile compounds from complex streams, providing extremely high enrichment and separation factors for target compounds, even in single step operations. This selectivity toward given solutes results from molecular interactions that these solutes may establish with the membrane material which, in fact, operates as a solid extractant in which sorption and desorption occur simultaneously at the feed or membrane and membrane or permeate interfaces. As regards the removal or recovery of volatile compounds from feed streams, the final product may be the processed feed stream (now free from the removed volatile compounds), or the recovered compounds, or even both. In all cases, the final product must exhibit the high quality required for food and nutraceutical applications, being totally free from solvent trace residues which may occur when traditional extraction is used.

The second major advantage of using organophilic PV for the removal or recovery of volatile compounds from complex agro-industrial streams results from the fact that the energy involved in the membrane-based operation is relatively low, when compared to other separation techniques (evaporation, distillation). In organophilic PV, the energy expenditure is only associated with the permeating species, which minimizes energy expenses with the bulk solvent. As the feed stream is commonly processed at room temperature, if necessary, the character and bioactivity of the compounds present in the feed stream and in the recovered permeate are kept.

These two features—no trace contamination with solvents and operation at room temperature—allow for the use of the label "natural" in the products obtained, which represents a major advantage, and added market value, for many products in the food, nutraceutical and cosmeceutical segments.

Organophilic PV for processing of agro-industrial streams has been mainly investigated with the following aims:

1. Removal of volatile compounds from feed streams, either because they are undesirable (e.g., off-flavors) or because the removed compounds will be added at a later stage of the overall process.
2. Recovery of valuable volatile products, for use as natural complex "extracts," or for fractionation in order to obtain selected compounds or fractions.

The first situation has been barely discussed in the literature, and, interestingly, the most frequent references are related to the removal of off-flavors for analytical purposes. A study published recently (Gomez-Ariza et al. 2006) is a good example of the use of organophilic PV for determination of off-flavors in wine. Reports on the removal of aroma compounds at an early process stage to be later reintroduced into the same matrix, or in other process streams, are also scarce. One of the few examples is a work (Crespo 2005) developed at Compal, a Portuguese fruit juice producing company, in which an integrated approach involving the removal of valuable aroma compounds from original fruit juice matrices prior to their pasteurization, followed by the later addition of the aroma concentrate to the corresponding fruit juice, was used. This procedure allows for minimizing aroma deterioration due to the pasteurization step and also the potential formation of off-flavors during the heating period. The increasing interest in the production of partially dealcoholized beverages has led to recent research projects aiming at removing ethanol with a minimal loss of intrinsic flavor compounds. With this objective, RO has been integrated with organophilic PV, thus allowing for the recovery and reintegration of relevant aromas into the final product.

In contrast, the recovery of valuable volatile products by PV has been extensively discussed in the literature since the pioneer works of Boddeker et al. (1990), Bengtsson et al. (1992), and Beaumelle et al. (1992). A significant amount of work has been devoted to understand the fundamental problems of the recovery of volatile compounds (in most cases aromas) from aqueous and hydroalcoholic media using model solutions (Börjesson et al. 1996, Tan et al. 2005, García et al. 2008), although increasing attention has been given to the recovery of aroma compounds from complex natural streams, such as wine-must (Schäfer et al. 1999, Schäfer 2002, Schäfer and Crespo 2007), fruit juices (Pereira et al. 2006), agro-industrial effluents (Souchon et al. 2002), and other natural matrixes (Cuperus 1998).

One of the most relevant problems of aroma recovery from dilute aqueous and hydroalcoholic media results from the difficulty to alleviate external mass transfer limitations, particularly when aromas with a high affinity towards the membrane are involved, such as ethyl hexanoate, isoamyl acetate and hexyl acetate toward PDMS membranes (see Figure 5.7). In this case, the selective potential of the membrane can be only fully explored if extremely effective fluid dynamics conditions are used in the feed compartment (situation 1 in Figure 5.7). Otherwise, under poor fluid dynamics conditions, the separation becomes controlled by transport through the concentration boundary layer (situation 2 in Figure 5.7), and the membrane intrinsic selectivity is lost (Schäfer 2002).

Additionally, the potential of using organophilic PV for the recovery of aroma compounds from biocatalytic processes (enzymatic and microbial) has also been explored, either as a postreaction process or as an *in situ*, integrated operation for recovery of valuable aromas from on-going bioconversions (Ehrenstein et al. 2005, Etschmann et al. 2005). This later approach offers the possibility for recovering the aroma compounds of interest and simultaneously relieves their potential inhibitory effect over the biocatalytic system.

Considering that the global flavoring market was valued at $4.80 billion in 2005 and is expected to reach $6.22 billion in 2012 (Foodnavigator 2006), and bearing in mind the growing consumers demand for "natural" products, PV will certainly play an important role in replacing traditional evaporative techniques as well as

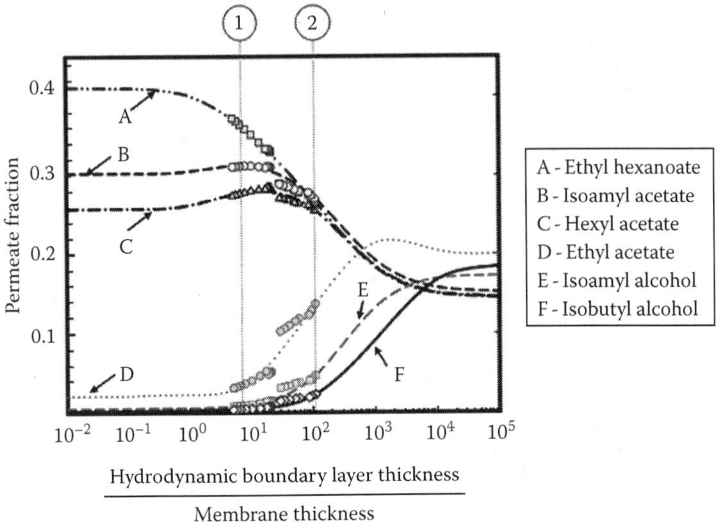

FIGURE 5.7 Effect of fluid dynamic conditions on the overall organophilic pervaporation selectivity for processing of different aroma compounds in a hydroalcoholic solution using a PDMS membrane is shown. In situation 1, the overall process is controlled by the membrane selectivity; in situation 2, the overall process becomes controlled by the liquid boundary layer at the membrane interface, due to poor fluid dynamics. As explained, the compounds with a high affinity toward the membrane are the ones strongly affected by this effect (note the evolvement of their partial contribution to the overall permeate composition). (Adapted from Schäfer, T., Recovery of wine-must aroma by pervaporation, PhD dissertation, Universidade Nova de Lisboa, Lisbon, Portugal, 2002.)

extraction-based processes in the near future. Still, there are relevant technical limitations that remain a challenge for scientists and engineers:

1. Membranes with low water permeability and a high selectivity for specific target solutes are necessary (the diversity of membranes available for organophilic PV is rather scarce).
2. Module design should be improved in order to optimize external mass transfer conditions in the liquid feed stream with high energy-efficiency, minimizing limitations due to concentration polarization of the target compounds with high affinity to the selective membrane.
3. Efficient approaches for capturing the permeating target compounds, using less energy-intensive solutions such as integrated encapsulation techniques, should be developed as an alternative to current condensation strategies.

5.4.2 Nanofiltration

NF is a membrane-based unit operation particularly adequate for the retention and/or fractionation of small organic compounds in the range of 100–1000 Da. As previously mentioned, it may also be used for a near complete retention of multivalent ions (rejections typically above 99%, depending on the operating conditions used) and for a partial retention of monovalent ions (rejections usually in the order of 70%). Due to these features, NF has rapidly gained acceptance in several industrial sectors, where it is not only used for the most basic operations, such as water softening for heating and steam generation, but also in rather complex situations, such as catalyst recovery and reuse, and recovery and fractionation of bioactive compounds from complex streams.

Particularly in the agro-industry field, NF has been used for partial demineralization of streams with a high salt content (monovalent ions), assuring high solvent fluxes and an effective retention of valuable low molecular weight organic compounds. This operation is rather interesting and has deserved a close attention in milk whey (Suárez et al. 2006, Cuartas-Uribe et al. 2007) and beet sugar (Gyura et al. 2002, Hinkova et al. 2002, Gyura et al. 2005) demineralization processes.

Considering its rejection characteristics, NF has been also proposed for the recovery and fractionation of bioactive molecules with molecular weight below 1 kDa from complex media. This approach has been followed for the recovery of biologically active oligosaccharides from milk (Sarney et al. 2000, Russo 2007), using a combination of enzymatic treatment of defatted milk with beta-galactosidase and NF. This approach was also shown to be suitable for the recovery of oligosaccharides from caprine milk (Martinez-Ferez et al. 2006). A similar concept has been used for the recovery of bioactive peptides from different sources, including the ones derived from proteins present in bovine colostrums, milk and cheese whey (Lapointe et al. 2003, Groleau et al. 2004, Ting et al. 2007). The most powerful strategies result from the combination of UF and NF steps (Butylina et al. 2006), enabling the production of diverse fractions with specific compositions and corresponding biological activity. NF offers an additional potential for recovering peptides and fractionating them according to their molecular weight, geometry, and charge (or dipole moment). It is important to bear in mind that size exclusion is not the sole mechanism to be explored in order to obtain the target selectivity. By adjusting the physicochemical properties of the media to be processed and selecting a membrane with adequate surface chemistry, membrane-solute Coulombic forces and Van de Waals interactions may be explored favorably.

One of the most promising areas for the use of NF in the agro-industry is the recovery of valuable bioactive compounds from process streams or residues, usually regarded as low-value products and, sometimes, even as an environmental burden (Nwuha 2000). One recent example is the production of natural extracts from olive subproducts, which are rich in a number of bioactive compounds such as hydroxytyrosol (the most potent natural antioxidant compound identified so far) (Nunes da Ponte et al. 2006). An extremely interesting target could also be the production of natural extracts from grape pomace residues, which are rich in a large diversity of well-recognized bioactive compounds such as resveratrol (Mestel 2008); the recovery of phytosterols from residues of vegetable oil processing industry; and the recovery of diluted organic compounds from waste streams of pulp and paper industry, in line with the new bio-refinery concepts. This list of emerging applications is not exhaustive and a large number of potential applications can be foreseen.

In all cases, the common feature derives from a very simple observation: in most plants the outer skin acts as a first protective barrier against environmental and external aggressions and, therefore, the skin is usually extremely rich in compounds that assure this protective role, such as molecules with an antioxidant activity. Additionally, the plant seeds are extremely rich in compounds that have to assure the initial development of a new organism. As a consequence, process streams and residues containing skin and seed compounds can be extremely rich in valuable bioactive molecules, representing a great source if the right technology is used to recover them. Nowadays, an additional challenge is to develop processes that not only isolate and/or purify but also meet the necessary requirements for an eco-efficient recovery of such compounds.

At this stage, two different approaches may be followed: the recovery of pure compounds or, alternatively, the recovery of a complex profile of compounds, recognized as "natural extracts." The first approach is appropriate if the final aim is the isolation of a pure compound, namely for pharmaceutical applications. In this case, further purification steps using chromatographic techniques may be needed. The second approach involves the production of "natural extracts," which may exhibit extremely high biological activities due to synergetic effects resulting from their complex composition. In this case, the extracts obtained by NF may not have to go through additional purification steps, if NF assures the desired quality and biological activity. Extracts are used as additives in the food industry and in new rapidly evolving market segments such as nutraceuticals and cosmeceuticals. In all cases, it is important that the NF step contributes for the exclusion of undesirable compounds, namely pesticides and heavy metals (which may accumulate in several plants).

In a significant number of applications, the target compound(s) are recovered in the NF permeate stream, while large compounds with undesirable bioactivity, pesticides, and heavy metals are retained by the membrane. As an example, a scheme for the recovery of hydroxytyrosol in a NF permeate followed by a RO concentration step is presented in Figure 5.8. In other situations, the compound(s) of interest is (are) retained in a retentate stream. This happens, for instance, in the recovery of the steryl esters formed by enzymatic esterification of phytosterols, which are present in deodistillate residues produced during vegetable oil processing. In this case, acyl glycerides, unreacted free fatty acids and sterols, as well as contaminating pesticides, permeate through the membrane. For this particular application, solvent-resistant NF membranes have to be used, which requires a better insight into the factors controlling the solvent flux in this newer and rapidly developing class of membranes (Santos et al. 2007b).

5.4.3 Electrodialysis

A comprehensive review referring to the historical development of ED and covering a number of its possible applications in dairy, sugar, wine, and other food industry branches has been published recently (Fidaleo and Moresi 2006). Different examples of using ED for the recovery of charged bio-products and/or removal of the frequently encountered product inhibition in acetic, lactic, and other organic and/or amino acids producing fermentation processes in microbial biotechnology have

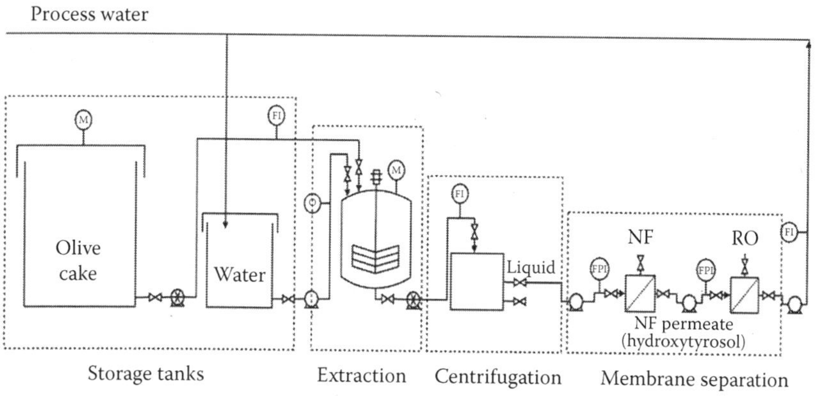

FIGURE 5.8 Process diagram for the production of hydroxytyrosol-rich natural extracts from olive oil solid residues after an aqueous extraction step. Hydroxytyrosol and other low molecular weight compounds are first recovered in the permeate stream of the NF process, after which a RO step is applied for concentrating hydroxytyrosol in the RO retentate stream. The permeate obtained in the RO step is recycled as process water, thus reducing the overall water demand of the process line.

been reviewed previously (Velizarov 1999). Concerning more recent research in the area of these classical ED applications, Mier et al. (2008) demonstrated the possibility of producing high-purity casein (95%) by its separation from milk using ED with bipolar ion exchange membranes, while Vera et al. (2007a,b, 2009) investigated the acidity reduction of fruit juices in different ED configurations, including mono- and bipolar ion exchange membranes, at both laboratory and preindustrial scales. An ED treatment for desalting of a scallop broth (scallop is one of the most popular seafood), which contains taste-active and other valuable components, such as taurine, glycine, and arginine, has been investigated by Atungulu et al. (2007). The high NaCl levels and the presence of some heavy metals still limit the utilization of this broth in the food processing industry, if compared to the soybean sauce. A remarkable decrease of more than 96% in the original NaCl content was found, which confirms the good potential of ED for performing desalting tasks. However, further work is still needed on the part of heavy metals due to their tendency to form complexes, e.g., with some proteins present, which leads to their lower migration rates through the ion exchange membranes.

Turning to studies evaluating the applicability of ED for the recovery of selected bioactive compounds (from model and/or real media), several selected examples will be briefly discussed.

Bipolar membrane ED was studied for producing a free L-ascorbic acid (Vitamin C) from its sodium L-ascorbate salt (Yu et al. 2002) in order to replace the conventional acidification method, which is based on the use of a strong inorganic acid (H_2SO_4) in a medium containing a toxic organic solvent (methanol), from which the insoluble Na_2SO_4 formed is then removed. A feed aqueous solution of sodium L-ascorbate was circulated in the space between a cation exchange membrane and the cation exchange layer of a bipolar membrane. Due to the applied electrical potential, Na^+

ions were transferred from this solution across the cation exchange membrane to the cathode, while the H^+ ions generated from the bipolar membrane (see Section 5.3.3) took their place, thus reducing the solution pH and forming free L-ascorbic acid. The authors consider this process economically feasible since, according to their industrial experience, the bipolar membrane ED process would be economical if its energy consumption were below 1 kWh/kg of L-ascorbic acid produced, while the actual electrical energy consumption was calculated to be 0.7 kWh/kg of L-ascorbic acid in the experiment performed (Yu et al. 2002).

ED has been also investigated as a possible way for the selective recovery of negatively charged polyphenols (precursors for bioactive phenolic compounds) from an aqueous tobacco extract used as a feed (diluate) solution (Bazinet et al. 2005). An aqueous NaCl solution was used as the concentrate solution. The migration degrees of chlorogenic acid, rutin, and scopoletin to the concentrate solution were determined in parallel with the total demineralization rates. Even though high total demineralization degrees (up to 77%) were observed as expected, the maximum polyphenol migration degrees were much lower: 28.7% for chlorogenic acid, 18.8% for scopoletin, and 10.3% for rutin. However, since the operating conditions in this study were not optimized, the ED treatment might be considered potentially interesting for the recovery of these and, most probably, other bioactive compounds from tobacco and/or other aqueous plant extracts.

In another study from the same group (Labbé et al. 2005), the potential of ED for recovering green tea catechins and caffeine from an aqueous extract, used as the feed (diluate) solution, was investigated. However, no statistically significant migration of the target compounds through the membranes was documented. This behavior could most probably be attributed to the rather high molecular weights of the target compounds (e.g., epigallocatechin gallate, which is the primary catechin in green tea, has a molecular weight of 458.4 Da). As already mentioned (Section 3.3), ED has limited potential for separating charged molecules with a molecular weight higher than 350 Da. For caffeine (molecular weight of 194.19 Da), the reason of its retention in the feed solution is less clear. It might probably be associated with the fact that the green tea solution pH significantly decreased during this ED treatment (from 5.5 to 2.5), which might have favored neutralization or formation of positively charged catechin species that were repealed by the anion exchange membranes.

The possibility of integrating ED with other membrane separation techniques has also been investigated. For example, Poulin et al. (2006) tested a hybrid membrane technique for separating bioactive peptides from a β-lactoglobulin hydrolysate using a UF membrane stacked in an ED cell, during a batch recirculation process step, to separate acidic (anionic) from basic (cationic) peptides and noncharged peptides. As it might be expected, pH was again a very important parameter controlling the direction of peptide migration but, surprisingly, not for all of the peptides studied.

5.5 FUTURE TRENDS AND PERSPECTIVES

Membrane separations are regarded as particularly suitable for the recovery of bioactive compounds because they can operate under mild conditions of temperature, pressure and stress, without involving the use of any mass agents such as solvents, avoiding product contamination and preserving the biological activity of the compounds recovered. The large variety of membrane materials available, as well as the diversity of membrane processes developed, underlines one of the strengths of membrane separations: the possibility of designing and fine-tuning the membrane and the membrane process for a specific task.

The use of membranes for the recovery of small bioactive molecules is expected to grow significantly within the next few years, in particular in what refers to the recovery of high added value compounds that have impact on human health. Some of the recovery processes will demand membranes that are more stable in hydro-alcoholic media, or even in organic media compatible with the final product and its use, exhibiting simultaneously a sharp cut-off behavior. The discovery of small molecules with desirable bioactive properties, present in natural matrixes such as plants and marine products, will boost the need for recovery processes regarded as clean and sustainable and allow for the use of the label "natural" in the final product. Membrane techniques fit this demand perfectly due to the mild conditions under which they operate.

REFERENCES

Atungulu, G., Koide, S., Sasaki, S., and Cao, W. 2007. Ion-exchange membrane mediated electrodialysis of scallop broth: Ion, free amino acid and heavy metal profiles. *J. Food Eng.* 78:1285–1290.

Ayala-Bribiesca, E., Araya-Farias, M., Pourcelly, G., and Bazinet, L. 2006. Effect of concentrate solution pH and mineral composition of a whey protein diluate solution on membrane fouling formation during conventional electrodialysis. *J. Membr. Sci.* 280:790–801.

Baker, R.W. 2004. *Membrane Technology and Applications*, 2nd Edition. Chichester, U.K.: John Wiley & Sons Ltd.

Bazinet, L., DeGrandpré, Y., and Porter, A. 2005. Electromigration of tobacco polyphenols. *Sep. Purif. Technol.* 41:101–107.

Beaumelle, D., Marin, M., and Gibert, H. 1992. Pervaporation of aroma compounds in water-ethanol mixtures: Experimental analysis of mass transfer. *J. Food Eng.* 16:293–307.

Bengtsson, E., Trägardh, G., and Hallström, B. 1992. Concentration of apple juice aroma from evaporator condensate using pervaporation. *LWT—Food Sci. Technol.* 25:29–34.

Böddeker, K.W. 1994. Recovery of volatile bioproducts by pervaporation. In *Membrane Processes in Separation and Purification*, Eds. J.C. Crespo and K.W. Böddeker, pp. 195–205, Dordrecht, the Netherlands: Kluwer Academic Publishers.

Boddeker, K.W., Bengtson, G., and Bode, E. 1990. Pervaporation of low volatility aromatics from water. *J. Membr. Sci.* 53:143–158.

Börjesson, J., Karlsson, H.O.E., and Tragardh, G. 1996. Pervaporation of a model apple juice aroma solution: Comparison of membrane performance. *J. Membr. Sci.* 119:229–239.

Butylina, S., Luque, S., and Nyström, M. 2006. Fractionation of whey-derived peptides using a combination of ultrafiltration and nanofiltration. *J. Membr. Sci.* 280:418–426.

Crespo, J. 2005. Compal, Portugal; Internal report.

Cuartas-Uribe, B., Alcaina-Miranda, M.I., Soriano-Costa, E., and Bes-Piá, A. 2007. Comparison of the behavior of two nanofiltration membranes for sweet whey demineralization. *J. Dairy Sci.* 90:1094–1101.

Cuperus, F.P. 1998. Membrane processes in agro-food state-of-the-art and new opportunities. *Sep. Purif. Technol.* 14:233–239.

Ehrenstein, U., Hennig, T., Kabasci, S., Kreis, P., and Górak, A. 2005. Pervaporation supported synthesis of natural aromatics through lipase catalyzed transesterification. *Chem.-Ing. Tech.* 77:1551–1556.

Etschmann, M.M.W., Sell, D., and Schrader, J. 2005. Production of 2-phenylethanol and 2-phenylethylacetate from L-phenylalanine by coupling whole-cell biocatalysis with organophilic pervaporation. *Biotechnol. Bioeng*. 92:624–634.

Fidaleo, M. and Moresi, M. 2006. Electrodialysis applications in the food industry. *Adv. Food Nutr. Res*. 51:265–360.

Foodnavigator. 2006. Health trends to drive flavour innovation in 2006. http://www.foodnavigator.com/Financial-Industry/Health-trends-to-drive-flavour-innovation-in-2006 (accessed November 18, 2008).

Fritzmann, C., Löwenberg, J., Wintgens, T., and Melin, T. 2007. State-of-the-art of reverse osmosis desalination. *Desalination* 216:1–76.

García, V., Diban, N., Gorri, D., Keiski, R., Urtiaga, A., and Ortiz, I. 2008. Separation and concentration of bilberry impact aroma compound from dilute model solution by pervaporation. *J. Chem. Technol. Biotechnol.* 83:973–982.

Ghidossi, R., Veyret, D., and Moulin, P. 2006. Computational fluid dynamics applied to membranes: State of the art and opportunities. *Chem. Eng. Process*. 45:437–454.

Gilron, J., Daltrophe, N., and Kedem, O. 2006. Trans-membrane pressure in nanofiltration. *J. Membr. Sci*. 286:69–76.

Gomez-Ariza, J.L., Garcia-Barrera, T., Lorenzo, F., and Beltran, R. 2006. Use of multiple headspace solid-phase microextraction and pervaporation for the determination of off-flavours in wine. *J. Chromatogr. A* 1112:133–140.

Groleau, P.E., Lapointe, J.-F, Gauthier, S.F., and Pouliot, Y. 2004. Effect of aggregating peptides on the fractionation of β-LG tryptic hydrolysate by nanofiltration membrane. *J. Membr. Sci*. 234:121–129.

Gyura, J., Seres, Z., and Eszterle, M. 2005. Influence of operating parameters on separation of green syrup colored matter from sugar beet by ultra- and nanofiltration. *J. Food Eng.* 66:89–96.

Gyura, J., Seres, Z., Vatai, G., and Molnár, E.B. 2002. Separation of non-sucrose compounds from the syrup of sugar-beet processing by ultra- and nanofiltration using polymer membranes. *Desalination* 148:49–56.

Hagmeyer, G. and Gimbel, R. 1998. Modelling the salt rejection of nanofiltration membranes for ternary ion mixtures and for single salts at different pH values. *Desalination* 117:247–256.

Helfferich, F. 1962. *Ion Exchange*, New York: McGraw-Hill.

Hinkova, A., Bubník, Z., Kadlec, P., and Pridal, J. 2002. Potentials of separation membranes in the sugar industry. *Sep. Purif. Technol.* 26:101–110.

Ho, W.S.W. and Sirkar, K.K. 1992. *Membrane Handbook*. New York: Van Nostrand Reinhold.

Katchalsky, A. and Curran, P.F. 1965. *Nonequilibrium Thermodynamics in Biophysics*. Cambridge, MA: Harvard University Press.

Kitts, D.D. 1994. Bioactive substances in food: Identification and potential uses. *Can. J. Physiol. Pharmacol*. 72:423–434.

Kris-Etherton, P.M., Hecker, K.D., Bonanome, A., et al. 2002. Bioactive compounds in foods: Their role in the prevention of cardiovascular disease and cancer. *Am. J. Med.* 113:71S–88S.

Labbé, D., Araya-Farias, M., Tremblay, A., and Bazinet, L. 2005. Electromigration feasibility of green tea catechins. *J. Membr. Sci*. 254:101–109.

Lapointe, J.-F, Gauthier, S.F., Pouliot, Y., and Bouchard, C. 2003. Effect of hydrodynamic conditions on fractionation of β-lactoglobulin tryptic peptides using nanofiltration membranes. *J. Membr. Sci*. 212:55–67.

Levenstein, R., Hasson, D., and Semiat, R. 1996. Utilization of the Donnan effect for improving electrolyte separation with nanofiltration membranes. *J. Membr. Sci.* 116:77–92.

Lim, H.S., Han, J.T., Kwak, D., Jin, M., and Cho, K. 2006. Photoreversibly switchable superhydrophobic surface with erasable and rewritable pattern. *J. Am. Chem. Soc.* 128:14458–14459.

Liu, S.X., Peng, M., and Vane, L. 2004. CFD modeling of pervaporative mass transfer in the boundary layer. *Chem. Eng. Sci.* 59:5853–5857.

Martinez-Ferez, A., Rudloff, S., Guadix, A., et al. 2006. Goats' milk as a natural source of lactose-derived oligosaccharides: Isolation by membrane technology. *Int. Dairy J.* 16:173–181.

Mestel, R. 2008. Resveratrol to stop aging? *Los Angeles Times* June 4, 2008. http://latimes-blogs.latimes.com/booster_shots/2008/06/its-long-been-s.html (accessed November 18, 2008).

Mier, M.P., Ibañez, R., and Ortiz, I. 2008. Influence of process variables on the production of bovine milk casein by electrodialysis with bipolar membranes. *Biochem. Eng. J.* 40:304–311.

Moulin, P., Veyret, D., and Charbit, F. 2001. Dean vortices: Comparison of numerical simulation of shear stress and improvement of mass transfer in membrane processes at low permeation fluxes. *J. Membr. Sci.* 183:149–162.

Mulder, M. 1996. *Basic Principles of Membrane Technology*, 2nd Edition. Dordrecht, the Netherlands: Kluwer Academic Publishers.

Nagasubramanian, K., Chlanda, F.P., and Liu, K.J. 1977. Use of bipolar membranes for generation of acid and base: An engineering and economic analysis. *J. Membr. Sci.* 2:109–124.

Nayak, A., Liu, H., and Belfort, G. 2006. An optically reversible switching membrane surface. *Angew. Chem.-Int. Ed.* 45:4094–4098.

Néel, J. 1991. Introduction to pervaporation. In *Pervaporation Membrane Separation Processes*, Ed. R.Y.M. Huang, pp. 1–110. Amsterdam, the Netherlands: Elsevier.

Nunes da Ponte, M.A., Santos, J.L.C., Figueiredo, A.M., Morgado, A.N., Duarte, C.M., and Crespo, J.G. 2006. Method of obtaining a natural hydroxytyrosol-rich concentrate from olive tree residues and sub-products using clean technologies. World Patent WO2007013032, filled July 25, 2006, and issued February 1, 2007.

Nwuha, V. 2000. Novel studies on membrane extraction of bioactive components of green tea in organic solvents: Part I. *J. Food Eng.* 44:233–238.

Oren, Y., Freger, V., and Linder, C. 2004. Highly conductive ordered heterogeneous ion-exchange membranes. *J. Membr. Sci.* 239:17–26.

Pereira, C.C., Ribeiro, C.P., Jr., Nobrega, R., and Borges, C.P. 2006. Pervaporative recovery of volatile aroma compounds from fruit juices. *J. Membr. Sci.* 274:1–23.

Poulin, J.-F, Amiot, J., and Bazinet, L. 2006. Simultaneous separation of acid and basic bioactive peptides by electrodialysis with ultrafiltration membrane. *J. Biotechnol.* 123:314–328.

Ritchie, S.M.C. and Bhattacharyya, D. 2002. Membrane-based hybrid processes for high water recovery and selective inorganic pollutant separation. *J. Hazardous Mater.* 92:21–32.

Russo, C. 2007. A new membrane process for the selective fractionation and total recovery of polyphenols, water and organic substances from vegetation waters (VW). *J. Membr. Sci.* 288:239–246.

Santos, J.L.C., de Beukelaar, P., Vankelecom, I.F.J., Velizarov, S., and Crespo, J.G. 2006. Effect of solute geometry and orientation on the rejection of uncharged compounds by nanofiltration. *Sep. Purif. Technol.* 50:122–131.

Santos, J.L.C., Geraldes, V., Velizarov, S., and Crespo, J.G. 2007a. Investigation of flow patterns and mass transfer in membrane module channels filled with flow-aligned spacers using computational fluid dynamics (CFD). *J. Membr. Sci.* 305:103–117.

Santos, J.L.C., Hidalgo, A.M., Oliveira, R., Velizarov, S., and Crespo, J.G. 2007b. Analysis of solvent flux through nanofiltration membranes by mechanistic, chemometric and hybrid modelling. *J. Membr. Sci.* 300:191–204.

Sarney, D.B., Hale, C., Frankel, G., and Vulfson, E.N. 2000. A novel approach to the recovery of biologically active oligosaccharides from milk using a combination of enzymatic treatment and nanofiltration. *Biotechnol. Bioeng*. 69:461–467.

Schäfer, T. 2002. Recovery of wine-must aroma by pervaporation, PhD dissertation, Universidade Nova de Lisboa, Lisbon, Portugal.

Schäfer, T. and Crespo, J.G. 2007. Study and optimization of the hydrodynamic upstream conditions during recovery of a complex aroma profile by pervaporation. *J. Membr. Sci.* 301:46–56.

Schäfer, T., Bengtson, G., Pingel, H., Böddeker, K.W., and Crespo, J.P.S.G. 1999. Recovery of aroma compounds from a wine-must fermentation by organophilic pervaporation. *Biotechnol. Bioeng*. 62:412–421.

Souchon, I., Pierre, F.X., Athes-Dutour, V., and Marin, M. 2002. Pervaporation as a deodorization process applied to food industry effluents: Recovery and valorisation of aroma compounds from cauliflower blanching water. *Desalination* 148:79–85.

Strathmann, H. 2000. Introduction to membrane science and technology, University of Colorado, Boulder, CO; Internal report.

Suárez, E., Lobo, A., Álvarez, S., Riera, F.A., and Álvarez, R. 2006. Partial demineralization of whey and milk ultrafiltration permeate by nanofiltration at pilot-plant scale. *Desalination* 198:274–281.

Tan, S., Li, L., Xiao, Z., Wu, Y., and Zhang, Z. 2005. Pervaporation of alcoholic beverages—The coupling effects between ethanol and aroma compounds. *J. Membr. Sci.* 264:129–136.

Ting, B.P.C.P., Gauthier, S.F., and Pouliot, Y. 2007. Fractionation of β-lactoglobulin tryptic peptides using spiral wound nanofiltration membranes. *Sep. Sci. Technol.* 42:2419–2433.

Vane, L.M. and Alvarez, F.R. 2002. Full-scale vibrating pervaporation membrane unit: VOC removal from water and surfactant solutions. *J. Membr. Sci.* 202:177–193.

Velizarov, S. 1999. Electric and magnetic fields in microbial biotechnology: Possibilities, limitations, and perspectives. *Electro- Magnetobiol.* 18:185–212.

Vera, E., Sandeaux, J., Persin, F., Pourcelly, G., Dornier, M., and Ruales, J. 2007a. Deacidification of clarified tropical fruit juices by electrodialysis. Part I. Influence of operating conditions on the process performances. *J. Food Eng*. 78:1427–1438.

Vera, E., Sandeaux, J., Persin, F., et al. 2007b. Deacidification of clarified tropical fruit juices by electrodialysis. Part II. Characteristics of the deacidified juices. *J. Food Eng.* 78:1439–1445.

Vera, E., Sandeaux, J., Persin, F., Pourcelly, G., Dornier, M., and Ruales, J. 2009. Deacidification of passion fruit juice by electrodialysis with bipolar membrane after different pretreatments. *J. Food Eng*. 90:67–73.

Wang, W., and de Mejia, E.G. 2005. A new frontier in soy bioactive peptides that may prevent age-related chronic diseases. *Comp. Rev. Food Sci. Food Saf.* 4:63–78.

Wesselingh, J.A. and Krishna, R. 2000. *Mass Transfer in Multicomponent Mixtures*. Delft, the Netherlands : Delft University Press.

Xu. T. 2005. Ion exchange membranes: State of their development and perspective *J. Membr. Sci*. 263:1–29.

Yu, L., Lin, A., Zhang, L., and Jiang, W. 2002. Large scale experiment on the preparation of vitamin C from sodium ascorbate using bipolar membrane electrodialysis. *Chem. Eng. Commun*. 189:237–246.

6 Recent Advances in Fruit-Juice Concentration Technology

Cláudio P. Ribeiro, Jr., Paulo L.C. Lage, and Cristiano P. Borges

CONTENTS

6.1 INTRODUCTION

Fruits and vegetables contain many health-promoting factors such as fiber and large amounts of minerals, vitamins, flavonoids, and phenolic acids. Epidemiological evidence suggests that a diet rich in fruits and vegetables can help in the protection of coronary heart disease, cancer, Alzheimer's, and other chronic diseases (Margetts and Buttriss, 2003; Dai et al., 2006; Ruxton et al., 2006). Most fruits, however, are rather fragile or perishable, deteriorating due to the action of heat or during transportation, which restrains both their commercialization and their application in the food industry. Another relevant aspect is the seasonal character of some fruits, which limits their production to some months of the year. The industrial processing of fruits into juices and pulps constitutes an alternative for enabling their production, storage, and transportation, which, in turn, can increase their application in the food industry. Furthermore, juices can be consumed more conveniently than whole fruits, which certainly plays an important role, since many people, especially those with a busy lifestyle, would rather take a supplement or pill than consume more fruits or vegetables to have a healthier diet (Mullen et al., 2007).

The commercial production of fruit juices started in Europe at the beginning of the twentieth century, when the Boehi process for storing apple juice under CO_2 and refrigeration was developed. In the United States, industrialized juices (tomato, grape, and citrus) began to be sold around 1929. During World War II, industrial juice processing grew considerably, and the market kept expanding in the following years with the application of new technologies (Moyer, 1980).

Nowadays, juice processing is a well established and rather important economic activity. In 1998, the total world market for fruit juices was greater than \$31 billion (Neves et al., 1999), a value that grew to \$52.5 billion in 2005 and is estimated to reach \$59.4 billion in 2010 (Datamonitor, 2007). As shown in Figure 6.1, orange juice is, by far, the most important product in this sector, followed by apple juice. From 2002 to 2005, the world production of orange juice was always greater than 2.2×10^6 metric tons. It is mainly concentrated in Brazil, which, within the period considered in Figure 6.1, was responsible for 51%–62% of the total production (USDA, 2007). The importance of the fruit-juice industry is also reflected in the data for total fruit production given in Figure 6.2. The large consumption of orange and other citrus juices renders these fruits the most produced in the world. Apples, associated with the second most important fruit juice, exhibit a production almost as large as grapes, which owe a large share of their market to wine production. Moreover, there is a

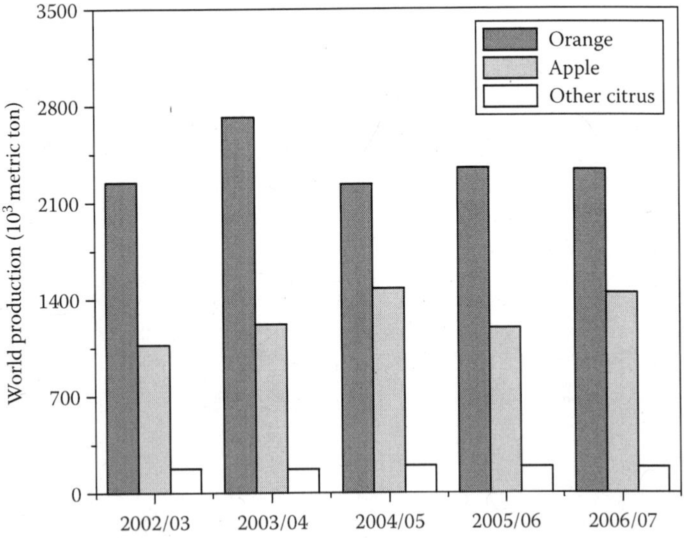

FIGURE 6.1 World production of some fruit juices, 2001–2005. (Data from USDA, Production, Supply and Distribution (PSD) Reports On-line, Foreign Agriculture Service, Washington, DC, 2007, http://www.fas.usda.gov.)

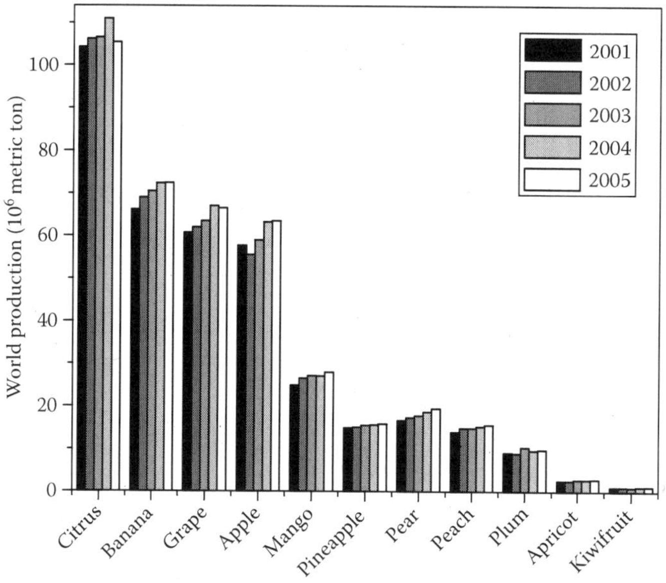

FIGURE 6.2 Global commercial production of selected fruit, 2001–2005. (Data from Pollack, S. and Perez, A., *Fruit and Tree Nuts Situation and Outlook Yearbook*, USDA, Washington, DC, 2006, http://www.ers.usda.gov.)

considerable difference between the production of apples and mangos, the next fruits in the series. It is interesting to note that bananas, mainly consumed as fresh fruit, constitute an exception, with a production that is secondary only to citrus.

The composition of a number of fruit juices is shown in Table 6.1. The processing steps in industrial juice production vary depending on the fruit, but usually include cleaning, extraction, clarification, and pasteurization (Moyer, 1980; Bates et al., 2001; Mosshammer et al., 2006). Due to the high water content of fruit juices, a concentration step is usually introduced to increase their solids content to 50–65 wt%. With the resulting decrease in the weight and volume of products, packaging, transportation, and storage costs are reduced. Furthermore, concentration provides microbiological stability, increasing the shelf life of fruit juices, as a result of an increase in osmotic pressure and a reduction in water activity. In addition, juice concentrates are used as ingredients in many products such as ice creams, fruit syrups, jellies, alcoholic beverages, soft drinks, and fruit-juice beverages.

Whenever used, concentration is a crucial step in juice processing on account of its direct effect upon the quality of the final product and, therefore, upon its acceptance by customers. Evaporation, probably the oldest concentration method, constitutes the dominant technology at the industrial scale (Ramteke et al., 1993; Jariel et al., 1996). Since juices are heat sensitive, operation is carried out under vacuum, in multiple-effect tubular or plate units. Four to seven effects are commonly installed and most plants are featured with an evaporator type known as TASTE (acronym for Thermally Accelerated Short Time Evaporator), which can be briefly described as a falling-film, long-tube, vertical evaporator. A detailed description of TASTE and

TABLE 6.1
Composition of Fruit Juices (per 100 g)

Fruit Juice	Water (%)	Energy (kJ)	Protein (%)	Fat (%)	Ash (%)	Fiber (%)	Carbohydrates (%)	Vitamin C (mg)	Ca (mg)	Fe (mg)	Mg (mg)	K (mg)	P (mg)
Acerola	94.30	96	0.40	0.30	0.20	0.3	4.8	1600.0	10	0.50	12	97	9
Apple	87.93	197	0.06	0.11	0.22	0.1	11.68	0.9	7	0.37	3	119	7
Apricot	86.62	201	0.63	0.04	0.37	1.6	12.34	4.9	12	0.30	10	165	20
Blackberry	90.90	158	0.30	0.60	0.40	0.1	7.80	11.3	12	0.48	21	135	12
Cranberry	87.13	194	0.39	0.13	0.15	0.1	12.20	9.3	8	0.25	6	77	13
Grape	90.00	163	0.50	0.10	0.20	0.1	9.20	38.0	9	0.20	12	162	15
Grapefruit	90.00	163	0.50	0.10	0.20	0.1	9.20	38.0	9	0.20	12	162	15
Lemon	90.73	105	0.38	0.00	0.26	0.4	8.63	46.0	7	0.03	6	124	6
Lime	90.79	104	0.42	0.07	0.31	0.4	8.42	30.0	14	0.09	8	117	14
Orange	88.30	188	0.70	0.20	0.40	0.2	10.40	50.0	11	0.20	11	200	17
Passion fruit	84.50	60	0.67	0.18	—	0.2	14.45	18.20	4	0.36	17	278	25
Peach	87.49	184	0.63	0.03	0.27	1.3	11.57	3.6	6	0.27	7	128	17
Pear	86.47	209	0.34	0.07	0.19	1.6	12.94	1.6	9	0.29	7	96	12
Pineapple	86.37	220	0.36	0.12	0.28	0.2	12.87	10.0	13	0.31	12	130	8
Plum	84.02	243	0.51	0.02	0.30	0.9	15.15	2.8	10	0.34	8	154	15
Pomegranate	83.65	—	1.03	—	0.32	—	14.6	—	24.5	2.21	5	333	6
Prune	81.24	297	0.61	0.03	0.68	1.0	17.45	4.1	12	1.18	14	276	25
Tangerine	88.90	180	0.50	0.20	0.30	0.2	10.10	31.0	18	0.20	8	178	14
Tomato	93.90	73	0.76	0.05	1.05	0.4	4.24	18.3	10	0.43	11	229	18

Sources: Adapted from U.S. Department of Agriculture, *National Nutrient Database*. Agricultural Research Service, Beltsville, MD, http://www.nal.usda.gov (accessed January 12, 2008); Modified from Al-Maiman, A.A. and Ahmad, D., *Food Chem.*, 76, 437, 2002. With permission.

other evaporator types is given by Standiford (1997). Evaporation installations in the citrus industry are rather large. A typical citrus concentrate plant evaporates 40–50 ton h^{-1} of water (Chen and Hernandez, 1997). Operating temperatures can vary from 105°C–90°C, in the first effect, to 40°C–45°C in the last effect, for a total residence time of 6–8 min (Koeseoglu et al., 1990).

Although evaporation is considered to be the most economical and technologically developed concentration method, it does exhibit some disadvantages when applied to fruit juices. Even under vacuum, operating temperatures are still high enough to bring about chemical changes in some juice components, leading to color degradation, loss of nutritional characteristics, and the development of a "cooked" taste. For example, lipids and ascorbic acid can be oxidized, amino acids and sugars can undergo the Maillard browning reaction, and pigments, especially anthocyanin, carotenoids, and chlorophyll, can be degraded (Toribo and Lozano, 1986; Lozano and Ibarz, 1997; Mikkelsen and Poll, 2002; Kato et al., 2003; Maskan, 2006). On the heat-transfer surfaces, since operating temperatures are even higher, considerable fouling is observed and periodic cleaning is required (Moyer, 1980; Ramteke et al., 1993).

Moreover, due to their high volatility in aqueous solutions, most of the so-called aroma compounds are lost to the vapor phase during evaporation (Nisperos-Carriedo and Shaw, 1990; Ramteke et al., 1990; Lin et al., 2002). Aroma is formed by a mixture of hundreds of different organic compounds present at very low concentrations, typically mg L^{-1} or μg L^{-1} levels. It has been demonstrated that some volatile components of aroma are effective to influence or even determine the consumption of a given food product (Reineccius, 2006). In the literature, more than 6000 different aroma components have already been identified, which are distinguished as esters, alcohols, aldehydes, ketones, carboxylic acids, hydrocarbons, amines, mercaptans, terpenes, ethers, phenols, lactones, etc. (Bomben et al., 1973). Table 6.2 shows the number of volatile aroma components already identified in some fruits.

TABLE 6.2
Number of Volatile Aroma Components Already Identified in Some Fruits

Fruit	Number of Identified Components	Reference
Apple	356	Rufino (1996)
Banana	225	Shiota (1993)
Black currant	150	Varming et al. (2004)
Orange	203	Rufino (1996)
Papaya	262	Rufino (1996)
Passion fruit	193	Werkhoff et al. (1998)
Peach	110	Narain et al. (1990)
Pineapple	157	Umano et al. (1992)
Strawberry	303	Zabetakis and Holden (1997)

Source: Modified from Pereira, C.C. et al., *J. Membr. Sci.*, 274, 1, 2006. With permission.

To avoid the resulting impairment of the flavor quality of the final product, these aroma compounds stripped during evaporation must be recovered and then added back to the concentrated juice before packing. The current industrial options are distillation, partial condensation, or a combination thereof (Bomben et al., 1973; Karlsson and Trägårdh, 1997). Due to the high dilution of the target components in the feed stream, commercial aroma recovery plants are somewhat complex. Distillation is conducted under vacuum, either in tray or packed columns, to avoid thermal degradation of the aroma compounds. In partial condensation, the first condenser operates at a higher temperature and produces a water-rich liquid stream. The remaining aroma-rich vapor is sent to a total condenser with a low operating temperature. In both cases, significant losses can be observed due to stripping in the vent gases, and the inclusion of expensive gas washing and condensation systems in the plant may be required (Mannheim and Passy, 1975).

In an attempt to overcome these disadvantages associated with traditional evaporation, the development of alternative concentration techniques has drawn considerable attention. Despite the intense research in this area, an alternative juice concentration process that can overcome all shortcomings of traditional vacuum evaporation and, at the same time, achieve the same total solids content at comparable costs, remains to be developed. In Sections 6.2 and 6.3, a critical review of some the different proposals found in the literature is provided, with emphasis on the most studied ones. In total, five techniques are discussed, namely direct-contact evaporation, Refractance Window®* evaporation, reverse osmosis (RO), osmotic distillation (OD), and membrane distillation (MD). Aroma recovery is addressed whenever it was taken into account in the specific processes under discussion, but techniques specially developed for aroma recovery are beyond the scope of the present chapter. Further information on aroma recovery can be found in the recent reviews of Pereira et al. (2006) and Nongonierma et al. (2006).

6.2 ADVANCED THERMAL TECHNIQUES

There are some alternatives to traditional vacuum evaporation for a heat-transfer-based removal of water from solutions. In the following sections, two of these alternatives are discussed: direct-contact evaporation and Refractance Window evaporation.

6.2.1 Direct-Contact Evaporation

The main feature of a direct-contact evaporator (DCE) is the absence of intervening walls separating the solution from the heating fluid. In a liquid–liquid DCE, the feed streams are two immiscible liquids and the dispersed phase, comprised by drops, contains the species to be evaporated. Alternatively, in a gas–liquid DCE, a superheated gas sparged as bubbles into the continuous phase is used as heating fluid to bring about vaporization (Ribeiro and Lage, 2005).

* Registered trademark of MCD Technologies Inc., Tacoma, WA.

The concept of gas–liquid direct-contact evaporation dates back to the end of the nineteenth century, and since the installation of the first commercial unit in the United States in 1935 (Burdick et al., 1949), it has been successfully applied to a wide variety of solutions (Swindin, 1949; Cronan, 1956; Weisman, 1961). The lack of heat-transfer walls confers many advantages on DCEs compared to traditional shell-and-tube units. Firstly, as a result of a reduced residence time and the bubbling-driven mixing of the liquid, DCEs exhibit higher heat-transfer efficiency, with typical values usually greater than 95% (Watson, 1966). Moreover, a gradual reduction in the heat-transfer efficiency due to fouling is not observed. Installations are more compact, with lower capital investment, as well as lower operating and maintenance costs (Cronan, 1956; Ownen and Moggio, 1955).

As illustrated in Figure 6.3, a gas–liquid DCE basically comprises a liquid column through which a superheated gas is bubbled. The sparger, responsible for bubble formation, is normally a perforated plate or a set of perforated pipes located at the bottom of the column. Most of the energy received by the liquid is usually transferred by the superheated gas during both bubble formation and ascension, but the sparger can also play an important or even dominant role depending upon the adopted operating conditions (Ribeiro and Lage, 2004a). Energy can be transferred either as sensible heat, leading to a temperature rise, or as latent heat, causing vaporization and thereby creating a mass flux from the surface to the interior of the bubbles. The amount of a volatile component that can be transferred to the dispersed phase is proportional to its saturation pressure at the gas–liquid interface (Campos and Lage, 2001; Ribeiro et al., 2005a). Consequently, the higher the temperature of the solution, the larger the amount of energy transferred to the liquid as latent heat. Eventually, an equilibrium

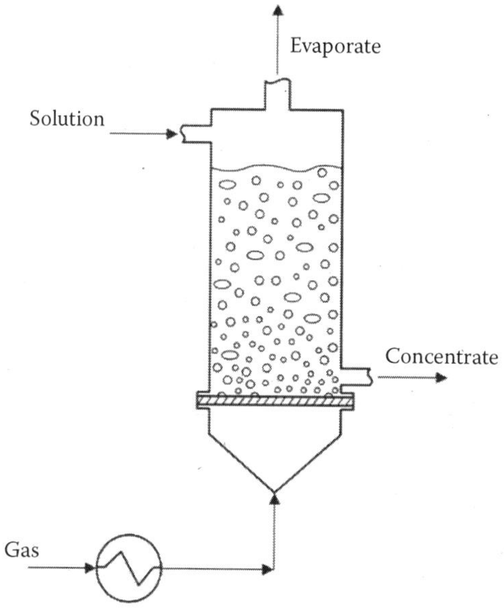

FIGURE 6.3 Schematic representation of a gas–liquid DCE. (From Ribeiro, C.P. and Lage, P.L.C., *Chem. Eng. Technol.*, 28, 1081, 2005. With permission.)

temperature is reached, at which almost all the energy is used in the vaporization, being the sensible heat fraction only responsible for compensating the heat losses of the unit.

Though originally developed and still only commercially adopted for concentrating highly fouling and/or corrosive solutions, gas–liquid direct-contact evaporation represents a promising alternative for the treatment of heat-sensitive solutions. Owing to the presence of noncondensable gases in the evaporate stream, solvent vaporization can be achieved at temperatures quite below its boiling point at the operating pressure. According to Kurz and Güthoff (1988), the equilibrium temperature of the continuous phase is normally 10°C–30°C lower than the boiling point of the liquid. Furthermore, the vigorous mixing produced by gas bubbling maintains a uniform liquid temperature, thereby preventing localized overheating. Apparently, Luedicke et al. (1979) were the first* to recognize the advantages of DCEs for concentrating heat-sensitive solutions. Working with a pilot unit, 43.2 cm in diameter, these authors concentrated fish hydrolysate from 6 to 30 wt% solids without any overheating or fouling problems, with a maximum liquid temperature equal to 76.1°C for operation under atmospheric pressure. Favorable results for the concentration of whey and milk were subsequently reported by Lovellsmith (1982) and Zaida et al. (1986), respectively.

The application of a DCE for the concentration of fruit juices was recently investigated by Ribeiro et al. (2004a), who adopted aqueous solutions of sucrose and sucrose/ethyl acetate as model liquids. Bench-scale runs at atmospheric pressure demonstrated that the initial 12 wt% sucrose concentration could be raised to 68 wt% for an equilibrium liquid temperature smaller than 67°C. However, ethyl acetate, the model aroma compound, was completely stripped by the gas even before the final equilibrium temperature was achieved.

To recover the stripped aroma compounds, considering the shortcomings of traditional aroma recovery techniques, the authors evaluated, in a subsequent work (Ribeiro et al., 2004b), the feasibility of applying vapor permeation, a membrane-based technique (Baker, 2004). Experiments carried out with ethyl acetate solutions at room temperature using commercial poly(dimethyl siloxane) (PDMS) hollow-fiber membranes evidenced that, in spite of the low concentration of this ester in the gas stream, a degree of recovery as high as 98% could be achieved. For a given membrane area, the degree of recovery of the ester decreased as the gas flow rate was raised because of the resulting drop in the residence time of the gaseous stream in the membrane module (Ribeiro and Borges, 2004), and operation under lower gas flow rates was recommended to reduce capital costs. This aspect complicates the treatment of a DCE exhaust stream in the membrane module, since the vaporization rate in a DCE increases with the gas superficial velocity† (Iyer and Chu, 1971; Kawasaki and Hayakawa, 1972; Ribeiro and Lage, 2004b) and operation at high gas

* A previous study on the application of a DCE in the food industry was published by E.L. Durkee, E. Lowe, K. Baker, and J. W. Burgess (*J. Food Sci.* 38: 507–11, 1978), but in this case the technique was chosen based on the scaling potential of the solution to be treated (spent cucumber pickle brine).

† In a DCE, the gas superficial velocity is defined as the ratio of the inlet gas flow rate to the cross-section area of the unit.

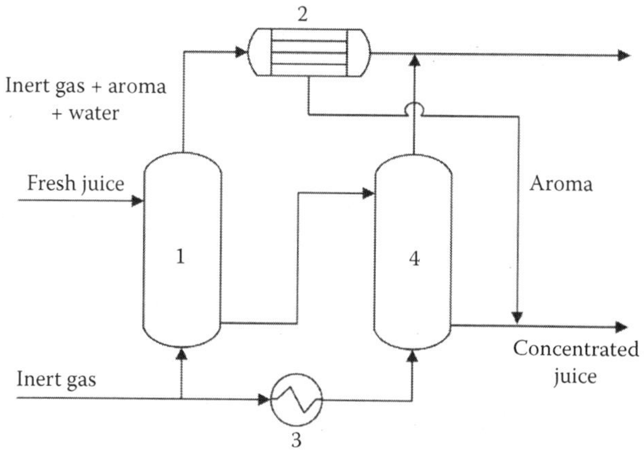

FIGURE 6.4 Two-step fruit-juice processing route proposed by Ribeiro et al. (2005b): 1, bubble-column stripper; 2, membrane module; 3, heat exchanger; 4, DCE. (From Ribeiro, C.P. et al., *Ind. Eng. Chem. Res.*, 44, 6903, 2005. With permission.)

flow rates are therefore preferred. Besides, operation under vacuum in the evaporator would require a larger membrane area for aroma recovery, for the partial pressure difference of the components across the membrane constitutes the driving force for mass transfer in vapor permeation.

To overcome these difficulties related to the coupling of a DCE and a vapor permeation module, Ribeiro and coworkers (Lage et al., 2005; Ribeiro et al., 2005b,c) proposed the two-step fruit-juice processing route schematically shown in Figure 6.4. In the first step, aroma compounds are stripped by an inert gas in a bubble column and subsequently recovered by vapor permeation in a membrane module. The aroma-depleted juice is then sent to concentration in a DCE. This route enables the individual optimization of gas flow rate and other operating parameters (such as temperature, pressure, and sparger type) for aroma recovery and juice concentration. It was tested at laboratorial scale with a synthetic juice, that is, an aqueous solution of sucrose (11.2 wt%), ethyl acetate (1000 mg L^{-1}) and ethyl butyrate (1000 mg L^{-1}). Even though the operating conditions were by no means optimized, degrees of recovery around 90% were already observed for the aroma compounds with the PDMS membrane. For the concentration step, some of the results reported by Ribeiro et al. (2005b) are reproduced in Figure 6.5. First of all, even though the operating pressure was 104.2 kPa (1.03 atm), the liquid temperature was always smaller than 66°C. Simulation results (Ribeiro et al., 2005c) indicated that, if pressure were reduced to 14 kPa, a value still greater than the one associated with the last evaporation effect in industrial units, the liquid temperature would not surpass 29°C. As shown in Figure 6.5a, a final sucrose concentration of 77 wt% was achieved, a value which is even greater than the current industrial standard of 50–65 wt%.

It is important to highlight that the viscosity of fruit juices increases rapidly with their solids content (Hernandez et al., 1995; Bailey et al., 2000; Nindo et al., 2005; Cassano et al., 2007), which brings about a significant drop in the heat-transfer coefficient for traditional shell-and-tube evaporators (Schwarz and Penn, 1948), whose

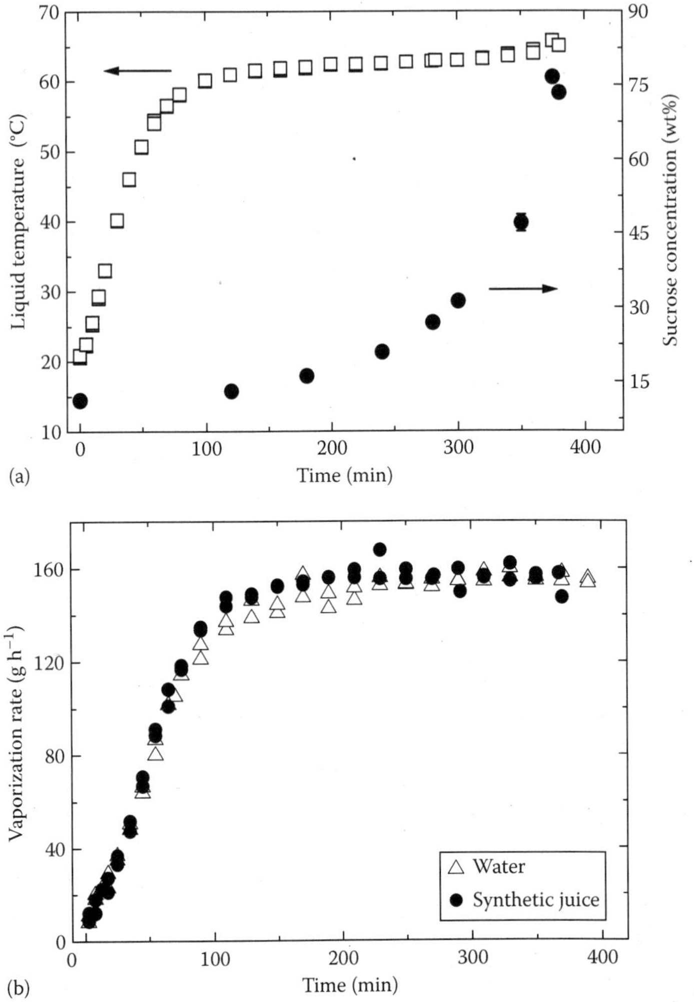

FIGURE 6.5 Concentration of a synthetic fruit juice in a DCE operating in semibatch mode: (a) liquid temperature and sucrose concentration; (b) vaporization rate. (Adapted from Ribeiro, C.P. et al., *Ind. Eng. Chem. Res.*, 44, 6888, 2005. With permission.)

heat-transfer efficiency is consequently reduced as concentration proceeds. In fact, it is precisely this reduction in heat-transfer efficiency that sets the current industrial limit of 65 wt% for the solids content of juice concentrates. On the other hand, in the case of the DCE, despite the continuous increase in the viscosity of the solution, there was no reduction whatsoever in the evaporation rate. As evidenced in Figure 6.5b, the evaporation rates observed by Ribeiro et al. (2005b) during the concentration of their synthetic juice were equal to those associated with the vaporization of distilled water at the same operating conditions for bubbling heights as small as 80 mm, in a clear demonstration of the high heat-transfer efficiency of DCEs. The authors emphasize that the value of 77 wt% for the solids content should not be regarded as a limit, since no operating problem was verified and the experiment was only interrupted because of the low bubbling height value as concentration proceeded in batch operation.

The influence of the sparger on the performance of a DCE applied in the concentration of sucrose and sucrose–ethyl acetate solutions was addressed by Ribeiro et al. (2007). Bubble coalescence was suppressed by both solutes and, as a result, the available interfacial area for heat and mass transfer in the unit increased considerably compared to the one obtained with distilled water upon a reduction in the orifice diameter of the sparger for all gas flow rates considered. Nonetheless, the increase in the interfacial area was not translated into an enhancement of the vaporization rate, indicating that all thermal energy in the superheated gas had already been transferred to the liquid with the smaller area.

These results obtained with synthetic juices are certainly promising, but further investigations with real fruit juices are required to verify whether the lower operating temperatures during concentration in a DCE will in fact be associated with higher sensory quality.

6.2.2 Refractance Window Evaporation

In the so-called Refractance Window evaporator (RWE), hot process water (95°C–98°C) and the solution to be concentrated, separated by a very thin (about 0.2 mm) transparent plastic sheet, flow concurrently in a inclined channel. A countercurrent flow of cool air over the solution exhausts the evaporated water, as shown in Figure 6.6. The solution is recirculated in the evaporator until the desired degree of concentration is reached. Even though operation takes place under atmospheric pressure, product temperatures seldom reach 70°C, and high solids concentration can be achieved without fouling problems (Nindo et al., 2004).

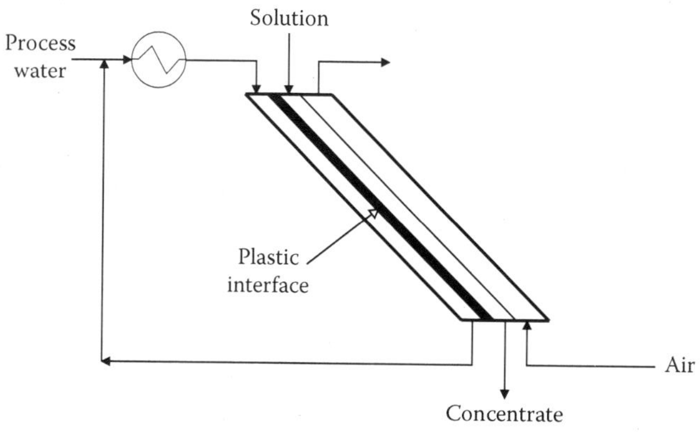

FIGURE 6.6 Schematic representation of a RWE. (Adapted from Nindo, C.I. et al., *Lebensm-Wiss. Technol.*, 40, 1000, 2007. With permission.)

The performance of a RWE with a surface area of 6.64 m^2 operating with tap water and reconstituted raspberry and blueberry juices (10 wt% solids) was investigated by Nindo et al. (2004). The flow rates of all fluids, including the cool air stream, were observed to influence the evaporation rate, whereas similar results were obtained for three different tilt angles of the inclined surface (24°, 30°, and 37°). The average product temperature during the concentration of the reconstituted juices was 65°C. For the highest hot water flow rate considered (6.8 kg s^{-1}), overall heat-transfer coefficients ranged from 619 to 733 W m^{-2} K^{-1}, values comparable to those observed in the last effects of multiple-effect vacuum evaporation systems, but much smaller than those associated with the initial effects, namely, 1500–2000 W m^{-2} K^{-1} (Chen and Hernandez, 1997).

In a subsequent work, Nindo et al. (2007) analyzed the changes in color and vitamin C content of cranberry and blueberry juices, respectively, during their concentration in a RWE. As concentration proceeded, the color of cranberry juice changed from intense red to bluer tones, and no significant difference was determined in the color (defined by the hue angle) of juices reconstituted (6°Brix) from concentrates (65°Brix) obtained in the RWE and in a industrial four-effect falling-film evaporator. With regard to vitamin C, its content in blueberry juice decreased by 32%–48%, depending on the temperature of the juice during concentration, which ranged from 55.5°C to 59.0°C and was observed to increase linearly with the inlet temperature of hot water. Higher product temperatures led to lower contents of vitamin C in the concentrate, but the numbers compared favorably with the 70% loss verified in an industrial falling-film evaporator operating at 68°C.

As highlighted by Nindo et al. (2004, 2007) themselves, rather than a substitute for the large multiple-effect evaporation systems of the citrus industry, the RWE would represent an alternative for small to medium-scale processors, who cannot afford the large investments associated with multiple-effect vacuum evaporators and therefore choose recirculation-type evaporators, which tend to expose juices to high temperatures over long periods of time, leading to low-quality products. The technology is commercially available, and some units have already been installed in Australia.

6.3 MEMBRANE-BASED TECHNIQUES

A membrane can be defined as a permeable barrier between two phases through which different species are transferred at different rates. It can be solid or liquid, organic or inorganic, porous or dense, isotropic or anisotropic. In every membrane-based unit operation, the feed stream is always split into two new streams: the permeate, enriched in those species that preferentially permeate the membrane, and the residue or retentate, which has a higher concentration of the compounds preferentially retained by the membrane. The permeate flux, defined as the permeate flow rate per membrane area, represents an important performance parameter, together with a quantitative measurement of the membrane selectivity, whose actual definition depends on the unit operation under consideration.

The selective permeation of species through a membrane constitutes a mass-transfer process and is hence driven by a chemical potential gradient for each species across the membrane. Depending on how this gradient is established and on the characteristics of the membrane used, different types of membrane-based unit operations are distinguished. We shall restrict our discussion to three of these operations: RO, OD, and MD. Insofar as juice concentration is concerned, water is always the main component in the permeate stream.

6.3.1 Reverse Osmosis

RO is one of the so-called pressure-driven membrane operations, in which a pressure difference across the membrane generates the chemical potential gradient required for mass transfer. Microfiltration (MF), ultrafiltration (UF), and nanofiltration (NF) are also included in this group. Their main features are listed in Table 6.3. The basic difference among these unit operations is the size of the species that they can retain, which stems from morphological differences in their membranes. As one goes from MF to RO, the mean porous size of the membrane decreases, and thus smaller species can be retained, but at the expense of larger pressure differences and lower values of average permeate flux. In fact, RO membranes are dense, with no pores, and transport occurs through volume elements created by thermal motion of the polymer chains that constitute the membrane. These free-volume elements constantly appear and disappear on about the same timescale of the motions of the permeants across the membrane (Baker, 2004). Apart from NF, which is somewhat recent, all other pressure-driven membrane operations are already industrially mature. In the specific case of RO, the combined market of membrane modules and processing systems was about $400 million worldwide in 1998 (Baker, 2004). Water desalination accounts for most of the installed industrial plants, whose global market in 2005 was estimated at $1.9 billion, with a prediction to exceed $3 billion by 2010 (Puffett and Drewes, 2007).

TABLE 6.3
Overview of Pressure-Driven Membrane Operations

Unit Operation	Nominal Pore Diameter (Å)	Operating Pressure (bar)	Permeability (Lh^{-1} m^{-2} bar^{-1})	Materials Rejected
MF	1,000–100,000	0.5–2	>1,000	Suspended particles and bacteria
UF	10–1,000	1–7	10–1,000	Colloids and macromolecules
NF	5–10	4–25	1.5–30	Multivalent ions and relatively small organics
RO	1–5	10–80	0.05–1.5	All solutes and suspended material

Source: Habert, A.C. et al., *Membrane Separation Processes*, Rio de Janeiro, 2006 (in Portuguese). With permission.

In the specific case of juice concentration, the first advantage of RO over traditional evaporation is an improvement in the quality of the final product, since thermal damages are minimized and loss of aroma and flavor is reduced. Additionally, unlike thermal methods, RO does not involve phase change for water removal, which leads to a more efficient energy use and thus lower operating costs (Dale et al., 1982; Robe, 1983; Molinari et al., 1995). Besides, RO units are more compact than falling-film evaporators. Studies on RO for juice concentration started in the late 1960s with Merson and Morgan (1968), just some years after the breakthrough for industrial application of RO, that is, the development of anisotropic cellulose acetate membranes by Loeb and Sourirajan (1963). Since then, a variety of juices has already been concentrated by RO at laboratory scale, including acerola, apple, camu-camu, grape, grapefruit, kiwi, orange, passion fruit, pear, pineapple, tomato, and watermelon. Comprehensive reviews of the earlier studies on the subject are available (Girard and Fukumoto, 2000; Jiao et al., 2004) and the main focus here will be on more recent reports.

The transport mechanism of RO was intensely debated in the 1960s and early 1970s, but now the solution–diffusion model is widely accepted (van der Bruggen et al., 2003; Baker, 2004). According to this model, permeants first dissolve in the membrane material and then diffuse through it down a concentration gradient. Separation is brought about by differences in solubilities and diffusivities of the compounds in the membrane. As detailed by Wijmans and Baker (1995), the solution–diffusion model leads to the following expression for the one-dimensional mass flux of a general component i (J_i^{m}) across a RO membrane:

$$J_i^{\mathrm{m}} = \frac{D_{\mathrm{m},i}\zeta_i}{l}\left\{c_i^{\mathrm{F}} - c_i^{\mathrm{P}}\exp\left[\frac{-\upsilon_i\left(P^{\mathrm{F}}-P^{\mathrm{P}}\right)}{RT}\right]\right\} \tag{6.1}$$

where

$D_{\mathrm{m},i}$ and ζ_i are the diffusion coefficient and solubility of component i in the membrane
l is the membrane thickness
c_i and υ_i are the mass concentration and molar volume of component i, respectively
T is the temperature
R is the universal gas constant
P is the pressure

The superscripts F and P refer to the feed and permeate streams, respectively.

In the case of water, considering that no flux is obtained when the applied hydrostatic pressure difference across the membrane ($\Delta P^{\mathrm{M}} = P^{\mathrm{F}} - P^{\mathrm{P}}$) equals the osmotic pressure difference ($\Delta\pi$), Equation 6.1 can be written as:

$$J_{\mathrm{w}}^{\mathrm{m}} = \frac{D_{\mathrm{m,w}}\zeta_{\mathrm{w}}c_{\mathrm{w}}^{\mathrm{F}}}{l}\left\{1 - \exp\left[\frac{-\upsilon_{\mathrm{w}}\left(\Delta P^{\mathrm{M}} - \Delta\pi\right)}{RT}\right]\right\} \tag{6.2}$$

which can be further simplified to

$$J_{\mathrm{w}}^{\mathrm{m}} = \frac{D_{\mathrm{m,w}}\zeta_{\mathrm{w}}c_{\mathrm{w}}^{\mathrm{F}}\upsilon_{\mathrm{w}}\left(\Delta P^{\mathrm{M}} - \Delta\pi\right)}{lRT} = A^{*}\left(\Delta P^{\mathrm{M}} - \Delta\pi\right) \tag{6.3}$$

since, under normal RO conditions, the term $-\upsilon_{\mathrm{w}}(\Delta P^{\mathrm{M}} - \Delta\pi)/(RT)$ is small, and the simplification $1 - \exp(x) \to x$ as $x \to 0$ applies. Equation 6.3 is the common expression found in RO studies, both for water desalination and juice concentration, and the constant A^* is known as the water permeability constant.

With regard to solutes, due to the small value of the term $\upsilon_i(P^{\mathrm{F}} - P^{\mathrm{P}})/(RT)$, the exponential term in Equation 6.1 is close to one, which leads to the simplification:

$$J_i^{\mathrm{m}} = \frac{D_{\mathrm{m},i}\zeta_i}{l}\left(c_i^{\mathrm{F}} - c_i^{\mathrm{P}}\right) = B^{*}\left(c_i^{\mathrm{F}} - c_i^{\mathrm{P}}\right) \tag{6.4}$$

where B^* is known as the permeability constant of solute i. Equation 6.4 is commonly found in RO studies on water desalination, but seldom utilized as far as juice concentration is concerned.

For each solute i, the separation efficiency is expressed in terms of a rejection coefficient, $\Re_i$, defined as:

$$\Re_i = \left(1 - \frac{c_i^{\mathrm{P}}}{c_i^{\mathrm{F}}}\right) \times 100 \tag{6.5}$$

An ideal membrane would give a solute-free permeate and therefore rejection coefficients equal to 100% for all solutes, whereas a completely ineffective membrane gives $c_i^{\mathrm{P}} = c_i^{\mathrm{F}}$ and $\Re_i = 0$.

An analysis of Equations 6.3 and 6.4 indicates that temperature, membrane material, applied hydrostatic pressure, and solute concentration (via osmotic pressure) are some of the operating parameters that can influence RO performance. Another important parameter, not evident from Equation 6.3, is the feed flow rate, or, more precisely, the Reynolds number on the feed side.

The solute concentration and the applied hydrostatic pressure (or transmembrane pressure), in particular, have a profound influence on the water flux due to their direct effect upon the driving force. For a fixed ΔP^{M}, there is always a sheer drop in the water flux as the juice becomes more and more concentrated. This is illustrated in Figure 6.7 with literature water flux data plotted as a function of the solids content in the feed for different juices and operating conditions. Each data set was normalized by the flux value at the beginning of the run ($J_{\mathrm{w,0}}^{\mathrm{m}}$). This flux decrease is a direct consequence of the fast increase in the osmotic pressure of juices with the total solids concentration (Girard and Fukumoto, 2000) and the resulting drop in the driving force. As anticipated by Equation 6.3, for a given ΔP^{M} and fruit juice, regardless of the membrane used, there is a maximum solids concentration that can be achieved, whose respective osmotic pressure equals ΔP^{M} and therefore $J_{\mathrm{w}}^{\mathrm{m}} = 0$. This trend is

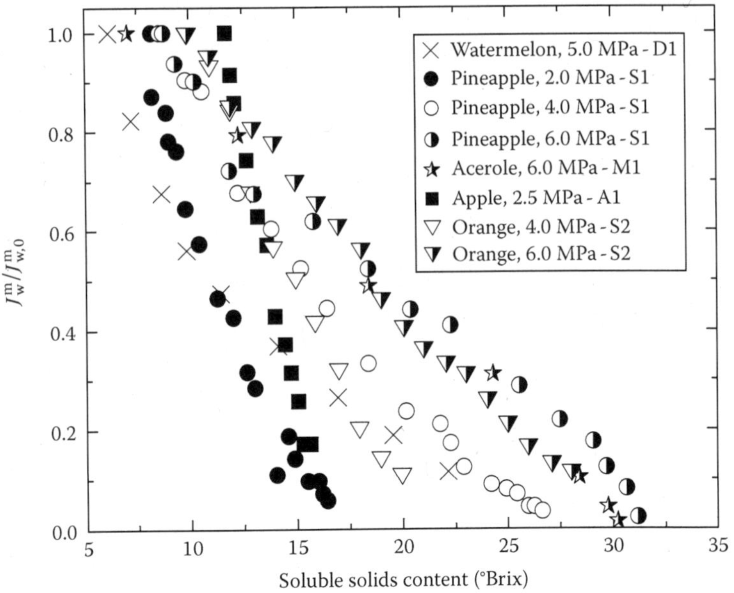

FIGURE 6.7 Evolution of water flux during the concentration of some fruit juices by RO using different membranes. For each data set, feed stream and adopted transmembrane pressure are specified. [Based on data from Álvarez, S. et al., *Ind. Eng. Chem. Res.*, 41, 6156, 2002 (A1); D1—From Das Gupta, D.K. and Jayarama, K.S., *J. Sci. Ind. Res.*, 55, 966, 1996 (D1); M1—From Matta, V.M. et al., *J. Food Eng.*, 61, 477, 2004 (M1); S1—From Sá, I.S. et al., *Braz. J. Food Technol.*, 6, 53, 2003 (S1); S2—From Singh, N.G.I. and Eipeson, W.E., *J. Food Sci. Technol.*, 37, 363, 2000 (S2)].

also perceived in Figure 6.7. It actually represents the main limitation of RO, as the required operating pressures are too high (>20 MPa) to reach the solids contents of 50–65 wt% associated with traditional vacuum evaporation. In practice, RO is economically limited to concentrations between 25 and 30°Brix and should be viewed, therefore, as a preconcentration step.

Apart from a higher limit for the solids concentration, an increase in ΔP^M leads to a higher water flux for a given fruit juice–membrane pair. For a constant solute concentration (which is experimentally achieved by recirculating both retentate and permeate streams), the effect is initially linear as predicted by Equation 6.3 (Álvarez et al., 1997, 2002; Jesus et al., 2007), but it starts to level off at higher pressures, as demonstrated, for instance, by Kozák et al. (2008), Sá et al. (2003), and Voit et al. (2006), and for black currant, pineapple, and tomato juices, respectively. A limit value is eventually achieved, with no changes in the water flux upon further increases in ΔP^M. Recent power consumption data (Voit et al., 2006) demonstrate that, despite the higher pump power requirements, operation at higher pressures reduces the total energy cost in view of the lower processing time to reach a given degree of concentration. Furthermore, contrary to water, solutes do not exhibit a pressure dependence in their flux equation, and, consequently, they become progressively more dilute in the permeate upon an increase in the operating pressure. This explains why, for the same membrane, the rejection coefficient of all solutes grows upon an increase in the operating pressure (Álvarez et al., 1998, 2002; Rodrigues et al., 2004).

The departure from a linear dependence of flux on ΔP^M is mainly a result of a phenomenon known as concentration polarization, which is intrinsic to all membrane operations. It refers to the development of a concentration gradient between the vicinity of the membrane and the liquid bulk. The concentration of rejected solutes increases close to the membrane as water permeates, which generates a diffusive flux of solutes back to the liquid bulk, and since solute is continuously brought to the membrane surface by convective transport driven by water permeation, an equilibrium eventually develops. The higher concentration in the vicinity of the membrane leads to an effective osmotic pressure higher than the value predicted based on the concentration in the liquid bulk, and the linear dependency of flux on ΔP^M is no longer observed. The smaller the Reynolds number (*Re*) on the feed side, the smaller the degree of mixing due to fluid flow and, thus, the higher the extent of this concentration gradient. The concentration polarization effect is usually accounted for with the aid of the film theory and a correlation to predict the mass-transfer coefficient on the feed side (Álvarez et al., 1997, 1998). Though normally not relevant in water desalination (Baker, 2004), concentration polarization can play an important role in juice concentration by RO, as evidenced in the numerical study of orange juice concentration presented by Curcio et al. (2001), who applied the boundary-layer theory and numerically solved a 1D model for simultaneous momentum and mass transfer in a flat channel with a permeable and a nonpermeable wall (the membrane and the module cover, respectively). The juice was represented as non-Newtonian fluid, the influence of apparent viscosity and solids concentration on the diffusion coefficient was accounted for, and the flux was computed with Equation 6.3 with both ΔP^M and $\Delta\pi$ expressed as a function of the axial coordinate. Depending on the water permeability constant A^*, the solutes concentration at the membrane surface reach up to seven times the value associated with the liquid bulk. Moreover, for $Re = 2000$, the results indicated that a fivefold increase in the value of A^* could still lead to lower flux values at the final part of the module due to the severe concentration polarization effect.

When the concentration of rejected solutes becomes sufficiently high, they start to precipitate on the membrane forming a gel layer, which gives rise to a permanent loss of flux known as fouling. Fouling is a major concern in all pressure-driven membrane operations. It requires periodic cleaning of the membrane and is the main cause of membrane replacement. RO concentration of cloudy fruit juices is usually associated with a high degree of fouling, mainly due to pectin and polysaccharides (Girard and Fukumoto, 2000; Jiao et al., 2004; Rektor et al., 2007). Apart from the substantial drop in flux, Medina and Garcia (1988) verified that excessive fouling can also result in a considerable loss of important aroma compounds, probably due to the entrapment of these compounds in the gel layer. Consequently, an initial juice clarification step by enzymatic treatment, usually coupled with either MF or UF, has become almost a standard procedure, with significant fouling reduction (Bottino et al., 2002; Matta et al., 2004; Gomes et al., 2005). Enzymatic treatment hydrolyzes pectin and other molecules such as starch, cellulose, and hemicellulose, causing them to flocculate and eventually settle. This enables their removal prior to RO concentration, which reduces fouling on the RO membrane, and decreases the viscosity of the juice, thereby enhancing RO performance. Coupling of enzymatic treatment with UF or MF, in turn, leads to a considerable reduction in the amount of enzyme used and to an improvement in the yield of the process (Koeseoglu et al., 1990; Álvarez et al., 2000). However, depending on the adopted membrane and module, fouling can become a problem in the UF/MF step, as observed for instance by Voit et al. (2006) for tomato juice in plate-and-frame units.

With regard to the operating temperature, a first glance at Equation 6.3 may lead to the prediction of a decrease in the water flux with increasing operating temperature. Nevertheless, in practice, the opposite trend is verified (Álvarez et al., 1997; Singh and Eipeson, 2000; Voit et al., 2006), which stems from the exponential increase of the diffusion coefficient in the membrane with temperature. Though initially very significant, this effect starts to level off as the solids concentration in the juice increases and the osmotic pressure influence on flux progressively dominates. Naturally, the increase in the diffusion coefficient is valid for all compounds, which means an increase in the flux of all solutes across the membrane (see Equation 6.4), or, in other words, a drop in their rejection coefficients, especially for some aroma compounds (Pearce and Bullen, 1995; Álvarez et al., 2002).

There has not been, to the authors' knowledge, any attempt to develop a RO membrane specifically for juice concentration. Commercial RO membranes, originally developed for water desalination, have been used in all cases. The current choices are either aromatic polyamides (Álvarez et al., 1997, 2000, 2002; Singh and Eipeson, 2000; Pozderovic et al., 2007) or the "thin-film composites" prepared by interfacial polymerization on the surface of a porous support (Das Gupta and Jayarama, 1996; Sá et al., 2003; Matta et al., 2004; Gomes et al., 2005). Although still commercially available, the first generation of RO membranes, based on cellulose acetate, was proven to give lower fluxes and higher losses of aroma and other juice constituents (Girard and Fukumoto, 2000; Jiao et al., 2004), and thus is not considered for juice concentration anymore.

The superior quality of reconstituted juices prepared by RO concentrates has been confirmed by sensory analysis. A high level of acceptability was already reported in the case of watermelon (Das Gupta and Jayarama, 1996), apple (Álvarez et al., 2000), and acerola (Gomes et al., 2005) juices. In the specific case of orange juice, Jesus et al. (2007) compared reconstituted juices (12°Brix) from concentrates (30°Brix) obtained by RO and thermal evaporation at 50°C. The characteristic aroma of the juice was found better preserved in the RO juice, even though its taste was found more acidic, which stem from a change in the sugar/acid ratio due to the different rejection coefficients of sugars and acids with the adopted membrane.

In conclusion, it can be said that, after 40 years since the pioneering work of Merson and Morgan (1968), RO has definitely proven its adequacy and advantages as a preconcentration technique for fruit juices. Industrially, it is already applied in some plants in the United States to remove more than half the total amount of water during apple and pear juice concentration (Duxbury, 1992), as well as in a plant in Italy to preconcentrate tomato pulp juice to 8.5% prior to further concentration by evaporation (Voit et al., 2006). Furthermore, commercial use for preconcentration of orange juice has also been reported (Girard and Fukumoto, 2000).

6.3.2 Osmotic Distillation and Membrane Distillation

OD and MD are two similar unit operations in which the solution to be concentrated, that is, the feed stream, is separated from a solvent-receiving phase by a microporous, nonwettable membrane. The two liquid phases are contacted through the gas (mostly air) entrapped within the membrane, and a chemical potential difference for the solvent is established between these two liquid phases to bring about mass transfer. The difference between the two unit operations lies in method used to obtain this chemical potential difference. In OD, an osmotic agent is dissolved in the receiving phase to reduce the chemical potential of the solvent compared to the feed, whereas in MD the feed temperature is raised to increase the chemical potential of its solvent compared to the solvent-receiving phase, which is normally (but not exclusively) water.* Since the liquids cannot enter the pores, the solvent (and any other volatile species in the feed stream) must vaporize on the feed side, where its chemical potential is higher, diffuse as vapor through the pores of the membrane, and then condense in the receiving phase.

In the specific case of fruit juices, since the solvent is water, the membranes must be hydrophobic. Studies on MD started in the early 1960s focusing on water desalination (Lawson and Lloyd, 1997), which remains its most promising application, and Kimura et al. (1987) were probably the first to report the concentration of liquid foods by MD. Research on OD, on the other hand, started in the late 1980s and has its genesis in the concentration of heat-sensitive solutions (Lefebvre, 1985, 1987). The technique has also been called osmotic evaporation, membrane evaporation, direct OD, isothermal MD, and gas membrane extraction. Although OD has still not reached commercial scale, it is by far the most studied technique for fruit-juice concentration in the recent literature, with a steadily increasing number of publications.

The large interest in OD is justified by its advantages over traditional evaporation and other membrane technologies. Operation can take place at room temperature and atmospheric pressure, preventing the thermal damage of heat-sensitive components and minimizing the loss of aroma compounds. In contrast to RO, OD is not limited by osmotic pressure, and total solids contents higher than 60 wt% are normally achieved by the choice of suitable operating conditions (Jiao et al., 2004; Nagaraj et al., 2006a). Additionally, OD is associated with lower energy consumption, since neither heating nor high feed pressures are required. Besides, the porous membranes adopted in OD are generally less expensive than dense composite RO membranes (Vaillant et al., 2005). The potential of OD for concentrating fruit juices has been demonstrated on laboratory scale for a variety of fruits, including melon (Vaillant et al., 2005), kiwifruit (Cassano et al., 2004; Cassano and Drioli, 2007), and cactus pear (Cassano et al., 2007), which are particularly difficult to concentrate by vacuum evaporation on account of their high sensitivity to heat-driven nonenzymatic browning. Additionally, successful operation at pilot-plant scale, with membrane areas

* This MD configuration is formally known as direct-contact MD to differentiate it from other possible configurations involving, for instance, a sweeping gas flow or vacuum on the permeate side. However, since it is the only configuration considered so far for juice concentration, it will be simply referred to as MD in this text. Details on other MD configurations and their uses can be found elsewhere (Curcio and Drioli, 2005; Khayet, 2008).

larger than 10 m², has also been reported (Hogan et al., 1998; Vaillant et al., 2001, 2005; Ali et al., 2003; Versari et al., 2004; Bui and Nguyen, 2005; Cisse et al., 2005; Kozák et al., 2008).

Although the advantage of reaching total solids contents higher than 60 wt% is also associated with MD, the technique is thermally driven, and for a given permeate flux, a product of inferior sensorial quality compared to OD can be theoretically expected. This is probably the reason why studies on fruit-juice concentration by MD are less common. Feed temperature is normally kept below 50°C for juices in an attempt to reach a compromise between product quality and flux magnitude (Calabrò et al., 1994; Gunko et al., 2006; Rektor et al., 2006), whereas values higher than 70°C are common in water desalination and wastewater treatment applications (El-Bourawi et al., 2006; Khayet, 2008).

6.3.2.1 Permeate Flux: Qualitative and Quantitative Aspects

One of the major limitations which still prevent commercial application of OD and MD for juice concentration is the low permeate flux associated with these techniques in comparison to RO and other pressure-driven membrane operations. Typical OD and MD water fluxes range from 0.5 to 7 kg m^{-2} h^{-1}, whereas values of 10–50 kg m^{-2} h^{-1} are common in the concentration of juices by RO. As a result, many authors have investigated the dependence of permeate flux on operating variables for OD and MD, and a summary of their main conclusions is provided in Table 6.4. Upon analyzing Table 6.4, one should bear in mind that it was elaborated based on the conceptual definitions of OD and MD, which require, for example, isothermal operation in OD (this explains why a feed temperature increase is not considered for OD in the referred table).

Most of the trends listed in Table 6.4 can be directly reasoned using the phenomenological equation below, which shows a linear relation between volumetric flux (J^V, m³ m^{-2} h^{-1}) and its driving force, that is, the partial pressure (p_i) difference between the membrane interfaces:

$$J_i^V = K_i\left(p_i^F - p_i^{RF}\right) = K_i\left(a_i^F P_i^{sat}\big|_{T=T^F} - a_i^{RF} P_i^{sat}\big|_{T=T^{RF}}\right) \tag{6.6}$$

where

- a_i and P_i^{sat} are the activity in the liquid and the saturation pressure of component i, respectively
- K_i is an overall mass-transfer coefficient and the superscript RF refers to the receiving phase

It will be demonstrated, though, that the apparent simplicity of Equation 6.6 is rather misleading, since the value of K_i is a complex function of operating conditions, membrane characteristics, and solvent properties. In fact, K_i can seldom be properly predicted without some degree of experimental input.

According to Equation 6.6, the total solids content of the feed stream will affect the driving force through its influence on water activity. As concentration proceeds, water activity on the feed side decreases, and water flux drops accordingly, as exemplified by the literature data presented in Figure 6.8. This effect is initially small

TABLE 6.4
Qualitative Effect of the Individual Increase of Different Operating Parameters on Transmembrane Flux in MD and OD

Parameter	Trend in Flux		References for Sample Data
	MD	OD	
Solids concentration in the feed	↓	↓	Hongvaleerat et al. (2008), Kimura et al. (1987), Lukanin et al. (2003), Versari et al. (2004)
Operating temperature	n. c.	↑	Bui et al. (2004), Gostoli et al. (1999)
Feed temperature	↑	n. c.	Bui and Nguyen (2006), Rektor et al. (2006)
Feed velocity	↱	↱	Alves et al. (2004), Bui and Nguyen (2006), Courel et al. (2000a), Gunko et al. (2006)
Temperature on permeate side	↓	n. c.	Calabrò et al. (1994), Laganà et al. (2000)
Osmotic agent concentration	n. c.	↑	Cassano et al. (2003), Warczok et al. (2007)
Fluid velocity on permeate side	↱	↱	Alves et al. (2004), Khayet et al. (2005), Laganà et al. (2000), Nagaraj et al. (2006b)

Note: ↑, increase; ↓, decrease; ↱, increase towards asymptotic value; n. c., not considered.

up to concentrations of about 40°Brix because of the low osmotic activity of sugars and other juice solutes (Hogan et al., 1998; Mansouri and Fane, 1999; Courel et al., 2000a; Hongvaleerat et al., 2008), but boundary-layer effects become progressively more important due to the exponential increase in viscosity with solids content, decreasing the value of K_i, and rather significant reductions in water flux can occur. Boundary-layer effects, also responsible for the dependence of flux on feed and permeate flow rates, are detailed in Section 6.3.2.3. It is important to note that, in comparison to RO, the relative drop in OD and MD fluxes due to increasing solids content in the feed is much less pronounced, as evidenced by a comparison between Figures 6.7 and 6.8.

As previously explained, the chemical potential difference for mass transfer in MD is created by forcing a temperature difference between the liquids in contact with the membrane. Therefore, any changes in this temperature difference will bring about a variation in flux. In view of the exponential increase of saturation pressure with temperature, at a given receiving-phase temperature (T^{RF}), water flux increases exponentially with feed temperature (Calabrò et al., 1994; Bui and Nguyen, 2006). In fact, for the feed temperatures considered in desalination studies ($T^F > 60°C$), MD fluxes higher than 20 kg m^{-2} h^{-1} are normally reported (Schofield

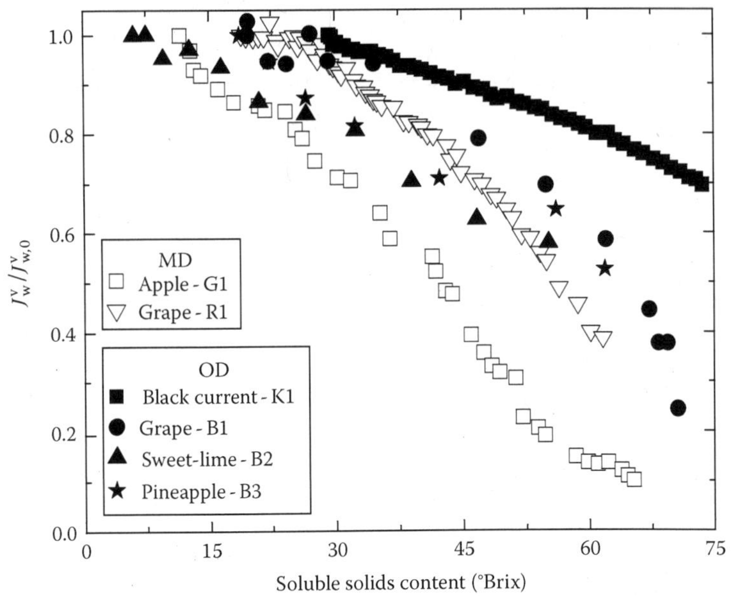

FIGURE 6.8 Evolution of water flux during the concentration of different fruit juices by membrane distillation (MD, open symbols) and osmotic distillation (OD, filled symbols) under different operating conditions. [Based on data from Bailey, A.F.G. et al., *J. Membr., Sci.*, 164, 195, 2000 (B1); Babu, B.R. et al., *J. Membr. Sci.*, 272, 58, 2006 (B2); Babu, B.R. et al., *J. Membr. Sci.*, 322, 146, 2008 (B3); Gunko, S. et al., *Desalination*, 190, 117, 2006 (G1); Kozák, A. et al., *Chem. Eng. Process.*, 47, 1171, 2008 (K1); Rektor, A. et al., *Desalination*, 191, 44, 2006 (R1)].

et al., 1990a; Bonyadi and Chung, 2007). Following the same reasoning, it follows that, if T^F is kept constant and T^{RF} increases, water flux drops, an effect whose magnitude decreases with increasing feed temperature (Gunko et al., 2006). An interesting consequence of the exponential dependence of P^{sat} on T is the fact that, for a fixed temperature difference $T^F - T^{RF}$, the permeate flux increases with the mean temperature $(T^F + T^{RF})/2$ (Izquierdo-Gil et al., 1999a; Khayet et al., 2005; Gunko et al., 2006). With regard to OD, since the process is isothermal, there is one less degree of freedom, with the temperature of both streams being varied at the same time. Once again, an exponential increase in flux with increasing temperature is observed (Gostolli et al., 1999; Narayan et al., 2002; Ali et al., 2003; Bui et al., 2004; Thanedgunbaworn et al., 2007a).

Bearing in mind the similarities between MD and OD, and considering that Equation 6.6 is a general expression that is simplified in different ways to be applied for each individual operation, one may wonder why the driving force for mass transfer has to be necessarily generated by a difference in either temperature or concentration, instead of both. It did not take long for researchers to realize this possibility and attempt to enhance water flux by applying a temperature difference between feed and osmotic agent streams in OD (Lefebvre, 1985; Godino et al., 1995; Courel et al., 2000a; Bui and Nguyen, 2006), or by adding an osmotic agent to the receiving phase in MD (Laganà et al., 2000; Tomaszewska, 2000; Wang et al., 2001). The unit operation, in this case, is formally referred to as osmotic membrane distillation (OMD) to highlight the fact that the driving force is a combination of temperature and activity differences, even though such conceptual distinction is not always correctly made in the literature (Gryta, 2005a).

Small temperatures differences in OMD can bring about significant improvements in flux. For instance, during the concentration of a model aqueous solution (6 wt% of sucrose and 1.5 wt% of ascorbic acid) using $CaCl_2$ 4.0 mol L^{-1} as receiving phase, Rodrigues et al. (2004) verified an increase of about 61% in the average J_w^m value compared to isothermal operation (that is, OD) when the temperature of the receiving phase was reduced from 30°C to 20°C. Another example is the work of Bélafi-Bakó and Koroknai (2006), in which increases of 10°C–20°C in the feed temperature resulted in increments of 26%–168% in J_w^V for pure water and two sucrose solutions (20 and 45 wt%), when $CaCl_2$ (3.5 and 6.0 mol L^{-1}) at 25°C was used as receiving phase. In addition, experimental data demonstrating the higher permeate flux for OMD in the case of real juices are also available (Bélafi-Bakó and Koroknai, 2006; Nagaraj et al., 2006b; Hongvaleerat et al., 2008). As discussed by Koroknai et al. (2006), for a given temperature and activity difference, the driving force in OMD is actually higher than the sum of the individual values associated with OD and MD alone, so the combination leads to a synergistic effect. Nonetheless, experimental data have revealed that, on account of boundary-layer effects, the actual gain in flux for a given temperature difference is not always equal to the prediction based solely on the driving force, and it decreases with increasing solids content in the feed (Courel et al., 2000a; Bélafi-Bakó and Koroknai, 2006).

A distinct feature of OD is the presence of an osmotic agent in the receiving phase, also referred to as strip solution. From Equation 6.6, it is clear that an increase in the concentration of the osmotic agent should cause an increase in the water flux, a well-documented trend that stems from the higher driving force brought about by the reduction in a_w in the brine (Deblay, 1992; Vaillant et al., 2001; Cassano et al., 2004; Babu et al., 2006; Cassano and Drioli, 2007). Intuitively, solutes of high water solubility and low equivalent weight (that is, high osmotic activity) are the natural candidates for osmotic agents. However, other properties must also be taken into account to ensure the safe, effective, and economic use of a given osmotic agent in OD, especially for food-related applications. Generally, a suitable solute should be (Michaels and Johnson, 1995; Hogan et al., 1998; Celere and Gostoli, 2004):

1. Chemically stable in solution and nonvolatile at all temperatures to which it is likely to be exposed during processing
2. Nonwetting and nondestructive to the hydrophobic membrane
3. Nontoxic to humans or animals
4. Devoid of detectable taste or smell in aqueous solutions
5. Noncorrosive toward the materials of construction of all process equipment to which it comes into contact
6. Substantially incapable of forming precipitates when exposed to volatile components of the feed solution
7. Commercially available in large scale and at a low cost

At a large-scale application, the diluted strip solution will have to be reconcentrated, so that it can be recycled and reused in the OD unit to reduce costs. Conventional evaporation is the most likely choice for this reconcentration, in which case the solute should also exhibit a large positive temperature coefficient of solubility in aqueous solution to maximize evaporative water removal and minimize, at the same time, the possibility of crystallization inside the evaporator. Other reconcentration methods have also been mentioned in the literature, including RO, pervaporation, solar evaporation, and electrodialysis (Petrotos and Lazarides, 2001).

Calcium chloride is, by far, the most used solute in OD studies, probably due to its low cost and exceptional ability to reduce water activity, followed by sodium chloride and magnesium chloride. Magnesium sulfate (Lefebvre, 1985) and lithium chloride (Albrecht et al., 2005) have also been tested. However, as pointed out by Michaels and Johnson (1995), none of these salts meet all requirements previously mentioned. In the case of NaCl, the driving force available is limited because of its low solubility ($a_w > 0.7$ at saturation at room temperature), and this salt has a detectable taste in aqueous solution. Apart from this last disadvantage, salts containing calcium and magnesium have the additional risk of formation of insoluble precipitates, particularly with compounds commonly found in process water of liquid food products such as carbon dioxide, sulfate, fluoride, and phosphates. Moreover, halide salts generally have the disadvantage of being highly corrosive to many common materials of construction, including stainless steel 304 and 316, as evidenced by the data in Table 6.5.

As an alternative, Michaels and Johnson (1995) proposed the use of potassium salts of phosphoric (H_3PO_4) and pyrophosphoric ($H_4P_2O_7$) acids, which, according to

TABLE 6.5
Corrosion-Pit Statistics on Stainless Steel 304 (SS304) and 316 (SS316)

	Number of Pits per cm²		Average Pit Diameter (μm)	
Brine	**SS304**	**SS316**	**SS304**	**SS316**
NaCl	2360	2040	48	37
$CaCl_2$	1570	1260	46	48
K_2HPO_4/KH_2PO_4	630	470	22	38
CH_3COOK	470	310	40	62
$K_4P_2O_7/H_4P_2O_7$	470	160	22	38
K_2HPO_4	470	160	19	48
$K_4P_2O_7$	310	310	24	24
K_2HPO_4/H_3PO_4	160	310	29	29

Source: Modified from Michaels, A.S. and Johnson, R. U.S. Patent 5824223, 1995.

Note: Samples were exposed to different brines in trials conducted at (96±2)°C for 453 h with coupons having a surface area of 8.6–8.8 cm² and a mass of 11.3–12.2 g.

the authors, meet all requirements for an osmotic agent in OD. First of all, such salts are normally present in biological fluids, being, thus, safe for food use when present in low concentrations. For a given mole fraction in the strip solution, permeate fluxes obtained in OD experiments with pure water as feed for these salts compared favorably with or were even superior to those related to $CaCl_2$. Furthermore, in immersion corrosion tests, these salts caused significantly fewer numbers of pits per square centimeter in both stainless steel 304 and 316 compared to halide salts (see Table 6.5), evidencing their low corrosivity. In a latter work, Narayan et al. (2002) found little different in the water flux obtained with K_2HPO_4 (4 mol L^{-1}) and $CaCl_2$ (5 mol L^{-1}) during the concentration of sugarcane juice by OD.

Concerned with corrosion and scaling problems involved in the reconcentration of salt solutions by conventional evaporation, Gostoli and coworkers (Gostoli et al., 1999; Celere and Gostoli, 2004, 2005; Versari et al., 2004) have investigated the use of organic solvents, namely glycerol and propylene glycol, as osmotic agents in OD. These two solutes have a similar ability to reduce a_w, which is higher than the one associated with NaCl but still smaller than the one obtained with $CaCl_2$. Initial tests with pure water in a commercial hollow-fiber module showed that glycerol and propylene glycol gave similar fluxes that were more than 60% larger than those obtained with NaCl. However, when orange juice was used as feed, the pressure drop was too large with these solutes, resulting in membrane wetting and contamination of the feed (Gostoli et al., 1999). Operation with a plate-and-frame module (1.5 m^2 of membrane area) solved the problem for glycerol, and orange juice could be concentrated up to 60°Brix, but propylene glycol was still detected in the feed. The use of glycerol was then tested in two bigger plate-and-frame modules (3.1 and 12.4 m^2 of membrane area) by Versari et al. (2004) for the concentration of sucrose solutions and grape juice. No evidence of membrane wetting was found, and the red wine produced from the grape juice concentrated by OD was preferred in the sensory analysis. Nevertheless, the high viscosity of concentrated glycerol solutions (>50 wt%) led to significant reductions in the value of K_i (Equation 6.6) owing to boundary-layer effects. In subsequent works, these effects were studied in detail by Celere and Gostoli (2004, 2005), both for hollow-fiber and plate-and-frame modules. To overcome the problem of high viscosity, the use of a glycerol (42.74 wt%)–NaCl (14.52 wt%) solution as stripping agent was suggested, whose a_w value is equal to that of a glycerol 70 wt% solution. Propylene glycol was tested again, but its use as osmotic agent was discouraged. Careful pressure drop control was required to avoid membrane wetting, and even when it did not occur, propylene glycol diffused back to the feed stream due to its not negligible volatility.

6.3.2.2 Transport Mechanism in the Membrane

Transport of gases and vapors through microporous membranes has been extensively studied, and three different mechanisms have been used in the literature for the development of theoretical models to predict the performance of MD and OD membranes. Usually, a membrane of uniform and noninterconnected cylindrical pores is assumed, and the transport is described by Knudsen diffusion, ordinary molecular diffusion, or a combination of both often summarized as the dusty gas model. Further details on these equations can be found elsewhere (Kunz et al., 1996; Lawson and Lloyd, 1997; Khayet, 2008).

The Knudsen number, *Kn*, defined as the ratio of the mean molecular free path of the diffusing molecule to the diameter of the pore (d_p^*), can be used as a theoretical criterion to determine which mechanism prevails. For small pores, $Kn > 10$, diffusing molecules tend to collide more frequently with the pore walls than with other molecules, and Knudsen diffusion is the controlling mechanism. On the other hand, if the pore diameter is relatively large, $Kn < 0.01$, the collisions between gas molecules themselves are more frequent, and ordinary molecular diffusion is predominant. Between these two limits, both mechanisms coexist and can be combined as proposed in the dusty gas model. The mean free path of water vapor at 25°C and atmospheric pressure is 0.13 μm, which is comparable with the typical value of nominal pore size of membranes used in OD and MD (Gostolli, 1999).

These three models predict a linear relation between the water flux and the partial pressure difference, as shown in Equation 6.6, and the differences lie in the formula for calculating the proportionality constant, that is, the membrane permeability, as a function of its structural properties (thickness, pore diameter, porosity, and tortuosity). In particular, the Knudsen model predicts a dependence of membrane permeability on pore diameter, whereas, according to the ordinary diffusion mechanism, membrane permeability does not depend on this parameter. In an attempt to determine the prevailing mechanism, different authors have compared experimental membrane mass-transfer coefficients (ratio of permeability to thickness), k^M, with values calculated by each model. The results, however, are not conclusive, with some authors in favor of Knudsen diffusion (Calabrò et al., 1994; Pena et al., 1998; Courel et al., 2000b), others supporting molecular diffusion (Gostoli et al., 1987; Kimura et al., 1987; Gostoli and Bandini, 1995; Celere and Gostoli, 2005; Gunko et al., 2006), and a third group using the dusty gas model to include both mechanisms (Alves et al., 2004; Alves and Coelhoso, 2004; Babu et al., 2006). In most cases, the predictions are very far from the experimental values, representing, at best, a crude estimation.

This apparent contradiction and the poor predictive character of the models can be reasoned bearing in mind that, in the vast majority of these studies, the average pore size and porosity provided by the membrane manufacturer were used in the equations. Commercial membranes are always anisotropic materials, with a selective skin and a support to provide mechanical strength. Barbe et al. (1998) found significant differences between manufacturer-specified nominal pore diameters and mean pore diameters on the surface determined by image analysis for nine different membranes. As shown by Courel et al. (2001), the structural properties of each layer of a composite, anisotropic membrane are very different from the parameters given by the manufacturer, which actually refer to the whole composite material. This is a very important issue, since the support layer has much larger pores and the aqueous solutions can usually penetrate it, which implies not only a difference between the thickness provided by the manufacturer and the effective membrane thickness for gas transport in OD or MD, but also an additional resistance to mass transfer, that is, diffusion in the stagnant liquid within the wetted pores (Romero et al., 2003). As an additional example to prove the point made here, the study of Khayet and Matsuura (2003) can be mentioned. These authors applied different techniques to obtain surface and bulk pore sizes of various laboratory-made flat-sheet and hollow-fiber membranes, and the results evidenced that surface pores were larger than those in the membrane bulk. In the case of flat-sheet membranes, for instance, the pores at the bottom surface were 3.7–9.8 times larger than those at the top membrane surface, which, in turn, were 2.1 times larger than the bulk pore sizes determined by gas permeation tests. Further evidence of the inadequacy of the nominal pore size as a characteristic length for gas transport models in OD and MD is the fact that examples of commercial membranes made of the same material, with the same nominal pore diameter, but with different liquid penetration pressures (see Section 6.3.2.4) can be found (Gostoli, 1999; Izquierdo-Gil et al., 1999b; Bui et al., 2004; Bui and Nguyen, 2006).

Another important aspect to consider is the fact that no value is provided by the manufacturer for tortuosity, which appears in the equations of all three models. In fact, this parameter is quite difficult to determine with accuracy, and the standard procedure is either to specify a value equal to 2.0 or to assume a given transport mechanism and fit tortuosity based on the experimental k^M. One should also bear in mind that all commercial membranes have a somewhat wide pore size distribution, whose effect is not included in any of the models.* Thus, unless all structural parameters of a given membrane are known with accuracy, and preferentially for a membrane with a very narrow pore size distribution, the distinction between transport mechanisms in OD and MD solely based on comparisons between predicted and experimental k^M values will remain inconclusive. It ought to be emphasized, though, that, because of the complex morphology of the applied membranes, from the practical point of view, this distinction is of limited interest, as none of the models are completely predictive, and to obtain reliable k^M values, at least one parameter has to be fitted based on the experimental k^M value for the membrane under consideration. Needless to say, as far as fitting is concerned, under relevant operating conditions for fruit-juice concentration, a particular k^M for given membrane can be obtained by all three models with different structural parameters. Therefore, in practice, k^M is always experimentally determined.

An alternative reasoning for determining the transport mechanism was discussed by Hogan et al. (1998). These authors pointed out that the equation researchers have adopted to estimate the mean free path of water vapor, derived from the kinetic theory of gases, is only valid for pure gases, and due to the presence of air in the pores of the membranes during MD and OD, the actual mean free path is much smaller. Therefore, unless the stagnant air trapped within the membrane pores is removed, ordinary molecular diffusion will be the dominant transport mechanism. Schofield et al. (1990b) predicted that complete deaeration of feed and permeate would give a sevenfold increase in k^M, which would be offset by a fivefold decrease in driving force resulting from boundary-layer effects. The predicted flux increase was experimentally observed in MD runs with pure water and different air pressures. With regard to OD, Hogan et al. (1998) worked with polypropylene

* Some attempts to include the effect of pore size distribution in the model have been proposed in the literature, as reviewed by Khayet (2008), but these did not result in a significant improvement in the predictive character of the model.

hollow fibers with a nominal pore diameter of 0.03 μm and verified that water fluxes were two to three times higher in the absence of air in the pores, that is, upon removal of dissolved air from feed and strip solution. As previously mentioned, according to the molecular diffusion mechanism, k^M must be independent of pore radius, a trend that has already been reported for both MD (Kimura et al., 1987; Schofield et al., 1987; Bui et al., 2007) and OD (Mansouri and Fane, 1999; Narayan et al., 2002; Brodard et al., 2003; Babu et al., 2006; Nagaraj et al., 2006b) when no special procedure was used to ensure the absence of air in the pores of the membrane. Thus, based on this reasoning and all these experimental results in its favor, it can be concluded that ordinary molecular diffusion is the predominant transport mechanism in MD and OD membranes at the operating conditions relevant for fruit-juice concentration.

6.3.2.3 Boundary-Layer Effects

Similar to what was previously discussed for RO, during OD, the concentration of rejected solutes on the feed side increases close to the membrane as water permeates, and a diffusional boundary layer is established. Additionally, due to condensation of the permeated water, the concentration of osmotic agent in the vicinity of the membrane is reduced compared to its value in the bulk of the strip solution, leading to another diffusional boundary layer. Upon analyzing Equation 6.6, one concludes that a higher concentration of solids on the feed side of the membrane (compared to its value in the bulk) and a lower concentration of osmotic agent on the strip side (also compared to the bulk) will lead to a lower driving force and thus, to a permeate flux that is lower than the one predicted based on the bulk concentrations. This diffusional-boundary-layer effect is known in the membrane literature as concentration polarization. It is also present in the case of MD, but since water is normally used as receiving phase, only the feed side contribution is observed.

Water permeation through the membrane in OD or MD implies evaporation on the feed–membrane interface, and, consequently, the corresponding latent heat must be provided. Initially, this is solely provided as sensible heat from the bulk feed, and thus the feed temperature decreases in the vicinity of the membrane compared to its value in the bulk, leading to a thermal boundary layer. Conversely, at the downstream surface of the membrane, energy is released as water vapor condenses into the receiving phase, increasing the temperature of the liquid close to the membrane compared to its value in the bulk, forming another thermal boundary layer. Therefore, even though the liquid streams fed to the membrane unit in OD are isothermal, a temperature difference is observed between the upstream and downstream sides of the membrane in both MD and OD. As a consequence, not only mass but also heat is transferred through the membrane in these operations.

Based on Equation 6.6, it follows that, similar to what occurs in the case of concentrations, the presence of these thermal boundary layers will decrease the driving force for mass transfer, producing a lower permeate flux compared to the value expected based upon the inlet temperatures of each liquid stream. In the membrane literature, this thermal-boundary-layer effect is usually referred to as temperature polarization, and it is also present in pervaporation, another membrane-based unit operation which involves phase transition.* Schematic representations of these diffusional and thermal boundary layers are easily found in the literature and will not be repeated here. As good examples, the illustrations provided by Alves and Coelhoso (2004) for OD, and by Bui et al. (2007) for MD, can be mentioned. Though not always clear in such representations, it is important to remember that the thickness of the diffusional and thermal boundary layers on a given side of the membrane is likely to be different.

The importance of boundary-layer effects in OD and MD was quickly recognized, and there has been a large amount of studies on this subject, which is not surprising, since such effects reduce water flux, whose low value is normally described as a disadvantage to be overcome toward industrial application. The traditional way of experimentally evidencing concentration and temperature polarization on a given side of the membrane is to keep a constant driving force for mass transfer and measure J_w^V (or J_w^m) as a function of Re for the liquid stream on that side, the latter being varied by changes in either the feed flow rate (continuous or closed-loop systems) or the stirring speed (batch systems) (Calabrò et al., 1994; Godino et al., 1995; Courel et al., 2000a; Narayan et al., 2002; Bui et al., 2004; Khayet et al., 2005; Babu et al., 2006, 2008; Bui and Nguyen, 2006; Gunko et al., 2006; Nagaraj et al., 2006b; Alves and Coelhoso, 2007; Thanedgunbaworn et al., 2007a). An asymptotic increase in J_w^V as a function of the varied parameter confirms that boundary-layer effects are present in the region where J_w^V is a function of Re. Because the thickness of the boundary layers decreases with increasing Re, their contribution to the total mass-transfer resistance becomes progressively less important as Re grows, up to the point at which they are negligible compared to the membrane resistance, and J_w^V does not depend on Re anymore. Some authors (Versari et al., 2004; Celere and Gostoli, 2004) have adopted an alternative approach, in which Re is kept constant, and J_w^V is measured for increasing values of driving force obtained by varying the concentration of the osmotic agent. For low driving forces, boundary-layer effects are not significant, and a liner relation between J_w^V and $(p_w^F - p_w^{RF})$ is verified, that is, K_w is constant and equal to k_w^M. However, as J_w^V grows, the thickness of the boundary-layers increases, and their resistance to mass transfer acquires importance, reducing the value of K_w and breaking the linear relation between J_w^V and $(p_w^F - p_w^{RF})$. An interesting experiment to show temperature polarization during MD was designed by Albrecht et al. (2005). Pure water was used on both sides of the membrane and the steady-state flux was measured as a function of the applied temperature difference, $T^F - T^{RF}$, which was both positive and negative. Choosing one direction to define a positive flux, the authors then plotted the values of $(p_w^F - p_w^{RF})$ estimated from the bulk temperatures as a function of the measured flux. If there were no temperature polarization, the point of zero flux should be associated with a zero driving force, but the linear regression line fitted to the data identified $J_w^V = 0$ for $(p_w^F - p_w^{RF}) = 150\,\text{Pa}$, which corresponds to a temperature difference of 1.3 K.

* The basic principles of pervaporation and its potential applications in the food industry are discussed in Chapter 5.

The changes in water flux due to concentration and temperature polarization can be rather significant. For instance, during the concentration of clarified passion fruit juice by OD in a pilot-scale unit with an effective membrane area of 10.2 m^2, Vaillant et al. (2001) continuously collected a 60 wt% concentrate over several hours with a constant flux of 0.49 kg h^{-1} m^{-2}, but this value dropped by almost 20% when the feed circulation flow was deliberately decreased, reducing the tangential velocity from 0.24 to 0.09 m s^{-1}. Babu et al. (2006), in turn, observed a 42% increase in J_w^V when the velocity of the feed stream (55 wt% sweet-lime juice) was raised from 0.94 to 3.75 mm s^{-1} in a module having a membrane area of 120 cm^2. On account of temperature polarization, a dependence of J_w^m on feed flow rate, with differences of up to 30% in relation to the asymptotic value, can be observed with pure water as feed stream in OD, as demonstrated by Gostoli (1999) and Celere and Gostoli (2002, 2004, 2005) for different osmotic agents. With regard to MD, in their study of orange juice concentration in a module with a membrane area of 20.42 cm^2, Calabrò et al. (1994) tested three temperature differences ($T^F - T^{RF}$ = 15°C, 20°C, and 25°C) and verified that J_w^m increased by 79%–143% as the feed and receiving-phase flow rates were simultaneously increased from 2.0 to 5.0 kg min^{-1}, the effect being more pronounced for smaller ($T^F - T^{RF}$) values. Working with apple juice as feed and a module with 490 cm^2 of membrane area, Gunko et al. (2006) observed a 12% increase in J_w^V for a feed containing 12 wt% of total solids upon changing the juice velocity from 0.37 to 0.58 m s^{-1}. When the total solids content of the juice reached 25 wt%, J_w^V became 25% higher for the same velocity change. Different factors can be pointed out to explain the differences among the relative changes in water flux verified in the cited examples, such as module design, feed temperature, and membrane characteristics, but the provided numbers should make it clear that boundary-layer effects must always be analyzed and taken into account.

A particularly important aspect of such effects during the concentration of fruit juices by OD, which equally applies to MD, was well illustrated by the data of Alves et al. (2004). Operating under previously determined hydrodynamic conditions to minimize boundary-layer effects in their module (effective internal area of 0.16 m^2), these authors reported a constant K_w (Equation 6.6) during the concentration of a 12 wt% sucrose solution by OD, up to a sucrose concentration of about 40 wt% in the feed stream. From this point on, as operation proceeded and sucrose concentration in the feed increased, K_w began to decrease rapidly due to the increasing contribution of the boundary layers to the overall mass-transfer resistance, which reached up to 55%. Similar behavior has been observed by other authors with real juices (Cassano et al., 2003, 2004, 2007; Bélafi-Bakó and Koroknai, 2006; Cassano and Drioli, 2007; Koroknai et al., 2008), and, in all cases, the points at which K_w starts to decrease and juice viscosity (μ) starts to increase rapidly (normally spanning two orders of magnitude) are almost identical. This is easy to understand bearing in mind that, as previously discussed, the thickness of all boundary layers is a function of *Re*, whose value depends on μ^{-1}. Furthermore, Hogan et al. (1998) state that, in view of the anomalously high viscosity values (and, frequently, non-Newtonian rheologic behavior) of highly concentrated juices, a "viscous fingering" phenomenon can take place as juices pass through a membrane-bounded channel. This phenomenon refers to the channeling of dilute, low-viscosity juice through the centers of channels bounded by membrane coated with essentially stationary films of concentrate with very high viscosity, leading to reduction in juice residence time in the membrane contactor and little access of channeled juice to the membrane surface, with substantial reduction in water flux. One clever way of minimizing these problems is to reach the target solids content of 60 wt% or higher in stages, operating with at least two membrane modules in series, a strategy whose advantage has been clearly demonstrated in a number of recent studies (Vaillant et al., 2001; Rodrigues et al., 2004; Bui and Nguyen, 2005; Cisse et al., 2005; Hongvaleerat et al., 2008).

The application of an acoustic field was suggested by Nagaraj et al. (2002) as a means of reducing boundary-layer effects in OD. Experiments were carried out with pure water and sugarcane juice as feed in a flat membrane cell (effective area of 15.9 cm^2) placed over an ultrasonic transducer (1.2 MHz), using both NaCl (5 mol L^{-1}) and $CaCl_2$ (5.3 mol L^{-1}) as receiving phases. In all cases, the average water flux was higher in the runs with acoustic field, which was ascribed to mild circulation currents induced by the acoustic field. For pure water as feed, the flux increase varied between 35% and 98%, depending on the adopted membrane and strip agent, whereas, for sugarcane juice, values as high as 204% were reported.

In Equation 6.6, the driving force was written only in terms of bulk conditions with the aid of an overall mass-transfer coefficient, K, for the sake of simplicity. Based on the aspects discussed so far, it follows that K must account for the membrane resistance and the effects of concentration and temperature polarization, which is accomplished based on the resistances-in-series approach. However, due to the presence of a diffusional and a thermal boundary layer, the final equation to actually calculate the volumetric flux in terms of mass-transfer resistances is different from Equation 6.6, with saturation pressures evaluated at the temperatures on the surface of the membrane (Alves et al., 2004; Bui et al., 2005; Babu et al., 2008):

$$J_i^{\mathrm{V}} = K_i^{\mathrm{OV}}\left(a_{i,\mathrm{b}}^{\mathrm{F}} P_{i,\mathrm{m}}^{\mathrm{sat,F}} - a_{i,\mathrm{b}}^{\mathrm{RF}} P_{i,\mathrm{m}}^{\mathrm{sat,RF}}\right) \tag{6.7}$$

where the subscripts b and m refer to evaluation of the corresponding variable in the bulk of the liquid and at the liquid–membrane interface, respectively.

To make it clear that Equation 6.6 only aids understanding but is not necessarily suitable for modeling purposes, a different symbol was used for the overall mass-transfer coefficient in Equation 6.7, whose value is given by

$$\frac{1}{K_i^{\mathrm{OV}}} = \frac{P_{i,\mathrm{m}}^{\mathrm{sat,F}}}{k_i^{\mathrm{F}}} + \frac{1}{k_i^{\mathrm{M}}} + \frac{P_{i,\mathrm{m}}^{\mathrm{sat,RF}}}{k_i^{\mathrm{RF}}} \tag{6.8}$$

where k_i^{F} and k_i^{RF} are the mass-transfer coefficients for component i in the diffusional boundary layers of the feed and receiving phases, respectively, both defined using activity differences as the driving force for mass transfer. In the development of Equation 6.8, all mass-transfer coefficients were defined in terms of the same area, which is not the only option in the case of a hollow-fiber membrane. An alternative definition of K^{OV} for hollow fibers is given by Alves and Coelhoso (2007), with each mass-transfer coefficient defined in terms of the physical area associated with its corresponding flux.

Equations 6.7 and 6.8 are valid for OD, MD, and OMD with the relevant simplifications depending on the composition of each liquid phase. For example, if pure water is adopted as receiving phase in MD, there is no diffusional boundary-layer in the receiving phase, and the last term of Equation 6.8 vanishes. In an alternative approach, with the aid of the Clausius–Clapeyron equation, an efficiency coefficient for the thermal effect can be introduced into Equation 6.6, as detailed elsewhere (Gostoli, 1999; Courel et al., 2000b; Celere and Gostoli, 2002; Thanedgunbaworn et al., 2007b). Reasoning that small differences between inlet and outlet temperatures of each liquid phase were observed in their OD experiments due to short contact times, some authors (Celere and Gostoli, 2004; Nagaraj et al., 2006b; Thanedgunbaworn et al., 2007a) have used simplified forms of Equations 6.7 and 6.8 to describe or analyze their data, assuming the same temperature on both sides of the membrane, that is, neglecting *a priori* temperature polarization effects.

The values of k_i^{F} and k_i^{RF} are either experimentally determined (Narayan et al., 2002; Versati et al., 2004; Celere and Gostoli, 2005) or estimated based on empirical correlations in terms of dimensionless numbers. Even though, at laboratory scale, both flat-sheet and hollow-fiber membranes can be used, the latter are regarded as the geometry of choice in the case of a commercial application of OD or MD for juice concentration due to the large area requirements anticipated based on the performance of current membranes. Therefore, only correlations for hollow-fiber modules will be discussed here. For the flat-sheet configuration, the reader is referred to the works of Babu et al. (2006, 2008), Courel et al. (2000b), Kimura et al. (1987), Nagaraj et al. (2006b), and Phattaranawik et al. (2003a).

A hollow-fiber module bears strong resemblance with a shell-and-tube heat exchanger. It basically comprises a bundle of fibers randomly packed into a shell. For the solution inside the fibers, the classical equations derived for heat transfer during fluid flow inside a cylindrical pipe, recently reviewed by Mengual et al. (2004) and Curcio and Drioli (2005), can be successfully used with the aid of the heat and mass-transfer analogy (Celere and Gostoli, 2004; Thanedgunbaworn et al., 2007a,b).

On the other hand, on the shell side, correlations originally developed for heat exchangers are not appropriate due to the packing randomness of the fibers inside the module. Working a module containing 85 fibers ($A = 0.04\,\mathrm{m}^2$, $L = 250\,\mathrm{mm}$), inside which pure water flowed, Celere and Gostoli (2004) tested the correlation proposed by Gostoli and Gatta (1980) based on data from a hollow-fiber dialyzer. A good agreement between predicted and experimental k values was observed with a 70 wt% glycerol solution as receiving phase, whereas the predicted values were well above the experimental ones in the case of a 40 wt% $CaCl_2$ solution.

Bui et al. (2005) used $J_{\mathrm{w}}^{\mathrm{m}}$ data related to glucose solutions (30–60 wt%) to assess a total of six correlations available in the literature for calculating k on the shell side of hollow-fiber modules (Yang and Cussler, 1986; Prasad and Sirkar, 1988; Costello et al., 1993; Gawronski and Wrzesinska, 2000; Wu and Chen, 2000; Lipnizki and Field, 2001). The data were obtained by Bui et al. (2004) in two different modules (A = 106.0 and $104.5\,\mathrm{cm}^2$, $L = 180\,\mathrm{mm}$) with $CaCl_2$ 45 wt% as strip solution under a variety of operating conditions ($25 \leq T \leq 45°C$, $5.8 \leq Re_{fiber} \leq 225.9$, $11.4 \leq Re_{shell} \leq 73.1$, $4.6 \leq Gz \leq 12.7$). For each correlation, the flux at a given experimental condition was used together with the estimated mass-transfer coefficients to compute the corresponding concentrations at the surface of the membrane on the feed and permeate sides. A similar procedure was adopted to account for the temperature polarization effect, and the water partial pressure difference across the membrane ($p_{w,m}^{F} - p_{w,m}^{RF}$), was computed for each experimental condition based on the different correlations. Finally, the correlations were evaluated according to the linearity between the experimental fluxes and the computed driving forces, that is, the ($p_{w,m}^{F} - p_{w,m}^{RF}$) values. The correlations of Prasad and Sirkar (1988) and Yang and Cussler (1986) resulted in unacceptable changes of concentrations and temperatures at the membrane surface, leading, in some cases, to negative ($p_{w,m}^{F} - p_{w,m}^{RF}$) values. The models of Lipnizki and Field (2001) and Gawronski and Wrzesinska (2000) were considered unsuitable due to the poor linearity between J_{w}^{m} and its driving force. The model of Wu and Chen (2000), in turn, performed well only for one of the modules. With coefficients of determination (R-sq) higher than 0.99 for both modules, the correlation of Costello et al. (1993), shown in Equation 6.9, was found to be the best option to represent the data set:

$$Sh = (0.53 - 0.58\varphi)Re^{0.53}Sc^{0.33} \qquad (6.9)$$

where φ is the packing density of the fibers in the module, defined as the ratio of the cross-sectional area occupied by the fibers to the total cross-sectional area of the shell

$$\varphi = n_{fibers}\left(\frac{d_{fiber}^{out}}{d_{shell}^{in}}\right)^2 \qquad (6.10)$$

where

d_{fiber}^{out} is the outside diameter of the fibers
d_{shell}^{in} is the inside diameter of the shell

It ought to be emphasized that none of the correlations considered by Bui et al. (2005) was originally developed based on OD data.

In a recent study, Thanedgunbaworn et al. (2007b) performed OD experiments with several hollow-fiber modules of packing densities ranging from 0.306 to 0.612 and used their data to develop the following empirical correlation for k on the shell side of the module in the laminar regime:

$$Sh = (-0.4575\varphi^2 + 0.3993\varphi - 0.0475)Re^{(4.0108\varphi^2 - 4.4296\varphi + 1.5585)}Sc^{0.33} \qquad (6.11)$$

All experiments were carried out at 35°C with water as feed flowing inside the fibers and $CaCl_2$ 44 wt% as strip solution on the shell side. The effective membrane area in the modules varied from 53.60 to 168.47 cm^2. The values of Sh predicted by Equation 6.11 were always within 10% of the experimental values. Thanedgunbaworn et al. (2007b) also used their experimental k_{w}^{RF} values to assess the performance of six literature correlations (Dahuron and Cussler, 1988; Prasad and Sirkar, 1988; Costello et al., 1993;

Viegas et al., 1998; Gawronski and Wrzesinska, 2000; Wu and Chen, 2000), none of which could suitably predict the experimental data, including the correlation of Costello et al. (1993), previously recommended by Bui et al. (2005).

Alves and Coelhoso (2007) have also addressed the estimation of mass-transfer coefficients in hollow-fibers modules. Using a module with an effective area of 0.16 m² ($\varphi = 0.37$, $n_{\text{fibers}} = 400$, $L = 20$ cm), these authors performed OD experiments at 25°C using water and sucrose solutions (20 and 45 wt%) as feed streams and $CaCl_2$ solutions (2.8, 3.5, and 6.0 mol L^{-1}) as receiving phases, varying the flow rate on the shell ($0.7 \le Re_{\text{shell}} \le 45.0$) and tube sides ($0.35 \le Re_{\text{fiber}} \le 41.9$). The experimental results were used to fit the three mass-transfer coefficients in Equation 6.8, and the following dimensionless correlations were then obtained for k values in the boundary layers:

$$Sh_{\text{shell}} = 15.4 Re_{\text{shell}}^{0.92} Sc_{\text{shell}}^{1/3} \left(\frac{d_{\text{eq}}}{L} \right) \tag{6.12}$$

$$Sh_{\text{fiber}} = 2.66 Re_{\text{fiber}}^{0.25} Sc_{\text{fiber}}^{1/3} \left(\frac{d_{\text{fiber}}^{\text{in}}}{L} \right)^{1/3} \tag{6.13}$$

An average deviation for the proposed correlations was not reported. However, a concentration experiment of a sucrose solution, from 12 to 60 wt% was carried out, and the fluxes predicted based on k values from Equations 6.12 and 6.13 were shown to agree with the experimental data.

Apart from the mass-transfer coefficients, the saturation pressures at the upstream and downstream surfaces of the membrane are also necessary to calculate K^{OV} and the permeate flux, which requires an analysis of heat transfer in OD and MD to estimate the solution temperature in these regions. Heat transfer in these operations occurs by two mechanisms. Firstly, there is latent heat transfer accompanying vapor flux, and secondly there is heat transfer by conduction across the membrane. Adding these two contributions, the total heat flux, Θ, is given by (Courel et al., 2000b; Romero et al., 2003; Bui et al., 2005; Thanedgunbaworn et al., 2007a):

$$\Theta = \sum_{i=1}^{s} J_i^V \rho_{l,i} \Delta H_i^{\text{vap}} + \frac{\lambda^{\text{M}}}{\delta} \left(T_{\text{m}}^{\text{F}} - T_{\text{m}}^{\text{RF}} \right) \tag{6.14a}$$

$$\delta = \begin{cases} l & \text{flat sheet} \\ 0.5 d_{\text{fiber}}^{\text{in}} \ln \left(d_{\text{fiber}}^{\text{out}} / d_{\text{fiber}}^{\text{in}} \right) & \text{hollow fiber} \end{cases} \tag{6.14b}$$

where

- s is the total number of permeated components
- $\rho_{l,i}$ is the density of component i in the liquid state
- ΔH_i^{vap} is the enthalpy of vaporization of component i
- λ^{M} is the membrane thermal conductivity

Experimental values of λ^M are available in the literature for some commercial membranes used in MD and OD (Schofield et al., 1987; Izquierdo-Gil et al., 1999a,b; García-Payo and Izquierdo-Gil, 2004; Celere and Gostoli, 2005). In most studies, however, λ^M has been computed from the individual conductivities of the polymer (λ_{pol}) and the gas (air) trapped inside the pores (λ_g), using the traditional equation for conduction through several bodies in parallel:

$$\lambda^M = \varepsilon\lambda_g + (1-\varepsilon)\lambda_{pol} \quad (6.15)$$

where is ε the membrane porosity. Schofield et al. (1987) reported that Equation 6.15 provided a good estimation of λ^M for a commercial polypropylene membrane (Enka $d_p^* = 0.1\,\mu m$). However, García-Payo and Izquierdo-Gil (2004) and Phattaranawik et al. (2003a) reported that Equation 6.15 considerably overestimated the value of λ^M, providing a review of other available equations to estimate the thermal conductivity of porous media.

At steady-state, the total heat flux across the membrane is equal to the heat flux in the thermal boundary layers, $h^F(T_b^F - T_m^F)$ and $h^{RF}(T_m^{RF} - T_b^{RF})$. Therefore, the temperatures at the membrane surfaces can be evaluated by the following equations (Thanedgunbaworn et al., 2007a; Babu et al., 2008):

$$T_m^F = \frac{(\lambda^M/\delta)[T_b^{RF} + (h^F/h^{RF})T_b^F] + h^F T_b^F - \sum_{i=1}^{s} J_i^V \rho_{l,i} \Delta H_i^{vap}}{\lambda^M/\delta + h^F[1 + \lambda^M/(\delta h^{RF})]} \quad (6.16a)$$

$$T_m^{RF} = \frac{(\lambda^M/\delta)[T_b^F + (h^{RF}/h^F)T_b^{RF}] + h^{RF} T_b^{RF} + \sum_{i=1}^{s} J_i^V \rho_{l,i} \Delta H_i^{vap}}{\lambda^M/\delta + h^{RF}[1 + \lambda^M/(\delta h^F)]} \quad (6.16b)$$

where h^F and h^{RF} are the heat-transfer coefficients in thermal boundary layers of the feed and receiving phases, respectively, whose values are traditionally estimated by the same correlations previously discussed for k using the heat and mass-transfer analogy. An analysis of Equations 6.7, 6.8, and 6.16 reveals that heat and mass fluxes in OD and MD are intrinsically related and must hence be simultaneously determined. All equations presented here were derived neglecting the concentration and temperature changes along the module, which are not necessarily small. More elaborate mathematical treatments to account for such changes are discussed elsewhere (Gostoli, 1999; Chernyshov et al., 2003; Alves and Coelhoso, 2007; Cheng et al., 2008; Teoh et al., 2008).

One important aspect to highlight is the fact that the heat flux by conduction in OD and MD has opposite directions. In MD, to generate the driving force for mass transfer, the feed stream is fed at a temperature higher than that of the receiving phase, that is, $T^F > T^{RF}$. Therefore, both contributions in Equation 6.14a have the same sign. It should be noted, however, that the conductive heat flux in this case actually represents a loss, as it uses part of the available mass-transfer driving force to reduce the temperature difference between the liquid streams. Under operating conditions relevant for water desalination, Fane et al. (1987) found that 20%–50% of the total heat transferred in a MD unit was lost by conduction. In extreme cases, when heat conduction across the membrane is too great, Gostoli et al. (1987) and Findley (1967) demonstrated that flux can be considerably reduced and even reversed (from permeate to feed side). Under operating conditions likely to be used for juice concentration, Bui et al. (2007) verified that the energy efficiency of their MD unit was always lower than 50%, and it could be as low as 2.1%. These authors, in particular, pointed out that the application of MD in liquid food concentration may be challenged by the problem of significant heat loss. Because such application requires operation at low feed temperature (<40°C), a competitive mass flux can only be achieved by applying a considerable temperature difference across the membrane, which increases heat loss by conduction. This problem does not occur in the case of water desalination, for which MD was originally conceived. In view of the exponential increase of P^{sat} with temperature, the higher feed temperatures allowed in water desalination enable a large partial pressure difference to be created by a relatively small temperature difference across the membrane.

On the other hand, in OD, as previously explained, the temperature difference is created as a result of the mass flux, and $T_m^{RF} > T_m^F$. Consequently, the two contributions to the heat flux in Equation 6.14a have opposite directions in OD, and the conductive heat across the membrane acts against the temperature difference created by the boundary-layer effect. Gostoli (1999) pointed out that, for industrial modules, in which the feed inlet temperatures are fixed and the bulk values evolve along the module, an asymptotic temperature difference across the membrane is eventually reached, at which the heat flux due to mass transfer is exactly balanced by the conductive flux across the membrane, and therefore $\Theta = 0$. This asymptotic temperature difference was experimentally determined by Gostoli (1999) and Celere and Gostoli (2002, 2005) for different sets of operating conditions. As discussed by these authors, operation at the condition of asymptotic temperature difference is especially useful for studying thermal boundary-layer effects in OD, since the temperature values at the membrane surface are known. Additionally, an experimental value for λ^M can also be obtained.

6.3.2.4 Membranes and Modules

Most MD and OD studies so far have been performed using commercial microporous membranes prepared from hydrophobic polymers such as polyethylene, polypropylene (PP), polytetrafluorethylene (PTFE), and poly(vinylidene difluoride) (PVDF), which are available both as flat sheets and hollow fibers. In general, the nominal pore diameter of these membranes ranges from 0.1 to 1.0 μm, their porosity from 50% to 80%, and their thickness is between 10 and 300 μm. Tables with manufacturer specifications for membranes commonly used in OD and MD can be found elsewhere (Kunz et al., 1996; Lawson and Lloyd, 1997; Khayet et al., 2005; Khayet, 2008).

Despite their widespread use, these commercial membranes are actually made for MF rather than MD or OD. Even though most of the required features are met by such membranes, their morphology has not been optimized for MD or OD purposes, and there is definitely room for improvement. It should be noted, however, that the ideal membrane parameters are not all the same for these two unit operations. An OD membrane should be highly porous and as thin as possible, since flux is directly proportional to porosity and inversely proportional to membrane thickness. In addition, it should have as high a λ^M value as possible to maximize the amount of latent heat supplied by conduction across the membrane, which minimizes thermal boundary-layer effects (Nagaraj et al., 2006a) and leads to operation close to isothermal conditions. On the other hand, for MD, since heat conduction across the membrane represents a loss, λ^M should be as low as possible. Although it may seem that a minimum thickness is also the best strategy for an ideal MD membrane, one has to bear in mind that heat loss by conduction inevitably increases with decreasing thickness, and the net result is the existence of an optimum thickness depending on the other properties of the membrane as well as on hydrodynamic conditions in the module (El-Bourawi et al., 2006; Bonyadi and Chung, 2007). As for porosity, the requirement is the same as in the case of OD, not only to maximize mass flux, but also to reduce λ^M, since the thermal conductivity of air ($0.024\,W\,m^{-1}\,K^{-1}$) is about one order of magnitude smaller than the typical values associated with hydrophobic polymers (0.1–$0.3\,W\,m^{-1}\,K^{-1}$).

Though still incipient, research on the preparation of membranes specifically designed for MD and OD has already started. In his recent review of the MD literature, Khayet (2008) pointed out that less than 8% of the articles published on MD between 1982 and 2005 were related to membrane preparation. The three main alternatives investigated for flux enhancement through membrane design are:

1. Study of relevant variables (polymer concentration, type of nonsolvent, exposure time prior to precipitation, use of additives, etc.) to optimize membrane morphology in the phase inversion process (Tomaszewska, 1996; Khayet and Matsuura, 2001).
2. Utilization of alternative hydrophobic polymers (Fujii et al., 1992; Feng et al., 2004; Albrecht et al., 2005).
3. Formation of hydrophobic–hydrophilic composite membranes by coating of hydrophilic supports (Cheng and Wiersma, 1980, 1982), surface modification methods (Kong et al., 1992; Wu et al., 1992; Khayet et al., 2005), and coextrusion spinning (Bonyadi and Chung, 2007).

The results are promising, with examples of taylor-made membranes that performed better in terms of flux than the commercial MF membranes in MD experiments with pure water and dilute NaCl solutions as feed (model feed for desalination applications). None of these taylor-made membranes, however, has been tested for fruit-juice concentration so far. The interested reader is referred to the recent reviews of Khayet (2008) and Curcio and Drioli (2005) for further details on this topic.

For a successful MD or OD operation, it is essential that neither the feed nor the receiving phase enters the pores of the membrane. Therefore, the pressure difference across the membrane, ΔP^M, must be smaller than the capillary penetration pressure of the liquid in the membrane pores, ΔP_{cap}. The value of the latter depends on the membrane material, pore size, and surface morphology, variables whose relationship is given by the Laplace–Young equation:

$$\Delta P_{cap} = -\frac{B\beta_l \cos\theta_c}{d_p^*} \quad (6.17)$$

in which

- β_l is the liquid–vapor surface tension
- θ_c is the contact angle
- B is a geometric factor determined by pore structure ($B = 1$ for a cylindrical pore)

Equation 6.17 is very useful to reason the factors that affect ΔP_{cap} but it cannot be used for accurate predictive purposes in the case of MD and OD membranes due to the complexity of their morphology (pore size distribution, variety of pore shapes, surface roughness effect on θ_c). Therefore, in practice, the values of ΔP_{cap} must be experimentally determined. Typical ΔP_{cap} values for water range from 100 to 400 kPa (Schneider et al., 1988; Bui et al., 2007; Khayet, 2008).

At laboratory scale, operation with $\Delta P^M < \Delta P_{cap}$ is relatively simple for fresh juices and the typical solutions used as receiving phase in OD and MD. However, Bui and Nguyen (2005) have shown that, at pilot scale, operating problems can arise as concentration progresses and the feed reaches the point of steep increase in viscosity with increasing solids content. The high viscosity leads to a considerable pressure drop along the length of the membrane module on the feed side, and eventually dictates the maximum feed flow rate that can be adopted. This is a very important aspect, since it suggests that, for concentrated solutions, a minimization of boundary-layer effects on the feed side by operation with sufficiently high flow rates is not possible. The value of *Re*, which as discussed in Section 6.3.2.3 is directly linked to the boundary-layer effects, is only a linear function of fluid velocity, whereas pressure drop grows with the second power of the same velocity. In a recent study, Hongvaleerat et al. (2008) demonstrated that pressure drop issues can dictate the maximum flow rate even at laboratory-scale modules ($A = 50\,cm^2$).

The particular composition of fruit juices creates a long-term challenge for hydrophobic membranes during concentration by OD and MD. Fruit juices contain many nonvolatile surface-active solutes, such as proteins, emulsified oil droplets, and colloidal hydrogel particles. Such solutes not only reduce the value of β_l compared to pure water, but more importantly, they may concentrate and precipitate, or adsorb upon, the surface of the membrane, reducing its hydrophobicity, and eventually leading to liquid penetration into the pores, which is referred to as membrane wet-out. Citrus juices, in particular, are a great concern in this regard, as they contain peel oils and other highly lipophilic flavor components that can easily promote wetting of hydrophobic surfaces (Hogan et al., 1998; Mansouri and Fane, 1999). The real impact of such effects can only be seen in long-term experiments, which, to the authors' knowledge, are still lacking insofar as juice concentration by OD and MD is concerned. Nonetheless, the significance of this matter is already illustrated by the results of Gryta (2005b) for the long-term performance of a MD desalination unit (A = 889 cm^2) equipped with PP membranes (Accurel®* PP S6/2). Within a 3-year period, stable water flux was verified for operation with RO permeate as feed, whereas the use of tap water as feed was enough to bring about a rapid decline of separation efficiency due to the partial membrane wet-out caused by deposition of $CaCO_3$ on the membrane surface.

To avoid this solute-driven wet-out of membranes, the standard strategy adopted in the literature is to coat the microporous hydrophobic membranes with a thin, dense hydrogel film. The details of the technique are well described in the patent of Michaels (1999), even though all examples are related to the preparation of the coated membranes without a single OD or MD permeation experiment. In the case of flat-sheet membranes, an even simpler idea was described in a previous patent (Michaels, 1997), which claimed to solve the problem by laying a semipermeable, hydrophilic barrier film, such as a dialysis or UF membrane, on top of the hydrophobic membrane and mounting the assembly in the module. Once again, permeation data to confirm the efficiency of such procedure were not presented.

Results confirming the efficiency of the hydrogel method were presented by Mansouri and Fane (1999), who coated three commercial MF membranes with poly(vinyl alcohol) (PVOH) crosslinked with either maleic acid or glutaraldehyde. Original and coated membranes were tested in OD runs at 25°C with $CaCl_2$ (24.5 and 36 wt%) as receiving phase and water, sucrose solutions, and limonene-containing (0.2, 0.5, and 1.0 wt%) feeds. Due to its high equilibrium water content (59.7–72.1 wt%), the PVOH layer offered little additional resistance to water flux, and within the experimental error, J_w^m values were the same for coated and uncoated membranes in the runs with pure water and sucrose solutions. Uncoated membranes promptly wetted out even with the smallest limonene concentration adopted, whereas PVOH-coated membranes enabled stable operation over 24 h with all oily feeds considered.

Instead of PVOH, Xu et al. (2004) applied sodium alginate as the coating material, which was tested with a commercial PTFE membrane. The coating procedure was somewhat simpler than the one adopted by Mansouri and Fane (1999) and could be done in the same cell where the OD runs were latter performed, even though the position of the cell had to be changed from horizontal (OD runs) to vertical (membrane coating). Water flux in OD runs with deionized water and water/orange oil (0.2, 0.4, 0.8, and 1.2 wt%) mixtures were measured at 23°C with $CaCl_2$ (40 wt%) as receiving phase. Wet-out of the uncoated membrane occurred within 2.0 min of operation with the feed containing 0.2 wt% of orange oil. Coated membranes, on the other hand, enabled safe operation for all water/orange oil mixtures, with a J_w^m value that was less than 5% smaller than the one obtained with the uncoated membrane and pure water as feed. In a durability test, a coated membrane retained hydrophobicity after 72 h of contact with the feed containing 1.2 wt% of orange oil. In a subsequent work, Xu et al. (2005a) included a cationic surfactant (myristyltrimethylammonium

* Registered trademark of Membrana GmbH, Wuppertal, Germany.

bromide) in the coating solution to increase the adhesion of the hydrophilic coating on the hydrophobic surface. The coating was shown to offer protection against membrane wet-out by both orange oil and sodium dodecylbenzene sulfonate (SDB), as well as to assist in protecting the membrane against fouling by components of whole milk. Alginic acid–silica hydrogel films were also shown by the same authors to provide membrane protection against wet-out by orange oil and SDB (Xu et al., 2005b).

Bowser (2000) proposed that high free-volume polymers, such as poly (1-trimethylsilyl-1-propyne) and some amorphous copolymers of perfluoro-2,2,-dimethyl-1,3-dioxole, could be used instead of hydrogels as the thin dense layer to coat the porous membrane. The idea was tested in OD runs with pure water and sucrose solutions (10 wt%) with and without limonene, using commercial PP hollow fibers ($d_p^* = 0.04\,\mu m$) and Teflon®* AF 2400 as the coating material. Although the coating effectively protected the PP membrane against wet-out by limonene, the water flux for a coating thickness of 1.3 μm was 40% lower for pure water compared to the value associated with the uncoated membrane, and the reduction was even higher in the case of the sucrose solution (drop of almost 56% in J_w^V). Nonetheless, in the case of water desalination by MD, Li and Sirkar (2004) demonstrated that, by proper design of the porous support and membrane module, PP membranes with a silicone fluoropolymer coating applied by plasmapolymerization technology could provide water fluxes similar to those associated with RO.

In spite of the predominance of polymeric membranes for OD and MD applications, the possibility of using ceramic membranes has been recently investigated. Originally hydrophilic, ceramic membranes necessarily require some treatment to be made hydrophobic. Brodard et al. (2003) prepared hydrophobic inorganic membranes by grafting siloxane compounds on α-alumina tubular supports. These membranes were tested in OD runs at four different temperatures ($25 \leq T \leq 38°C$) with pure water as feed and $CaCl_2$ (50 wt%) as brine. The conductivity on the feed side was continuously monitored and remained constant during all experiments, confirming the hydrophobicity of the membrane. In a subsequent work from the same group (Gabino et al., 2007), a contact angle of 141° was reported for this membrane, and its performance was kept after 10 cleaning cycles with acidic (HNO_3 1%), basic (NaOH 1%), and sanitizing agents (200 ppm Cl^-). However, when a commercial, surfactant-containing cleaning agent (Ultrasil®†-10) was tested, a drop in water flux was verified.

Despite its importance, the membrane is not the only aspect to be considered, as it must be properly contacted with feed and receiving phases to work. This contact takes place in the so-called membrane modules. Flat-sheet membranes can be used in either plate-and-frame or spiral-wound modules, whereas the hollow-fiber geometry requires a module design which bears the same name. The main features of each kind of module are well described in the literature (Mulder, 1991; Baker, 2004; Habert et al., 2006), and we shall limit our attention here to one particular aspect, the

* Registered trademark of E. I. du Pont de Nemours & Company, Inc., Wilmington, DE.

† Registered trademark of Ecolab Inc., St. Paul, MN.

packing density, which is defined as the ratio of membrane area in the module to its total volume. Typical packing densities for plate-and-frame and spiral-wound modules are equal to 400–600 and 800–1000 m^2 m^{-3}, respectively, whereas, for hollow-fiber modules, values of about 9200 m^2 m^{-3} are normally obtained (Mulder, 1991). In view of the relatively low fluxes associated with current OD and MD membranes, a large membrane area is anticipated for a commercial application, which asks for as high a packing density as possible. Thus, there is a consensus among researchers that hollow-fiber modules will be the configuration of choice for a juice concentration plant by OD or MD. Indeed, apart from Versari et al. (2004), who chose a plate-and-frame configuration, all other authors have utilized hollow-fiber modules in their pilot-plant studies (Hogan et al., 1998; Vaillant et al., 2001, 2005; Ali et al., 2003; Bui and Nguyen, 2005; Cisse et al., 2005; Kozák et al., 2008).

In a hollow-fiber module, the porous fibers are randomly arranged in bundles that are then placed inside a shell and potted. The flow on the shell-side of the module, that is, outside the fibers, is known to be prone to severe concentration polarization effects due to flow maldistribution, as illustrated for instance by the simulation results of Zhongwei et al. (2003). Probably in an attempt to minimize such effects, some authors have fed the fruit juice (or the adopted model feed stream) into the bore of the fibers (tube side) (Vaillant et al., 2001, 2005; Bui et al., 2004; Bui and Nguyen, 2005; Cisse et al., 2005). However, in view of the small diameter of the fibers (<1.0 mm), viscous fingering is hard to minimize with feed flow inside them (Hogan et al., 1998), and pressure drop can become rather significant for concentrated streams (Alves and Coelhoso, 2004). Therefore, the best alternative is to keep the feed stream on the shell side. Both countercurrent and cocurrent arrangements have been used. Those in favor of the former use the analogy with other unit operations to justify a higher overall driving force for mass transfer along the fibers. Nevertheless, Deblay (1992) highlights that the cocurrent configuration minimizes the transmembrane pressure gradient along the fibers, a fact that should not be neglected, especially in a scenario where successful operation implies keeping $\Delta P^M < \Delta P_{cap}$. In the case of MD, the cocurrent configuration also reduces conductive heat loss. To guarantee that, in the event of membrane failure, feed liquid leaks into the receiving phase and not *vice versa*, OD units are normally operated with a slightly higher pressure on the feed side of the module.

It has been demonstrated that significant water flux enhancement can be obtained for both OD and MD when the juice is previously clarified by MF or UF (Calabrò et al., 1994; Bailey et al., 2000; Lukanin et al., 2003). In fact, just like in the case of RO, a MF/UF pretreatment of the fresh juice, with or without enzymes, is now a standard procedure in OD/MD studies (Shaw et al., 2001; Vaillant et al., 2001, 2005; Cassano et al., 2003, 2004, 2007; Rodrigues et al., 2004; Bui and Nguyen, 2005; Cisse et al., 2005; Gunko et al., 2006; Rektor et al., 2006; Cassano and Drioli, 2007; Galaverna et al., 2008; Hongvaleerat et al., 2008; Koroknai et al., 2008; Kozák et al., 2008). The improved performance is attributed to a reduction in juice viscosity due to the removal of proteins and other biopolymers, with a resulting decrease in boundary-layer effects for the same operating conditions compared to untreated juice. Additionally, removal of these biopolymers prior to concentration prevents them from depositing on the surface of the membrane, reducing the risk of membrane wet-out and, at the same time, minimizing membrane fouling. Furthermore, the MF/UF pretreatment yields a particle-free stream to the OD/MD contactor, preventing clogging of the tiny flow channels in the hollow-fiber module.

If the highly viscous stream has to flow on the shell side, special care must be taken to optimize hydrodynamic conditions and reduce boundary-layer effects. As previously discussed, increasing feed flow rate is only of limited help on account of pressure-drop limitations, and design solutions are required. One effective way of promoting mixing in the vicinity of the fibers, even in laminar flow, is to force the liquid stream to flow normal to the long axis of the fibers, the latter arranged in a relatively closely spaced array to assure uniform flow of fluid to all fibers. This is the so-called cross-flow configuration, which can provide for an efficient displacement of the boundary layer around each fiber and an efficient micromixing of fluid elements (Yang and Cussler, 1986; Wickramasinghe et al., 1992). In the case of water desalination by MD, the advantages of the cross-flow configuration have been clearly demonstrated by Sirkar and coworkers (Li and Sirkar, 2004; Song et al., 2007, 2008) both at laboratory and pilot-plant scales. At commercial scale, an intermediate flow configuration between pure tangential and cross flows can be obtained in the hollow-fiber contactors called Liqui-Cel®,* which are manufactured by Membrana-Charlotte (Charlotte, NC) with nominal membrane areas up to 220 m^2. A schematic representation of these contactors is shown in Figure 6.9. The units are provided with PP fibers woven into a fabric that is spirally wound around a perforated tube through which the liquid flowing on the shell side enters and leaves the module. This central perforated tube and the shell are baffled to force a change in fluid flow direction and prevent most of the liquid in the shell to flow tangential to the fiber axis. According to Hogan et al. (1998), even with low mass flow rates (pressure drops in the 20–50 kPa range), water fluxes in OD experiments with juices concentrates containing as much as 70 wt% solids were within one-half to two-thirds of those predicted based on data for pure water as feed when these Liqui-Cel contactors were used.

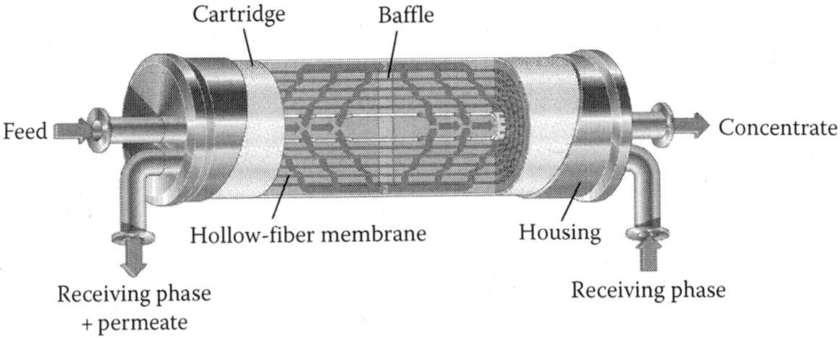

FIGURE 6.9 Schematic representation of a commercial hollow-fiber contactor that provides an intermediate configuration between pure tangential and cross flows on the shell side. (Courtesy of Membrana-Charlotte, Charlotte, NC.)

* Registered trademark of Membrana-Charlotte, a division of Celgrad LLC, Charlotte, NC.

The utilization of spacers and other turbulence promoters is an option to improve fluid mixing in membrane modules commonly used in UF and MF, but it remains little investigated insofar as MD and OD applications are concerned. The efficiency of spacers in enhancing J_w^m (up to 60% increase) during water desalination by MD using flat-sheet membranes has been reported by a few authors (Martínez-Díez et al., 1998; Phattaranawik et al., 2001, 2003a,b; Martínez and Rodríguez-Maroto, 2006), but it was only very recently that a study concerning hollow-fiber modules was presented. Using commercial PP membranes, Teoh et al. (2008) prepared different hollow-fiber modules with baffles (window, helical, helical + braided), sieve spacers (1–3), and different fiber configurations (braided, twisted). These modules (90–100 fibers per module, $L = 15\,cm$) were tested in MD experiments with a 3.5 wt% NaCl solution as feed and deionized water as receiving phase. The options tested by the authors led to improvements in the value of J_w^m ranging from 4% to 47%, with the use of baffles being particularly successful. It remains to be seen what improvements such strategy can provide for the viscous solutions involved in juice concentration by OD and MD, in which case pressure drop increase will also have to be taken into account.

6.3.2.5 Product Quality

In view of the mild operating conditions associated with OD, one can theoretically expect a fruit-juice concentrate with an overall quality superior to the one associated with the product of traditional vacuum evaporation. This theoretical expectation has indeed been confirmed by a considerable amount of experimental data related to different juices.

First of all, the vitamin C content, an important nutritional characteristic, is only slightly decreased. Retention values of vitamin C higher than 96% are reported in the literature for passion fruit (Vaillant et al., 2001), camu-camu (Rodrigues et al., 2004), kiwifruit (Cassano et al., 2003; Cassano and Drioli, 2007), melon (Vaillant et al., 2005), orange (Cisse et al., 2005), cactus pear (Cassano et al., 2007), blood orange (Galaverna et al., 2008), and pineapple (Babu et al., 2008; Hongvaleerat et al., 2008) juices, all concentrated to 55°Brix or higher by OD. In some of these works, the vitamin C retention for thermal evaporation was also measured, and the values obtained were substantially lower (passion fruit, 11%; kiwifruit, 11%; orange, 59%; blood orange, 69.4%). In addition to vitamin C, total acidity and sugars content also show very similar values in the original juice and in the juice reconstituted from OD concentrates (Vaillant et al., 2001, 2005; Cisse et al., 2005; Cassano et al., 2007). In particular, contrary to what occurs in RO, the acidity/sugar ratio of the original juice is not changed upon concentration by OD. All these results are reasoned considering that, in order to permeate trough the OD membrane, any compound must vaporize, which clearly does not happen to sugars, vitamin C or the other organic acids of the juice under the adopted operating conditions. In the case of vitamin C, the small losses are believed to be related to oxidation by exposition to the oxygen entrapped within the pores of the membrane (Cisse et al., 2005).

The dietary value of fruit juices is directly related to their antioxidant capacity (Cassano et al., 2003; Mullen et al., 2007). Bermúdez-Soto and Tomás-Barberán (2006), for instance, demonstrated a linear correlation between antiradical activity and total phenolic, anthocyanin, and flavonol contents in juice samples. As a result, the total antioxidant activity (TAA) is an important quality parameter for juices. In the case of kiwifruit juice (9.4°Brix), Cassano and Drioli (2007) observed that the TAA remained constant and independent of the total solids content during concentration by OD up to 66.6°Brix. In addition, the contribution of vitamin C to the TAA also did not change. On the other hand, when concentration was performed by thermal evaporation, the TAA was reduced to about 50% of its value in the clarified juice, even when the final solids content was only 20°Brix. Additionally, the contribution of vitamin C to the TAA was reduced by 72% compared to the clarified juice. In a subsequent work, Cassano et al. (2007) reported the same trend of constant TAA for the concentration of cactus pear juice by OD (from 13.0 to 58.0°Brix). Furthermore, upon applying OD to concentrate chokeberry, red currant, and cherry juices, which are amongst the most important sources of antioxidant compounds within the group of red fruits, Koroknai et al. (2008) observed that the reconstituted juices (from concentrates with at least 62.4°Brix) retained more than 97% of the TAA related to the fresh juices.

In their analysis of an integrated membrane process for blood orange juice (12.6°Brix) concentration, comprised of a clarification step by UF, a preconcentration step by RO (21.4°Brix), and a final concentration to 60.6°Brix by OD, Cassano et al. (2003) verified a slight decrease (13.6%) in the TAA after the RO step, without further changes in the subsequent OD operation. When the clarified juice was directly concentrated by OD, removing the RO step, the final TAA value, though slightly higher (11.0% reduction), was not very different from the one obtained in the RO/OD configuration. In both cases, the reduction was smaller than the one associated with traditional thermal evaporation (20.8% for a 56.3°Brix concentrate). In order to better understand these findings, Galaverna et al. (2008) used high-performance liquid chromatography to quantify several antioxidant components of blood orange juice before and after the different concentration routes. In the UF–RO–OD process, no significant variations for hydroxycinnamic acids and flavanones were observed, whereas a reduction of about 23% was measured for anthocyanins, particularly during the RO step. A slight decrease (ca. 15%) for vitamin C, the main antioxidant component of this particular juice, was also observed, mainly during the UF step (ca. 10%). When the RO step was removed, hydroxycinnamic acids and flavanones remained unaffected, and the reductions in the anthocyanins and vitamin C contents were somewhat smaller (ca. 22% and 11%, respectively). Finally, in the case of the traditional thermally evaporated juice, for a final concentration of 60.6°Brix, the TAA content decreased by almost 26%, and the contents of all antioxidants components underwent reductions: 36% for anthocyanins, 55% for hydroxycinnamates, ca. 30% for vitamin C, and ca. 23% for flavanones.

As highlighted in Section 6.1, one clear disadvantage of vacuum evaporation for juice concentration is the color degradation resulting from thermally driven undesired reactions, such as pigments degradation and Maillard browning reactions. Absence of color changes during juice concentration by OD has been confirmed in the literature (Bui and Nguyen, 2005; Cisse et al., 2005; Vaillant et al., 2001, 2005; Hongvaleerat et al., 2008; Kozák et al., 2008) both with sensory analysis (apple, black currant, passion fruit) and with the parameter L^* of the Hunter color system (pineapple, melon, camu-camu, orange). Galaverna et al. (2008) stated that, despite the reduction in the anthocyanins content, the bright red color characteristic of blood orange juice was perfectly preserved with their UF–RO–OD process. Betaxanthins and betacyanins, important pigments found in cactus pear juice, were monitored by Cassano et al. (2007) during concentration by OD, and no significant changes in the levels of these compounds were noticed during processing.

Although the dietary value and appearance items discussed so far are rather relevant, the flavor quality of the final juice concentrate plays a central role in its acceptance in the market. The loss of important aroma compounds, as well as the development of a "cooked flavor," is well known and documented for concentration by traditional vacuum evaporation, leading to a reconstituted juice of inferior quality. In the case of orange juice, which is by far the most important product in this sector as shown in Figure 6.1, two practical results of these aspects can be mentioned. First, in the United States, it is not unusual to find the label "not from concentrate" on the package of retailed products as an indicative of superior quality. On the other hand, in countries like Brazil, where the costumer is used to the flavor of fresh orange juice, the market for the concentrated product is rather limited (Tribess and Tadini, 2001).

In view of their hydrophobicity, many aroma compounds have very high activity coefficients in aqueous solutions (Sorrentino et al., 1986; Sancho et al., 1998; Van Ruth and Villeneuve, 2002), being therefore volatile in such systems. Thus, contrary to sugars, proteins, and other juice components, aroma compounds can be transported across the OD membrane, and at least some degree of loss is anticipated. Shaw et al. (2001) utilized headspace gas chromatography to quantify such losses during a threefold concentration of orange and passion fruit juices by OD using a PP membrane ($A = 10.3\,m^2$). A total of 35 aroma compounds were considered for orange juice, with an average aroma retention of 66%–69%. In the case of passion fruit juice, 22 compounds were analyzed, and the average aroma retention was 61%. Reconstituted juices were submitted to sensory evaluation using an untrained taste panel. For both fruits, a triangle difference test showed a difference at the 99.9% confidence level between the initial and concentrated samples. A trained panel, on the other hand, judged the initial orange juice to be only slightly stronger in six flavor characteristics in comparison to the concentrated product.

Barbe et al. (1998) investigated the transport of water and seven aroma compounds (ethanol, 3-methylbutanal, ethyl acetate, ethyl hexanoate, α-pinene, β-myrcene, limonene) during the OD of a synthetic solution and two fruit juices (grape and orange) using nine different commercial membranes (four PP, two surface-modified PVDF, and three PTFE membranes). For all feeds, the greatest aroma loss per unit water removal occurred for the membrane having the smallest pore diameters at the surface (measured by image analysis), and this loss was progressively reduced as the surface pore diameter increased. This result was reasoned in terms of the differences in the additional resistances to mass transfer in the diffusional boundary layers on the upstream and downstream sides of the membrane arising from different degrees of liquid intrusion into the membrane pores. The aroma to water flux ratios obtained with grape and orange juices were one and two orders of magnitude higher than those measured with the synthetic solution, but no attempt to explain such trend was made by the authors.

One important aspect related to aroma losses in OD was pointed out by Ali et al. (2003). Working with a PP membrane ($A = 10.2\,m^2$) and a synthetic juice ($160\,g\,L^{-1}$ sucrose, $40\,g\,L^{-1}$ citric acid) containing four aroma compounds (hexyl acetate, ethyl butyrate, hexanol, and benzaldehyde—all at an initial concentration of $4–6\,mg\,kg^{-1}$), these authors verified that the concentration dropped significantly for all aroma compounds in the first hour of operation, a decrease that could not be explained by a reduction in the mass-transfer driving force or by modifications of the physical properties of the solution. Considering that both the membrane and the aroma compounds are hydrophobic, Ali et al. (2003) assumed that such behavior could be related to aroma adsorption by the membrane. To verify such hypothesis, in a new trial, the membrane was previously contacted with a solution containing the same aroma compounds at $6\,mg\,kg^{-1}$, being the OD experiment started with a new solution immediately after draining the installation. In this case, the losses measured for all compounds were significantly lower. The influence of this preconditioning of the membrane on the measured losses of aroma compounds was later confirmed by Cisse et al. (2005) in OD concentration experiments (from 11.8 to 62.0°Brix) with orange juice. After membrane preconditioning, average losses were strongly reduced for all classes of aroma compounds considered but aldehydes, as shown in Figure 6.10.

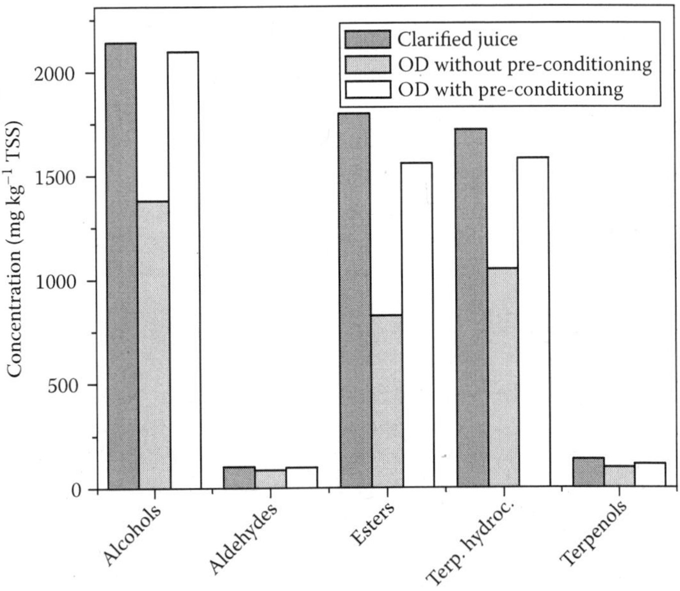

FIGURE 6.10 Comparison of the aroma content in clarified and reconstituted orange juices concentrated by OD in a pilot-plant unit with and without preconditioning of the polypropylene hollow-fiber membrane. Aroma compounds were analyzed by gas chromatography with a mass spectrometer detector and included three alcohols, four aldehydes, three esters, six terpenic hydrocarbons, and five terpenols. (Data from Cisse, M. et al., *Int. J. Food Sci. Technol.*, 40, 105, 2005.)

There is, therefore, some loss of aroma compounds during concentration of fruit juices by OD. Nonetheless, sensory analysis results have appeared in the literature suggesting that the consequent effect on juice quality may not be very drastic. For instance, in spite of the changes in the aroma profile shown in Figure 6.10, Cisse et al. (2005) reported that a panel of 25 highly trained orange juice tasters did not notice significant difference (at 95% confidence level) between the initial clarified juice and the one reconstituted from OD concentrate. On the other hand, juices reconstituted from concentrates obtained by OD and traditional vacuum evaporation were significantly recognized as different, with a better "fresh orange juice" aroma profile attributed to the OD concentrate. Vaillant et al. (2001) submitted their OD concentrate (60°Brix) of passion fruit juice to quality tests with highly trained tasters, who did not notice significant differences (95% confidence level) between fresh, clarified juice, and clarified reconstituted juice in a triangular difference test. The same tasters could easily recognize the difference between fresh juice and juice reconstituted from concentrate obtained by traditional vacuum evaporator. In another study, Bui and Nguyen (2005) asked 15 panelist to rank fresh and reconstituted (from 48°Brix OD concentrate) apple juice samples according to their color, aroma, flavor, and overall acceptability. For all items, no significant difference (95% confidence level) between samples was found. Grape juice was also concentrated (65°Brix) in the same OD unit and analyzed by the same test. A little change in the color of the reconstituted juice was identified by the panelists, but aroma and flavor were difficult to distinguish between the fresh and reconstituted samples, with no significant difference in acceptability. A comparison of fresh, OD-treated, and commercial black currant juices was recently presented by Kozák et al. (2008). Sensory analysis was performed by trained panelists, who evaluated the juice samples according to their color intensity, transparent ability, odor intensity, sweet flavor intensity, acidic flavor intensity, black currant flavor intensity, and general impression. The commercial juice was refused by the panel because it was only sweet, with neither the black currant nor the typical acidic flavors. It also had a higher transparency due to its low black currant content. With regard to the OD-treated juice, a certain extent of loss in odor intensity and black currant flavor intensity was recognized, but the other properties were similar to those of the fresh juice.

Considering now the MD operation, to the authors' knowledge, there has been only one study in which the quality of a juice concentrate obtained by MD was somehow accessed. Using commercial PVDF membranes, Calabrò et al. (1994) carried out a threefold concentration of orange juice by MD and measured the retention of sugars and organic acids, both of which were equal to 100%. The color and flavor of concentrated juice were described as "satisfactory," but the authors themselves highlighted that no formal sensory test was performed. Based on all the theory discussed in the previous sections, it is reasonable to expect a performance similar to the one observed in OD as far as the retention of nonvolatile compounds is concerned, and the results of Calabrò et al. (1994) further confirm that. The biggest question to be addressed, however, is how the heat-sensitive compounds and the volatile solutes, such as aroma compounds, will be affected. Because of the higher temperature on the feed side, one does expect differences between OD and MD in this case. Color and sensory analysis of juices concentrated by MD will have to be conducted to provide the definitive answer to this question.

6.3.2.6 Comparative Studies

Considering the many similarities between OD and MD, the choice between these two operations as the right alternative for a commercial application concerning juice concentration is not very straightforward. From the energy point of view, OD has the advantage of being isothermal. However, it requires an osmotic agent in the receiving phase, which raises the issue of poor consumers' perception due to the use of "chemicals," not to mention the costs of regenerating the receiving phase and the potential corrosion problems involved in such task. These disadvantages are not associated with MD, since cold water constitutes the receiving phase.

The performances of OD and MD with two different PVDF membranes for the concentration of glucose solutions were compared by Bui and Nguyen (2006). Three operating parameters (feed temperature, feed velocity, and feed concentration) were varied, one at a time, and the corresponding effects on J_w^m for each unit operation were compared. Based on their data, the authors concluded that MD was more sensitive to boundary-layer effects than OD, since the former was more affected by feed velocity and feed concentration than the latter. Nevertheless, such conclusion is actually not necessarily supported by the data, which were obtained with the feed flowing inside the fibers in the OD runs but on the shell side in the case of MD. Thus, equality of operating conditions did not translate into equality of hydrodynamic conditions in the module for the two operations, and, as discussed in Section 6.3.2.4, boundary-layer effects are harder to minimize on the shell side of hollow-fiber modules. Besides, in MD, when the feed flow rate is varied and the membrane area is kept constant, the drop in the driving force for mass transfer due to heat conduction decreases (Kimura et al., 1987; Cheng et al., 2008), and flux has to increase, even if the boundary-layer resistance remains the same.

Another experimental comparison between MD and OD in the same experimental unit was presented by Alves and Coelhoso (2006), who investigated both water and aroma fluxes during the concentration of a sucrose solution (12 wt%) and synthetic juices (18 mg L^{-1} citral, 18 mg L^{-1} ethyl butyrate in water, with and without 45 wt% sucrose). In order to minimize aroma adsorption effects (Section 6.3.2.5), the flat-sheet module used for the synthetic juice had a considerably smaller area (34 cm^2) than the hollow-fiber module used to concentrate the sucrose solution ($A = 0.16\,m^2$, $L = 20\,cm$). OD was conducted at 25°C with $CaCl_2$ 4.9 mol L^{-1} as receiving phase, whereas, in MD, pure water was the receiving phase and a temperature difference was chosen ($T^F = 34°C$, $T^{RF} = 23°C$) to give the same driving force as in OD. In spite of this, J_w^V values were more than twice as high in OD compared to MD, an effect that was reasoned in terms of a pronounced temperature polarization effect in the case of MD on the shell side of the module. Though important, temperature polarization alone is not likely to be responsible for all the flux differences in these tests. As highlighted by Alves and Coelhoso (2006) themselves, inlet and outlet temperatures for all streams were different in both OD and MD, demonstrating a variation of the driving force for mass transfer along the module. Because vapor pressure is an exponential function of temperature, the use of an arithmetic mean of inlet and outlet temperatures to calculate the driving force, as performed by the authors, could be a crude approximation to estimate the contribution of boundary-layers effects depending on the extent of the temperature difference, which unfortunately is not provided by Alves and

Coelhoso (2006). Even if the reasoning presented by the authors can be questioned, the experimental results themselves are rather important, since they demonstrate that for a given module and a theoretical driving force estimated based on inlet conditions, OD and MD do not give the same water flux.

Equally important, or perhaps even more relevant, are the findings of Alves and Coelhoso (2006) related to the flux of aroma compounds. In all experiments, the ratio of aroma to water flux was at least twice as high in MD compared to OD, and the difference increased when sucrose was present in the feed stream. For example, at the end of an 8h MD run without sucrose in the feed ($T^F = 33.6°C$), the receiving phase contained 51% and 49% of the initial mass of citral and ethyl butyrate in the feed, respectively, whereas, for OD ($T^F = 25.1°C$), only 12% of the initial mass of each aroma compound was transferred to the receiving phase. Furthermore, there was no significant difference in the extent of aroma loss for MD when the driving force was created by cooling the receiving phase ($T^{RF} = 9.1°C$) and keeping the feed at room temperature. As pointed out by the authors, the considerably smaller temperature difference between both sides of the membrane and the "salting out effect," that is, the significant increase in the activity coefficient of hydrophobic compounds in aqueous solutions upon addition of salts, are responsible for the higher retention of aroma per amount of water removal in OD. The same reasoning was previously applied by Hogan et al. (1998) to justify the low losses of aroma compounds during OD.

6.4 CONCLUDING REMARKS

A definitive substitute for traditional vacuum evaporation as a concentration method for fruit juices has still not been developed, and further research on this topic is necessary to achieve such goal. In this chapter, five of the many methods already proposed in the literature have been reviewed: direct-contact evaporation, Refractance Window evaporation, RO, OD, and MD.

Direct-contact evaporation has shown great potential in experiments with multicomponent solutions, as it allows water vaporization at temperatures considerably lower than its boiling point at a given operating pressure. Aroma compounds, however, are still lost and must be recovered. The technique should be tested with real juices to verify whether the lower operating temperatures will actually lead to better sensory quality. This technique has been used at industrial scale for many years, but never to concentrate heat-sensitive solutions.

For small- to medium-scale processors, who usually adopt recirculation-type evaporators instead of multiple-effect vacuum systems on account of the large investments associated with the latter, Refractance Window evaporation is commercially available and can be an alternative to improve the sensory and nutritional quality of fruit-juice concentrates. Nonetheless, it cannot replaced the large-scale TASTE systems of the citrus industry.

Membrane-based techniques have been intensely investigated for juice concentration because of their mild operating conditions. RO, a well-established industrial operation with an enormous global market for water desalination, can only be used as a preconcentration technique, as it is not able to economically reach solids contents higher than 25–30°Brix. However, in this range of solids contents, it is less energy-intensive than multiple-effect vacuum evaporation and it provides concentrates with rather superior sensory and nutritional quality. It has already found commercial application in the fruit-juice industry, and its market share in this segment is very likely to grow in the future.

OD constitutes the most studied technique for juice concentration in the recent literature. Operating at room temperature and atmospheric pressure, it can reach and even surpass the solids contents associated with traditional vacuum evaporation. After more than 20 years of research on this topic, it is already proven that the technique is feasible and leads to a juice concentrate with organoleptic properties very close to those of the fresh juice. Besides, the effects of the main operating variables are already well understood. Industrial application has not been achieved yet, but many pilot-plant studies have been reported. Low permeate fluxes, lack of data for long-term operation, and absence of cost and energy consumption estimates compared to traditional vacuum evaporation are some of the issues to be addressed if the technique is to reach commercial scale. Future studies should focus on membrane and module designs specifically for this operation, as well as on long-term performance and economic analysis.

MD, the third membrane-based unit operation investigated in the literature for juice concentration, is similar to OD and can also reach the target solids content of the juice industry. However, it is thermally driven. Originally conceived for water desalination, it can lead to water permeation fluxes that are comparable to those of RO for that particular application when the feed temperature is higher than 70°C. However, as fruit juices are heat-sensitive, operating temperatures must be kept below 40°C, which leads to lower fluxes and energy efficiency problems in view of increased heat loss by conduction through the membrane. Quality of MD fruit-juice concentrates has not been well investigated in the literature, but recent evidence of higher loss of aroma compounds compared to OD has been found in experiments with multicomponent aqueous solutions. Even if the technique seems to be less appropriate than OD for a commercial application in the fruit industry, the two operations are very similar, and future researchers should definitely benefit from the vast literature on MD for water desalination, which has been up to now largely ignored. In particular, advances reported for the latter case, such as taylor-made membranes with optimized morphology and cross-flow modules, could be easily tested for juice concentration.

REFERENCES

Albrecht, W., Hilke, R., Kneifel, K., Weigel, T., and Peinemann, K.-V. 2005. Selection of microporous hydrophobic membranes for use in gas/liquid contactors: An experimental approach. *J. Membr. Sci.* 263: 66–76.

Ali, F., Dornier, M., Duquenoy, A., and Reynes, M. 2003. Evaluating transfers of aroma compounds during the concentration of sucrose solutions by osmotic distillation in a batch-type pilot plant. *J. Food Eng.* 60: 1–8.

Al-Maiman, A. A. and Ahmad, D. 2002. Changes in physical and chemical properties during promegranate (*Punica granaum* L.) fruit maturation. *Food Chem.* 76: 437–441.

Álvarez, V., Álvarez, S., Riera, F. A., and Álvarez, R. 1997. Permeate flux prediction in apple juice concentration by reverse osmosis. *J. Membr. Sci.* 127: 25–34.

Álvarez, S., Riera, F. A., Álvarez, R., and Coca, J. 1998. Permeation of apple aroma compounds in reverse osmosis. *Sep. Purif. Technol.* 14: 209–220.

Álvarez, S., Riera, F. A., Álvarez, R., Coca, J., Cuperus, F. P., and Bouwer, S. Th. 2000. A new integrated membrane process for producing clarified apple juice and apple juice aroma concentrate. *J. Food Eng.* 46: 109–125.

Álvarez, S., Riera, F. A., Álvarez, R., and Coca, J. 2002. Concentration of apple juice by reverse osmosis at laboratory and pilot-plant scales. *Ind. Eng. Chem. Res.* 41: 6156–6164.

Alves, V. D. and Coelhoso, I. M. 2004. Effect of membrane characteristics on mass and heat transfer in the osmotic evaporation process. *J. Membr. Sci.* 228: 159–167.

Alves, V. D. and Coelhoso, I. M. 2006. Orange juice concentration by osmotic evaporation and membrane distillation: A comparative study. *J. Food Eng.* 74: 125–133.

Alves, V. D. and Coelhoso, I. M. 2007. Study of mass and heat transfer in the osmotic evaporation process using hollow fiber membrane contactors. *J. Membr. Sci.* 289: 249–257.

Alves, V. D., Boroknai, B., Bélafi-Bakó, K., and Coelhoso, I. M. 2004. Using membrane contactors for fruit juice concentration. *Desalination* 162: 263–270.

Babu, B. R., Rastogi, N. K., and Raghavarao, K. S. M. S. 2006. Mass transfer in osmotic membrane distillation of phycocyanin colorant and sweet-lime juice. *J. Membr. Sci.* 272: 58–69.

Babu, B. R., Rastogi, N. K., and Raghavarao, K. S. M. S. 2008. Concentration and temperature polarization effects during osmotic membrane distillation. *J. Membr. Sci.* 322: 146–153.

Bailey, A. F. G., Barbe, A. M., Hogan, P. A., Johnson, R. A., and Sheng, J. 2000. The effect of ultrafiltration on the subsequent concentration of grape juice by osmotic distillation. *J. Membr. Sci.* 164: 195–204.

Baker, R. W. 2004. *Membrane Technology and Applications*, 2nd edn. Chichester, U.K.: John Wiley & Sons.

Barbe, A. M., Bartley, J. P., Jacobs, A. L., and Johnson, R. A. 1998. Retention of volatile organic flavor/fragrance components in the concentration of liquid foods by osmotic distillation. *J. Membr. Sci.* 145: 67–75.

Bates, R. P., Morris, J. R., and Crandall, P. G. 2001. *Principles and Practices of Small- and Medium-Scale Fruit Juice Processing*. FAO Agricultural Services Bulletin 146. http://www.fao.org/ag/ags/subjects/en/harvest/docs/ags_bulletins/principles_and_practices.pdf (accessed March 23, 2008).

Bélafi-Bakó, K. and Koroknai, B. 2006. Enhanced water flux in fruit juice concentration: Coupled operation of osmotic evaporation and membrane distillation. *J. Membr. Sci.* 269: 187–193.

Bermúdez-Soto, M. J. and Tomás-Barberán, F. A. 2006. Evaluation of commercial red fruit juice concentrates as ingredients for antioxidant functional juices. *Eur. Food Res. Technol.* 219: 133–141.

Bomben, J. L., Bruin, S., Thijssen, H. A. C., and Merson, R. L. 1973. Aroma recovery and retention in concentration and drying of foods. *Adv. Food Res.* 20: 1–111.

Bonyadi, S. and Chung, T. S. 2007. Flux enhancement in membrane distillation by fabrication of dual layer hydrophilic-hydrophobic hollow fiber membranes. *J. Membr. Sci.* 306: 134–146.

Bottino, A., Capannelli, G., Turchini, A., Della Valle, P., and Trevisan, M. 2002. Integrated membrane processes for the concentration of tomato juice. *Desalination* 148: 73–77.

Bowser, J. J. 2000. Osmotic distillation process. WO Patent 0112304, filled July 28, 2000, and issued Feb. 22, 2001.

Brodard, F., Romero, J., Belleville, M. P., Sanchez, J., Combes-James, C., Dornier, M., and Rios, G. M. 2003. New hydrophobic membranes for osmotic evaporation process. *Sep. Purif. Technol.* 32: 3–7.

Bui, A. V. and Nguyen, H. M. 2005. Scaling up of osmotic distillation from laboratory to pilot plant for concentration of fruit juices. *Int. J. Food Eng.* 1: 1–18. http://www.bepress.com/ijfe/vol1/iss2/art5.

Bui, V. A. and Nguyen, M. H. 2006. The role of operating conditions in osmotic distillation and direct contact membrane distillation—A comparative study. *Int. J. Food Eng.* 2: 1–12. http://www.bepress.com/ijfe/vol2/iss2/art1.

Bui, V. A., Nguyen, M. H., and Muller, J. 2004. A laboratory study on glucose concentration by osmotic distillation in hollow fiber module. *J. Food Eng.* 63: 237–245.

Bui, V. A., Nguyen, M. H., and Muller, J. 2005. Characterization of the polarizations in osmotic distillation of glucose solutions in hollow fiber module. *J. Food Eng.* 68: 391–402.

Bui, V. A., Nguyen, M. H., and Muller, J. 2007. The energy challenge of direct contact membrane distillation in low temperature concentration. *Asia-Pac. J. Chem. Eng.* 2: 400–406.

Burdick, E. M., Anderson, C. O., and Duncan, W. E. 1949. Application of submerged combustion to processing of citrus waste products. *Chem. Eng. Prog.* 45: 539–544.

Calabrò, V., Jiao, B. L., and Drioli, E. 1994. Theoretical and experimental study on membrane distillation in the concentration of orange juice. *Ind. Eng. Chem. Res.* 33: 1803–1808.

Campos, F. B. and Lage, P. L. C. 2001. Modeling and simulation of direct contact evaporators. *Braz. J. Chem. Eng.* 18: 277–286.

Cassano, A. and Drioli, E. 2007. Concentration of clarified kiwifruit juice by osmotic distillation. *J. Food Eng.* 79: 1397–1404.

Cassano, A., Conidi, C., Timpone, R., D'Avella, M., and Drioli, E. 2007. A membrane-based process for the clarification and the concentration of the cactus pear juice. *J. Food Eng.* 80: 914–921.

Cassano, A., Drioli, E., Galaverna, G., Marchelli, R., Di Silvestro, G., and Cagnasso, P. 2003. Clarification and concentration of citrus and carrot juices by integrated membrane processes. *J. Food Eng.* 57: 153–163.

Cassano, A., Jiao, B., and Drioli, E. 2004. Production of concentrated kiwifruit juice by integrated membrane process. *Food Res. Int.* 37: 139–148.

Celere, M. and Gostoli, C. 2002. The heat and mass transfer phenomena in osmotic membrane distillation. *Desalination* 147: 133–138.

Celere, M. and Gostoli, C. 2004. Osmotic distillation with propylene glycol, glycerol and glycerol-salt mixtures. *J. Membr. Sci.* 229: 159–170.

Celere, M. and Gostoli, C. 2005. Heat and mass transfer in osmotic distillation with brines, glycerol and glycerol-salt mixtures. *J. Membr. Sci.* 257: 99–110.

Chen, C. S. and Hernandez, E. 1997. Design and performance evaluation of evaporation. In: *Handbook of Food Engineering Practice*, eds. K. J. Valentas, E. Rostein, and R. P. Singh, pp. 211–252, Boca Raton, FL: CRC Press.

Cheng, D. Y. and Wiersma, S. J. 1980. Composite membrane for a membrane distillation system. U.S. Patent 4316772, filled Feb. 4, 1980, and issued Feb. 23, 1982.

Cheng, D. Y. and Wiersma, S. J. 1982. Composite membrane for a membrane distillation system. U.S. Patent 4419242, filled Feb. 18, 1982, and issued Dec. 6, 1983.

Cheng, L. H., Wu, P. C., Kong, C. K., and Chen, J. 2008. Spatial variations of DCMD performance for desalination through countercurrent hollow fiber modules. *Desalination* 234: 323–334.

Chernyshov, M. N., Meindersma, G. W., and de Haan, A. B. 2003. Modeling temperature and salt concentration distribution in membrane distillation feed channel. *Desalination* 157: 315–324.

Cisse, M., Vaillant, F., Perez, A., Dornier, M., and Reynes, M. 2005. The quality of orange juice processed by coupling crossflow microfiltration and osmotic evaporation. *Int. J. Food Sci. Technol.* 40: 105–116.

Costelo, M. J., Fane, A. G., Hogan, P. A., and Scholfield, R. W. 1993. The effect of shell side hydrodynamics on the performance of axial flow hollow fiber modules. *J. Membr. Sci.* 80: 1–11.

Courel, M., Dornier, M., Herry, J.-M., Rios, G. M., and Reynes, M. 2000a. Effect of operating conditions on water transport during the concentration of sucrose solutions by osmotic distillation. *J. Membr. Sci.* 170: 218–289.

Courel, M., Dornier, M., Rios, G. M., and Reynes, M. 2000b. Modeling of water transport in osmotic distillation using asymmetric membrane. *J. Membr. Sci.* 173: 107–122.

Courel, M., Tronel-Peyroz, E., Rios, G. M., Dornier, M., and Reynes, M. 2001. The problem of membrane characterization for the process of osmotic distillation. *Desalination* 140: 15–25.

Cronan, C. S. 1956. Submerged combustion flares anew. *Chem. Eng.* 63: 163–167.

Curcio, E. and Drioli, E. 2005. Membrane distillation and related operations—A review. *Sep. Purif. Rev.* 34: 35–86.

Curcio, S., Calabrò, V., Iorio, G., and Cindio, B. 2001. Fruit juice concentration by membranes: Effect of rheological properties on concentration polarization phenomena. *J. Food Eng.* 48: 235–241.

Dahuron, L. and Cussler, E. L. 1988. Protein extraction with hollow fibers. *AIChE J.* 34: 130–136.

Dai, Q., Borenstein, A. R., Wu, Y., Jackson, J. C., and Larson, E. B. 2006. Fruit and vegetable juices and Alzheimer's disease: The *Kame* project. *Am. J. Med.* 119: 751–759.

Dale, M. C., Okos, M. R., and Nelson, P. 1982. Concentration of tomato products: Analysis of energy saving process alternatives. *J. Food Sci.* 47: 1853–1858.

Das Gupta, D. K. and Jayarama, K. S. 1996. Studies on the membrane concentration of watermelon juice. *J. Sci. Ind. Res.* 55: 966–970.

Datamonitor. 2007. Juices: Global Industry Guide—Highlights. http://www.just-drinks.com/store/product.aspx?id=52552&lk=s (accessed March 14, 2008).

Deblay, P. 1992. Process for at least partial dehydration of an aqueous composition and devices for implementing the process. U.S. Patent 5832365, filled Oct. 21, 1992, and issued Jan. 17, 1995.

Duxbury, D. D. 1992. Apple, pear juice concentrates 100% pure. *Food Process.* 53: 78–79.

El-Bourawi, M. S., Ding, Z., Ma, R., and Khayet, M. 2006. A framework for better understanding membrane distillation separation process. *J. Membr. Sci.* 285: 4–29.

Fane, A. G., Schofield, R. W., and Fell, C. J. D. 1987. The efficient use of energy in membrane distillation. *Desalination* 64: 231–243.

Feng, C., Shi, B., Li, G., and Wu, Y. 2004. Preparation and properties of microporous membrane from poly(vinylidene fluoride-*co*-tetrafluoroethylene) (F2.4) for membrane distillation. *J. Membr. Sci.* 237: 15–24.

Findley, M. E. 1967. Vaporization through porous membranes. *Ind. Eng. Chem. Process Des. Dev.* 6: 226–230.

Fujii, Y., Kogoshi, S., Iwatani, H., and Aoyama, M. 1992. Selectivity and characteristics of direct contact membrane distillation type experiment. I. Permeability and selectivity through dried hydrophobic fine porous membranes. *J. Membr. Sci.* 72: 53–72.

Gabino, F., Belleville, M. P., Preziosi-Belloy, L., Dornier, M., and Sanchez, J. 2007. Evaluation of the cleaning of a new hydrophobic membrane for osmotic evaporation. *Sep. Purif. Technol.* 55: 191–197.

Galaverna, G., Di Silvestro, G., Cassano, A., Sforza, S., Dossena, A., Drioli, E., and Marchelli, R. 2008. A new integrated membrane process for the production of concentrated blood orange juice: Effect on bioactive compounds and antioxidant activity. *Food Chem.* 106: 1021–1030.

García-Payo, M. C. and Izquierdo-Gil, M. A. 2004. Thermal resistance technique for measuring the thermal conductivity of thin microporous membranes. *J. Phys. D: Appl. Phys.* 37: 3008–3016.

Gawronski, R. and Wrzesinska, B. 2000. Kinetics of solvent extraction in hollow fiber contactors. *J. Membr. Sci.* 168: 213–222.

Girard, B. and Fukumoto, L. R. 2000. Membrane processing of fruit juices and beverages: A review. *Crit. Rev. Biotechnol.* 20: 109–175.

Godino, M. P., Pena, L., Zarate, J. M. O., and Mengal, J. I. 1995. Coupled phenomena membrane distillation and osmotic distillation through a porous hydrophobic membrane. *Sep. Sci. Technol.* 30: 993–1011.

Gomes, E. R. S., Mendes, E. S., Pereira, N. C., and Barros, S. T. D. 2005. Evaluation of the acerola juice concentrated by reverse osmosis. *Braz. Arch. Biol. Technol.* 48: 175–183.

Gostoli, C. 1999. Thermal effects in osmotic distillation. *J. Membr. Sci.* 163: 75–91.

Gostoli, C. and Bandini, S. 1995. Gas membrane extraction of ethanol by glycols: Experiments and modeling. *J. Membr. Sci.* 98: 1–12.

Gostoli, C. and Gatta, A. 1980. Mass transfer in a hollow fiber dialyzer. *J. Membr. Sci.* 6: 133–148.

Gostoli, C., Cervellti, A., and Zardi, G. 1999. Gas membrane extraction: A new technique for the production of high quality juices. *Fruit Process.* 8: 417–421.

Gostoli, C., Sarti, G. C., and Matulli, S. 1987. Low temperature distillation through hydrophobic membranes. *Sep. Sci. Technol.* 22: 855–872.

Gryta, M. 2005a. Osmotic MD and other membrane distillation variants. *J. Membr. Sci.* 246: 145–156.

Gryta, M. 2005b. Long-term performance of membrane distillation process. *J. Membr. Sci.* 265: 153–159.

Gunko, S., Verbych, S., Bryk, M., and Hilal, N. 2006. Concentration of apple juice using direct contact membrane distillation. *Desalination* 190: 117–124.

Habert, A. C., Borges, C. P., and Nobrega R. 2006. *Membrane Separation Processes.* Rio de Janeiro: E-papers (in Portuguese).

Hernandez, E., Chen, C. S., Johnson, J., and Carter, R. D. 1995. Viscosity changes in orange juice after ultrafiltration and evaporation. *J. Food Eng.* 25: 387–396.

Hogan, P. A., Canning, R. P., Peterson, P. A., Johnson, R. A., and Michaels, A. S. 1998. A new option: Osmotic distillation. *Chem. Eng. Prog.* 94: 49–61.

Hongvaleerat, C., Cabral, L. M. C., Dornier, M., Reynes, M., and Ningsanond, S. 2008. Concentration of pineapple juice by osmotic evaporation. *J. Food Eng.* 88: 548–552.

Iyer, P. A. and Chu, C. 1971. Submerged combustion. *Desalination* 9: 19–31.

Izquierdo-Gil, M. A., García-Payo, M. C., and Fernández-Pineda, C. 1999a. Direct contact membrane distillation of sugar aqueous solutions. *Sep. Sci. Technol.* 34: 1773–1801.

Izquierdo-Gil, M. A., García-Payo, M. C., and Fernández-Pineda, C. 1999b. Air gap membrane distillation of sucrose aqueous solutions. *J. Membr. Sci.* 155: 291–307.

Jariel, O., Reynes, M., Courel, M., Durand, N., Dornier, M., and Deblay, P. 1996. Comparison of some fruit juice concentration techniques. *Fruits* 51: 437–450 (in French).

Jesus, D. F., Leite, M. F., Silva, L. F. M., Modesta, R. D., Matta, V. M., and Cabral, L. M. C. 2007. Orange (*Citrus sinensis*) juice concentration by reverse osmosis. *J. Food Eng.* 81: 287–291.

Jiao, B., Cassano, A., and Drioli, E. 2004. Recent advances on membrane processes for the concentration of fruit juices: A review. *J. Food Eng.* 63: 303–324.

Karlsson, H. O. E. and Trägårdh, G. 1997. Aroma recovery during beverage processing. *J. Food Eng.* 34: 159–178.

Kato, T., Shimoda, M., Suzuki, J., Kawaraya, A., Igura, N., and Hayakawa, I. 2003. Changes in the odors of squeezed apple juice during thermal processing. *Food Res. Int.* 36: 777–785.

Kawasaki, J. and Hayakawa, T. 1972. Direct-contact mass and heat transfer between vapor and liquid with change of phase. *J. Chem. Eng. Jpn.* 5: 119–124.

Khayet, M. 2008. Membrane distillation. In: *Advanced Membrane Technology and Applications*, eds. N. N. Li, A. G. Fane, W. S. W. Ho, and T. Matsuura, pp. 297–369. Hoboken, NJ: John Wiley & Sons.

Khayet, M. and Matsuura, T. 2001. Preparation and characterization of poly (vinylidene fluoride) membranes for membrane distillation. *Ind. Eng. Chem. Res.* 40: 5710–5718.

Khayet, M. and Matsuura, T. 2003. Determination of surface and bulk pore sizes of flat-sheet and hollow-fiber membranes by atomic force microscopy, gas permeation, and solute transport methods. *Desalination* 158: 57–64.

Khayet, M., Mengual, J. I., and Matsuura, T. 2005. Porous hydrophobic/hydrophilic composite membranes: Application in desalination using direct contact membrane distillation. *J. Membr. Sci.* 252: 101–113.

Kimura, S., Nakao, S.-I., and Shimatani, S.-I. 1987. Transport phenomena in membrane distillation. *J. Membr. Sci.* 33: 285–298.

Koeseoglu, S. S., Lawhon, J. T., and Lusas, E. W. 1990. Use of membranes in citrus juice processing. *Food Technol.* 44: 90–97.

Kong, Y., Lin, X., Wu, Y., Chen, J., and Xu, J. 1992. Plasma polymerization of octafluorocyclobutane and hydrophobic microporous composite membranes for membrane distillation. *J. Appl. Polym. Sci.* 46: 191–199.

Koroknai, B., Csanádi, Z., Gubicza, L., and Bélafi-Bakó, K. 2008. Preservation of antioxidant capacity and flux enhancement in concentration of red fruit juices by membrane processes. *Desaltination* 228: 295–301.

Koroknai, B., Gubicza, L., and Bélafi-Bakó, K. 2006. Coupled membrane process applied to fruit juice concentration. *Chem. Pap.* 60: 399–403.

Kozák, A., Bánvölgyi, S., Vincze, I., Kiss, I., Békássy-Molnár, E., and Vatai, G. 2008. Comparison of integrated large scale and laboratory scale membrane processes for the production of black currant juice concentrate. *Chem. Eng. Process.* 47: 1171–1177.

Kunz, W., Benhabiles, A., and Ben-Aim, R. 1996. Osmotic evaporation through macroporous hydrophobic membranes: A survey of current research and applications. *J. Membr. Sci.* 121: 25–36.

Kurz, G. and Güthoff, H. 1988. Submerged combustion gathers a variety of applications. *Process Eng.* 68: 31–32.

Laganà, F., Barbieri, G., and Drioli, E. 2000. Direct contact membrane distillation: Modeling and concentration experiments. *J. Membr. Sci.* 166: 1–11.

Lage, P. L. C., Borges, C. P., and Ribeiro, C. P. 2005. Combined vapor-permeation direct-contact-evaporation route for concentrating fruit juices. Brazilian Patent BRPI0501787-4, filled May 20, 2005, and issued Jan. 16, 2007 (in Portuguese).

Lawson, K. W. and Lloyd, D. R. 1997. Membrane distillation. *J. Membr. Sci.* 124:1–25.

Lefebvre, M. 1985. Method of performing osmotic distillation. U.S. Patent 4781837, filed Nov. 21, 1985, and issued Nov. 1, 1988.

Lefebvre, M. S. M. 1987. Osmotic distillation process and semipermeable barriers therefor. Australian Patent AU607142, filled May 05, 1987, and issued Feb. 28, 1991.

Li, B. and Sirkar, K. K. 2004. Novel membrane and device for direct contact membrane distillation-based desalination process. *Ind. Eng. Chem. Res.* 43: 5300–5309.

Lin, J., Rouseff, R. L., Barros, S., and Naim, M. 2002. Aroma composition changes in early season grapefruit produced from thermal concentration. *J. Agric. Food Chem.* 50: 813–839.

Lipnizki, F. and Field, R. W. 2001. Mass transfer performance for hollow fiber modules with shell-side axial feed flow: Using an engineering approach to develop a framework. *J. Membr. Sci.* 193: 195–208.

Loeb, S. and Sourirajan, S. 1963. Sea water demineralization by means of an osmotic membrane. *Adv. Chem. Ser.* 28: 117–132.

Lovellsmith, J. E. R. 1982. Submerged combustion for the concentration of whey. *N. Z. J. Dairy Sci. Technol.* 17: 161–170.

Lozano, J. E. and Ibarz, A. 1997. Color changes in concentrated fruit pulp during heating at high temperatures. *J. Food Eng.* 63: 303–324.

Luedicke, A. H., Hendrickson, B., and Pigott, G. M. 1979. A method for the concentration of proteinaceous solutions by submerged combustion. *J. Food Sci.* 44: 1469–1473.

Lukanin, O. S., Gunko, S. M., Bryk, M. T., and Nigmatullin, R. R. 2003. The effect of content of apple juice biopolymers on the concentration by membrane distillation. *J. Food Eng.* 60: 275–280.

Mannheim, C. H. and Passy, N. 1975. Aroma recovery and retention in liquid foods during concentration and drying—Part I: Processes. *Process Biotechnol.* 10: 3–10.

Mansouri, J. and Fane, A. G. 1999. Osmotic distillation of oily feeds. *J. Membr. Sci.* 153: 103–120.

Margetts, B. and Buttriss, J. 2003. Epidemiology linking consumption of plant foods and their constituents with health. In: *Plants: Diet and Health*, eds. G. Goldberg, pp. 49–64, Oxford, U.K.: Blackwell Publishing.

Martínez, L. and Rodríguez-Maroto, J. M. 2006. Characterization of membrane distillation modules and analysis of mass flux enhancement by channel spacers. *J. Membr. Sci.* 274: 123–137.

Martínez-Díez, L., Vázquez-González, M. I., and Florido-Díaz, F. J. 1998. Study of membrane distillation using channel spacers. *J. Membr. Sci.* 144: 45–56.

Maskan, M. 2006. Production of pomegranate (*Punica granatum* L.) juice concentrate by various heating methods: Color degradation and kinetics. *J. Food Eng.* 72: 218–224.

Matta, V. M., Moretti, R. H., and Cabral, L. M. C. 2004. Microfiltration and reverse osmosis for clarification and concentration of acerola juice. *J. Food Eng.* 61: 477–482.

Medina, B. G. and Garcia, A. 1988. Concentration of orange juice by reverse osmosis. *J. Food Process Eng.* 10: 217–230.

Mengual, J. I., Khayet, M., and Godino, M. P. 2004. Heat and mass transfer in vacuum membrane distillation. *Int. J. Heat Mass Transf.* 47: 865–875.

Merson, R. L. and Morgan, A. I. 1968. Juice concentration by reverse osmosis. *Food Technol.* 22: 631–634.

Michaels, A. S. 1997. Osmotic distillation process using a membrane laminate. U.S. Patent 5938928, filled Sep. 18, 1997, and issued Aug. 17, 1999.

Michaels, A. S. 1999. Membrane laminates and methods for their preparation. WO Patent 9940997, filled Feb. 11, 1999, and issued Aug. 19, 1999.

Michaels, A. S. and Johnson, R. 1995. Methods and apparatus for osmotic distillation. U.S. Patent 5824223, filled Nov. 8, 1995, and issued Oct. 20, 1998.

Mikkelsen, B. B. and Poll, L. 2002. Decomposition and transformation of aroma compounds and anthocyanins during black currant juice processing. *J. Food Sci.* 67: 3447–3455.

Molinari, R., Gagliardi, R., and Drioli, E. 1995. Methodology for estimating saving of primary energy with membrane operations in industrial processes. *Desalination* 100: 125–137.

Mosshammer, M. R., Stintzing, F. C., and Carle, R. 2006. Cactus pear fruits (*Opuntia* spp.): A review of processing technologies and current uses. *J. Prof. Assoc. Cactus Dev.* 8: 1–25.

Moyer, J. C. 1980. Fruit juices. In: *Encyclopedia of Chemical Technology*, eds. H. F. Mark, D. F. Othmer, C. G. Overberger, and G. T. Seaborg, vol. 11, pp. 300–316. New York: John Wiley & Sons.

Mulder, M. 1991. *Basic Principles of Membrane Technology*. Dordrecht, the Netherlands: Kluwer.

Mullen, W., Marks, S. C., and Crozier, A. 2007. Evaluation of phenolic compounds in commercial fruit juices and fruit drinks. *J. Agric. Food Chem.* 55: 3148–3157.

Nagaraj, N., Patil, B. S., and Biradar, P. M. 2006a. Osmotic membrane distillation—A brief review. *Int. J. Food Eng.* 2: 1–22. http://www.bepress.com/ijfe/vol2/iss2/art5.

Nagaraj, N., Patil, G., Babu, B. R., Hebbar, U. H., Raghavarao, K. S. M. S., and Nene, S. 2006b. Mass transfer in osmotic membrane distillation. *J. Membr. Sci.* 268: 48–56.

Narain, H., Hsieh, T. C. Y., and Johnson, C. E. 1990. Dynamic headspace concentration in gas-chromatography of volatile flavor compounds in peach. *J. Food Sci.* 55: 1303–1307.

Narayan, A. V., Nagaraj, N., Hebbar, H. U., Chakkaravarthi, A., Raghavarao, K. S. M. S., and Nene, S. 2002. Acoustic field-assisted osmotic membrane distillation. *Desalination* 147: 149–156.

Neves, M. F., Neves, E. M., and Val, A. M. 1999. Sweet perspectives of worldwide orange juice consumption. *Gazeta Mercantil*, November 22, section "Interior Paulista" (in Portuguese).

Nindo, C. I., Powers, J. R., and Tang, J. 2007. Influence of refractance window evaporation on quality of juices from small fruits. *Lebensm-Wiss. Technol.* 40: 1000–1007.

Nindo, C. I., Tang, J., Powers, J. R., and Bolland, K. 2004. Energy consumption during refractance window evaporation of selected berry juices. *Int. J. Energy Res.* 28: 1089–1100.

Nindo, C. I., Tang, J., Powers, J. R., and Singh, P. 2005. Viscosity of blueberry and raspberry juices for processing applications. *J. Food Eng.* 69: 343–350.

Nisperos-Carriedo, M. O. and Shaw, P. E. 1990. Comparison of volatile flavor components in fresh and processed orange juices. *J. Agric. Food Chem.* 38: 1048–1052.

Nongonierma, A., Cayot, P., Le Quere, J. L., Springett, M., and Voilley, A. 2006. Mechanisms of extraction of aroma compounds from foods using adsorbents: Effect of various parameters. *Food Rev. Int.* 22: 51–94.

Owen, J. J. and Moggio, W. A. 1955. Submerged combustion evaporation of neutral sulphite spent cooking liquor. *TAPPI J.* 38: 144–147.

Pearce, S. and Bullen, J. D. 1995. Juice and juice aroma concentrate production using reverse osmosis under controlled conditions. U.K. Patent Application GB2289857, filled Feb. 06, 1995.

Pena, L., Paz Godino, M., and Mengual, J. I. 1998. A method to evaluate the net membrane distillation coefficient. *J. Membr. Sci.* 143: 219–233.

Pereira, C. C., Ribeiro, C. P., Nobrega, R., and Borges, C. P. 2006. Pervaporative recovery of aroma compounds from fruit juices. *J. Membr. Sci.* 274: 1–23.

Petrotos, K. B. and Lazarides, H. N. 2001. Osmotic concentration of liquid foods. *J. Food Eng.* 49: 201–206.

Phattaranawik, J., Jiraratananon, R., Fane, A. G., and Halim, C. 2001. Mass flux enhancement using spacer filled channels in direct contact membrane distillation. *J. Membr. Sci.* 187: 193–201.

Phattaranawik, J., Jiraratananon, R., and Fane, A. G. 2003a. Heat transport and membrane distillation coefficients in direct contact membrane distillation. *J. Membr. Sci.* 212: 177–193.

Phattaranawik, J., Jiraratananon, R., and Fane, A. G. 2003b. Effects of net-type spacers on heat and mass transfer in direct contact membrane distillation and comparison with ultrafiltration studies. *J. Membr. Sci.* 217: 193–206.

Pollack, S. and Perez, A. 2006. *Fruit and Tree Nuts Situation and Outlook Yearbook*, Market and Trade Economics Division. Washington, DC: USDA. http://www.ers.usda.gov (accessed November 20, 2007).

Pozderovic, A., Moslavac, T., and Pichler, A. 2007. Influence of processing parameters and membrane type on permeate flux during solution concentration of different alcohols, esters, and aldehydes by reverse osmosis. *J. Food Eng.* 78: 1092–1102.

Prasad, R. and Sirkar, K. K. 1988. Dispersion-free solvent extraction with microporous hollow-fiber modules. *AIChE J.* 34: 177–188.

Puffett, J. and Drewes, J. 2007. Membranes part I: A process approach to control and prevent biofouling in pressure-driven polyamide membrane systems. *Ultrapure Water J.* 24(5): 33–42.

Ramteke, R. S., Eipeson, W. E., and Patwardhan, M. V. 1990. Behavior of aroma volatiles during the evaporative concentration of some tropical fruit juices and pulps. *J. Sci. Food Agric.* 50: 399–405.

Ramteke, R. S., Singh, N. I., Rekha, M. N., and Eipeson, W. E. 1993. Methods for concentration of fruit juices: A critical evaluation. *J. Food Sci. Technol.* 30: 391–402.

Reieccius, G. 2006. *Flavor Chemistry and Technology*. Boca Raton, FL: Taylor & Francis.

Rektor, A., Kozák, A., Vatai, G., and Bekassy-Molnar, E. 2007. Pilot plant RO-filtration of grape juice. *Sep. Purif. Technol.* 57: 473–475.

Rektor, A., Vatai, G., and Békássy-Molnár, E. 2006. Multi-step membrane processes for the concentration of grape juice. *Desalination* 191: 446–453.

Ribeiro, C. P. and Borges, C. P. 2004. Using pervaporation data in the calculation of vapor permeation hollow-fibre modules for aroma recovery. *Braz. J. Chem. Eng.* 21: 629–640.

Ribeiro, C. P. and Lage, P. L. C. 2004a. Direct-contact evaporation in the homogeneous and heterogeneous bubbling regimes—Part I: Experimental analysis. *Int. J. Heat Mass Transf.* 47: 3825–3840.

Ribeiro, C. P. and Lage, P. L. C. 2004b. Direct-contact evaporation in the homogeneous and heterogeneous bubbling regimes—Part II: Dynamic simulation. *Int. J. Heat Mass Transf.* 47: 3841–3854.

Ribeiro, C. P. and Lage, P. L. C. 2005. Gas-liquid direct-contact evaporation: A review. *Chem. Eng. Technol.* 28: 1081–1107.

Ribeiro, C. P., Borges, C. P., and Lage, P. L. C. 2004a. Concentration of synthetic fruit juices by direct-contact evaporation. *Paper Presented at the 31st Brazilian Congress on Particulate Systems*, Uberlândia, MG.

Ribeiro, C. P., Borges, C. P., and Lage, P. L. C. 2005a. Modeling of direct-contact evaporation using a simultaneous heat and multicomponent mass transfer model for superheated bubbles. *Chem. Eng. Sci.* 60: 1761–1772.

Ribeiro, C. P., Borges, C. P., and Lage, P. L. C. 2005b. A new route combining direct-contact evaporation and vapor permeation for obtaining high-quality fruit-juice concentrates. Part I: Experimental analysis. *Ind. Eng. Chem. Res.* 44: 6888–6902.

Ribeiro, C. P., Borges, C. P., and Lage, P. L. C. 2005c. A new route combining direct-contact evaporation and vapor permeation for obtaining high-quality fruit-juice concentrates. Part II: Modeling and simulation. *Ind. Eng. Chem. Res.* 44: 6903–6915.

Ribeiro, C. P., Borges, C. P., and Lage, P. L. C. 2007. Sparger effects during the concentration of synthetic fruit juices by direct-contact evaporation. *J. Food Eng.* 79: 979–988.

Ribeiro, C. P., Lage, P. L. C., and Borges, C. P. 2004b. A combined gas-stripping vapour permeation process for aroma recovery. *J. Membr. Sci.* 238: 9–19.

Robe, K. 1983. Hyperfiltration methods for preconcentrating juice save evaporation energy. *Food Process.* 44: 100–101.

Rodrigues, R. B., Menezes, H. C., Cabral, L. M. C., Dornier, M., Rinos, G. M., and Reynes, M. 2004. Evaluation of reverse osmosis and osmotic evaporation to concentrate camu-camu juice (*Myrciaria dubia*). *J. Food Eng.* 63: 97–102.

Romero, J., Rios, G. M., Sanchez, J., Bocquet, S., and Savedra, A. 2003. Modeling heat and mass trasnfer in osmotic evaporation process. *AIChE J.* 49: 300–308.

Rufino, J. R. M. 1996. Recovery of aroma compounds by pervaporation. MSc dissertation, Federal University, Rio de Janeiro, Brazil (in Portuguese).

Ruxton, C. H. S., Gardner, E. J., and Walker, D. 2006. Can pure fruit and vegetable juices protect against cancer and cardiovascular disease too? A review of the evidence. *Int. J. Food Sci. Nutr.* 57: 249–272.

Sá, I. S., Cabral, L. M. C., and Matta, V. M. 2003. Concentration of pineapple juice by membrane processes. *Braz. J. Food Technol.* 6: 53–62 (in Portuguese).

Sancho, M. F., Rao, M. A., and Downing, D. L. 1998. Infinite dilution activity coefficients of apple juice aroma compounds. *J. Food Eng.* 34: 145–158.

Schneider, K., Holz, W., and Wollbeck, R. 1988. Membranes and modules for transmembrane distillation. *J. Membr. Sci.* 39: 25–42.

Schofield, R. W., Fane, A. G., and Fell, C. J. D. 1987. Heat and mass transfer in membrane distillation. *J. Membr. Sci.* 33: 299–313.

Schofield, R. W., Fane, A. G., Fell, C. J. D., and Macon, R. 1990a. Factors affecting flux in membrane distillation. *Desalination* 77: 279–294.

Schofield, R. W., Fane, A. G., Fell, C. J. D., and Macon, R. 1990b. Gas and vapor transport through microporous membranes. II. Membrane distillation. *J. Membr. Sci.* 53: 173–185.

Schwarz, H. W. and Penn, F. E. 1948. Production of orange juice concentrate and powder. *Ind. Eng. Chem.* 40: 938–944.

Shaw, P. E., Lebrun, M., Dornier, M., Ducamp, M. N., Courel, M., and Reynes, M. 2001. Evaluation of concentrated orange and passion fruit juices prepared by osmotic evaporation. *Lebnsm.-Wiss. Technol.* 34: 60–65.

Shiota, H. 1993. New esteric components in the volatiles of banana fruit (*Musa sapientum* L.). *J. Agric. Food Chem.* 41: 2056–2062.

Singh, N. G. I. and Eipeson, W. E. 2000. Concentration of clarified orange juice by reverse osmosis. *J. Food Sci. Technol.* 37: 363–367.

Song, L., Li, B., Sirkar, K. K., and Gilron, J. L. 2007. Direct contact membrane distillation-based desalination: Novel membranes, devices, larger-scale studies, and a model. *Ind. Eng. Chem. Res.* 46: 2307–2323.

Song, L., Ma, Z., Liao, X., Rosaraju, P. B., Irish, J. R., and Sirkar, K. K. 2008. Pilot plant studies of novel membranes and devices for direct contact membrane distillation-based desalination. *J. Membr. Sci.* 323: 257–270.

Sorrentino, F., Voilley, A., and Richon, D. 1986. Activity coefficients of aroma compounds in model food systems. *AIChE J.* 32: 1988–1993.

Standiford, F. C. 1997. Evaporators. In: *Perry's Chemical Engineering Handbook*, 7th edn., eds. R. H. Perry, D. W. Green, and J. O. Maloney, 11-107–11-118. New York: McGraw-Hill.

Swindin, N. 1949. Recent developments in submerged combustion. *Trans. Inst. Chem. Eng.* 27: 209–211.

Teoh, M. M., Bonyadi, S., and Chung, T.-S. 2008. Investigation of different hollow fiber module designs for flux enhancement in the membrane distillation process. *J. Membr. Sci.* 311: 371–379.

Thanedgunbaworn, R., Jiraratananon, R., and Nguyen, M. H. 2007a. Mass and heat transfer analysis in fructose concentration by osmotic distillation process using hollow fiber module. *J. Food Eng.* 78: 126–135.

Thanedgunbaworn, R., Jiraratananon, R., and Nguyen, M. H. 2007b. Shell-side mass transfer of hollow fiber modules in osmotic distillation process. *J. Membr. Sci.* 290: 105–113.

Tomaszewska, M. 1996. Preparation and properties of flat-sheet membranes from poly (vinylidene fluoride) for membrane distillation. *Desalination* 104: 1–11.

Tomaszewska, M. 2000. Concentration and purification of fluosilicic acid by membrane distillation. *Ind. Eng. Chem. Res.* 39: 3038–3041.

Toribo, J. L. and Lozano, J. E. 1986. Heat induced browning of clarified apple juice at high temperatures. *J. Food Sci.* 51: 172–175, 179.

Tribess, T. B. and Tadini, C. C. 2001. Minimally processed orange juice: An alternative to expand the orange juice market in Brazil. *Paper Presented at the III International Conference on Agri-Food Chain Economics and Management*, Ribeirão Preto, SP, Brazil (in Portuguese).

Umano, K., Hagi, Y., Nakahara, K., Shoji, A., and Shibamoto, T. 1992. Volatile constituents of green and ripened pineapple (*Ananas comosus* L.). *J. Agric. Food Chem.* 40: 599–603.

USDA 2007. Production, Supply and Distribution (PSD) Reports On-Line. Washington, DC: Foreign Agriculture Service. http://www.fas.usda.gov (accessed November 20, 2007).

USDA 2008. *National Nutrient Database*. Beltsville, MD: Agricultural Research Service. http://www.nal.usda.gov (accessed January 12, 2008).

Vaillant, F., Cisse, M., Chaverri, M., Perez, A., Viquez, M. D. F., and Mayer, C. D. 2005. Clarification and concentration of melon juice using membrane processes. *Innov. Food Sci. Emerg. Technol.* 6: 213–220.

Vaillant, F., Jeanton, E., Dornier, M., O'Brien, G. M., Reynes, M., and Decloux, M. 2001. Concentration of passion fruit juice on an industrial pilot scale using osmotic evaporation. *J. Food Eng.* 47: 195–202.

Van der Bruggen, B., Vandecasteele, C., Van Gestel, T., Doyen, W., and Leysen, R. 2003. A review of pressure-driven membrane processes in wastewater treatment and drinking water production. *Environ. Prog.* 22: 46–56.

Van Ruth, S. M. and Villeneuve, E. 2002. Influence of β-lactoglobulin, pH and presence of other aroma compounds on the air/liquid partition coefficients of 20 aroma compounds varying in functional group and chain length. *Food Chem.* 79: 157–164.

Varming, C., Petersen, M. A., and Poll, L. 2004. Comparison of isolation methods for the determination of important aroma compounds in black currant (*Ribes nigrum* L.) juice, using nasal impact frequency profiling. *J. Agric. Food Chem.* 52: 1647–1652.

Versari, A., Ferrarini, R., Tornielli, G. B., Parpinello, G. P., Gostoli, C., and Celotti, E. 2004. Treatment of grape juice by osmotic evaporation. *J. Food Sci.* 69: 422–427.

Viegas, R. M. C., Rodriguez, M., Luque, S., Alvarez, J. R., Coelhoso, I. M., and Crespo, J. P. S. G. 1998. Mass transfer correlations in membrane extraction: Analysis of Wilson-plot methodology. *J. Membr. Sci.* 145: 129–142.

Voit, D. C., Santos, M. R., and Paul Singh, R. 2006. Development of a multipurpose fruit and vegetable processor for a manned mission to Mars. *J. Food Eng.* 77: 230–238.

Wang, Z., Zheng, F., and Wang., S. 2001. Experimental study of membrane distillation with brine circulated in the cold side. *J. Membr. Sci.* 183: 171–179.

Warczok, J., Gierszewska, M., Kujawski, W., and Güell, C. 2007. Application of osmotic membrane distillation for reconcentration of sugar solutions from osmotic dehydration. *Sep. Purif. Technol.* 57: 425–429.

Watson, B. R. 1966. The combustion heating of liquids. *Metals Eng. Quart.* 6: 56–60.

Weisman, W. I. 1961. Submerged combustion equipment. *Ind. Eng. Chem.* 53: 708–712.

Werkhoff, P., Guntert, M., Krammer, H., Sommer, J., and Kaulen, J. 1998. Vacuum headspace method in aroma research: Flavor chemistry in yellow passion fruits. *J. Agric. Food Chem.* 46: 1076–1093.

Wickramasinghe, S. R. J., Semmens, M. J., and Cussler, E. L. 1992. Mass transfer in various hollow fiber geometries. *J. Membr. Sci.* 69: 235–250.

Wijmans, J. G. and Baker, R. W. 1995. The solution-diffusion model: A review. *J. Membr. Sci.* 107: 1–21.

Wu, J. and Chen, V. 2000. Shell-side mass transfer performance of randomly packed hollow fiber modules. *J. Membr. Sci.* 172: 59–74.

Wu, Y., Kong, Y., Lin, X., Liu, W., and Xu, J. 1992. Surface-modified hydrophilic membranes in membrane distillation. *J. Membr. Sci.* 72: 189–196.

Xu, J. B., Lange, S., Bartley, J. P., and Johnson, R. A. 2004. Alginate-coated microporous PTFE membranes for use in the osmotic distillation of oily feeds. *J. Membr. Sci.* 240: 81–89.

Xu, J. B., Bartley, J. P., and Johnson, R. A. 2005a. Application of sodium alginate-carrageenan coatings to PTFE membranes for protection agains wet-out by surface-active agents. *Sep. Sci. Technol.* 40: 1067–1081.

Xu, J. B., Spittler, D. A., Bartley, J. P., and Johnson, R. A. 2005b. Alginic acid-silica hydrogel coatings for the protection of osmotic distillation membranes against wet-out by surface-active agents. *J. Membr. Sci.* 260: 19–25.

Yang, M. C. and Cussler, E. L. 1986. Designing hollow-fiber contactors. *AIChE J.* 32: 1910–1916.

Zabetakis, I. and Holden, M. A. 1997. Strawberry flavor: Analysis and biosynthesis. *J. Agric. Food Chem.* 74: 421–434.

Zaida, A. H., Sarma, S. C., Grover, P. D., and Heldman, D. R. 1986. Milk concentration by direct-contact heat exchange. *J. Food Process Eng.* 9: 63–79.

Zhongwei, D., Liying, L., and Runyu, M. 2003. Study on the effect of flow maldistribution on the performance of the hollow fiber modules used in membrane distillation. *J. Membr. Sci.* 215: 11–23.

7 Encapsulation Technologies for Modifying Food Performance

Maria Inês Ré, Maria Helena Andrade Santana, and Marcos Akira d'Ávila

CONTENTS

7.1 INTRODUCTION

The food industry faces serious challenges in the twenty-first century. Consumers are demanding more from the foods they eat. Now foods must not only taste good and aid immediate nutrition, but also assist in mitigating disease, provide clear health benefits, and help to reduce health care costs. To meet these demands, food manufacturers must prepare safe, healthy, and convenient foods that are of good value and of great taste. The global functional foods market is a dynamic and growing segment of the food industry, being expected to represent 5% of the total global food market in 2010 (Market Research, 2004).

To date, a number of national authorities, academic bodies, and the industry have proposed definitions for functional foods. Although the term "functional food" has already been defined several times, there is no universally accepted definition for this emerging food category. Definition ranges from the very simple to the more complex (Siró et al., 2008): "foods that, by virtue of physiologically active components, provide health benefits beyond basic nutrition (International Life Sciences Institute, 1999);" "food similar in appearance to a conventional food, consumed as part of the usual diet, with demonstrated physiological benefits, and/or to reduce the risk of chronic disease beyond basic nutritional functions" (Health Canada, 1998).

Examples of functional foods that have potential benefits for health and whose market has grown tremendously include baby food, ready meals, snacks, soft drinks such as energy and sport drinks, meat products, and spreads.

A functional benefit is usually obtained by fortification with an active (functional) ingredient. Examples of functional ingredients are flavors, vitamins, minerals, enzymes, peptides, bioactive lipids, antioxidants, and probiotic microorganisms. Functional ingredients come in a variety of molecular and physical forms, with different polarities (polar, nonpolar, amphiphilic), molecular weights (low to high), and physical states (solid, liquid). They are rarely used directly in their pure form. Instead, they are often incorporated into some form of delivery system generated by encapsulation technologies.

A delivery system must perform a number of different roles. First, it serves as a vehicle for carrying the functional ingredient to the desired site of action. Second, it may have to protect the functional ingredient from chemical or biological degradation (e.g., oxidation) during processing, storage, and utilization; this maintains the functional ingredient in its active state. Third, it may have to be capable of controlling the release of the functional ingredient, either responding to specific environmental conditions that trigger release (e.g., pH, ionic strength, or temperature) or varying the release rate. Fourth, the delivery system has to be food-grade and also compatible with the physicochemical and qualitative attributes (i.e., appearance, texture, taste, and shelf life) of the final product (Weiss et al., 2006).

The characteristics of the delivery systems are one of the most important factors influencing the efficacy of functional ingredients in many industrial products. The delivery systems can be used to deliver a host of ingredients in a range of food formulations. The demand for encapsulation technologies is growing at around 10% annually, driven by both the increasing fortification with healthy ingredients and the consumer demand for novel products (Brownlie, 2007). Encapsulation technologies are attracting growing interest because they can decrease costs for food makers, particularly those using sensitive ingredients like probiotics, by reducing the need for preservatives. Encapsulation also allows manufacturers of food and beverages, as well as other consumer products, to add into their formulations ingredients that would be normally used in traditional processing.

7.2 ENCAPSULATION TECHNOLOGIES

Encapsulation is a topic of interest in a wide range of scientific and industrial areas, varying from pharmaceutics to agriculture, and from pesticides to enzymes. The development of these technologies is characterized by strong fundamental research and several industrial applications, demonstrated by the growing number of scientific papers and patent applications. The European network of patent databases (esp@cenet) has approximately 600 patent documents worldwide containing "microencapsulation" in the title or in the abstract, whereas the U.S. Patent and Trademark Office has approximately 160 patents and 140 applications with at least one claim containing the term "microencapsulation."* In the past 5 years, there have been 200–300 peer-reviewed articles per year on microencapsulation in the *Web of Science*®† database (Figure 7.1), in contrast to 1600 papers in the previous 20 years.

Encapsulation involves the coating or entrapment of a desired component (active or core material) within a secondary material (encapsulating, carrier, coating, shell, or wall material) to prevent or delay the release of the active or core ingredient until a certain

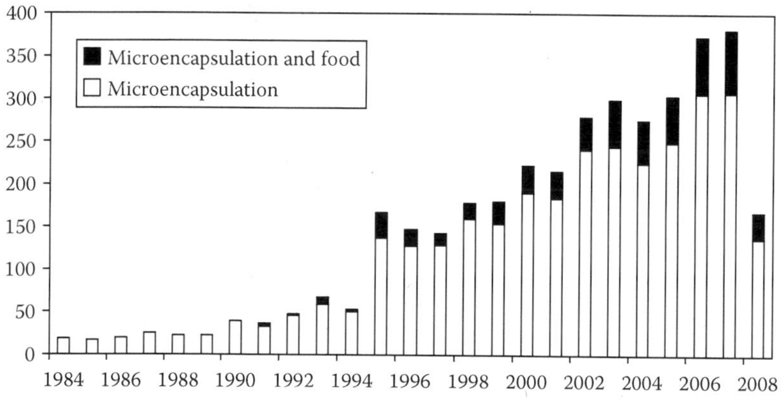

FIGURE 7.1 Evolution of the number of scientific publications in the Web of Science database (up to July 2008) on "microencapsulation" and "microencapsulation of food ingredients." Microencapsulation applied to food represents approximately 20% of the papers published in the microencapsulation area.

* Data related to a search performed on the U.S. Patent and Trademark Office: 1976–January 2009 (patents) and 2001–January 2009 (patent applications); and on the espac@cenet: 1970–January 2009.

† Registered trademark of Thomson Reuters, New York.

time or a set of conditions is achieved. Encapsulation can potentially offer numerous benefits to the materials being encapsulated. Various properties of active ingredients may be changed by encapsulation. For example, handling and flow properties can be changed by converting a liquid to a solid encapsulated form, and hygroscopic materials can be protected from moisture.

Encapsulation systems in food applications are typically used with at least one of the following purposes:

- To solve formulation problems arising from a limited chemical or physical stability of the active ingredient
- To overcome incompatibility between active ingredient and food matrix
- To control the release of a sensorial active compound
- To assist or enhance the absorption of a nutrient

Examples of food additives that may benefit from encapsulation and controlled release are flavors, minerals, and lipids, among others.

There are several items to consider when choosing or developing an encapsulated food ingredient. Each of these is very important to the success of the product. The molecular characteristics of the ingredient, its desired function, the type of coating, the required protection rate, the particle size, and the processing conditions must all be clearly defined. In this regard, there are some important questions that must be answered:

- What are the molecular structure, size, and charge of the ingredient?
- What is it that you are trying to achieve with the encapsulation?
- When or how do you want the core ingredient to be released?
- Do you want the coating to melt and release the core ingredient at a certain temperature? Do you want the coating to break away by mechanical action or dissolving in water?
- Is a change in pH such as in the gastrointestinal (GI) tract going to release the core? Or is there another way your core ingredient will be released?

A critical step in developing encapsulated food products is to determine the encapsulant formulation that meets the desired stability and release criteria. The GRAS (generally recognized as safe) encapsulating material must stabilize the core material, must not react with or deteriorate the active ingredient, and should release it under the specific conditions based on product application. In addition, delivery systems should be developed from inexpensive ingredients, since the additional costs associated with the encapsulation of the active ingredient should be overcome by its benefits, for example, improved shelf life, better bioavailability, or enhanced marketability.

The variety of encapsulating materials allows food producers to select compounds that work for water or fat-soluble food ingredients; dissolve, melt, or rupture to release core material; and provide textural characteristics to satisfy consumer palates. Commonly used encapsulating materials are carbohydrates due to their ability to absorb and retain flavors, cellulose (based on its permeability), gums (which offer good gelling properties and heat resistance), lipids (based on their hydrophobicity), and usually gelatin as a protein, which is nontoxic, inexpensive, and commercially available. Some combinations of these encapsulating agents are also commonly used.

New encapsulating materials for foods have not really emerged in recent years. However, a great effort has been made with respect to food proteins. Food proteins have been engineered as a range of new GRAS matrices with the potential to incorporate nutraceutical compounds and provide controlled release via the oral route. The advantages of food protein matrices include their high nutritional value, abundant renewable sources, and acceptability as naturally occurring food components degradable by digestive enzymes. In addition, food proteins can be used to prepare a wide range of matrices and multicomponent matrices in the form of hydrogels, micro- or nanoparticles, all of which can be tailored for specific applications in the development of innovative functional food products, as recently reviewed (Chen et al., 2006).

Encapsulation and microencapsulation are often used interchangeably when discussing the process technology. Microencapsulation is encapsulation at the microscale, producing delivery devices ranging from 1 to 1000 μm in size, generally less than 200 μm.

Delivery devices can have many morphologies, depending upon the materials and methods used in their preparation. In general, one can distinguish between two main groups of device architecture, depending on the way the core (solid or liquid) is distributed within the system (Figure 7.2):

1. A reservoir system, in which the core is largely concentrated near the center and enveloped by a continuous film (wall) of the encapsulating material
2. A matrix system, in which the core is finely dispersed throughout a continuous matrix of the encapsulating material

The active constituent/encapsulating material ratio is usually high in reservoir systems (between 0.70 and 0.95), whereas for matrix systems, this ratio is generally lower than 0.5 (more commonly between 0.2 and 0.35). The delivery devices defined in 1 are often referred to as *microcapsules* and those described in 2 are called *microspheres*.

Delivery devices do not necessarily have a spherical shape, as illustrated in Figure 7.2e. A great variety of shapes can be obtained when a solid core material is encapsulated by a shell. Particle size is an important characteristic of these structures because it is one of the many parameters that can be tailored to control release rates of encapsulated ingredients. However, the production of microcapsules often gives a certain particle size polydispersity. The active ingredient release kinetics depends on the particle size distribution. It is thus necessary to determine both the mean particle size and the size distribution for the targeted delivery.

The release of the core material from a delivery device may be programmed to be immediate, delayed, pulsatile, or prolonged over an extended period of time. In general, the release depends upon the architecture and the physical structure of the device, as well as upon the barrier properties of the encapsulating material used to form the system.

Delivery devices can be formulated to release the core material for food applications through a variety of mechanisms to meet product performance requirements,

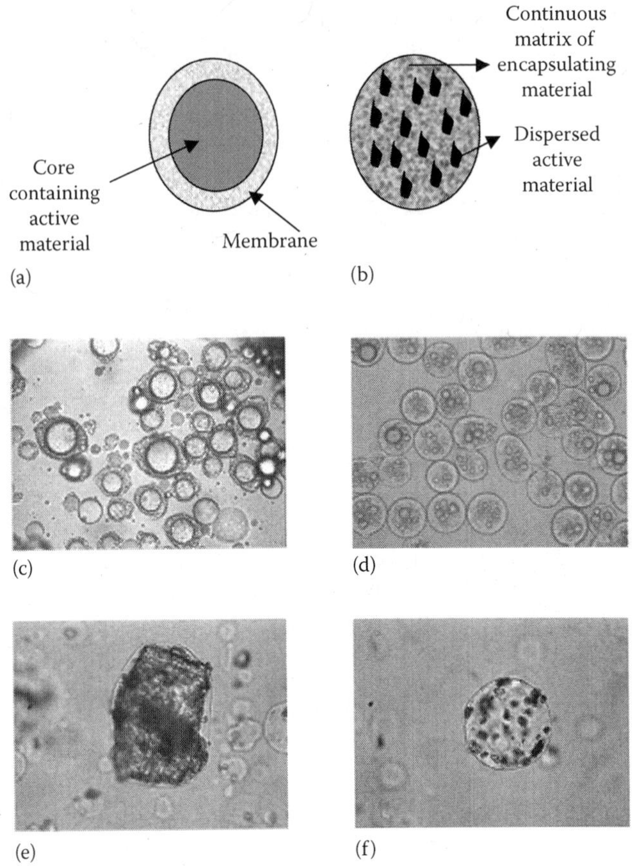

FIGURE 7.2 Schematic diagram of microcapsules morphology: (a) reservoir system (simple wall); (b) matrix system; (c) simple wall (liquid core); (d) multicore; (e) simple wall (solid and irregular core); and (f) matrix (solid core dispersed into the polymeric matrix).

which may be based on temperature or solvent effects, diffusion, degradation, or particle fractures. For example, the encapsulating material can be fractured by external or internal forces such as chewing (Figure 7.3). Melting of the core or the encapsulating material by means of an appropriate solvent or thermally is another

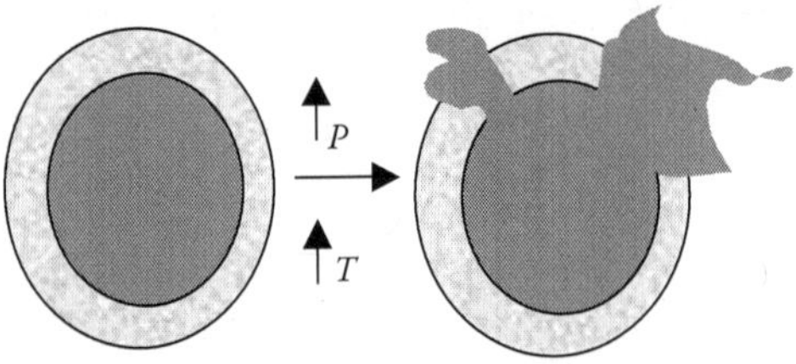

FIGURE 7.3 Release from reservoir device (microcapsule) fractured by mechanical forces.

way of controlling the release of the active ingredient. Solvent release is based upon the solubilization of the encapsulant (water is typically the solvent), followed by subsequent release of the encapsulated ingredient. Release may be regulated by controlling the dissolution rate of the encapsulant and pH effects. For example, coating materials can be selected to dissolve upon consumption, slowly or quickly in the acidic gastric medium, or only when a certain pH is reached. Thermal release is commonly used for fat capsules and occurs during baking. The release of the core material may be delayed until the proper temperature is reached, delaying a chemical reaction. For example, sodium bicarbonate is a baking ingredient that reacts with food acids to produce leavening agents, which give baked goods their volume and lightness of texture. To delay and control the leavening process, the sodium bicarbonate is encapsulated in a fat, which is solid at room temperature but melts at a temperature of about 50°C–52°C. Release of the active ingredient from microcapsules can be accomplished through biodegradation processes, if the encapsulating agent is sensitive to enzymatic actions. For example, lipid coatings may be degraded by the action of lipases.

Diffusion is another important mechanism in release into foods because it is dominant in controlled release from matrix systems (microspheres). Diffusion occurs when the active ingredient passes through the encapsulating material. This mechanism can occur on a macroscopic scale (as through pores in the matrix) or on a molecular level, by permeation through the structuring material. Examples of diffusion-release systems are shown in Figure 7.4.

The typical release profiles shown in Figure 7.4 for reservoir and matrix delivery systems may present variations: a burst effect due to the presence of some core

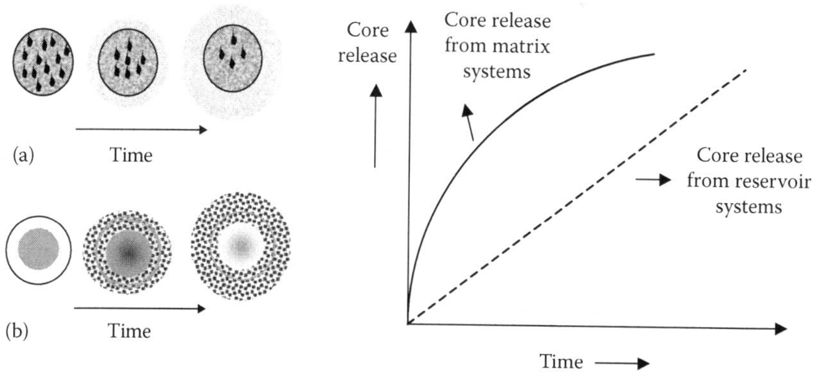

FIGURE 7.4 Examples of diffusion-release systems. (a) Diffusion from a typical matrix delivery system: Diffusion occurs when the active ingredient passes from the structuring matrix into the external environment. As the release continues, the rate normally declines with time since the active agent has a progressively longer distance to travel and therefore requires a longer diffusion time to release. (b) Diffusion from a reservoir system, whether the active core is surrounded by a film or membrane of an encapsulating material. The only structure effectively limiting the release of the core material is the encapsulating layer surrounding the core. Since this layer is uniform and has a constant thickness, the diffusion rate of the core can be kept fairly stable throughout the lifetime of the delivery system.

material too close to the external device surface for matrix devices, or a delayed time to start diffusion due to the diffusion of the core through the encapsulating layer of the reservoir device. Also, the physical state of the core material (dissolved or dispersed) defines the release kinetics. For example, a reservoir system in which the active core is not dissolved results in zero-order kinetics (constant flow), whereas it results in first-order kinetics (exponentially decreasing flow) if the core is dissolved in the encapsulated material.

Encapsulation of food ingredients can be achieved by physical, physicochemical, or chemical techniques (Shahidi and Han, 1993; Desai and Park, 2005a; Champagne and Fustier, 2007). The various encapsulation technologies allow product formulators to make delivery devices from less than one micrometer to several thousand micrometers in size. Each technology offers specific attributes, such as high production rates, large production volume, high product yield, and different capital and operating costs. Other process variables include degree of flexibility in the selection of the encapsulating material and differences in the device size and morphology. The selection of an encapsulation technique is governed by the properties (physical and chemical) of core and encapsulating materials and the intended application of food ingredients.

The material to be encapsulated can be solid or liquid. The principle of most encapsulation technologies is quite simple, combining three consecutive steps (Poncelet and Dreffier, 2007):

1. The active ingredient is mixed within an encapsulating material, in most cases a polymer solution.
2. The resultant formulation, in a liquid form, is dispersed into fine droplets by dripping (drop-by-drop), spraying, or emulsification. This step increases the liquid surface available for further transformations (evaporation, cooling, gelation, chemical reactions) and favors the generation of dispersed (liquid or solid) delivery structures.
3. The oil or aqueous droplets are stabilized in a third step. Oil droplets can be made of a molten material and can be transformed into solid particles by cooling. Depending on their formulation, aqueous droplets can be subjected to a number of solidification processes such as gelation, polymerization, or crystallization.

When the active ingredient is in a solid form (solid particles), encapsulation can also be achieved by spraying a coating solution onto the particles surface, by many different processes that will keep the particles in motion (e.g., agitated particles bed in a fluid bed or pan rotating bed), followed by a consecutive step of stabilization of the coating by solidification or membrane formation.

7.3 ENCAPSULATING SYSTEMS

Delivery systems can be solid or liquid, depending on the food matrix where they are introduced. Some examples of *solid systems* are spray-dried and gel microparticles, whereas *liquid systems* include liposomal and emulsified systems. Each of these delivery systems has its own specific advantages and disadvantages for encapsulation, protection, and delivery of food ingredients. These aspects are briefly discussed in the following, together with a description of the basic principles of each technique, its physicochemical characteristics, and the current challenges for its application in foods.

7.3.1 Spray-Dried Microparticles

7.3.1.1 Concept, Structure, and Properties

Spray drying (SD) is a preservation technique commonly used in the food industry, mainly for dairy products. By decreasing water content and water activity when converting liquids into powders, this technique increases the storage stability of products, minimizes the risk of chemical or biological degradations, and also reduces the storage and transport costs.

SD is a unique drying process since it involves both particle formation and drying. From a microstructural viewpoint, the formation of spray-dried powders involves the droplet formation from a liquid state followed by a solidification operation driven by solvent evaporation, as schematically represented in Figure 7.5.

Liquid atomization is a decisive stage in SD, defining the evaporation surface. It covers the process of liquid bulk breakup into millions of individual droplets forming a spray. To illustrate, consider the division of one liquid droplet with an initial diameter of 1 cm into N droplets of an equal final diameter of 100 μm. For the same liquid volume, this disintegration mechanism generates 10^6 droplets of 100 μm. The superficial area of the liquid, for example, the available surface to heat and mass transfer between the liquid and the drying air is thereby increased 100 times.

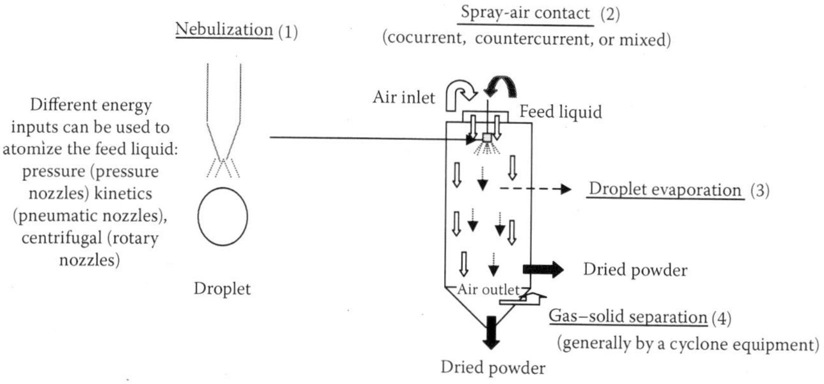

FIGURE 7.5 The main stages involved in the spray-drying process: 1, nebulization of feed liquid into small droplets (spray formation); 2, spray-air contact (mixing and flow), which can be made by several modes: cocurrent (the liquid is nebulized in the same direction as the airflow, as illustrated), countercurrent (liquid droplets and hot air flowing in opposite directions), or mixed flow (the liquid is sprayed upward and only remains in the hot zone for a short time); 3, droplet evaporation; and 4, separation of dried product from the air.

Liquid nebulization into small droplets can be carried out by kinetic pressure or centrifugal energy. Conventional atomizer or nebulizer devices include centrifugal or pneumatic nozzles and rotary discs. Selection of the atomizer configuration is one of most important choices in the SD design. It depends on the nature and viscosity of the liquid to be sprayed and has a significant effect on the size distribution of the final dry particles. As a general rule, an increase in the energy available for nebulization (i.e., rotary nozzle speed, nozzle pressure, or air-to-liquid ratio in a pneumatic nozzle) reduces the particles size (Masters, 1985). However, for pneumatic nozzles, there is an optimum air-to-liquid ratio, above which an increase in energy input does not increase the efficiency of the nozzle to disintegrate the liquid and represents a waste of energy (Ré et al., 2004). By modifying the atomization and the drying conditions, as well as the liquid formulation, it is possible to alter and control properties of spray-dried powders such as particle size, solubility, dispersibility, moisture content, hygroscopicity, flowability, and bulk density.

Spray-dried particles are produced as a very fine powder. Mean particle sizes range from few microns to several tens of microns, with a relatively narrow size distribution, which actually depends on the spray–air contact mode. Particles in the range of 1–50 μm in diameter are typically produced in the cocurrent mode, the most usual contact mode to dry thermosensitive products. Larger particles, ranging from 50 to 200 μm can be obtained in a countercurrent mode due to their agglomeration inside the drying chamber. Spray-dried particles may be spherical or present superficial indentations formed by shrinkage of the droplets during the early stages of the drying process due to the effect of a surface-tension-driven viscous flow (Masters, 1985). The interested reader is referred to Chapter 9 for further details on the principles of SD and its use as a preservation method in the food industry.

SD is used as a microencapsulation method because structured microparticles can be created starting from complex liquid mixtures comprising an active ingredient dissolved, dispersed, or emulsified with an encapsulating material in an aqueous or an organic solvent (water is used in most cases). An appropriate combination of the active ingredient with a range of encapsulating materials and other formulating ingredients, such as surfactants, can give the in-use properties of the encapsulated product. In a single operation, after nebulization of the liquid feed into the spray dryer, the active ingredient can be entrapped into the main component of the structure, for example, the encapsulating material. This process permits the isolation of microspheres or microcapsules depending on the initial formulation (Figure 7.6), as discussed by Ré (2006).

An important step in developing spray-dried microcapsules for food application is the choice of the encapsulating material. This greatly affects the encapsulation efficiency, appropriate thermal or dissolution release, mechanical strength, microcapsule stability, and compatibility with the food product. The criteria for this choice are mainly based on the solubility in water at an acceptable level, the film-forming and emulsifying properties of the limited number of food-grade materials available. Carbohydrates such as starches, maltodextrins, corn syrup solids, and gum arabic are used extensively in spray-dried encapsulations of food ingredients as the encapsulating material (Ré, 1998). Proteins such as whey proteins, soy proteins, and sodium caseinate have also been used (Gharsallaoui et al., 2007). However, the selection of

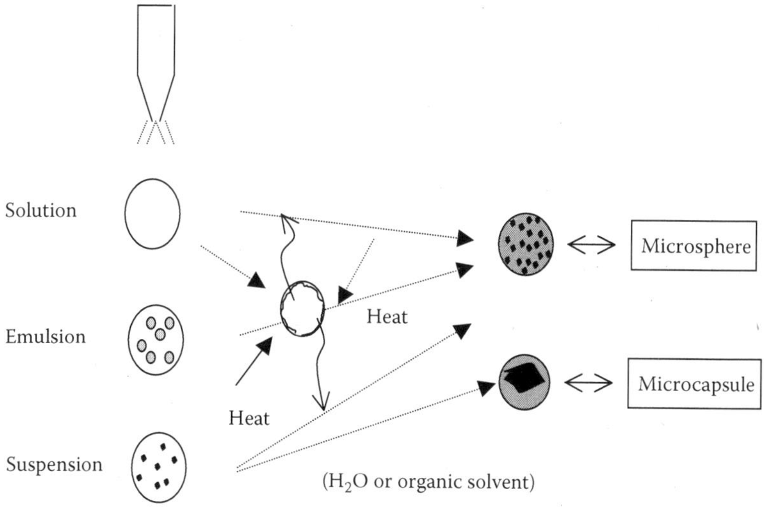

FIGURE 7.6 Architecture of spray-dried particles, depending on the initial formulation (solution, suspension, or emulsion). (From Ré, M.I., *Drying Technol.*, 24, 433, 2006. With permission.)

the encapsulating material for spray-dried microcapsules production has traditionally involved trial-and-error procedures. Firstly, a material is chosen and microcapsules are produced using different mass proportions of encapsulating material to active ingredient and various drying conditions. The spray-dried powders are then evaluated for encapsulation efficiency, stability under different storage conditions, and degree of protection provided to the core material, among other physical characterizations (particle size, powder density, powder flowability, etc.). This procedure is costly and time consuming. Some efforts have been made to develop quantitative methods for selecting the most suitable wall materials for spray-dried microcapsules, mainly for lipid encapsulation.

A method proposed by Matsuno and Adachi (1993) for screening the most suitable wall materials for lipid encapsulation is based on measurements of the drying rate of an emulsion as a function of its moisture content. A suitable material for this application should possess a high emulsifying activity, a high stability, a tendency to form a fine and dense network during drying, and, at the same time, should not permit lipid separation from the emulsion during dehydration. Because the isothermal drying rate is governed by the water diffusion rate, the drying rate may reflect the sample matrix characteristics, i.e., the finer and denser the matrix, the lower the drying rate. According to this method, a characteristic drying curve as a function of moisture content for a suitable group of encapsulating materials is presented in Figure 7.7 (type 1 curve). This curve has been interpreted in terms of the ability of the wall material to form a dense network. The drying rate decreases rapidly as the water content decreases, reflecting a rapid formation of a dense skin and a good protection of the core ingredient against oxygen transfer and possible deterioration. Some materials presenting this type of drying curves are maltodextrin and gum arabic,

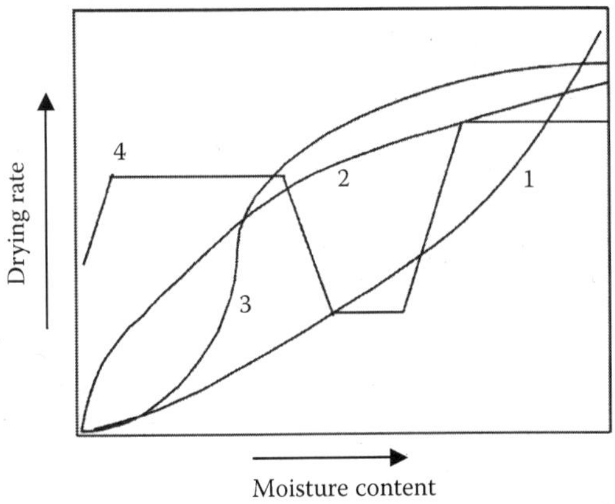

FIGURE 7.7 Schematic representation of isothermal drying curve for the selection of encapsulating materials for spray-dried microcapsules (Matsuno and Adachi method). Type 1 curve corresponds to materials that form fine, dense, two-dimensional skins immediate upon drying. Types 2, 3, and 4 curves correspond to materials that do not form dense skins at an early stage of drying. (From Perez-Alonso, C. et al., *Carbohydr. Polym.*, 53, 197, 2003. With permission.)

which are considered as the most suitable for microencapsulation by SD. According to this method, materials that do not form dense skins at early stages of drying are unsuitable for efficient lipid encapsulation (types 2, 3, and 4 curves). Characteristic type 2 materials are caseinate and albumin, type 3 materials are low molecular weight saccharides that do not crystallize readily like glucose, and type 4 materials are those that easily crystallize upon dehydration such as mannitol (Matsuno and Adachi, 1993). However, not all the materials showing an early decreasing rate, in which water evaporation is controlled by diffusion mechanisms, are suitable for lipid encapsulation when used alone. Therefore, it is desirable to determine an optimal combination of materials that will provide excellent emulsifying capacity and very low oxygen diffusion. In this respect, as analyzed by Pérez-Alonso et al. (2003), the Adachi and Masuno method does not allow for an effective discrimination between materials showing similarly shaped drying curves.

Another method, proposed by Pérez-Alonso et al. (2003), uses the activation energy of carbohydrate polymer blends dried isothermally, as a discriminating parameter for selecting the most suitable mixture as wall material for spray-dried microcapsules. The activation energy provides a measure of the necessary energy required for evaporating a mass of water from the material to be dried. This method requires the knowledge of the drop volume shrinkage of every conceivable blend, which can be achieved as follows: A drop of a blend constituent aqueous solution is put on a glass slide and micrographs of the *X*–*Y*, *X*–*Z*, and *Y*–*Z* planes of the drop are taken. The area of each plane is calculated and approximated to that of a sphere. These steps are repeated as the drops are dried isothermally at intervals of approximately 10% moisture content decrease (determined by drop mass loss). The experimental points are then fitted with a polynomial reported for each blend constituent and assumed additive volumes of the blend constituents to determine the drop shrinkage of studied blends.

Despite these efforts, screening for new wall materials, such as milled citrus fruit fibers as a potential replacement for maltodextrin-type carriers (Chiou and Langrish, 2007), is still mainly done by trial and error.

When selecting SD to produce an encapsulated food ingredient, one is generally looking for high production in a short time and for a product in a powder form. Spray-dried microparticles are commonly used to encapsulate flavors or lipids, and the release mechanism is generally linked to the dissolution of the encapsulating agents. The encapsulated ingredients are, in general, rapidly released due to the dissolution of the spray-dried structures at the time of consumption by dispersing the powder in a wet formulation (e.g., instant beverages). The challenge is how to encapsulate thermosensitive and volatile compounds by a drying process operation, avoiding thermal degradation and volatile losses during drying, and generating an encapsulated product in a powder form with good flowability, stability, and acceptable shelf life after drying.

7.3.2 Gel Microparticles

7.3.2.1 Concept, Structure, and Properties

Gel microparticles are formed from the concept of gelation as an encapsulation technology. Gelation is based on the formation of a solution, dispersion, or the emulsification of the core material in an aqueous solution containing a hydrophilic polymer (hydrocolloid) capable of forming a gel under an external action, either physical or chemical.

There are many techniques for physical gelation, whose use depends on whether the hydrocolloid can gel in water without additives (thermal gelation) or ions are required to aid gelation (ionotropic gelation). In thermal gelation, typically, a solution is made by dissolving a hydrocolloid in powder form in water at high temperature and then cooling to room temperature. As the solution cools, enthalpically stabilized chain helices may form from segments of individual chains, leading to a three-dimensional network (Burey et al., 2008). Examples of such systems are gelatin and agar.

Ionotropic gelation occurs via cross-linking of hydrocolloid chains with ions, generally cation-mediated gelation of negatively charged polysaccharides. Examples of such systems are alginate, carrageenan, and pectin. A typical example is the formation of alginate beads by dropping an alginate solution into a bath containing calcium chloride to form the insoluble calcium-alginate. There are two main methods by which ionotropic gelation can be done, namely external and internal gelation. External gelation involves the introduction of a hydrocolloid solution into an ionic solution, with gelation occurring via diffusion of ions from outside into the hydrocolloid solution. This method is the easiest and most often used one for encapsulation by ionotropic gelation. However, it can often cause inhomogeneous gelation of gel particles due to its diffusion-based mechanism, which constitutes a drawback. Surface gelation often occurs prior to core gelation, and the former can inhibit the latter, leading to gel particles with firm outer surfaces and soft cores (Chan et al., 2006). Internal gelation overcomes the main disadvantage of the external gelation method, as it requires the dispersion of ions prior to their activation to cause gelation of hydrocolloid particles. This usually involves the addition of an inactive form of the ion that will cause cross-linking of the hydrocolloid, which is then activated, for example, by a change in pH, after the ion dispersion is sufficiently complete. Internal gelation is particularly useful in alginate systems, which can gel rapidly and may become inhomogeneous if gelation occurs before adequate ion dispersion has occurred (Poncelet et al., 1992; Chan et al., 2002). For example, in the production of alginate particles by external gelation, the alginate solution is extruded as droplets into a solution of a calcium salt. For internal gelation, an insoluble calcium salt is added to the alginate–drug solution and the mixture extruded into oil (Liu et al., 2002). The latter is acidified to bring about the release of Ca^{2+} from the insoluble salt for cross-linking with the alginate. Despite their homogeneity, internal gelated matrices may be more permeable, resulting in lower encapsulation efficiencies and faster release rates (Vanderberg and De La Noüe, 2001), which may also be overcome by manipulating the pH of the medium and the amount of calcium salt used.

Whatever the technique used (thermal or ionotropic gelation), gel particles are generally formulated in a two-step procedure involving a droplet formation and hardening. The droplet formation step determines the mean size and the size distribution of the resulting gel particles. In the following, the main procedures used for droplet formation—droplet extrusion, nebulization (spray), and emulsification—are described.

7.3.2.1.1 Droplet Extrusion

Extrusion denotes feeding the hydrogel solution, typically containing the active material to be encapsulated, through a single or plurality of pathways directly into the continuous gelation bath. Henceforth, for simplicity, the hydrogel solution containing the active material to be encapsulated (generally an aqueous dispersion) is referred to only as "hydrogel" solution.

In the droplet extrusion technique, also referred to as the drop method, hydrogel solutions are extruded through a small tube or needle (Figure 7.8a), permitting the formed droplets to freely fall into a gelation bath. The droplets may be cross-linked by addition of an appropriate cross-linker to the receiving solution. The size of the

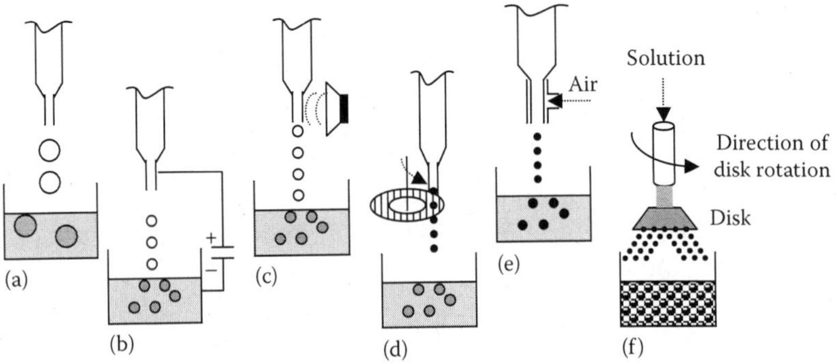

FIGURE 7.8 Schematic representation of the different techniques for drop formation employed in gelation: (a) droplet extrusion; (b) electrostatic dripping; (c) laminar jet breakup; (d) jet-cutting; (e) jet nebulizer; and (f) disk nebulizer.

droplets, and thus the size of the subsequent gel particles, depends upon the diameter of the needle, the flow rate of the solution, its viscosity, and the concentration of the ionic solution. Typical gel particle sizes obtained using the conventional syringe-drop method are 0.5–6 mm, and on the scale of hundreds of microns, if modified techniques suitable for large-scale processes are used to disperse the hydrocolloid solution into droplets (Burey et al., 2008). The main difference among the reported techniques for a mass production of small narrowly dispersed or monosized hydrocolloid gel particles lies in the way the drops are formed, i.e., electrostatic dripping (Figure 7.8b), jet breakup through mechanical vibrations (Figure 7.8c), jet-cutting (Figure 7.8d), and jet and rotating disc atomizers (Figure 7.8e and f, respectively).

The electrostatic technique uses an electric potential difference to pull the droplets from a needle tip (Figure 7.8b). The electrostatic potential difference is established between the needle feeding the solution and the gelling bath. In the absence of an electric field, a droplet forming on a needle tip will grow until its mass is large enough to escape the surface tension at the needle–droplet interface. With the introduction of an electric field, a charge is induced on the droplet surface. Mutual charge repulsion results in an outwardly directed force acting downward on the forming droplet. The additional electrostatic force pulls the droplet from the needle tip at a much lower mass and hence size. Capsule size may be controlled by adjusting the magnitude of the voltage. The higher the voltage, the higher the electrical force pulling the droplet, and therefore, the smaller the obtained droplet and, consequently, the capsule. The literature reports different capsule ranges that can be achieved using this method, for example, 40–2500 μm according to Burey et al. (2008) or 50–800 μm according to Poncelet and Dreffier (2007).

In the laminar jet breakup technique, a laminar flow of the hydrocolloid solution is converted into a succession of identical droplets by the action of an ultrasound vibrating nozzle (Figure 7.8c). Jet-cutting is another method, in which the fluid is pressed through a nozzle in the form of a liquid jet. This jet is cut into uniform cylindrical segments by a means of a rotating cutting tool (Figure 7.8d). Due to surface tension effects, these segments form spherical beads while falling down. The diameter of the resulting bead is determined by the number of cutting wires, the number of rotations of the cutting tool, and the mass flow through the nozzle, which, in turn, depends on both the nozzle diameter and the fluid velocity.

7.3.2.1.2 Nebulization

The dispersion of the hydrocolloid solution may also be achieved by jet nebulizers using, for example, a coaxial air stream that pulls droplets from a needle tip into a gelling bath (Figure 7.8e). Small quantities of gel particles ranging in size down to around 400 μm are achieved by this method (Herrero et al., 2006). Droplet formation is aided by the shear energy of gas flow, used to overcome the viscous and surface tension forces of the fluid. The viscosity and surface tension of a liquid being nebulized can thus alter the properties of the aerosol generated (Figure 7.8 f).

7.3.2.1.3 Emulsification

In the emulsion technique, solutions are mixed and dispersed into a nonmiscible phase. For food applications, vegetable oils are used as the continuous phase. In some cases, emulsifiers are added to form a better emulsion, since such chemicals lower the surface tension, resulting in smaller droplets. After emulsion formation, gelating and/or membrane formation is initiated by cooling and/or addition of a gelling agent to the emulsion, or by introducing a cross-linker. In a last step, the gel particles formed are washed to remove oil (Chan et al., 2002).

Stirring is the most straightforward method to generate droplets of a dispersed phase in a continuous phase to produce an emulsion. In the simplest approach, the continuous phase is poured into a vessel and stirred by an impeller (Figure 7.9a). The dispersed phase is then added, dropwise or all at once, under agitation at a sufficient speed to reach the desired droplet size. The final droplet size of the liquid–liquid dispersion in stirred vessels depends on parameters such as the physicochemical characteristics of the two phases (e.g., viscosity, interfacial tension, and stabilizer concentration), the preparation conditions of the emulsion (e.g., temperature, addition order of the components), and the stirring system (e.g., shear rate, design of the stirrer, and containing vessel). Hydrocolloid solution droplet sizes normally range from 0.2 to 80 μm, although they can be as large as 5000 μm, and gel particles can range from 10 to 3000 μm, as summarized in a recent review (Burey et al., 2008).

A number of approaches have been proposed for a continuous emulsification and improved control of product size distribution: static mixers, membrane emulsification, and microchannel emulsification. For the last two, emulsions are produced by extruding a liquid through many individual pores or microchannels.

Static mixers consist of a series of geometric mixing elements fixed within a pipe. The particular arrangement of these mixing elements repeatedly splits and recombines the stream of fluid passing through the tube (Figure 7.9b). Recombination

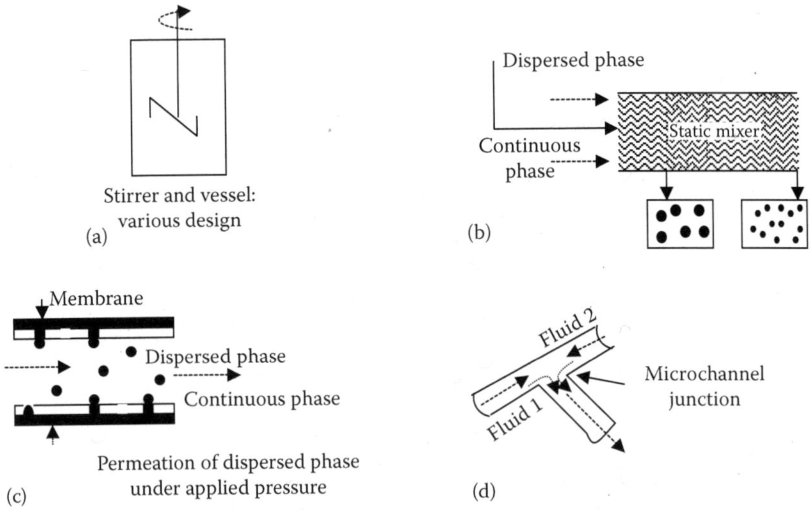

FIGURE 7.9 Schematic diagram of different emulsification processes to produce gel particles: (a) mechanical stirring; (b) static mixing; (c) membrane emulsification; and (d) microchannel emulsification.

occurs through impingement of the substreams, creating turbulence and inducing back mixing. Static mixers installed in tubes allow continuous production, and have already been used to produce gel particles (Belyaeva et al., 2004).

Membrane emulsification is a relatively new method for the preparation of spherical particles with a highly uniform size distribution, and it can be used to extrude a hydrocolloid solution into a gelation bath (Oh et al., 2008). The method involves the use of a porous membrane with a highly uniform pore size. The dispersed phase is pressed through the membrane pores, whereas the continuous phase flows along the membrane surface. Droplets grow at pore outlets until they detach upon reaching a certain size (Figure 7.9c). A low-pressure drop is applied to force the dispersed phase to permeate through the microporous membrane into the continuous phase. Details on such methods using membrane emulsification are summarized in several review papers (Joscelyne and Trägardh, 2000; Vladisavljevic and Williams, 2005). The distinguishing feature is the fact that the resulting droplet size is primarily controlled by the choice of the membrane and not by the generation of turbulent droplet breakup. The technique is highly attractive given its simplicity, potentially lower energy demands, need for less surfactant, and the resulting narrow droplet-size distributions. It is applicable to both oil-in-water (O/W) and water-in-oil (W/O) emulsions, and it has been recently applied to prepare microgel particles with a uniform size distribution (Wang et al., 2005; Zhou et al., 2007).

Microfluidic methods have recently been mentioned as an emerging technology for the production of monosized gel particles (Amici et al., 2008). Typically, the term microfluidics refers to the manipulation of fluids in systems or devices having a network of chambers and reservoirs connected by channels, whose typical cross-sectional dimensions range from 1.0 to 500 μm, the so-called "microchannels." Materials such as silicon, cofired ceramic, glass, quartz, and polymers have been explored for the fabrication of microfluidic systems (Nam-Trung and Zhigang, 2005). Due to the large surface-to-volume ratio and low inertial forces encountered at the microscale, highly precise and specific flow manipulation and control can be achieved by appropriate microfluidic design.

Emulsions are produced in microfluidic devices when two immiscible fluids, flowing in two separate microchannels, are forced through a microchannel junction, and the flow of one fluid (usually the fluid that wets the channel surface) breaks the flow of the other to form microdroplets (Figure 7.9d). Drop formation is reproducible, and the drop size can be regulated by operating on factors such as flow rates, fluid viscosities, and surfactant concentration. Gel formation in microfluidic systems has been reported, including, for instance, the thermal gelation of κ-carrageenan (Walther et al., 2005) and agarose (Xu et al., 2005). Ionotropic gelation can also be achieved in microchannels through both the internal and the external approaches (Huang et al., 2007; Amici et al., 2008).

Alginate microspheres have been extensively used as delivery system because they are very easy to prepare, the process is very mild, and virtually any ingredient can be encapsulated (Vladisavljevic and Williams, 2005). That is the reason why they are chosen to exemplify the use of the membrane and channel emulsification processes to produce gel particles.

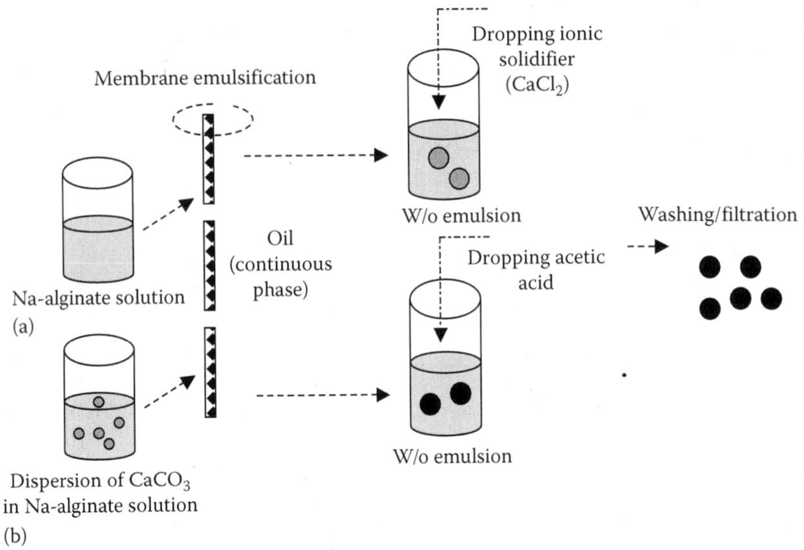

FIGURE 7.10 Illustration of the production of alginate microgels by membrane emulsification: (a) external gelation and (b) internal gelation.

Figure 7.10 illustrates the preparation of Ca-alginate gel particles by membrane emulsification by two different approaches. In the first one, an aqueous Na-alginate solution is extruded through a hydrophobic membrane into an oil phase to form a W/O emulsion. The droplets are then gelled by adding a $CaCl_2$ solution, and the gelled particles can be colleted by filtration. This process was used by Weiss et al. (2004), enabling the formation of uniform gel particles with an adjustable mean diameter (20–300 μm). In the second approach, Ca-alginate gel particles can be produced by internal gelation; in brief, a dispersion of an aqueous Na-alginate solution containing $CaCO_3$ is extruded through a microporous membrane into an oil phase. The gelation is initiated by the addition of acetic acid into the emulsion, dissolving $CaCO_3$ to release Ca^{2+} and form Ca-alginate. Using the second approach, Liu et al. (2003) obtained Ca-alginate gel particles with a mean size of 55 μm (with a coefficient of variation of 27%), using a nickel membrane with a pore size of 2.9 μm.

Figure 7.11 illustrates the preparation of Ca-alginate gel particles by microchannel emulsification. To produce alginate gel particles by external gelation, an alginate aqueous solution flows through a flow-focusing channel, and an alginate droplet is formed from the balance of interfacial and viscous drag forces resulting from the continuous (oil) phase flowing past the alginate solution. It immediately reacts with an adjacent $CaCl_2$ drop that is extruded into the main flow channel by another flow-focusing channel located downstream in relation to the site of the alginate drop creation. This procedure has been used in the literature, generating monosized alginate beads within a range of 50–200 μm, depending on flow conditions (Hong et al., 2007). To produce alginate drops in a microfluidic system by internal gelation (Amici et al., 2008), two aqueous streams, one acidic and one containing alginate and calcium carbonate, can merge immediately prior to entering a channel where a continuous

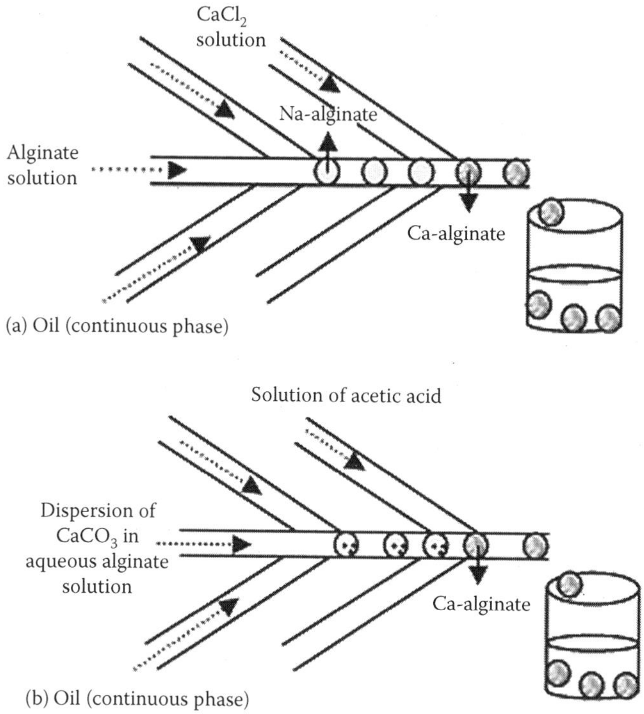

FIGURE 7.11 Illustration of the microfluidic production of alginate microgels by (a) internal gelation and (b) external gelation.

flow of an oil phase breaks the flow of the aqueous phase to form microdroplets (Figure 7.11b). Several variations of both approaches have been developed in the literature by using new microfluidic systems (Liu et al., 2006; Choi et al., 2007).

7.3.3 Liposomal Systems

7.3.3.1 Basic Principles

Liposomes are conceptually biomimetic model systems. They allow studies of the lipid matrix of biomembranes, as well as the investigation of membrane embedded proteins and certain fundamental aspects of organelles in a biomimicking environment, outside the living cell. Structurally, liposomes are self-assembled colloidal particles in which an aqueous nucleus is enclosed by one or several concentric phospholipid bilayers. Phospholipids are amphiphilic molecules, which in aqueous solution form energy-favorable structures as a result of hydrophilic and hydrophobic interactions. Depending on the size and the number of bilayers, liposomes are classified as multilamellar vesicles (MLVs) or large and small unilamellar vesicles (LUVs and SUVs). Other classifications for larger liposomes include the plurilamellar vesicles (PLVs), when nonconcentric bilayers enclose various aqueous compartments, and oligolamellar vesicles (OLVs), in which small vesicles are included in the structure of a large vesicle. The size of unilamellar liposomes may vary from 20 to 500 nm

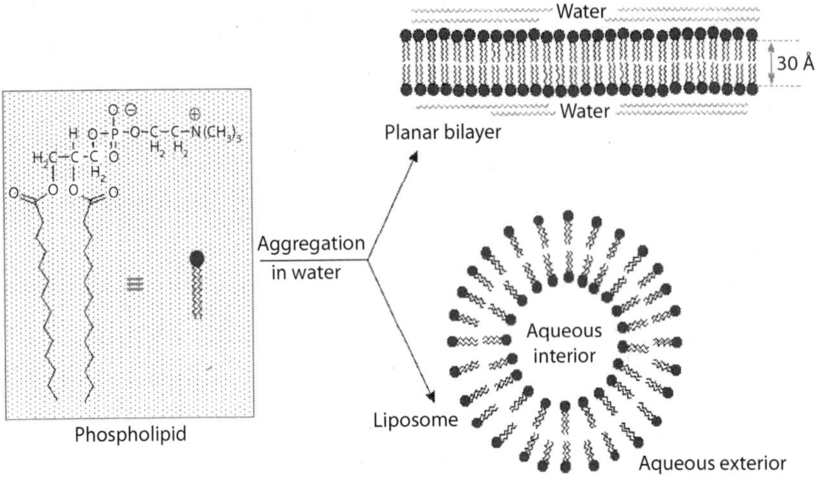

FIGURE 7.12 Phospholipid aggregation in planar bilayer and vesiculation in liposomal structure.

approximately, and the thickness of one lipid bilayer is about 4 nm. Due to the presence of hydrophilic and hydrophobic domains in the structure, liposomes are able to encapsulate water-soluble molecules in the aqueous nucleus, water-insoluble within the lipid membrane, or amphiphiles between the two domains. Therefore, liposomes are a powerful solubilizing system for a wide range of compounds.

Figure 7.12 shows schematically the phospholipid aggregation in bilayer, and the vesiculation, which occurs in excess of water, forming the liposomal structure. Due to their structure, chemical composition, and size, all of which can be well controlled by preparation processes, liposomes exhibit several properties that are useful in a large range of applications. The most important properties are colloidal size, bilayer phase behavior, mechanical properties, permeability, and charge density. Liposomes also allow surface modifications through the attachment of ligands and bounded or grafted polymers. Figure 7.13 shows the various modifications of liposome surfaces.

For historical reasons, when no physical or chemical surface modification is introduced, liposomes are called conventional to be distinguished from the surface-modified liposomes.

Liposomes can be made entirely from naturally occurring substances, and are therefore biodegradable and nonimmunogenic. In some cases, the introduction of synthetic lipids is useful for specific characteristics such as stability and charge. The lipid composition, colloidal characteristics, and surface modifications also allow liposomes to have functional properties, such as stability, controlled release of the encapsulated molecules, specific targeting, and controlled pH and temperature sensitivity.

From the thermodynamic point of view, liposomes are not stable structures, and so they cannot form themselves spontaneously. To produce liposomes, some energy from extrusion, homogenization, or sonication must be dissipated into the system. Subsequently, after formation, liposomes can aggregate, fuse, and form larger structures that eventually settle out of the liquid. Thermodynamically stable systems,

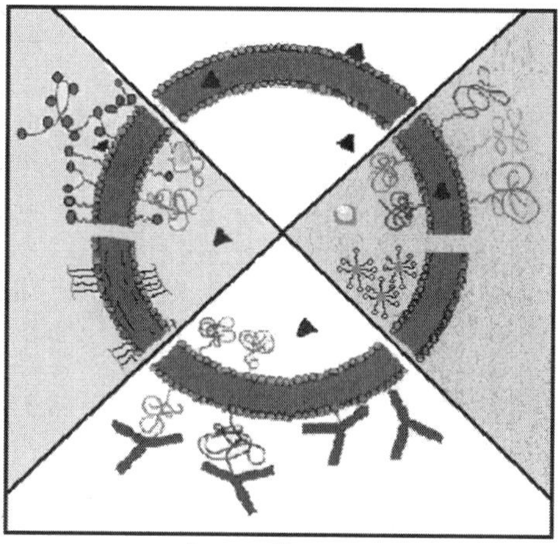

FIGURE 7.13 Surface modifications in liposomes.

such as micelles, stay at the same phase forever. The precursor phase of liposomes is the symmetric bilayer composed of self-assembled phospholipid molecules with comparable areas of the polar and nonpolar portions. Liposomes are formed when, at high concentration, the self-assembled molecules form a long-range ordered liquid-crystalline lamellar phase, which diluted in excess water is dispersed in stable colloidal particles. These particles retain the short-range symmetry of their original parent phase.

An extension of the Tanford's (1980) treatment of the shapes of micelles, based on the molecular-shape analysis and the concept of the shape parameter (ratio between the nonpolar and polar cross-section areas), was introduced by Israelachvili (1991) for a qualitative understanding of the topology of lipid aggregates with different lipid compositions. However, the liposome models are approximate, and a rigorous thermodynamic analysis fails because liposomes are not at the thermodynamic equilibrium. The rigorous analysis would yield a very narrow size distribution, which is never observed in practice, as well as the spontaneous formation, while a high-energy process is typically needed to produce liposomes.

Helfrich (1973) explained the instability of liposomes by the bending elasticity. Symmetric membranes prefer to be flat (spontaneous curvature equal to zero), and energy is required to curve them. Various factors, such as asymmetrically changes by ionization or insertion of molecules like surfactants, change the spontaneous curvature to nonzero values, but the spontaneous formation does not produce stable liposomes for drug encapsulation. Stable liposomes for drug-delivery must be in a kinetically trapped and thermodynamically unstable state. As a consequence of that, liposomes maintain their integrity in spite of changes in the environment like dilution or *in vivo* administration. Micelles and microemulsions, which are thermodynamically stable systems disintegrate, aggregate, or change phase under perturbations of the medium. The vesiculation accumulates an excess of free energy around

10–50 kT (1 kT = 4.11×10^{-21} J at T = 298 K) in the curvature of the liposomes, which generate instabilities promoting fusion and disintegration, or may become bioavailable upon vesicle fusion with cell membranes.

Phase transition is one of the most important properties of liposomes. It happens when lipid bilayers change from a solid-ordered phase, at low-temperature, to a fluid-disordered phase, above the phase transition temperature. Bilayers can also undergo transitions into different liquid-crystalline phases, such as hexagonal or micellar. The phase transitions can be triggered by physical or chemical factors, resulting in different effects on the liposomes. Thus, transitions from gel to liquid-crystalline phase caused by temperature or ionic strength changes cause liposome leakage.

Physicochemical stability determines the shelf life stability of liposomes. By optimizing the size distribution, pH, and ionic strength, as well as by the addition of antioxidants and chelant agents, the stability of liposomes can be preserved for years. They can be stored in a frozen or in dried form, but cryoprotectants have to be added to prevent fusion. Electrostatic and steric stabilization also reduce fusion and disintegration by freezing. Biological *in vivo* stabilization is obtained by reducing the interactions of liposomes with macromolecules, blood protein, and disintegrating enzymes, as well as adverse pH conditions. Therefore, *in vivo* stability of liposomes depends on the route of administration. Steric stabilization is generally provided by the protective coating from the grafting of the liposome surface with inert hydrophilic polymers such as polyethylene glycol (Lasic, 1995) or hyaluronic acid (Eliaz and Szoka, 2001).

Since their discovery in the 1960s and the first studies where they were used as a drug delivery system in the 1970s, liposomes have been used as an encapsulation system in a myriad of applications ranging from material science to analytical chemistry, food, and medicine.

Numerous books and reviews describe the various physicochemical and biological aspects of liposomes, as well as their construction and characterization. This literature has been written by scientists who composed the scientific base of the design and construction of liposomes for the various applications (Gregoriadis, 1988; Lasic, 1993; Lasic and Papahadjopoulus, 1998).

7.3.3.2 Applications in Foods

Gomez-Hens and Fernandez-Romero (2006) depict the number of articles and patents published in the 1990–2004 period regarding the use of the liposomal systems in five areas: drug and gene delivery systems, biochemical and biotechnological applications, cosmetics, nutrition, and foods. Figure 7.14 shows the evolution of articles and patents from 2004 to 2008. In both cases, the tendency is the same: there have been significant advances in the applications of liposomes in the biomedical and pharmaceutical industries related to therapeutic drugs, opposed to their applications in foods, which are presently at an early stage of development.

The applications of liposomes in foods fulfill similar requirements to their applications in pharmaceuticals, and have been focused in the following categories: formulation aid, processing, preservation, stabilizer, nutritional supplement, and nutraceutical carrier.

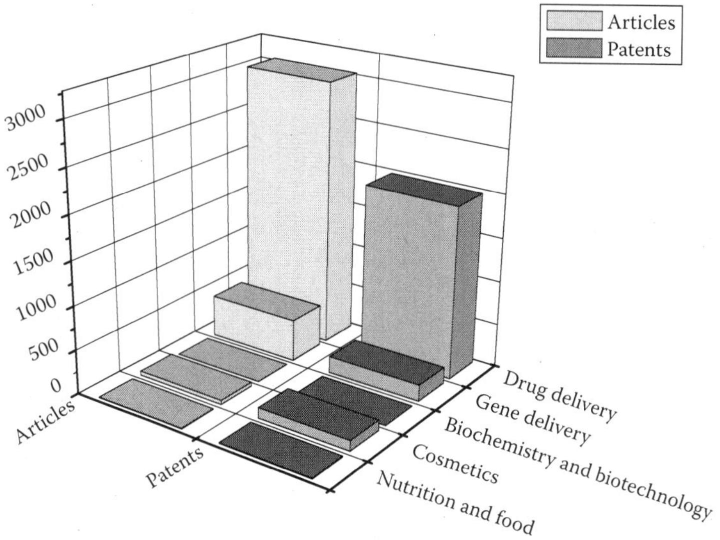

FIGURE 7.14 Articles and patents published on liposomal delivery systems in the 2004–2008 period (last year incomplete). (Courtesy of SciFinder Scholar®, American Chemical Society, Columbus, OH.)

Liposomes aid in formulation because they are powerful in solubilizing non-water-soluble compounds, enhancing their bioavailability. Furthermore, liposomes entrap hydrophilic molecules into their interior, and hydrophobic molecules into their lipophilic membrane, encapsulating various nutritional molecules. The first example of liposomes in foods is human milk, which has been studied for years. Electron microscopic studies show the presence of liposomes along with emulsion droplets and casein micelles (Roger and Anderson, 1998). Liposomes, as a microstructural component of breast milk, may play an important role in enhanced nutrient absorption, colloidal stability, and immunogenicity (Keller, 2001).

In the processing of foods, liposomes accelerate cheese ripening and increase the yield in bioconversion through uniform distribution of hydrophilic enzymes in hydrophobic medium.

Applications of liposomes in cheese ripening were developed by the 1980s (El Soda, 1986). Enhancement of proteolysis by encapsulated cyprosins was evident 24 h after manufacture of Manchego cheese. Addition of encapsulated cyprosins to milk perceptibly accelerated the development of flavor intensity in experimental cheese through 15 days of age without enhancing bitterness (Picon et al., 1996). The capability of neutral and charged liposomes to entrap the proteolytic enzyme neutrase, and the stability of the preparation, were evaluated in the ripening of Saint-Paulin cheese milk (Alkhalaf et al., 1989).

Liposomes also promote sustained release of antimicrobial peptides assuring protection of the formulation. Liposome-entrapped nisin retained higher activity against *Listeria innocua* and improved stability in cheese production, proving to be a powerful inhibitor in the growth of *L. innocua* in cheese, while not preventing the detrimental effect of nisin on the actual cheese-ripening process. Coencapsulation of calcein and nisin, and calcein and lysozyme demonstrated that production and optimization of stable nanoparticulate aqueous dispersions of polypeptide antimicrobials for microbiological stabilization of food products depend on selection of suitable lipid–antimicrobial combinations (Benech et al., 2002; Were et al., 2003).

Besides solubilization, the encapsulation capability of liposomes protects labile compounds from chemical degradation, light oxidation during storage, and harmful compounds from the environment. The stabilizing capability of liposomes also preserves the taste and flavors of foods during processing and storage. Liposomes improve the nutritional effects of foodstuff by entrapping nutritionally important compounds, such as vitamins, polyunsaturated fatty acids, minerals, and antioxidants.

Health benefits of nutrients by changing their kinetics of release and enhancing their bioavailability are obtained by liposome encapsulation. Recently, immunoliposomes, which are liposomes containing antibodies for site-specific targeting, have been studied for nutrient targeting regulation. A useful model involving the leptin protein and immunoliposomes was used to illustrate the nutrient regulation of the endocrine system (Xianghua and Zirong, 2006).

Liposome as a carrier matrix in foods has become an attractive system, because they can be constructed entirely from acceptable edible compounds (food-grade ingredients), like proteins and carbohydrates. Lecithin is the main natural phospholipid, routinely extracted from nutrients, such as egg yolks and soybeans. Additionally, the phospholipids in the liposome matrix are also versatile nutraceuticals for functional foods. The benefits are for the brain, liver, and blood circulation. Phosphatidylcholine is a highly effective nutraceutical for recovery of the liver following toxic or chronic viral damage. It has exceptional emulsifying properties, which the liver draws to produce the digestive bile fluid. The lung and intestinal lining cells use phosphatidylcholine to make the surfactant coating essential for their gas and fluid exchange functions. Phosphatidylcholine exhibits potentially lifesaving benefit against pharmaceutical and deathcap mushroom poisoning, alcohol-damaged liver, and the chronic hepatitis B. Phosphatidylserine establishes its benefits for higher brain functions such as memory, learning and words recall, mood elevation, and action against stress. Phosphatidylserine also has a salutary revitalizing effect on the aging brain, and may also be helpful to children with cognitive and mood problems. The fast access of glycerophosphocholines to the human brain and its capacity to sharpen mental performance also makes it well suited for drink formulations. The nutraceutical properties of phospholipids are described extensively by Kidd (2001). Therefore, the product value comes from the health benefits of the phospholipids associated with the benefits of the selected nutrient. This combined phospholipid–nutrient approach is suited to produce chewable tablets, confectionery products, cookies, granulates, spreads, bars, and emulsified or purely aqueous phase beverages.

Although liposomes carrying nutrients are ingested via the GI system, the oral route also offers a way through the sublingual mucosal membranes. In the first case, the adverse conditions of the environment (low pH of the stomach, surfactant action of bile salts, and the presence of lipases) destabilize the conventional liposome formulation. The sublingual route avoids the first pass liver clearance and metabolism offering a direct uptake of nutrients into the bloodstream through the mucosal membranes. Additionally, the sublingual administration avoids swallowing difficulties from the ingestion of tablets or large capsules by old people or children, in addition to being an alternative for personal preference.

The performance of the sublingual administration of nutrients has been demonstrated using the CoEnzyme Q10 in spray formulation, compared to the powder formulation in hard gelatin capsules. The results showed increased bioavailability of 100% for CoQ10 over endogenous levels with the sublingual spray compared to 50% increase over baseline levels with a two-piece gelatin capsule as measured by area under the curve in 24h (Gibaldi, 1991). Additionally, the time of onset of the spray formulation administration was shorter than the capsule one, adding benefits to the treatments which require immediate onset, like the cardiogenic supplement CoQ10, diet aids, pain treatment, fever, or insomnia (Rowland and Towzer, 1995). Keller (2001) listed some products that have been formulated using this novel oral liposomal delivery system.

The main issue in liposome encapsulation for food industry is the scaling up of the processes at an acceptable cost. Methods of liposome formation now exist that do not make use of sonication (Batzri and Korn, 1973; Kirby and Gregoriadis, 1984; Zhang et al., 1997) or of any organic solvents (Frederiksen et al., 1997; Zheng et al., 1999), and allow the continuous production of microcapsules on a large scale. Nowadays, liposome encapsulation can also become a routine process in the food industry (Gregoriadis, 1987; Kirby and Law, 1987; Kim and Baianu, 1991; Gouin, 2004).

The great advantage of liposomes over other encapsulation technologies is the stability imparted by liposomes to water-soluble materials in high water activity application: spray-dried, extruded, and fluidized beds impart great stability to food ingredients in the dry state but release their content readily in high water activity application, giving up all protective properties.

Microfluidization techniques have been shown to be an effective and solvent-free continuous method for the production of liposomes with high encapsulation efficiency. The method can process a few hundred liters per hour of aqueous liposomes on a continuous basis. The process has been reported in the literature (Vuillemard, 1991; Maa and Hsu, 1999; Zheng et al., 1999).

Multitubular systems represent a scalable version of the Bangham method. It is adequate to prepare liposomes for food application, due to its simplicity and easy scaling up (Tournier et al., 1999; Carneiro and Santana, 2004; Latorre et al., 2007).

Dry liposomes circumvent the drawback of liposome stability in large-scale production, storage, and shipping of encapsulated food ingredients. Freeze drying the liposome suspension can only be carried out by high price encapsulated ingredients in a niche market, due to the considerable cost of large-scale freeze-drying processes. Moreover, not all liposome formulations can be freeze dried, and the reconstitution of the wet formulation is not always straightforward and usually requires complex steps and processes (Lasic, 1993).

These problems are reduced when the bioactive ingredient is incorporated in a lipid matrix by spray drying and, subsequently, liposomes are made through mechanical stirring (Alves and Santana, 2004). The operational conditions modulate the crystallinity of the lipid matrix and the efficiency of incorporation of the bioactive compound. Conventional or special mechanical stirrers can be used to adjust the size and distribution of liposomes.

Supercritical fluid offers another attempt to avoid the use of organic solvent in the production of liposomes. Basically, the process involves the solubilization of the phospholipids under supercritical condition, followed by the release of the supercritical mixture into an aqueous phase containing the dissolved active ingredient, which results in the formation of liposomes containing the active ingredient in their aqueous cores. Although this method is scientifically interesting, the encapsulation efficiency reported so far is limited to 15% (Frederiksen et al., 1997), which would have to be dramatically increased for this technique to become interesting from an industrial point of view.

Liposomes can also be associated with other technologies. Many hydrophilic and hydrophobic compounds, including various vitamins and antioxidants, can be dispersed in other matrices by liposomes encapsulation.

7.3.4 Emulsified Systems

7.3.4.1 Concept, Structure, and Properties

Emulsions are defined as a mixture of two immiscible liquids wherein one phase is dispersed in the other in the form of droplets. Food emulsions usually contain an aqueous phase (polar), an oil phase (apolar), and surfactants that are added to stabilize the system by reducing the interfacial tension between dispersed and continuous phases. When oil droplets are present in the emulsion they are called O/W emulsions. On the other hand, when the aqueous phase is dispersed, the emulsion is called W/O, and sometimes it is referred to as inverse emulsion. Figure 7.15 shows a schematic picture of O/W and W/O emulsions stabilized by surfactant molecules. These molecules are adsorbed on the interface and oriented according to the type of emulsion. In the case of O/W emulsions, the hydrophilic head is "dissolved" in the continuous phase, whereas in W/O emulsions it is in the dispersed phase. Formation of O/W or W/O emulsions depends on the migration of surfactant molecules to the interface, which stabilize the droplets, and coalescence, leading to the destruction

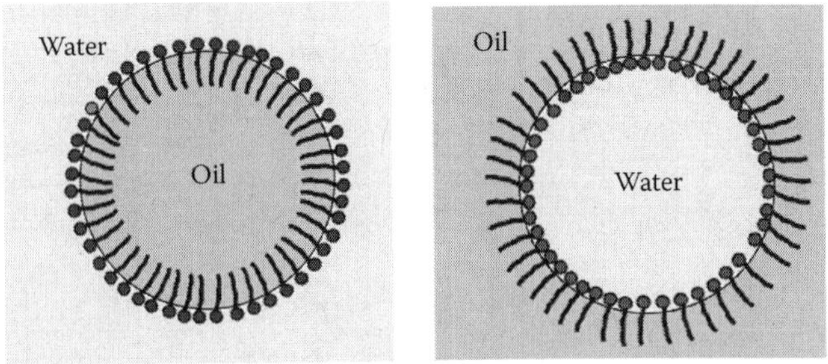

(a) (b)

FIGURE 7.15 Schematic picture of (a) O/W and (b) W/O emulsions stabilized by surfactant molecules.

of the droplets. Thus, this is a competing process and the phase that presents higher coalescence rates will become the continuous phase (Evans and Wennerstrom, 1994). The rates of migration and coalescence depend on the components chemical structure, leading to specific surfactant conformations when adsorbed on the oil–water interface. One practical parameter is the hydrophilic–lipophilic balance (HLB) of the surfactant, which is a number that relates the number of hydrophobic and hydrophilic groups in a surfactant molecule. In general, surfactants with HLB from 3 to 9 tend to stabilize W/O emulsions, whereas molecules with HLB higher than 9 tend to stabilize O/W emulsions. Therefore, it is possible to prepare emulsions with droplet volume fractions up to 90% provided that an adequate surfactant or a mixture of surfactants is used. Sometimes, emulsions are also stabilized by adding high molecular weight components, such as long chain polymers and proteins, which adsorb on the interface acting as a surfactant. Small solid particles also tend to adsorb on the oil–water interface and can act as surfactant in an emulsion.

Another type of emulsion that has gained interest in food applications is the so-called multiple emulsion, which is basically an emulsion contained in a droplet. For example, a water-in-oil-in-water (W/O/W) emulsion means a multiple emulsion of water droplets inside an oil droplet that is dispersed in a continuous water phase. Recently, potential industrial applications in encapsulating active food components were recognized. One of the main difficulties in applying multiple emulsions is their low stability, which limits the applicability when prolonged stability and release are necessary (Muschiolik, 2007). Figure 7.16 shows a schematic representation of a W/O/W emulsion.

The fact that the dispersed phase can be composed of either oil or water shows that emulsions can be used to encapsulate both lipophilic and hydrophilic bioactives (Flanagan and Singh, 2006). Food systems based on emulsions are recognized to have great potential in delivering functional components such as omega-3, β-carotene, fatty acids, phytosterols, and antioxidants, among others (McClements et al., 2007). The main preparation methods are based on addition and stirring components, mechanical mixing, homogenization, and heating, which make these systems suited for industrial scale-up.

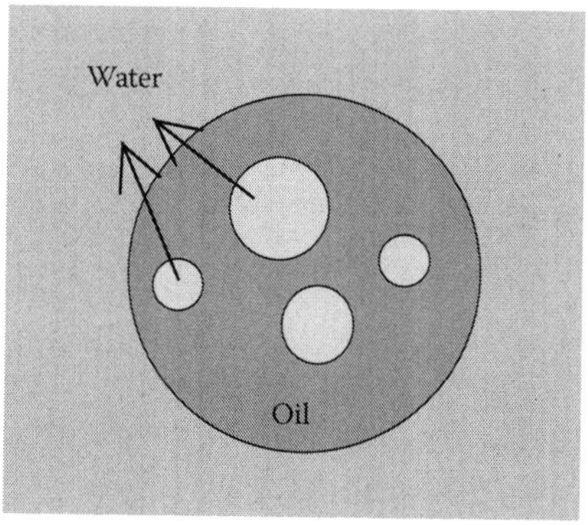

FIGURE 7.16 Schematic representation of a W/O/W emulsion.

Emulsified systems can be classified according to their thermodynamic stability and their droplets size. Macroemulsions (or simply emulsions) are metastable systems, i.e., the system is not in thermodynamic equilibrium, and it will breakdown into two distinct phases if sufficient time is allowed. However, emulsions that keep their kinetic stability for periods of months or years can be prepared by using appropriate components and amounts (McClements et al., 2007). This is the most common type of emulsion, and it is found in many food systems such as milk and salad dressing. Macroemulsions are usually polydisperse, with droplet sizes in the range of 1–100 μm. The main destabilization mechanisms in macroemulsions are droplets creaming, flocculation, and coalescence.

Microemulsions, on the other hand, are thermodynamically stable emulsions. Therefore, at a given temperature, pressure, and composition, these systems keep their morphological characteristics and are not affected by the destabilization mechanisms cited above. Microemulsions are usually monodisperse, with droplets in a nanoscale range (10–100 nm) (Flanagan and Singh, 2006). Thermodynamic stability is achieved by the proper choice of the components, as well as their proportions, leading to a negative overall free energy of mixing (Evans and Wennerstrom, 1994). Usually, large amounts of surfactants are required, and different surfactants and cosurfactants are generally used. In microemulsions, surfactants are important because they not only decrease the interfacial tension between the oil and water phases, but also affect the energy balance of the system through the formation of self-assembled structures in the continuous phase, such as micelles. Furthermore, the chemical structure of the surfactant affects the interface spontaneous curvature, which is an important factor in determining the droplets size, as well as the type of emulsion.

Microemulsions exhibit different phase behavior under equilibrium, which is classified using the Winsor classification system. A Winsor I system means that the microemulsion coexists with an oil-rich region. When there is a water-rich region present in the system, the microemulsion is said to be a Winsor II system. In case there are both oil-rich and water-rich regions coexisting with the microemulsion, one speaks of a Winsor III system. Finally, a Winsor IV system means that there is no phase coexistence and only the microemulsified phase is observed. This phase behavior is desired for food delivery systems and, as said earlier, depends on the proper choice of system components, as well as temperature conditions (Flanagan and Singh, 2006). Figure 7.17 shows a schematic representation of the different microemulsion regimes.

Depending on its composition, a single-phase microemulsified system can also exhibit different morphologies. The three main structures are O/W, W/O, and bicontinuous. The latter is a structure in which both oil and water exist as a continuous phase, but all three structures have a surfactant monolayer in the interface separating both phases. These structures are shown in Figure 7.18. Usually, an O/W microemulsion is formed when oil concentration is low, and a W/O microemulsion is formed when water concentration is low. Bicontinuous systems are formed when the amounts of water and oil are similar (Lawrence and Rees, 2000).

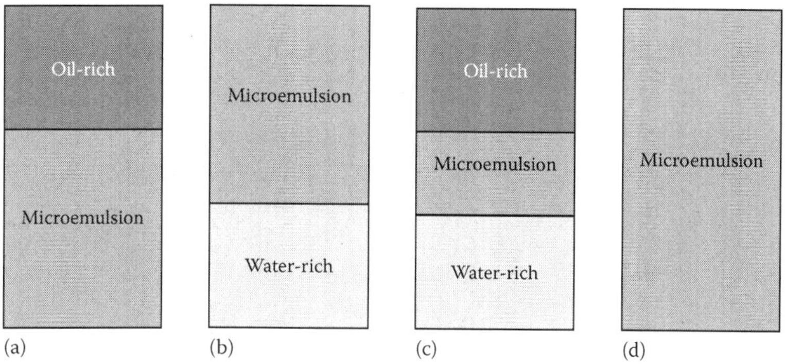

FIGURE 7.17 Schematic representation of the microemulsion regimes: (a) Winsor I; (b) Winsor II; (c) Winsor III; and (d) Winsor IV.

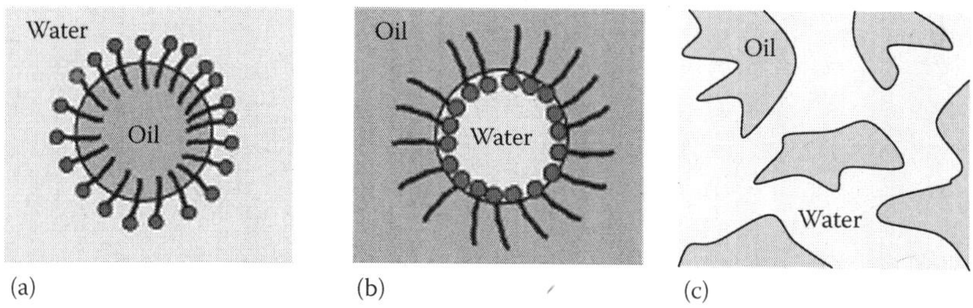

FIGURE 7.18 Schematic representation of the main single-phase microemulsion systems: (a) O/W; (b) W/O; and (c) bicontinuous.

Microemulsions as food systems have great potential, which can be attested in patented products (Bauer et al., 2002; Allgaier et al., 2004; Chanamai, 2007). Incorporation of proteins in microemulsions might also have impact in food applications in the future (Rohloff et al., 2003). Studies on microemulsions applied to the pharmaceutical field might also be of interest when seeking food applications, since biocompatible components are used in this field. Studies in this area have been summarized in review articles (Lawrence and Rees, 2000; Rane and Anderson, 2008).

Recently, metastable emulsions with nanosized droplets have started to receive attention due to their technological potential in pharmaceutical and food industries (Solans et al., 2005), and their fundamental properties have been studied (Mason et al., 2006b). Such systems are called nanoemulsions or miniemulsions. Basically, it is an emulsion with droplet sizes in the range of 50–200 nm. Nanoemulsions have the same physical appearance as a microemulsion, i.e., they have droplets in the nanoscale range and usually exhibit transparency and low viscosity. Although they can lose stability through coalescence and flocculation, the main destabilization mechanism is Ostwald ripening, due to the high Laplace pressure of the droplets (Porras et al., 2004).

Nanoscale emulsions have gained technological interest because the transport efficiency of functional components in emulsion food systems is increased when droplets are in the nanoscale (Spernath and Aserin, 2006). In addition, these emulsions are transparent, and they have lower viscosity when compared to conventional emulsions, which make them suitable for use in beverages, for example. In recent years, considerable research effort was made to understand the physical properties, preparation, phase behavior, and stability of micro- and nanoemulsions.

Nanoemulsions are metastable structures. This fact confers to nanoemulsions advantages and disadvantages for functional food applications in comparison with microemulsions. Main advantages are that nanoemulsions do not require the use of large amounts of surfactants and there is a wider range of possibilities of combination of different components for a given system (Solans et al., 2005). Furthermore, concentrated systems can be prepared, and their rheological properties can be explored for different food applications (Mason et al., 2006b). The main disadvantages are the limited kinetic stability of the system, which has to be monitored to keep the desired properties for a sufficient period of time for a given application. This is an important factor to determine the shelf life of food products based on nanoemulsions.

Although the system is metastable, creaming stability is highly enhanced due to the small droplet sizes, which leads to homogeneous systems even for low-viscosity continuous phases. It has been reported that kinetically stable nanoemulsions from flocculation and coalescence can be prepared (Solans et al., 2005; Mason et al., 2006a). In fact, the main destabilization mechanism in nanoemulsions is Ostwald ripening, due to high Laplace pressures of droplets, which is significantly higher when compared to conventional emulsions.

7.3.4.2 Preparation and Characterization of Microemulsions

Microemulsions are prepared by adding the proper amounts of the components, which form the microemulsion after a given period of time. This is the great advantage of microemulsion preparation when compared to conventional emulsification methods, since the preparation does not require the input of high amounts of energy in the system. However, microemulsion systems can lose their characteristic morphology with variations in temperature and composition. Therefore, the range of the parameters that maintain the microemulsion characteristics has to be determined to define the applicability range of a given system.

Preparation methods for microemulsions consist essentially in adding and mixing the components to the system in different ways and conditions to form a microemulsion. Usually, a single surfactant is not sufficient to decrease the interfacial tension up to the point at which spontaneous emulsification occurs. Thus, one or more cosurfactants are used, which are usually amphiphilic molecules with different HLB of the main surfactant, or alcohols, but their applicability can be limited as food systems due to toxicity issues. Although the system to be formed is thermodynamically favorable, i.e., the microemulsion is formed spontaneously, it is usually necessary to overcome kinetic energy barriers. The main preparation methods are the low energy emulsification and the phase inversion temperature (PIT) method. The first method can be achieved by (1) adding water in a mixture of oil and surfactant; (2) adding oil in a mixture of water and surfactant; and (3) by mixing all components together at once. It has been reported in the literature that the order of ingredient addition can play a significant role in the formation of the microemulsion (Flanagan and Singh, 2006).

In the PIT method, an initial emulsion, for example, a W/O emulsion, is heated up to a temperature, called PIT, for which the interfacial tension between the oil and water phases reaches a minimum. At this point there is an inversion, and an O/W emulsion is formed. The system is then cooled while stirring and a stable microemulsion is formed. Sometimes high-pressure homogenization is used to prepare microemulsions, but this method is limited due to the high heat dissipation involved.

Characterization of microemulsions requires the construction of phase diagrams, which can be done by titration methods (Lawrence and Rees, 2000). Ternary or pseudoternary phase diagrams are usually built by varying the amount of the components and observing the phase behavior of the system to identify the region where a clear and isotropic emulsion is formed, i.e., a Winsor IV system. In general, pseudoternary diagrams are found since microemulsion systems usually contain cosurfactants. Figure 7.19 shows a schematic pseudoternary phase diagram of a microemulsion system, indicating the phase behavior for a given composition. Each axis corresponds to the volume or mass fraction of each component or group of components, which are usually water, oil, and surfactant/cosurfactant.

The microstructure (or nanostructure) characterization of microemulsions can be performed by using dynamic light scattering (DLS) to determine droplet sizes. Scanning electron microscopy (SEM), small-angle x-ray scattering (SAXS), small-angle neutron scattering, and nuclear magnetic resonance (NMR) can be used to determine other structural features such as the presence of worm-like reverse micelles and other liquid-crystalline phases.

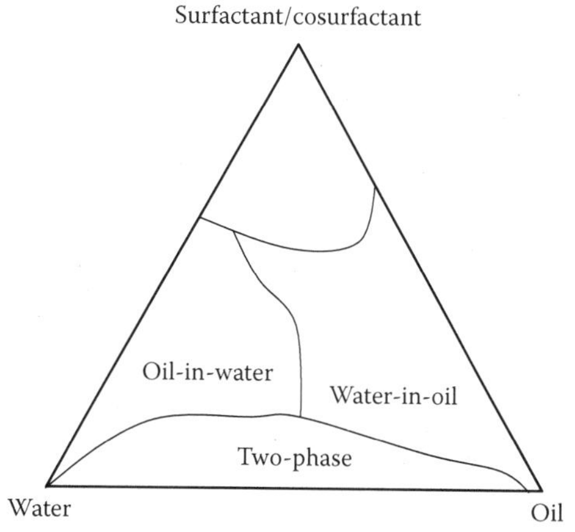

FIGURE 7.19 Schematic pseudoternary diagram showing regions of two-phase and microemulsion regimes.

Microstructure and dynamical behavior of microemulsions strongly affect the macroscopic properties of microemulsions. Therefore, it is important to characterize the macroscopic properties of a determined system. In food applications, it is very important to characterize the rheological properties of microemulsions, which can be performed using conventional rheometers. Techniques such as conductive and dielectric properties measurements can be used to determine the type of microemulsion formed (W/O or O/W) and monitor percolation, phase inversion, and other structural and dynamical features. Optical features of microemulsions are important for food applications since it is necessary for the system to be visually appealing when considering a commercial functional food system (Flanagan and Singh, 2006).

7.3.4.3 Preparation and Characterization of Nanoemulsions

Nanoemulsions can be prepared by low-energy methods based on PIT, which is similar to the one described in Section 7.3.4.2, but in this case the system, when cooled, keeps its morphology in a nonequilibrium state (Förster et al., 1995; Izquierdo et al., 2004, 2005; Morales et al., 2003). Other low-energy methods resulting in nanoemulsions are the phase inversion composition and autoemulsification methods. The former is similar to the PIT method, but the phase inversion occurs by modifying the composition of the system, leading to a kinetically stable nanoemulsion (Forgiarini et al., 2001; Porras et al., 2004). The latter is based on the dilution of an initial stable microemulsion, usually at the bicontinuous phase, resulting in a nanoemulsion (Pons et al., 2003; Wang et al., 2007, 2008).

Preparation using low-energy methods results in nanoemulsions lying in a region close to thermodynamic equilibrium. Therefore, their properties are very similar to a Winsor IV microemulsion system, and a distinction between this type of nanoemulsion and a thermodynamic stable microemulsion has been contested by some research groups (Mason et al., 2006b). However, stability of nanoemulsions prepared by low-energy methods, showing their nonequilibrium character, has been studied and summarized in review articles, and such emulsions have gained wide acceptance due to their potential in applications in food systems and pharmaceuticals (Tadros et al., 2004; Solans et al., 2005; Gutiérrez et al., 2008).

Another method that has gained importance is the preparation under high shear. In the past, this method was limited due to the lack of high-pressure homogenizers that were able to generate shear rates high enough to break droplets in to nanometer sizes. Therefore, most nanoemulsions prepared using this method were limited to laboratory research using homebuilt homogenizers. Recently, affordable high-pressure homogenizers with the capacity to generate high shear, such as those based on microchannel flows, appeared in the market. Consequently, the industrial interest in nanoemulsions is expected to increase. This method is particularly interesting for food applications, since most of the conventional food emulsions are prepared in this way, and high-pressure homogenization operations are suited for scale-up when industrial production rates are desired.

Nanoemulsions prepared by high shear methods form metastable systems far from equilibrium, and despite the comparable droplet sizes, their characteristics are not similar to microemulsions. The main advantage of these nanoemulsions is the fact that kinetically stable systems can be prepared using considerably lower amounts of surfactants, while exhibiting physical properties similar to the counterpart microemulsions, such as low viscosity and transparency (Mason et al., 2006b). Therefore, nanoemulsions prepared under high shear can be advantageous if surfactant cost is an important issue when developing food delivery systems. Also, it is possible to develop concentrated systems, which can exhibit unique rheological properties that could be explored in food applications.

The main morphological parameters in nanoemulsions are their average droplet size, droplet size distribution, and droplet volume fraction. Usually, droplet sizes are measured using DLS, and the stability of the system can be monitored by size measurements in a time interval. NMR can be used to measure droplet sizes and droplet interactions in concentrated systems, and it has been used in stability studies of emulsified systems. It can also be used when in situ measurements are required, and it is able to characterize concentrated systems, the emulsion type (W/O or O/W), and flow and mixing properties, although this technique has more widespread use in characterizing conventional emulsions, but it is feasible for applications in nanoemulsions. Morphological aspects of concentrated nanoemulsions have been studied using small-angle neutron scattering (Mason et al., 2006a). Rheological and optical properties are important when seeking food applications. Therefore, the dependence of viscosity and other viscoelastic properties on droplet sizes and droplet concentrations are important in nanoemulsion characterization.

Reports of nanoemulsions applied as drug delivery systems can be found in the literature (Solans et al., 2005). However, few studies are found concerning applications in food systems, but this number can be expected to grow due to the increase in the availability of homogenization systems of ultrahigh shear based on microchannel flow and other recently developed preparation methods (Kentish et al., 2008; Yuan et al., 2008). Furthermore, fundamental research on nanoemulsion formation through high shear has recently appeared (Meleson et al., 2004). Reviews on the fundamentals and potential applications of nanoemulsions can be found in the literature (Solans et al., 2005; Mason et al., 2006b; Gutiérrez et al., 2008). Recently, a patent was filed claiming production of W/O nanoemulsion for food applications (Del Gaudio et al., 2007).

7.4 APPLICATIONS OF ENCAPSULATED ACTIVE INGREDIENTS IN FOODS

Encapsulated ingredients are used in many food applications. They can be incorporated in beverages, dairy products, baked products, and manufactured meats, including infant and other specialized formulations. Examples of food products in which encapsulated ingredients can be incorporated are UHT milk, cheese, ice cream, margarines, muesli bars, yoghurt, infant foods, dietetic food supplements, spreads, health drinks, mayonnaise, baked products, and breakfast cereals.

Encapsulation may be used to deliver traditional active ingredients, such as flavors, vitamins, minerals, sweeteners, and antioxidants, or relatively novel ones, such as probiotic microorganisms. SD and gel microparticles, liposome and emulsified systems, which are under focus in this chapter, have been used for some of these applications. Their functionality as delivery systems is discussed in Sections 7.4.1 through 7.4.5.

7.4.1 Flavors

One of the largest food applications is the encapsulation of flavors. Flavors can be among the most valuable ingredients in any food formula. Even small amounts of some aromatic substances can be expensive, and because they are usually sensitive and volatile, preserving them is often a top concern of food manufacturers. Encapsulating these high-cost materials can result in cost savings, as the loss through storage and processing is limited.

Encapsulation of flavors has been attempted and commercialized using many different methods such as SD, spray chilling or spray cooling, extrusion, coacervation, and molecular inclusion. Among them, SD is the widely used commercial process in large-scale production of encapsulated flavors and volatiles. One of the reasons is the large-scale capacity of production in a continuous mode. Microencapsulation of flavors by SD presents the challenge of removing water by evaporation, while retaining substances that are much more volatile than water, which is the case for most organic compounds. However, SD microcapsules with a high retention of aromatic compounds can be obtained, due to the phenomenon known as selective diffusion (Thijssen, 1971). This concept is based on the fact that the diffusion of water in concentrated solutions behaves differently from the diffusion of other substances. According to this concept, favorable conditions for obtaining high volatiles retention can be created in SD due to the rapid decrease of the water concentration at the drying droplet surface in contact with the hot drying air. Once the droplet surfaces have dried sufficiently, selective diffusion comes into effect because the diffusion coefficients of organic compounds in the surface region become much lower than that of water. When volatile substances are encapsulated, successful microencapsulation relies on achieving high retention of the core material during processing and storage. Many studies have been carried out on the influence of encapsulating material compositions and the operating conditions on the encapsulation efficiency and on the controlled release of encapsulated flavors. Most frequently used carriers include carbohydrates, gums, and food proteins. Each group of materials has certain advantages and disadvantages. For this reason, many coatings are actually composite formulations. Thorough reviews of the technology have been published (Madene et al., 2006), some of them, with special emphasis (Ré, 1998), are dedicated to the encapsulation of flavors by SD (Gharsallaoui et al., 2007).

A subject of increasing interest in this area concerns the development of alternative and inexpensive polymers that may be considered as natural, like gum arabic (good emulsifying properties), and could encapsulate flavors with a good efficiency. For example, sugar-beet pectin (Drusch, 2007) has been regarded as an alternative emulsifying encapsulating material for flavors.

Another area of interest to optimize the encapsulation efficiency of food flavors and oils by SD is the submicronization of the droplets oil of the emulsion. It has been well documented that emulsion droplet size has a pronounced effect on the encapsulation efficiency of different core materials by SD (Jafari et al., 2008). The findings clearly show that reducing emulsion size can result in encapsulated powders with higher retention of volatiles and lower content of unencapsulated oil at the surface of powder particles. The presence of oil on the surface of the powder particles is the most undesirable property of encapsulated powders, and it has been pointed out as a frequent problem with the quality of spray-dried products. This surface oil not only deteriorates the wettability and dispersability of the powder, but it is also readily susceptible to oxidation and to the development of rancidity.

Much of the work in this area has been done in emulsions having a droplet size of more than 1 μm, and the application of submicron (nano) emulsions in encapsulation of oils and flavors is relatively new in the literature. Some works have been carried out to determine the influence of submicron emulsions produced by different emulsification methods on encapsulation efficiency and to investigate the encapsulated powder properties after SD for different emulsion droplet sizes and surfactants. The process has been referred to as "nanoparticle encapsulation" since a core material in nanosize range is encapsulated into a matrix of micron-sized powder particles (Jafari et al., 2008). This area of research is developing. Some patents were filed in the past describing microemulsion formulations applied to flavor protection (Chung et al., 1994; Chmiel et al., 1997) and applications in flavored carbonated beverages (Wolf and Havekotte, 1989). However, there is no clear evidence on how submicron or nanoemulsions can improve the encapsulation efficiency and stability of food flavors and oils into spray-dried powders.

SD of emulsions has been traditionally used for the encapsulation of flavors; however, novel encapsulation and delivery properties can be achieved by encapsulating flavors into liposomes. Because liposomes have the ability to carry hydrophilic and fat-based flavors, they protect them from degradation during processing and storage and also increase the longevity of the flavor in the system where they are being used. Therefore, their use in the beverage industry has been widespread (Reineccius, 1995).

The bioflavor compounds of blue cheese, obtained from fermentation of *Aspergillus* spp., were encapsulated in soy lecithin liposomes and spray-dried to obtain the powder form by Santana et al. (2005). A sensory evaluation was performed, by adding the liposome-bioflavor powder in a base of light cream cheese, which was spread on toasts. Flavor intensity, acceptance by the consumers, and purchasing intention were the tests done in the sensory evaluation. The results showed that the encapsulation maintained the characteristic flavor of blue cheese and the product was classified by the consumers as acceptable. The dried liposome-stabilized flavor was useful to add in foods and to be kept in storage.

7.4.2 Antioxidants

There is a growing demand for delivery of antioxidants through functional foods with the concomitant challenge of protecting their bioactivity during food processing and subsequent passage through the GI tract. Antioxidants such as lycopene, and β-carotene can be encapsulated by SD. Blends of sucrose and gelatin have been successfully used in encapsulation of lycopene (Shu et al., 2006) and blends of maltodextrin and starches can be used to encapsulate β-carotene (Loksuwan, 2007).

The product quality has been analyzed with respect to the retention of the antioxidant activity of the spray-dried powder. Polyphenolic compounds present in several extracts (grape seed, apple polyphenolic extract, or olive-leaf) were also encapsulated by SD in protein–lipid emulsions (Kosaraju et al., 2008) and chitosan (Kosaraju et al., 2006).

The common carotenoids in fruits and vegetables (licopen, lutein, zeaxantin, astaxantin, and β-criptoxantina) are used as ingredients in foods. They substitute artificial colorants and are also functional ingredients due to their pro-vitamin A activity, apart from the fact that they act as antioxidants (Fernandez-Garcia et al., 2007). The potential market of water-dispersible carotenoids is broad, including ice creams, soups, disserts, meat products, and animal foods (Delgado-Vargas et al., 2000). Nevertheless, carotenoids lose color under oxidation, as they suffer isomerization easily under heat, acidic pH, or light exposition. The necessity of protection as well as the lipophilic or amphiphilic nature of carotenoids makes them attractive to liposome encapsulation. However, the approach of the studies of carotenoids in liposomes only focuses upon their oxidant activity and interactions into the lipid bilayer (Socaciu et al., 2000, 2002; Kostecka-Gugala et al., 2003; Gruszecki and Strzalka, 2005; Jemiola-Rzeminka et al., 2005; McNulty et al., 2007; Sujak et al., 2007). Applications of carotenoids in liposomes are intended to increase their longevity in the foods, as well as to increase their oral bioavailability due to the presence of lipids as coadjuvant in the formulation (Fernandez-Garcia et al., 2007; Parada and Aguilera, 2007).

Reports of conventional O/W emulsions used to encapsulate lycopene and β-carotene are found in the literature (Ribeiro et al., 2006; Santipanichwong and Suphantharika, 2007). Stability of emulsified lycopene was evaluated after its incorporation in liquid food matrices (Ribeiro et al., 2003). Microemulsions have been applied to increase the efficiency of antioxidants such as ascorbic acid (Moberger et al., 1987). Furthermore, microemulsions have successfully been used for lycopene solubilization (Spernath et al., 2002). Applications of microemulsions to increase stability of antioxidants have been reported (Moberger et al., 1987; Yi et al., 1991).

7.4.3 Omega-3 Fatty Acids

Omega-3 acids are considered essential to human health, but cannot be manufactured by the human body and must therefore be obtained from food. These acids are naturally present in most fishes and certain plant oils such as soybean and canola, which are foods that people rarely consume in large quantities. Moreover, the direct addition of omega-3 fatty acids to many foods is prevented due to some characteristics (fishy flavors, readily oxidized), which together reduce the sensory acceptability of foods containing fatty acids, limit shelf life, and potentially reduce the bioavailability of the acids. Encapsulation responds to the challenges of omega-3 fatty acid delivery and extends the reach of its health benefits.

These acids can be encapsulated by SD. The success of the encapsulation is based on the ability of the spray-dried powder to provide, firstly, a good retention of these compounds within the structures and, secondly, a good oxidative stability. The goal is to find the appropriate carriers to create good barrier properties as shown in the literature. For example, glucose syrup was used in combination with proteins such as whey protein isolate or soy protein isolate (Rusli et al., 2006) or with sugar beet pectin (Drusch et al., 2007) to encapsulate fish oil, leading to a product that was more stable to oxidation than bulk oils. Other formulation compositions such as a blend of modified celluloses (methylcellulose and hydroxypropylmethylcellulose) with good emulsifying properties and maltodextrin (Kolanowski et al., 2004) have also been tested to encapsulate fish oil. Another strategy recently developed was based on the production of multilayer membranes of lecithin and chitosan around the oil droplets of the O/W emulsion before SD (Klinkesorn et al., 2006). Despite research being in progress, the influence of various process variables on oil oxidation during the emulsifying and drying stages is still not well known.

Encapsulation of omega-3 fatty acids using O/W emulsions was recently reported in the literature (Lee et al., 2006). The potential for emulsion-based delivery systems of omega-3 molecules in different types of foods such as yogurts, milk, and ice cream has been recognized (McClements et al., 2007). However, emulsion technology to encapsulate these molecules is difficult, requiring the development of antioxidant technologies to stabilize the system due to the complex oxidation reaction of omega-3 molecules (McClements and Decker, 2000).

Omega-3 fatty acids can also be protected against oxidation when encapsulated within liposomes (Haynes et al., 1991; Wallach and Mathur 1992).

7.4.4 Vitamins and Minerals

Encapsulating vitamins and minerals offers several benefits. It increases their stability when exposed to air, heat, or moisture. Many of these micronutrients are often destroyed in the baking process. Loss through processing and storage is prevented, resulting in the ability to use less of these products and thus save cost. Finally, many of these micronutrients have undesirable flavors or odors, which can be masked, keeping the micronutrients available to be absorbed in the GI tract.

Fortifying foods with minerals and vitamins is becoming more and more common. Mineral deficiency is one of the most important nutritional problems in the world. The best method to overcome this problem is to make use of an external supply, which may be nutritional or supplementary, like the fortification of foods with highly bioavailable mineral sources. Major interests of mineral encapsulation are linked to the fact that this technique enables to reduce mineral reactions with other ingredients, when they are added to dry mixes to fortify a variety of foods, and it can also incorporate time-release mechanisms of the minerals into the formulations. For example, iron is the most difficult mineral to add to foods and ensure adequate absorption, and iron bioavailability is severely affected by interactions with food ingredients (e.g., tannins, phytates, and polyphenols). Additionally, iron catalyses the oxidative degradation of fatty acids and vitamins (Schrooyen et al., 2001).

Liposomes have been used to encapsulate bioactive vitamins and minerals. Milk enriched with ferrous sulfate encapsulated in liposomes enabled an increase in the iron concentration compared to free iron. The encapsulated ferrous sulfate was stable to heat sterilization (100°C, 30 min) and storage at 4°C for 1 week.

Furthermore, liposomes provided the same bioavailability as the free sulfate, adding the advantage of being coated with a phospholipid membrane, which kept the iron from contacting with the other components of food, thus preventing undesirable interactions (Boccio et al., 1997; Uicich et al., 1999; Lysionek et al., 2000, 2002; Shuqin and Shiying, 2005).

Orange juice, cereals, and even candies are fortified with vitamins and minerals such as vitamin C and calcium. Vitamin C, also known as ascorbic acid or ascorbate, is added extensively to many types of foods for two quite different purposes: as a vitamin supplement to reinforce dietary intake of vitamin C, and as an antioxidant, to protect the sensory and nutritive quality of the food itself. Encapsulation of vitamin C improves and broadens its applications in the food industry. Spray-dried structures encapsulating vitamin C can be produced by using several carriers as encapsulating materials, among them, methacrylate copolymers named Eudragit®.* The resulting delivery systems were able to offer a controlled release at different pH values due to the Eudragit characteristics (Esposito et al., 2002). Chitosan, a hydrophilic polysaccharide also used as a dietary food additive, was used to encapsulate vitamin C by SD (Desai and Park, 2005b). Chitosan was cross-linked with a nontoxic cross-linking agent, tripolyphosphate.

SD was also used to formulate calcium microparticles using cellulose derivatives and polymethacrylic acid as encapsulating materials (Oneda and Ré, 2003), to modify the dissolution rate of calcium from calcium citrate and calcium lactate. Microparticulate systems with incorporated time-release mechanism were obtained to modify the calcium release from these commercial salts used in fortification of diet.

Liposomes composed of phosphatidylcholine, cholesterol, and DL-α-tocopherol improved shelf life of vitamin C from a few days up to 2 months, especially in the presence of common food components which normally speed up decomposition, such as copper ions, ascorbate oxidase, and lysine (Kirby et al., 1991). Calcium lactate was also encapsulated in lecithin liposomes, in this case to prevent undesirable calcium–protein interactions (Champagne and Fustier, 2007). The liposomal calcium levels of fortified soymilk were equivalent to those found in cow's milk. A synergistic effect of coencapsulation of vitamins A and D in liposomes promoted calcium absorption in the GI tract (Champagne and Fustier, 2007).

W/O emulsions based on olive oils were also used to encapsulate vitamin C (Mosca et al., 2008). Solubilization of vitamin E in microemulsions based on polyoxyethylene (POE) surfactants was reported in the past (Chiu and Yang, 1992). Phase behavior studies were conducted on microemulsions based on different oil phases such as limonene, medium-chain triglycerides (MCT), short-chain alcohols, polyols, and different surfactants (Garti et al., 2001; Papadimitriou et al., 2008; Zhang et al., 2008). In addition, the phase behavior of microemulsions prepared with food grade components based on lecithin has been investigated (Patel et al., 2006), showing potential for applications in encapsulating vitamins and minerals.

* Registered Trademark of Rohm GmbH & Co. KG, Darmstadt, Germany.

7.4.5 Probiotics

According to the Food and Agriculture Organization (FAO) of the United States and the World Health Organization (WHO), a probiotic is a live microorganism which, when administered in adequate amounts, confers a health benefit to the host. The FAO/WHO Expert Consultation lists benefits that had substantial support from peer-reviewed publication of human studies (FAO/WHO, 2001).

Probiotics may consist of a single strain or a mixture of several strains. Most common are lactic acid bacteria from the *Lactobacillus* and *Bifidobacterium* genera. Species of bacteria and yeasts used as probiotics include *Bifidobacterium bifidum*, *B. breve*, *Lactobacillus casei*, *L. acidophilus*, *Saccharomyces boulardii*, and *Bacillus coagulan*, among others (Champagne et al., 2005).

Dietary supplements containing viable probiotic microorganisms (referred herein as probiotics) are increasing in popularity in the marketplace as their health benefits become recognized. Probiotics are sensitive to various environmental conditions such as pH, moisture, temperature, and light. When these conditions are not properly controlled, the product viability (measured in colony forming units or CFU), and therefore its efficacy, can be substantially reduced. The viability and stability of probiotics have both been a marketing and technological challenge for industrial producers.

Losses of microorganisms occur during manufacture and during the product shelf life. In addition, probiotics with good characteristics for effectiveness against disease and other conditions may not have good survival characteristics during transit through the GI tract. The probiotic cultures encounter gastric juices in the stomach ranging from pH 1.2 (on an empty stomach) through pH 5.0. These cultures stay in the stomach from around 40 min up to 5 h. In the stomach and the small intestine, these probiotics also encounter bile salts, and hydrolytic and proteolytic enzymes, which are able to kill them. These cultures are able to grow or survive only when they reach higher pH regions of the gastrointestine. During this transit, probiotics also have to compete with resident bacteria for space and nutrients. In addition, they have to avoid being flushed out of the tract by normal peristaltic action and being killed by antimicrobials produced by other microorganisms. Ability to adhere to surfaces, such as intestinal mucosal layer and the epithelial cell walls of the gut, is an important characteristic of a probiotic. The term "colonization" is used, and it means that the microorganism has mechanisms that enable it to survive in a region of the gastrointestine on an ongoing basis.

Intensive research efforts have been focused on protecting the viability of probiotic cultures both during product manufacture and storage, and through the gastric transit until the target site is reached. Protection may be achieved by several ways, among them, encapsulation.

At least five encapsulation methods have been investigated to protect probiotics: spray coating, SD, extrusion droplets, emulsion, and gel-particle technologies, including spray chilling. Several reviews in the literature are dedicated to probiotic (Mattila-Sandholm et al., 2002; Champagne et al., 2005) or, more specifically, to probiotic encapsulation (Kailasapathy, 2002; Krasaekoopt et al., 2003; Anal and Singh, 2007). In the specific case of fermented dairy products, a through discussion of both probiotics and prebiotics is given in Chapter 20.

SD is rarely considered for cell immobilization because of the high mortality resulting from simultaneous dehydration and thermal inactivation of microorganisms. Despite this limitation, several works have been evaluated SD as a process for encapsulating probiotics. Technical alternatives have been proposed to increase thermal resistance of the microorganisms during the dehydration process, such as the proper adjustment and control of the inlet and outlet drying temperatures (O'Riordan et al., 2001), the use of complex thermoprotector (prebiotic) carbohydrates as encapsulating materials (Rodriguez-Huezo et al., 2007), or even a previous encapsulation of the microorganisms by another technique, as proposed by Oliveira et al. (2007). These authors encapsulated probiotics (*B. lactis* and *L. acidophilus*) in a casein–pectin complex formed by complex coacervation, and the wet encapsulated microorganisms were dried by SD.

It has been demonstrated that a variety of probiotic cultures can be protected via encapsulation by SD in a variety of carriers, including whey protein (Picot and Lacroix, 2004), a matrix of gelatin, soluble starch, skim milk, or gum arabic (Lian et al., 2003), and cellulose phthalate, which is an enteric release pharmaceutical compound (Favaro-Trindade and Grosso, 2002), and a matrix of protective colloids (whey protein isolate, mesquite gum, and maltodextrin in a 17:17:66 ratio) associated to aguamiel as a prebiotic thermoprotector (Rodriguez-Huezo et al., 2007). In fact, the protective effect exerted by SD encapsulation against stressful conditions (the environment in the food product, during storage, and during the passage through the stomach or intestinal tract) may vary with the carriers or encapsulating materials and the microorganisms but, in all cases, the thermal resistance of strains is a critical parameter that should always be taken into consideration if SD is the intended method for encapsulation (Picot and Lacroix, 2004; Su et al., 2007).

Most of the literature reported on the encapsulation of probiotics has investigated the use of gel particles for improving their viability in food products and intestinal tract. The bacterial cells are dispersed into the hydrocolloid solution before gelation.

Entrapping probiotic bacteria in gels with ionic cross-linking is typically achieved with polysaccharides (alginate, pectin, and carrageenan). By far, the most commonly used material for this purpose is alginate, and the most commonly reported encapsulation procedure is based on the calcium-alginate gel formation. The droplets form gel spheres instantaneously (sodium-alginate in calcium chloride) entrapping the cells in a three-dimensional lattice of ionically cross-linked alginate. The success of this method is due to the gentle environment it provides for the entrapped material, cheapness, simplicity, and its biocompatibility (Anal and Singh, 2007). In the past 10 years, there have been 93 peer-reviewed articles on "probiotics encapsulation" (Web of Science database), out of which 47 especially on "encapsulation using alginate."

Various researchers have studied factors affecting the gel particles characteristics and their influence on the encapsulation of probiotics, such as concentrations of alginate and $CaCl_2$, timing of hardening of the gel particles, and cell concentrations (Chandramouli et al., 2004). Most of them have shown that probiotics can be protected in calcium-alginate beads, what is generally demonstrated by an increase in the survival of bacteria under different harsh conditions, compared with free microorganisms. Alginates also demonstrate easy release of the encapsulated bacteria when suspended in an alkaline buffer. However, the degree of protection might depend on the gel particle size, suggesting that these microorganisms should be encapsulated within a specific gel particle size range. For example, very large calcium-alginate beads (>1 mm) can negatively affect the textural and sensorial properties of food products in which they are added (Hansen et al., 2002), whereas reduction of the sphere size to less than 100 μm would be advantageous for texture considerations, allowing direct addition of encapsulated probiotics to a large number of foods. However, it has been demonstrated that particles smaller than 100 μm do not significantly protect the probiotics in simulated gastric fluid, compared with free cells (Hansen et al., 2002). One limitation for cell loading in small particles is also the large size of microbial cells, typically 1–4 μm, or particles freeze-dried culture (more than 100 μm). On the other hand, there are evidences in the literature that calcium-alginate gel particles with mean diameters of 450 μm (Chandramouli et al., 2004) and 640 μm (Shah and Ravula, 2000) could protect probiotics from adverse gastric conditions. In the latter case, the particles, after being freeze dried, also protected the viability of the microorganisms in fermented frozen dairy desserts. In fact, Chandramouli et al. (2004) found an optimal particle size of 450 μm for the calcium-alginate gel particles to protect the cells (*L. acidophilus*), when testing gel particles of different sizes (200, 450, and 1000 μm).

The composition of the alginate also influences bead size (Martinsen et al., 1989). Alginates are heterogeneous groups of polymers, with a wide range of functional properties. Alginates with a high content of guluronic acid blocks (G blocks) are preferable for capsules formation because of their higher mechanical stability and better tolerance to salts and chelating agents.

In addition to the reports of benefits of encapsulation in protecting probiotics against the stressful conditions of the GI tract, there is increasing evidence that the procedure is helpful in protecting the probiotic cultures destined to be added to foods. For example, encapsulation technologies have been used satisfactorily to increase the survival of probiotics in high acid fermented products such as yoghurts (Krasaekoopt et al., 2003), including Ca-alginate gel particles. Other reported food vehicles for delivery of encapsulated probiotic bacteria are cheese, ice cream, and mayonnaise (Kailasapathy, 2002).

Despite the suitability of alginate as entrapment matrix material, this system has some limitation due to its low stability in the presence of chelating agents such as phosphate, lactate, and citrate. The chelating agents share affinity for calcium and destabilize the gel (Kailasapathy, 2002). Special treatments, such as coating the alginate particles, can be applied to improve the properties of encapsulated gel particles. Coated beads not only prevent cell release but also increase mechanical and chemical stability. It has been reported that cross-linking with cationic polymers, coating with other polymers, mixing with starch, and incorporating additives can improve stability of beads (Krasaekoopt et al., 2003). For example, alginate can be coated with chitosan, a positively charged polyamine. Chitosan forms a semipermeable membrane around a negatively charged polymer such as alginate. This membrane, like alginate, does not dissolve in the presence of Ca^{2+} chelators or antigelling agents, and thus enhances the stability of the gel and provides a barrier to cell release (Krasaekoopt et al., 2004, 2006; Urbanska et al., 2007).

Various other polymer systems have been used to encapsulate probiotic microorganisms. κ-Carrageenan (Adhikari et al., 2003), gellan gum, gelatin, starch, and whey proteins (Reid et al., 2007) have also been used as gel encapsulating systems for probiotics. An increasing interest in developing new compositions of gel particles to improve the viability of the probiotic microorganisms to harsh conditions (thermotolerance, acid-tolerance, etc.) is marked by the more recent researches reported in the literature. Some of these systems include alginate plus starch (Sultana et al., 2000), alginate plus methylcellulose (Kim et al., 2006), alginate plus gellan (Chen et al., 2007), alginate–chitosan–enteric polymers (Liserre et al., 2007), alginate-coated gelatin (Annan et al., 2008), gellan plus xanthan (McMaster et al., 2005), κ-carrageenan with locust bean gum (Muthukumarasamy et al., 2006), and alginate plus pectin plus whey proteins (Guerin et al., 2003). In some cases, systems have been developed not only to provide better probiotic viability but also to deliver a prebiotic synergy (Iyer and Kailasapathy, 2005; Crittenden et al., 2006).

Improving the number of possibilities to encapsulate probiotics is a important tool even because, in recent years, the consumer demand for non-dairy-based probiotic products has increased (Prado et al., 2008), and the application of probiotic cultures in nondairy products represents a great challenge, because they may represent new hostile environment for probiotics (heat-processed foods, storage at room temperature, more acid foods like fruit juices, etc.).

Emulsified systems have also been investigated to protect probiotics. Incorporation of *L. acidophilus* in a W/O/W emulsion was recently reported and the protective effect of the probiotic in a low pH environment was evaluated (Shima et al., 2006). Lactic acid bacteria were encapsulated in sesame oil emulsions and, when subjected to simulated high gastric or bile salt conditions, a significant increase in survival rate was observed (Hou et al., 2003).

7.5 ENCAPSULATION CHALLENGES

The challenges in developing an encapsulated food ingredient commercially viable depend on selecting appropriate and food grade (GRAS) encapsulating materials, selecting the most appropriate process to provide the desired size, morphology, stability, and release mechanism, and economic feasibility of large-scale production, including capital, operating, and other miscellaneous expenses, such as transportation and regulatory costs.

However, the development of any encapsulation technique must not be treated as an isolated operation but as part of an overall process starting with ingredient production followed by processes, including encapsulation, right through to liberation and utilization of the ingredient. Furthermore, a selection has to be made between batch, semicontinuous, and continuous encapsulation processes, resulting in a difficult choice for process designers. Cost is often the main barrier of the implementation of encapsulation, and multiple benefits are generally required to justify the cost of encapsulation. Indeed, in the food industry, regulations with respect to ingredients, processing methods, and storage conditions are tight, and the price margin is much lower than in, for example, the pharmaceutical industry.

This procedure is something of an art, as Asajo Kondo asserts in *Microcapsule Processing and Technology* (Kondo, 1979):

> Microencapsulation is like the work of a clothing designer. He selects the pattern, cuts the cloth, and sews the garment in due consideration of the desires and age of his customer, plus the locale and climate where the garment is to be worn. By analogy, in microencapsulation, capsules are designed and prepared to meet all the requirements in due consideration of the properties of the core material, intended use of the product, and the environment of storage.

Encapsulation technology remains something of an art, although firmly grounded in science. Combining the right encapsulating materials with the most efficient production process for any given core material and its intended use requires extensive scientific knowledge of all the materials and processes involved and a good feel for how materials behave under various conditions.

Continuing research is clearly necessary to improve and extend the technology to the encapsulation to a wide variety of beneficial ingredients. Researchers are investigating the next generation of encapsulation technologies, including

- The development of *new, natural food materials and encapsulated products* that can be used by food manufacturers, among them, nonproteinaceous materials to eliminate allergens, that protect the encapsulated ingredients while they travel through the body to a targeted site in GI tract
- The increase in the range of *processing techniques*, with special interest for processes producing in continuous mode with high productivity
- The potential use of *coencapsulation methodologies*, where two or more bioactive ingredients can be combined to have a synergistic effect
- The *targeted delivery* of bioactives to various parts of the GI tract
- The trial of new ways of *incorporating bioactives into foods* with minimal loss of bioactivity and without compromising the quality of the food that is used as a delivery vehicle
- The understanding of the self-assembly and stabilization of *nanoemulsions* during food processing

These developments will give food manufacturers new opportunities to produce a greater variety of innovative functional foods that promote the health and well being of consumers.

REFERENCES

Adhikari, K., Mustapha, A., and Grun, I.U. 2003. Survival and metabolic activity of microencapsulated *Bifidobacterium longum* in stirred yogurt. *J. Food Sci.* 68:275–280.

Alkhalaf, W., El Soda, M., Gripon, J.-C., and Vassal, L. 1989. Acceleration of cheese ripening with liposomes-entrapped proteinase: Influence of liposomes net charge. *J. Dairy Sci.* 72:2233–2238.

Allgaier, J., Willner, L., Richter, D., Jakobs, B., Sottmann, T., and R. Strey. 2004. Method for increasing the efficiency of surfactants with simultaneous suppression of lamellar mesophases and surfactants with an additive added thereto, U.S. Patent 2004054064-A1, filed Aug. 19, 2003, and issued Mar. 18, 2004.

Alves, G.P. and Santana, M.H.A. 2004. Phospholipid dry powders produced by spray drying processing: Structural, thermodynamic and physical properties. *Powder Technol.* 145:141–150.

Amici, E., Tetradis-Meris, G., Pulido de Torres, C., and Jousse, F. 2008. Alginate gelation in microfluidic channels. *Food Hydrocolloid.* 22:97–104.

Anal, A.K. and Singh, H. 2007. Recent advances in microencapsulation of probiotics for industrial applications and targeted delivery. *Trends Food Sci. Technol.* 18:240–251.

Annan, N.T., Borza, A.D., and Hansen L.T. 2008. Encapsulation in alginate-coated gelatin microspheres improves survival of the probiotic *Bifidobacterium adolescentis* 15703T during exposure to simulated gastro-intestinal conditions. *Food Res. Int.* 41:184–193.

Batzri, S. and Korn, E.D. 1973. Single bilayer liposomes prepared without sonication. *Biochim. Biophys. Acta* 298:1015–1019.

Bauer, K., Neuber, C., Schmid, A., and K.M. Voelker. 2002. Oil in water microemulsion. U.S. Patent 6426078-B1, filed Feb. 26, 1998, and issued Jul. 30, 2002.

Belyaeva, E., Della Valle, D., Neufeld, R.J., and Poncelet, D. 2004. New approach to the formulation of hydrogel beads by emulsification/thermal gelation using a static mixer. *Chem. Eng. Sci.* 59:2913–2920.

Benech, R.-O., Kheadr, E.E., Lacroix, C., and Fliss, I. 2002. Antibacterial activities of nisin Z encapsulated in liposomes or produced in situ by mixed culture during cheddar cheese ripening. *Appl. Environ. Microbiol.* 68:5607–5619.

Boccio, J.R., Zubillaga, M.B., Caro, R.A., Gotelli, C.A., and Weill, R. 1997. A new procedure to fortify fluid milk and dairy products with high bioavailable ferrous sulfate. *Nutr. Rev.* 55:240–246.

Brownlie K. 2007. Marketing perspective of encapsulation technologies in food applications. In *Encapsulation and Controlled Release Technologies in Food Systems*, ed. J.M. Lakkis. Ames, IA: Blackwell Publishing Professional, pp. 213–233.

Burey, P., Bhandari, B.R., Howes, T., and Gidley, M.J. 2008. Hydrocolloid Gel particles: Formation, characterization, and application. *Crit. Rev. Food Sci. Nutr.* 48:361–377.

Carneiro, A.L. and Santana, M.H.A. 2004. Production of liposomes in a multitubular system useful for scaling-up of processes. *Prog. Colloid Polym. Sci.* 128:273–277.

Champagne, C.P. and Fustier, P. 2007. Microencapsulation for the improved delivery of bioactive compounds into foods. *Curr. Opin. Biotechnol.* 18:184–190.

Champagne, C.P., Gardner, N.J., and Roy, D. 2005. Challenges in the addition of probiotic cultures to foods. *Crit. Rev. Food Sci. Nutr.* 45:61–84.

Chan, L., Lee, H., and Heng, P. 2002. Production of alginate microspheres by internal gelation using an emulsification method. *Int. J. Pharm.* 241:259–262.

Chan, L.W., Lee, H.Y., and Heng, P.W.S. 2006. Mechanisms of external and internal gelation and their impact on the functions of alginate as a coat and delivery system. *Carbohydr. Polym.* 63:176–187.

Chanamai, R. 2007. Microemulsions for use in food and beverage products. U.S. Patent 087104-A1, filed Oct. 6, 2006, and issued Apr. 19, 2007.

Chandramouli, V., Kailasapathy, K., Peiris, P., and Jones, M. 2004. An improved method of microencapsulation and its evaluation to protect *Lactobacillus* spp. in simulated gastric conditions. *J. Microbiol. Methods* 56:27–35.

Chen, L., Remondetto, G.E., and Subirade, M. 2006. Food protein-based materials as nutraceutical delivery systems. *Trends Food Sci. Technol.* 17:262–283.

Chen, M.J., Chen, K.N., and Kuo, Y.T. 2007. Optimal thermotolerance of *Bifidobacterium bifidum* in gellan-alginate microparticles. *Biotechnol. Bioeng.* 98:411–419.

Chiou D. and Langrish, T.A.G. 2007. Development and characterisation of novel nutraceuticals with spray drying technology. *J. Food Eng.* 82:84–91.

Chiu, Y.C. and Yang, W.L. 1992. Preparation of vitamin E microemulsion possessing high resistance to oxidation. *Coll. Surf.* 63:311–322.

Chmiel, O., Traitler, H., and K. Vopel. 1997. Food microemulsion formulations. WO Patent 96/23425, filed Jan. 24, 1996, and issued Aug. 8, 1996.

Choi, C.H., Jung, J.H., Rhee, Y.W., Kim, D.P., Shim, S.E., and Lee, C.S. 2007. Generation of monodisperse alginate microbeads and in situ encapsulation of cell in microfluidic device. *Biomed. Microdevices* 6:855–862.

Chung, S.L., Tan, C.-T., Tuhill, I.M., and L.G. Scharpf. 1994. Transparent oil-in-water microemulsion flavor or fragrance concentrate, process for preparing same, mouthwash or perfume composition containing said transparent microemulsion concentrate, and process for preparing same. U.S. Patent 5283056, filed Jul. 1, 1993, and issued Feb. 1, 1994.

Crittenden, R., Weerakkody, R., Sanguansri, L., and Augustin, M. 2006. Symbiotic microcapsules that enhance microbial viability during nonrefrigerated storage and gastrointestinal transit. *Appl. Environ. Microbiol.* 72:2280–2282.

Del Gaudio, L., Lockhart, T.P., Belloni, A., Bortolo, R., and Tassinari, R. 2007. Process for the preparation of water-in-oil and oil-in-water nanoemulsions. WO Patent 2007/112967-A1, filed Mar. 28, 2007, and issued Oct. 11, 2007.

Delgado-Vargas, F., Jimenez, A.R., and Paredes-Lopez, O. 2000. Natural pigments: Carotenoids, anthocyanins and betalains—Characteristics, biosynthesis, processing and stability. *Crit. Rev. Food Sci. Nutr.* 40:173–189.

Desai, K.G. and Park, H.J. 2005a. Recent developments in microencapsulation of food ingredients. *Drying Technol.* 23:1361–1394.

Desai, K.G. and Park, H.J. 2005b. Encapsulation of vitamin C in tripolyphosphate cross-linked chitosan microspheres by spray drying. *J. Microencapsul.* 22:179–192.

Drusch, S. 2007. Sugar beet pectin: A novel emulsifying wall component for microencapsulation of lipophilic food ingredients by spray-drying. *Food Hydrocoll.* 21:1223–1228.

Drusch, S., Serfert, Y., Scampicchio, M., Schmidt-Hansberg, B., and Schwarz, K. 2007. Impact of physicochemical characteristics on the oxidative stability of fish oil microencapsulated by spray-drying. *J. Agric. Food Chem.* 55:11044–11051.

El Soda, M. 1986. Acceleration of cheese ripening: Recent advances. *J. Food Prot.* 49:395–399.

Eliaz, R.E. and Szoka, F.C. Jr., 2001. Liposome-encapsulated doxorubicin targeted to CD44: A strategy to kill CD44-overexpressing tumor cells. *Cancer Res.* 61:2592–2601.

Esposito, E., Cervellati, F., Menegatti, E., Nastruzzi, C., and Cortesi, R. 2002. Spray dried Eudragit microparticles as encapsulation devices for vitamin C. *Int. J. Pharm.* 242:329–334.

Evans, D.F. and Wennerstrom, H. 1994. *The Colloidal Domain—Where Physics, Chemistry, Biology and Technology Meet.* New York: Wiley-VCH.

FAO/WHO. 2001. Evaluation of health and nutritional properties of powder milk and live lactic acid bacteria. Food and Agriculture Organization of the United Nations and World Health Organization Expert Consultation Report. Cordoba, Argentina. Available at: http://www.who.int/foodsafety/publications/fs_management/en/probiotics.pdf, accessed Feb. 16, 2009.

Favaro-Trindade, C.S. and Grosso, C.R.F. 2002. Microencapsulation of L-acidophilus (La-05) and B-lactis (Bb-12) and evaluation of their survival at the pH values of the stomach and in bile. *J. Microencapsul.* 19:485–494.

Fernandez-Garcia, E., Minguez-Mosquera, M.I., and Perez-Galvez, A. 2007. Changes in composition of the lipid matrix produce a differential incorporation of carotenoids in micelles. Interaction effect of cholesterol and oil. *Innov. Food Sci. Emerg. Technol.* 8:379–384.

Flanagan, J. and Singh, H. 2006. Microemulsions: A potential delivery system for bioactives in food. *Crit. Rev. Food Sci. Nutr.* 46:221–237.

Forgiarini, A., Esquena, J., Gozales, C., and Solans, C. 2001. Formation of nanoemulsions by low-energy emulsification methods at constant temperature. *Langmuir* 17:2076–2083.

Förster, T., Rybinski, W.V., and Wadle, A. 1995. Influence of microemulsion phases on the preparation of fine-disperse emulsions. *Adv. Coll. Int. Sci.* 58:119–149.

Frederiksen, L., Anton, K., van Hoogevest, P., Keller, H.R., and Leuenberger, H. 1997. Preparation of liposomes encapsulating water-soluble compounds using supercritical CO_2. *J. Pharm. Sci.* 86:921–928.

Garti, N., Yaghnur, A., Leser, M.E., Clement, V., and Watzke, H.J. 2001. Improved oil solubilization in oil/water food grade microemulsions in the presence of polyols and ethanol. *J. Agric. Food Chem.* 49:2552–2562.

Gharsallaoui, A., Roudaut, G., Chambin, O., Voilley, A., and Saurel, R. 2007. Applications of spray-drying in microencapsulation of food ingredients: An overview. *Food Res. Int.* 40:1107–1121.

Gibaldi, M. 1991. *Biopharmaceutics Clinical Pharmacokinetics*, 4th edn. Philadelphia, PA: Lea & Febiger.

Gomez-Hens, A. and Fernandez-Romero, J.M. 2006. Analytical methods for the control of liposomal delivery systems. *Trends Anal. Chem.* 25:167–177.

Gouin, S. 2004. Microencapsulation: Industrial appraisal of existing technologies and trends. *Trends Food Sci. Technol.* 15:330–347.

Gregoriadis, G. 1987. Encapsulation of enzymes and other agents liposomes. In *Chemical Aspects in Food Enzymes*, ed. A.J. Andrews. London, U.K.: Royal Society of Chemistry.

Gregoriadis, G., ed. 1988. *Liposomes as Drug Carriers*. Chichester, U.K.: John Wiley & Sons Ltd.

Gruszecki, W.I. and Strzalka, K. 2005. Carotenoids as modulators of lipid membrane physical properties. *Biochim. Biophys. Acta* 1740:108–115.

Guerin, D., Vuillemard, J.C., and Subirade, M. 2003. Protection of bifidobacteria encapsulated in polysaccharide-protein gel beads against gastric juice and bile. *J. Food Prot.* 66:2076–2084.

Gutiérrez, J.M., González, C., Maestro, A., Solè, I., Pey, C.M., and Nolla, J. 2008. Nano-emulsions: New applications and optimization of their preparation. *Curr. Opin. Colloid Interface Sci.* 13:245–251.

Hansen, L.T., Allan-Wojtas, P.M., Jin, Y.L., and Paulson, A.T. 2002. Survival of Ca-alginate microencapsulated *Bifidobacterium* spp. in milk and simulated gastrointestinal conditions. *Food Microbiol.* 19:35–45.

Haynes, L.C., Levine, H., and Finley, J.W. 1991. Liposome composition for stabilization of oxidizable substances. U.S. Patent 5015483, filed Sep. 02, 1989, and issued May 14, 1991.

Helfrich, W. 1973. Elastic properties of lipid bilayers: Theory and possible experiments. *Z. Naturforsch.* 28C:693–703.

Health Canada. 1998. Policy Paper—Nutraceuticals/functional foods and health claims on foods. Available at http://www.hc-sc.gc.ca/fn-an/label-etiquet/claims-reclam/nutra-funct_foods-nutra-fonct_aliment-eng.php, accessed Feb. 16, 2009.

Herrero, E.P., Martin Del Valle, E.M., and Galan, M.A. 2006. Development of a new technology for the production of microcapsules based in atomization processes. *Chem. Eng. J.* 117:137–142.

Hong, J.S., Shin, S.J., Lee, S.H., Wong, E., and Cooper-White, J. 2007. Spherical and cylindrical microencapsulation of living cells using microfluidic devices. *Korea-Aust. Rheol. J.* 19:157–164.

Hou, R.C.W., Lin, M.Y., Wang, M.M.C., and Tzen, J.T.C. 2003. Increase of viability of entrapped cells of *Lactobacillus delbrueckii* ssp bulgaricus in artificial sesame oil emulsions. *J. Dairy Sci.* 86:424–428.

Huang, K.S., Lai, T.H., and Lin, Y.C. 2007. Using a microfluidic chip and internal gelation reaction for monodisperse calcium alginate microparticles generation. *Front. Biosci.* 12:3061–3067.

International Life Sciences Institute. 1999. Safety assessment and potential health benefits of food components based on selected scientific criteria. ILSI North America Technical Committee on Food Components for Health Promotion. *Crit. Rev. Food Sci. Nutr.* 39:203–316.

Israelachvili, J.N. 1991. *Intramolecular and Surface Forces*. New York: Academic Press.

Iyer, C. and Kailasapathy, K. 2005. Effect of co-encapsulation of probiotics with prebiotics on increasing the viability of encapsulated bacteria under in vitro acidic and bile salt conditions and in yogurt. *J. Food Sci.* 70:18–23.

Izquierdo, P., Esquena, J., Tadros, T.F. et al. 2004. Phase behavior and nano-emulsion formation by the phase inversion temperature method. *Langmuir* 20:6594–6598.

Izquierdo, P., Feng, J., Esquena, J. et al. 2005. The influence of surfactant mixing ratio on nano emulsion formation by the pit method. *J. Colloid Interface Sci.* 285:388–394.

Jafari, S.M., Assadpoor, E., Bhandari, B., and He, Y. 2008. Nano-particle encapsulation of fish oil by spray drying. *Food Res. Int.* 41:172–183.

Jemiola-Rzeminka, M., Pasenkiewicz-Gierula, M., and Strzalka, K. 2005. The behaviour of β-carotene in the phosphatidylcholine bilayer as revealed by a molecular simulation study. *Chem. Phys. Lipids* 135:27–37.

Joscelyne, S.M. and Trägardh, G. 2000. Membrane emulsification—A literature review. *J. Membr. Sci.* 169:107–117.

Kailasapathy, K. 2002. Microencapsulation of probiotic bacteria: Technology and potential applications. *Curr. Issues Intest. Microbiol.* 3:39–48.

Keller, B.C. 2001. Liposomes in nutrition. *Trends Food Sci. Technol.* 12:25–31.

Kentish, S., Wooster, T.J., Ashokkumar, M., Balachandran, S., Mawson, R., and Simons, L. 2008. The use of ultrasonics for nanoemulsion preparation. *Innovat. Food Sci. Emerg. Technol.* 9:170–175.

Kidd, P.M. 2001. Phospholipids, nutrients for life. *Total Health*, 23:Sept.–Oct. issue.

Kim, H.H.Y. and Baianu, I.C. 1991. Novel liposome microencapsulation techniques for food applications. *Trends Food Sci. Technol.* 2:55–61.

Kim, C.J., Jun, S.A., and Lee, N.K. 2006. Encapsulation of *Bacillus polyfermenticus* SCD with alginate-methylcellulose and evaluation of survival in artificial conditions of large intestine. *J. Microbiol. Biotechnol.* 16:443–449.

Kirby, C.J. and Gregoriadis, G. 1984. Dehydration-rehydration vesicles: A simple method for high yield drug entrapment in liposomes. *Biotechnology* 2:979–984.

Kirby, C.J. and Law, B. 1987. Development in the microencapsulation of enzymes in food technology. In *Chemical Aspect of Food Enzymes*, ed. A.T. Andrews. London, U.K.: Royal Society of Chemistry.

Kirby, C.J., Whittle, C.J., Rigby, N., Coxon, D.T., and Law, B.A. 1991. Stabilization of ascorbic acid by microencapsulation in liposomes. *Int. J. Food Sci. Technol.* 26:437–449.

Klinkesorn, U., Sophanodora, P., Chinachoti, P., Decker, E.A., and McClements, J. 2006. Characterization of spray-dried tuna oil emulsified in two-layered interfacial membranes prepared using electrostatic layer-by-layer deposition. *Food Res. Int.* 39:449–457.

Kolanowski, W., Laufenberg, G., and Kunz, B. 2004. Fish oil stabilisation by microencapsulation with modified cellulose. *Int. J. Food Sci. Nutr.* 55:333–343.

Kondo, A. 1979. *Microcapsule Processing and Technology*. New York: Marcel Dekker, Inc..

Kosaraju, S.L., D'ath, L., and Lawrence A. 2006. Preparation and characterisation of chitosan microspheres for antioxidant delivery. *Carbohydr. Polym.* 64:163–167.

Kosaraju, S.L., Labbett, D., Emin, M., Konczak, L., and Lundin, L. 2008. Delivering polyphenols for healthy ageing. *Nutr. Diet.* 65:48–52.

Kostecka-Gugala, A., Latowski, D., and Strzalka, K. 2003. Thermotropic phase behaviour of alpha-dipalmitoylphosphatidylcholine multibilayers is influenced to various extents by carotenoids containing different structural features—Evidence from differential scanning calorimetry. *Biochim. Biophys. Acta* 1609:193–202.

Krasaekoopt, W., Bhandari, B., and Deeth, H. 2003. Evaluation of encapsulation techniques of probiotics for yoghurt. *Int. Dairy J.* 13:3–13.

Krasaekoopt, W., Bhandari, B., and Deeth, H. 2004. The influence of coating materials on some properties of alginate beads and survivability of microencapsulated probiotic bacteria. *Int. Dairy J.* 14:737–743.

Krasaekoopt, W., Bhandari, B., and Deeth, H.C. 2006. Survival of probiotics encapsulated in chitosan-coated alginate beads in yoghurt from UHT and conventionally treated milk during storage. *Lwt—Food Sci. Technol.* 39:177–183.

Lasic, D.D. 1993. *Liposomes: From Physics to Applications*. New York: Elsevier.

Lasic, D.D. 1995. Pharmacokinetics and antitumor activity of anthracyclines precipitated in sterically (stealth) liposomes. In *Stealth Liposomes*, eds. D.D. Lasic and F.J. Martin. Boca Raton, FL: CRC Press.

Lasic, D.D. and Papahadjopoulos, D., eds. 1998. *Medical Applications of Liposomes*. New York: Elsevier.

Latorre, L.G., Carneiro, A.L., Rosada, R.S., Silva, C.L., and Santana, M.H.A. 2007. A mathematical model describing the kinetic of cationic liposome production from dried lipid films adsorbed in a multitubular system, *Braz. J. Chem. Eng.* 24:1–10.

Lawrence, M.J. and Rees, G. 2000. Microemulsion-based media as novel drug delivery systems. *Adv. Drug Deliv. Rev.* 45:89–121.

Lee, S., Hernandez, P., Djordjevic, D. et al. 2006. Effect of antioxidants and cooking on stability of n-3 fatty acids in fortified meat products. *J. Food Sci.* 71:C233–C238.

Lian, W.C., Hsiao, H.C., and Chou, C.C. 2003. Viability of microencapsulated bifidobacteria in simulated gastric juice and bile solution. *Int. J. Food Microbiol.* 86:293–301.

Liserre, A.M., Ré, M.I., and Franco, B.D.G.M. 2007. Microencapsulation of *Bifidobacterium animalis* subsp. *lactis* in modified alginate-chitosan beads and evaluation of survival in simulated gastrointestinal conditions. *Food Biotechnol.* 21:1–16.

Liu, X.D., Yu, X.W., Zhang, Y., Xue, W.M. et al. 2002. Characterization of structure and diffusion behaviour of Ca-alginate beads prepared with external or internal calcium sources. *J. Microencapsul.* 10:775–782.

Liu, X.D., Bao, D.C., Xue, W. et al. 2003. Preparation of uniform calcium alginate gel beads by membrane emulsification coupled with internal gelation. *J. Appl. Polym. Sci.* 87:848–852.

Liu, K., Ding, H.J., Chen, Y., and Zhao, X.Z. 2006. Shape-controlled production of biodegradable calcium alginate gel microparticles using a novel microfluidic device. *Langmuir* 22:9453–9457.

Loksuwan, J. 2007. Characteristics of microencapsulated beta-carotene formed by spray drying with modified tapioca starch, native tapioca starch and maltodextrin. *Food Hydrocoll.* 21:928–935.

Lysionek, A.E., Zubillaga, M.B., Sarabia, M.I. et al. 2000. Study of industrial microencapsulated ferrous sulfate by means of the prophylactic-preventive method to determine its bioavailability. *J. Nutr. Sci. Vitaminol.* 6:125–129.

Lysionek, A.E., Zubillaga, M.B., Salgueiro, M.J., Pineiro, A., Caro, R.A., Weill, R., and Boccio, J.R. 2002. Bioavailability of microencapsulated ferrous sulfate in powdered milk produced from fortified fluid milk: A prophylactic study in rats. *Nutrition* 18:279–281.

Maa, Y.F. and Hsu, C. 1999. Performance of sonication and microfluidization for liquid-liquid emulsification. *Pharm. Dev. Technol.* 4:233–240.

Madene, A., Jacquot, M., Scher, J., and Desobry, S. 2006. Flavour encapsulation and controlled release—A review. *Int. J. Food Sci. Technol.* 41:1–21.

Market Research. 2004. Global Market Review of Functional Foods—Forecasts to 2010. http://www.marketresearch.com, accessed July 20, 2008.

Martinsen, A., Skjak-Braek, C., and Smidsrod, O. 1989. Alginate as immobilization material. I. Correlation between chemical and physical properties of alginate gel beads. *Biotechnol. Bioeng.* 33:79–89.

Mason, T.G., Graves, S.M., Wilking, J.N., and Lin, M.Y. 2006a. Extreme emulsification: Formation and structure of nanoemulsions. *Condens. Matter Phys.* 9:193–199.

Mason, T.G., Wilking, J.N., Meleson, K., Chang, C.B., and Graves, S.M. 2006b. Nanoemulsions: Formation, structure, and physical properties. *J. Phys.: Condens. Matter* 18:635–666.

Masters, K. 1985. *Spray Drying Handbook*. New York: Halsted Press.

Matsuno, R. and Adachi, S. 1993. Lipid encapsulation technology—Techniques and applications to food. *Trends Food Sci. Technol.* 4:256–261.

Mattila-Sandholm, T., Myllarinen, P., Crittenden, R., Mogensen, G., Fondén, R., and Saarela, M. 2002. Technological challenges for future probiotic foods. *Int. Dairy J.* 12:173–182.

McClements, D.J. and Decker, E.A. 2000. Lipid oxidation in oil-in-water emulsions: Impact of molecular environment on chemical reactions in heterogeneous food systems. *J. Food Sci.* 65:1270–1282.

McClements, D.J., Decker, E.A., and Weiss, J. 2007. Emulsion-based delivery systems for lipophylic bioactive components. *J. Food Sci.* 72:109–124.

McMaster, L.D., Kokott, S.A., Reid, S.J., and Abratt, V. 2005. Use of traditional African fermented beverages as delivery vehicles for *Bifidobacterium lactis* DSM 10140. *Int. J. Food Microbiol.* 102:231–237.

McNulty, H.P., Byun, J., Lockwood, S.F., Jacob, R.F., and Mason, R.P. 2007. Differential effects of carotenoids on lipid peroxidation due to membrane interactions: X-ray diffraction analysis. *Biochim. Biophys. Acta* 1768:167–174.

Meleson, K., Graves, S., and Mason, T.G. 2004. Formation of concentrated nanoemulsions by extreme shear. *Soft Mat*er 2:109–123.

Moberger, L., Larsson, K., Buchheim, W., and Timmen, H. 1987. A study on fat oxidation in a microemulsion system. *J. Disper. Sci. Technol.* 8:207–215.

Morales, D., Gutierrez, J.M., Garcia-Celma, M.J., and Solans, Y.C. 2003. A study of the relation between bicontinuous micro emulsions and oil/water nano-emulsion formation. *Langmuir* 19: 7196–7200.

Mosca, M., Ceglie, A., and Ambrosone, L. 2008. Antioxidant dispersions in emulsified olive oils. *Food Res. Int.* 41:201–207.

Muschiolik, G. 2007. Multiple emulsions for food use. *Curr. Opin. Colloid Interface Sci.* 12:213–220.

Muthukumarasamy, P., Allan-Wojtas, P., and Holley, R.A. 2006. Stability of *Lactobacillus reuteri* in different types of microcapsules. *J. Food Sci.* 71:20–24.

Nguyen, N.T. and Zhigang, W. 2005. Micromixers—A review. *J. Micromech. Microeng.* 15:R1–R6.

Oh, J.K., Drumright, R., Siegwart, D.J., and Matyjaszewski, K. 2008. The development of microgels/nanogels for drug delivery applications. *Prog. Polym. Sci.* 33:448–477.

Oliveira, A.C., Moretti, T.S., Boschini, C., Baliero, J.C.C., Freitas, O., and Favaro-Trindade, C.S. 2007. Stability of microencapsulated B lactis (Bl 01) and L acidophilus (LAC 4) by complex coacervation followed by spray drying. *J. Microencapsul.* 24:685–693.

Oneda, F. and Ré, M.I. 2003. The effect of formulation variables on the dissolution and physical properties of spray-dried microspheres containing organic salts. *Powder Technol.* 130:377–384.

O'Riordan, K., Andrews, D., Buckle, K., and Conway, P. 2001. Evaluation of microencapsulation of a *Bifidobacterium* strain with starch as an approach to prolonging viability during storage. *J. Appl. Microbiol.* 91:1059–1066.

Papadimitriou, V., Pispas, V., Syriou, S. et al. 2008. Biocompatible microemulsions based on limonene: Formulation, structure, and applications. *Langmuir* 24:3380–3386.

Parada, J. and Aguilera, J.M. 2007. Food microstructure affects the bioavailability of several nutrients. *J. Food Sci R: Conc. Rev. Hypoth. Food Sci.* 72:21–32.

Patel, N., Schmid, U., and Lawrence, M. J. 2006. Phospholipid-based microemulsions suitable for use in foods. *J. Agric. Food Chem.* 54:7817–7824.

Pérez-Alonso, C., Báez-González, J.G., Beristain, C.I., Vernon-Carter, E.J., and Vizcarra-Mendonza, M.G. 2003. Estimation of the activation energy of carbohydrate polymers blends as selection criteria for their use as wall material for spray-dried microcapsules. *Carbohydr. Polym.* 53:197–203.

Picon, A., Serrano, C., Gaya, P., Medina, M., and Nunhez, M. 1996. The effect of liposome-encapsulated cyprosins on manchego cheese ripening. *J. Dairy Sci.* 79:1694–1705.

Picot, A. and Lacroix, C. 2004. Encapsulation of bifidobacteria in whey protein-based microcapsules and survival in simulated gastrointestinal conditions and in yoghurt. *Int. Dairy J.* 14:505–515.

Poncelet, D. and Dreffier, C. 2007. Les methods de microencapsulation de A à Z (ou presque). *In Microencapsulation: Des sciences aux technologies*, eds. T. Vandamme, D. Poncelet, and P. Subra-Paternault. Paris, France: Tec&Doc (Editions), pp. 23–33.

Poncelet, D., Lencki, R., Beaulieu, C., Halle, J.P., Neufeld, R.J., and Fournier, A. 1992. Production of alginate beads by emulsification internal gelation. I. Methodology. *Appl. Microbiol. Biotechnol.* 38:39–45.

Pons, R., Carrera, I., Caelles, J., Rouch, J., and Panizza, P. 2003. Formation and properties of miniemulsions formed by microemulsions dilution. *Adv. Colloid Interface Sci.* 106:129–146.

Porras, M., Solans, C., Gonzalez, C., Martinez, A., Guinart, A., and Gutierrez, J.M. 2004. Studies of formation of W/O nano-emulsions. *Colloids Surf. A* 249:115–118.

Prado, F.C., Parada, J.L., Pandey, A., and Soccol, C.R. 2008. Trends in non-dairy probiotic beverages. *Food Res. Int.* 41:111–123.

Rane, S.S. and Anderson, B.D. 2008. What determines drug solubility in lipid vehicles: Is it predictable? *Adv. Drug Deliv. Rev.* 60:638–656.

Ré, M.I. 1998. Microencapsulation by spray drying. *Drying Technol.* 16:1195–1236.

Ré, M.I. 2006. Formulating drug delivery systems by spray drying. *Drying Technol.* 24:433–446.

Ré, M.I., Messias, L.S., and Schettini, H. 2004. The influence of the liquid properties and the atomizing conditions on the physical characteristics of the spray-dried ferrous sulfate microparticles. Paper presented at the *Annual International Drying Symposium*, Campinas, Brazil, August 2004, 1174–1181.

Reid, A.A., Champagne, C.P., Gardner, N., Fustier, P., and Vuillemard, J.C. 2007. Survival in food systems of *Lactobacillus rhamnosus* R011 microentrapped in whey protein gel particles. *J. Food Sci.* 72:31–37.

Reineccius, G.A. 1995. Liposomes for controlled release in the food industry. In *Encapsulation and Controlled Release of Food Ingredients, American Chemical Society Symposium Series* 590, eds. S.J. Risch and G.A. Reineccius. Washington, DC: American Chemical Society, pp. 113–131.

Ribeiro, H.S., Ax, K., and Schubert, H. 2003. Stability of lycopene emulsions in food systems. *J. Food Sci.* 68:2730–2734.

Ribeiro, H.S., Guerrero, J.M.M., Briviba, K., Rechkemmer, G., Schuchmann, H.P., and Schubert, H. 2006. Cellular uptake of carotenoid-loaded oil-in-water emulsion in colon carcinoma cell in vitro. *J. Agric. Food Chem.* 54:9366–9369.

Rodriguez-Huezo, M.E., Duran-Lugo, R., Prado-Barragan, L.A. et al. 2007. Pre-selection of protective colloids for enhanced viability of *Bifidobacterium bifidum* following spray-drying and storage, and evaluation of aguamiel as thermoprotective prebiotic. *Food Res. Int.* 40:1299–1306.

Roger, J.A. and Anderson, K.E. 1998. The potential of liposomes in oral drug delivery. *Crit. Rev. Ther. Drug. Carrier Syst.* 15:421–481.

Rohloff, C.M., Shimek, J.W., and Dungan, S.R. 2003. Effect of added alpha-lactalbumin protein on the phase behavior of AOT-brine-isooctane systems. *J. Colloid Interface Sci.* 261:514–523.

Rowland, M. and Towzer, T.N. 1995. *Clinical Pharmacokinetics: Concepts and Applications*, 3rd edn. Baltimore, MD: Williams & Williams.

Rusli, J.K., Sanguansri, L., and Augustin, M.A. 2006. Stabilization of oils by microencapsulation with heated protein-glucose syrup mixtures. *J. Am. Oil Chem. Soc.* 83:965–997.

Santana, M.H.A., Martins, F., and Pastore, G.M. 2005. Processes for stabilization of bioflavors through encapsulation in cyclodextrins and liposomes. *Brazilian Patent Application*, PI 0403279-9 A, filed Dez. 2005 (in Portuguese).

Santipanichwong, R. and Suphantharika, M. 2007. Carotenoids as colorants in reduced-fat mayonnaise containing spent brewer's yeast beta-glucan as fat replacer. *Food Hydrocoll.* 21:565–574.

Schrooyen, P.M.M., van der Meer, R., and De Krif, C.G. 2001. Microencapsulation: Its application in nutrition. *Proc. Nutr. Soc.* 60:475–479.

Shah, N.P. and Ravula, R. 2000. Microencapsulation of probiotic bacteria and their survival in frozen fermented dairy desserts. *Aust. J. Dairy Technol.* 55:139–144.

Shahidi, F. and Han, X.Q. 1993. Encapsulation of food ingredients. *Crit. Rev. Food Sci. Nutr.* 33:501–547.

Shima, M., Morita, Y., Yamashita, M., and Adachi, S. 2006. Protection of *Lactobacillus acidophilus* from the low pH of a model gastric juice by incorporation in a W/O/W emulsion. *Food Hydrocoll.* 20:1164–1169.

Shu, B., Yu, W., Zhao, Y., and Liu, X. 2006. Study on microencapsulation of lycopene by spray-drying. *J. Food Eng.* 76:664–669.

Shuqin, X. and Shiying, X. 2005. Ferrous sulfate liposomes: Preparation, stability and application in fluid milk. *Food Res. Int.* 38:289–296.

Siró, I., Kálpona E., Kálpona B., and Lugasi A. 2008. Functional food. Product development, marketing and consumer acceptance—A Review. *Appetite* 51:456–467.

Socaciu, C., Bojarski, P., Aberle, L., and Diehl, H.A. 2002. Different ways to insert carotenoids into liposomes affect structure and dynamics of bilayer differently. *Biophys. Chem.* 99:1–15.

Socaciu, C., Jessel, R., and Diehl, H.A. 2000. Competitive carotenoid and cholesterol incorporation into liposomes: Effects on membrane phase transition, fluidity, polarity and anisotropy. *Chem. Phys. Lipids* 106:79–88.

Solans, C., Izquierdo, P., Nolla, J., Azemar, N., and Garcia-Celma, M.J. 2005. Nano-emulsions. *Curr. Opin. Colloid Interface Sci.* 10:102–110.

Spernath, A. and Aserin, A. 2006. Microemulsions as carriers for drugs and nutraceuticals. *Adv. Colloid Interface Sci.* 128–130:47–64.

Spernath, A., Yaghmur, A., Aserin, A., Hoffman, R.E., and Garti, N. 2002. Food-grade microemulsions based on nonionic emulsifiers: Media to enhance lycopene solubilization. *J. Agric. Food Chem.* 50:6917–6922.

Su, L.C., Lin, C.W., and Chen, M.J. 2007. Development of an oriental-style dairy product coagulated by microcapsules containing probiotics and filtrates from fermented rice. *Int. J. Dairy Technol.* 60:49–54.

Sujak, A., Strzalka, K., and Gruszecki, W.I. 2007. Thermotropic phase behaviour of lipid bilayers containing carotenoid pigment canthaxanthin: A differential scanning calorimetry study. *Chem. Phys. Lipids* 145:1–12.

Sultana, K., Godward, G., Reynolds, N., Arumugaswamy, R., Peiris, P., and Kailasapathy, K. 2000. Encapsulation of probiotic bacteria with alginate-starch and evaluation of survival in simulated gastrointestinal conditions and in yoghurt. *Int. J. Food Microbiol.* 62:47–55.

Tadros, T., Izquierdo, R., Esquena, J., and Solans, C. 2004. Formation and stability of nano-emulsions. *Adv. Colloid Interface Sci.* 108:303–318.

Tanford, C. 1980. *The Hydrophobic Effect: Formation of Micelles and Biological Membranes.* New York: Wiley-Interscience.

Thijssen, H.A.C. 1971. Flavour retention in drying preconcentrated food liquids, *J. Appl. Chem. Biotechnol.* 21:372–376.

Tournier, H., Schneider, M., and C. Guillot. 1999. Liposomes with enhanced entrapment capacity and their use in imaging. U.S. Patent 5,980,937, filed Aug. 12, 1997 and issued Nov. 9, 1999.

Uicich, R., Pizarro, F., Almeida, C. et al. 1999. Bioavailability of microencapsulated ferrous sulfate in fluid cow milk: Studies in human beings. *Nutr. Rev.* 19:893–897.

Urbanska, A.M., Bhathena, J., and Prakash, S. 2007. Live encapsulated *Lactobacillus acidophilus* cells in yogurt for therapeutic oral delivery: Preparation and in vitro analysis of alginate-chitosan microcapsules. *Can. J. Physiol. Pharmacol.* 85:884–893.

Vanderberg, G.W. and De La Noüe, J. 2001. Evaluation of protein release from chitosan-alginate microcapsules produced using external or internal gelation. *J. Microencapsul.* 18:433–441.

Vladisavljevic, G.T. and Williams, R.A. 2005. Recent developments in manufacturing emulsions and particulate products using membranes. *Adv. Colloid Interface Sci.* 113:1–20.

Vuillemard, J.-C. 1991. Recent advances in the large-scale production of lipid vesicles for use in food products: Microfluidization. *J. Microencapsul.* 8:547–562.

Wallach, D.F.H. and Mathur, R. 1992. Method of making oil filled paucilamellar lipid vesicles. U.S. Patent 5160669, filed Oct. 16, 1990, and issued Nov. 03, 1992.

Walther, B., Cramer, C., Tiemeyer, A. et al. 2005. Drop deformation dynamics and gel kinetics in a co-flowing water-in-oil system. *J. Colloid Interface Sci.* 286:378–386.

Wang, L., Li, X., Zhang, G., Dong, J., and Eastoe, J. 2007. Oil-in-water nanoemulsions for pesticide formulations. *J. Colloid Interface Sci.* 314:230–235.

Wang, L., Mutch, K.J., Eastoe, J., Heenan, R.K., and Dong, J. 2008. Nanoemulsions prepared by a two-step low-energy process. *Langmuir* 24:6092–6099.

Wang, L.Y., Ma, G.H., and Su, Z.G. 2005. Preparation of uniform sized chitosan microspheres by membrane emulsification technique and application as a carrier of protein drug. *J. Controlled Release*, 106:62–75.

Weiss, J., Kobow, K., and Muschiolik, G. 2004. Preparation of microgel particles using membrane emulsification. Abstracts of Food Colloids, Harrogate, U.K., B-34.

Weiss, J., Takhistov, P., and McClements, D.J. 2006. Functional materials in food nanotechnology. *J. Food Sci.* 71:107–116.

Were, L.M., Bruce, B.D., Davidson, P.M., and Weiss, J. 2003. Size, stability, and entrapment efficiency of phospholipid nanocapsules containing polypeptide antimicrobials. *J. Agric. Food Chem.* 51:8073–8079.

Wolf, P.A. and Havekotte, M.J. 1989. Microemulsions of oil in water and alcohol. U.S. Patent 4,835,002, filed Jul. 10, 1987, and issued May 30, 1989.

Xianghua, Y. and Zirong, X. 2006. The use of immunoliposome for nutrient target regulation (a review). *Crit. Rev. Food Sci. Nutr.* 46:629–638.

Xu, S., Nie, Z., Seo, M. et al. 2005. Generation of monodisperse particles by using microfluidics: Control over size, shape, and composition. *Angew. Chem. Int. Ed.* 44:724–728.

Yi, O.S., Han, D., and Shin, H.K. 1991. Synergistic antioxidative effects of tocopherol and ascorbic-acid in fish oil lecithin water-system. *J. Am. Oil Chem. Soc.* 68:881–883.

Yuan, Y., Gao, Y., Zhao, J., and Mao, L. 2008. Characterization and stability evaluation of beta-carotene nanoemulsions prepared by high pressure homogenization under various emulsifying conditions. *Food. Res. Int.* 41:61–68.

Zhang, H., Feng, F., Li, J. et al. 2008. Formulation of food-grade microemulsions with glycerol monolaurate: Effects of short-chain alcohols, polyols, salts and nonionic surfactants. *Eur. Food Res. Technol.* 226:613–619.

Zhang, L., Liu, J., Lu, Z., and Hu, J. 1997. Procedure for preparation of vesicles with no leakage from water-in-oil emulsion. *Chem. Lett.* 8:691–692.

Zheng, S., Alkan-Onyuksel, H., Beissinger, R.L., and Wasan, D.T. 1999. Liposome microencapsulations without using any organic solvent. *J. Disper. Sci. Technol.* 20:1189–1203.

Zhou, Q.Z., Wang, L.Y., Ma, G.H., and Su, Z.G. 2007. Preparation of uniform-sized agarose beads by microporous membrane emulsification technique. *J. Colloid Interface Sci.* 311:118–127.

8 Perspectives of Fluidized Bed Coating in the Food Industry

Frédéric Depypere, Jan G. Pieters, and Koen Dewettinck

CONTENTS

8.1 INTRODUCTION

Microencapsulation is a process in which a pure active ingredient or a mixture is coated with or entrapped within a protecting material or system. As a result, useful and otherwise unusual properties may be conferred to the microencapsulated ingredient, or unuseful properties may be eliminated from the original ingredient (Shahidi and Han 1993).

The microencapsulated ingredient is also referred to as core, active, ingredient, fill, encapsulant, internal phase, or payload. The encapsulating material can also be called capsule, wall, coat, coating, envelope, covering, membrane, carrier, or shell (Kanawjia et al. 1992, Gibbs et al. 1999).

Originally developed in sectors such as those involved in the production of carbonless carbon paper and the pharmaceutical industry, microencapsulation is now increasingly applied in the food industry to tune, time, or enhance the effect of functional ingredients and additives (Dewettinck and Huyghebaert 1999). In contrast to high cost tolerating encapsulation applications in the areas of pharmacy, cosmetics, or health, microencapsulation in the food industry should be considered a large volume operation whereby production costs have to be minimized. Although food ingredient microencapsulation was originally considered a rather high-priced custom route to solving unique problems, today's increased production volumes and well-developed, cost-effective preparation techniques and materials have resulted in a significant increase in the number of microencapsulated food products (DeZarn 1995).

Encapsulated ingredients are used in prepared foods, fortified foods, nutritional mixes, seasonings, fillings, desserts, dry mix puddings, teas, dry mix beverages, and dairy mixes. As can be seen from this list, microcapsules are generally constituents of a larger food system and must function within that system. Consequently, a number of performance requirements must be fulfilled by the microcapsule, taking into account the limited number of encapsulating materials accepted for food applications and feasible microencapsulation methods for the industry (Versic 1988).

Table 8.1 summarizes the variety of constituents that have successfully been encapsulated so far. Additionally, it is possible to incorporate more than one food ingredient or additive in one microcapsule. As can be observed from Table 8.1, a host of food core materials can be transformed into microcapsules provided the right encapsulation process and encapsulating material are selected and appropriate manufacturing conditions are applied.

Conversely, the number of materials that can be applied to realize microencapsulation in the food industry is rather limited owing to the constraint of selecting the encapsulating material from a list of U.S. Food and Drug Administration (FDA)-approved, generally recognized as safe (GRAS) materials (Kanawjia et al. 1992). However, the FDA continuously adds new GRAS materials to the list, allowing

TABLE 8.1
Encapsulated Ingredient and Additive Types in the Food Industry (Not Comprehensive)

Acidulants	Enzymes	Microorganisms	Spices
Amino acids	Flavors	Minerals	Sweeteners
Antioxidants	Gases	Preservatives	Vitamins
Colorants	Leavening agents	Redox agents	Water
Dough conditioners	Lipids and oils	Salts	Yeast

TABLE 8.2
Applicable Food Coating Materials (Not Comprehensive)

Category	Examples
Proteins	Caseinate, albumin, gelatine, soy, gluten, zein[a]
Polysaccharides–hydrocolloids	(modified) starches, maltodextrins, β-cyclodextrins, alginate, gum arabic, pectin/polypectate, carrageenan, agarose
Cellulose derivatives	Methylcellulose, ethylcellulose[a], carboxymethylcellulose (CbMC)
Fats and fatty acids	(hydrogenated) vegetable oils[a], mono-/di-/tri-glycerides[a]
Waxes	Shellac[a], beeswax[a]

[a] Water-insoluble coatings.

researchers to revisit unsolved problems of the past. Table 8.2 lists some important examples of coating materials for microencapsulation in the food industry, including a distinction between water-soluble and water-insoluble coatings.

Microencapsulation can be applied for a variety of reasons, but the general underlying philosophy is always to add value to a conventional food ingredient or additive. Furthermore, it is also the source of totally new ingredients with matchless properties (Gouin 2004). Further details on the principles and possible benefits of microencapsulation, as well as on the release mechanisms of encapsulated ingredients, can be found in Chapter 7. The referred chapter also addresses the variety of microencapsulation processes that have been developed, modified, and improved over the years to adapt the microencapsulation process to different objectives.

Specifically for the food powder industry, it is acknowledged that spray drying (SD) and fluidized bed coating (FBC) are the most common microencapsulation methods (Janovsky 1993). As stated above, food products are ruled by stringent regulations of safety and cost, limiting the number of coating materials that can be used for microencapsulation. The strict cost regulations are also the reason why expensive microencapsulation techniques cannot be applied. This explains why the majority of food products are encapsulated by SD or FBC in large batch or continuous fluidized bed (FB) systems.

8.2 FLUIDIZED BED COATING

8.2.1 Basic Principles

Film coating techniques are characterized by the deposition of a uniform film onto the surface of a substrate. In the production of coated particles, the rotating pan and the FB are most frequently used.

In the rotating pan, the coating solution is sprayed onto particles, set in motion by rotating the drum or pan, and dried by the supply of hot gas. This method has been in use for a long time and has been modified with several improvements, the most recent being focused on fluidization of particles inside the drum (Litster and Sarwono 1996). However, its reproducibility is shown to be rather low, making this technology suitable only for producing particles for which no high coating uniformity is required.

Conversely, the FB technology is recognized for its good mixing and optimal heat and mass transfer, which are related to the bubble characteristics of the fluidizing air (Senadeera et al. 2000). One of the main reasons for the success of this technology in the food industry is that an FB allows a large number of unit operations to be performed within the same piece of equipment, either separately or sequentially.

Originally developed for drying, the FB technology is widely used in typical food processing applications such as coating, agglomeration and granulation, SD, explosion puffing, freezing and cooling, freeze drying, classification, blanching, and cooking (Shilton and Niranjan 1993, Becher and Schlünder 1997).

Also termed air suspension coating, FBC is accomplished by suspending solid core particles in an upward-moving air current, which serves both as a heating or cooling medium and as a momentum carrier. The bed of particles is supported by drag forces, exerted by the air flow, and acquires the characteristics of a boiling liquid; hence the term fluidization. Suspending each particle, thereby exposing the entire particle surface area to the air stream, results in optimal convective heat transfer.

The coating, which may be dissolved in a volatile solvent or applied in a molten state, is atomized through nozzles into the coating chamber. Owing to fast and effective evaporation of the coating liquid, the coating material is deposited as a thin layer on the surface of the suspended particles. Taking advantage of the evaporative efficiency of the fluid bed, it is even possible to coat water-sensitive products with aqueous coating materials (Jones 1988).

Depending on the configuration, the solution can be sprayed from the top (top-spray configuration), from the bottom of the FB (bottom-spray), or from tangentially positioned nozzles submerged inside the particle bed (tangential- or rotary-spray). As each method has a different effect on the final film quality, they are discussed separately. However, the underlying coating principle is the same for each option, as shown in Figure 8.1.

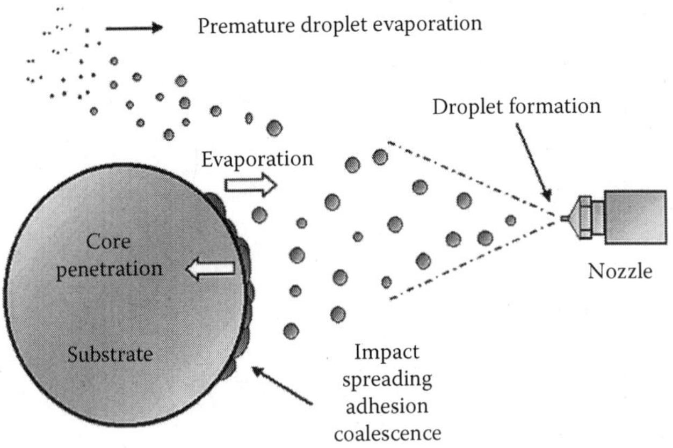

FIGURE 8.1 Principles of the FBC process.

The working zone of a FBC unit is the coating zone around the atomization nozzle (DeZarn 1995). With every particle passing through the coating zone, coating material is applied, until the desired surface coverage is reached. Droplet formation, impact, spreading, adhesion, coalescence, and evaporation, as illustrated in Figure 8.1, are occurring almost simultaneously during the process (Guignon et al. 2002). When dealing with porous particles, penetration of the core by water or solvents should be avoided. This can be accomplished using the evaporative efficiency of a FB, although care should be taken to avoid premature evaporation of the coating solution.

When the coating solution droplets do not manage to reach and attach to the surface of a fluidized core particle (e.g., when the traveling distance of the droplets toward the particles is too long), they are subject to premature evaporation and drying, in a process similar to the one observed in the SD operation. Although light spray-dried particles are entrained by the air stream and removed from the FB by elutriation, the heavier fines remain inside the bed. The latter may collide with other spray-dried particles (fines agglomeration) or may be captured by the fluidized core particles, causing surface imperfections and eventually leading to ball growth (Nienow 1995, Maronga 1998). These side-effect phenomena are shown in Figure 8.2.

If, however, the droplets of the atomized coating solution manage to collide and successfully adhere to the core particles, the latter become wetted. Depending on the inlet air temperature and absolute humidity, the core particles may be individually coated and the coating layered uniformly, or, in the case of excessive wetting, liquid bridges are formed between the core particles, resulting in the formation of large, wet clumps, and eventually, collapse and defluidization of the bed.

Besides wet quenching, dry quenching can occur with moderately wetted core particles. This happens when the adhesion forces between many connected individual particles are too strong. When a small number of particles are joined by solid bridges, agglomeration is likely to occur (Nienow 1995, Maronga 1998). In Figure 8.3, the different stages of coating and agglomeration processes are illustrated.

Finally, Maronga (1998) also describes the possibility of fragmentation of dried liquid bridges with weak adhesive forces, leading to particle coating.

From these possible processes, it is obvious that efficient process control, leading to particle coating and avoiding the above side-effects, is of crucial importance.

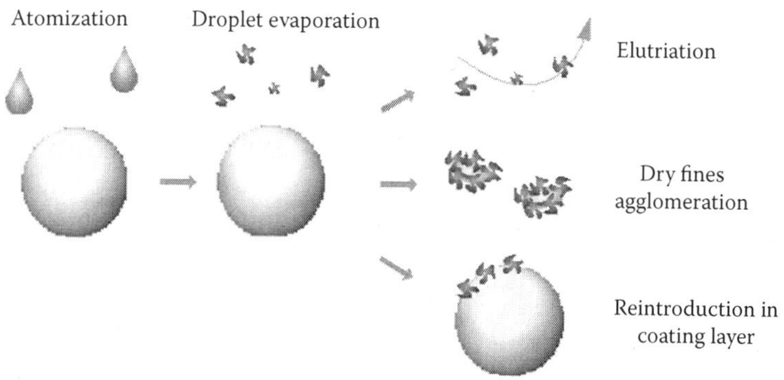

FIGURE 8.2 Coating solution spray-drying side-effect.

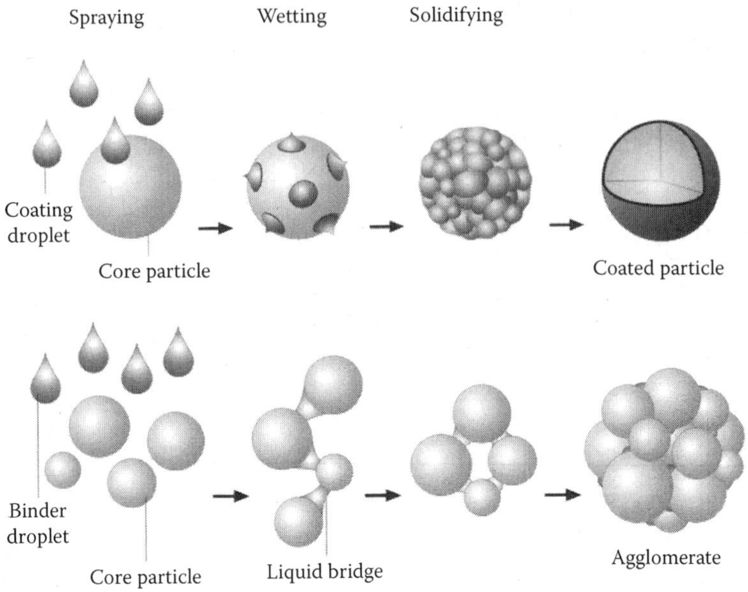

FIGURE 8.3 Stages of coating and agglomeration processes. (Modified from Glatt GmbH, Germany).

8.2.2 Batch Fluidized Bed Coating

The principle and possible applications of the three basic processing options of the batch FB are discussed below.

Particles, as small as 100 µm, have been coated using top-spray FBC (Jones 1994). Attempts have been made to coat smaller particles, but agglomeration was found to be almost unavoidable because of nozzle limitations and the tackiness of most coating materials. Moreover, these small particles tend to be carried away in the exhaust air (Jackson and Lee 1991). The bottom-spray system has also been used successfully to coat particles as small as 100 µm and attempts to coat smaller particles may lead to the same difficulties as encountered in the top-spray configuration. Finally, the tangential-spray FB has been used to coat particles of at least 250 µm (Jones 1994). Compared with the previous configurations, this design is more susceptible to adhesion of particles on the upper product container wall owing to static electricity; hence, coating of smaller and lighter particles with this system is discouraged.

8.2.2.1 Top-Spray Fluidized Bed Coating

Figure 8.4 shows a schematic representation of the top-spray FB configuration. Both the top- and bottom-spray methods are characterized by a conical product container and expansion chamber, resulting in a more vigorous fluidization pattern and a decrease in velocity as the particles move upward in the expansion chamber.

In a top-spray FB, air is introduced through a uniform air distribution plate (also called "distributor" or "grid plate") and the resulting fluidization pattern has long been denoted as random and unrestricted (Jones 1994). However, using particle

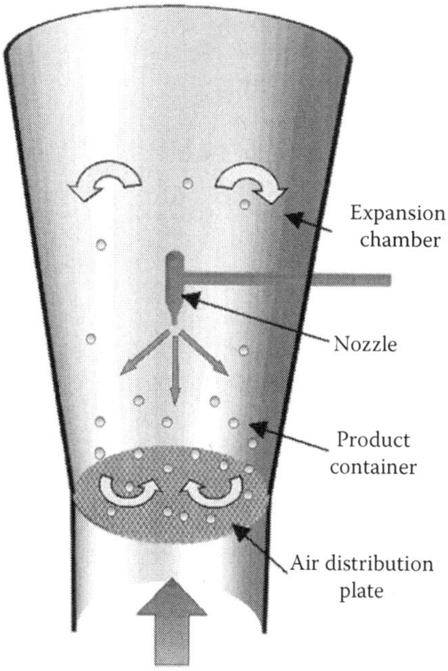

FIGURE 8.4 Principle of top-spray FB processing.

tracking, Depypere (2005) have recently shown that, overall, particles follow a clear circulation pattern: upwards in the center and downwards along the walls. Particular to the top-spray method, the nozzle is positioned above the particle bed, and the coating liquid is sprayed into the fluidized core particles, countercurrently to the air flow. In this configuration, controlling each droplet's travel distance and time before impinging on a core particle is impossible, possibly leading to quite severe premature evaporation of the coating liquid. Although the resulting imperfect coating is unsuitable for sustained or controlled release purposes, satisfactory results may be obtained for taste masking.

With top-spray coating, the use of any organic solvent is discouraged. However, it is the system of choice when lipid coating or hot-melt coating is envisaged (Rácz et al. 1997, Achanta et al. 2001). Using heated and insulated nozzles, molten materials such as hydrogenated oils and waxes are sprayed against a stream of cool air, causing the molten wall material to solidify around the core particle. The degree of protection offered by the coating, and hence the coating quality, is related to the application rate and the congealing rate. A product bed temperature that is too low results in premature congealing prior to complete spreading and, consequently, pores and defects can occur in the coating. Conversely, if the product bed temperature is too close to the melting point of the coating, the result is a significant increase in the viscous drag of the bed, favoring particle–particle agglomeration (Joszwiakowski et al. 1990). The use of cold rather than heated air and the application of 100% coating material result in short processing time and low energy consumption, making this process economically feasible for many food applications (Sinchaipanid et al. 2004).

8.2.2.2 Bottom-Spray Fluidized Bed Coating

Bottom-spray film coating is accomplished by means of the "Wurster" system, originally developed by Wurster (1959), and it differs in many aspects from the top-spray configuration. A scheme of the bottom-spray FB is given in Figure 8.5.

An open-ended cylindrical inner partition, for which the distance above the air distribution plate (partition height) can be varied, divides the product container into an inner area or high-velocity zone and an outer area, also called annulus or down-bed section.

Compared with the top-spray configuration, the Wurster-based FBC system does not contain any simple FB regions in the traditional sense. Several researchers recognize four distinct regions, each with different controlling parameters (Christensen and Bertelsen 1997).

In the bottom-spray configuration, the number and the diameter of the holes of the air distribution plate are different between the inner and outer sections, and so most free area is provided in the center section, just below the partition. In this way, a high-velocity zone is created inside the partition, separating the core particles and transporting them past the nozzle, which is mounted at the bottom and sprays the coating liquid upwards. After passing the nozzle, the coated particles enter the expanded area, slow down and fall back into the outer section of the product container, while continuously being dried by the upwardly flowing air. The particles in the down-bed region remain sufficiently fluidized to allow them to continuously move toward the distribution plate and to enter the horizontal transport region, where they are drawn back into the high-velocity air stream. This cycling process continues until the desired coating level has been achieved.

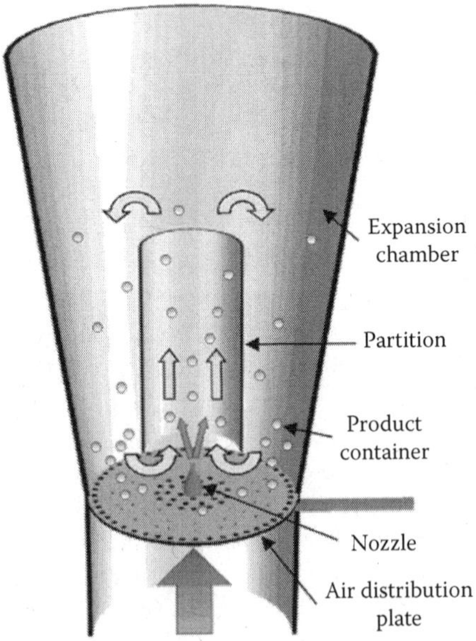

FIGURE 8.5 Principle of bottom-spray ("Wurster") FB processing.

Compared with that from the top-spray configuration, the resulting coating film is very uniformly applied, which makes bottom-spray FBC an excellent tool to apply reproducible, controlled release protection barriers. It is the system of choice when enteric coating systems are envisaged (Singiser and Lowenthal 1961, Mehta et al. 1986). The superior coating quality obtained is a result of three essential features:

- The spraying is effected concurrently, i.e., from below the product bed into the direction of the core particle movement and process air flow, making the distance between the spray nozzle and the core particles extremely short.
- Efficient heat exchange is allowed, as very high particle separation forces are generated in the inner partition.
- The core particles follow a reproducible repetitive cycling motion from the inner partition to the down-bed region and back, so they are exposed at very regular time intervals to a uniform spray with droplets of uniform spreadability, leading to a coating film of uniform thickness (Eichler 1989).

Recently, the Wurster high-speed processing insert, patented by Jones (1995), was introduced with the aim of offering larger batch volumes, higher throughputs, and the ability to coat finer particles down to 50 μm, without the prevalence of agglomeration. The modification consisted of shielding the spray nozzle by mounting a second cylindrical partition, connected with the air distribution plate, inside the inner partition area (Jones 1994). The upper end of the inner tubular partition prevents the premature entrance of particles into the spray nozzle area and shields the initial spray pattern until the spray pattern has fully developed. This allows the droplet density to decrease upon contact with the particles of the FB, resulting in a more evenly wetted particle surface and thereby preventing excessive particle agglomeration. Accordingly, higher spray rates can be achieved, leading to a decrease in overall coating process time.

A relatively new spraying technology in the Wurster configuration is dry powder coating, whereby fine particles together with a plasticizer are sprayed onto core particles. This technology allows coating of the core particles without the use of heated air. In this way, heat-sensitive materials can be processed in a convenient way. Other advantages are the low energy requirements and the short processing times (Ivanova et al. 2005). Following the coating step, the microcapsules are often heat-treated to allow better coalescence of the fine particles on the core particles (Pearnchob and Bodmeier 2003).

8.2.2.3 Tangential-Spray (Rotor) Fluidized Bed Coating

The latest developed configuration in FBC is the tangential-spray or rotary-spray set-up, depicted in Figure 8.6. Unlike the two previous configurations, the product container and expansion chamber are cylindrically shaped and the air distribution plate is replaced with a rotating disk with adjustable height.

As a result, three forces determine the fluidization pattern, best described as a spiraling helix. The combination of centrifugal forces, an upward fluidizing air flow, and gravity effects results in a rapidly tumbling FB. Tangentially immersed in the powder bed, a nozzle is positioned to spray the coating liquid concurrently with the

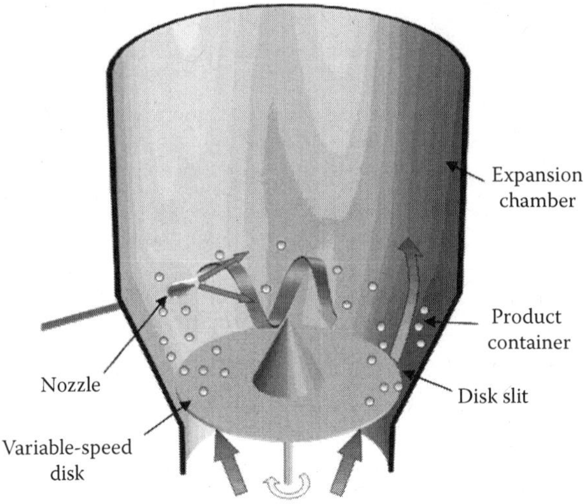

FIGURE 8.6 Principle of tangential-spray (rotary-spray) FB processing.

particle flow. As the three main physical criteria are the same, the film coating performance of the tangential-spray system can be compared with that of the bottom-spray configuration. It can be expected that a rotor-applied film has at least the same quality as a Wurster-applied film. However, its use with friable substrates is discouraged, since compared with the other systems, the tangential-spray system exerts the greatest mechanical stress (Jones 1988).

Alternatively, in a patent by Jones et al. (1992), the use of a rotor-spray FB for layering powder onto core particles is described. Compared with an original tangential-spray FB system, the coating liquid delivery system is characterized by an annular powder outlet encompassing the coating liquid outlet. In this way, the coating liquid and powder can be sprayed simultaneously onto the core particles. To accomplish immobilization of the applied powder on the substrate surface by liquid bridges, the rotating disk slit width is usually narrow and the heating air volume and temperature are relatively low.

8.2.2.4 Design Modifications and Possibilities for the Food Industry

During recent decades, a number of design modifications on a variety of the process components have been adopted to further improve the coating process. These range from a turning bottom discharge system, which is the most effective discharge method in batch configurations, to advanced dynamic filter systems, which continuously recycle product particles into the process area.

In a patent by Hüttlin (1990), the conventional air distribution plate is replaced with a series of slanted plates, directing the incoming air into a controlled toroidal motion pattern. Concurrent with the resulting air flow, built-in three-component spray nozzles deliver the coating solution directly into the fluidized product, ensuring that the liquid is delivered at the place of highest product velocity. The optimized nozzle design with a microclimate around the nozzle tip eliminates caking and blockage, thereby increasing efficiency and minimizing the need for nozzle maintenance.

TABLE 8.3
Comparison of Batch FB Configurations

Parameter	Top-Spray	Bottom-Spray	Tangential-Spray
Air distribution plate	Homogeneous	Heterogeneous	N/A
Spray modus	Countercurrent	Concurrent	Concurrent
Nozzle position relative to particle bed	Above to top submerged	Bottom submerged	Tangentially submerged
Fluidization pattern	Least controlled, irregular circulation	Controlled, regular coating intervals	Controlled, spiraling helix
Min. particle size (μm)	100	100 (50)[a]	250
Max. batch size (kg)	1500	600	500
Coating quality	Porous to dense, imperfect	Uniform, excellent	Dense, excellent
Advantages	Simplicity, cost efficiency	Wide application range	Spheronization
Disadvantages	Spray-drying losses	Tedious set up and cleaning	Unsuitable for friable substrates
Aqueous particle coating	Good	Superior	Excellent
Hot-melt coating	Superior	Good	Can be done
Powder layering	Not recommended	Can be done	Superior

[a] High-speed Wurster.

Table 8.3 summarizes the main characteristics of the three batch FB configurations. Evaluating the advantages and disadvantages, it can be concluded that the aqueous particle coating quality resulting from bottom- and tangential-spray systems exceeds the performance of the top-spray system. However, it should be emphasized that a universal definition of coating quality does not exist, as its judgment depends heavily on the intended application (Finch 1993). For specific applications, the microcapsule surface may not need to be completely covered to obtain the desired coating quality (Rümpler and Jacob 1998). Because of the specific needs and constraints of the food industry, and taking into account the high versatility, relatively large batch size and relative simplicity of the conventional top-spray configuration, Dewettinck and Huyghebaert (1999) felt that this system had the greatest possibility of being introduced successfully in the food industry.

8.2.3 Continuous Fluidized Bed Coating

A recent engineering novelty in FB technology is the development of continuous systems, with little down time and operating times up to 8000 h/year. Particularly for the food industry, in which large volume throughputs and lower costs are required, the introduction of a continuous coating process offers considerable advantages and opportunities (Rümpler and Jacob 1998). For example, scale-up problems, an important issue in batch systems (Jones 1985, Leuenberger 2001, Knowlton et al. 2005), are avoided.

Figure 8.7 shows a schematic diagram of a continuous top-spray FB coater. The core material is introduced at one end of the unit and the processed product is discharged continuously at the other end of the bed. The particle mixing behavior is dependent on the product throughput and on the geometric design of the FB. Compared with a nearly ideal mixing behavior in batch systems, the residence time distribution of the particles inside the processing unit is a very important characteristic for continuous FBC systems (Teunou and Poncelet 2002). The division of the inlet air plenum into multiple chambers allows the use of different inlet air temperatures or velocities along the FB. After the transport of the core particles by the dosing unit, they are conveyed within the device by means of the fluidizing air alone. As specially designed top-spray or bottom-spray nozzles can be positioned, where appropriate for a particular application, it is possible to realize different unit operations within one continuous FB unit, for instance coating in the first chamber, followed by drying and cooling in subsequent parts of the processing chamber. As the process can be equipped for handling solvent or lipid coating materials and as the amount of fluid sprayed from the nozzle and the distance between the nozzles can be adjusted, the continuous system can be adapted to specific requirements, while assuring proper covering capability and maintaining the flexibility needed for large product ranges (Dewettinck and Huyghebaert 1999).

A similar continuous multi-cell particle coating process is described in a patent by Liborius (1997). In this application, particles are pneumatically conveyed through the coating application while controlled recirculation of particles occurs within each cell, and a controlled spray is applied to the core particles in order to produce a substantially uniform coating distribution.

One of the latest developments in continuous powder coating is the spouted bed (SB) technology (e.g., Glatt ProCell®*). In Figure 8.8, a classical SB is compared with the Glatt ProCell configuration. In classical SB systems (Mathur and Epstein 1974), the spouting air is centrally introduced through a nozzle, rather than

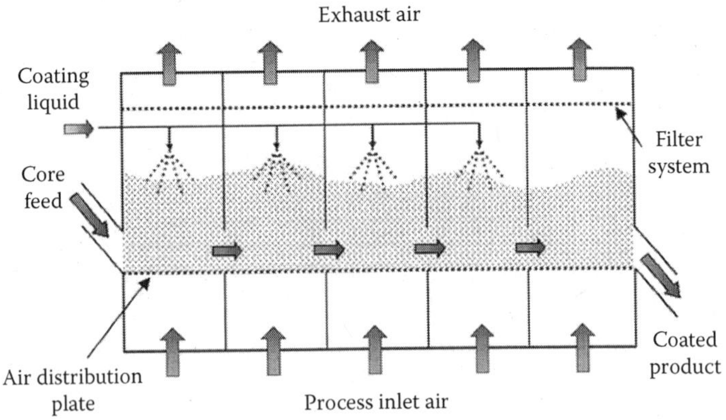

FIGURE 8.7 Continuous top-spray FBC.

* Registered trademark of Glatt Ingenieur Technik-GmbH, Weimar, Germany.

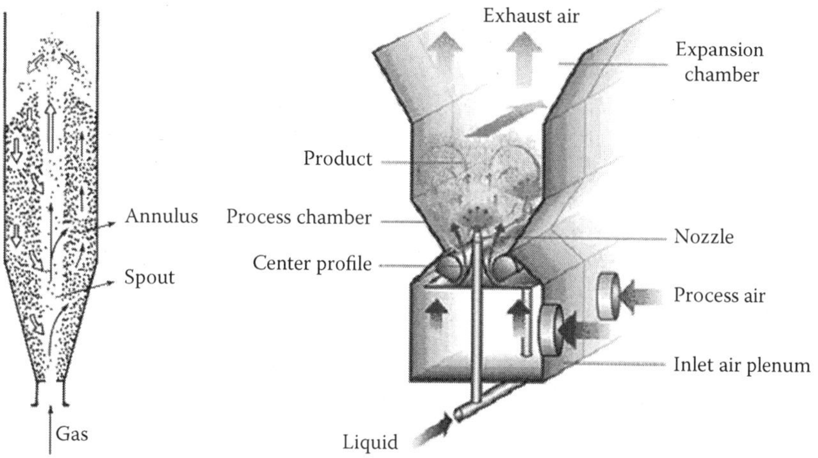

FIGURE 8.8 Comparison between a classical SB (left) and the Glatt ProCell SB (right). (Left: Modified from Kunii, D. and Levenspiel, O., *Fluidization Engineering*, Butterworth-Heinemann, Boston, MA, 1991. Right: Modified from Glatt GmbH, Germany.)

uniformly through a distributor plate as in FBs. The particle flow and the spouting air are concurrent in the spout and countercurrent in the annulus, which makes up the major portion of the SB. Although spouting can be beneficially achieved with particles larger than 1 mm, significant progress has been made in the spouting of finer particles. An example of this progress is the Glatt ProCell technology, in which the spouting air enters the process chamber through slots in the side wall. Owing to a significant increase in the cross section of the process chamber toward the top, a sharp decrease of the spouting velocity and a controlled airflow pattern are obtained. Coating can be accomplished by nozzles positioned in the top- or the bottom-spray configuration. Among the reported advantages of this technology are:

- The possibility to handle fine (down to 50 μm) and sticky particles
- Gentle drying of temperature-sensitive products
- Process convenience (e.g., short processing times, easy cleaning, easy access)
- Excellent coating quality

8.3 ISSUES AND PROBLEMS IN FOOD POWDER COATING TECHNOLOGY

8.3.1 Process and Coating Material Selection

Although firmly grounded in research, microencapsulation, and more specifically FBC technology, is sometimes considered more as art than science.

The selection of an appropriate coating material and the most efficient and cost-effective coating process for a given core material with its intended use, stability and storage requires extensive scientific knowledge of all applicable materials and processes involved, as well as a good estimation of the behavior of materials under a variety of conditions. As in the decision making process for many other industrial applications, economic considerations must be balanced against microcapsule quality and performance requirements (Balassa and Fanger 1971).

From the previous discussion, it should be clear that there is not a single method to successfully encapsulate all kinds of possible core materials. Considering only FBC, the choice remains between batch, semi-continuous, and continuous applications. According to the choice of core and coating material, solvent or lipid-based microencapsulation will be involved. Some general aspects dealing with process and material selection are pointed out in the following paragraphs.

Before considering the desired microcapsule properties, the aim of microencapsulation should be clearly defined. The process conditions are substantially different when envisaging improvement of powder handling characteristics such as dust reduction and increase in free-flowing characteristics on the one hand and reproducible controlled release on the other (Dziezak 1988).

The objectives of the coating operation are critical for the choice of a suitable coating material and microcapsule type (reservoir or matrix). A coating material that is nonreactive with both the core and the food system to which the microcapsules will be added should be chosen.

Furthermore, a compromise must be made between the coating solution composition (e.g., plasticizer, solvent, or binder addition), the legislation, and the process operations (Teunou and Poncelet 2002). Care should be taken that the microcapsule is tailored to the processing conditions it is subject to, to eliminate premature release of its contents.

Also, the desired mechanism of release is to a great extent a determinant of the choice of a proper coating wall material. Core release should be studied and optimized against application parameters such as pH, temperature, and pressure.

The overall cost of polymer coat and coating process should be justified in terms of improved performance (Arshady 1993). As a rule of thumb, Spooner (1994) argued more than 20 years ago that microencapsulation at least doubles the cost of any product. On the other hand, in a more recent review by Teunou and Poncelet (2002), a significant reduction in coating operation cost is observed when moving from a batch Wurster to a continuous FB. Although it may be more appropriate to compare a similar FB spray configuration (batch versus continuous top-spray), it is to be expected that costs may decrease owing to increased process volumes and more effective encapsulation techniques. Additionally, cost savings derived from increased productivity, improved yields, extended product shelf life, and more consistent quality products should be taken into account (Spooner 1994). A more updated rule of thumb was recently given by Gouin (2004), who approached the economic aspect of microencapsulated foodstuffs from a customer point of view. Bearing in mind that functional ingredients are used at low levels (1%–5%) in foodstuffs, the maximum cost for a microencapsulation process in the food industry is roughly estimated at €0.1 per kg food product. Although FB encapsulation is not inexpensive, its adaptability makes it cost-effective for many ingredient applications.

To facilitate the selection of a suitable coating material, further research into the behavior of biopolymers during encapsulation processes (e.g., film-forming properties) and into the behavior of food-grade materials for controlled release purposes may be suggested.

8.3.2 Core Particle Selection in Fluidized Bed Coating

Concerning FBC, powders can be classified into four groups according to their fluidization properties at ambient conditions (Geldart 1972, 1973). The Geldart classification of powders, shown in Figure 8.9 (density difference between the particle and the gas as a function of the mean particle diameter), is widely used in all fields of powder technology. The influence of particle size on the flow of fluidized powders is further explained in Gilbertson and Eames (2003).

The transition from type C to A powders is marked by a dotted line, as research on this distinction is still ongoing. Type A (aeratable, e.g., catalysts) and B (sand-like, e.g., salt) powders, characterized by a moderate mean particle size and density, may be easily fluidized and are ideal to coat in conventional FB systems. Conversely, small Geldart C (cohesive, e.g., starch, flour) particles are not only difficult to fluidize but also very prone to agglomeration because of their high cohesiveness. For coatings of relatively large and heavy type D (spoutable, e.g., grains, peas) particles, SB systems are usually used. Particularly for powder type D, it is difficult to obtain good fluidization, as slugging and channeling are likely to occur. To obtain uniform fluidization with type C and D powders, the FB features have to be extended with mechanical agitation, vibration, centrifugal forces, internal baffles or by adding free-flowing agents (e.g., silica) (Senadeera et al. 2000, Mawatari et al. 2001).

Gas–solid contact and solids mixing determine the quality of fluidization. In the literature, a number of procedures for establishing the quality and nature of fluidization are discussed (Kai et al. 1991, Saxena et al. 1993). For instance, monitoring of a time series of pressure fluctuations in the FB can be used to detect changes in the fluidization quality (e.g., the onset of defluidization) (Schouten and van den Bleek

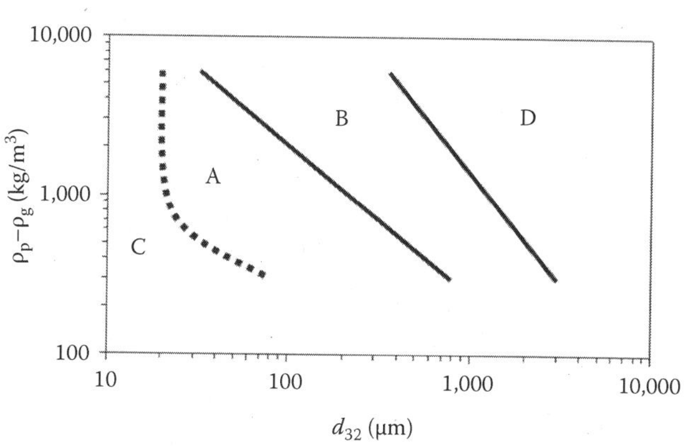

FIGURE 8.9 Geldart powder classification (ρ_p: particle density, ρ_g: gas density, d_{32}: Sauter mean particle diameter).

1998). Furthermore, pressure fluctuation signals can be used to characterize and distinguish the flow behavior of different classes of particles in the Geldart diagram (Bai et al. 1999).

With respect to the fluidization capacity of powders, Guignon et al. (2002) mentioned that this is not significantly affected by coating operations, as the amount of coating material applied usually leads to a very small variation in particle size and density. On the other hand, the injected liquid in the FB spreads over the particles and is known to thereby increase the core particle cohesivity and decrease the bed fluidity (McDougall et al. 2005). As a result, a change in particle motion may be expected, as recently demonstrated by Depypere et al. (2009).

Relevant particle forces in fluidization research have been reviewed by Seville et al. (2000). Liquid bridges, in particular, are interesting from a practical point of view, since their magnitude can be adjusted by altering the amount of free liquid and its properties. By taking into account inter-particle forces, one can distinguish between powder groups (Makkawi and Wright 2004), similar to the classification of Geldart (1973). It is postulated that by influencing the inter-particle forces, the fluidization behavior of the powder may shift from one powder group to another. The effect of adding small quantities of nonvolatile liquid to a FB of Geldart B particles has been the scope of earlier work and contradictory results were obtained. Contrary to the general assumption that adding liquid shifts the powder behavior toward groups A and C (Seville and Clift 1984), Makkawi and Wright (2004) reported a shift away from group A behavior.

An intrinsic problem in dealing with most food powders is their relatively wide particle size distribution, which may even undergo changes during FB processes due to attrition and/or agglomeration (Al-Zahrani 2000). For each particle size, a minimum fluidization velocity (incipient fluidization) and terminal settling velocity (lower limit for particle entrainment) can be defined (Shilton and Niranjan 1993, Motte and Molodtsof 2001). Considering that the FB operates at air velocities between these two described velocities, it can be derived that for powders with a wide particle size distribution, the behavior of the particles in the FB is far from being homogeneous. Furthermore, between these two velocities, a number of fluidization types can be distinguished (Jones 1994).

The key tool for improving the coating quality obtained with food powder coating in a continuous FB is to increase the probability of collisions between coating solution droplets and fluidized core particles (Teunou and Poncelet 2002). Particularly for the top-spray configuration, it is observed that when introducing polydispersed core particles, smaller cores have a higher chance of being coated than their larger counterparts. Therefore, Maronga and Wnukowski (2001) introduced a segregation factor to account for the probability of passing through the coating zone for particles of different sizes.

Besides particle size and density, another factor determining the feasibility of encapsulating a core material using FB technology is particle shape, with a perfect sphere being most favorable to uniform encapsulation. In practice, the maximum and minimum core particle size are determined by the maximum air flow capacity (hence, turbine capacity) of the FB and the porosity of the exit air filter, respectively (DeZarn 1995). However, it should be taken into account that for smaller particles, the surface area to be coated increases substantially, and agglomeration becomes almost unavoidable.

Moreover, the desired release mechanism determines to a great extent the core particle requirements. When sustained release, relying on film thickness and quality, is envisaged, surface area, integrity, and porosity are essential core particle features, and very stringent requirements for raw materials have to be fulfilled (Jones 1994).

8.3.3 Problems Encountered in Fluidized Bed Coating

8.3.3.1 Agglomeration and Premature Evaporation

Under some circumstances, FBC processes tend to produce low yields of encapsulated product. Particularly with the top-spray configuration, side-effects such as premature evaporation of the coating solution and particle agglomeration may occur, resulting in unexpectedly low yields.

In most cases, selection of input variables to produce high-quality coated solids conflicts with the selected input variables for optimal yield. Previous research has illustrated that wet film properties rather than coating solution properties determine the tendency to agglomerate. Glass transition phenomena have been mentioned as causing alterations in rheological properties upon evaporation of the coating solution droplets (Dewettinck et al. 1998, Bhandari and Howes 1999). Even though the agglomeration phenomenon benefits from particle stickiness, it is considered a major problem affecting product quality and yield during powder handling, drying, and coating operations. Despite the recent attempt to correlate glass transition measurements with stickiness (Adhikari et al. 2005), there is still a need for accurate, simpler, and cheaper techniques to characterize stickiness behavior of food powders (Boonyai et al. 2004).

Side-effect agglomeration is usually prevented by increasing the kinetic energy of the particles, by increasing the drying or evaporative capacity of the inlet air and/or by lowering the bed moisture content. While the two former modifications can be accomplished by higher inlet air flow rates, a change in air distribution plate and vessel design, increased process temperatures, etc., the latter (lowering the bed moisture content) somewhat contradicts the time- and cost-saving aspects the food industry has to deal with (Dewettinck and Huyghebaert 1999). Higher flow rates of temperature- and humidity-controlled air increase the drying capacity with the potential of suppressing side-effect agglomeration, but at the same time they increase the probability of premature evaporation of the coating solution.

In contrast to FB drying and agglomeration operations, the exhaust air from a coating unit is usually not saturated, explaining the stronger potential of the latter for premature evaporation of the coating solution (Jones 1985). Minimization of dried coating material, which is collected by the filter system at the top of the vessel and should be considered a loss, should be one of the primary aims of the food technologist.

From previous research, it could be concluded that coating losses can be reduced drastically by choosing appropriate processing conditions. These involve not only evaporative capacity but also, among others, spray nozzle factors, such as atomization pressure and nozzle position, and coating solution characteristics. Increasing the atomization pressure, for example, was shown not only to decrease the size of the atomized droplets, but also to increase the droplet velocity and to decrease the FB temperature (Dewettinck and Huyghebaert 1998).

The distance the droplets have to travel toward the fluidized particles is an important factor determining the degree of premature evaporation of the coating solution. In all three possible FB configurations, the nozzle is positioned to minimize droplet travel distance (Jones 1994). Actually, this is an issue only in the top-spray configuration, where the relative position of the nozzle above the bed can be adjusted. In bottom- and tangential-spray coating systems, the binary nozzles are already submerged inside the FB.

As previously mentioned, equipment modifications are performed, for example, in the high-speed Wurster configuration, to prevent excessive wetting of the particles in the vicinity of the spray nozzle. Furthermore, the use of metal filters and high throughput nozzles is known to reduce losses due to premature evaporation. To avoid both unwanted phenomena, the ratio of core particle diameter to droplet size should be adequately controlled and must be at least 10 (Jones 1994, Guignon et al. 2002).

8.3.3.2 Film Coating Operation and Core Penetration

As indicated earlier, owing to the high number of phenomena occurring almost simultaneously during film coating, the application of a thin coating layer (5–10 μm) to a solid should be considered a challenge (Jones 1994).

First, the applicable concentration of the coating solution is limited to the range in which the solution remains sprayable and easy to dry. Coating solution characteristics such as viscosity, surface tension, and tackiness are of paramount importance to achieve a uniform, high-quality coating (Shavit and Chigier 1995). Improper control of the coating viscosity, for instance, may result in a porous and voluminous film, characterized by holes and incorporated spray-dried droplet material. This so-called orange peel effect can typically be noticed when top-spray FBC is carried out incorrectly (Eichler 1989).

A particular problem when dealing with food powders is their inherent porosity, possibly affecting core penetration by compounds of the atomized coating solution. Hemati et al. (2003) defined a nongrowth period during which the coating solution is deposited inside the pore volume. To prevent this, the application of an initial very fine coating layer, by a combination of a reduced spray rate and a higher inlet air temperature, may be suggested. After this initial step, resulting in the formation of a thin barrier against core penetration, the fluidization parameters may be adjusted to the desired processing values.

8.3.3.3 Other Problems

When dealing with food powders of small particle size, development of static electricity may pose a serious problem to FB and SD processing. The small scale effects may be visible through particles tending to stick to each other, leading to agglomeration, and adherence to the vessel walls, leading to process losses. On a larger scale, the danger of explosions must be emphasized. The factors causing electrostatic charging and the mechanisms of reducing charge accumulation are still poorly understood (Park et al. 2002). However, as adding moisture is a general method of dispelling static electricity, a quantity of moisture in the inlet air is usually helpful (Jones 1994).

Certain materials are far from being beneficial toward encapsulation and require various surface modifications. For instance, hygroscopic, sticky, cohesive, and self-agglomerating particles, or particles that are prone to mechanical stress, friability or heat, all need to be processed in a specific way. However, this does not necessarily mean that the use of the FB system as such is ruled out. For example, a proper choice of vessel geometry, vessel wall material and air distribution system may minimize coating attrition problems caused by particle–particle and particle–wall collisions (Guignon et al. 2002).

One of the key goals of every batch microencapsulation is to produce, batch by batch, uniform product quality and morphology (Eichler 1989). The continuous operating mode is evaluated according to its ability to match the coating quality of the batch mode. With respect to this, the process controllability is of tremendous importance.

However, the FB systems discussed are characterized by a vast number of input variables, including operating conditions, core and coating material properties, and environmental variables. One of the major issues is to assess how these input variables affect microcapsule properties such as morphology, size, coating thickness, and uniformity.

Furthermore, better understanding of the relationships between microcapsule properties and microcapsule functionality, such as release profile, stability, and processability, is needed.

For instance, in FBC, when using inlet air of relatively low temperature (50°C–90°C), the effect of changing environmental variables such as temperature and relative humidity of ambient air may become quite pronounced. This can result in a change in drying capacity, implying changes in film density and porosity, and hence changes in the release profile. This undesirable "weather effect" can be overcome by dehumidifying the inlet air or by humidification through additional spraying. Alternatively, adjustments within predetermined limits of the spray rate, inlet air temperature and velocity provide, to some extent, the means for controlling the thermodynamic operation point (TOP) (Dewettinck et al. 1999).

To evaluate the suitability of a set of operating conditions, accurate monitoring of the final microcapsule properties such as surface characteristics, coating layer thickness, and uniformity must be accomplished. A survey of possible methods for assessing the amount of coating deposited on core particles is given in the literature (Maronga and Wnukowski 2001). These methods include weight measurements, sieve analyses, dissolution and release profiles, scanning electron microscopy (SEM), etc. For food powders, which have a wide particle size distribution, only measurements including a sufficiently large number of individual particles should be considered appropriate.

8.3.4 Control and Modeling of Fluidized Bed Process

Despite the widespread use of microencapsulated ingredients in the manufacture of food products, details of coating as a unit operation are not fully understood. Particularly, the fine tuning of microcapsules for optimum performance requires a thorough understanding of the polymer properties and processing involved in microencapsulation. Among other future developments, both real-time measurements and efficient predictive modeling capacity will contribute to improved understanding of the coating operation, allowing better process control to be developed.

Currently, several methods have been established to assess microcapsule properties such as particle size, particle shape, coating uniformity, and coating functionality. As these methods require extensive manipulations and dedicated analyzing equipment, they can be performed only off-line. However, to establish process control systems that are able to detect and counteract the onset of process failure and can ensure constant output quality in both batch and continuous encapsulation processes, real-time measurements are necessary. Distinction can be made between in-line measurements, in which the sample interface is located in the process stream, and on-line measurements, in which the sample is transferred automatically to the analyzer. Furthermore, measurements can be invasive or noninvasive, the latter being preferred (Rantanen 2000).

Various authors have described new measurement methods that are in an experimental or an early commercialized stage for in-line or on-line measurements in FB processes. These measurement methods include laser diffraction for in-line droplet and particle size measurement (Yuasa et al. 1999), infrared spectroscopy for the quantification of moisture content (Watano et al. 1993, Rantanen et al. 2001), and optically based methods for particle size and particle morphology measurement (Watano 2001). However, these methods offer the possibility of assessing a wide variety of parameters and consequently generate enormous amounts of data. The challenge will be to identify the most relevant parameters in real-time measurements. A variety of data processing techniques, such as neural networks, fuzzy logic, or principal component analysis can be used (Eerikäinen et al. 1993, Haley and Mulvaney 1995, Thyagarajan et al. 1998, Jia et al. 2000). The most important advantage of the real-time measurement techniques is that the generated measurement results can be used immediately for process control.

Process controllers could be implemented as feedback, feedforward, or predictive controllers (Haley and Mulvaney 1995). Feedforward controllers are particularly interesting, for instance in continuous coating operations. Feedforward compensation measures unanticipated disturbances in process inputs, such as changes in particle size distribution or food powder composition, and performs corrective actions before these disturbances can affect the process. However, to be successful, feedforward control requires an accurate model of how disturbances affect the actual coating operation.

One of the first approaches in FB control of the TOP consisted in a reduction of the numerous variables to two instruments: the actual temperature difference within the product and the actual process air flow rate (Alden et al. 1988, Eichler 1989). Control of the former can be improved by in-line temperature measurements over the whole product bed instead of registering only one product bed temperature. Definition of the process air flow rate is a weak point, as reliable monitoring systems are rather expensive.

During the last decade, attempts have been made to effect process control in a variety of ways, such as on–off and proportional integral derivative (PID) humidity control using infrared moisture sensors (Watano et al. 1993).

Dewettinck et al. (1999) developed a thermodynamic computer model, Topsim, able to calculate and effectuate predictive control of the steady-state TOP of a FB process. The Topsim model can also be used to develop control strategies for countering the weather effect.

Larsen et al. (2003) mention a new process control strategy for aqueous film coating, based on in-process calculation of the degree of utilization of the potential evaporation energy and the relative humidity of the outlet air. While controlling these two outlet air properties and controlling the product temperature by regulating the inlet air temperature, the maximum coating liquid spray rates are envisaged.

In the patent literature (Tondar et al. 2002), a method and a device function are described for the monitoring and control of a FBC unit by determining the product moisture in a contact-free manner using electromagnetic radiation in the high frequency or microwave range. Taking into account the product temperature, the total product moisture is held in a predetermined range via a control circuit, regulating the spray rate, inlet air temperature, and volume flow.

In two review publications (Guignon et al. 2002, Teunou and Poncelet 2002), process modeling and optimization of FBC are discussed in more detail. The majority of established models were developed using a black box approach and their parameters were adjusted from close correlation to experimental data. In most cases, generalization and extrapolation of obtained results to other conditions does not seem feasible, as they describe the particular behavior of the core component A with coating material B, under process conditions C, using equipment D. Therefore, the authors (Guignon et al. 2002) conclude that with the established knowledge, to date, it is not appropriate to use only one model to describe all the phenomena appearing in the FBC process.

8.4 CONCLUSIONS

Microencapsulation in the food industry is not only a technology that adds value to a conventional ingredient or additive, but it is also often the source of totally new ingredients with matchless properties. Hereby, the food technologist is challenged to select on the one hand an economically feasible encapsulation technique and, on the other, to opt for a suitable coating material from a relatively short list of food grade approved substances.

With respect to FBs, no other microencapsulation technology can apply as broad a range of coating materials. Recent improvements in the coating process and the advent of continuous FB systems aimed at increasing production volumes and making the process more cost efficient. This, together with the numerous possibilities microencapsulation offers the food technologist, explains the increasing interest in food ingredient microencapsulation.

Despite the developments in online control instruments and process simulation tools, microencapsulation is in many cases still considered more an art than science. Particularly in industry, often empirical know-how ("art") instead of science is relied on, which so far has limited the possibilities to further improve product quality.

While a uniform product quality and morphology is primordial to every batch FB encapsulation process, continuous processes are also evaluated according to their ability to match the coating quality of batch processes. With respect to product quality, process controllability is of utmost importance. To achieve this, a full characterization of all the partial aspects of the FBC process is a prerequisite.

REFERENCES

Achanta, A.S., Adusumilli, P.S., James, K.W., and Rhodes, C.T. 2001. Hot-melt coating: Water sorption behavior of excipient films. *Drug Dev. Ind. Pharm.* 27:241–250.

Adhikari, B., Howes, T., Lecomte, D., and Bhandari, B.R. 2005. A glass transition temperature approach for the prediction of the surface stickiness of a drying droplet during spray drying. *Powder Technol.* 149:168–179.

Alden, M., Torkington, P., and Strutt, A.C.R. 1988. Control and instrumentation of a fluidized-bed drier using the temperature-difference technique. I. Development of a working model. *Powder Technol.* 54:15–25.

Al-Zahrani, A.A. 2000. Particle size distribution in a continuous gas-solid fluidized bed. *Powder Technol.* 107:54–59.

Arshady, R. 1993. Microcapsules for food. *J. Microencapsul.* 10:413–435.

Bai, D., Grace, J.R., and Zhu, J.X. 1999. Characterization of gas fluidized beds of Group C, A and B particles based on pressure fluctuations. *Can. J. Chem. Eng.* 77:319–324.

Balassa, L.L. and Fanger, G.O. 1971. Microencapsulation in the food industry. *Crit. Rev. Food Technol.* 2:245–263.

Becher, R.D. and Schlünder, E.U. 1997. Fluidized bed granulation: Gas flow, particle motion and moisture distribution. *Chem. Eng. Process.* 36:261–269.

Bhandari, B.R. and Howes, T. 1999. Implication of glass transition for the drying and stability of dried foods. *J. Food Eng.* 40:71–79.

Boonyai, P., Bhandari, B., and Howes, T. 2004. Stickiness measurement techniques for food powders: A review. *Powder Technol.* 145:34–46.

Christensen, F.N. and Bertelsen, P. 1997. Qualitative description of the Wurster-based fluid-bed coating process. *Drug Dev. Ind. Pharm.* 23:451–463.

Depypere, F., Pieters, J.G., and Dewettinck, K. 2009. PEPT visualization of particle motion in a tapered fluidized bed-coater. *J. Food Eng.* 93:324–336.

Dewettinck, K. and Huyghebaert, A. 1998. Top-spray fluidized bed coating: Effect of process variables on coating efficiency. *Lebensm.-Wiss. Technol.—Food Sci. Technol.* 31:568–575.

Dewettinck, K. and Huyghebaert, A. 1999. Fluidized bed coating in food technology. *Trends Food Sci. Technol.* 10:163–168.

Dewettinck, K., Deroo, L., Messens, W., and Huyghebaert, A. 1998. Agglomeration tendency during top-spray fluidized bed coating with gums. *Lebensm.-Wiss. Technol.—Food Sci. Technol.* 31:576–584.

Dewettinck, K., De Visscher, A., Deroo, L., and Huyghebaert, A. 1999. Modeling the steady-state thermodynamic operation point of top-spray fluidized bed processing. *J. Food Eng.* 39:131–143.

DeZarn, T.J. 1995. Food ingredient encapsulation—An overview. In *Encapsulation and Controlled Release of Food Ingredients*, Eds. S.J. Risch and G.A. Reineccius, pp. 74–86. Washington, DC: American Chemical Society. *ACS Symposium Series*, nr. 590.

Dziezak, J.D. 1988. Microencapsulation and encapsulated ingredients. *Food Technol.* 42:136–153.

Eerikäinen, T., Linko, P., Linko, S., Siimes, T., and Zhu, Y.H. 1993. Fuzzy logic and neural network applications in food science and technology. *Trends Food Sci. Technol.* 4:237–242.

Eichler, K. 1989. *Principles of Fluid Bed Processing*. Binzen, Germany: Glatt GmbH.

Finch, C.A. 1993. Industrial microencapsulation: Polymers for microcapsule walls. In *Encapsulation and Controlled Release*, Eds. D.R. Karsa and R.A. Stephenson, pp. 1–11. Cambridge, U.K.: The Royal Society of Chemistry.

Geldart, D. 1972. The effect of particle size and size distribution on the behaviour of gas fluidized beds. *Powder Technol.* 6:201–215.

Geldart, D. 1973. Types of gas fluidization. *Powder Technol.* 7:285–292.

Gibbs, B.F., Kermasha, S., Alli, I., and Mulligan, C.N. 1999. Encapsulation in the food industry: A review. *Int. J. Food Sci. Nutr.* 50:213–224.

Gilbertson, M.A. and Eames, I. 2003. The influence of particle size on the flow of fluidised powders. *Powder Technol.* 131:197–205.

Gouin, S. 2004. Microencapsulation: Industrial appraisal of existing technologies and trends. *Trends Food Sci. Technol.* 15:330–347.

Guignon, B., Duquenoy, A., and Dumoulin, E.D. 2002. Fluid bed encapsulation of particles: Principles and practice. *Drying Technol.* 20:419–447.

Haley, A.T. and Mulvaney, S.J. 1995. Advanced process control techniques for the food industry. *Trends Food Sci. Technol.* 6:103–110.

Hemati, M., Cherif, R., Saleh, K., and Pont, V. 2003. Fluidized bed coating and granulation: Influence of process-related variables and physicochemical properties on the growth kinetics. *Powder Technol.* 130:18–34.

Hüttlin, H. 1990. Fluidized bed apparatus for the production and/or further treatment of granulate material. U.S. patent 4,970,804, filed October 17, 1989, and issued November 20, 1990.

Ivanova, E., Teunou, E., and Poncelet, D. 2005. Encapsulation of water sensitive products: Effectiveness and assessment of fluid bed dry coating. *J. Food Eng.* 71:223–230.

Jackson, L.S. and Lee, K. 1991. Microencapsulation and the food industry. *Lebensm.-Wiss. Technol.—Food Sci. Technol.* 24:289–297.

Janovsky, C. 1993. Encapsulated ingredients for the baking industry. *Cereal Foods World* 38:85–87.

Jia, F., Martin, E.B., and Morris, A.J. 2000. Non-linear principal components analysis with application to process fault detection. *Int. J. Syst. Sci.* 31:1473–1487.

Jones, D. 1994. Air suspension coating for multiparticulates. *Drug Dev. Ind. Pharm.* 20:3175–3206.

Jones, D.M. 1985. Factors to consider in fluid-bed processing. *Pharm. Technol.* 9:50–62.

Jones, D.M. 1988. Controlling particle size and release properties—Secondary processing techniques. In *Flavor Encapsulation*, Eds. S.J. Risch and G.A. Reineccius, pp. 158–175. Washington, DC: American Chemical Society. *ACS Symposium Series*, nr. 370.

Jones, D.M. 1995. Fluidized bed with spray nozzle shielding. U.S. patent 5,437,889, filed August 9, 1993, and issued August 1, 1995.

Jones, D.M., Hirschfeld, P.F.F., and Nowak, R. 1992. Apparatus and method for producing pellets by layering powder onto particles. U.S. patent 5,132,142, filed March 19, 1991, and issued July 21, 1992.

Joszwiakowski, M.J., Jones, D.M., and Franz, R.M. 1990. Characterization of a hot-melt fluid bed coating process for fine granules. *Pharm. Res.* 7:1119–1126.

Kai, T., Murakami, M., Yamasaki, K.I., and Takahashi, T. 1991. Relationship between apparent bed viscosity and fluidization quality in a fluidized bed with fine particles. *J. Chem. Eng. Jpn.* 24:494–500.

Kanawjia, S.K., Pathania, V., and Singh, S. 1992. Microencapsulation of enzymes, microorganisms and flavours and their applications in foods. *Indian Dairyman* 44: 280–287.

Knowlton, T.M., Karri, S.B.R., and Issangya, A. 2005. Scale-up of fluidized-bed hydrodynamics. *Powder Technol.* 150:72–77.

Kunii, D. and Levenspiel, O. 1991. *Fluidization Engineering*. Boston, MA: Butterworth-Heinemann.

Larsen, C.C., Sonnergaard, J.M., Bertelsen, P., and Holm, P. 2003. A new process control strategy for aqueous film coating of pellets in fluidised bed. *Eur. J. Pharm. Sci.* 20:273–283.

Leuenberger, H. 2001. New trends in the production of pharmaceutical granules: Batch versus continuous processing. *Eur. J. Pharm. Biopharm.* 52:289–296.

Liborius, E. 1997. Continuous multi-cell process for particle coating providing for particle recirculation in the respective cells. U.S. patent 5,648,118, filed August 23, 1995, and issued July 15, 1997.

Litster, J.D. and Sarwono, R. 1996. Fluidized drum granulation: Studies of agglomerate formation. *Powder Technol.* 88:165–172.

Makkawi, Y.T. and Wright, P.C. 2004. Tomographic analysis of dry and semi-wet bed fluidization: The effect of small liquid loading and particle size on the bubbling behaviour. *Chem. Eng. Sci.* 59:201–213.

Maronga, S. 1998. On the optimization of the fluidized bed particulate coating process. PhD dissertation, Royal Institute of Technology, Stockholm, Sweden.

Maronga, S.J. and Wnukowski, P. 2001. Growth kinetics in particle coating by top-spray fluidized bed systems. *Adv. Powder Technol.* 12:371–391.

Mathur, K.B. and Epstein, N. 1974. *Spouted Beds*. New York: Academic Press.

Mawatari, Y., Koide, T., Tatemoto, Y., Takeshita, T., and Noda, K. 2001. Comparison of three vibrational modes (twist, vertical and horizontal) for fluidization of fine particles. *Adv. Powder Technol.* 12:157–168.

McDougall, S., Saberian, M., Briens, C., Berruti, F., and Chan, E. 2005. Effect of liquid properties on the agglomerating tendency of a wet gas-solid fluidized bed. *Powder Technol.* 149:61–67.

Mehta, A.M., Valazza, M.J., and Abele, S.E. 1986. Evaluation of fluid-bed processes for enteric coating systems. *Pharm. Technol.* 10:46–56.

Motte, J. and Molodtsof, Y. 2001. Predicting transport velocities. *Powder Technol.* 120:120–126.

Nienow, A.W. 1995. Fluidised bed granulation and coating: Applications to materials, agriculture and biotechnology. *Chem. Eng. Commun.* 139:233–253.

Park, A.H., Bi, H.S., and Grace, J.R. 2002. Reduction of electrostatic charges in gas-solid fluidized beds. *Chem. Eng. Sci.* 57:153–162.

Pearnchob, N. and Bodmeier, R. 2003. Coating of pellets with micronized ethylcellulose particles by a dry powder coating technique. *Int. J. Pharm.* 268:1–11.

Rácz, I., Dredán, J., Antal, I., and Gondár, E. 1997. Comparative evaluation of microcapsules prepared by fluidization atomization and melt coating process. *Drug Dev. Ind. Pharm.* 23:583–587.

Rantanen, J. 2000. Near-infrared reflectance spectroscopy in the measurement of water as a part of multivariate process monitoring of fluidised bed granulation process. PhD dissertation, University of Helsinki, Helsinki, Finland.

Rantanen, J., Räsänen, E., Antikainen, O., Mannermaa, J.P., and Yliruusi, J. 2001. In-line moisture measurement during granulation with a four-wavelength near-infrared sensor: An evaluation of process-related variables and a development of non-linear calibration model. *Chemom. Intell. Lab. Syst.* 56:51–58.

Rümpler, K. and Jacob, M. 1998. Continuous coating in fluidized bed. *Int. Food Market. Tech.* 12:41–43.

Saxena, S.C., Rao, N.S., and Tanjore, V.N. 1993. Diagnostic procedures for establishing the quality of fluidization of gas-solid systems. *Exp. Thermal Fluid Sci.* 6:56–73.

Schouten, J.C. and van den Bleek, C.M. 1998. Monitoring the quality of fluidization using the short-term predictability of pressure fluctuations. *AIChE J.* 44:48–60.

Senadeera, W., Bhandari, B.R., Young, G., and Wijesinghe, B. 2000. Methods for effective fluidization of particulate food material. *Drying Technol.* 18:1537–1557.

Seville, J.P.K. and Clift, R. 1984. The effect of thin liquid layers on fluidisation characteristics. *Powder Technol.* 37:117–129.

Seville, J.P.K., Willett, C.D., and Knight, P.C. 2000. Interparticle forces in fluidisation: A review. *Powder Technol.* 113:261–268.

Shahidi, F. and Han, X.Q. 1993. Encapsulation of food ingredients. *Crit. Rev. Food Sci. Nutr.* 33:501–547.

Shavit, U. and Chigier, N. 1995. The role of dynamic surface tension in air assist atomization. *Phys. Fluids* 7:24–33.

Shilton, N.C. and Niranjan, K. 1993. Fluidization and its applications to food processing. *Food Struct.* 12:199–215.

Sinchaipanid, N., Junyaprasert, V., and Mitrevej, A. 2004. Application of hot-melt coating for controlled release of propanolol hydrochloride pellets. *Powder Technol.* 141:203–209.

Singiser, R.E. and Lowenthal, W. 1961. Enteric filmcoats by the air-suspension coating technique. *J. Pharm. Sci.* 50:168–170.

Spooner, T.F. 1994. Encapsulation worth looking into, but check out economics. *Milling Baking News*, April, 50–51.

Teunou, E. and Poncelet, D. 2002. Batch and continuous fluid bed coating—Review and state of the art. *J. Food Eng.* 53:325–340.

Thyagarajan, T., Shanmugam, J., Panda, R.C., Ponnavaikko, M., and Rao, P.G. 1998. Artificial neural networks: Principle and application to model based control of drying systems—A review. *Drying Technol.* 16:931–966.

Tondar, M., Luy, B., and Prasch, A. 2002. Method for monitoring and/or controlling a granulation, coating and drying process. U.S. Patent 6,383,553, filed September 24, 1999, and issued May 7, 2002.

Versic, R.J. 1988. Flavor encapsulation—An overview. In *Flavor Encapsulation*, Eds. S.J. Risch and G.A. Reineccius, pp. 1–6. Washington, DC: American Chemical Society. *ACS Symposium Series*, nr. 370.

Watano, S. 2001. Direct control of wet granulation by image processing system. *Powder Technol.* 117:163–172.

Watano, S., Harada, T., Terashita, K., and Miyanami, K. 1993. Development and application of a moisture control system with IR moisture sensor to aqueous polymeric coating process. *Chem. Pharm. Bull.* 41:580–585.

Wurster, D.E. 1959. Air suspension technique of coating drug particles, a preliminary report. *J. Am. Pharm. Assoc. (Baltim)* 48:451–454.

Yuasa, H., Nakano, T., and Kanaya, Y. 1999. Suppression of agglomeration in fluidized bed coating. II. Measurement of mist size in a fluidized bed chamber and effect of sodium chloride addition on mist size. *Int. J. Pharm.* 178:1–10.

9 Spray Drying and Its Application in Food Processing

Huang Li Xin and Arun S. Mujumdar

CONTENTS

9.1 INTRODUCTION

Spray drying is a one-step continuous processing operation that can transform feed from a fluid state into a dried form by spraying the feed into a hot drying medium. The product can be a single particle or agglomerates. The feed can be a solution, a paste, or a suspension. This process has become one of the most important methods for drying liquid foods to powder form.

The main advantages of spray drying are the following:

- Product properties and quality are more effectively controlled.
- Heat-sensitive foods, biologic products, and pharmaceuticals can be dried at atmospheric pressure and low temperatures. Sometimes inert atmosphere is employed.
- Spray drying permits high-tonnage production in continuous operation and relatively simple equipment.
- The product comes into contact with the equipment surfaces in an anhydrous condition, thus simplifying corrosion problems and selection of materials of construction.
- Spray drying produces relatively uniform, spherical particles with nearly the same proportion of nonvolatile compounds as in the liquid feed.
- As the operating gas temperature may range from 150°C to 750°C, the efficiency is comparable to that of other types of direct dryers.
- The principal disadvantages of spray drying are as follows:
- Spray drying generally fails if a high bulk density product is required.
- In general it is not flexible. A unit designed for fine atomization may not be able to produce a coarse product, and vice versa.
- For a given capacity, evaporation rates larger than other types of dryers are generally required due to high liquid content requirement. The feed must be pumpable. Pumping power requirement is high.
- There is a high initial investment compared to other types of continuous dryers.
- Product recovery and dust collection increases the cost of drying.

The development of the process has been closely associated with the dairy industry. The use of spray drying in the dairy industry dates back to around 1800, but it was not until 1850 that it became possible to dry milk on industrial scale. Since then, this technology has been developed and expanded to cover a large food group which is now successfully spray-dried. Over 20,000 spray dryers are estimated to be in use commercially, at present, to agro-chemical products, biotechnology products, fine and heavy chemicals, dairy products, foods, dyestuffs, mineral concentrates, and pharmaceuticals in evaporation capacities ranging from a few kg per hour to 50 tons/h (Mujumdar 2000).

9.2 PRINCIPLES OF SPRAY DRYING

9.2.1 Spray Drying Fundamentals

A conventional spray dryer flow sheet in its most simplified form is shown in Figure 9.1. It consists of the following four essential components:

- Air heating system and hot-air distribution system
- Feed transportation and atomization
- Air and spray contacts and drying
- Dried particles collection system

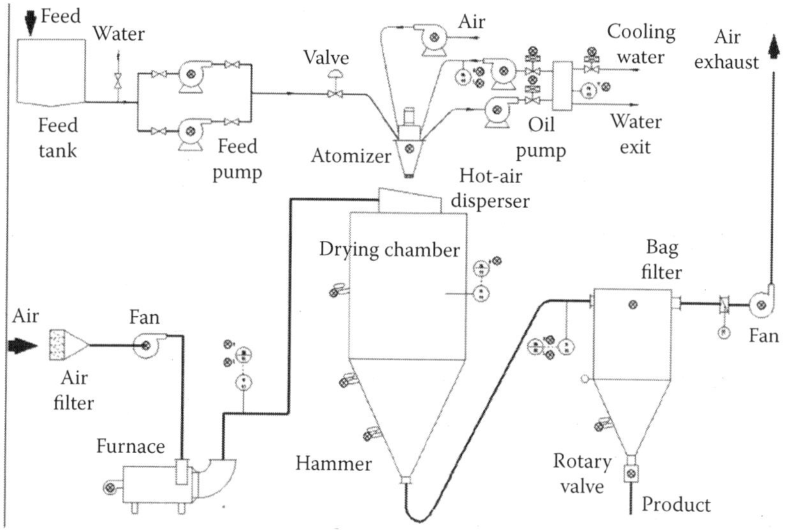

FIGURE 9.1 A typical basic spray dryer flow diagram.

From Figure 9.1, it is seen that the process operates in the following way:

The liquid is pumped from the product feed tank to the atomization device, e.g., a rotary disc atomizer, pressure nozzle, pneumatic nozzle, or ultrasonic nozzle, which is usually located in the air distributor at the top of the drying chamber. The drying air is drawn from the atmosphere through a filter by a supply fan and is passed through the air heater, e.g., oil furnace, electrical heater, and steam heater, to the air distributor. The droplets produced by the atomizer meet the hot air and the evaporation takes place, cooling the air in the meantime. After the drying of the droplets in the chamber, the majority of the dried product falls to the bottom of the chamber and entrains in the air. Then they pass through the bag filter for separation of the dried particles from air. The particles leave the bag filter at the bottom via a rotary valve and are collected or packed later. The exhausted air is discharged to the atmosphere via the exhaust fan.

This process shown in Figure 9.1 is the one generally used in industrial spray drying. It is called open-cycle process. Its main feature is that the air is drawn from the atmosphere, passed through the heating system and the drying chamber, and then exhausted to the atmosphere.

Some foodstuffs have to be prepared in organic solvents rather than water to prevent oxidation of one or more of the active ingredients. In these applications, a closed-cycle spray drying system using an inert gas, such as nitrogen, is typically used. This is a special system. Another application is the drying of flammable and toxic materials

In closed-cycle spray drying plants, the atomized droplets are contacted by hot nitrogen in the spray drying chamber and processed into a free flowing powder like any other food formulation. Dried product is discharged from the drying chamber and the cyclone, but the spent drying gas must be introduced into a condensation system.

The solvent evaporated in the drying chamber has to be condensed and recovered. The off-gases from the condensation tower are then reheated in an indirect heater for being reused in the drying chamber. The process is shown in Figure 9.2.

Depending on how the drying medium and droplets produced by the atomizer are contacted, three basic air-droplet contacting configurations can be identified, i.e., cocurrent flow, countercurrent flow, and mixed flow.

In the cocurrent flow configuration, the liquid spray and air pass through the drying chamber in the same direction, although spray–air movement in reality is far from cocurrent in initial contact. This type of contact is commonly used in a centrifugal atomization spray dryer. It can lead to product temperatures lower than those obtained by the other two flow patterns (Masters 1991).

In a countercurrent flow system, the spray and hot drying air enter the drying chamber at the opposite ends of the dryer. It typically produces high bulk density powders.

The mixed-flow spray drying system is a combination of the previous two systems. The droplets may contact the drying medium in the same and opposite direction in one drying chamber. It is usually found in spray dryers fitted with pressure nozzle or pneumatic nozzle. It is sometimes used to obtain agglomerated powders in a small drying chamber. It should be noted that the product must be nonheat sensitive.

In spray drying of foods, the one-stage system like that shown in Figure 9.1 is the most common choice. It can be used with several different atomization, air-spray contact and process layout arrangements.

Two-stage spray dryer systems are sometimes called Spray-Fluidizers, since a fluid bed is installed at the cone of the spray dryer chamber. Alternatively, a vibrated fluid bed (VFB) is installed at the bottom of the drying chamber. It can produce

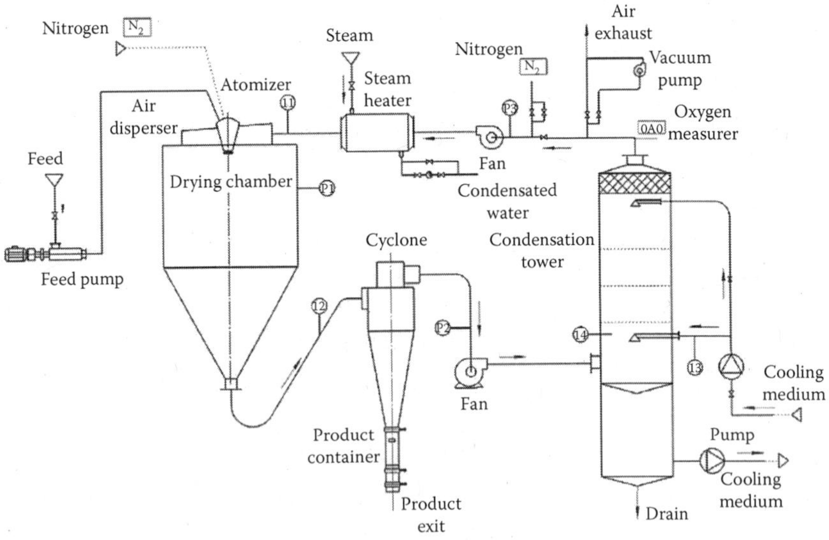

FIGURE 9.2 Layout of a closed-cycle spray drying system.

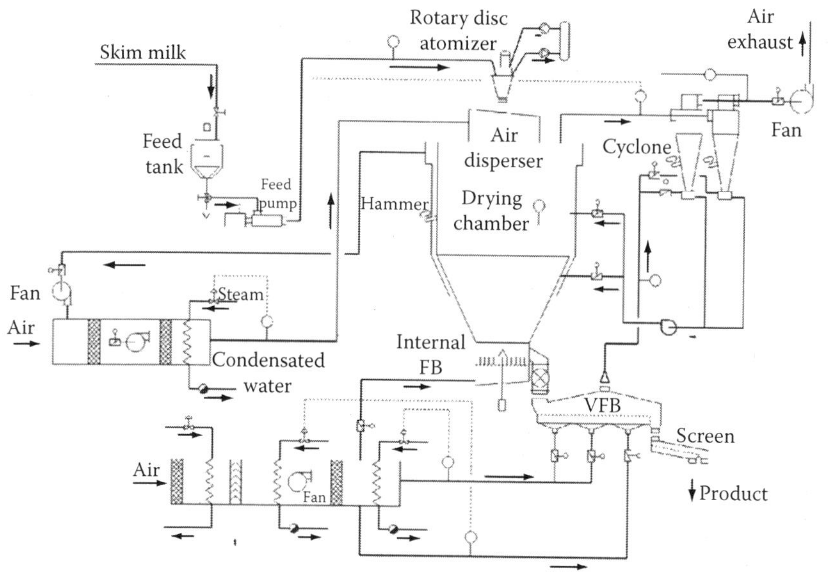

FIGURE 9.3 A schematic diagram of a typical three-stage spray drying system.

instantly soluble products, such as instant coffee, milk, cocoa, etc., by agglomeration of the product. It is ideal to handle heat sensitive products.

The three-stage spray dryer system includes a fluid bed and a vibrated fluid bed dryer or agglomerator together with the spray dryer. It is usually used to produce an agglomerated product by spraying the viscous feed at the beginning of VFB. The process is shown in Figure 9.3. Huang et al. (2001) reported that such a system can save about 20% energy compared to a single stage spray dryer. In this case, a spray dryer is used to evaporate the surface moisture in the drying chamber, and then the moist powder is further dried in the fluid bed installed at the cone chamber of spray dryer. Finally the powder leaves the integrated fluid bed to enter a VFB for final drying and cooling. In some applications, the VFB dryer is replaced by a belt dryer.

The main advantages of a three-stage system are

- Improved agglomeration of particles
- Less thermal degradation of product
- Increased thermal efficiency
- Low product packaging temperature
- Easy to add new operations, such as, spray coating, agglomeration, etc.

9.2.2 Spray Drying Components

From Figure 9.1, it is clear that spray drying consists of the following four essential stages, i.e., heating of the drying air and its distribution, feed transportation and atomization, contacting of hot air and spray for drying of spray and recovery of dried products (final air cleaning and dried product handling).

Although the physical design, operation mode, handling of feedstock, and product requirements can be diverse, each stage must be carried out in all spray dryers. The formation of a spray and its effective contacting with the heated air are the key characteristic features of spray drying.

9.2.2.1 Atomization

Since the choice of the atomizer is very crucial, it is important to note the key advantages and limitations of different atomizers (centrifugal, pressure, and pneumatic atomizers). Other atomizers, e.g., ultrasonic atomizer, can also be used in spray dryers (Bittern and Kissel 1999) but they are expensive and have rather low capacity. Although different atomizers can be used to dry the same feedstock, the final product properties (bulk density, particle size, flowability, etc.) are quite different and hence a proper selection is necessary.

9.2.2.1.1 Centrifugal Atomizer

The centrifugal atomizer is sometimes called a rotary wheel or disk atomizer. This is a spinning disk assembly with radial or curved vanes, which rotate at high velocities (7,000–50,000 rpm) with wheel diameters of 5–50 cm. The feed is delivered near the center, spreads between the two plates, and is accelerated to high linear velocities before it is thrown off the disk in the form of thin sheets, ligaments or elongated ellipsoids. However, the subdivided liquid immediately attains a spherical shape under the influence of surface tension.

The atomizing effect is dependent upon the centrifugal force generated by rotation of the disk; it also depends upon the frictional influence of the external air. The liquid is continuously accelerated to the disk rim by the centrifugal force produced by the disk rotation. Thus the liquid is spread over the disk internal surface and discharged horizontally at a high speed from the periphery of the disk.

Masters (1991) noted that Equation 9.1 appears to be the most suitable one for the prediction of the mean droplet size generated by a rotating disc atomizer.

$$d_{32} = \frac{1.4 \times 10^4 (\dot{M}_l)^{0.24}}{(u_{\text{atom}} d_{\text{atom}})^{0.83} (nH_{\text{atom}})^{0.12}} \tag{9.1}$$

where

d_{32} is the Sauter mean diameter (μm)
$\dot{M}_l$ is the mass liquid feed rate (kg/h)
u_{atom} is the atomizer wheel speed (rpm)
d_{atom} is the atomizer wheel diameter (m)
n is the number of channels in the wheel
H_{atom} is the atomizer wheel vane height (m)

Centrifugal atomizers have less tendency to become clogged, which is a great advantage. For this reason, they are preferred for spray drying of nonhomogeneous foods. Its advantages are summarized below:

- Handles large feed rates with single wheel or disk
- Suited for abrasive feeds with proper design
- Has negligible clogging tendency
- Change of wheel rotary speed to control the particle size distribution
- More flexible capacity (but with changes in powder properties)

The limitations associated with this type of atomizer are

- Higher energy consumption compared to pressure nozzles
- More expensive to be manufactured than the other nozzles
- Broad spray pattern requires large drying chamber diameter

9.2.2.1.2 *Pressure Nozzle*

High pressure nozzles are alternative atomizing systems in which a fluid acquires a high-velocity tangential motion while being forced through the nozzle orifice. Orifice sizes are usually in the range of 0.5–3.0 mm. The fluid emerges with a swirling motion in a cone shaped sheet, which breaks up into droplets. Greater pressure drop across the orifice produces smaller droplets.

Keey (1991) and Masters (1991) introduced the following correlations to predict the mean droplet diameter produced by pressure nozzle for the commercial conditions

$$d_{32} = \frac{2774 Q_l^{0.25} \mu_l}{\Delta P^{0.5}} \tag{9.2}$$

where
Q_l is the volumetric feed rate (mL/s)
μ_l is the feed viscosity (MPa s)
ΔP is the operating pressure (kPa)

Its advantages are

- Simple, compact, and cheap
- No moving parts
- Low energy consumption

Its limitations are

- Low capacity (feed rate for single nozzle)
- High tendency to clog
- Erosion can change spray characteristics

9.2.2.1.3 *Pneumatic Nozzle*

The pneumatic nozzle is an atomizer with internal or external mixing of gas and liquid. Here atomization is accomplished by the interaction of the liquid with a second fluid, usually compressed air. Such a design permits air or steam to break up the stream of liquid into a mist of fine droplets. Neither the liquid nor the air requires very high pressure, with 200–450 kPa being typical. The particle size is controlled by varying the ratio of the compressed air flow to that of the liquid. The mean spray size produced by pneumatic nozzle atomization follows the relation (Masters 1991)

$$d_{32} = \frac{A_1}{(u_{\text{rel}}^2 \rho_g)^{\alpha}} + A_2 \left(\frac{\dot{M}_g}{\dot{M}_l} \right)^{-\beta} \tag{9.3}$$

where

The exponents α and β are function of nozzle design
A_1 and A_2 are constants involving nozzle design and liquid properties
u_{rel} is the relative velocity between gas and liquid (m/s)
$\dot{M}_g$ and $\dot{M}_l$ are mass flow rate of compressed air and feed, respectively (Masters 1991)

Its advantages are

- Simple, compact, and cheap
- No moving parts
- Handle the feedstocks with high viscosity
- Produce products with very small size particle

Its limitations are

- High energy consumption
- Low capacity (feed rate)
- High tendency to clog

9.2.2.1.4 *Ultrasonic Nozzle*

Ultrasonic nozzles are designed to specifically operate from a vibration energy source. In ultrasonic atomization, a liquid is subjected to a sufficiently high intensity of ultrasonic field that splits it into droplets, which are then ejected from the liquid-ultrasonic source interface into the surrounding air as a fine spray (Rajan and Pandit 2001). A number of basic ultrasonic atomizer types, like capillary wave, standing wave, bending wave, fountain, vibrating orifice, and whistle, etc. exist.

Rajan and Pandit (2001) assessed the impact of various physicochemical properties of liquid, its flow rate, the amplitude and frequency of ultrasonic, and the area and geometry of the vibrating surface on the droplet size distribution. A correlation was proposed to predict the droplet size formed using an ultrasonic atomizer taking into consideration the effect of liquid flow rate and viscosity. The droplet size distribution from an ultrasonic nozzle follows a log-normal distribution (Berger 1998).

Its advantages are

- Simple and compact
- No moving parts
- Droplets with narrow size distribution

Its limitations are

- Low capacity (feed rate)
- High tendency to clog

9.2.2.2 Air and Spray Contact and Droplet Drying

Spray and hot air contact determines the evaporation rate of volatiles in the droplet, droplet trajectory, droplet residence time in the drying chamber, and the deposit in the chamber wall. It also influences the morphology of particles and product quality. So, apart from the selection of atomizers, the drying chamber and air disperser selection are other important factors in spray drying. They determine the air flow pattern in the drying chamber. Several authors (e.g., Gauvin et al. 1975, Crowe 1980, Oakley 1994, Kieviet 1997, Langrish and Kockel 2001, Huang et al. 2003) worked in this area, but the amount of published data on spray–air contact is still limited and is mainly applicable to small-scale spray dryers. Experimental measurements are very difficult to make in an operating spray dryer.

9.2.2.2.1 Hot Air Distribution

Indeed, the hot air distribution is one of the crucial points in a spray drying system design. Today, there are three types of hot air distributors which can be found in spray drying systems in the food industry, i.e., rotating air flow distributor, plug air flow distributor, and central pipe air distributor, A schematic representation of each of these systems is shown in Figure 9.4.

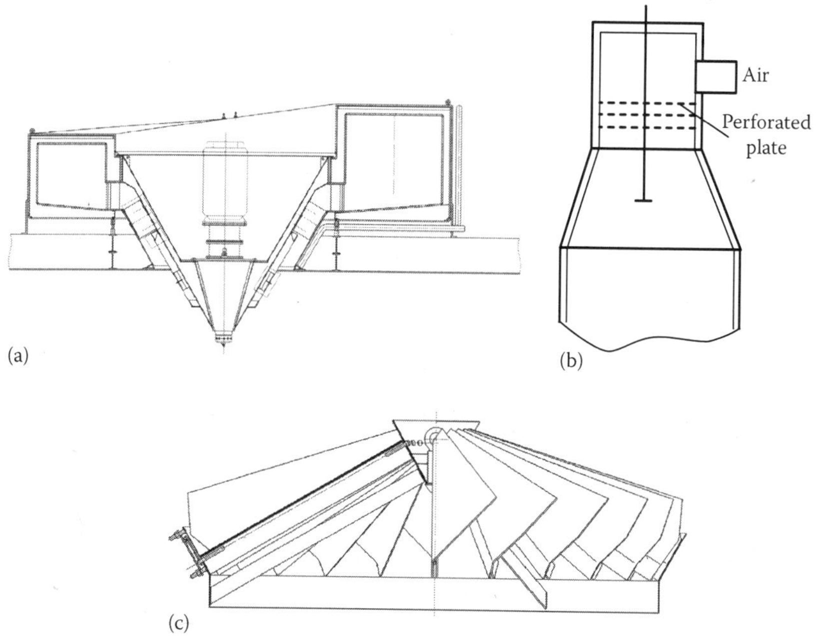

FIGURE 9.4 Hot air dispersers for spray drying in food products: (a) rotating distributor; (b) plug flow distributor; (c) central pipe distributor.

The rotating air flow distributor (Figure 9.4a) is installed at the top of the drying chamber. At this condition, the hot air coming from the heater enters tangentially into a spiral-shaped house where it is distributed radially by the distributed guide vanes and led downward over the second set of guide vanes. The second set of guide vanes is used to make the distributed air rotate by the adjustment of the vanes. This type of distributor is usually used for the spray drying system in which the rotary disc atomizer or nozzle atomizer is installed at the center of the distributor.

The plug flow air distributor (Figure 9.4b) is also installed at the top of the drying chamber. In this condition, the hot air enters radially from the distributor side and is distributed by air passing through the mesh or perforated plate. In order to make air distributed evenly, the mesh or plate is always arranged in two or three layers. Typically, the nozzle atomizer is used for this hot air distributor.

The central pipe air distributor (Figure 9.4c) is installed at the hot air pipe located at the center of the drying chamber. Such a distributor is usually used for the spray dryer, which operates at high inlet temperatures.

9.2.2.2.2 *Drying Chamber Selection and Design*

Main heat and mass transfer between droplets and drying medium takes place in the drying chamber. The drying chamber designs are directly related to the results of droplets drying.

In spray drying market, various designs of the drying chamber can be seen. The cylinder-on-cone chamber is commonly used. According to the product properties, the conical angle is adjusted within 40°–60°. Small angle will help the dried powder leave the chamber by gravity. But it has not necessarily been optimized. Huang et al. (2003) studied three new types of drying chamber designs, i.e., pure conical, lantern and hourglass, comparing their performance to one of the cylinder-on-cone design under the same spray drying operation conditions. They found that pure conical and a lantern geometry can also be used as the spray dryer chamber.

Masters (2002) has suggested a simplified method to design the drying chamber. Huang et al. (1997) reported an empirical correlation for the volumetric drying intensity of the drying chamber in a centrifugal atomization spray dryer as follows:

$$\frac{\dot{M}_w}{V_{\text{dryer}}} = \frac{(T_{\text{in}} + 273)^{3.4287}}{(T_{\text{out}} + 273)^{3.34}} \tag{9.4}$$

where

$\frac{\dot{M}_w}{V_{\text{dryer}}}$ is the volumetric drying intensity (kg/m^3) viz., evaporation rate per unit drying chamber volume

T_{in} and T_{out} are the inlet drying air temperature and outlet air temperature (°C)

Another important parameter for the drying chamber design is the droplets residence time in the drying chamber. The typical particle residence time in the drying chamber listed in Table 9.1 was suggested by Mujumdar (2000).

TABLE 9.1
Residence Time Requirements for Spray Drying of Various Products

Residence Time in Chamber	Recommended for
Short (10–20 s)	Fine, nonheat sensitive products, surface moisture removal, nonhygroscopic
Medium (20–35 s)	Fine-to-coarse spray (d_{vs} = 180 μm), drying to low final moisture
Long (>35 s)	Large powder (200–300 μm); low final moisture, low temperature operation for heat sensitive products

Source: Mujumdar, A.S. Dryers for particulate solids, slurries, and sheet-form materials, in *Mujumdar's Practical Guide to Industrial Drying*, Devahastion, S., Ed., Exergex Corp., Brossard, Quebec, Canada, 2000. With permission.

In the spray drying market, the horizontal box type drying chambers are also found in food industries. In such a design, spray is generated horizontally in a box. The contacts and heat and mass transfer between spray and drying medium are within the box. At the bottom of the box, usually a forced powder removal system, e.g., screw, is installed. This is necessary for the system to remove the dried product so that the heat sensitive material, e.g., food, is not degraded. Huang (2005) suggested that a fluidized bed can be fitted at the bottom of such a box chamber. It may increase the drying capacity for the system, as well as remove the dried particles.

9.2.2.3 Other Components of the Spray Drying System

9.2.2.3.1 Air Heating System

There are two types of air heaters which can be used in a spray drying system, e.g., direct air heaters and indirect heaters. Direct air heaters, such as direct gas or oil fired furnace, can be used whenever the contact between combustion gas and spray is acceptable. When products of combustion of fossil fuels cannot contact with the spray, an indirect heater, such as indirect steam air heater, indirect gas or oil fired heater, is recommended. Interested readers can find more details about it in the literature (Matsers 1991).

9.2.2.3.2 Dried Particle Collection System

The dried products that are entrained in the exhaust air from the drying chamber must be separated and collected. Generally, there are two main types of collectors, i.e., dry collectors and wet collectors. Dry collectors include cyclones, bag filters, and electrostatic precipitators, whereas wet collectors include wet scrubbers, wet cyclones, and spray towers.

Dry collectors are often used as the first stage collector in a spray dryer. Due to the high cost and maintenance of electrostatic precipitator, it is less often used. Cyclones are first chosen due to their low cost and low maintenance requirements. But the relatively low collection efficiency (90%–98%) is not enough in some cases. Under this condition, a bag filter is used as the second collection stage or a wet collector follows. The limitations of a bag filter are its maintenance and cleaning difficulties. The bags rupture will lead to loss of products. Recently, Niro company in the United Kingdom brought out a new type of bag filter which could be cleaned-in-place (CIP) for better economy, quieter running, higher yields, and a greatly reduced chance of cross-contamination. Their new filter, the SANCIP, can be used with a cyclone in series or as the only means of powder separation and environmental control. SANCIP makes it possible to reduce water and chemical consumption and features a purge system that allows CIP of the entire air assembly.

Addition of wet collectors means additional cost. The scrubber liquid needs to be retreated. So the selection of collection method is dependent on the product value and environmental regulations. See Mujumdar (1995) for further details on product collection methods.

9.2.2.3.3 System Control

Spray dryers can be controlled manually or automatically. No matter what control method is used, the outlet temperature from the drying chamber is always controlled or monitored. It usually determines the residual moisture within the product. Based on the control of outlet temperature, two basic control systems (A and B) can be considered.

- System A: It maintains the outlet temperature by adjusting the feed rate. It is particularly suitable for centrifugal atomization spray dryers. This control system usually has another control loop, i.e., controlling the inlet temperature by regulating air heater.
- System B: It maintains the outlet temperature by regulating the air heater and maintaining the constant spray rate. This system can be particularly used for nozzle spray dryers because varying spray rate will result in the change of the droplet size distribution for pressure or pneumatic nozzle.

9.3 MODELING AND SIMULATION OF SPRAY DRYERS USING COMPUTATIONAL FLUID DYNAMICS

Although spray drying systems are widely found and used in the industries, their design is still based on empirical methods and experience. Pilot tests are necessary for each spray drying system. Therefore, more systematic studies must be carried out on spray formation and air flow and heat and mass transfer for optimizing and controlling the drying mechanisms to achieve the highest quality of the powder produced. This process–product association requires a more complex model, which must predict not only the material drying kinetics as a function of the spray drying (SD) operation variables, but also changes in the powder properties during drying in order to quantify the end-product quality. Such combination can be established by introducing into the SD operation, model empirical correlations for predicting the most important product quality requirements (statistical approach) or by describing mechanisms of changes in the material properties during drying (kinetic approach).

Fortunately, computer and software technology are under constant development, which makes the mathematical modeling of spray dryers possible. Such a model is used to predict the droplet or particle movement and the evaporation or drying of droplets in a spray dryer. Two kinds of numerical models, i.e., one-way coupling and two-way coupling, have appeared in the literature (Crowe et al. 1977). In the one-way coupling model, it is assumed that the condition of the drying medium is not affected by the spray or evaporating droplets, although the droplet or particle characteristics change due to the evaporation and drying process. In order to improve the model to simulate spray drying, taking into account heat and mass transfer between spray and drying medium, the two-way coupling approach was developed. This model considers the interaction, e.g., heat, mass, and momentum transfer, between the two phases, i.e., droplets and drying medium. Arnason and Crowe (1993) and Crowe et al. (1998) summarized this approach as shown in Figure 9.5.

On the other hand, the models can also be categorized in terms of the geometry, i.e., one-dimensional (1D), two-dimensional (2D), and three-dimensional (3D).

Crowe et al. (1977) proposed an axi-symmetric spray drying model called Particle-Source-In-Cell model (PSI-Cell model). This model includes two-way mass, momentum, and thermal coupling. In this model, the gas phase is regarded as a continuum (Eulerian approach) and is described by pressure, velocity, temperature, and humidity fields. The droplets or particles are treated as discrete phases which are characterized by velocity, temperature, composition, and the size along trajectories (Lagrangian approach). The model incorporates a finite difference scheme for both the continuum and discrete phases. The authors used this PSI-Cell model to simulate a cocurrent spray dryer. But no experimental data were compared with it. More details can be found in the work by Crowe et al. (1977).

Papadakis and King (1988a,b) used this PSI-Cell model to simulate a spray dryer and compare their predicted results with limited experimental results associated with a lab-scale spray dryer. They have shown that the measured air temperatures at various levels below the roof of the spray drying chamber were well predicted by the computational fluid dynamics (CFD) model. Negiz et al. (1995) developed a program to simulate a cocurrent spray dryer based on the PSI-Cell model. Straatsma et al. (1999) developed a drying model, named NIZO-DrySim, to simulate aspects of drying processes in the food industry. It can simulate the gas flow in a 2D spray dryer chamber and calculate the particle trajectories.

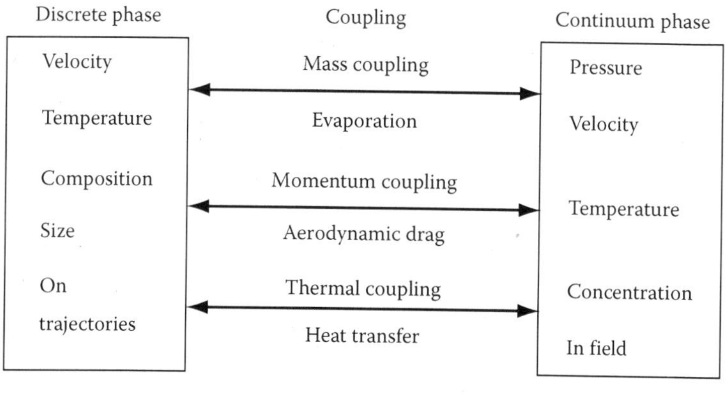

FIGURE 9.5 Two-way coupling between discrete and continuum phases.

Livesley et al. (1992) and Oakley and Bahu (1990) found that numerical simulations using the k–ε turbulence model are useful for simulating the measured particle sizes and mean axial velocities in industrial spray dryers.

Oakley and Bahu (1990) reported a 3D simulation using the CFD code FLOW3D which is an implementation of the PSI-Cell model. They proposed that additional research needs to be done to verify the performance of their model. This model was used by Goldberg (1987), who predicted the trajectories of typical small, medium, and large droplets of water in a spray dryer with a 0.76 m diameter chamber with 1.44 m height. But in the open literature, most of the studies were carried out in small scale spray dryers. For example, Langrish and Zbicinski (1994) carried out an experiment in a 0.779 m^3 spray dryer.

Kieviet (1997) carried out the measurement of air flow patterns and temperature profiles in a cocurrent pilot spray dryer (diameter 2.2 m). FLOW3D was used to model such a spray dryer and the results showed that experimental data agree qualitatively with the predicted ones, corroborating needs of improvements on measurement techniques and on turbulence model considerations.

Fletcher et al. (2003) used a commercial CFD software (CFX) to simulate the full-scale spray dryers. Results obtained for the three scales (laboratory, pilot, and industrial) SD in two configurations (tall and short height to diameter chambers) showed the possibility of exploring the effect of the operational variables on flow stability, wall deposition, and product quality.

Cakaloz et al. (1997) studied a horizontal spray dryer to dry α -amylase. However, it is observed that in the flow pattern in Cakaloz et al. design is not optimal for spray drying since the main air inlet is located at a corner of the chamber. This arrangement makes the spray more likely to hit the top wall unless designed carefully.

Verdurmen et al. (2004) proposed an agglomeration model to be included into CFD models to predict agglomeration process in a spray dryer. However, this is still under investigation.

Huang and Mujumdar (2007) were the first to investigate a spray dryer fitted with a centrifugal atomizer using a CFD model. In their model, they model the rotary disk atomization into the disk side point injection which is the same as the holes in the disk.

9.3.1 Governing Equations for the Continuous Phase

For any fluid, its flow must obey the conservation of mass and momentum. These conservation equations can be found in standard fluid dynamic literature (for incompressible gas) (Bird et al. 1960).

The general form of the continuity equation for mass conservation is

$$\frac{\partial \rho}{\partial t} + \frac{\partial(\rho u_i)}{\partial x_i} = M_{\mathrm{m}} \tag{9.5}$$

The source term M_{m} is the mass added to the continuous phase, coming from the dispersed phase due to droplet evaporation. The general form of the equation for momentum conservation is

$$\frac{\partial(\rho u_i)}{\partial t}+\frac{\partial(\rho u_i u_j)}{\partial x_i}=-\frac{\partial P}{\partial x_j}+\frac{\partial \tau_{ij}}{\partial x_i}+\rho g_i+M_{\mathrm{F}} \tag{9.6}$$

This is in accordance with Newton's law (mass times acceleration = sum of forces) where the first term on the left-hand side of Equation 9.6 is dedicated to the rate of increase of momentum per unit volume and the second term is the momentum increase or decrease per unit volume due to convection. The first term on the right-hand side of Equation 9.6 is the pressure force on a fluid element per unit volume, the second term is the viscous force on a fluid element per unit volume and the third term is the gravitational force on a fluid element per unit volume. The last term M_{F} is the momentum source term.

For Newtonian fluids, the components of the stress tensor τ_{ij} in Equation 9.6 can be written as

$$\tau_{ij}=\mu\left(\frac{\partial u_i}{\partial x_j}+\frac{\partial u_j}{\partial x_i}\right)-\frac{2}{3}\delta_{ij}\frac{\partial u_1}{\partial x_1}$$

with the fluid viscosity μ and the volume dilation term with the "Kronecker" delta:

$$\delta_{ij}=\begin{cases}1 & \text{for} \quad i=j\\ 0 & \text{for} \quad i\neq j\end{cases}$$

The general form of the energy equation is

$$\frac{\partial(\rho c_{\mathrm{p}} T)}{\partial t}+\frac{\partial(\rho c_{\mathrm{p}} u_i T)}{\partial x_i}=\frac{\partial}{\partial x_i}[\lambda\frac{\partial T}{\partial x_i}-\overline{\rho u_i' T'}]+M_h \tag{9.7}$$

where the first term on the left-hand side of Equation 9.7 is dedicated to the rate of increase of energy per unit volume and the second term is the energy increase/decrease per unit volume due to convection. The first term on the right-hand side of Equation 9.7 is energy on a fluid element per unit volume and the last term M_{h} is the energy source term.

9.3.2 Governing Equations for the Particle

Based on the solution obtained for the flow field of the continuous phase, using an Euler–Lagrangian approach we can obtain the particle trajectories by solving the force balance for the particles taking into account the discrete phase inertia, aerodynamic drag, gravity g_i and further optional user-defined forces F_{xi}.

$$\frac{\mathrm{d}u_{\mathrm{p}i}}{\mathrm{d}t}=C_{\mathrm{D}}\frac{18\mu}{\rho_{\mathrm{p}}d_{\mathrm{p}}^2}\frac{\mathrm{Re}}{24}(u_i-u_{\mathrm{p}i})+g_i\frac{\rho_{\mathrm{g}}-\rho}{\rho_{\mathrm{g}}}+F_{xi} \tag{9.8}$$

with particle velocity $u_{\mathrm{p}i}$ and fluid velocity u_i in direction i, particle density ρ_{p}, gas density ρ_{g}, particle diameter d_{p}, and relative Reynolds number

$$Re = \frac{\rho d_{\rm p} \left| u_{\rm p} - u_{\rm g} \right|}{\mu} \tag{9.9a}$$

and drag coefficient

$$C_{\rm D} = a_1 + \frac{a_2}{Re} + \frac{a_3}{Re^2} \tag{9.9b}$$

where a_1 to a_3 are constants (FLUENT 2007).

Two-way coupling allows for interaction between both phases by including the effects of the particulate phase on the fluid phase. In order to simplify the model and computation, the particles are usually assumed to be fully dispersed, i.e., they are not interacting with each other. The particle trajectory is updated in fixed intervals (so-called length scales) along the particle path. Additionally, the particle trajectory is updated each time the particle enters a neighboring cell.

9.3.3 Turbulence Models

In turbulent flows, the instantaneous velocity component u_i is the sum of a time-averaged (mean) value $\bar{u}_i$ and a fluctuating component u_i' as shown in Equation 9.10.

$$u_i = \bar{u}_i + u_i' \tag{9.10}$$

These fluctuations need to be accounted for in the above illustrated Navier–Stokes equation

$$\frac{\partial}{\partial t}(\rho u_i) + \frac{\partial}{\partial x_j}(\rho u_i u_j) = -\frac{\partial P}{\partial x_i} + \frac{\partial}{\partial x_j}\left[\mu\left(\frac{\partial u_i}{\partial u_j} + \frac{\partial u_j}{\partial x_i} - \frac{2}{3}\delta_{ij}\frac{\partial u_l}{\partial x_l}\right)\right] + \frac{\partial}{\partial x_j}(-\rho\overline{u_i'u_j'}) \tag{9.11}$$

Compared with Equation 9.6, Equation 9.11 contains the term $-\rho\overline{u_i'u_j'}$, the so-called Reynolds stress, which represents the effect of turbulence and must be modeled by the CFD code. Limited computational resources restrict the direct simulation of these fluctuations, at least for the moment. Therefore the transport equations are commonly modified to account for the averaged fluctuating velocity components. Three commonly applied turbulence modeling approaches have been used in the CFD model of spray drying system, i.e., k–ε model (Launder and Spalding 1972, 1974), RNG k–ε model (Yakhot and Orszag 1986), and a Reynolds stress model (RSM) (Launder et al. 1975).

The standard k–ε model focuses on mechanisms that affect the turbulent kinetic energy. Robustness, economy, and reasonable accuracy over a wide range of turbulent flows explain its popularity in industrial flow and heat transfer simulations. The RNG k–ε model was derived using a rigorous statistical technique (called Re-Normalization Group theory). It is similar in form to the standard k–ε model, but the effect of swirl on turbulence is included in the RNG mode enhancing the accuracy for swirling flows.

The RSM solves transport equations for all Reynolds stresses and the dissipation rate ε and therefore does not rely on the isotropic turbulent viscosity μ_t. This makes the RSM suitable to predict even swirling flows, however, the major drawback of this model is the computational effort needed to solve its equations. For 3D-simulations, seven additional transport equations must be solved (six for the Reynolds stresses and one for ε). However, the RSM is highly recommended if the expected flow field is characterized by anisotropy in the Reynolds stresses as is the case with swirling flows, e.g., cyclones or spray drying with tangential inlet ducts. Crowe et al. (1980) emphasized that the k–ε model is not suitable for swirl flow problems.

9.3.4 Heat and Mass Transfer Models

In general, there are two drying rate periods, i.e., constant drying rate period (CDRP) and falling drying rate period (FDRP) during droplet drying. CDRP is controlled by mass transfer between the drying medium and the droplet. But FDRP is controlled by the mass diffusion within the droplets or particles.

The heat transfer between the droplet and the hot gas is updated according to the heat balance relationship given as follows

$$M_p c_p \frac{dT_p}{dt} = hA_p(T_g - T_p) + \frac{dM_p}{dt}\Delta H^{vap} \tag{9.12}$$

where

M_p is the mass of the particle (kg)
c_p is heat capacity of the particle (J kg^{-1} K^{-1})
A_p is the surface area of the particle (m^2)
T_g is the local temperature of the hot medium (K)
h is the convective heat-transfer coefficient (W m^{-2} K^{-1})
ΔH^{vap} is the latent heat (J kg^{-1})
$\frac{dM_p}{dt}$ is the evaporation rate (kg s^{-1})

The heat transfer coefficient is evaluated using the correlation of Ranz and Marshall (1952a,b):

$$Nu = \frac{hd_p}{\lambda_g} = 2.0 + 0.6Re^{1/2}Pr^{1/3} \tag{9.13}$$

where

d_p is the particle diameter (m)
λ_g is the thermal conductivity of the hot medium (Wm^{-1} K)
Re is the Reynolds number based on the particle diameter and the relative velocity (Equation 9.9a)
Pr is the Prandtl number of the hot medium ($Pr = c_p\mu_g/\lambda_g$)

The mass transfer rate is given by

$$J_i^{\mathrm{m}} = k(c_{i,\mathrm{sur}} - c_{i,\mathrm{g}}) \tag{9.14}$$

where
- J_i^{m} is the mass flux of vapor (kg m^{-2} s^{-1})
- k is the mass transfer coefficient (m s^{-1})
- $c_{i,\mathrm{sur}}$ is the vapor concentration at the droplet surface (kg m^{-3})
- $c_{i,\mathrm{g}}$ is the vapor concentration in the bulk gas (kg m^{-3})

The concentrations $c_{\mathrm{w,sur}}$ and $c_{\mathrm{w,g}}$ are defined as

$$c_{i,\mathrm{sur}} = MW_i \frac{P_i^{\mathrm{sat}}(T_{\mathrm{p}})}{RT_{\mathrm{p}}} \tag{9.15}$$

$$c_{i,\mathrm{g}} = MW_i y_i \frac{P_{\mathrm{op}}}{RT_{\mathrm{g}}} \tag{9.16}$$

where
- P_i^{sat} is the vapor pressure at the particle surface and corresponding temperature (Pa)
- R is the universal gas constant (8.314 J mol^{-1} K^{-1})
- y_i and MW_i are the local bulk mole fraction and the molecular weight of species i
- P_{op} is the operating pressure (Pa)

The mass transfer coefficient in Equation 9.17 is calculated from the Sherwood correlation (Ranz and Marshall 1952a,b):

$$Sh = \frac{kd_{\mathrm{p}}}{D_{\mathrm{m}}} = 2.0 + 0.6Re^{1/2}Sc^{1/3} \tag{9.17}$$

where
- D_{m} is the diffusion coefficient of vapor in the bulk (m^2 s^{-1})
- Sc is the Schmidt number ($Sc = \frac{\mu}{\rho D_{\mathrm{m}}}$)

9.3.5 Solver

In order to get the numerical results, the above equations must be solved in series. In general, the solution step of a CFD problem is carried out in two steps:

- Discretization: Integration of the governing equations for conservation of mass and momentum, and other scalars (e.g., turbulence parameters) on a cell (=control volume) yielding a set of mathematical expressions for the dependent variables, such as velocity, pressure, etc.

- Linearization: The above-obtained set of mathematical expressions has to be linearized and solved to update the dependent variables in the control volume (cells).

Starting with an initial guessed solution—provided by the user—this solution procedure is repeated until the preset convergence criteria are met and a final solution is obtained via an iterative process. The solution history can be monitored by plotting the sum of the residuals for each dependent variable at the end of each iteration. For a converged solution, the residuals should be a small value (so-called round off).

9.3.6 Typical Simulation Results Using CFD Model

Here is an example using CFD model by Huang (2005). Good agreement was obtained through comparison of its simulated results with the measurements. In the simulation, a pressure nozzle is used in this spray dryer. The feed is skim milk. The physical properties of skim milk are selected in constant values (Holman 1976). The properties of air in the simulation are varied with its temperature (Li et al. 1978). Different mesh sizes are designed and selected to get the mesh-independent results. In Table 9.2, the drying chamber dimension, main operating parameters, and values for simulation are summarized.

Figure 9.6 shows the typical results, e.g., air velocity vector, air streamline, temperature contour, and particles trajectories. From Figure 9.6a and b, it is seen that there is a strong recirculation zone in the drying chamber. It is also noted that there is a nonuniform velocity distribution in the core region of the chamber. The velocity is reduced as the air flows downwards in the chamber.

The temperature contours (Figure 9.6c) show that temperatures in the central core region vary significantly. This is due to the intense heat and mass transfer that occurs during the initial contact between the spray and drying air caused by the high

TABLE 9.2
Drying Chamber Dimension and Main Operating Parameters for Spray Drying

Drying Chamber			
Cylindrical Height (m)	Cylindrical Diameter (m)	Cone Height (m)	Cone Angle (°)
2.0	2.215	1.725	60

Main Operating Parameters				
Air Flow Rate (kg/s)	Air Temperature (°C)	Feed Rate (kg/h)	Droplet Size (μm)	Droplet Velocity (m/s)
0.336	195	50	10–138	59
Feed Solid content (%)	Turbulent Kinetic Energy (m^2/s^2)	Energy Dissipation Rate (m^2/s^3)		
42.5	0.027	0.3740		

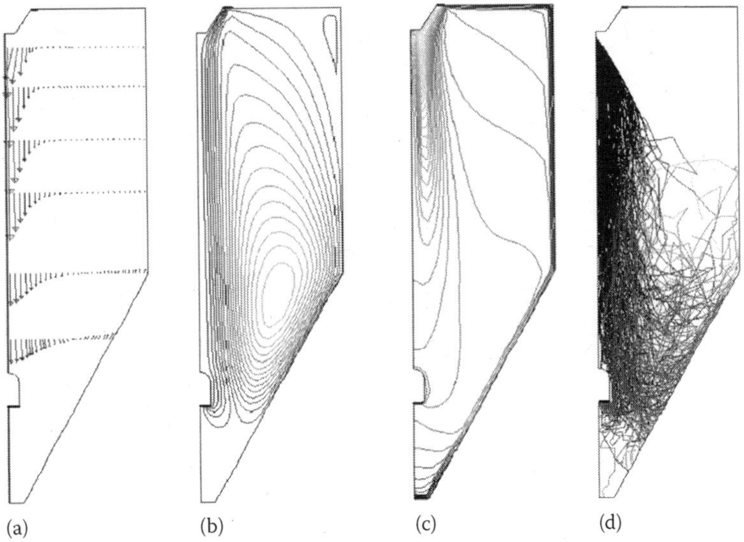

FIGURE 9.6 Simulation results using a CFD model. (a) Velocity vector. (b) Streamline. (c) Temperature contours. (d) Particle trajectory.

relative velocity between these two phases and the large temperature driving force. Only a minor radial variation of air temperature is found in the remaining volume except for the region very near the chamber wall.

From Figure 9.6d of the particle trajectories, it is easy for the user to find the particle deposit position in the chamber. This information is useful for the designer as well as the spray dryer user to know the place of the deposit in the chamber.

9.4 SPRAY DRYING APPLICATIONS IN THE FOOD INDUSTRY

Most food processing companies use spray dryers to produce powdered products. Spray drying has the ability to handle heat sensitive foods with maximum retention of their nutritive content. The flexibility of spray dryer design enables powders to be produced in the various forms required by consumer and industry. This includes agglomerated and nonagglomerated powders having precise particles size distribution, residual moisture content, and bulk density. As examples, spray drying of milk, tomato juice, tea extracts, and coffee is discussed.

9.4.1 Spray Drying of Milk

Milk is one of the most nutritious foods. It is rich in high quality protein providing all 10 essential amino acids. It contributes to total daily energy intake, as well as essential fatty acids, immunoglobulins, and other micronutrients. Commercially available milk can be classified into two major groups: liquid milk and dried or powdered milk. Due to long shelf life of the powdered milk, it is more popular in our daily life.

In general, there are two ways of spray drying milk, i.e., one-stage spray drying system with pneumatic conveying system (shown in Figure 9.7) and multistage spray

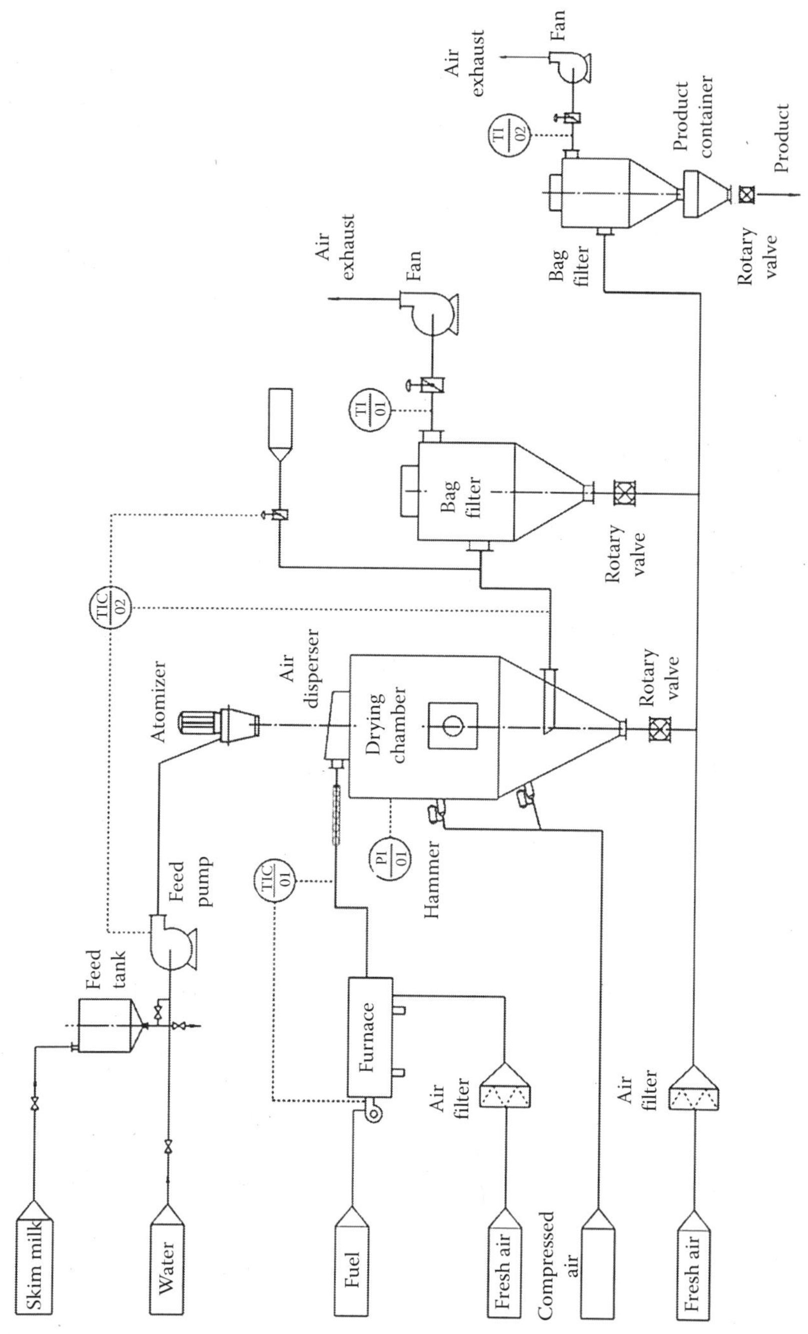

FIGURE 9.7 Conventional spray dryer with pneumatic conveying system.

drying system with or without inner static fluid bed (IFB) and external vibrated fluid bed (VFB) dryer or cooler (shown in Figure 9.3). This is the basic dairy plant featuring a cocurrent drying chamber with either rotary disc or nozzle atomization. The drying chamber can be either a standard conical design or has a static fluid bed integrated into the chamber base.

The advantage of the multistage spray drying system can be summarized as follows: (1) higher capacity per unit drying air; (2) better economic performance, since low outlet temperature can be used; (3) better product quality, e.g., good solubility, good flowability, and high bulk density of product. Here, the pneumatic conveying system is replaced by the internal fluid bed and vibrated fluid bed.

Both semi-instant skim milk and instant, agglomerated skim milk can be produced in MSD. Since the internal static fluid bed (IFB) can be used as the second drying stage and the main agglomeration device, lower outlet temperature can be obtained. The energy consumption is reduced significantly as well. At this process, the final stage (VFB) is usually used as a cooler. If the multistage system only includes SD and VFB, VFB is used as both dryer and cooler.

The performance of various designs for producing skim milk powders was compared by Masters (1991) and Westergaard (1994). Their results showed that MSD gives minimum energy consumption per unit product. The powders produced by MSD are agglomerated and free flowing and with low bulk density.

9.4.2 Spray Drying of Tomato Juice

Fresh, ripe tomatoes are soaked in a vat and then transported by a rolling conveyor to a spray-washing vat. Following washing, the tomatoes are manually sorted and crushed in a chopper to obtain the pulp. If the "hot break" is used, the tomatoes must be heated at 85°C–90°C prior to crushing. Seeds and skin need to be removed prior to refinement of particle size. Since tomato puree which contains skin and seeds produce a more dryable product, it is recommended that the seeds are grinded to 325 μm in order to pass through a specific mesh screen. When a "cold break" is used, tomato pulp is held for a few seconds in order to obtain pectin decomposition, which will provide an easier-to-spread paste. However, the powder obtained from the hot break is more desirable. The juice is concentrated in a multieffect evaporator and this product goes to a feeding tank to be pumped to the spray drying (Masters 1976). The spray drying system is similar to that shown in Figure 9.3. The IFB and VFB system attached to the spray dryer plays an important role in the drying process of tomato juice. The cooling of drying chamber wall is sometimes necessary since it has some low melting-point ingredients in the tomato powder.

Cold break tomato pastes are spray dried at higher concentrations than hot break pastes. For moderate drying air temperatures, the intake of cool air is controlled in order to maintain a temperature between 40°C and 50°C. The tomato heavy paste goes to a rotary atomizer that contains several vanes. The paste is sprayed into a stream of hot air (140°C–150°C) and then cooled. Droplets of 120–250 μm are desirable for tomatoes with 28% solids constituents. When low drying temperatures are used during spray drying, the system has slow evaporation rates and very long chambers are required. Usually the chambers are tens of meter high to increase the droplet falling and drying time. Droplet expansion takes place very slowly because of the low drying temperature. However, the volatile compounds present in the paste will be retained as well as the quality of tomato solids. Tomato powder comes out of the drying chamber with 10% moisture content which is later reduced by using a fluidized bed attached to the base of the dryer chamber. The tomato powder is packaged in an air-conditioned packing room (Masters 1976).

Lumpiness decreases as cooling powder increases. However, the maximum powder temperature to obtain a lump-free product during storage depends upon the type of the tomato. If the product is utilized within few months and it is atmosphere packaged in dry air at low temperature, the product will suffice. In order to prevent lumpiness noncaking agents can be used. Since tomato powder cannot contain more than 2% moisture content, nitrogen or carbon dioxide atmosphere packaging is the most appropriated. For this reason low moisture content of dried fruit juices is required for storage. Food gel silica gel and other noncaking additives can be utilized to prevent caking.

9.4.3 Spray Drying of Tea Extracts

There are different kinds of tea, such as green tea, black tea, oolong tea, toasted tea, brown rice tea, etc. Here, we take green tea as an example. First of all, an extract solution from tea leaves is obtained by pouring water onto tea leaves, heat-treating, and filtering it. It is preferable to obtain this solution at a relatively low temperature, for example, in the range of 10°C–40°C. By a low-temperature extraction process, components for bitterness and astringency are not extracted from tea leaves to a great extent, whereas savory components can be extracted effectively. The separation of an extract solution and tea leaves may be carried out by pressing and centrifugal separation. Thus, the elution of components for bitterness and astringency can be prevented.

The extract solution of 3%–5% solid content obtained according to the above method is concentrated by means of multieffect tube-evaporators or a reverse osmosis membrane. This concentration by reverse osmosis has the advantages of the flavor not disappearing and the deterioration due to heat is small, since mild operating temperatures are employed. It is preferable to concentrate the solution at a temperature of about 25°C, since water flux can be too low at lower temperatures. The solids content in the extract solution is increased from 10% to 30%.

The tea extracts at a concentration of about 20%–25% are stored in a tank for spray drying. A pressure nozzle spray drying system is usually selected. The process is similar to that shown in Figure 9.2. When the tea extracts are transported into the pressure nozzle, it is separated into millions of fine droplets at an operating pressure of 2.7–3.5 MPa. The inlet air temperature is usually at 180°C–220°C. A cooled and dehumidified air must be used to transport and cool the powder products exiting from the drying chamber and cyclone base. Otherwise, too high packing temperature may degrade the tea powder quality.

9.4.4 Spray Drying of Coffee

Coffee is one of the world's most popular beverages. The production of coffee powder usually consists of roasting, grinding, extraction, spray drying, and agglomeration. The production of soluble coffee is a typical example that shows the need of this new drying process definition. Consumers are more demanding about the instant coffee quality requiring similar flavor and aroma of the regular coffee. To become more competitive in the international market, the coffee producers must enhance the quality of their soluble coffee. The traditional process can be categorized into three basic steps: (1) drying of green coffee beans; (2) roasting and grinding of these beans; and (3) extracting and drying of the coffee liquor. To improve the energy efficiency and to avoid environmental pollution, one more step has been added to this process corresponding to the treatment of the coffee sludge for reusing as a fuel (sludge must be dried and, if feasible, gasified). Roasting of green beans needs to be improved to enhance coffee flavor. Both, new gas-particle contactor equipment and grain propriety variable measurable on line (different from color) are required to save energy and to control precisely the roasting operation time, increasing the end-product quality. The spray drying operation must be optimized as a function of the energy consumption and the powder quality. Changes in the solid material properties should be described together with the specific drying mechanisms observed in the spray chamber to overcome the loss of some volatile compounds responsible for the coffee flavor and aroma. This loss can be minimized by reducing the length of the spray formation region and decreasing faster water concentration into droplets until the critical value dictated by crust formation. Such considerations lead to work in a nozzle tower spray dryer (high height to diameter ratio with a pressure nozzle atomizer type) with a concurrent and a more concentrated coffee liquor. There is also the feasibility to use a secondary cooling air flow at the spray dryer base, as shown in Figure 9.3 (Huang and Mujumdar 2007).

Generally, the preconcentrated coffee extract (40%–50% solid content) is atomized by high pressure nozzles in a cocurrent flow drying chamber. A direct air heater is usually used. The drying operates at the inlet and outlet temperatures of 220°C–240°C and 105°C–115°C. Sometimes the second dehumidified and cool air is induced into the chamber cone to further cool the coffee powder. The exhaust air and fine coffee powder exit from the outlets in the middle of the chamber bustle. The upper chamber may be insulated and lower cone un-insulated. The powder from the drying chamber may be further dried and cooled in VFB dryer and cooler. The collected powder from the cyclone may be conveyed to the VFB or return to the atomization zone for agglomeration. The nonagglomerated coffee powder with 3% residual moisture normally has a particle size in the range 100–400 μm.

9.4.5 Spray Drying of Eggs

Egg is the most nutritious natural product. Eggs are rich in protein, vitamins, and minerals. During the last three decades, the poultry industry in China has made remarkable progress and grown into an organized and highly productive industry. Dried egg powder can be stored and transported at room temperatures. It is quite stable and has long shelf life. The manufacture of egg powder is an important segment of egg consumption. Nowadays there are some plants of eggs powder production with a suitable capacity across the world.

Manufacture of dried egg powder starts with breaking of eggs and removing egg-shells. After removal of shells, the mixture is filtered and stored in tanks at about 4°C. Before it is spray-dried, it is taken to a tubular heater wherein it is heated to about 65°C for 8–10 min. The conventional spray drying system with a pressure nozzle is usually selected. The centrifugal atomization is also found in some factories. The spray drying process is similar to that shown in Figure 9.1. The operating drying temperature is 180°C–200°C. Recently, at least two U.S. companies produce horizontal spray dryers, especially for heat-sensitive foods, such as eggs (Rogers 2008, FES 2008).

9.5 SUMMARY

In this chapter, a summary of the fundamentals of spray drying, selection of spray dryers, and the use of spray dryers in the food industry is provided. Spray dryers, both conventional and innovative, will continue to find increasing applications in various industries; almost all industries need or use or produce powders starting from liquid feedstocks. Therefore, although it is very difficult to generate rules for the selection of spray dryers in the food area because of numerous possible exceptions and new developments, it is important for the users of spray dryers to understand the typical and main characteristics of spray drying in the food industry. Despite advances in modeling of spray dryers, it is important to carry out careful pilot tests and evaluate the quality parameters that cannot yet be predicted with confidence. Further advances in mathematical modeling of sprays and spray dryers are needed before confident design and scale-up can be carried out using models.

REFERENCES

Arnason, G. and Crowe, C. T. 1986. Assessment of numerical models for spray drying. In *Drying'86*, Ed. A. S. Mujumdar. New York: Hemisphere Publishing Corp.

Berger, H. L. 1998. *Ultrasonic Liquid Atomization: Theory and Application*. New York: Partridge Hill Publishers.

Bird, R. B., Stewart, W. E., and Lightfoot, E. N. 1960. *Transport Phenomena*. New York: John Wiley & Sons Inc.

Cakaloz, T., Akbaba, H., Yesugey, E. T., and Periz, A. 1997. Drying model for α-amylase in a horizontal spray dryer. *J. Food Eng*. 31:499–510.

Crowe, C. T. 1980. Modeling spray air contact in spray drying systems. In *Advances in Drying*, Ed. A. S. Mujumdar, Vol. 1, pp. 63–99. New York: Hemi-sphere.

Crowe, C. T., Sharma, M. P., and Stock, D. E. 1977. The particle-source-in-cell (PSI-Cell) model for gas-droplet flows. *J. Fluid Eng*. 9:325–332.

Crowe, C., Sommerfeld, M., and Tsuji, Y. 1998. *Multiphase Flows with Droplets and Particles*. Boca Raton, FL: CRC Press.

FES company 2008. http://www.fesintl.com/htmfil.fld/sprydryh.htm

Fletcher, D., Guo, B., Harvie, D. et al. 2003. What is important in the simulation of spray dryer performance and how do current CFD models perform? *Proceedings of the 3rd International Conference on CFD in the Minerals and Process Industries*, Melbourne, Australia, www.cfd.com.au.

FLUENT Manual 2007. http://www.fluent.com.

Gauvin, W. H., Katta, S., and Knelman, F. H. 1975. Drop trajectory predictions and their importance in the design of spray dryers. *Int. J. Multiphase Flow*, 1:793–816.

Goldberg, J. E. 1987. Prediction of spray dryer performance, PhD thesis, University of Oxford, Oxford, U.K.

Huang, L. X. 2005. Simulation of spray drying using computational fluid dynamics, PhD thesis, National University of Singapore, Kent Ridge, Singapore.

Huang, L. X., Kumar, K., and Mujumdar, A. S. 2003. Use of computational fluid dynamics to evaluate alternative spray chamber configurations. *Drying Technol.* 21:385–412.

Huang, L. X. and Mujumdar, A. S. 2007. Simulation of an industrial spray dryer and prediction of off-design performance. *Drying Technol.* 25:703–714.

Huang, L. X., Tang, J., and Wang, Z. 1997. Computer-aided design of centrifugal spray dryer, *J. Nanjing Forestry Univ.* 21(add.): 68–71 (in Chinese).

Huang, L. X., Wang, Z., and Tang, J. 2001. Recent progress of spray drying in China, *Chem. Eng. (China)* 29:51–55 (in Chinese).

Holman, J. P. 1976. *Heat Transfer*. New York: McGraw-Hill.

Keey, R. B. 1991. Private communication by Dr. Masters. In *Spray Drying Handbook*, 5th edn., p. 243, New York: John Wiley & Sons Inc.

Kieviet, F. G. 1997. Modeling quality in spray drying, PhD thesis, Endinhoven University of Technology, the Netherlands.

Langrish, T. A. G. and Kockel, T. K. 2001. The assessment of a characteristic drying curve for milk powder for use in computational fluid dynamics modeling. *Chem. Eng. J.* 84:69–74.

Langrish, T. A. G. and Zbicinski, I. 1994. The effect of air inlet geometry and spray cone angle on the wall deposition rate in spray dryers, *Trans. I. Chem. E.* 72:420–430.

Launder, B. E., Reece, G. J., and Rodi, W. 1975. Progress in the development of a reynolds-stress turbulence closure. *J. Fluid Mech.* 68:537–566.

Launder, B. E. and Spalding, D. B. 1972. *Lectures in Mathematical Models of Turbulence*. London, U.K.: Academic Press.

Launder, B. E. and Spalding, D. B. 1974. The numerical computation of turbulent flows. *Comput. Methods Appl. Mech. Eng.* 3:269–289.

Li, Y. K., Mujumdar, A. S., and Douglas, W. J. M. 1978. Coupled heat and mass transfer under a laminar impinging jet. In *Proceedings of the First International Symposium on Drying*, Ed. A. S. Mumudar, pp. 175–184, Montreal, Canada: McGill University, Canada.

Livesley, D. M., Oakley, D. E., Gillespie, R. F. et al. 1992. Development and validation of a computational model for spray-gas mixing in spray dryers. In *Drying's 92*, Ed. A. S. Mujumdar, pp. 407–416, New York: Hemisphere Publishing Corp.

Masters, K. 1976. *Spray Drying Handbook*, 1st edn., London, U.K.: George Godwin Ltd.

Masters, K. 1991. *Spray Drying Handbook*, 5th edn., New York: John Wiley & Sons.

Masters, K. 2002. *Spray Drying in Practice*. Denmark: SprayDryConsult International ApS.

Mujumdar, A. S. 1995. Superheated steam drying. In *Handbook of Industrial Drying*, 2nd edn., Ed. A. S. Mujumdar, pp. 1071–1086, New York: Marcel Dekker.

Mujumdar, A. S. 2000. Dryers for particulate solids, slurries and sheet-form materials. In *Mujumdar's Practical Guide to Industrial Drying*, Ed. S. Devahastion, pp. 37–71, Brossard, Quebec, Canada: Exergex Corp.

Negiz, A., Lagergren, E. S., and Cinar, A. 1995. Mathematical models of cocurrent spray drying, *Ind. Eng. Chem. Res.* 34:3289–3302

Oakley, D. 1994. Scale-up of spray dryers with the aid of computational fluid dynamics, *Drying Technol.*, 12:217–233.

Oakley, D. E. and Bahu, R. E. 1990, Spray/gas mixing behavior within spray dryers, *7th International Symposium Drying*, in *Drying'91*, ed. A. S. Mujumdar and I. Filkova, pp. 303–313, Amsterdam, the Netherlands: Elsevier.

Papadakis, S. E. and King, C. J. 1988a. Air temperature and humidity profiles in spray drying, part 1: Features predicted by the particle source in cell model. *Ind. Eng. Chem. Res.* 27:2111–2116.

Papadakis, S. E. and King, C. J. 1988b. Air temperature and humidity profiles in spray drying, part 2: Experimental measurements. *Ind. Eng. Chem. Res.* 27:2116–2123.

Rajan, R. and Pandit, A. B. 2001. Correlations to predict droplet size in ultrasonic atomization. *Ultrasonics* 39:235–255.

Ranz, W. E. and Marshall, W. R. 1952a. Evaporation from drops. *Chem. Eng. Prog.* 48:141–146.

Ranz, W. E. and Marshall, W. R. 1952b. Evaporation from drops. *Chem. Eng. Prog.* 48:173–180, Rogers company 2008. http://www.cerogers.com/html/horizontal_dryer.html

Straatsma, J., Houwelingen, G. V., Steenbergen, A. E. et al. 1999. Spray drying of food products: 1. Simulation model. *J. Food Eng.* 42:67–72.

Verdurmen, R. E. M., Menn, P., Ritzert, J. et al. 2004. Simulation of agglomeration in spray drying installations: The EDECAD project. *Drying Technol.* 22:1403–1462.

Westergaard, V. 1994. *Milk Powder Technology: Evaporation and Spray Drying*, Denmark: Niro A/S.

Yakhot, V. and Orszag, S. A. 1986. Renormalization group analysis of turbulence: i. basic theory, *J. Sci. Comput.* 1:1–51.

10 Superheated-Steam Drying Applied in Food Engineering

Somkiat Prachayawarakorn
and Somchart Soponronnarit

CONTENTS

10.1 INTRODUCTION

The primary aim of food drying is to preserve the product, as a decrease in the product moisture content can prevent the growth of microorganisms and enzymatic reactions. However, drying may have an adverse effect on physical, chemical, and nutritional values of food products. The success or failure of a drying process depends on the product quality obtained after drying and also on the efficiency of the process.

An idea of using superheated steam instead of hot gas (air, combustion, or flue gases) for drying was initially introduced in Germany in 1908 as reported by Douglas (1994), but its application was limited to only a few industries. Later, its use has widely been acknowledged in many applications, i.e., paper (Svensson, 1980; Douglas, 1994), foods (Iyota et al., 2001; Taechapairoj et al., 2003; Tang et al., 2005), coal (Chen et al., 2000), and wood (Pang and Dakin, 1999). The product quality after drying is usually better in superheated-steam drying (SSD) than in hot-air drying, particularly in terms of less shrinkage and higher product porosity.

A superheated-steam dryer is normally designed as a closed loop in which exhaust steam may be either reused or employed in other processes resulting in net energy savings. The temperature of superheated steam used under atmospheric pressure generally varies between 100°C and 150°C. The high temperature restricts its application to foods that are not sensitive to heat. Mostly, the applications of SSD are related to starch-based products such as parboiled rice, noodles, potato chips, and durian chips (Iyota et al., 2001; Markowski et al., 2003; Taechapairoj et al., 2004; Jamradloedluk et al., 2007).

In this chapter, the basic concepts of SSD, drying characteristics, mathematical modeling of the process operation, and quality consideration of the selected food products are outlined and discussed.

10.2 ADVANTAGES AND LIMITATIONS OF SUPERHEATED STEAM DRYING

When water in a liquid state is contained in a closed vessel and heated under a given pressure, its temperature increases up to the boiling point as shown in Figure 10.1. At this point, water is called saturated liquid (shown by letter a in Figure 10.1). Upon continued heating, the temperature would remain constant at that saturation value and water would continually vaporize. During this heating period, the system contains two phases, liquid and vapor. After the last droplet of water vaporizes, there is only water vapor in the system, which is called saturated steam (shown by letter b in Figure 10.1). From a practical viewpoint, saturated steam cannot be used to dry any materials since the steam is wet. To obtain dry or superheated steam, saturated steam

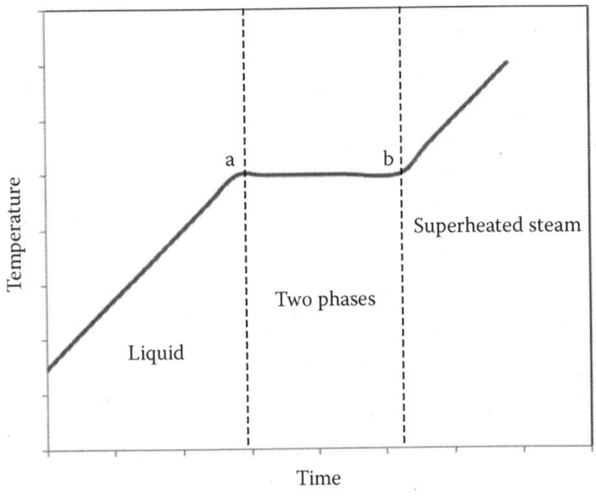

FIGURE 10.1 Illustration of changing states of water during heating.

needs to be further heated by flowing it through a heat exchanger. As a result, the temperature of steam increases beyond its saturation temperature. The temperature difference between superheated and saturated steam is called degree of superheat.

Using superheated steam in an industrial process can lead to substantial energy savings if the vapor evaporating from the product being dried is condensed and if the latent heat of condensation can be recovered and used in other unit operations. This benefit makes superheated steam a more attractive drying medium for industrial use, especially if the energy cost for drying is a major proportion of the total production cost. In addition, the use of superheated steam for drying foods has advantages to both consumers and industry as detailed by many researchers (Lane and Stern, 1956; Shibata and Mujumdar, 1994; Tang and Cenkowski, 2000; Deventer and Heijmans, 2001; Soponronnarit et al., 2006):

1. Use of superheated steam as the drying medium leads to an oxygen-free drying system, which prevents oxidative reactions, resulting in an improvement of the product quality.
2. SSD allows pasteurization, sterilization, and deodorization of food products during drying. In addition, products are partially cooked, with possible favorable changes in their textural properties.
3. SSD may reduce processing time and processing steps. For example, in producing parboiled rice by a conventional process, the process consists of three main stages, namely, soaking, steaming, and drying. However, steaming and drying can be combined into one stage with the use of SSD, thereby allowing a significantly shorter processing time.
4. Higher drying rates are obvious in both constant and falling rate periods if the product can be dried at a temperature above the inversion temperature since the temperature difference between particle and fluid stream is larger in superheated steam than in hot air, resulting in higher heat flux from the gas to the particle surface. More details about the inversion temperature will be given in Section 10.3. The higher drying rates would increase the dryer performance, which leads to a reduction in equipment size or an increase in drying capacity.
5. SSD may have high thermal efficiency if the exhaust steam can be used elsewhere in the process.
6. SSD enhances physical food qualities as it leads to less shrinkage and high product porosity due to evolution of moisture inside the product during drying.

However, SSD also has the following limitations:

1. There is unavoidable condensation at the beginning of drying because raw materials are generally fed into the system at ambient temperature. This results in an increase in the moisture content of materials by approximately 2%–3%, resulting in an increase in the drying time by 10%–15%. In addition to the increase in the moisture content, condensation may interrupt the drying system, particularly in the case of a fluidized-bed dryer. Condensation of steam causes particle surface to be very wet and hence the particles would

agglomerate, resulting in more difficulty in fluidizing them. To alleviate this problem, raw materials need to be warmed up before being fed into the dryer.

2. An SSD system is more complicated and requires higher investment for insulation and auxiliary heating than in the case of a hot-air drying system. Shut down and start up also take longer for a superheated-steam dryer than for a hot-air dryer.
3. Application of SSD for heat-sensitive materials is unfavorable. Products that may suffer degradation at higher temperatures cannot be dried in superheated steam. However, it may be possible to use a two-stage drying technique, i.e., SSD followed by hot-air drying to alleviate this shortcoming. For example, drying of chicken meat using this two-stage technique appears to produce a higher quality dried product, i.e., product with less brown color and higher rehydration ability (Nathakaranakule et al., 2007).

10.3 FUNDAMENTALS OF DRYING IN SUPERHEATED STEAM

During drying, moisture from a drying product vaporizes into the drying medium, which then carries the water vapor away. The drying rate remains constant as long as the evaporated moisture of the material is available at its surface. During this constant drying rate period, heat transfer controls the moisture evaporation rate. In the case of hot-air drying, the moisture evaporation rate can be calculated by:

$$\frac{\mathrm{d}M_{\mathrm{w}}}{\mathrm{d}t} = \frac{hA\left(T_{\text{drying medium}} - T_{\text{samp sur}}\right)}{\Delta H_{\mathrm{w}}^{\mathrm{vap}}} \tag{10.1}$$

where

$\mathrm{d}M_{\mathrm{w}}/\mathrm{d}t$ is the drying rate (kg water s^{-1})
h is the heat transfer coefficient (W $\mathrm{m}^{-1}\cdot\mathrm{K}^{-1}$)
A is the surface area of material in contact with the drying medium (m^2)
$\Delta H_{\mathrm{w}}^{\mathrm{vap}}$ is the latent heat of evaporation of moisture (J kg^{-1})
T is the temperature (K)

The temperature at sample surface is equal to the wet bulb temperature of hot air. However, when hot air is replaced by superheated steam, $T_{\text{samp sur}}$ in Equation 10.1 is replaced by the saturation (boiling) temperature of steam at the dryer operating pressure. Equation 10.1 is derived based on the assumptions that other modes of heat transfer from the gas medium to the solid sample, such as sensible heat effects and heat losses from solid sample, are neglected.

According to Equation 10.1, the evaporation rate of moisture into superheated steam and hot air may have different values, depending on the temperature difference between the sample surface and the bulk gas/vapor phase, and on the heat transfer coefficient of the drying medium. When the drying temperature is lower than the so-called inversion temperature, the evaporation rate of moisture into hot air is higher than that into superheated steam as illustrated in Figure 10.2. On the other hand, the moisture evaporation rate into a superheated-steam environment becomes

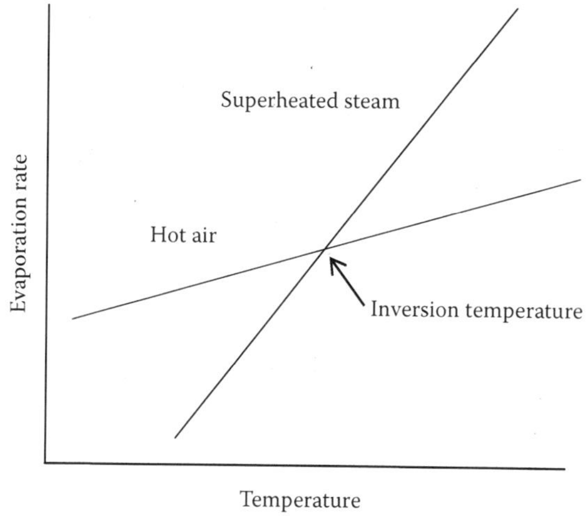

FIGURE 10.2 Variation of the evaporation rates with drying temperatures.

higher than that associated with a hot-air environment at temperatures beyond the inversion temperature.

Below the inversion temperature, the higher moisture evaporation rate in hot-air drying is related to the heat-transfer coefficient and the temperature difference between the sample surface and bulk gas phase. Considered at the same drying condition, i.e., a given gas velocity and a fixed operating pressure, both the heat-transfer coefficient and the temperature difference are higher for hot air than for superheated steam, being the latter variable more important. When the drying temperature reaches the inversion value, the temperature difference in Equation 10.1 is, in turn, smaller in hot-air drying than in SSD. However, this result can be compensated by the higher heat transfer coefficient for hot-air dying, and, hence, the drying rates in SSD and hot-air drying are equal at the inversion temperature. From this point on, the temperature difference is significantly larger for SSD. Moreover, the larger temperature difference in SSD is more dominant than the heat-transfer coefficient in hot-air drying. Thus, it provides higher heat flux and subsequently higher moisture evaporation rate in SSD.

The first experimental investigation to explore the inversion phenomenon was carried out by Yoshida and Hyōdō (1970). They reported an inversion temperature between 160°C and 176°C for the water–air countercurrent flow in a wetted-wall column. Chow and Chung (1983a,b) later presented numerical results of the evaporation of water into superheated steam and dry air for laminar (Chow and Chung, 1983a) and turbulent (Chow and Chung, 1983b) forced convection over a flat plate. Their analysis showed an inversion temperature of 250°C for laminar flow and 190°C for turbulent flow. Schwartze and Bröcker (2002) theoretically studied the evaporation of water into superheated steam, dry air, and humid air. They found the inversion temperature of 199°C for turbulent flow in a wetted-wall column.

The above findings are the results of evaporation rate, which is governed by convective transport phenomena. Several reports have shown that the inversion temperature was also found in the case of porous food drying, in which internal moisture movement mainly governs overall transport mechanisms. The inversion temperature in this case varies from material to material. Prachayawarakorn et al. (2002), for example, dried shrimp in superheated steam and hot air and found the inversion temperature to be between 140°C and 150°C. For drying of potato slices, however, there existed an inversion temperature between 145°C and 160°C for the first drying stage, in which the moisture content was above 2.6 dry basis (d.b.) and between 125°C and 145°C during the last drying stage, when the moisture content was below 2.6 d.b. (Tang and Cenkowski, 2000).

10.3.1 Drying Characteristic Curves

During air-drying, the moisture content of the material decreases continually with time. An illustrative example of this phenomenon is shown in Figure 10.3. With superheated steam, however, the drying characteristic curve is different from that found with hot air; there is a gain in the moisture content in the early stage of drying. This increase in the moisture content is caused by the condensation of steam. However, for longer drying times, the drying curve becomes similar to the one in hot air.

Condensation occurs when superheated steam contacts the material surface, which is at a lower temperature. As a result, superheated steam releases its energy and changes its phase from vapor to liquid. The amount of steam condensed depends on the degree of superheat, which is defined as the difference between the actual

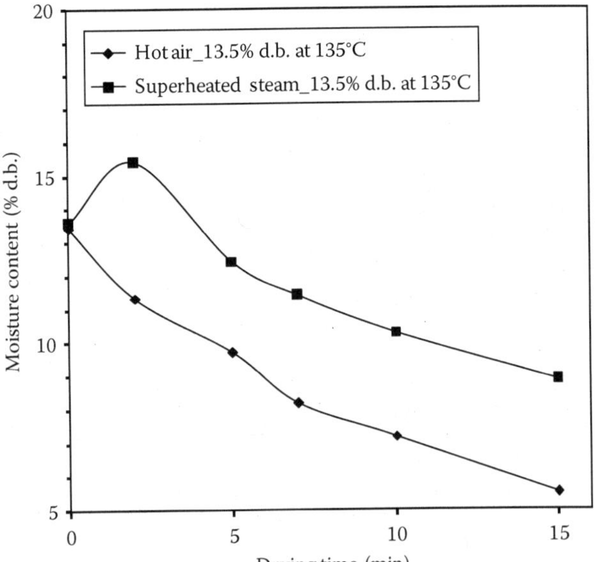

FIGURE 10.3 Drying curves of soybean in superheated-steam and hot-air fluidized bed. (From Prachayawarakorn, S. et al., *LWT-Food Sci. Technol.*, 39, 773, 2006. With permission.)

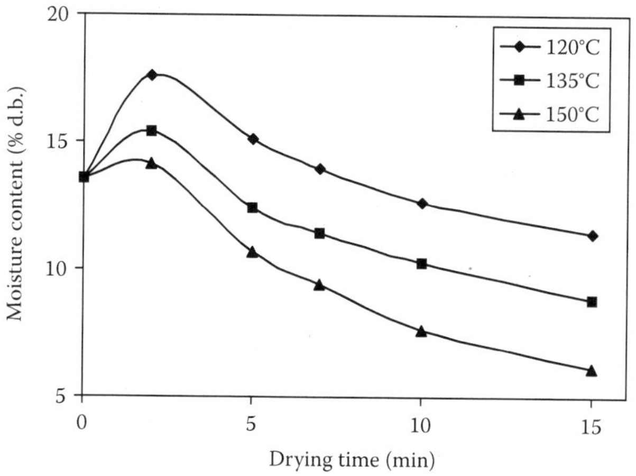

FIGURE 10.4 Change of moisture content of soybean at three different inlet steam temperatures. (From Prachayawarakorn, S. et al., *Drying Technol.*, 22, 2105, 2004. With permission.)

superheated-steam temperature and its saturation value at the dryer operating pressure. A higher degree of superheat obviously leads to a smaller amount of steam condensation, and thus to a smaller increase in the moisture content of the material. The effect of the degree of superheat on the moisture uptake can be seen in Figure 10.4.

Condensation of steam may render the dryer operation unstable, particularly in the case of a fluidized-bed dryer. Condensation and formation of liquid around the feed stream of food material leads to a difficulty in maintaining a desired level of fluidization. In such cases, solid preheating may be a useful option because the amount of steam condensation would then decrease.

Nevertheless, steam condensation does not have only the disadvantages as mentioned earlier. Actually, it sometimes provides several advantages. James et al. (2000) compared three different treatment methods for decontaminating lamb carcasses, namely, steam condensation, hot-water immersion, and chlorinated-hot-water immersion. All three treatments significantly reduced aerobic plate counts on the carcasses. There was no significant difference between steam and hot-water treatments; both treatments reduced the counts by approximately 1 $\log_{10}$ CFU cm^{-2}. In all cases, no significant differences were found in the evaluation of lean appearance, color appearance, odor, and overall acceptability of treated and untreated carcasses after 48 h of chilling and chilled storage. However, the use of steam seemed to have the greatest potential for industrial application since legislation and consumer concern would limit the use of chemical substances such as chlorine in decontamination applications, whereas the use of hot water may present problems in filtering, cleaning, and disposal stages.

In addition, steam condensation helps with gelatinization of starch. It has been reported that gelatinization of many starches is more complete in superheated steam than in hot air, in particular on the material surface, from which the moisture can easily be removed. Figure 10.5, for instance, shows morphologies of durian slices dried with superheated steam and hot air at 150°C (Jamradloedluk et al., 2007).

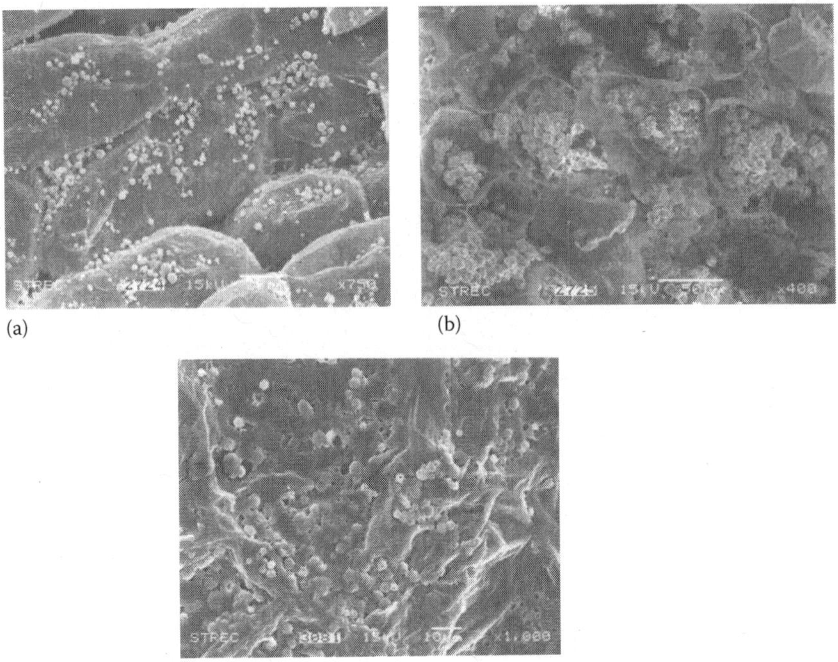

(a) (b) (c)

FIGURE 10.5 Scanning electron micrographs of durian. (a) Surface of raw sample. (b) Surface of sample dried with hot air. (c) Surface of sample dried with superheated steam. (From Jamradloedluk, J. et al., *J. Food Eng.*, 78, 203, 2007. With permission.)

As shown in Figure 10.5a, durian starch granules are spherical with an average diameter of 1–3 μm. When the samples were dried with superheated steam, the starch granules disappeared as shown in Figure 10.5c, indicating that starch was gelatinized. In contrast, ungelatinized starch granules still appeared when durian was subjected to hot-air drying as shown in Figure 10.5b. The gelatinization of starch granules on the material surface led to a smoother surface, produced a transparent layer, and provided a glossy product.

10.3.2 Moisture Diffusivity

Moisture diffusivity is a fundamental parameter for analyzing, designing, and optimizing a drying system. When a material is being dried with hot air, moisture inside it moves through interfacial void spaces, evaporating and reaching the surface. Moisture is then transported away to the flowing stream on account of the moisture concentration difference between the thin, air boundary layer on the material surface and the bulk air stream. Inside the material, the transport of moisture in the falling rate period may occur by several mass transfer mechanisms, such as Knudsen diffusion, molecular diffusion, and capillary flow. All the drying mechanisms, as well as the effect of the porous structure of the solid material, are lumped together into an effective (apparent) diffusion coefficient. This effective moisture diffusivity can be determined from an experimental drying curve under a constant drying temperature, using classical analytical solutions of the diffusion equation, given by Crank (1975).

In SSD, the transport of moisture is driven by the pressure difference between the material surface and bulk stream and there is no mass transfer resistance on the gas side. During drying, the pressure at the material surface is equal to the saturation pressure at material temperature, and the pressure in the bulk stream is equal to the operating pressure. Moisture that exists inside the material starts to be removed when the material temperature reaches the saturation temperature of steam. The mechanisms of moisture movement inside the material in SSD are possibly similar to those in hot-air drying as mentioned above.

The morphology of the food material undergoing SSD may also be different from that of the material undergoing hot-air drying. Differences in such morphologies, in turn, affect the effective moisture diffusivity. Figure 10.6 shows experimental values and trends of the effective moisture diffusivity for foods dried by superheated steam and hot air, indicating lower values of this parameter when superheated steam is used (Poomsa-ad et al., 2002; Taechapairoj et al., 2004; Uengkimbuan et al., 2006; Jamradloedluk et al., 2007). The lower moisture diffusivity is due to the physically dense structures of food materials created during SSD. This led consequently to a lower drying rate in superheated steam compared to hot air.

10.3.3 Mathematical Modeling

This section describes a mathematical model that can be used to predict the temperature and moisture content of a material, especially cereal grain, in a batch superheated-steam dryer, operating near the atmospheric pressure. The model is first derived based on the fundamental transport equations for a drying particle, together with energy and mass balances for the dryer (fluidized bed), neglecting heat transfer by radiation, particle shape or size deformation (disregarding particle shrinkage or growth), and temperature gradients inside the particle. Superheated steam is injected into the bed of particles at constant inlet conditions (i.e., temperature, pressure, and mass flow rate). It is assumed that the inlet steam mass flow rate insures the bed-fluidization regime, and also that bubble formation and flow do not directly affect drying operation in SSD as they do in hot-air fluidized-bed dryers. Note that the free-bubbling regime assures a well mixing of superheated steam, but bubbles pass through the bed as an inactive phase, without the establishment of any concentration gradient to bring about water vapor transport, contrary to what occurs in hot-air drying. During drying, the wet particle with initially low temperature comes into contact with superheated steam. This naturally leads to steam condensation and an increase in both the particle moisture content and temperature. This initial condensation period is described in the mathematical model as the heating-up period. After the particle temperature reaches the saturation temperature of 100°C at an atmospheric pressure, condensation stops. Once the period of initial condensation is over, moisture evaporation starts. During the evaporation period, drying is presumably divided into two subperiods: constant drying-rate period and falling rate periods.

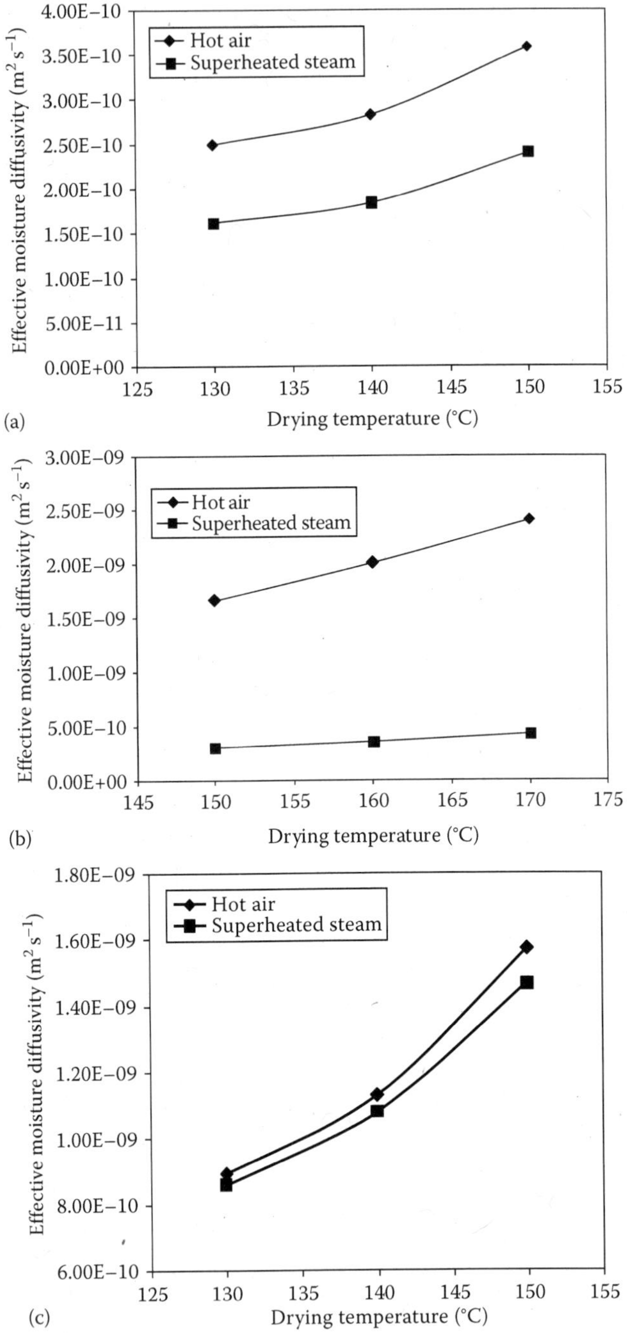

FIGURE 10.6 Comparison of effective moisture diffusivities of food materials dried with superheated steam and hot air. (a) Durian slice. (From Jamradloedluk, J. et al., *J. Food Eng.*, 78, 201, 2007. With permission.) (b) Rice. (From Taechapairoj, C. et al., *Drying Technol.*, 22, 729, 2004; and Poomsa-ad, N. et al., *Drying Technol.* 20, 202, 2002. With permission.) (c) Pork slice. (From Uengkimbuan, A. et al., *Drying Technol.* 24, 1666, 2004. With permission.)

10.3.3.1 Heating-Up Period

During the heating-up period, it is assumed that the condensed water forms a layer of liquid uniformly distributed over the entire external surface of each particle in the bed. This layer acts as a resistance to heat transfer from steam to particle surface. The rate of heat transfer, q, to particles into the bed can be described by:

$$q = h_{\text{fl}} A_{\text{pfl-bed}} \left(T_{\text{st}}^{\text{sat}} - T_{\text{p}} \right) \qquad (10.2)$$

with $A_{\text{pfl-bed}}$ representing the total heat transfer area of particles into the bed, including the thickness of the liquid film covering them.

This rate of heat transfer is equal to the rate of heat released by steam plus the cooling of the condensate, which is expressed by (Holman, 1997):

$$q = \dot{M}_{\text{cond}} \left\{ c_{\text{p,st}} \left(T_{\text{st}}\big|_{\text{in}} - T_{\text{st}}^{\text{sat}} \right) + \Delta H^{\text{vap}} + \frac{3}{8} c_{\text{p,w}} \left(T_{\text{st}}^{\text{sat}} - T_{\text{p}} \right) \right\} \qquad (10.3)$$

or

$$q = \dot{M}_{\text{cond}} \, \Delta H^{\text{vap}} \qquad (10.4)$$

Equation 10.3 is obtained by assuming a linear temperature distribution in the condensed liquid film and a laminar flow of the liquid film. It is also assumed that, in the bed, the condensed water in contact with particles has their temperature. The condensation of steam onto a spherical particle is schematically shown in Figure 10.7.

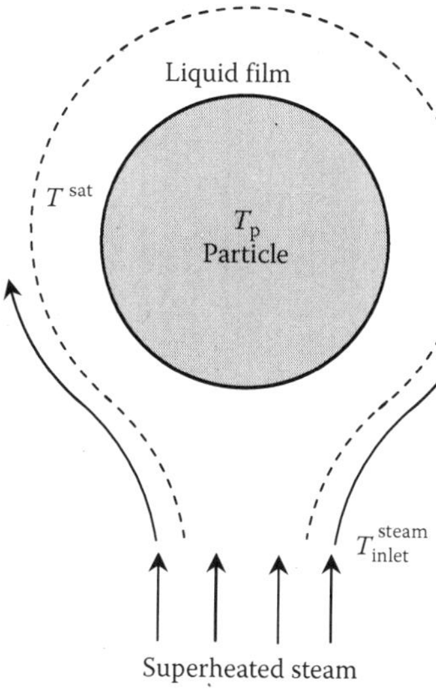

FIGURE 10.7 Liquid film condensation onto the surface of a particle.

Thus, based on these assumptions, the condensation rate of steam, $\dot{M}_{\text{cond}}$, can be calculated. The value of $\dot{M}_{\text{cond}}$ calculated with Equation 10.3 may not be precisely accurate since the thin liquid film may not be stagnated in reference to the fluidized particles and there are strong interactions amongst the solid particles, resulting in less uniformity of liquid film covering the entire particle surface. When the condensation rate is higher than the mass flow rate of steam entering the bed, one would expect that the steam flowing through the bed be completely condensed. In this case, the warming up of particles can be calculated using Equation 10.3, with $\dot{M}_{\text{cond}}$ replaced by the mass flow rate of steam. This situation may occur if the temperature of particles that are fed into the fluidized-bed dryer is rather low. Hence, the particles in the bed cannot fluidize.

For an energy balance of the solid phase, heat released from steam condensation can warm up both the particles and the condensed water, thus:

$$q = \left[M_{\text{ss}} \left(c_{\text{p,s}} + \bar{m}\, c_{\text{p,w}} \right) + M_{\text{flw}} c_{\text{p,w}} \right] \frac{\text{d}T_{\text{p}}}{\text{d}t} \tag{10.5}$$

where

c_{p} is the specific heat capacity
$\bar{m}$ is the average moisture content of particles
M_{flw} is the mass of condensed water covering the particle surface

The initial condition for solving Equation 10.5 is the particle temperature being equal to the ambient air temperature.

The heat transfer coefficient of the film condensation can be calculated from an empirical equation for an isolated sphere (Holman, 1997):

$$h_{\text{fl}} = 0.815 \left[\frac{\rho_{\text{w}} \left(\rho_{\text{w}} - \rho_{\text{st}} \right) g \Delta H^{\text{vap}} \lambda_{\text{st}}^{3}}{\mu_{\text{st}} d_{\text{p}} \left(T_{\text{st}}^{\text{sat}} - T_{\text{p}} \right)} \right]^{\frac{1}{4}} \tag{10.6}$$

The physical properties of steam and of condensed water are evaluated at the film temperature, $T_{\text{fl}} = \dfrac{T_{\text{st}}^{\text{sat}} + T_{\text{p}}}{2}$. Using Equation 10.6 to calculate the condensation heat transfer coefficient in the fluidized bed may lead to some errors since the fluidized particles change the gas fluid dynamics and subsequently the formation of boundary layers around the particles, which are not similar to the boundary layer around a single sphere.

The resulting steam condensation in the bed increases the moisture content of the particles, and this can be mathematically described by a differential mass balance for moisture in the particle. Assuming the food particle to be spherical with a constant volume, and neglecting convective effects, this mass balance (in spherical coordinates) is then expressed by

$$\frac{\partial m}{\partial t} = D_{\text{m}}^{\text{eff}} \left[\frac{\partial^2 m}{\partial r^2} + \frac{2}{r} \frac{\partial m}{\partial r} \right] \tag{10.7}$$

where D_m^{eff} is the effective diffusion coefficient (m^2s^{-1}), assumed to be independent of the particle moisture content and temperature dependent. The value of D_m^{eff} during sorption of condensed water can be determined directly from soaking experiments at a given temperature. Initial and boundary conditions associated with Equation 10.7 are given by

$$m(r,t) = m_0, \quad \text{at } t = 0 \tag{10.8}$$

$$\left.\frac{\partial m}{\partial r}\right|_{r=0} = 0, \quad \text{at } t > 0 \tag{10.9}$$

$$m\left(\frac{d_p}{2}, t\right) = m_{eq,c} \quad \text{at } t > 0 \tag{10.10}$$

Equations 10.9 and 10.10 imply, respectively, that the minimum moisture content is at the center of the particle and that the surface moisture of the particle is in equilibrium with the condensed water. $m_{eq,c}$ is the moisture content at the particle surface, which is supposed to equilibrate with the condensed water covering the surface. The value of $m_{eq,c}$ can be determined by soaking the particles in hot water at the boiling-point temperatures until the particle mass does not change. The average moisture content of the particle, $\bar{m}$, is given by

$$\bar{m}(t) = \frac{4\pi}{V_p}\int_0^{d_p/2} m(r,t)\, r^2 dr \tag{10.11}$$

where V_p is the particle volume (m^3). The amount of the condensed water sorbed by all particles in the bed at time t, $M_{sor}(t)$, is

$$M_{sor}(t) = M_{ss}\left[\bar{m}(t) - \bar{m}(t - \Delta t)\right] \tag{10.12}$$

where the time interval, Δt, should be infinitesimal (i.e., tends to zero).

The amount of the remaining condensate in the bed at time t, M_{flw}, is thus:

$$M_{flw}(t) = \sum_{i=0}^{t}\left(\dot{M}_{cond}(i)\Delta t - M_{sor}(i)\right) \tag{10.13a}$$

Initial condensation of steam enables faster development of particle temperature. However, subsequent condensation rate decreases, whereas the rate of water, sorbed by particles, increases. These opposite behavior might result in the sorption rate being faster than the condensation rate. If such a case occurs, the gain in moisture by particles would come from the water condensation at the present time plus the remaining condensed water on the particle surface, resulting in a smaller amount of condensed water into the bed. In this case, the remaining water at the present time is calculated by Equation 10.13a, which can be rewritten as

$$M_{\text{flw}}(t) = \dot{M}_{\text{cond}}(t)\Delta t - M_{\text{sor}}(t) + \sum_{i=0}^{t-\Delta t} M_{\text{flw}}(i) \tag{10.13b}$$

As there is no available condensed water left at the particle surface before condensation stopped, the increase in moisture content cannot be calculated directly by Equation 10.7. Therefore, at this time t_{H} (time just before condensation stop), $M_{\text{flw}}(t_{\text{H}}) = 0$, $T_{\text{p}}(t_{\text{H}}) = T_{\text{st}}^{\text{sat}}$, and $M_{\text{sor}}(t) = M_{\text{ss}}\left[\bar{m}(t_{\text{H}}) - \bar{m}(t_{\text{H}} - \Delta t)\right]$. Then, upon dividing Equation 10.13b by M_{ss} and using these equations at $t = t_{\text{H}}$, one obtains:

$$\bar{m}(t_{\text{H}}) = \bar{m}(t_{\text{H}} - \Delta t) + \frac{\dot{M}_{\text{cond}}(t_{\text{H}})\Delta t}{M_{\text{ss}}} + \sum_{i=0}^{t_{\text{cond}}-\Delta t} \frac{M_{\text{fw}}(i)}{M_{\text{ss}}} \tag{10.14}$$

From Equation 10.14, one may expect the occurrence of the constant drying rate period when the average particle moisture content is higher than the critical moisture content. Accordingly, the drying period should be divided into two: constant drying rate and the falling rate.

10.3.3.2 Drying Period

A mathematical description of the temperature changes of steam and particles within the bed is derived from an energy balance in the solid and gas phases. The solid-phase energy balance can generally be written as

$$hA_{\text{p-bed}}\left(T_{\text{st}}\big|_{\text{bed}} - T_{\text{p}}\right) = \left(-M_{\text{ss}}\frac{d\bar{m}}{dt}\right)\left[\Delta H^{\text{vap}} + c_{\text{p,st}}\left(T_{\text{st}}\big|_{\text{bed}} - T_{\text{p}}\right)\right] + M_{\text{ss}}\left(c_{\text{p,s}} + \bar{m}\,c_{\text{p,w}}\right)\frac{dT_{\text{p}}}{dt} \tag{10.15}$$

Equation 10.15 represents the energy used for vaporizing water and heating up this vapor to the steam temperature inside the bed, $T_{\text{st}}|_{\text{bed}}$, (indicating that the steam within the bed is in a gaseous state), as well as for heating up the bed particles. When the particle temperature reaches the steam saturation value and drying starts in the constant rate regime, the last term on the right hand side of Equation 10.15 is neglected. However, this term is relevant for the falling rate period.

As observed from the experiments carried out in a fluidized bed (Soponronnarit et al., 2006), while the material is being dried, the temperature of steam inside the bed dryer, $T_{\text{st}}|_{\text{bed}}$, is lower than the one at the inlet, $T_{\text{st}}|_{\text{in}}$, but $T_{\text{st}}|_{\text{bed}}$ is insignificantly different from the exhaust steam temperature, $T_{\text{st}}|_{\text{out}}$. Therefore, at least for shallow fluidized-bed dryers, the gas phase can be considered to flow in perfect mixing inside the dryer chamber (i.e., $T_{\text{st}}|_{\text{bed}} = T_{\text{st}}|_{\text{out}}$), with its energy balance written as

$$\dot{M}_{\text{st,in}}\,c_{\text{p,st}}\left(T_{\text{st}}\big|_{\text{in}} - T_{\text{ref}}\right) - \dot{M}_{\text{st,out}}\,c_{\text{p,st}}\left(T_{\text{st}}\big|_{\text{bed}} - T_{\text{ref}}\right) = hA_{\text{p-bed}}\left(T_{\text{st}}\big|_{\text{bed}} - T_{\text{p}}\right) + \Gamma_{\text{st-amb}}A_{\text{wall}}\left(T_{\text{st}}\big|_{\text{bed}} - T_{\text{amb}}\right) \tag{10.16}$$

The first and second terms on the left-hand side of Equation 10.16 represent the energy carried by steam that, respectively, enters and leaves the drying chamber. This energy change is equivalent to the heat supplied for drying and heating particles plus the heat lost to the environment, whose temperature, T_{amb}, is known. In Equation 10.16, $\Gamma_{\text{st-amb}}$ represents the overall heat transfer coefficient between the steam inside the dryer and the surroundings, and A_{wall} is the contact area between the dryer walls and the surroundings, adopted as reference in the calculation of $\Gamma_{\text{st-amb}}$. It is assumed that the dryer internal wall is only in contact with steam, and particles therefore cannot lose their energy directly to the surroundings.

The gas-phase mass balance is now considered. The mass flow rate of steam leaving the drying chamber at any instant, $\dot{M}_{\text{st,out}}$, can be calculated by

$$\dot{M}_{\text{st,out}} = \dot{M}_{\text{st,in}} + \left(-M_{\text{ss}} \frac{\mathrm{d}\overline{m}}{\mathrm{d}t} \right) \tag{10.17}$$

Equation 10.17 represents an evaporation operation ($\mathrm{d}\overline{m}/\mathrm{d}t < 0$) leading to the mass flow rate of steam at the outlet being higher than that at the inlet under normal operation. Condensation, on the other hand, decreases the steam mass flow rate within the bed ($\mathrm{d}\overline{m}/\mathrm{d}t > 0$).

10.3.3.2.1 Constant-Rate Drying Period

As mentioned earlier, when the period of condensation finishes, the particle temperature reaches the steam saturation temperature ($T_{\text{p}} = T_{\text{st}}^{\text{sat}}$) and drying starts with the evaporation of unbound water from particle. Since only free water is removed in this case, drying occurs at a constant rate dictated by external conditions (steam flow rate, temperature, and pressure). This constant-rate drying period can be divided into two subperiods: the first one concerning the evaporation of the condensed water film covering particles, and the second one regarding the unbound water on the surface of particles.

During the first subperiod, one can consider that the interface at which water evaporates moves with decreasing film thickness until reaching the particle surface. Therefore, the moisture profile inside the particle remains unchanged until all condensed water has been removed from the surface. The temperatures of both the particle and the condensed water remain equal to $T_{\text{st}}^{\text{sat}}$ and Equation 10.15 is reduced to

$$h_{\text{fl}} A_{\text{pfl-bed}} \left(T_{\text{st}}\big|_{\text{bed}} - T_{\text{st}}^{\text{sat}} \right) = \left(-\frac{\mathrm{d}M_{\text{flw}}}{\mathrm{d}t} \right) \left[\Delta H^{\text{vap}} + c_{\text{p,st}} \left(T_{\text{st}}\big|_{\text{bed}} - T_{\text{st}}^{\text{sat}} \right) \right] \tag{10.18a}$$

After the condensed water film has been removed, unbound water starts to be evaporated at particle surface, which marks the beginning of the second drying subperiod. In this case, Equation 10.18a is changed to Equation 10.18b:

$$hA_{\text{p-bed}} \left(T_{\text{st}}\big|_{\text{bed}} - T_{\text{st}}^{\text{sat}} \right) = \left(-M_{\text{ss}} \frac{\mathrm{d}\overline{m}}{\mathrm{d}t} \right) \left[\Delta H^{\text{vap}} + c_{\text{p,st}} \left(T_{\text{st}}\big|_{\text{bed}} - T_{\text{st}}^{\text{sat}} \right) \right] \tag{10.18b}$$

where h, the heat transfer coefficient between particle and steam, is determined from Ranz and Marshall (1952). The changes in internal moisture content of particle, and therefore the vaporization rate, can be calculated using Equation 10.7, with the following initial and boundary conditions:

$$m(r,t) = m(r,t_1), \quad \text{at } t = t_1 \tag{10.19}$$

$$\left.\frac{\partial m}{\partial r}\right|_{r=0} = 0, \quad \text{at } t > t_1 \tag{10.20}$$

$$-\frac{6D_{\mathrm{m}}^{\mathrm{eff}}}{d_{\mathrm{p}}}\left.\frac{\partial m}{\partial r}\right|_{r=\frac{d_{\mathrm{p}}}{2}} = N_{\mathrm{c}}, \quad \text{at } t > t_1 \tag{10.21}$$

where

- t_1 is the time at the end of condensed water removal period (s)
- N_c is the constant drying rate (kg evaporated water/s·kg dry matter), which can be determined experimentally

Equation 10.19 represents the existing moisture gradients at the beginning of the second subperiod of constant-rate drying. As mentioned earlier, this moisture profile inside the particle is the one obtained at the end of the condensed water removal period, since during the first subperiod of constant-rate drying only the condensed water film covering the particle is removed.

For both subperiods, the energy balance for the gas phase (steam) is still similar to Equation 10.16, with T_p replaced by T_{st}^{sat}.

10.3.3.2.2 *Falling-Rate Drying Period*

After the moisture content of the particle drops below the critical moisture content, further reduction of moisture occurs in the falling-rate drying period. The critical moisture content of the particle can be determined from the experiment and this value is an input parameter in the model. Equation 10.7 still describes water diffusion inside particle, now with the following initial and boundary conditions:

$$m(r,t) = m(r,t_2), \quad \text{at } t = t_2 \tag{10.22}$$

$$\left.\frac{\partial m}{\partial r}\right|_{r=0} = 0, \quad \text{at } t > t_2 \tag{10.23}$$

$$m\left(\frac{d_{\mathrm{p}}}{2}, t\right) = m_{\mathrm{eq}}, \quad \text{at } t > t_2 \tag{10.24}$$

where

- t_2 is the drying time at the end of the constant-rate drying period (s)
- m_{eq} is the equilibrium moisture content, whose value is the moisture content of the particle in equilibrium with superheated steam at a given temperature and pressure

In SSD near the atmospheric pressure, the value of m_{eq} is assumed to be zero since the high drying temperature of 100°C is used The calculations of the steam bed temperature, the particle temperature, and the steam flow still follow Equations 10.15 through 10.17, with the particle temperature increasing with time from the steam saturation temperature to one close to the steam bed temperature ($=T_{st}|_{out}$).

In Figure 10.8, the predicted and experimental data of moisture content vs. time for drying paddy in a superheated-steam fluidized-bed dryer at 150°C are compared. These results show that the prediction of the time evolution of the paddy moisture content from the mathematical model presented is very closed to experimental data. The calculations also indicated that the period of steam condensation in the bed of particles being fluidized depends on the steam velocity, being shorter at higher steam velocities. More specifically, the condensation periods were around 2–2.5 s for the steam velocities of 1.3 and 1.5 u_{mf} (corresponding to the superficial velocities of 2.6–3 m/s). For the thin layer dryer, in which superheated steam flows across material and the material is not fluidized, condensation takes longer than in the fluidized-bed dryer. Pronyk et al. (2004) investigated the drying characteristics of foodstuffs using a superheated-steam thin-layer dryer at velocities between 0.25 and 0.35 ms^{-1} and steam temperatures from 125°C to 165°C. It was found that the condensation time was in the range of 6–7 s for Asian noodle before drying started.

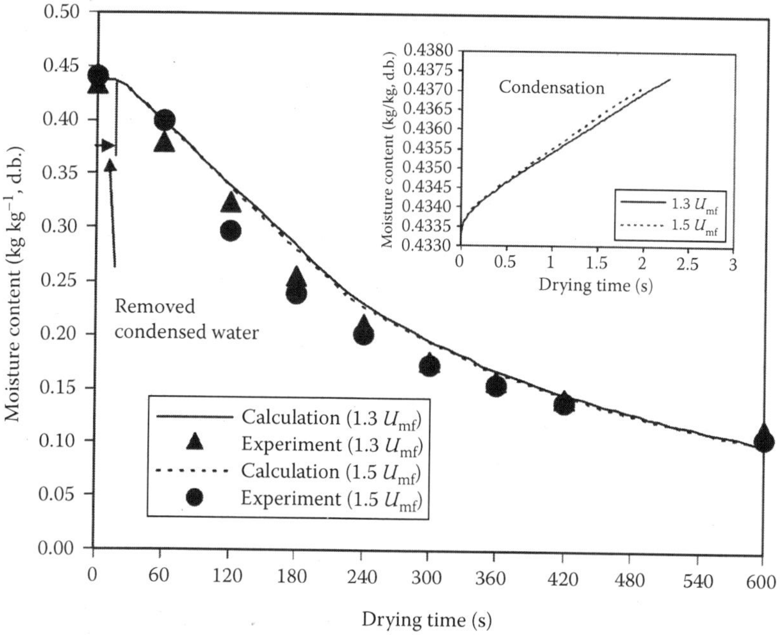

FIGURE 10.8 Comparison between predicted and experimental data of moisture content of paddy at superficial velocities of 1.3 and 1.5 times minimum fluidization velocity and inlet steam temperature of 150°C. (From Soponronnarit, S. et al., *Drying Technol.*, 24, 1462, 2006. With permission.)

10.4 APPLICATIONS OF SUPERHEATED STEAM DRYING TO FOOD MATERIALS

10.4.1 Parboiled Rice

Parboiled rice offers some advantages over unparboiled rice, such as strengthening of kernel integrity, high milling yield, and decrease in solid loss after cooking. Other characteristics of parboiled rice are its firmer and less sticky texture. A process for producing parboiled rice consists essentially of three steps including soaking of paddy, steaming, and drying to the predetermined moisture content. Figure 10.9 shows the conventional production process of parboiled rice that is currently operated in some countries. This process generally takes 3–4 h for the steps of soaking and drying to a moisture content of 22% d.b. A hot-air fluidized-bed dryer is used to dry paddy to moisture content of 39% d.b. (28.5% w.b.). Then, paddy is tempered (see No. 5) and dried again in a fluidized-bed dryer to 21% d.b.

Recently, however, Taechapairoj et al. (2004) found that drying of paddy in a superheated-steam fluidized bed could give a rice texture similar to the parboiled rice; milling yield was also noted to be higher. The moisture content of paddy after drying should not be lower than 18% d.b., otherwise the head rice yield would be very low. The kinetics of rice-starch gelatinization during SSD could suitably be explained by a zeroth-order reaction rate whose rate constant, N_k (s^{-1}), was related to the bed depth and drying temperature in the following form:

$$N_k = -3.0340\times10^1 + 3.1920H_{bed} - 1.0984\times10^{-1}H_{bed}^2 - \exp\left(-\frac{1.5186\times10^3}{T}\right) \tag{10.25}$$

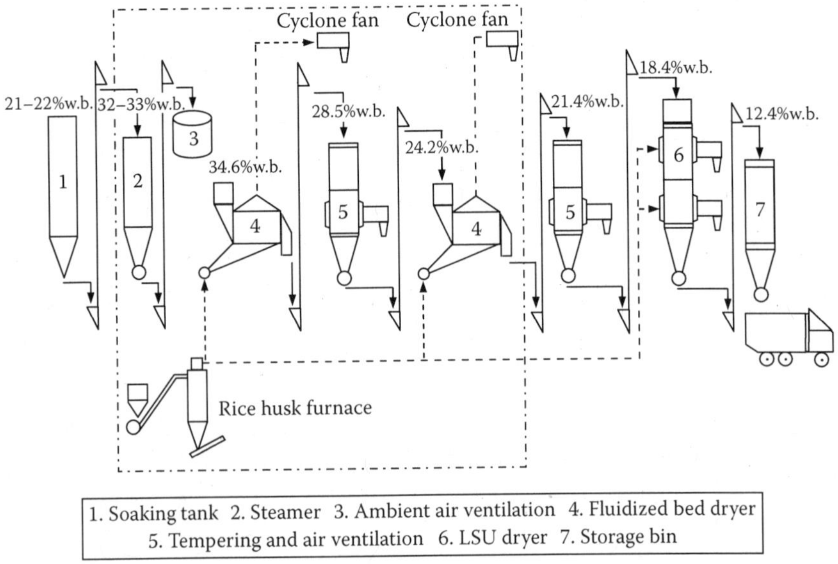

FIGURE 10.9 Conventional parboiled rice process. (From Rordprapat, W. et al., *J. R. Inst. Thailand*, 30, 377, 2005. With permission.)

where

H_{bed} is the bed depth (cm)

T is the drying temperature (K)

The values of correlation coefficient and absolute mean error associated with Equation 10.25 were 0.97 and 3.66, respectively. Complete gelatinization of rice starch could be established within 5–6 min at steam temperatures of 150°C–160°C, and within 4 min at a steam temperature of 170°C.

The findings of parboiled rice characteristics in SSD would make a significant progress in the parboiling process because superheated steam itself can act as both steaming and drying media at the same time, thereby reducing steps in the parboiling process. With superheated steam, steaming and drying can be combined into one step. Thus, all the units from No. 2 to No. 5, shown in Figure 10.9 by the dash-dot line (Rordprapat et al., 2005), could be replaced by a single superheated-steam dryer. Additionally, the processing time would be much shorter.

Soponronnarit et al. (2006) successfully fabricated and tested a pilot-scale, continuous superheated-steam fluidized-bed dryer, with a capacity of $100 \, kg \, h^{-1}$. A cyclonic rice husk furnace was used as a heating source to generate steam for the dryer. A schematic diagram of the referred pilot-scale dryer is shown in Figure 10.10.

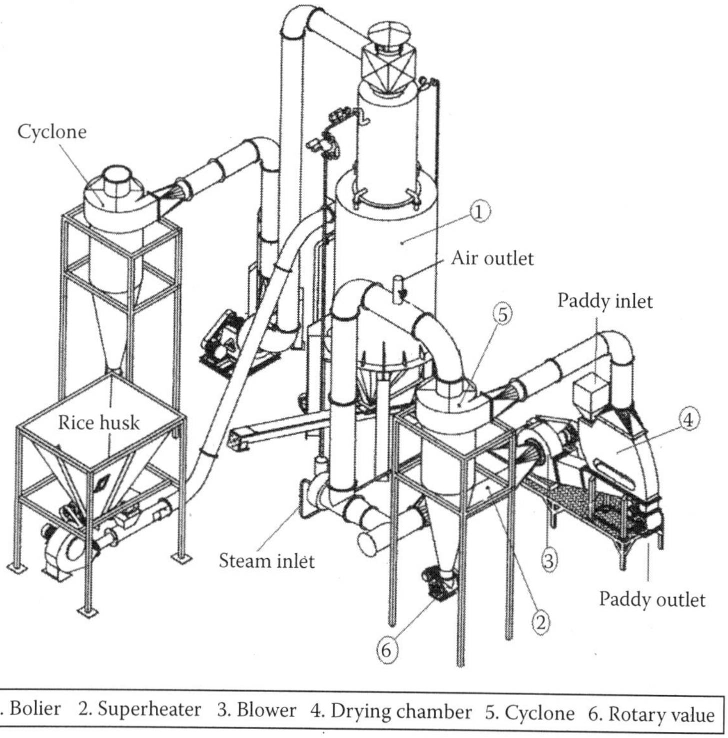

FIGURE 10.10 Schematic diagram of a pilot-scale superheated-steam fluidized bed. (From Soponronnarit, S. et al., *Drying Technol.*, 24, 1462, 2006. With permission.)

TABLE 10.1
Paddy Qualities (Chainat 1 Variety) Soaked at 70°C for 7–8h and Dried at Different Inlet Steam Temperatures

Drying Condition	Feed Rate ($kg h^{-1}$)	Moisture Content (d.b.)	Head Rice Yield (%)	Whiteness	White Belly (%)	Hardness (N)	Water Adsorption (g Water/g Rice)
After soaking	—	0.456 ± 0.008	56.6 ± 1.1[a]	39.0 ± 0.2[a]	5.7 ± 0.6[*a]	37.9 ± 1.3[a]	3.51 ± 0.07[a]
128°C	106	0.290 ± 0.008	63.5 ± 0.6[b]	32.2 ± 0.8[b]	0[b]	48.6 ± 0.9[b]	2.82 ± 0.05[b]
144°C	98	0.23 ± 0.004	66.9 ± 0.6[c]	29.4 ± 0.5[c]	0[b]	51.9 ± 0.6[c]	2.19 ± 0.01[c]
160°C	120	0.218 ± 0.004	67.9 ± 0.6[c]	28.0 ± 0.6[c]	0[b]	55.0 ± 1.3[d]	3.99 ± 0.05[d]

Source: Soponronnarit, S. et al., *Drying Technol.*, 24, 1462, 2006. With permission.
Note: Same letters on the same column indicate that values are insignificantly different at $p < 0.05$.
* Chalky grains.

The equipment was also installed in and demonstrated to some parboiled rice factories. Table 10.1 shows the paddy quality after drying with superheated steam. Before drying, paddy was soaked with hot water at 70°C for 7–8h. The head rice yield of the reference paddy, which was dried in shade, was 56.6%, and the white belly, representing the incomplete gelatinization, was 5.6%. After drying, the head rice yield was improved, being in the range of 63%–68%, and the white belly was not observed. Moreover, the value of hardness of dried parboiled rice significantly increased while less water was adsorbed.

Figure 10.11 shows the pasting viscosity of rice flour after SSD. It can be seen that the peak viscosity, final viscosity, and setback viscosity of dried rice flour are lower than those of the reference rice flour. The lowering of setback viscosity implies a firmer texture of rice flour. The trend of changing pasting viscosity of rice flour dried in SSD was similar to that of commercial parboiled rice obtained from the conventional process, but the values themselves are lower. Note that the extent to which the viscosity of rice flour obtained from the superheated-steam treatment is lower depends on the drying temperature. The pasting curve of flour made from rice dried at a temperature of 128°C is comparable to that of the commercially produced parboiled rice flour.

10.4.2 Soybean Meal

Full-fat soybean or soybean prior to oil extraction is used as a feedstuff because of its high-oil and high-quality protein contents. However, the presence of biologically active compounds in raw full-fat soybean such as trypsin inhibitors, hemagglutinins,

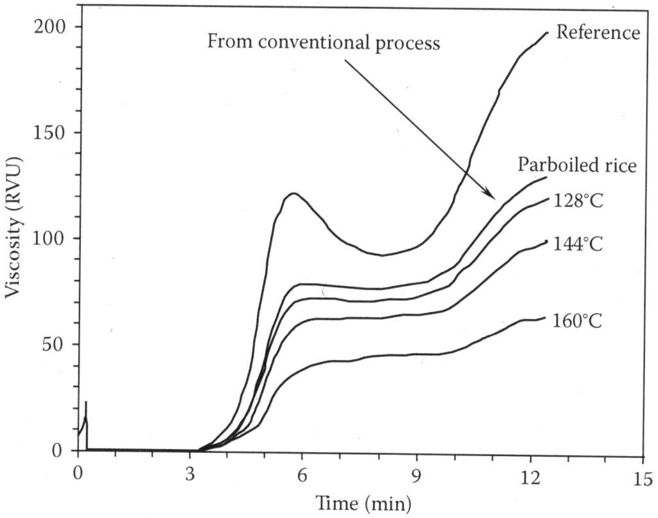

FIGURE 10.11 Pasting viscosities of rice flour samples dried at different steam temperatures. (From Soponronnarit, S. et al., *Drying Technol.*, 24, 1466, 2006. With permission.)

lectins, and saponins limits the utilization of its nutritive values (Hensen et al., 1987; Liener, 1994), resulting in compromised health and performance of nonruminants and immature ruminants. To eliminate these antinutritional factors, heat treatment is needed. Trypsin inhibitors are the main parameter used to control the quality of soybean meal. However, direct measurement of trypsin inhibitors is not frequently practiced. Instead, the urease activity is usually measured to indicate the activity of trypsin inhibitors because their inactivation rate is equal to that of the urease enzyme (Baker and Mustakas, 1973). The residual urease activity for adequately treated soybean is in the range of 10% and 20%; a value below 10% indicates overheating.

Heat-treatment methods frequently used are, for example, cooking, microwave, and roasting (Raghavan and Harper, 1974; Hensen et al., 1987; Stewart et al., 2003). Prachayawarakorn et al. (2006) studied the treatment of full-fat soybean using superheated steam and hot-air fluidized-bed dryers. It was found that both heating media could eliminate the urease enzyme, but at different rates, although the medium temperature used was the same. Insufficient inactivation was obvious with hot air at 120°C for soybean with initial moisture content of 13.5% d.b. As shown in Figure 10.12, the residual urease activity remained steadily at 40% although an extended drying period was applied. As the initial moisture content of soybean increased to 19.5% d.b., however, sufficient inactivation was noted at a heating time of over 25 min. In contrast to hot air, inactivation of the urease enzyme could be achieved at a superheated-steam temperature of 120°C; the levels of residual activity were below 20% for soybean with initial moisture contents of 13.5% and 19.5% d.b. when heated for 7 and 5 min, respectively. The thermal inactivation of urease followed a modified first-order reaction model, which is expressed by

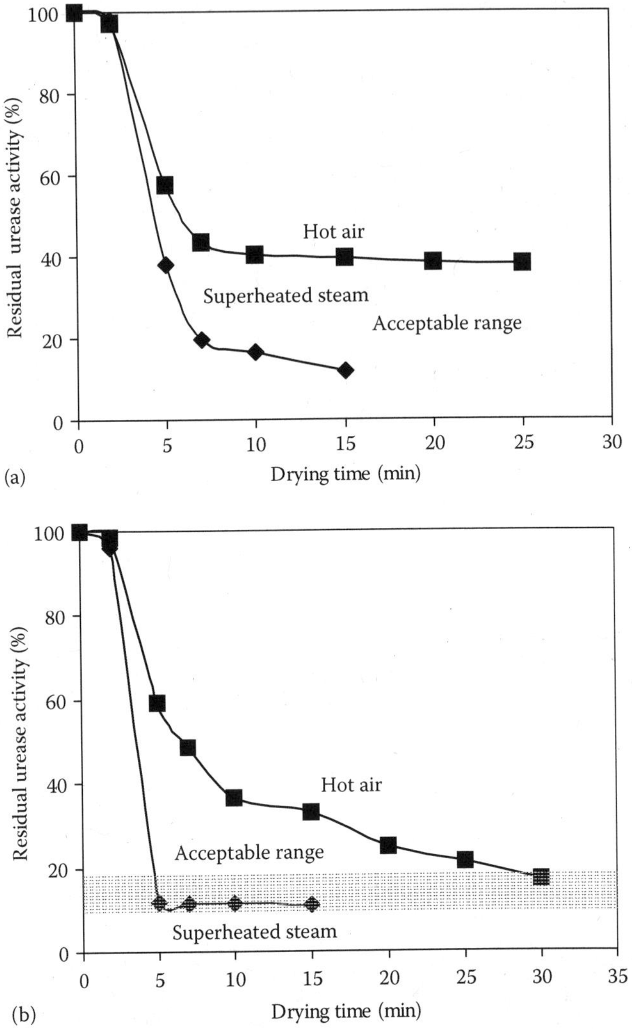

FIGURE 10.12 Apparent inactivation kinetics of urease inactivation present in soybean during thermal treatments with superheated steam and hot air at 120°C: (a) M_{in} = 13.5 % d.b. and (b) M_{in} = 19.5% d.b. (From Prachayawarakorn, S. et al., *LWT-Food Sci. Technol.*, 39, 775, 2006. With permission.)

$$\frac{a_{ur}(t) - a_{ur}\big|_{eq}}{a_{ur}\big|_{in} - a_{ur}\big|_{eq}} = \exp(-N_k t) \tag{10.26}$$

where

$a_{ur}(t)$ is the residual urease activity at time t (%)
$a_{ur}\big|_{in}$ is the urease activity at the beginning of heat treatment (%)
$a_{ur}\big|_{eq}$ is the residual urease activity at infinite time (%)
N_k is the apparent constant kinetic rate (min^{-1})
t is the drying time (min)

The apparent constant kinetic rate could be expressed as a function of the operating parameters, i.e., heating temperature and moisture content, as

$$N_k = \left[-1.01\times10^4 + 7.50\times10^{-2}T^2 - 7.66\times10^4 m_{in} + 1.89\times10^4 m_{in}^2 + \frac{2.62\times10^7 m_{in}}{T} \right] \times \exp\left(-\frac{5.35\times10^{-1}}{m_{in}} - \frac{1.51\times10^3}{T} \right) \quad (10.27)$$

In addition, the expression for $a_{ur}|_{eq}$ is given by

$$a_{ur}|_{eq} = \left(-4.7523\times10^{-2} + 7.1742\times10^{-2} m_{in}^2 + 1.2200\times10^{-4} T\right) \times \exp\left(\frac{3.2273\times10^3 - 9.7815\times10^2 m_{in}}{T} \right) \quad (10.28)$$

Equations 10.27 and 10.28 both presented a correlation coefficient equal to 0.95, and root mean square errors of 0.347 and 1.47, respectively.

Prachayawarakorn et al. (2006) recommended that a fluidized-bed dryer should be operated at 135°C–150°C for hot air and below 135°C for superheated steam to eliminate the urease enzyme present in soybean. This could simultaneously preserve the nutritional qualities, protein solubility, and lysine content. Under such temperature ranges, the use of superheated steam as the heating/drying medium resulted in higher protein solubility of treated sample than the use hot-air drying when applied to the dried soybean (13.5% d.b.), as can be seen in Table 10.2. In addition, the protein solubility of soybean was similar to that treated by an industrial micronizer (Wiriyaumpaiwong et al., 2004). For the moist soybean, the heating medium types did not affect the quality since the urease enzyme was inactivated over a short period of time before protein denaturation largely occurred.

10.4.3 Snack Foods

Snack foods, e.g., potato chips, banana chips, durian chips, are generally commercially produced by deep-fat frying. The development of a porous structure provides snacks with better quality, especially in terms of texture, compared with air-dried products. Snack products typically have a high oil content and cannot be kept for a long period of time due to possible lipid oxidation, leading to rancidity. In addition, consumers who are concerned with their health may dislike these products. The product with low fat content is an interesting commodity to the consumer.

High-temperature drying could be an alternative to frying to produce low-fat snack products. During high-temperature drying, the product moisture vaporizes and expands rapidly, thereby allowing the formation of pores. Li et al. (1999) studied superheated-steam-impingement drying of tortilla chips and founded that a higher steam temperature resulted in more pores, coarser appearance, and higher modulus of deformation of tortilla chips. In addition, the superheated steam-dried tortilla chips had fewer but

TABLE 10.2
Protein Solubility and Lysine Content of Soybean Treated with Superheated Steam and Hot Air at Various Temperatures

Temperature (°C)		m_{in} (% d.b.)	Heating Time (min)	$m(t)$ (% d.b.)	Urease Activity (%)	Protein Solubility (%)	Lysine Content ($mg\,g^{-1}$ Soybean)
Raw soybean					100	94.28	2.9–3.1
Hot air	120	13.5	50	3.9	37	71.1	N/A
		19.5	30	6.9	19	70.98	3.0
	135	13.5	5	9.7	15	78.68	3.1
		19.5	5	14.4	13	84.33	N/A
	150	13.5	5	8.2	16	74.17	N/A
		19.5	2	15.5	10	87.86	N/A
Superheated steam	120	13.5	7	13.9	16	85.66	2.8
		19.5	5	18.3	12	85.2	2.7
	135	13.5	5	12.4	11	71.76	2.7
		19.5	5	16.5	9	82.25	2.7
	150	13.5	5	10.6	12	54.94	2.8
		19.5	2	18.9	11	83.84	N/A

Source: Prachayawarakorn, S. et al., *LWT-Food Sci. Technol.*, 39, 776, 2006. With permission.
N/A, not available.

larger pores than the air-dried product. Similar results were also found for dried durian chips (Jamradloedluk et al., 2007). In spite of the different morphologies of products obtained from the two drying media, textural properties such as hardness and stiffness were not significantly different. In the case of some particular fruits such as banana, however, though pores could be formed during high-temperature drying and the texture of the final product might be acceptable, the color of the product normally becomes brown when its moisture content is reduced below 80% d.b. (Prachayawarakorn et al., 2008). In addition, some pores might collapse during the early drying period, resulting in shrinkage of the product and subsequent less crispiness and hard texture. To avoid shrinkage, the structure should be rigid so that it can resist stresses generated during drying. This can be achieved by drying the product at temperatures below the glass transition temperature, and the freezing drying technique is normally applied.

Another technique that provides less material shrinkage involves puffing, by which the moisture inside the material is rapidly vaporized. Vapor expansion inside the product then creates voids or ruptures the existing structure, leading to a more porous dried product. Boualaphanh et al. (2008) studied the textural properties of banana after puffing with superheated steam at different conditions; their experimental results are shown in Figure 10.13, indicating that puffing temperature and moisture content of banana before puffing strongly affected the textural properties and shrinkage of the product. The sample shrunk least, approximately 10%–15%, when the moisture content of banana before puffing ranged between 20% and 25% d.b.

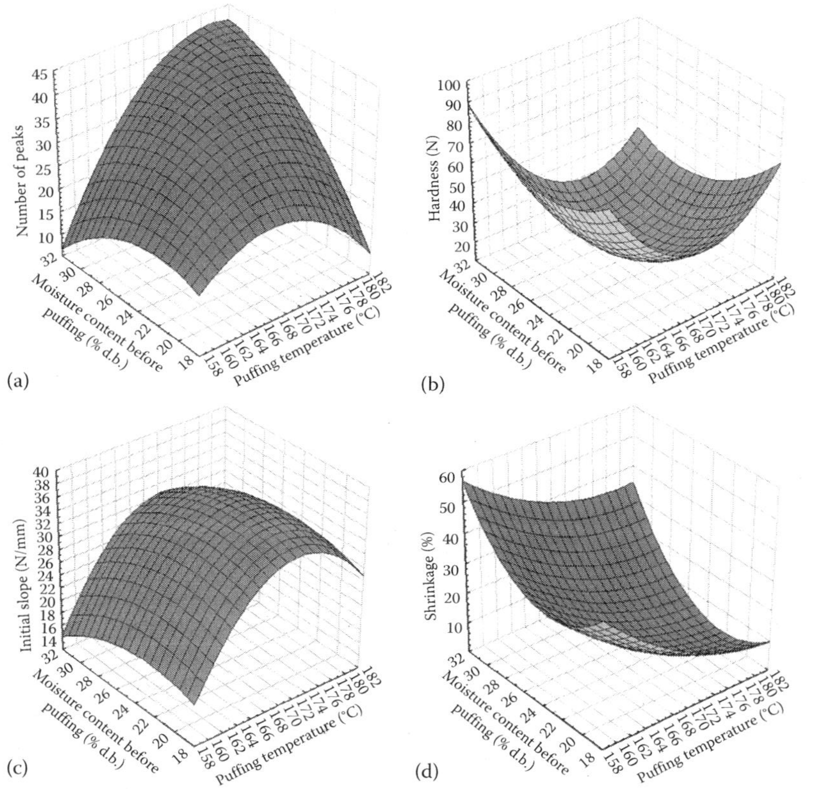

FIGURE 10.13 Quality attributes of banana puffed at different conditions (2 min puffing time). (a) Number of peaks. (b) Hardness. (c) Initial slope. (d) Shrinkage. (From Boualaphanh, K. et al., Optimization of the superheated steam puffing of banana, in *Proceedings of the 9th Thai Society of Agricultural Engineering Conference*, Chiang Mai, Thailand, January 31–February 1, 2008, CR2-16 [in Thai]. With permission.)

and the puffing temperature was in the range of 170°C–180°C. Under these puffing conditions, the color of the sample fell into the color group of grayed-orange 163C, and the textural property values were in ranges of 35 ± 5 N/mm for the initial slope, 37 ± 5 for the number of peaks and 33 ± 11 N for the hardness. These quality attributes were similar to those of the commercially vacuum fried product; the textural property values of the commercial product were 48 ± 8 N for the hardness, 42 ± 5 for the number of peaks and 45 ± 16 for the initial slope.

Figure 10.14 shows the morphology of puffed banana, indicating the very large voids within the banana sample and the small pores at the exterior. The produced large pores imply the vaporization and expansion of moisture during superheated-steam puffing.

However, a drawback of superheated steam became evident when it was applied to pop amaranth seeds (Iyota et al., 2005). Amaranth, a well-known product in the market as a source of food and high nutritional components, can be consumed after popping. The volume expansion after popping is an important parameter; a high volume expansion produces softer and more appealing texture. Figure 10.15 shows the volume expansion ratio of popped amaranth seeds at different temperatures. In hot air, the maximum expansion ratio reached 7.7 at 260°C. A decrease in expansion ratio and browning occurred at higher temperature of 290°C. In the case of superheated steam, the expansion ratio was approximately 10% lower than that obtained in hot air. This occurred because the seed coats were apparently softened by steam condensation.

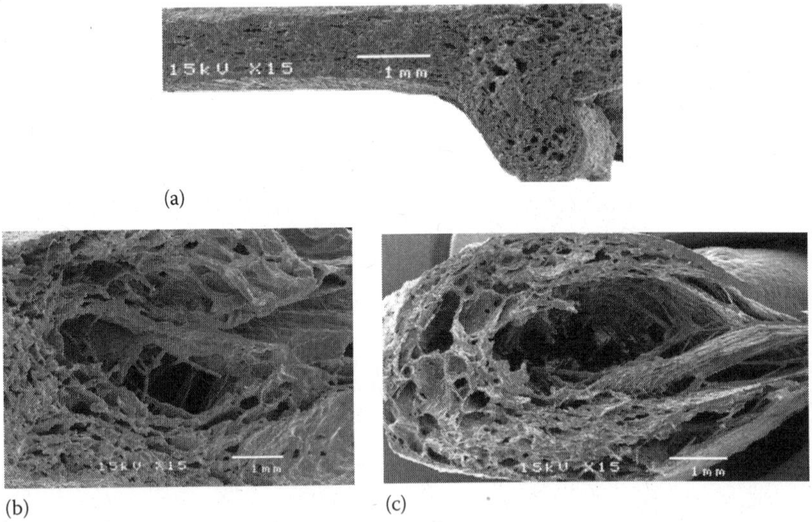

FIGURE 10.14 Morphologies of banana at different puffing temperatures. (a) No puffing. (b) Puffing temperature of 170°C. (c) Puffing temperature of 180°C. (From Boualaphanh, K. et al., Optimization of the superheated steam puffing of banana, in *Proceedings of the 9th Thai Society of Agricultural Engineering Conference*, Chiang Mai, Thailand, January 31–February 1, 2008, CR2-16 [in Thai]. With permission.)

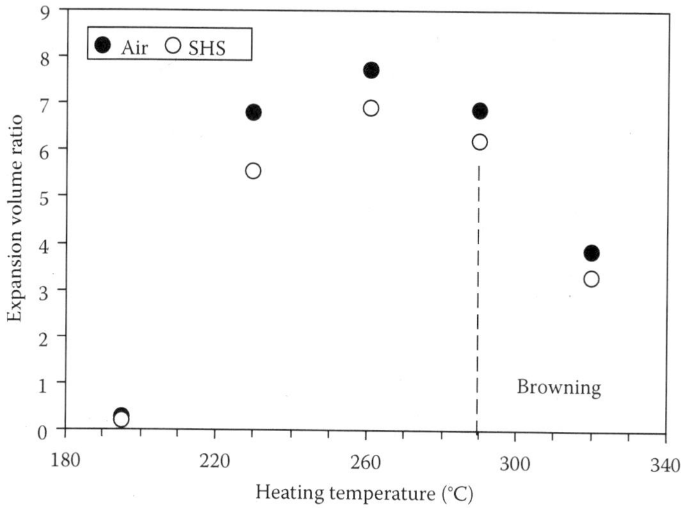

FIGURE 10.15 Effect of heating temperature on volume expansion ratio of amaranth seed in hot air and superheated steam (SHS) (initial moisture content of 13% d.b.). (From Iyota, H. et al., *Drying Technol.*, 23, 1287, 2005. With permission.)

10.5 CONCLUDING REMARKS

SSD is a feasible alternative to hot-air drying for a number of food products. The condensation of steam onto the material surface provides a rapid temperature rise, and this causes the food quality obtained from SSD to be different from that related to hot-air drying. The physical appearance of foods is better in SSD than in the hot-air drying, particularly when SSD is applied to starchy foods. Superheated-steam dried products such as tortilla, potato, and durian chips have a smoother surface and are glossier since the starch present in them can form a gel network to a larger extent in SSD than in the case of hot-air drying. Paddy dried by superheated steam has physicochemical and physical properties similar to those of parboiled rice. A temperature of 150°C is recommended for the dryer operation to produce parboiled rice.

In addition to the drying of starch-based foods, superheated steam can also be applied to eliminate the antinutritional factors present in legumes. However, the treatment of soybean with superheated steam should be done at temperatures below 135°C to preserve the protein solubility at the level that is required for feed meal.

A mathematical model for SSD that includes both steam condensation and drying has been derived based on the transport equations for drying a single particle, together with the energy and mass balances in the dryer. It can predict the change of moisture content of product during SSD relatively well. The mathematical model developed for the fluidized-bed dryer can be extended to other dryer types, with some modifications in the governing equations.

ACKNOWLEDGMENTS

The authors express their sincere appreciation to the Thailand Research Fund and Commission on Higher Education for the financial support of projects for more than 10 years.

REFERENCES

Baker, E.C. and Mustakas, G.C. 1973. Heat inactivation of trypsin inhibitor, lipoxgygenase and urease in soybeans: Effect of acid and base additives. *J. Am. Oil Chem. Soc.* 50:137–141.

Boualaphanh, K., Prachayawarakorn, S., and Soponronnarit, S. 2008. Optimization of the superheated steam puffing of banana. In *Proceedings of the 9th Thai Society of Agricultural Engineering Conference*. Chiang Mai, Thailand, January 31–February 1, 2008, CR2-16 (in Thai).

Chen, Z., Wu, W., and Agarwal, P.K. 2000. Steam-drying of coal. Part 1. Modeling the behavior of a single particle. *Fuel* 79:961–973.

Chow, L.C. and Chung, J.N. 1983a. Evaporation of water into a laminar stream of air and superheated steam. *Int. J. Heat Mass Transfer* 26:373–380.

Chow, L.C. and Chung, J.N. 1983b. Water evaporation into a turbulent stream of air, humid air or superheated steam. In *Proceedings of the ASME National Heat Transfer Conference*, No. 83-HT-2 ASME, New York.

Crank, J. 1975. *The Mathematics of Diffusion*, 2nd ed. Oxford, NY: Clarendon Press.

Deventer, H.C. and Heijmans R.M.H. 2001. Drying with superheated steam. *Drying Technol.* 19:2033–2045.

Douglas, W.J.M. 1994. Drying paper in superheated steam. *Drying Technol.* 12:1341–1355.

Hensen, B.C., Flores, E.S., Tanksley, Jr., T.D., and Knabe, D.A. 1987. Effect of different heat treatments during processing of soybean meal on nursery and growing pig performance. *J. Anim. Sci.* 65:1283–1291.

Holman, J.P. 1997. *Heat Transfer*, 8th ed. New York: McGraw-Hill.

Iyota, H., Konishi, Y., Inoue, T., Yoshida, K., Nishimura, N., and Nomura, T. 2005. Popping of amaranth seeds in hot air and superheated steam. *Drying Technol.* 23:1273–1287.

Iyota, H., Nishimura, N., Onuma, T., and Nomura, T. 2001. Drying of sliced raw potatoes in superheated steam and hot air. *Drying Technol.* 19:1411–1424.

James, C., Thornton, J.A., Ketteringham, L., and James, S.J. 2000. Effect of steam condensation, hot water or chlorinated hot water immersion on bacterial numbers and quality of lamb carcasses. *J. Food Eng.* 43:219–225.

Jamradloedluk, J., Nathakaranakule, A., Soponronnarit, S., and Prachayawarakorn, S. 2007. Influences of drying medium and temperature on drying kinetics and quality attributes of durian chip. *J. Food Eng.* 78:198–205.

Lane, A.M. and Stern, S. 1956. Application of superheated-vapor atmospheres to drying. *Mech. Eng.* 78:423–426.

Li, Y.B., Seyed-Yagoobi, J., Moreira, R.G., and Yamsaengsung, R. 1999. Superheated steam impingement drying of tortilla chips. *Drying Technol.* 17:191–213.

Liener, I.E. 1994. Implications of antinutritional components in soybean foods. *Crit. Rev. Food Sci. Nutr.* 34:31–67.

Markowski, M., Cenkowski, S., Hatcher, D.W., Dexter, J.E., and Edwards, N.M. 2003. The effect of superheated steam dehydration kinetics on textural properties of Asian noodles. *Trans. ASAE* 46:389–395.

Nathakaranakule, A., Kraiwanichkul, W., and Soponronnarit, S. 2007. Comparative study of different combined superheated steam drying techniques for chicken meat. *J. Food Eng.* 80:1023–1030.

Pang, S. and Dakin, M. 1999. Drying rate and temperature profile for superheated steam vacuum drying and moist air drying of softwood lumber. *Drying Technol.* 17:1135–1147.

Poomsa-ad, N., Soponronnarit, S., Prachayawarakorn, S., and Terdyothin, A. 2002. Effect of tempering on subsequent drying of paddy using fluidization. *Drying Technol.* 20:195–210.

Prachayawarakorn, S., Prachayawasin, P., and Soponronnarit, S. 2006. Heating process of soybean using hot-air and superheated-steam fluidized-bed dryer. *LWT-Food Sci. Technol.* 39:770–778.

Prachayawarakorn, S., Soponronnarit, S., Wetchacama, S., and Jaisut, D. 2002. Desorption isotherms and drying characteristics of shrimp in superheated steam and hot air. *Drying Technol.* 20:669–684.

Prachayawarakorn, S., Tia, W., Plyto, N., and Soponronnarit, S. 2008. Drying kinetics and quality attributes of low-fat banana slices dried at high temperature. *J. Food Eng.* 85:509–517.

Pronyk, C., Cenkowski, S., and Muir, W.E. 2004. Drying foodstuffs with superheated steam. *Drying Technol.* 22:899–916.

Raghavan, G.S.V. and Harper, J.M. 1974. Nutritive value of salt-bed roasted soybean for broiler chicks. *Poultry Sci.* 53:547–553.

Ranz, W. and Marshall, W. 1952. Evaporation from drops-Part I. *Chem. Eng. Prog.* 48, 141.

Rordprapat, W., Nathakaranakule, A., Tia, W., and Soponronnarit, S. 2005. Development of a superheated-steam-fluidized-bed dryer for parboiled rice. *J. R. Inst. Thailand.* 30:363–378 (in Thai).

Schwartze, J.P. and Bröcker, S. 2002. A theoretical explanation for the inversion temperature. *Chem. Eng. J.* 86:61–67.

Shibata, H. and Mujumdar, A.S. 1994. Steam drying technologies: Japanese R&D. *Drying Technol.* 12:1485–1524.

Soponronnarit, S., Prachayawarakorn, S., Rordprapat, W., Nathakaranakule, A., and Tia, W. 2006. A superheated-steam fluidized-bed dryer for parboiled rice: Testing of a pilot-scale and mathematical model development. *Drying Technol.* 24:1457–1467.

Stewart, O.J., Raghavan, G.S.V., Orsat, V., and Golden, K.D. 2003. The effect of drying on unsaturated fatty acids and trypsin inhibitor activity in soybean. *Process Biochem.* 39:483–489.

Svensson, C. 1980. Steam drying of pulp. In *Drying '80*, vol. 2, Ed. A.S. Mujumdar, pp. 301–307. New York: Hemisphere.

Taechapairoj, C., Dhuchakallaya, I., Soponronnarit, S., Wetchacama, S., and Prachayawarakorn, S. 2003. Superheated steam fluidised bed paddy drying. *J. Food Eng.* 58:67–73.

Taechapairoj, C., Prachayawarakorn, S., and Soponronnarit, S. 2004. Characteristics of rice dried in superheated steam fluidized-bed. *Drying Technol.* 22:719–743.

Tang, Z. and Cenkowski, S. 2000. Dehydration dynamics of potatoes in superheated steam and hot air. *Can. Agr. Eng.* 42:43–49.

Tang, Z., Cenkowski, S., and Izydorczyk, M. 2005. Thin-layer drying of spent grains in superheated steam. *J. Food Eng.* 67:457–465.

Uengkimbuan, N., Soponronnarit, S., Prachayawarakorn, S., and Nathkarankule, A. 2006. A comparative study of pork drying using superheated steam and hot air. *Drying Technol.* 24:1665–1672.

Wiriyaumpaiwong, S., Soponronnarit, S., and Prachayawarakorn, S. 2004. Comparative study of heating processes for full-fat soybeans. *J. Food Eng.* 65:371–382.

Yoshida, T. and Hyōdō, T. 1970. Evaporation of water in air, humid air, and superheated steam. *Ind. Eng. Chem. Process Des. Dev.* 9:207–214.

11 Drying of Tropical Fruit Pulps: An Alternative Spouted-Bed Process

Maria de Fátima D. Medeiros, Josilma S. Souza, Odelsia L. S. Alsina, and Sandra C. S. Rocha

CONTENTS

11.1 INTRODUCTION

Fruit culture is recognized as one of the important agriculture potentials in Brazil, which stands out as a large producer of a wide variety of fruits, ranging from tropical fruits to those considered cold climate fruits, such as apples, pears, and peaches. Out of the total annual production, it is estimated that 14% (about 5 million tons) constitutes little exploited tropical fruits (FAO 2005), such as umbu (*Spondias tuberosa*), hog plum (*Spondia lutea*), red mombin (*Spondia purpurea*), hog plum mango (*Spondias dulcis*), soursop (*Annona muricata*), sapodilla (*Manilkara achras*), and mangaba (*Hancornia speciosa*), among others. Tropical fruits are tasty, aromatic and, in addition to being hydrating, are energetic and rich in vitamins and mineral salts, mainly calcium, iron, and phosphorous. Despite the significant fruit consumption in Brazil, whether in their natural form, or as juices (the best form to maximize their nutrients) or prepared as sweets, jams, compotes, ice creams, etc., and the surge in agribusiness and exports, there is still a high level of fruit wastage, mainly in cyclically produced seasonal fruits.

To increase fruit life, without altering its nutritive and sensory characteristics, new fruit processing and preprocessing technologies have been developed and introduced into the agribusiness sector. These technologies aim at avoiding wastage, increasing consumption during the out-of-season period, and exploiting fruits as raw materials in the manufacture of industrialized foods such as candies, sweets, baby food, ice creams, etc. Dry fruit consumption has increased significantly in recent years, mainly in health foods, such as granola, enriched cereals, and whole wheat breads. However, most drying techniques used require long exposure to heat, with losses in thermosensitive nutrients and irreversible changes in physical and chemical characteristics. As a consequence, the rehydration process does not regenerate all the natural characteristics of dried fruits (Rahman and Pereira 1999, Fellows 2000).

Some cultivated fruits, mainly in the northeast of Brazil, such as hog plum, umbu, red mombin, and surinam cherry (*Eugenia uniflora*), among others, are very acidic and juicy, with a low pulp or pit ratio, which renders them unsuitable for dry fruit production. These fruits, whose consumption is mainly in the form of juice and ice cream, are depulped and commercialized frozen, requiring large storage and transportation space. Conservation by freezing results in high energy costs and adds no value to the product.

The development of powdery fruit as a postharvest processing option ensures a product with low water content, greater stability, and prolonged storage under ambient temperature conditions. Among the techniques used in the fruit powder production are lyophilization, foam-mat drying (FMD), encapsulation of juices by cocrystallization with sucrose, and spray drying (SD), as well as fluidized-bed (FB) and spouted-bed (SB) drying with inert particles. Lyophilization is a complicated, costly process. Though already studied as an alternative for obtaining dried fruit (Righetto 2003, Marques et al. 2007), this technique is used more in the drying of heat-sensitive products of high commercial value, such as medicines, dry extracts, etc. SD is considered a viable alternative in fruit powder production. As shown in Chapter 9, SD produces high amounts of solids and offers fundamental advantages over traditional drying methods, since the nutritional characteristics of the product are maintained owing to the short contact time of the raw material with the heating gases inside the dryer. However, in SD, as well as in cocrystallization (Astolfi-Filho et al. 2005) and FMD processes (Soares et al. 2001), additives and/or adjuvants are added to the fruit juice formulation as emulsifiers, thickeners, and wall or encapsulating materials. Generally, wall material is used in the formulation to retain volatile ingredients by microencapsulation, and also to avoid degradation of the vitamins and caramelization of the sugars present in fruits. Despite the importance of adjuvants in the mentioned processes, their presence causes changes in the characteristic flavor of the fruit, modifying significantly the original fruit composition.

Fruit powder production with the minimum addition of adjuvants, using simpler low-cost drying techniques, has been the object of study in recent years, not only in Brazil but also in other countries (Lima et al. 1992, Martinez et al. 1995, Reyes et al. 1996, Medeiros et al. 2002). Drying of tropical fruit pulps in SB with inert particles has been extensively studied in Brazil on account of the excellent quality of the powdery products processed, the equipment simplicity, and easiness of operation, as well as its low costs of building, setup, and maintenance (Lima et al. 1998, 2000, Ramos et al. 1998, Medeiros et al. 2002, Souza et al. 2007). In this chapter a summary of the results obtained is presented. Emphasis is laid on three aspects: (1) the effects of the addition and composition of fruit pulps on the SB fluid dynamic behavior and on process performance; (2) the optimization of production as a function of the composition of processed pulp; and (3) the development of new products formulated from the mixture of fruit pulps with adjusted composition.

11.2 SPOUTED-BED DRYING WITH INERT PARTICLES

Mathur and Gishler developed in 1954 a solid–fluid contact device, the SB, useful for processing coarse particles that do not easily fluidize and are known as type D in the Geldart classification. One of the SB applications is the drying of pulps, pastes, slurries, and suspensions, fed over a spouting bed of inert particles. The conventional equipment, shown in Figure 11.1, consists of a drying chamber constituted by a cylindrical column with a conical base. A gas, usually air, is injected into the column by the inlet orifice. When gas velocities over certain values are used, the inert particles start a cyclic movement, ascending in the central region and falling along the outer cross section or annulus of the bed. Particles in

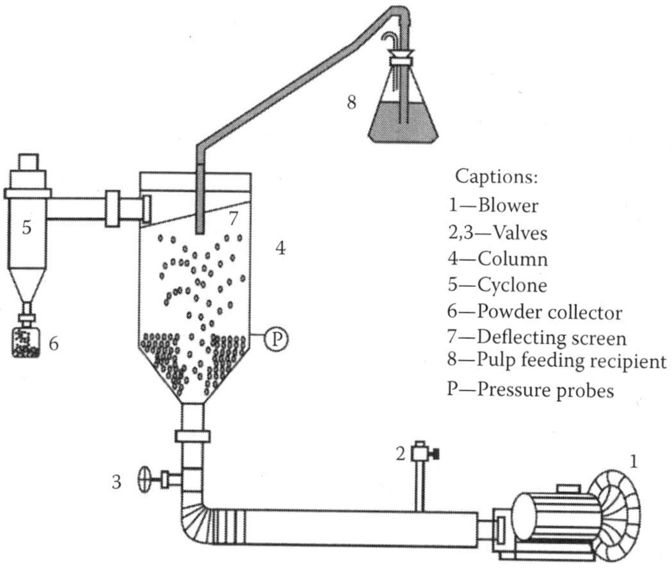

FIGURE 11.1 Scheme of the spouted-bed dryer with inert particles.

the spout, going in upward direction and after reaching the top of the bed under decelerated motion, form a fountain just above the annulus. The particle path in the annulus is directed toward the base, returning to the central channel or spout. This cyclic movement of particles creates three distinct regions characteristic of a SB: the spout—a dilute phase with high porosity; the annulus—a low porosity moving bed; and the fountain—a region above the bed where particles change the direction of their vertical motion from upward to downward. This gas–particle contact configuration induces high rates of heat and mass transfer between gas and particles, allowing efficient drying of pastes and suspensions (Mathur and Gishler 1955). The suspension feed is sprayed, dropped, or injected into the bed of inert particles, usually at the top or bottom of the SB column. Feeding can be continuous, batchwise, or intermittent. The liquid spreads on the surface of the particles and, after drying, forms a thin film covering them. The interparticle attrition and slippage cause the breakage of the film, resulting in a fine powder, which is entrained by air and collected at the cyclone.

SB hydrodynamics is usually represented by characteristic curves of pressure drop across the bed of particles versus the gas superficial velocity (Mathur and Epstein 1974). The main parameters to describe SB dynamics are (1) the minimum spouting superficial velocity, U_{ms}, or the minimum spouting flow rate, Q_{ms} ($=U_{ms} \times$ transversal area of empty column); (2) the minimum spouting pressure drop, ΔP_{ms}; (3) the maximum pressure drop across the bed of particles, ΔP_{max}; and (4) the stable spouting pressure drop, ΔP_{ssp}. These parameters are influenced by the presence of liquids or suspensions and by the powder produced during drying. Besides these effects, depending on the amount of suspension added into the bed, the SB behavior can become unstable and the spout can collapse. Another important parameter, especially in drying pastes and suspensions, is the solids circulation rate, i.e., the inert particle mass flow rate, since the spout stability is mainly governed by the particle path, velocity, and stickiness (cohesion between particles, adhesion to particles covered by liquid film, and also particle adhesion to wall).

11.3 DRYING OF FRUIT PULP IN SPOUTED BEDS

By the end of the 1980s and early the 1990s a number of studies were developed in order to determine the feasibility of the SB equipment to obtain fruit powder, especially tropical fruits, by drying their natural pulps using inert particles in the bed. The early studies of Kachan and coworkers with banana (Hufenussler and Kachan 1988) and tomato paste pointed to a good powder quality (Kachan and Chiapetta 1988, Kachan et al. 1988). The moisture level, the reconstitution time, and sensorial properties met, in general, under proper operational conditions, the required conservation standards, some of which have not been achieved by powder from rotary dryers. The authors have also observed the advantages in comparison with FB and SD products.

During the 1990s, Alsina and coworkers studied the drying of fruit pulps, mainly umbu and acerola in SB (Lima et al. 1992, Lima 1993, Lima and Alsina 1994, Alsina et al. 1995, 1996). Their first results referred to preliminary hydrodynamic tests of the equipment working in the conical region and to the influence of the operating temperature and the airflow rate on the powder moisture content, the thermal efficiency, and the process yield. The pulp influence on SB hydrodynamics was evidenced, showing a decrease in Q_{ms} in the presence of pulp. These studies were then widened to other varieties of tropical fruits, as hog plum and red mombin pulps, and to other SB column geometries, including the conventional cone-cylindrical one. The main results of the influence of some operating conditions, namely, the inlet air temperature, $T_g|_{in}$, the inert particle load, M_{inert}, and the airflow rate, Q, on the process performance and product quality are summarized below:

- The drying performance of fruit pulps in SBs is mainly affected by the spout regime stability, which limits the pulp feeding flow rate and, consequently, the dryer capacity.
- Powdery products have presented, in general, low moisture contents, in the range from 4 to 6 wt%, under stable operating conditions.
- As expected, drying at higher $T_g|_{in}$ produces powders with lower moisture contents, below 4 wt%. Nevertheless, this condition, although favorable to evaporation, can induce dark colored product and agglomerates, as in the case of acerola dried at 50°C and red mombin dried at temperatures close to 80°C.
- In terms of vitamin C preservation, there is an optimum range of $T_g|_{in}$, attributed to faster drying, which avoids degradation and that should be found for each pulp, i.e., the vitamin C preservation during drying in spouted beds depends in a large extent on the pulp composition. There is not a unique range of acceptable losses of vitamin C for all tropical fruits. For instance, in the drying of umbu pulp, vitamin C was found to increase from 596 mg/100 g (dry basis) in the powder obtained at 50°C to a maximum of 655 mg/100 g (dry basis) when operating at 75°C. This trend was attributed to the influence of two opposite effects: possible losses of vitamin C that increases with temperature balanced by high drying rates that preserves vitamin due to the low residence time.
- An increase in the inert load, M_{inert}, also increases the maximum allowable amount of pulp into the bed, above which the SB operation becomes unstable. In addition, no significant effects of M_{inert} on the powder moisture content and production have been observed for a given Q/Q_{ms} ratio.
- High airflow rates, i.e., high Q/Q_{ms} ratios, increase the solids circulation rate and, consequently, the powder entrainment can be improved. An increase in solids circulation implies shorter time for inert particles circulating in the bed. This low cyclic time of inert particles may explain the higher powder moisture content obtained at higher Q/Q_{ms} ratios, because, for cyclic times, the liquid film that covers the particle surface would still be partially wet when its breakage occurs.
- Rewetting inert particles (already coated by the liquid film) contributes to a low powder production and to accumulation of wet powder inside the bed.
- The bed collapse is always accompanied by abrupt changes in the SB flow regime. However, the stable spout regime is restored when pulp feeding is interrupted, and drying proceeds under stable conditions.

- Effects of rewetting particles, powder accumulation, and spout collapse can be minimized by an intermittent pulp feeding.
- The process yield, defined as the ratio between the produced powder and total solid content in the fed pulp (wt% d.b.), increases at higher $T_g|_{in}$ and Q.
- The powder moisture content, m_{pd}, increases and the process yield tends to decrease with increasing pulp feeding flow rate if all other operating conditions are kept constant.
- The thermal efficiency, defined as the ratio of the heat used for water evaporation to the total heat transferred by the gas, decreases with increasing $T_g|_{in}$ and Q, and increases with increasing pulp feeding flow rate. High airflow rates are necessary when drying fruit pulps to preserve the dynamic regime stability. Maintaining the pulps feeding constant, more energy is supplied to the system at higher airflow and temperature. Although enhancing heat and mass transfers, it results in losses to the environment and high exhausting air temperatures with consequent decrease in the thermal efficiency. Inverse behavior is verified upon increasing the pulps feeding to the bed for constant inlet airflow rate or temperature.
- The effective area of heat exchange, defined as the contact area between gas and particles per total bed volume = $(S_p/V_p)(1-\varepsilon)$, does not correspond to the total available surface area of inert particles per bed volume. There is a fraction of inert particles covered by a thin liquid film that modifies their surface area and volume, as well as the annular bed voidage. This fraction of wet particles has been determined as a direct function of the ratio between the pulp feeding flow rate and the total volume of inert particles in the bed.
- The adverse behavior observed for some fruit pulps in SBs, which could not be ascribed to their physical and rheological properties, suggests a possible influence on the drying performance of the pulp chemical composition and of the particles size distribution in the pulps suspensions.

Concerning this last result, Martinez et al. (1995) analyzed the drying of liquid foods, such as milk at different concentrations of fat and without fat, orange juice, carrots, and coffee, in a SB of inert polypropylene particles. Their main results point to the influence of the chemical compositions of the feed on drying performance, which is later discussed in Section 11.4.

Following these results, drying of hog plum, hog plum mango, umbu, red mombin, mango, and sweetsop in a SB of inert particles at fixed operating conditions was analyzed by Ramos et al. (1998) and Lima et al. (2000) aiming to evaluate the effects of the chemical composition of the pulp feed on process performance and product quality. In addition, the powder production kinetics was obtained during each pulp processing. The results showed that losses of vitamin C were up to 50% in relation to the original pulp, mainly for powders from the most acid fruits. However, the high aggregate values of the products may justify such losses in pulp processing. It should be noted that, in spite of the losses, the powdery products presented high vitamin C content due to overall concentration by dehydration. No significant changes in pH were observed and losses of acidity and sugars, which are inherent in processes involving thermal treatment, did not affect the product quality. Except for the hog plum powder, all other pulp powders presented moisture contents below 6%, which is considered suitable for conservation and storage. Regarding powder production, the best results were obtained using umbu and mango pulps, with efficiencies of 60% and 47%, respectively. In the case of sweetsop and hog plum mango drying, the SB fluid dynamic conditions were unstable, resulting in spout collapse. Instability, as described by these authors, was characterized by a low particle concentration in the fountain region and by significant variations in the pressure drop across the bed and in the outlet air temperature. Such a type of SB instability coincides with the one pointed out by Mujumdar (1989), who states that the unsteady regime, being a result of powder accumulation inside the bed of inert particles, is one of the disadvantages of the SB equipment.

Ramos et al. (1998) and Lima et al. (2000) also analyzed the influence of air and pulp feeding, as well as the inert type, on mango pulp drying in a SB. They used high-density PE (polyethylene), low-density PS (polystyrene), and PP (polypropylene) particles as inert. The amount of pulp added into the bed of inert particles was the most significant variable, causing powder retention in the SB and hazarding SB fluid dynamics, leading to the bed collapse at a specific value, the so-called maximum allowable pulp amount. Below this value, powder was steadily produced at a constant rate, reaching values around 60% of total solids feed. Important effects of the type of inert particles on drying operation were observed. Low-density PS particles showed the best results. Unsatisfactory results were obtained with PP particles due to adsorption of the mango pulp on their surface.

11.4 EFFECTS OF PULP COMPOSITION ON SB DRYING OPERATION AND PRODUCT QUALITY

Effects of the paste-like material feeding on SB fluid dynamic parameters and drying performance were initially studied in the 1980s (Pham 1983, Patel et al. 1986, Ré and Freire 1986, 1988, Kachan and Chiapetta 1988, Mujumdar 1989). Afterward, in the 1990s, drying pastes and suspensions in SBs with inert particles gained more interest, since this technique appeared as an alternative to SD, offering lower cost and resulting in products with similar or superior quality (Barret and Fane 1990, Schneider and Bridgwater 1990, 1993, Lima et al. 1992, Reyes and Massarani 1992, Alsina et al. 1996, Passos et al. 1997, Spitzner Neto and Freire 1997).

Patel et al. (1986) verified that upon adding water and glycerol to the SB of inert particles, there was a reduction of about 10% in Q_{ms}, which was attributed to the decrease in the number of particles entering the spout region in beds of wet particles. Lima et al. (1992) analyzed the influence of umbu pulp feeding on Q_{ms}, ΔP_{max}, and ΔP_{ssp}. Their findings showed that an increase in the mass of umbu pulp feeding resulted in a significant increase in ΔP_{max}, no significant variation in ΔP_{ssp}, and a significant decrease in Q_{ms}. However, Reyes and Massarani (1992) observed an increase in Q_{ms} when pure water or an aqueous alumina suspension was added into the bed of inert particles. This effect was more pronounced for the alumina suspension. By analyzing the fluid dynamic behavior of a SB with acerola pulp addition, Alsina et al. (1996) reported that Q_{ms} decreased and ΔP_{max} and ΔP_{ssp} increased as the mass of pulp fed into the bed was augmented in relation to M_{inert}. These results are in accordance with those of Lima and Alsina (1994) and Patel et al. (1986). In the work of Spitzner Neto and Freire (1997), variations in the bed dynamic behavior were analyzed as a function of the continuous addition of water (considered as "standard paste"). The results showed an increase of about 20% in Q_{ms} and a decrease of about 10% in ΔP_{ssp} for the conical bed column, whereas the ΔP_{ssp} decrease was about 20% for the conventional cone-cylindrical geometry.

The complex fluid dynamic behavior of the SB in the presence of paste-like materials, which seemed to result in different discrepant observations by different authors, is a function of the paste feeding mode (continuous or intermittent), its physicochemical properties, the bed geometry, and also the paste-inert wettability characteristics.

A critical analysis of the experimental results of Reyes (1993) and Lima et al. (1992) was presented by Passos et al. (1997), leading to the following conclusions: (1) a critical feed flow rate exists, above which the interparticle cohesion forces caused by liquid bridges formation are significant; (2) the cohesion forces depend on the paste and inert properties and are higher for particles having smaller sizes and sphericity; (3) when these forces are significant, Q_{ms} increases with the paste feeding flow rate until the bed collapses; and (4) Q_{ms} decreases when cohesion forces are negligible due to the formation of a thin layer of paste on the inert surface causing particles slipping.

Studies on drying of different pastes in SBs, such as foods of vegetable and animal origins, organic and inorganic chemical products, bioproducts, and medicines, emphasized the material characteristics as one of the factors that significantly influences the process. Ré and Freire (1988) accounted for a relation between the paste viscosity (vegetable extracts) and the mean dry particle diameter. Martinez et al. (1995) reported high powder retention into the SB during the drying of low fat milk; contrarily, for drying integral milk, the powder was easily carried out from the SB dryer by air. This different behavior was attributed to the high lipid content of integral milk. In that work, the authors also tested the drying of orange and carrot juices in SB and no powder was carried out from the dryer. Inert particles agglomerated, resulting in the interruption of the spouting regime.

Analyses of the drying behavior of different vegetable products have related particle adherence and the spout regime collapse because of the sticky characteristics of fruit pulps and vegetable juices due to the high sugar contents in these products (glucose, fructose, and sucrose). However, none of the available work in the literature could quantify and properly explain these interactions. Therefore, a systematic methodology was developed to analyze and quantify such interactions. These interactions, expressed as empirical correlations, will contribute to implement this type of dryer in the food industry for small-scale powder production.

11.4.1 Influence of Pulp Composition on SB Fluid Dynamics and Drying Performance

Following the previous work of Medeiros and coworkers on drying tropical fruit pulps in SBs of inert particles (Lima 1992, Lima et al. 1992, 1995, 1996, 1998,

2000) and also taking into account results from other authors (Martinez et al. 1995). Medeiros (2001) specially focused on the influence of the chemical composition of fruit pulps on fluid dynamics and drying operation in SBs. Since the pulp chemical composition directly affects the paste properties, it is expected to influence both drying performance and product quality. To quantify these effects, an experimental methodology has been developed.

11.4.1.1 Experimental Methodology

Different tropical fruit pulps (umbu, hog plum, hog plum mango, sweetsop, red mombin, acerola, and mango), without any addition of chemicals or water, were analyzed to identify and quantify their main constituents. They were characterized by the following contents: reducing and nonreducing sugars, fibers, starch, pectin, total solids, soluble solids, lipids, and water. Determinations of pH and citric acid percentage were also performed. All these analyses followed standard, well-established methods in the literature (Adolf Lutz Institute 1985). Starch and pectin were identified as significant components, apart from reducing sugars, lipids, and fibers, as shown in Table 11.1. Composition data of powders obtained from drying these pulps in a SB are also presented in Table 11.1 (Medeiros 2001).

Preliminary drying experiments showed the effects of reducing sugars, lipids, and fibers on the powder production efficiency. Therefore, the five significant constituents (reducing sugars, lipids, fibers, starch, and pectin) were defined as independent variables for analyzing the drying process of tropical fruit pulps in SB of inert particles. Natural mango pulp was adopted as a standard pulp and its composition could be modified, when needed, by adding known amounts of reducing sugars, starch, pectin, lipids, and/or fibers. A 2^{5-1} fractional factorial experimental design with three replicates at the central point was adopted to study the effect of the variables, as shown in Table 11.2 by the 6 first columns. Because the efficiency was defined in relation to the total solids in the fed pulp, the water content was no more an independent variable. For that reason, the pulps' compositions were parametrized in relation to the water content ($C_i^{\mathrm{w}} = C_i/C_{\mathrm{w}}$) in order to normalize, to get easier comparisons between the factor influences and looking for more generalized results. The maximum levels for the concentration of each component were the maximum ones found in tropical fruit pulps (as reported in Table 11.1 or in the literature), whereas the minimum levels corresponded to those of the mango pulp. Materials used to modify and control the pulp composition were glucose and fructose (as reducing sugars), soluble starch, citric pectin, olive oil (as lipids), fibers (extracted from the natural mango pulp), and distilled water.

Six different response variables were chosen to evaluate the process performance. The first three, namely, Q_{ms}, ΔP_{ssp}, and $\theta_{\text{p-fl}}$, the drained angle of repose for the pulp-wetted inert particles, are related to SB fluid dynamic behavior and instability. The other three variables, mainly used to optimize the pulp composition, were the drained angle of repose of the inert particles after drying, $\theta_{\text{p-dry}}$, the powder production efficiency, η_{pd}, and the retention of powder on the inert particle

TABLE 11.1
Chemical Composition of Fruit Pulps and Their Respective Powders

	Fruits													
	Acerola		Umbu		Hog Plum		Hog Plum Mango		Red Mombin		Sweetsop		Mango	
Analysis (wt%)	Pulp	Powder	Pulp	Powder	Pulp	Powder	Pulp	Powder	Pulp	Powder	Pulp	Powder	Pulp	Powder
m (w.b.)	91.6	8.3	89.2	4.0	88.7	9.8	85.8	5.0	79.7	7.1	75.5	5.6	82.4	4.3
RS	6.5	—	4.1	26.4	6.6	33.4	6.3	25.4	8.4	28.1	16.5	26.6	6.2	4.3
NRS	0.1	—	2.2	10.8	0.1	5.4	5.5	14.0	5.6	12.7	0.5	8.6	9.3	—
Lipids	0.3	—	2.3	—	1.4	—	0.9	—	0.8	—	1.3	—	2.2	—
Fibers	0.5	—	0.4	—	0.2	—	0.6	—	0.5	—	2.6	—	0.4	—
Pectin	0.8	—	0.8	—	0.6	—	—	—	0.5	—	—	—	0.5	—
Starch	0.3	—	1.6	—	0.6	—	—	—	4.2	—	—	—	0.4	—

TABLE 11.2
Results of Drying Modified Pulps according to the Experimental Design

Run No.	C^{w}_{sugar} (%)	C^{w}_{lipids} (%)	C^{w}_{fibers} (%)	C^{w}_{starch} (%)	C^{w}_{pectin} (%)	η_{pd} (%)	χ_{pd} (%)	Loss (%)	$\theta_{p\text{-}dry}$ (°)
1	7.67	0.77	0.50	0.52	1.82	14.06	68.95	16.99	24.5
2	19.95	0.77	0.50	0.52	0.67	0.00	97.07	2.93	31.0
3	7.67	6.84	0.50	0.52	0.67	22.10	22.64	55.26	23.0
4	19.95	6.84	0.50	0.52	1.82	0.00	23.05	76.95	20.5
5	7.67	0.77	2.01	0.52	0.67	12.19	75.02	12.79	31.5
6	19.95	0.77	2.01	0.52	1.82	0.00	94.89	5.11	26.5
7	7.67	6.84	2.01	0.52	1.82	27.77	9.09	63.14	21.0
8	19.95	6.84	2.01	0.52	0.67	0.00	58.37	41.63	20.0
9	7.67	0.77	0.50	4.66	0.67	20.83	49.08	30.09	21.0
10	19.95	0.77	0.50	4.66	1.82	9.11	58.14	32.75	26.0
11	7.67	6.84	0.50	4.66	1.82	49.65	19.87	30.48	25.0
12	19.95	6.84	0.50	4.66	0.67	6.69	33.07	60.23	23.0
13	7.67	0.77	2.01	4.66	1.82	26.70	37.29	36.01	27.5
14	19.95	0.77	2.01	4.66	0.67	5.89	66.12	27.99	27.5
15	7.67	6.84	2.01	4.66	0.67	27.57	12.56	59.87	22.0
16	19.95	6.84	2.01	4.66	1.82	23.88	6.72	69.40	19.5
17	13.81	3.81	1.26	2.59	1.24	16.09	16.69	67.22	23.0
18	13.81	3.81	1.26	2.59	1.24	15.86	18.58	55.56	22.0
19	13.81	3.81	1.26	2.59	1.24	19.49	11.05	69.46	20.0

surface (adhered or adsorbed on particle surface), χ_{pd}. The values of η_{pd} and χ_{pd} are defined as follows:

$$\eta_{pd} = \frac{M_{pd}(1 - m_{pd(wb)})}{M_{pp}(1 - m_{pp(wb)})} \times 100 \qquad (11.1)$$

$$\chi_{pd} = \frac{M_{pd\text{-}ret}}{M_{pp}(1 - m_{pp(wb)})} \times 100 \qquad (11.2)$$

The drained angle of repose of particles is assumed to be equal to the measured angle of slide (minimum angle to the horizontal of a flat inclined surface that will allow one particle layer to slide against another under its own weight). To measure the drained angle of repose, a representative sample of material was spread over a horizontal platform, in layers, until the predefined thickness was completed. The platform was then slowly inclined and the drained angle of repose measured when the first particle layer was observed to slide.

Besides the responses to the factorial experimental design (Q_{ms}, ΔP_{ssp}, $\theta_{p\text{-}fl}$, $\theta_{p\text{-}dry}$, η_{pd}, and χ_{pd}) the bed dynamic was followed every minute during the first 10 min after feeding the pulp into the bed. Air superficial velocity at the cyclone outlet (U_{cyc}) and the bed pressure drop were the parameters chosen to follow and analyze the dynamic changes due to pulp feeding into the bed.

To perform the tests, a stainless steel cone-cylindrical SB dryer with acrylic windows was used. Based on the scheme presented in Figure 11.1, the column conical base had an included angle of 60°, a height of 0.13 m, and a gas inlet orifice diameter of 0.03 m. The column cylindrical part was 0.72 m in height and 0.18 m in diameter. This SB column dryer was equipped with an air blower, an electrical heater, an airflow meter, thermocouples to measure and record the air temperature at different points inside the dryer, and a controller for ensuring a constant air inlet temperature. A Lapple cyclone, 0.10 m in diameter, was installed at the dryer outlet to collect the dried powder.

High-density PE particles ($d_p = 3.9$ mm, $\rho_p = 950$ kg/m^3, and $\phi_p = S_p/[\pi d_p^2] = 0.76$) were chosen as inert. The static bed porosity of these particles is 0.29 and their drained angle of repose is 19.5°. From ΔP vs. Q curves obtained for characterizing the SB of inert particles (without pulp), the minimum spouting conditions could be determined as $Q_{ms0} = 17.04 \times 10^{-3}$ m^3/s and $\Delta P_{ms0} = 670$ Pa, at $T_g|_{in} = 70$°C and $M_{inert} = 2.5$ kg.

Based on preliminary tests, the following operating conditions were selected for carrying out the experiments: $M_{inert} = (2.500 \pm 0.005)$ kg, $M_{pp} = (50 \pm 1)$ g, $t_{op} = 40$ min, $T_g|_{in} = (70 \pm 1)$°C and $Q/Q_{ms} = 1.25 \pm 0.05$. The ΔP vs. Q curves were obtained in the SB of inert with pulp according to the following procedure: (1) spouting the bed of inert particles by air at $Q/Q_{ms0} = 1.25$ and $T_g|_{in} = 70$°C until reaching the steady stable regime (i.e.,: $\Delta P_{ms0} = \Delta P_{ssp}$ = constant and $T_g|_{out}$ = constant); (2) feeding the pulp over the fountain region using a syringe, during about 1 min; (3) monitoring ΔP, Q, $T_g|_{out}$, annulus and fountain heights during and after feeding the pulp, until steady state was reestablished; (4) decreasing Q slowly, recording its value and the correspondent pressure drop, as well as any other parameters that were changing; and (5) registering as Q_{ms} the lowest Q value for which the fountain was still observed. For comparison, distilled water was also used as a standard liquid feeding to the SB of inert particles.

11.4.1.2 Influence of the Pulp Chemical Composition on Spouted-Bed Fluid Dynamics

Changes in the fluid dynamic variables are evident within a few minutes after injecting pulp into the bed of inerts. For almost all experiments, the SB flow regime stabilized about 10 min after injecting the pulp, during which time most of the pulp moisture was evaporated. Such behavior was not observed for pulp 2, in which case the SB was characterized by a diluted and lower-height fountain and serious trends to collapse the spout.

To better evaluate changes in the SB hydrodynamic behavior due to pulp feeding, the fluid dynamic variables were parameterized in relation to their initial values obtained before feeding the pulp. Pulps 1, 2, 3, 5, and 9 (see Table 11.2) were chosen to analyze the effect of the chemical composition on the SB stability and hydrodynamics, since these pulps exhibit the maximum concentration of each individual main component: pectin, reducing sugars, lipids, fibers, and starch.

Figures 11.2 and 11.3 illustrate, respectively, initial changes in the $\Delta P/\Delta P_0$ and $U_{cyc}/U_{cyc,0}$ ratios. One can infer that the presence of water results in a small increase in U_{cyc} and a sharp decrease in ΔP. An expansion of the annulus and an increase in the fountain height were also observed in this case. However, as drying proceeds, initial values of U_{cyc} and ΔP are gradually restored, even though changes in the

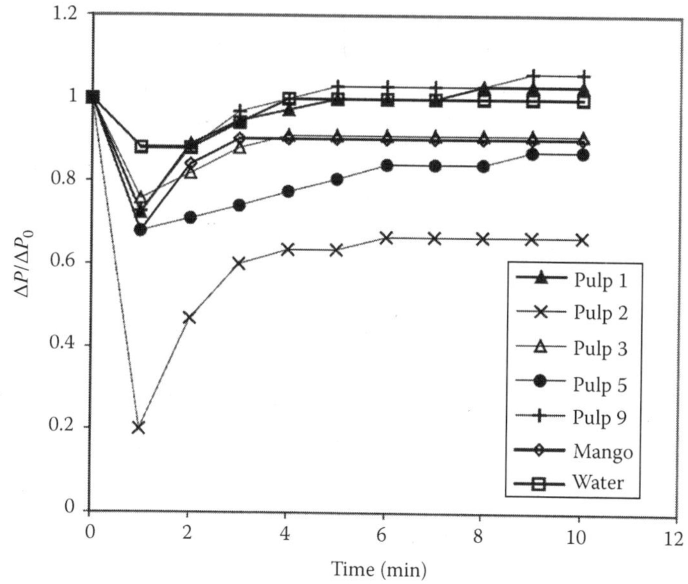

FIGURE 11.2 Influence of liquid feeding on the bed pressure drop: pulp 1 (the highest level of pectin), pulp 2 (the highest level of reducing sugars), pulp 3 (the highest level of lipids), pulp 5 (the highest level of fibers), pulp 9 (the highest level of starch), natural mango, and water.

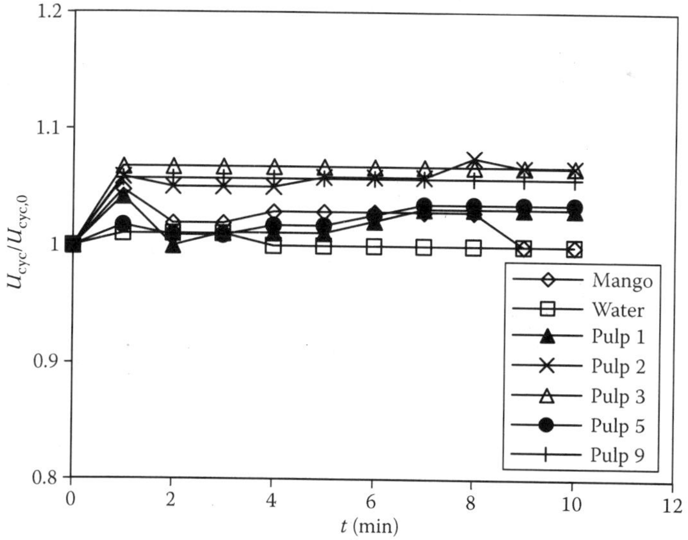

FIGURE 11.3 Influence of liquid feeding on the gas superficial velocity at the cyclone outlet. The legend for different pulps is the same as adopted in Figure 11.2.

fountain height and a little expansion of the annulus are still present. This fountain height behavior is the same as pointed out by Schneider and Bridgwater (1993) upon injecting water in a bed of glass beads. Similar behavior was also obtained with the addition of mango pulp. Immediately after feeding the mango pulp (see Figure 11.2), ΔP sharply decreases and U_{cyc} shows a tiny increase. As drying continues, both ΔP and U_{cyc} increase. However, the SB steady condition is attained at a smaller ΔP compared to ΔP_0. These observations confirm that, apart from the first instantaneous effects of water and mango pulp feeding, an influence of the mango pulp on the bed fluid dynamics is still present when practically all water has already been evaporated, which may be attributed to powder retention on the surface of the inert particles.

All other pulps also caused a sharp decrease in ΔP immediately after their addition into the bed of particles. According to Spitzner Neto and Freire (1997), this decrease is explained by the agglomeration that, together with the pulp viscosity, jeopardizes the particle circulation in the bed, increasing airflow rate in the spout region. With a greater resistance in the annulus region, the airflow rate and the voidage in the spout region increase, leading to a decrease in the pressure drop across the bed. The higher the fluid dynamic instability brought about by the pulp, the larger this reduction in ΔP. As shown in Figure 11.2, water leads to a pressure drop decrease of about 12% in relation to the dry bed, while, for the fruit pulps, the reduction ranges from 24% (for pulp 3 with maximum content of lipids) to 80% (for pulp 2 with maximum content of reducing sugars). This corroborates that the content of reducing sugars in pulps contributes more to instabilities on SB dynamics than the content of lipids. Excluding pulp 2, one can see that the recovery of the $\Delta P/\Delta P_0$ ratio is somewhat slower for pulp 5 (maximum of fibers), even though the $\Delta P/\Delta P_0$ value after 10 min is equal to one obtained for mango pulp and pulp 3.

From Figure 11.3, it is seen that, except in the case of water, the fluid dynamic regime tends to stabilize at a $U_{cyc}/U_{cyc,0}$ somewhat higher than 1. This trend shows the effect of powder adherence to the surface of the inert particles on SB fluid dynamics, since, during these first 10 min of operation, no powder has been carried out by air to the cyclone yet.

The initial values of Q/Q_{ms}, based on the bed of inert particles without pulp, were fixed for every experiment. As the experiments progressed, the bed fluid dynamics changed due to the presence of fruit pulp. As mentioned earlier in this chapter, the pressure drop, fountain high, solid circulation rate, and the structural characteristics of the bed are affected by the presence of pulp. The superficial velocity U_{cyc}, as well as the velocity distribution, changes consequently in order to adjust to these new loading conditions, even though no changes were made on the input valve aperture.

Pulp 2, with the highest reducing sugar content, shows oscillations in the $U_{cyc}/U_{cyc,0}$ ratio, which derive from an increase in the resistance to gas flow through the annulus region due to particle agglomeration observed during this test.

The SB fluid dynamic changes due to differences in pulp composition are apparently complex. For high lipid and starch contents, a somewhat higher $U_{cyc}/U_{cyc,0}$ ratio is associated with better fluid dynamic conditions, in which higher particle circulation rate and lower particle-wall adherence were observed during experiments. For pulp 2, however, the little increase in $U_{cyc}/U_{cyc,0}$ is a consequence of particle agglomeration, which creates a preferential path for airflow in the spout region, decreasing the $\Delta P/\Delta P_0$ ratio.

Similar behavior of increasing fountain height just after liquid addition was observed for all pulps, except for pulp 2. After feeding pulp 2, the fountain height decreased and oscillated (around 85% of the initial value) as drying proceeded. With regard to the annulus, its expansion, measured by the increase in height, was observed after feeding each one of 19 pulps, and the highest expansion recorded was for pulp 2. This annular bed expansion is explained by a decrease in the circulation rate, which causes a redistribution of particles inside the bed, with higher solids concentration in the annulus, leading to its expansion.

The values of Q_{ms} and $\theta_{p\text{-fl}}$ for all the pulps are shown in histograms in Figures 11.4 and 11.5, respectively, to provide a general view of the global results. The observed Q_{ms} values ranged from 13.0×10^{-3} m³/s to 19.6×10^{-3} m³/s, while $\theta_{p\text{-fl}}$ varied between 2.01 and 1.21. It is verified, however, that these parameters varied within a narrow range for all the fruits. The maximum Q_{ms} corresponds to pulp 2, whereas

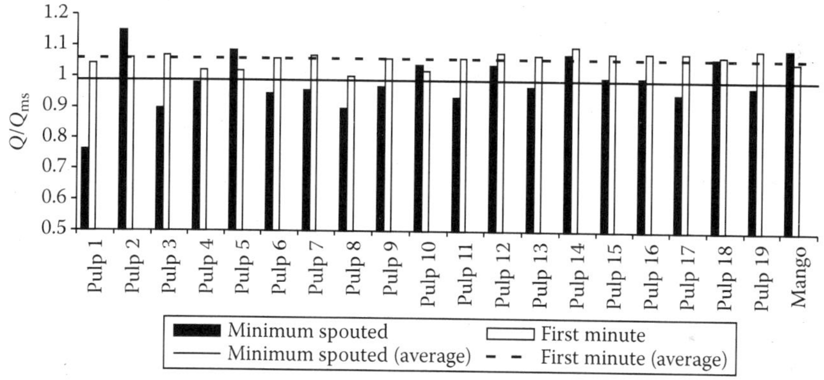

FIGURE 11.4 Minimum spouting flow rate. Runs with the modified pulps (1–19).

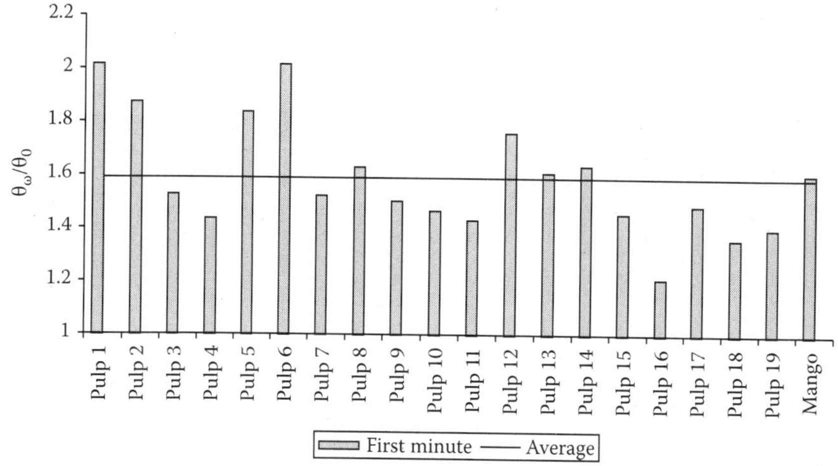

FIGURE 11.5 Angles of repose for the inert with the pulps (1–19) addition.

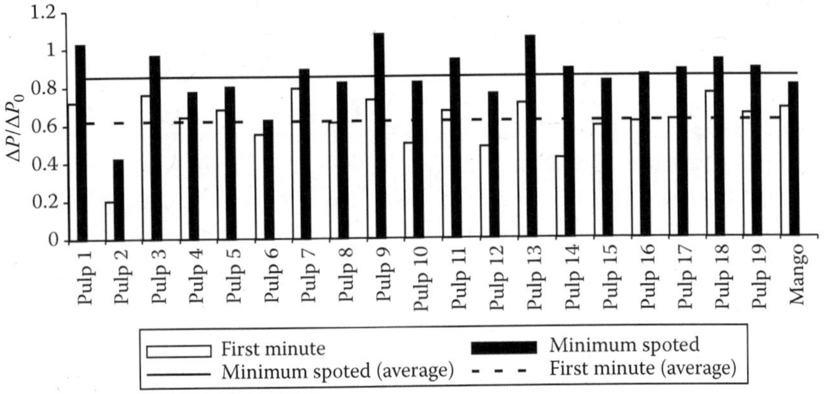

FIGURE 11.6 Pressure drop ratio at $t = 1$ min and at minimum spouting conditions, during drying. Runs with the modified pulps (1–19), and with the mango pulp.

$\theta_{p\text{-fl}}$ presented its minimal value in the case of pulp 1. The results suggest that high starch and lipid content could favor the bed flowability as it was expected due to the lubricant characteristic of these pulps (Martinez et al. 1995). Nevertheless, a more detailed statistical analysis would have to be made in order to confirm the observed trends.

In Figure 11.6, data of $\Delta P_{ms}/\Delta P_{ms0}$ are compared with $\Delta P/\Delta P_0$ measured at 1 min after pulp feeding for all paste-like materials tested. The effect of liquid addition on the pressure drop is the same for all tests, i.e., ΔP decreases in the presence of these liquids. Comparing the two different instants of the drying process, the behavior follows the same tendency for each pulp composition. Lower and higher pressure drops for the wet bed at 1 min after the pulp addition correspond to lower and higher stable spout pressure drop.

11.4.1.3 Drying Performance

The four last columns in Table 11.2 show the results obtained in relation to the drying parameters. The loss refers to the percent of powder mass retained in the equipment, i.e., adhered on walls, dispersed at the column outlet, and, eventually, lost in the cyclone.

As detailed by Medeiros (2001) and Medeiros et al. (2004), the results of the statistical analysis of the fractional factorial design showed that all the pulp components, except fibers, exert significant effect (at 95% confidence level) on η_{pd}. Reducing sugars cause a decrease in η_{pd} and their effect is the most significant one. Starch, pectin, and lipids favor η_{pd}, the starch concentration being the most influent. This result is in accordance with the dynamic results for reducing sugars, starch, and lipids concentrations.

As regards χ_{pd}, the same components present significant effects, and, as expected, such effects are opposite to those on η_{pd}.

Pulps having high concentrations of reducing sugars form highly adherent films on the surface of inert particles, and interparticle attrition and impacts are not sufficient to break these films as drying goes on (Martinez et al. 1995, Ramos et al. 1998). In addition, due to the sticky characteristics of reducing sugars, particle agglomeration occurs, compromising the regime stability and even leading to SB collapse. Lipid concentration followed by starch exerts the most important effects on χ_{pd}. The important negative effect of lipids on χ_{pd} is related to their lubricant characteristic, which interferes on the bed dynamics, enhancing the particles circulation and thus, facilitating breakage of the adherent film (Martinez et al. 1995).

For $\theta_{p\text{-}dry}$, only the lipid concentration had a significant effect (Medeiros 2001, Medeiros et al. 2004).

According to the results of the fractional factorial design, the fiber concentration does not significantly affect the drying performance. Therefore, this variable was excluded from the statistical analysis, and the experimental design could then be rearranged into a complete 2^4 factorial design with three replicates at the central point.

Based on this 2^4 complete factorial design, a predictive model was obtained for η_{pd} at a confidence level of 95%:

$$\begin{aligned}\eta_{pd} = 15.68 - 9.71\frac{(C^{w}_{sugar} - 13.81)}{6.14} + 4.31\frac{(C^{w}_{lipids} - 3.81)}{3.03} + 5.89\frac{(C^{w}_{starch} - 2.59)}{2.07}\\ + 3.49\frac{(C^{w}_{pectin} - 1.24)}{0.57} - 2.36\frac{(C^{w}_{sugar} - 13.81)}{6.14}\frac{(C^{w}_{lipids} - 3.81)}{3.03}\\ + 2.55\frac{(C^{w}_{starch} - 2.59)}{2.07}\frac{(C^{w}_{pectin} - 1.24)}{0.57}\end{aligned} \quad (11.3)$$

resulting in

$$\begin{aligned}\eta_{pd} = 17.40 - 1.10C^{w}_{sugar} + 3.17C^{w}_{lipids} + 0.17C^{w}_{starch} + 0.52C^{w}_{pectin}\\ -0.13C^{w}_{sugar}C^{w}_{lipids} + 2.16C^{w}_{starch}C^{w}_{pectin}\end{aligned} \quad (11.4)$$

where

$C^{w}_{i} = C_i/C_w$ is expressed in wt%

η_{pd} in mass percent of powder produced in relation to the solids content in the pulp feeding

As shown by Medeiros et al. (2002), the percentage of explained variation of η_{pd} predicted by Equation 11.4 is 93.73%. More details about the regression and residue analyses can be found in Medeiros (2001).

The correlation represented by Equation 11.4 was assessed using experimental data related to the drying of different natural tropical fruit pulps, as shown in Figure 11.7. Drying of red mombin pulp resulted in the only significant deviation for the η_{pd} prediction given by Equation 11.4. This deviation can be explained by the different

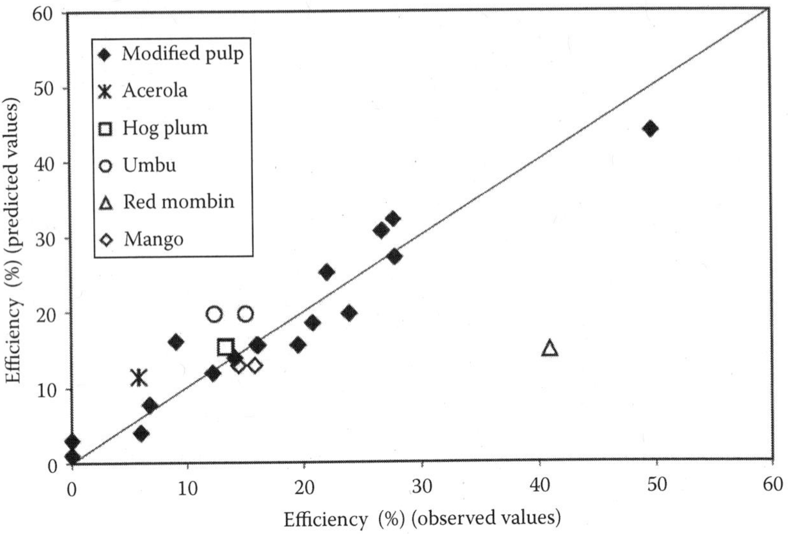

FIGURE 11.7 Comparison between experimental efficiency of powder production and the values predicted with Equation 11.3 for tests with natural fruit pulps and modified pulp.

drying conditions used in processing this pulp (lower Q and $T_g|_{in}$ with $Q/Q_{ms} = 1.05$ and $T_g|_{in} = 50°C$) and by agglomeration problems verified during this test.

The promising results obtained for modeling η_{pd} led to the proposition of an optimized pulp composition, which maximizes the efficiency in an amplified range of maximum and minimum concentrations of reducing sugars, lipids, starch, and pectin, now covering the range encountered in many tropical fruit pulps. As the mango pulp was the basis for the modified pulps compositions, concentrations of the components below the ones encountered in the standard mango pulp could not be tested in the experimental design. According to Equation 11.4, the maximum efficiency should occur for minimum sugar content and maximum lipids, starch, and pectin contents. Based on an optimization routine (Medeiros 2001), the optimized pulp, associated with a maximum for η_{pd} equal to 81%, would have the following composition: $C^{w}_{sugar} = 5.52$ wt%, $C^{w}_{lipids} = 14.69$ wt%, $C^{w}_{starch} = 4.93$ wt%, and $C^{w}_{pecti} = 2.78$ wt%.

A modified pulp having the optimized composition was prepared and the drying experiment was carried out at $Q/Q_{ms} = 1.22$ and $T_g|_{in} = 70°C$, with five intermittent pulp feedings into the bed of inert particles. The experimental results obtained for η_{pd} are shown in Figure 11.8. Excellent drying performance was reached in this experiment, with uniform powder production, and η_{pd} about 70%, reproducing the drying behavior obtained with the tropical fruit pulps, but showing significantly higher powder production rate and efficiency. This result for the optimized pulp composition has motivated the development of the next step of the research utilizing mixtures of fruits, taking advantage of the different natural pulp compositions to generate the optimized composition, with the help of some oil and starch additives. It is worth mentioning that powder mixtures of fruits have been showing very good market acceptance.

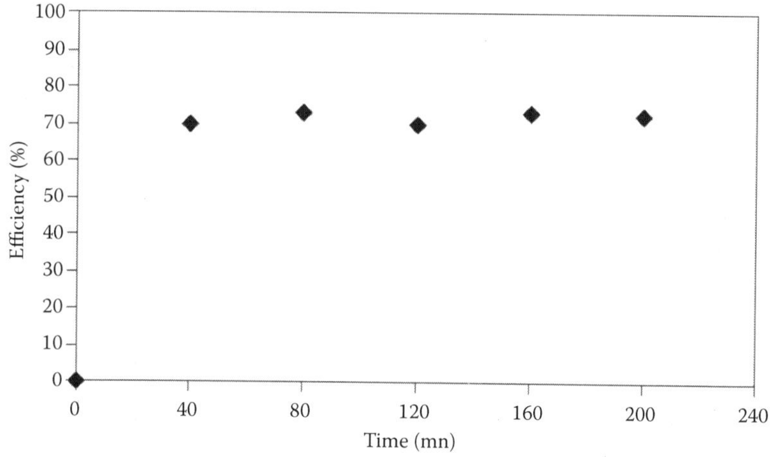

FIGURE 11.8 Efficiency of powder production in the tests with modified pulp.

11.5 DRYING OF MIXTURE OF FRUITS WITH ADDITIVES

The results of pulp drying with optimized composition indicate that fruit pulp drying in a SB is viable. However, with respect to the product, it is necessary to search for ways of optimizing the addition of adjuvant to maintain the sensory and nutritional quality of powdery fruit. Mixing pulps with the addition of specific components (starch, lipids, and pectin) in proportions that combine optimization of the drying process and functionality of the powder obtained, point to very favorable perspectives with respect to the use of the SB in the drying and production of powdery fruits.

The concern in obtaining nutritionally and sensorially acceptable products comes from the current challenge of the food industry to develop products with high nutritional value and sensory characteristics equal to, or greater than, those of natural or traditionally processed foods. The choice of a mixture of tropical fruit pulps is explained by the functionality of the mixture promoted by the synergy of the individual compositions. This powder mixture, with natural flavors, aromas, and functional components, may result in products of sensory and nutritional quality that will make their way into the market.

11.5.1 Definition of the Mixture Formulation and Drying Results

Preliminary drying tests of fruit mixtures (mango, hog plum, umbu, red mombin, and acerola) with the addition of cornstarch, pectin, and olive oil were performed in a SB to identify promising mixture formulations (Araújo et al. 2007, Souza et al. 2007). The mixture formulations were defined based on the composition of fruit pulps and on the additives to be incorporated to attain the optimized composition and to make SB drying feasible, as pointed out by Medeiros (2001). It is worth highlighting that, although the concentration of lipids identified as optimal for fruit pulp drying is above 6 wt%, the mixtures were formulated to contain 2 wt% of lipids due to their adverse effects for health. The best results were obtained in the drying of mango pulp (high fiber and carotenoid contents, with aroma preservation) mixed with red mombin (high starch content) and umbu (high lipid, vitamin C, and complex B contents). However, results from the sensory analysis carried out in yogurts with the addition of this dried powder were not satisfactory, owing to the characteristic flavor and aroma of olive oil (Araújo et al. 2007).

Considering the need to modify the lipid source, another experimental series was performed, involving the drying of mixtures constituted by red mombin, mango, and umbu pulps with the following additives: cornstarch or boiled green banana, as sources of starch, citric pectin, and different types of fat (linseed oil and wheat germ; olive and Brazilian nut oils; coconut milk, milk cream, and powdery palm fat). The codes used to identify these mixtures and their respective formulations are presented in Table 11.3.

Experimental conditions and the procedure used in the drying tests of these six fruit mixtures were similar to those developed and adopted in the drying tests of

TABLE 11.3
Code and Mixture Formulations in the Drying Runs with Fruit Pulp Mixtures

	Formulations	
Codes	**Fruit Pulps (wt%)**	**Additives (wt%)**
F01	Umbu pulp—30	Olive oil—1.3
	Red mombin pulp—30	Pictin—1.4
	Mango pulp—30	Cornstarch—1.4
		Water—6.0
F02	Umbu pulp—28.6	Cream milk—5.7
	Red mombin pulp—28.6	Pictin—1.3
	Mango pulp—28.6	Cornstarch—1.4
		Water—5.7
F03	Umbu pulp—28.6	Coconut milk—5.7
	Red mombin pulp—28.6	Pictin – 1.3
	Mango pulp—28.6	Cornstarch—1.4
		Water—5.7
F04	Umbu pulp—30	Brazilian nut oil—1.3
	Red mombin pulp—30	Pictin—1.4
	Mango pulp—30	Cornstarch—1.4
		Water—6.0
F05	Umbu pulp—29.5	Powdery palm fat—2.4
	Red mombin pulp—29.5	Pictin—1.5
	Mango pulp—29.5	Cornstarch—1.8
		Water—5.9
F06	Umbu pulp—27.5	Powdery palm fat—2.2
	Red mombin pulp—27.5	Pictin—1.4
	Mango pulp—27.5	Boiled green banana—8.3
		Water—5.5

modified pulps. The SB dryer and inert particles were also the same, but the mixture was fed into the bed of particles by a peristaltic pump connected to a twin-fluid atomizer, whose atomizing air was supplied by a low-power compressor. In each experiment, the mixture feeding was intermittent and divided into three loads of approximately 100 g for 20 min each, with a 15 min interval between feedings. Parameters measured during 105 min of drying operation were the same as those listed in Section 11.4.1.1 for the tests performed with modified pulps. In addition, all the six processed mixtures and their corresponding powders were characterized with respect to moisture, total acidity (% citric acid), soluble solids, and pH, following standard methods and procedures (Adolf Lutz Institute 1985). Measurements of the water activity of pulps and powders were performed in specific equipment.

Table 11.4 shows the physicochemical results of the six formulated mixtures. No significant variations in the moisture content, soluble solids, acidity, or pH of the mixtures were found. The same behavior was observed in the case of water activity.

No fluid dynamic instability occurred during the drying trials of these six formulated mixtures, with the pressure drop across the bed, annulus, and fountain heights stable over the whole course of the experiments.

Table 11.5 shows that the drying of the mixtures displays η_{pd} in the range from 37% to 52%. This is considered relatively high when compared to previous η_{pd} data

TABLE 11.4
Physicochemical Characterization of the Fruit Pulp Mixture

Mixtures	$m_{pp(wb)}$ (%)	Soluble Solids (%)	Total Acidity	pH	a_w
F01	82.9	11.33	1.06	3.27	0.986
F02	83.1	9.97	0.79	3.20	0.983
F03	83.4	10.73	0.83	3.25	0.985
F04	83.4	11.37	0.90	3.22	0.989
F05	82.2	13.07	0.86	3.28	0.985
F06	82.8	12.15	0.90	3.28	0.992

TABLE 11.5
Drying Performance in Runs with Fruit Pulp Mixtures

Mixtures	$m_{pd(wb)}$ (%)	η_{pd} (%)	χ_{pd} (%)
F01	4.6	38.7	30.4
F02	4.7	35.6	58.9
F03	8.1	44.7	14.3
F04	7.6	42.1	16.3
F05	6.3	52.3	37.9
F06	5.4	37.6	31.2

of umbu, mango, and red mombin pulps, drying in the same SB of inert particles (Medeiros 2001). It can be observed in Table 11.5 that different types of fat (F02 and F03) or of starch (F05 and F06) interfere in η_{pd}. This is likely due to the amount of powder retained on the surface of inert particles (high χ_{pd} in case of F02) or the amount of powder adhered to dryer walls (high value in case of F03 and low in case of F05). Note that tests with low η_{pd} and χ_{pd}, as F04, can be characterized by significant losses of powder due to its adherence to the dryer walls. Low powder moisture contents translate into products that exhibit good conservation and storage characteristics.

Figure 11.9 illustrates the powder collection as a function of time for the different mixtures. The mass of produced powder increases linearly with time, indicating that the production rates are practically constant in these runs, except for the run with F05, which shows an increased production rate from 90 min.

The curves shown in Figure 11.10 represent the evolution of η_{pd} during the drying operation. From these curves, it is clear that η_{pd} can be maintained practically at the same level during all drying operations by controlling the intermittence of pulp feeding. Therefore, by using an intermittent and well-controlled pulp feeding, powder can be produced continuously, overcoming problems of SB instability. This assures the feasible implementation of SBs of inert particles for drying paste-like materials, as pointed out by Passos et al. (2004) and Honorato (2007), especially in the food industry.

A sensory analysis performed for powders obtained from F01, F02, F03, F04, and F05 formulations revealed good sensory acceptance for F05 powder (Araújo et al. 2007). This result, along with the better dryer performance associated with this mixture, leads to the choice of palm fat as the lipid source.

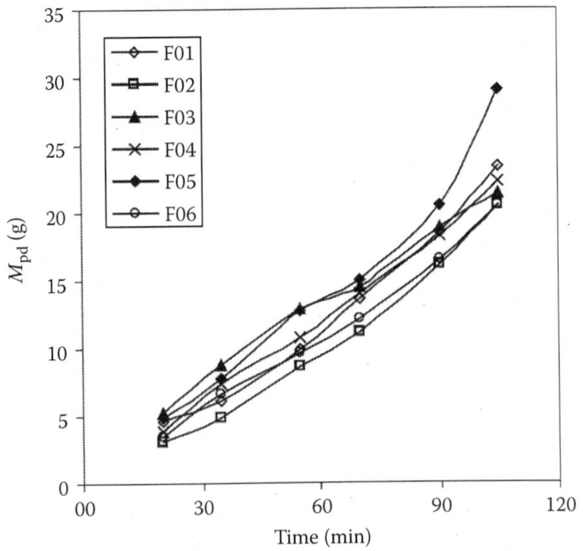

FIGURE 11.9 Kinetics of powder production in the tests with fruit pulp mixtures.

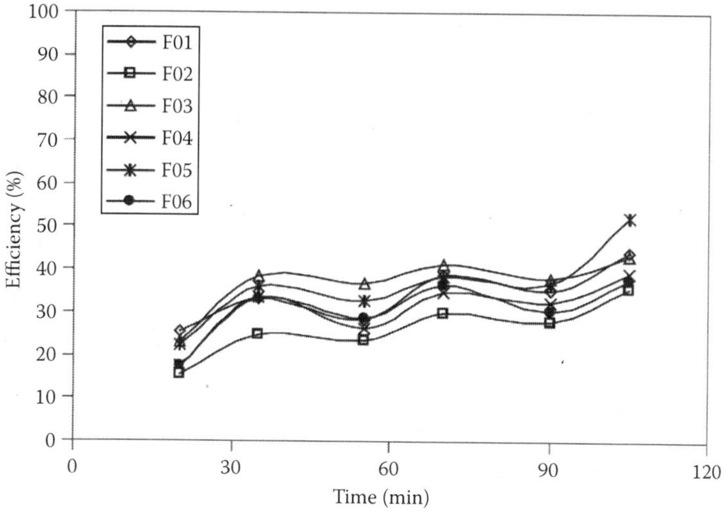

FIGURE 11.10 Efficiency of powder production during drying in the tests with fruit pulp mixtures.

11.5.2 Quality of Powdery Fruit Pulp Mixtures

To define the starch source, the quality of the powders obtained from mixtures F05 and F06 was evaluated for physical, physicochemical, and sensory characteristics. In addition to physical characterization and water activity measurements, the following powder properties were determined: apparent density, according to the methodology described by Birchal et al. (2005), flowability, and solubility (Moreira 2007); reconstitution time and laser granulometry. The powder apparent density was determined under two specific conditions, one of free packing (poured bulk density) and another of maximum compaction (tapped bulk density). The reconstitution of the mixture was achieved by powder rehydration, determining its reconstitution time at 30 s intervals.

Since these particular pulp mixtures come from acidic fruits, their concentrated powders have high acidity. Their low moisture content and water activity are characteristics of long shelf life. The mean time to obtain the reconstituted mixtures was 300 s for the powders of both F05 and F06 formulations. This result is similar to those found by Kachan et al. (1988) and Dacanal (2006) analyzing the reconstitution of tomato paste and dehydrated acerola juice, respectively. Both powders presented good solubility, 99.1% for F05 and 99.35 for F06, a range compatible with that found by Moreira (2007) and Cano-Chauca et al. (2005) for the extract obtained from acerola residue and for mango juice microencapsulated with maltodextrine, respectively, both processed in a spray dryer.

The ratio of the apparent density under maximum compaction to the one under free packing condition was 1.76 for the powder produced from F05 mixture, and 1.54 for the powder produced from F06 mixture. This result shows that both powders are cohesive, since this ratio is higher than 1.3, with F05 mixture giving the more cohesive powder. Angles of repose resulting from the flowability tests showed values below 45°, satisfactory fluidity, but not excellent. The particle average diameter was 255.60 μm for the powder from F05 mixture, and 317.81 μm for the powder from F06 mixture, compatible with the granulometry of the powders obtained by Dacanal (2006) and Moreira (2007).

11.5.3 Sensory Evaluation of Yogurt Prepared with a Mixture of Fruit Powder

Sensory analysis was performed with natural yogurt produced in the laboratory with the addition of either reconstituted powder mixtures (F05 and F06) or natural pulp mixtures.

Powders were reconstituted by dissolution in water, added at the proportion needed to recover the water content of the original mixture (approximately 12° Brix). The mixture was submitted to stirring at 200 rpm for 30 s intervals until complete dissolution was obtained. The reconstituted fruit pulp mixtures were added to the natural yogurt. The pulp/yogurt/sugar proportions on the labels of the industrialized yogurts enriched with fruit pulp were adopted. The samples were prepared 1 h before the test and served for sensory assessment at a temperature of (10 ± 2)°C.

The following attributes were analyzed: color, texture, appearance, flavor, and aroma, using the 9-point, nonstructured hedonic scale. The Acceptability Index (AI) was calculated considering a maximum score of 100% and mean score in percent. The product is considered acceptable when the AI is higher than 70%.

Sensory analysis was carried out with a panel of –25 untrained tasters, each one receiving three identified samples: the first, natural yogurt composed of the product in powder form with the addition of commercial cornstarch (F05); the second, a powder mixture with boiled green banana replacing the cornstarch (F06); and the third, a mixture of natural pulps (F07).

Figure 11.11 shows the mean values obtained for the samples assessed. The results demonstrate that sample acceptability was near or higher than 70% in most of the attributes evaluated. This result indicates that the production of the fruit pulp mixture in powder form is viable from the sensorial viewpoint; however, ways of improving the quality of the product should be investigated.

The data presented in Table 11.6 were generated from the analysis of variance and from Tukey's test, where equal letters on the same line mean that the samples are not significantly different in the attribute assessed. For the attributes of color, appearance, and texture, the samples showed no significant differences in acceptance ($p > 0.05$), although sample F01 had the highest means, in absolute terms, with respect to these attributes. The taste of the samples showed significant differences, and the lowest mean was associated with the yogurt containing the mixture of natural pulps, whereas sample F01 had the best acceptability.

Sample F07 obtained the best acceptability index for aroma, a finding that differed significantly from samples F05 and F06. Sample F07 was prepared only with natural red mombin, mango, and umbu pulps without additives such as fat, which has a characteristic aroma. It must also be considered that the powdery products were obtained from the drying process and submitted to heating, resulting in the loss of volatile components that identify the aroma and flavor of fruits.

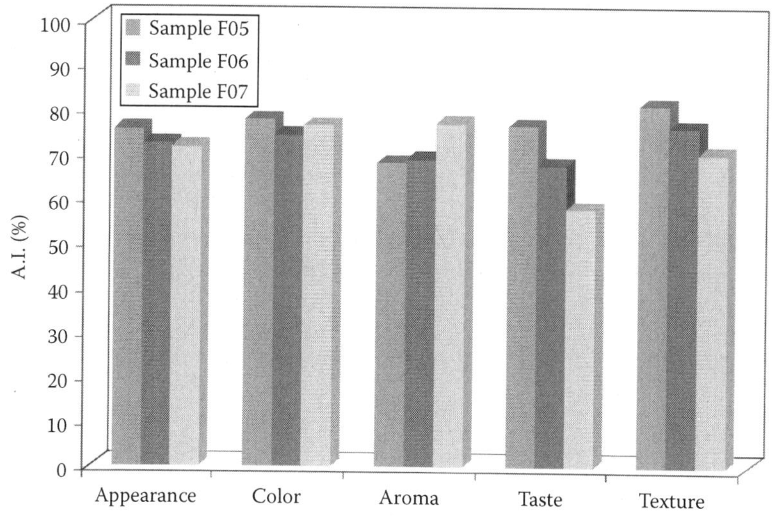

FIGURE 11.11 Frequency of the notes attributed to the panel in the sensory analyses of yogurt prepared with fruit pulp mixtures.

TABLE 11.6
Mean of Absolute Sensory Attributes Used in the Analysis of Yogurt with the F05, F06, and F07 Formulations

Attributes	Sample F05 (%)	Sample F06 (%)	Sample F07 (%)
Appearance	6.7[a]	6.7[a]	6.7[a]
Color	7.0[a]	6.8[a]	7.0[a]
Aroma	6.2[a]	6.4[a]	6.9[a]
Taste	6.9[a]	6.3[b]	5.9[c]
Texture	7.2[a]	7.1[a]	6.5[b]

Note: Same letters for attributes indicate that samples do not differ significantly ($p \leq 0.05$) among themselves.

Despite the loss of volatile components, sample F05 had a higher mean in the taste attribute. The low indices of acceptability with respect to taste of samples F06 and F07 may be related to the astringency of green banana and the acidity of the natural fruits, respectively. The loss of volatile components and the fruit mixture prevented the tasters from identifying the fruits used in the samples.

11.6 CONCLUDING REMARKS

This chapter provides a review on the application of spouted-bed dryers to the drying of tropical fruit pulps. It was shown that the fluid dynamics is affected by the presence of pulp but it is possible to operate in stable spout regime by controlling the feeding rate. High yield and efficiency can be attained (80% or higher), provided that proper operational conditions are chosen. Strong evidence about the influence of pulp composition showed that it is not favorable to process some kinds of fruits in spouted bed. Nevertheless, even in these cases, it is possible to operate with good yield and stability by adjusting the pulp composition close to the optimized value. In this context, the use of blend of pulps could be an interesting application. Some questions remain to be solved before the industrial production, especially the scaling up, and the conditions for a continuous operation, but the results showed that this process is technically feasible and leads to a good quality dry fruit powder product.

REFERENCES

Adolfo Lutz Institute. 1985. Analytical standards of Adolfo Lutz Institute. Physical and chemical methods for analyses of foods. São Paulo, vol. 1, 316 p. (in Portuguese).

Alsina, O. L. S., Lima, L. M. R., Morais, V. L. M., and Nóbrega, E. S. 1995. Study the solids circulation of conventional spouted bed for drying of West Indian Cherry paste. Paper presented at the 1st Ibero-American Food Engineering Congress, Campinas, Brazil.

Alsina, O. L. S., Morais, V. L. M., Lima, L. M. R., and Soares, F. H. L. 1996. Studies on the performance of the spouted bed dryer for the dehydration for West Indian Cherry pulp. In *Drying'96*, ed. A.S. Mujumdar. New York: Hemisphere Publishing Corp., pp. 867–872.

Araújo, V. P. U., Souza, J. S., Rocha, S. C. S., and Medeiros, M. F. D. 2007. Studies of drying mixtures of pulp of tropical fruit. Paper presented at the VII Brazilian Congress on Undergraduate Research in Chemical Engineering, São Carlos/SP, Brazil.

Astolfi-Filho, Z., Souza, A. C., Reipert, E. C. D., and Telis, V. R. N. 2005. Encapsulation of passion fruit juice by co-crystallization with sucrose: Crystallization kinetics and physical properties. *Sci. Technol. Foods* 25: 795–801 (in Portuguese).

Barret, N. and Fane, A. 1990. Drying liquid materials in a spouted bed. In *Drying'89*, ed. A.S. Mujumdar. New York: Hemisphere Publishing Corp., pp. 415–420.

Birchal, V. S., Passos, M. L., Wildhagen, G. R. S., and Mujumdar, A. S. 2005. Effect of spray-dryer operating variables on the whole milk powder quality. *Drying Technol.* 23: 611–636.

Cano-Chauca, M., Stringheta, P. C., Ramos, A. M., and Cal-Vidal, J. 2005. Effect of the carriers on the microstructure of mango powder obtained by spray drying and its functional characterization. *Innov. Food Sci. Emerg. Technol.* 6: 420–428.

Dacanal, G. C. 2006. Study of granulation of acerola juice dehydrated in fluidized bed. MSc dissertation, State University of Campinas, Campinas, Brazil (in Portuguese).

FAO. 2005. https://www.fao.org.br/publicacoes.asp (accessed November, 2007).

Fellows, P. J. 2000. *Food Processing Technology*, 2nd edn. Cambridge, U.K.: Ellis Horwood Ltd.

Honorato, G. C. 2007. Drying of Cefalotórax of the shrimp (Peneaus vannamei) for production of proteinic concentrate. PhD thesis, Federal University of Rio Grande do Norte, Natal, Brazil (in Portuguese).

Hufenussler, M. and Kachan, G. C. 1988. Drying of puree of banana in a spouted bed dryer. Paper presented at the XIII Brazilian Meeting on Porous Media, São Paulo, Brazil.

Kachan, G. C. and Chiapetta, E. 1988. Dehydration of tomato paste in a spouted bed dryer. Paper presented at the VIII Brazilian Congress on Chemical Engineering, São Paulo, Brazil.

Kachan, G. C., Taqueda, M. E., and Gunther, P. A. S. 1988. Characteristics of tomato powder, obtained by dehydration of the tomato paste in spouted bed. Paper presented at the VIII Brazilian Congress on Chemical Engineering, São Paulo, Brazil.

Lima, C. A. P. 1993. Drying of umbu pulp in spouted bed. MSc dissertation, Federal University of Paraíba, Campina Grande, Brazil (in Portuguese).

Lima, M. F. M. 1992. Dehydration of umbu pulp in spouted bed—Fluid dynamic and thermal analysis. MSc dissertation, Federal University of Paraíba, Campina Grande, Brazil (in Portuguese).

Lima, M. F. M. and Alsina, O. L. S. 1994. Dehydration of umbu pulp in spouted bed. Thermal studies. Paper presented at the European Congress on Fluidization, Las Palmas de Gran Canaria.

Lima, M. F. M., Almeida, M. M., Vasconcelos, L. G. S., and Alsina, O. L. S. 1992. Drying of umbu pulp in spouted bed: Characteristic curves. In *Drying'92*, ed. A.S. Mujumdar. New York: Hemisphere Publishing Corp., pp. 1508–1515.

Lima, M. F. M., Lima, L. M. O., Santos, E. M. B. D., and Santos, C. I. 1995. Influence of operating variables on seriguela pulp dehydration in spouted bed. Paper presented at the XXIII Brazilian Congress in Particulate Systems, Maringá, Brazil.

Lima, M. F. M., Lima, L. M. O., Santos, E. M. B. D., and Santos, C. I. 1996. Drying of cajá in spouted bed. Paper presented at the XI Brazilian Congress on Chemical Engineering, Rio de Janeiro, Brazil.

Lima, M. F. M., da Mata, A. L. M., Lima, L. M. F., and Moreno, M. T. S. 1998. Dehydration of beetroots (*Beta vulgaris* L.) in spouted bed. Paper presented at the XII Brazilian Congress on Chemical Engineering, Porto Alegre, Brazil.

Lima, M. F. M., Rocha, S. C. S., Alsina, O. L. S., Jerônimo, C. E. M., and da Mata, A. L. M. 2000. Influence of material chemical composition on the drying performance of fruits in spouted beds. Paper presented at the XIII Brazilian Congress of Chemical Engineering, Campinas, Brazil.

Marques L. G., Ferreira, M. C., and Freire, J. T. 2007. Freeze-drying of acerola (West Indian Cherry) (*Malpighia glabra L.*). *Chem. Eng. Proc.* 2: 451–457.

Martinez, O. L. A., Brennam, J. G., and Nirajam, K. 1995. Study of food drying in a fountain dryer with inert. Paper presented at the 1st Ibero American Food Congress, Campinas, Brazil.

Mathur, K. B. and Epstein, N. 1974. *Spouted Beds*. New York: Academia Press.

Mathur, K. B. and Gishler, P.E. 1955. A technique of contacting gases with coarse solid particles. *AICHE J.* 1: 157–171.

Medeiros, M. F. D. 2001. Influence of material chemical composition on spouted bed drying performance of fruit pulps. PhD thesis, State University of Campinas, Campinas, Brazil (in Portuguese).

Medeiros, M. F. D., Rocha, S. C. S., Alsina, O. L. S. et al. 2002. Drying of pulps of tropical fruits in spouted bed: Effect of composition on dryer performance. *Drying Technol.* 20: 855–881.

Medeiros, M. F. D., Alsina, O. L. S., Rocha, S. C. S., Jerônimo, C. E. M., and Medeiros, U. K. L. 2004. Drying of pastes in spouted beds: Influence of the paste composition on the material retention in the bed. Paper presented at the 14th International Drying Symposium (IDS 2004), Campinas/SP, Brazil.

Moreira, G. E. G. 2007. Production and characterization of the microencapsulated extract of agro-industrial residue of acerola. MSc dissertation, Federal University of Rio Grande do Norte, Natal, Brazil (in Portuguese).

Mujumdar, A. S. 1989. Spouted beds: Principles and recent developments. Paper presented at the XVII Brazilian Meeting on Porous Media, São Carlos, Brazil.

Passos, M. L., Massarani, G., Freire, J. T., and Mujumdar, A. S. 1997. Drying of pastes in spouted beds of inert particles: Design criteria and modeling. *Drying Technol.* 15: 605–627.

Passos, M. L., Trindade, A. L. G., d'Angelo, J. V. H., and Cardoso, M. 2004. Drying of black liquor in spouted bed of inert particles. *Drying Technol.* 22: 1041–1067.

Patel, K., Bridgwater, J., Baker, C. G. J., and Schneider, T. 1986. Spouting behavior of wet solids. In *Drying'86*, ed. A.S. Mujumdar. New York: Hemisphere Publishing Corp., pp. 183–189.

Pham, Q. T. 1983. Behavior of a conical spouted bed dryer for animal blood. *Can. J. Chem. Eng*. 61: 426–434.

Rahman, M. S. and Pereira, C. O. 1999. Drying and food preservation. In *Handbook of Food Preservation*, ed. M.S. Rahman. New York: Marcel Dekker, pp. 173–216.

Ramos, C. M. P., Lima, M. F. M., and Maria, Z. L. 1998. Obtaining dried fruit powder in a spouted bed dryer. *Rev. Bras. Eng. Quim*. 47: 33–36 (in Portuguese).

Ré, M. I. and Freire, J. T. 1986. Drying of animal blood in a spouted bed. Paper presented at the XIV Brazilian Meeting on Porous *Media*, Campinas, Brazil.

Ré, M. I. and Freire, J. T. 1988. Drying of pastelike materials in spouted beds. Paper presented at the 6th International Drying Symposium, Versailles, France.

Reyes, A. E. 1993. Drying of suspensions in a conical spouted bed. PhD thesis, Federal University of Rio de Janeiro, Rio de Janeiro, Brazil (in Portuguese).

Reyes, A. E. and Massarani, G. 1992. Hydrodynamics and evaporation of water in a conical spouted bed. Paper presented at the XX Brazilian Meeting in Porous Media, São Carlos, Brazil.

Reyes, A. E., Diaz, G., and Blasco, R. 1996. Experimental study of slurries on inert particles in spouted bed and fluidized bed dryers. In *Drying' 96*, ed. A.S. Mujumdar. New York: Hemisphere Publishing Corp., pp. 605–612.

Righetto, A. M. 2003. Physico-chemical characterization and stability of acerola juice microencapsulated by green spray and lyophilization. PhD thesis, State University of Campinas, Campinas, Brazil (in Portuguese).

Schneider, T. and Bridgwater, J. 1990. Drying of solutions and suspensions in spouted beds. In *Drying'89*, ed. A.S. Mujumdar. New York: Hemisphere Publishing Corp., pp. 421–425.

Schneider, T. and Bridgwater, J. 1993. The stability of wet spouted beds. *Drying Technol*. 11: 277–301.

Soares, E. C., Oliveira, G. S. F., Maia, G. A. et al. 2001. Dehydration the acerola pulp (*Malpighia emarginata*) by foam mat process. *Sci. Technol. Food* 21:164–170 (in Portuguese).

Souza Jr., F. E., Souza, J. S., Rocha, S. C. S., and Medeiros, M. F. D. 2007. Drying of mixtures of fruit pulp in spouted bed. Influence the addition of fats and properties pulp in the performance of the process. Paper presented at the VII Brazilian Congress on Undergraduate Research in Chemical Engineering, São Carlos, Brazil.

Spitzner Neto, P. I. and Freire, J. T. 1997. Study of pastes drying in spouted beds: Influence of the presence of the paste on the process. Paper presented at the XXV Brazilian Congress in Particulate Systems, São Carlos, Brazil.

12 Application of Hybrid Technology Using Microwaves for Drying and Extraction

Uma S. Shivhare, Valérie Orsat, and G. S. Vijaya Raghavan

CONTENTS

12.1 INTRODUCTION

An integral part of the universe, electromagnetic waves, characterized by their frequency and wavelength, radiate from all bodies above absolute zero temperature. Microwaves (MWs) are high-frequency electromagnetic waves generated by magnetrons and klystrons and are composed of an electric and a magnetic field. The frequency range for MWs is from 0.3 to 300 GHz, equivalent to wavelengths of 1 mm to 1 m. The science of MW owes its origin to the development of radar, which gained momentum during World War II. Some of the important applications of MWs include communication, navigation and radar, heating, and physical diathermy. The most commonly used frequencies for industrial, scientific, and medical band applications are 0.433, 0.915, and 2.45 GHz. The ability of MWs to penetrate and dissipate power as heat in dielectric materials, such as food products, has brought MW technology to the common household.

12.2 BASIC CONCEPTS

When a dielectric material is subjected to a MW field, part of the energy is transmitted, part is reflected, and part is absorbed by the material, where it is dissipated as heat. Heating is due to the "molecular friction" of permanent dipoles within the material as they try to reorient themselves within the oscillating electric field of the incident wave. The power generated in a material is proportional to the frequency of the source, the dielectric loss of the material, and the square of the field strength within it. The device in which a material is subjected to MW energy is known as an applicator or cavity. Considering all these features, it is possible to identify materials and processes that can use MW heating effectively and to understand MW-ingredient interaction mechanisms.

The interaction of an electric field with dielectric materials is due to the response of charged particles to the applied field (Metaxas and Meredith 1983; Schiffmann 1995). The electric field induces polarization by displacing electrons around the nuclei (electric polarization) or by causing the relative polarization of nuclei as a result of unequal charge distributions in molecules (atomic polarization). Some lossy materials, known as polar dielectrics, e.g., water, contain permanent dipoles due to the asymmetric charge distribution of unlike charge partners in a molecule. This charge distribution tends to realign under the influence of a changing electric field, giving rise to orientation polarization and it is known as dipolar polarization. In addition, polarization can also arise from a charge build-up at the interfaces between components in heterogeneous systems. Such polarization is termed interfacial or Maxwell–Wagner polarization. The relative contribution of these types of polarization to MW absorption depends on the MW frequency and on the temperature and composition of the dielectric material. Hasted (1973) has demonstrated that dipolar polarization is probably the most significant process occurring with dielectrics in the MW frequency range.

12.2.1 Dielectric Properties

Dielectric properties of biological materials are affected by the manner in which water molecules and ions are associated with constituents such as carbohydrates and proteins. The relative dielectric activity of free water present in the food products is high, whereas bound water exhibits relatively low dielectric activity (Drecareau 1985). The dielectric properties of a material are defined by Von Hippel (1954):

$$\in^* = (\in' - j\in'')\in_0 \qquad (12.1)$$

where

ϵ^* is the complex permittivity
ϵ' is the dielectric constant
ϵ'' is the dielectric loss factor
ϵ_0 is the permittivity of free space (=8.854 × 10^{-12} Fm^{-1})
j is the complex operator

The dielectric constant is a measure of the ability of a dielectric material to store electric energy, whereas the loss factor is a measure of the energy absorbed from the applied field. The larger the loss factor the more easily the material absorbs the incident MW energy. A material's dielectric properties are dependent upon its moisture content and temperature, as well as on the frequency of the field. Both the dielectric constant and the loss factor increase with moisture content but decrease with frequency. The dielectric constant increases with temperature but the temperature dependence of the loss factor is unpredictable and may either increase or decrease with frequency and moisture content, depending upon the range of each (Nelson 1979; Liao et al. 2001, 2002, 2003a,b).

12.2.2 Volumetric Heating

Dielectric heating is a volumetric process in which heat is generated inside food materials by the selective absorption of electromagnetic energy by water molecules. Volumetric power absorption and the rate of heat generation depend on the intensity and frequency of the field as well as on the dielectric properties of the material. The relationships, as expressed by Goldblith (1967) and Smith and Hui (2004), are

$$\mathrm{PW} = 55.63 \times 10^{-12}\,\mathrm{fr}\,\epsilon''\,E^2 V \tag{12.2}$$

$$\frac{\mathrm{d}T}{\mathrm{d}t} = \frac{\mathrm{PW}}{\rho V c_\mathrm{p}} = \frac{55.63 \times 10^{-12}\,\mathrm{fr}\,\epsilon''\,E^2}{\rho c_\mathrm{p}} \tag{12.3}$$

where

PW is the power dissipation (W)
fr is the frequency of the field (Hz)
E is the electric field strength (V cm^{-1})
V is the material volume (cm^3)
T is the material temperature (°C)
t is the time (s)
ρ is the material density (g cm^{-3})
c_p is the specific heat (J g^{-1} $°C^{-1}$)

The use of Equation 12.3 in the analysis of MW heating of food materials is limited because the dielectric loss factor, power absorption, and electric field strength within the material vary with temperature and moisture content during the course of irradiation, making it difficult to predict the temperature rise (Drecareau 1985; Richardson 2001). It is evident from Equation 12.2 that the power dissipated in a lossy material can be increased by either increasing the frequency or electric field strength, or both. MWs, being of higher frequency than radiofrequency waves, clearly cause greater dissipation of power density in a lossy material. The magnitude of dissipated power can also be increased by raising the strength of the electric field. However, there seems to be an upper limit to the strength of the electric field used, beyond which arcing may occur, leading to equipment and material damage and fire hazard (Metaxas and Meredith 1983; Perkin 1983).

12.2.3 Penetration Depth

Another important parameter in the dielectric processing of lossy materials is the penetration depth. It indicates to what depth the material can be effectively heated by the high-frequency field. The power absorbed per unit volume of material increases with frequency, whereas the depth of the layer in which the power is absorbed varies inversely with frequency. Power attenuation occurs as electromagnetic waves penetrate the lossy material; and the degree of penetration for a constant frequency depends on the dielectric properties. Penetration depth is defined as the distance in the material where the field strength of electromagnetic waves propagating into the surface has decreased by e^{-1} (36.8%) of the surface value. Mathematically, it is expressed as

$$\text{pd} = \frac{\lambda_{L0}\left(\epsilon'\right)^{0.5}}{2\pi\,\epsilon''} \tag{12.4}$$

where

pd is penetration depth (cm)
λ_{L0} is free space wavelength (cm)

The wavelength of the field greatly influences the penetration depth. The free space wavelength is a function of the frequency of the field; a lower frequency will result in higher values of λ_{L0} and a greater pd. At a constant frequency, the penetration depth increases with the dielectric constant but decreases with the material's loss factor. Both the dielectric constant and the loss factor decrease with a reduction in moisture content, but the penetration depth is more sensitive to the variation in loss factor (Equation 12.4). Hence, the penetration depth increases as the material's moisture content decreases during processing (e.g., drying).

12.3 MICROWAVE-ASSISTED DRYING

Drying is the oldest technique to preserve food materials. Advantages of drying include reduction in water activity, enhancement of shelf life, maintenance of flavor and nutritive value, and reduction in packaging and transportation costs. Conventional drying techniques involve transfer of thermal energy from the surroundings to the surface of food material. Heat is generally conducted slowly from the material's surface into the material itself, owing to the low thermal conductivity of food materials. Furthermore, as moisture transport occurs in the opposite direction as heat transfer, drying rates are slow, leading to long drying times. MWs offer the opportunity to drastically shorten drying times. The physical mechanisms involved in drying using MW are distinctly different from those of conventional drying techniques. MWs penetrate to the interior of material and generate heat through the absorption of electromagnetic radiation by dipolar molecules like water. The temperature of moisture-laden interior layers is thus higher than that of the surface. This results in a greater water vapor pressure differential between the center and the surface of the food material, allowing for rapid moisture transfer. MW drying is therefore rapid, more uniform, and energy efficient compared to conventional drying techniques. Because of these advantages, MW drying has been used in several industries including food, paper, textile, timber, and ceramics. Studies have shown that MW can be successfully applied in drying foods.

Major drawbacks limiting the application of MW in drying are nonuniformity of the electromagnetic field and low penetration depth. To overcome these limitations, a number of studies have investigated hybrid drying systems combining MW with conventional techniques. To heighten its efficiency, hot-air drying may be combined with MW for its efficient utilization (Shivhare 1991, Shivhare et al. 1991, Prabhanjan et al. 1995, Schiffmann 1995, Gunasekaran 1999, Andrés et al. 2004, Sunjka et al. 2004, Zhang et al. 2006).

12.3.1 Microwave for the Entire Duration of Drying

MWs can be used throughout the drying period along with air, and this technology may be applied at either constant or variable power levels in combination with heated or unheated air of varying velocities. As an example, the experimental setup used by Beaudry et al. (2003) for MW hot air drying of cranberries is shown in Figure 12.1.

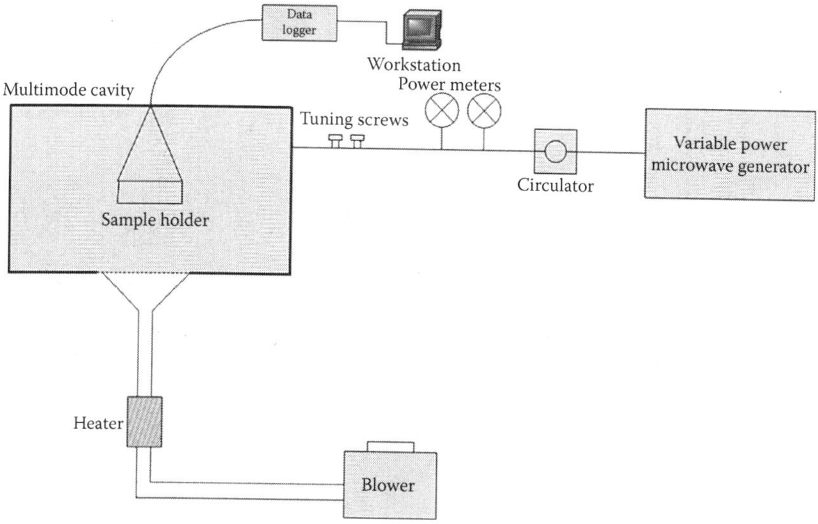

FIGURE 12.1 MW hot air drying.

Selection of the MW power level, temperature, and velocity of air are the important parameters in MW-assisted air-drying. Lower MW power levels are preferred to prevent over heating of solids. In a study on drying corn (*Zea mays* L.), Shivhare et al. (1992a) demonstrated that absorbed MW power levels exceeding 0.75 W g^{-1} resulted in discoloration and cracking of kernels. Absorbed MW power of 0.13 and 0.25 W g^{-1} used in drying resulted in 92%–99% germination of dried soybean (*Glycine max* (L.) Merr.) and corn, respectively, conforming to the standards for the certified grade (Shivhare et al. 1992a, 1993).

In combination drying, the relative contribution of air temperature in achieving the desired particulate temperature is insignificant or low compared to that of MW power (Shivhare et al. 1992a, Tulasidas 1995, Tulasidas et al. 1995, Andrés et al. 2004). Shivhare et al. (1992a) demonstrated that the drying air temperature did not significantly affect the drying kinetics of corn in a MW environment. A similar observation on the effect of varying air temperatures on MW drying of apples (*Malus domestica* Borkh.) has been reported by Andrés et al. (2004). The role of passing air may therefore be considered to be that of a moisture carrier, and ambient (unheated) air may be used in combination drying. The role of air velocity is particularly significant in controlling the temperature of solids in combination drying. Temperature of grains in combination drying decreases with increasing air flow velocity, and the effect is more pronounced when higher levels of MW power are used for drying (Shivhare 1991, Shivhare et al. 1991). In such cases, air acts as a cooling medium by maintaining the particulate temperature at safe limits.

Feng and Tang (1998) found that MW drying caused little discoloration of diced apples, whereas hot air drying resulted in severe discoloration. Similar results were reported by Maskan (2000) for the drying of bananas (*Musa acuminata* Colla), a fruit difficult to dry by traditional hot-air methods. Combining MW with hot air resulted in enhanced drying rates and substantial shortening (40%–89%) of drying time. Sharma and Prasad (2001) obtained good quality garlic (*Allium sativum* L.) cloves using combined MW air-drying method. They reported that the combination drying resulted in reduction of about 91% total drying time. For kiwifruits (*Actinidia deliciosa*) however, MW increased the rate of color deterioration and produced a brownish product (Maskan 2001). The finding implies the need for appropriate pretreatment, such as osmotic concentration in sugar syrup, to reduce the extent of discoloration.

Application of unvarying MW power level for the entire duration would obviously constantly increase the solid temperature, especially when higher power levels are used for drying. Furthermore, the solid temperature would increase appreciably when not enough moisture is present in grains to absorb incident MWs. This implies that MW power application should decrease with the progression of drying. This operation is referred to as variable MW power application (Shivhare et al. 1991). To investigate this approach, Shivhare et al. (1991) subjected corn to higher levels of MW power (0.50 and 0.75 W g^{-1}) for fixed periods in the early stages of drying and decreased the MV to lower levels (0.25 and 0.50 W g^{-1}, respectively) for the remaining duration of drying. In doing so, a moisture reduction equivalent to the one obtained under the continuous mode was achieved but with lesser loss of energy and better product quality.

12.3.2 Microwave during Final Stages of Drying

For some materials, removal of moisture present in the core during the final stages of drying is difficult and highly time consuming. This condition is exemplified by a solid with dry surface but wet core. Applying MW at this step generates internal heat and, therefore, a water vapor pressure, which forces the moisture to the surface, where it is readily removed (Prabhanjan et al. 1995). In such situations, MW can be used for finish drying. MW when applied at the final stage results in reduced drying time, higher thermal efficiency, and better product quality (Xu et al. 2004). MW finish drying of banana was investigated by Maskan (2000), who reported that this combined technique reduced the total drying time by about 64%, and the dried product had a lighter color and better rehydration characteristics than the hot-air dried product. Industrial application of MW finish drying to level off the moisture content in pasta, biscuits, cracker, snacks, and potato chips, has been reported by Osepchuk (2002).

12.3.3 Intermittent Microwave during Drying

Several studies have shown that higher levels of MW power enhance the drying rate but at the expense of greater heat loss through the airflow and deleterious effects on quality of the dried product. Loss of heat energy can be reduced if MW is used in pulsed mode for drying. An intermittent supply of MW, which can enhance thermal energy use and the quality of dried heat-sensitive products, is best applied when internal heat and mass transfer rates control the overall drying rate (Gunasekaran 1999).

For corn drying, Shivhare et al. (1992b) found that total drying time increased but the effective duration for which MW power was applied in pulsed mode was substantially lower than that under continuous operation. Pulsed MW operation resulted in reduction in the energy loss through the outlet air and in the overall energy requirement, which was lower than that when constant levels of MW power were used continuously (Sanga et al. 2002). Tulasidas et al. (1994) reported the drying of grapes (*Vitis vinifera* L.) using pulsed MW at selected power levels. They demonstrated that application of pulsed MW resulted in highly acceptable quality of raisins.

Mathematical models were proposed by Yang and Gunasekaran (2001, 2004) to predict the temperature distribution inside the solid material during pulsed MW heating based on Maxwell's equation and Lambert's law. They demonstrated that unevenness of temperature distribution obtained during continuous MW heating was dramatically reduced when pulsed MW heating was used. In a recent study, Gunasekaran and Yang (2007) further validated that pulsed MW heating should be preferred to continuous MW heating when sample temperature uniformity is critical. Chua and Chou (2005) compared the efficacy of intermittent-MW and -IR (infrared radiation) drying for potato (*Solanum tuberosum* L.) and carrot (*Daucus carota* L.). On the basis of the drying kinetics, they concluded that intermittent-MW drying was an effective method to reduce drying time and product color change in comparison with intermittent-IR drying or even convective drying.

12.3.4 Microwave with Fluidized or Spouted-Bed Drying

Nonuniform heating has been a major limitation in commercial application of MW processing. Various methods using mechanical means (Torringa et al. 2001) and pneumatic agitation (Feng and Tang 1998) have been developed to achieve uniform heating. Household MW ovens usually have a turntable to accomplish constant movement of the material being heated within the MW cavity. Pneumatic agitation can be achieved by fluidization of particles in the MW cavity (Salek-Mery 1986, Kudra 1989, Feng and Tang 1998). Salek-Mery (1986) combined fluidization and MW for drying grains. Application of MW in the fluidized wheat (*Triticum aestivum* L.) bed resulted in 50% higher drying rates than in the fluidized bed alone.

Spouted beds can be used for processing coarse particles that are difficult to fluidize in a conventional fluidized bed (Jumah and Raghavan 2001). This microwave with spouted-bed drying (MWSBD) technique provided much more uniform heating as indicated by more uniform temperature distribution and color in dried apple dices (Feng and Tang 1998). Total time of drying was greatly shortened, and the dried diced apple exhibited better reconstitution characteristics. In a subsequent study, Feng et al. (2001) developed a heat and mass transfer model to predict moisture, temperature, and pressure history and distribution for MWSBD of particulate materials. Nindo et al. (2003) used MWSBD to evaluate product quality and found that this technique produced asparagus (*Asparagus officinalis* L.) with good color and rehydration attributes. Enhanced retention of total antioxidant activity was achieved when asparagus were dried by MWSBD at a 2 W g^{-1} power level and 60°C hot air.

12.3.5 Microwave with Vacuum Drying

MW vacuum drying (MWVD) is an emerging technique involving incorporation of MW in a conventional vacuum dryer for heating and evaporation of moisture. MW energy is an efficient mechanism of energy transfer through vacuum and into the interior of the food. MWVD combines advantages of rapid volumetric heating and low-temperature evaporation of moisture with rapid moisture removal by vacuum (Pappas et al. 1999, Durance and Wang 2002, Mousa and Farid 2002, Sunjka et al. 2004). The schematic of an experimental setup for MWVD is shown in Figure 12.2.

While drying banana slices using MWVD, Drouzas and Schubert (1996) found that color, taste, aroma, and shape of the final product were all comparable to those of the freeze-dried product. Application of pulsed MWVD on cranberries (*Vaccinium oxycoccos* L.) was investigated by Yongsawatdigul and Gunasekaran (1996), who concluded that such a technique was more efficient than continuous MW heating. Lin et al. (1998) dried carrot slices by MWVD and compared their quality with the one of air and freeze-dried products. MW-vacuum-dried carrot slices had higher values of rehydration, α-carotene and vitamin C, lower density and color degradation, and softer texture than the air-dried product did. Furthermore, the MW-vacuum-dried carrot slices were rated as comparable to the freeze-dried product by the sensory panel with respect to color, texture, flavor, and overall acceptability, in both the dried and rehydrated state. Cui et al. (2004) reported that in MWVD of carrot slices, drying rate was strongly affected by MW power but only slightly affected by vacuum pressure.

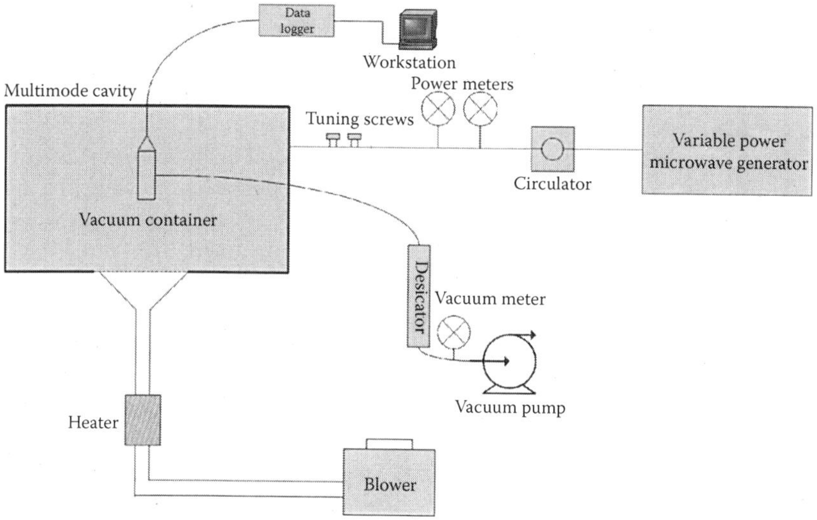

FIGURE 12.2 Schematic of MWVD.

Puffing of a product is achieved as a result of the greater difference between the water vapor pressure within the material and the vacuum chamber pressure. Puffiness is desirable as it increases rehydration characteristics of dehydrated fruits and vegetables (Lin et al. 1998, Sham et al. 2001). According to Erle (2005), combination of air drying and MW-vacuum puffing is being used in Germany and Poland for fruits and vegetables. Absence of air during MWVD may inhibit oxidation leading to better retention of color and nutrients. MWVD of tomatoes resulted in least energy consumption, while total drying time was reduced by more than 18 times as compared to conventional hot-air drying (Durance and Wang 2002).

12.3.6 Microwave with Freeze-Drying

Copson and Decareau (1957) were the first to use MW to accelerate freeze-drying of beefsteak and achieve good product quality. Hoover et al. (1966a,b) demonstrated that total drying times for beef were 3- to 13-fold less when MW was combined with freeze drying compared to freeze-drying alone. Several researchers (Litvin et al. 1998, Xu et al. 2008) have reported similar results by using this microwave-assisted freeze-drying (MWFD) technique.

When applied along with freeze-drying, MW may cause plasma discharge, a phenomenon which occurs when the electric field intensity in the vacuum chamber is above a threshold value. Ionization of the gases present in the vacuum chamber results in burning of the product surface and substantial energy loss. It is therefore necessary to control vacuum pressure and MW power levels to prevent the occurrence of this phenomenon. Mathematical models have been developed to describe and optimize the MWFD process (Ma and Peltre 1975a,b, Ang et al. 1977, Heng et al. 2007).

12.3.7 Microwave with Osmotic Drying

Combined osmotic and MW drying results in a more homogeneous heating of food by modification of its dielectric properties due to uptake of solids, thus reducing drying time and shrinkage, and improving rehydration characteristics (Erle and Schubert 2001, Torringa et al. 2001, Beaudry et al. 2003). Prothon et al. (2001) evaluated the effects of MW-air drying with and without osmotic pretreatment on apple. Apple cubes were pretreated in sugar syrup followed by MW-air drying at selected hot-air temperatures. The authors concluded that osmotic pretreatment before MW-air drying resulted in better product quality. However osmotic pretreatment lowered the drying rate and effective moisture diffusivity. Heredia et al. (2007) used osmotic solutions containing sugar, salt, and calcium lactate for osmotic treatment of cherry tomatoes (*Lycopersicon esculentum* Mill.) prior to MW-assisted-air-drying. They reported that an osmotic solution containing 27.5 wt% sugar, 10 wt% salt, and 2 wt% calcium lactate combined with MW-assisted-air-drying produced dried and intermediate moisture tomato products that were shelf stable and showed better quality than the traditional product.

12.3.8 Microwave with Vacuum and Osmotic Drying

MWVD of osmotically pretreated foods combines the benefits of both techniques and yields product of good quality in terms of color, taste, nutrients, structure, and volume (Venkatachalapathy and Raghavan 1999, Erle and Schubert 2001, Prothon et al. 2001, Raghavan and Silveira 2001, Beaudry et al. 2004, Heredia et al. 2007).

MWVD with and without osmotic pretreatment of carrot slices was investigated by Sutar (2008). Rehydration ratio, texture, shrinkage and density, color, sensorial attributes, as well as the vitamin C and β-carotene content of the product were used as quality indicators. Sutar (2008) found that the quality of MW vacuum-dried and combined osmotic MW vacuum-dried carrots was better than that of air-dried carrots, but inferior to that of the freeze-dried product.

12.4 MICROWAVE-ASSISTED EXTRACTION

Another important area in which the industrial application of MW has immense potential is in MW-assisted extraction (MWAE). Solvent extraction is a process of separation of solute(s) from a solid matrix using solvents. The rate and efficiency of extraction depends on many factors, including matrix characteristics, distribution of the target solute(s) in the matrix, solubility of the target and interfering-solutes, and temperature. Enhancement in the extraction rate, leading to a substantial reduction in extraction time, can be achieved by increasing the temperature and breakdown of the solid matrix. When MWs are used for extraction, these waves are absorbed by the material, mainly the water in the glandular and vascular systems, whose expansion in volume leads to explosion at the cellular level (Chen and Spiro 1994, Chemat et al. 2005). The solute(s) located in the cells can then diffuse easily into the surrounding solvent.

In 1879, von Soxhlet developed an extraction device based on repeated circulation of the solvent while accumulating the solute in a heated flask. The Soxhlet apparatus is commonly used because no filtration is required. The solid repeatedly comes in contact with fresh solvent, and both polar and nonpolar solvents can be used for extraction. Based on micrographs of caraway seed flakes obtained by scanning electron microscopy, Chemat et al. (2005) observed the complete absence of rupturing of the glands during hexane extraction. Limitations of the Soxhlet extraction device are that it uses large amounts of solvent, that the solvent must be evaporated to get the solute, and that it takes a long time to complete the extraction. Recently, pressurized fluid and supercritical fluid extraction (SCFE), which use an extractant at a higher temperature and pressure, have emerged as viable improvements over the traditional Soxhlet method. However, these methods are expensive. Extraction of solutes using MWs incorporates the advantages of these extraction methods while eliminating most of the disadvantages. The solvent consumption is reduced and extraction time is drastically decreased (approximately 10 min compared to about 2 h with SCFE and 16–48 h with Soxhlet). An additional advantage of MWAE is the enhancement of the number of samples handled, as several samples can be extracted simultaneously (Sparr Eskilsson and Björklund 2000). This is due to the fact that MWs heat the liquid–solid mixture directly, thereby accelerating the heating rate. The major drawback of MWAE appears to be the lack of selectivity as compared to SCFE, which results in the co-extraction of other solutes present in the solid.

Though application of MW for sample digestion was successfully demonstrated by Abu-Samra et al. (1975), the first study on sample extraction using MW appeared in 1986. Ganzler et al. (1986) used MWAE for extraction of various compounds (crude fat, gossypol, vicine, convicine, and organophosphate) from food, feed, and soil. These researchers demonstrated the effectiveness of MWAE over the conventional Soxhlet method in terms of considerably savings in time and energy.

Considerable research on MWAE has been carried out toward isolation of essential oils (Chen and Spiro 1994, Lucchesi et al. 2004, Chemat et al. 2006, Ferhat et al. 2006, Wang et al. 2006, Lucchesi et al. 2007), lipids (Simoneau et al. 2000; ElKhori et al. 2007), pectin (Fishman et al. 2006), pigments (Kiss et al. 2000), amino acids (Kovacs et al. 1998), antioxidants (Raghavan and Richards 2007), antinutritional compounds (Martín-Calero et al. 2007), pesticides (Chee et al. 1996), hydrocarbons (Chee et al. 1996), and trace elements (Konieczynski and Wesolowski 2007). A summary of these studies is presented in Table 12.1.

Based on the dielectric properties of the materials to be extracted and the solvents, MWAE may be classified as

- Mechanism I—the material is extracted in a pure solvent or a mixture of solvents of high dielectric loss, thus absorbing MWs strongly.
- Mechanism II—the material is extracted in a solvent or mixture of solvents with both high and low dielectric losses.
- Mechanism III—material exhibiting a high loss factor is extracted in low loss factor solvent(s) (Jassie et al. 1997, Sparr Eskilsson and Björklund 2000).

TABLE 12.1
Applications of MWAE in Food Products

Compound	Matrix	MWAE Features	Reference
Fat	Cocoa	Hexane, isopropanol, acetone, petroleum ether, water 460 s	ElKhori et al. (2007)
	Chocolate, cocoa products	Petroleum ether 1100 kPa, 100°C, 15 min	Simoneau et al. (2000)
	Cheese	Hexane 40 min	Garcia-Ayuso et al. (1999)
	Bakery products	Hexane	Luque-Garcia and Luque de Castro (2004)
Essential oils	Peppermint leaves	Hexane, carbon tetrachloride, toluene less than 60 s	Chen and Spiro (1994)
	Orange peel	SFME 30 min	Ferhat et al. (2006)
	Cardamom	Water 75 min	Lucchesi et al. (2007)
	Origanum vulgare L.	SFME 0.054 mLg^{-1} yield, 80% reduction compared to hydrodistillation	Bayramoglu et al. (2008)
Edible oil	Olive seed, rapeseed, soybean, sunflower	Hexane 20–25 min	Garcia-Ayuso and Luque de Castro (1999)
Volatile oils	Mentha piperita	Hexane and other alkanes	Paré et al. (1994)
Vicine, convicine, gossypol	Cotton seed, fava, beans, maize, meat, soybean, walnut, yeast	Methanol:water (1:1) 30 s	Ganzler et al. (1986)
Ergosterol	Fungal contaminations	35 s	Young (1995)
Azadirachtin-related limonoids	Neem seed	4.5 min	Dai et al. (1999)
Curcuminoids	Turmeric rhizomes	Methanol 80°C, 6 min	Kaufmann and Christen (2002)
Carotenoids	Paprika	Acetone, dioxane, ethanol, methanol, tetrahydrofuran	Kiss et al. (2000)
Antioxidants	Cranberry cake	Ethanol, methanol, acetone, water 125°C, 10 min	Raghavan and Richards (2007)
Pectin	Lime	140°C, 3 min	Fishman et al. (2006)
Phenolic compounds	Grape seed	Methanol:water (90:10) 73°C	Hong et al. (2001)
Aroma compounds	Pepper	5 min	Plessi et al. (2002)
Off flavor	Catfish	120°C, 25 min	Zhu et al. (1999)
Pesticide residue	Sunflower seeds	Dichloromethane 45 min	Prados-Rosales et al. (2003)

TABLE 12.1 (continued)
Applications of MWAE in Food Products

Compound	Matrix	MWAE Features	Reference
Trace elements	Canned foods	Hydrogen peroxide, nitric acid 23 min	Tuzen and Soylak (2007)
Calcium, iron, magnesium	Carrot, coconut water, milks	Nitric acid 250°C	Oliveira et al. (2000)
Copper, lead, zinc	Bovine lever	Nitric acid: perchloric acid (4:1) 1.5–3.0 min	Abu-Samra et al. (1975)
Mercury	Fish	Nitric acid, sulfuric acid	Barrett et al. (1978)
Cadmium, chromium, mercury, lead	Food packaging materials	Nitric acid, sulfuric acid, hydrogen peroxide, intermittent power	Perring et al. (2001)
Essential oils, calcium, copper, iron, potassium, magnesium, manganese, sodium, zinc	Medicinal plants, lavender	Steam, hydrogen peroxide, nitric acid 100°C, 10 min	Chemat et al. (2006)

MWAE can be carried out either at atmospheric pressure in open vessels or under elevated pressure using closed vessels. Typical pressures reached within closed-vessel systems range from 1,350 to 10,750 kPa (Lopez-Avila and Benedicto 1996). Extraction time decreases with increasing pressure, but the extent to which the pressure can be increased is influenced by the heat sensitivity of the material to be extracted.

Digestion of samples for determination of trace metals (e.g., mercury in foods) requires the destruction of samples by digestion, a commonly used technique which involves oxidation of organic matter by heating with strong acids. Besides the necessary long digestion times, this method demands constant supervision and requires a specialized fume hood to handle acid fumes safely. Moreover, a danger of explosion exists if established procedures are not strictly followed. These limitations can be overcome by the use of MWs. MW-assisted digestion is rapid and uses less mineral acid.

Extraction of crude fat, vitamins, pigments, essential oils, and antinutrients from foods is generally carried out using solvents at elevated temperatures for long periods of time. Consequently, degradation of the extracted compounds may take place. Due to its rapidity, MWAE is even suitable for extraction of heat-labile constituents from food matrices. Using MW, the degrading effects of high temperatures can be prevented. MWAE therefore offers great potential in food processing. A considerably shorter extraction process may also avoid the extraction of undesirable pigments and compounds that are located in tissue. The quality of the extract thus obtained is superior to that obtained under traditional methods.

The parameters affecting the MWAE process are the following: selection of solvent, solvent to sample ratio, temperature and time of extraction, and matrix characteristics including moisture content (Sparr Eskilsson and Björklund 2000).

Selection of an appropriate solvent is essential in achieving optimum efficiency in MWAE. Solvent selection criteria should include its dielectric properties, its interaction with the food matrix, and the solubility of the desired solute in the solvent (Sparr Eskilsson and Björklund 2000). The solvent should be a lossy dielectric and possess a strong but selective affinity toward the solute. Minor changes in solvent composition may enhance extraction considerably. For instance, while working on the extraction of phenolic compounds from grape seed, Hong et al. (2001) obtained significantly higher recovery when water was added to methanol at a volumetric ratio of 1:9. Raghavan and Richards (2007) evaluated the effect of different solvents (acetone, ethanol, methanol, water, and their mixtures) on MWAE of antioxidants from cranberry cake, and found 100% ethanol to be the most effective. Furthermore, the dielectric properties of the material extracted influence the solvent selection. For example, a MW-transparent solvent may be effectively used for MWAE of moist materials exhibiting high dielectric loss (Paré et al. 1991, 1994, Chen and Spiro 1994, Lucchesi et al. 2004). While studying MWAE of moist mint (*Mentha* sp.) leaves in the presence of hexane, an MW-transparent solvent, Paré et al. (1994) demonstrated that essential oils were extracted within 20 s. Moist glands and vascular system containing mint oil were selectively heated leading to rupture of the tissue, and the essential oil was released into the relatively cool organic solvent (hexane).

Some studies have established that an organic solvent is not required for MWAE of moist materials exhibiting high dielectric loss. Solvent-free MWAE (SFME) has been tested for release of essential oil from various parts of plants (Lucchesi et al. 2004, Ferhat et al. 2006). The materials extracted using SFME by Lucchesi et al. (2004) included basil, garden mint, and thyme. Due to high moisture content (80–95 wt% in wet basis), water present may have aided the extraction process. The SFME can also be effectively used for extraction of dried matrices. Wang et al. (2006) demonstrated the use of MW absorption media (iron carboxyl powder, active carbon powder, and graphite powder) for extraction of essential oils from dried spices and medicinal plants. The media essentially acted as the water-replacing agent and facilitated MW absorption during MWAE.

When a mixture of two solvents is used in MWAE, one is usually of higher dielectric loss than the other. In this case, the lossy solvent may achieve rapid heating, while the other relatively MW-transparent solvent may facilitate rapid extraction. Organic solvent mixtures of hexane–acetone, hexane–toluene, isooctane–acetone, and ethyl acetate–cyclohexane have been widely used by various researchers for MWAE of pesticides, hydrocarbons, herbicides, fungicides, esters, phenols, etc. (López-Avila et al. 1995, Chee et al. 1996, Punt et al. 1999, Eskilsson and Björklund 2000, Esteve-Turrillas et al. 2004). Some researchers have also explored use of water to replace one of the organic solvents for MWAE and obtained higher solute recoveries (Hong et al. 2001, ElKhori et al. 2007, Lucchesi et al. 2007).

Studies have shown that MWAE can be successfully used for extraction of various compounds at lower temperatures, and that solvent requirements for MWAE are much smaller than those for the Soxhlet (Kovacs et al. 1998). However, the amount must be adequate to ensure that the entire extract is immersed during extraction and to solubilize the solute extracted. Kovacs et al. (1998) studied MAE of free amino acids from foods and evaluated effects of varying amounts of extracts at selected temperatures (40°C, 50°C, 60°C, and 80°C) and time on the yield. Extraction yields obtained using MWAE were about 10% higher, while the extraction times were about one-third of those associated with the conventional technique. While investigating the effect of extract to sample ratio on yield, these researchers found greater yield for foods with higher protein (cheese) and fat (salami) contents. The relative composition of the amino acids was not affected by the extract volume however. Sample to extract ratio and temperature did not affect the extraction yield of amino acids from samples of plant origin (cauliflower, *Brassica oleracea* L. var. *botrytis*).

Hong et al. (2001), while using MWAE of phenolic compounds from grape seed, found that both power and time of extraction did not significantly affect the yield. Increased yield was attained by changing the polarity of the solvent by adding water (10 wt%) to methanol. Several researchers have demonstrated the suitability of MWAE for improving the conventional solvent extraction of lipids from food materials (García-Ayuso and Luque de Castro 2001, Luque-García and Luque de Castro 2004).

Dai et al. (1999) developed a method for the determination of total Azadirachtin-related limonoids in neem seed kernel extracts. Their results indicated the possibility of acceleration of the extraction process significantly when MW was applied. Furthermore, results showed that it took 36 h for the room temperature extraction, whereas the MWAE took only 4.5 min to achieve the same result. In their subsequent work, Dai et al. (2001) discovered that the selection of solvent plays a major role in affecting the efficiency and the selectivity of the MWAE process.

Bayramoglu et al. (2008) used SFME for the extraction of essential oil from *Origanum vulgare* L. and compared the effect of MW power level. Their results show that SFME offers significantly higher extraction yield (0.054 mLg^{-1}) when compared to hydrodistillation (0.048 mLg^{-1}). SFME also reduces the required extraction time by 80% at 622 W power level. There is no significant difference in physical properties of the essential oil obtained from both methods.

12.5 SUMMARY AND CONCLUSIONS

This chapter has shown that it is possible to develop unique drying methods using MW to achieve high quality drying. Time reduction is the major advantage associated with application of MW for drying. However, controlling the temperature during drying is important to achieve higher quality dried products. Most of the methods discussed in this chapter have been carried out on lab scale. Further research is necessary to successfully scale up the methods for large-scale industrial applications. MW freeze-drying can help in reducing the overall cost involved with freeze-drying while helping to produce dried products with excellent dehydration properties. Intermittent application of MW during drying offers an attractive alternative when heat sensitive products need to be dried.

MW assisted extraction not only dramatically reduces the extraction time; it also offers significant control over the extraction process. The selectivity of MW offers greater flexibility over conventional extraction methods. Solvents play a major role in affecting the efficiency of MW extraction process. Researchers have developed solvent free MW extraction methods, which could help in wider adaptation of MWAE.

ACKNOWLEDGMENT

The authors would like to express their sincere gratitude for the financial support offered by CIDA, NSERC, and FQRNT, which enabled this work.

REFERENCES

Abu-Samra, A., Morris, J. S., and Koirtyohann, S. R. 1975. Wet ashing of some biological samples in a microwave oven. *Anal. Chem.* 47:1475–1477.

Andrés, A., Bilbao, C., and Fito, P. 2004. Drying kinetics of apple cylinders under combined hot air-microwave dehydration. *J. Food Eng.* 63:71–78.

Ang, T. K., Ford, J. D., and Pei, D. C. T. 1977. Microwave freeze-drying of food: A theoretical investigation. *Int. J. Heat Mass Transfer* 20:517–526.

Barrett, P., Davidowski Jr. L. J., Penaro, K. W., and Copeland, T. R. 1978. Microwave oven-based wet digestion technique. *Anal. Chem.* 50:1021–1023.

Bayramoglu, B., Sahin, S., and Sumnu, G. 2008. Solvent-free microwave extraction of essential oil from oregano. *J. Food Eng.* 88:535–540.

Beaudry, C., Raghavan, G. S. V., and Rennie, T. J. 2003. Microwave finish drying of osmotically dehydrated cranberries. *Drying Technol.* 21:1797–1810.

Beaudry, C., Raghavan, G. S. V., Ratti, C., and Rennie, T. J. 2004. Effect of four drying methods on the quality of osmotically dehydrated cranberries. *Drying Technol.* 22:521–539.

Chee, K. K., Wong, M. K., and Lee, H. K. 1996. Microwave-assisted solvent elution technique for the extraction of organic pollutants in water. *Anal. Chim. Acta* 330:217–227.

Chemat, S., Aït-Amar, H., Lagha, A., and Esveld, D. C. 2005. Microwave-assisted extraction kinetics of terpenes from caraway seeds. *Chem. Eng. Process.* 44:1320–1326.

Chemat, F., Lucchesi, M. E., Smadja, J., Favretto, L., Colnaghi, G., and Visinoni, F. 2006. Microwave accelerated steam distillation of essential oil from lavender: A rapid, clean and environmentally friendly approach. *Anal. Chim. Acta* 555:157–160.

Chen, S. and Spiro, M. 1994. Study of microwave extraction of essential oil constituents from plant materials. *J. Microw. Power Electromagn. Energy* 29:231–241.

Chua, K. J. and Chou, S. K. 2005. A comparative study between intermittent microwave and infrared drying of bioproducts. *Int. J. Food Sci. Technol.* 40:23–39.

Copson, D. and Decareau, R. 1957. Microwave energy in freeze-drying process. *Food Res. Int.* 22:402–403.

Cui, Z.-W., Shi-Ying Xu, Z.-W., and Da-Wen Sun, Z.-W. 2004. Effect of microwave-vacuum drying on the carotenoids retention of carrot slices and chlorophyll retention of Chinese chive leaves. *Drying Technol.* 22:563–575.

Dai, J., Yaylayan, V. A., Raghavan, G. S. V., and Paré, J. R. 1999. Extraction and colorimetric determination of azadirachtin-related limonoids in neem seed kernel. *J. Agric. Food Chem.* 47:3738–3742.

Dai, J., Yaylayan, V. A., Raghavan, G. S. V., Paré, J. R. J., Liu, Z., and Bélanger, J. M. R. 2001. Influence of operating parameters on the use of the microwave-assisted process (MAP) for the extraction of azadirachtin-related limonoids from neem (*Azadirachta indica*) under atmospheric pressure conditions. *J. Agric. Food Chem.* 49:4584–4588.

Drecareau, R. 1985. *Microwaves in Food Processing Industries*. Orlando, FL: Academic Press.

Drouzas, A. E. and Schubert, H. 1996. Microwave application in vacuum drying of fruits. *J. Food Eng.* 28:203–209.

Durance, T. D. and Wang, J. H. 2002. Energy consumption, density, and rehydration rate of vacuum microwave- and hot-air convection-dehydrated tomatoes. *J. Food Sci.* 67:2212–2216.

ElKhori, S., Paré, J. R. J., Bélanger, J. M. R., and Pérez, E. 2007. The microwave-assisted process (MAPTM1): Extraction and determination of fat from cocoa powder and cocoa nibs. *J. Food Eng.* 79:1110–1114.

Erle, U. 2005. Drying using microwave processing. In *The Microwave Processing of Foods*, Eds. H. Schubert and M. Regier, pp. 142–152. Cambridge, U.K.: Woodhead Publications.

Erle, U. and Schubert, H. 2001. Combined osmotic and microwave-vacuum dehydration of apples and strawberries. *J. Food Eng.* 49:193–199.

Eskilsson, C. S. and Björklund, E. 2000. Analytical-scale microwave-assisted extraction. *J. Chromatogr. A* 902:227–250.

Esteve-Turrillas, F. A., Aman, C. S., Pastor, A., and de la Guardia, M. 2004. Microwave-assisted extraction of pyrethroid insecticides from soil. *Anal. Chim. Acta* 522:73–78.

Feng, H. and Tang, J. 1998. Microwave finish drying of diced apples in a spouted bed. *J. Food Sci.* 63:679–683.

Feng, H., Tang, J., Cavalieri, R. P., and Plumb, O. A. 2001. Heat and mass transport in microwave drying of porous materials in a spouted bed. *AIChE J.* 47:1499–1512.

Ferhat, M. A., Meklati, B. Y., Smadja, J., and Chemat, F. 2006. An improved microwave Clevenger apparatus for distillation of essential oils from orange peel. *J. Chromatogr. A* 1112:121–126.

Fishman, M. L., Chau, H. K., Hoagland, P. D., and Hotchkiss, A. T. 2006. Microwave-assisted extraction of lime pectin. *Food Hydrocoll.* 20:1170–1177.

Ganzler, K., Salgó, A., and Valkó, K. 1986. Microwave extraction: A novel sample preparation method for chromatography. *J. Chromatogr. A* 371:299–306.

García-Ayuso, L. E., and Luque de Castro, M. D. 2001. Employing focused microwaves to counteract conventional Soxhlet extraction drawbacks. *TrAC, Trends Anal. Chem.* 20:28–34.

García-Ayuso, L. E., and Luque de Castro, M. D. 1999. A multivariate study of the performance of a microwave-assisted soxhlet extractor for olive seeds. *Anal. Chim. Acta* 382: 309–316.

García-Ayuso, L. E., Velaseo, J., Do barganes, M. C., and Luque de Castro, M. D. 1999. Accelerated extraction of the fat content in cheese using a focussed microwave-assisted soxhlet device. *J. Agric. Food Chem.* 47:2308–2315.

Goldblith, S. 1967. Basic principles of microwaves and recent developments. *Adv. Food Res.* 15:277–301.

Gunasekaran, S. 1999. Pulsed microwave-vacuum drying of food materials. *Drying Technol.* 17:395–412.

Gunasekaran, S. and Yang, H.-W. 2007. Effect of experimental parameters on temperature distribution during continuous and pulsed microwave heating. *J. Food Eng.* 78:1452–1456.

Hasted, J. 1973. *Aqueous Dielectrics*. London, U.K.: Chapman and Hall Publications.

Heng, S., Hongmei, Z., Haidong, F., and Lie, X. 2007. Thermoelectromagnetic coupling in microwave freeze-drying. *J. Food Process. Eng.* 30:131–149.

Heredia, A., Barrera, C., and Andrés, A. 2007. Drying of cherry tomato by a combination of different dehydration techniques. Comparison of kinetics and other related properties. *J. Food Eng.* 80:111–118.

Hong, N., Yaylayan, V. A., Raghavan, G. S. V., Paré, J. R. J., and Bélanger, J. M. R. 2001. Microwave-assisted extraction of phenolic compounds from grape seed. *Nat. Prod. Rep.* 15:197–204.

Hoover, M., Markantonatos, A., and Parker, W. 1966a. Engineering aspects of using UHF dielectric heating to accelerate the freeze-drying of foods. *Food Technol.* 20:107–110.

Hoover, M., Markantonatos, A., and Parker, W. 1966b. UHF dielectric heating in experimental acceleration of the freeze-drying of foods. *Food Technol.* 20:103–107.

Jassie, L., Revesz, R., Kierstead, T., Hasty, E., and Matz, S. 1997. Microwave-assisted solvent extraction. In *Microwave Enhanced Chemistry. Fundamental, Sample Preparation, and Applications*, Eds. H. M. Kingston and S. J. Haswell, pp. 569–609. Washington, DC: American Chemical Society.

Jumah, R. Y. and Raghavan, G. S. V. 2001. Analysis of heat and mass transfer during combined microwave convective spouted-bed drying. *Drying Technol.* 19:485–506.

Kaufmann, B. and Christen, P. 2002. Recent extraction techniques for natural products: Microwave-assisted extraction and pressurized solvent extraction. *Phytochem. Anal.* 13:105–113.

Kiss, G. A. C., Forgács, E., Cserháti, T., Mota, T., Morais, H., and Ramos, A. 2000. Optimisation of the microwave-assisted extraction of pigments from paprika (*Capsicum annuum* L.) powders. *J. Chromatogr. A* 889:41–49.

Konieczynski, P. and Wesolowski, M. 2007. Total phosphorus and its extractable form in plant drugs. Interrelation with selected micro- and macroelements. *Food Chem.* 103:210–216.

Kovacs, A., Ganzler, K., and Sarkadi, L. S. 1998. Microwave assisted extraction of free amino acids from foods. *Z. Lebensm. Unters. Forsch. A* 207:26–30.

Kudra, T. 1989. Dielectric drying of particulate materials in a fluidized state. *Drying Technol.* 7:17–34.

Liao, X., Raghavan, G. S. V., and Yaylayan, V. A. 2001. Dielectric properties of alcohols (C1-C5) at 2450 MHz and 915 MHz. *J. Mol. Liquid* 94:51–60.

Liao, X., Raghavan, G. S. V., and Yaylayan, V. A. 2002. A novel way to prepare n-butylparaben under microwave irradiation. *Tetrahedron Lett.* 43:45–48.

Liao, X., Raghavan, G. S. V., Dai, J., and Yaylayan, V. A. 2003a. Dielectric properties of [alpha]-D-glucose aqueous solutions at 2450 MHz. *Food Res. Int.* 36:485–490.

Liao, X., Raghavan, G. S. V., Wu, G., and Yaylayan, V. A. 2003b. Dielectric properties of lysine aqueous solutions at 2450 MHz. *J. Mol. Liquid* 107:15–19.

Lin, T. M., Durance, T. D., and Scaman, C. H. 1998. Characterization of vacuum microwave, air and freeze dried carrot slices. *Food Res. Int.* 31:111–117.

Litvin, S., Mannheim, C. H., and Miltz, J. 1998. Dehydration of carrots by a combination of freeze drying, microwave heating and air or vacuum drying. *J. Food Eng.* 36:103–111.

Lopez-Avila, V. and Benedicto, J. 1996. Microwave-assisted extraction combined with gas chromatography and enzyme-linked immunosorbent assay. *TrAC, Trends Anal. Chem.* 15:334–341.

Lopez-Avila, V., Young, R., Benedicto, J., Ho, P., Kim, R., and Beckert, W. F. 1995. Extraction of organic pollutants from solid samples using microwave energy. *Anal. Chem.* 67:2096–2102.

Lucchesi, M. E., Chemat, F., and Smadja, J. 2004. An original solvent free microwave extraction of essential oils from spices. *Flavour Frag. J.* 19:134–138.

Lucchesi, M. E., Smadja, J., Bradshaw, S., Louw, W., and Chemat, F. 2007. Solvent free microwave extraction of *Elletaria cardamomum* L.: A multivariate study of a new technique for the extraction of essential oil. *J. Food Eng.* 79:1079–1086.

Luque-García, J. L. and Luque de Castro, M. D. 2004. Focused microwave-assisted Soxhlet extraction: Devices and applications. *Talanta* 64:571–577.

Ma, Y. H. and Peltre, P. R. 1975a. Freeze dehydration by microwave energy: Part I. Theoretical investigation. *AIChE J.* 21:335–344.

Ma, Y. H. and Peltre, P. R. 1975b. Freeze dehydration by microwave energy: Part II. Experimental study. *AIChE J.* 21:344–350.

Martín-Calero, A., Pino, V., Ayala, J. H., González, V., and Afonso, A. M. 2007. Focused microwave-assisted extraction and HPLC with electrochemical detection to determine heterocyclic amines in meat extracts. *J. Liquid Chromatogr. Relat. Technol.* 30:27–42.

Maskan, M. 2000. Microwave/air and microwave finish drying of banana. *J. Food Eng.* 44:71–78.

Maskan, M. 2001. Kinetics of colour change of kiwifruits during hot air and microwave drying. *J. Food Eng.* 48:169–175.

Metaxas, A. and Meredith, R. 1983. *Industrial Microwave Heating*. London, U.K.: Peter Peregrinus Publication.

Mousa, N. and Farid, M. 2002. Microwave vacuum drying of banana slices. *Drying Technol.* 20:2055–2066.

Nelson, S. O. 1979. RF and microwave dielectric properties of shelled, yellow-dent field corn. *Trans. ASAE* 22:1451–1457.

Nindo, C. I., Sun, T., Wang, S. W., Tang, J., and Powers, J. R. 2003. Evaluation of drying technologies for retention of physical quality and antioxidants in asparagus (*Asparagus officinalis*, L.). *Lebensm. Wiss. Technol.* 36:507–516.

Oliveira, C. C., Sartini, R. P., and Zagatto, E. A. G. 2000. Microwave-assisted sample preparation in sequential injection: Spectrophotometric determination of magnesium, calcium and iron in food. *Anal. Chim. Acta* 413:41–48.

Osepchuk, J. M. 2002. Microwave power applications. *IEEE Trans. Microw. Theory Tech.* 50:975–985.

Pappas, C., Tsami, E., and Marinos-Kouris, D. 1999. The effect of process conditions on the drying kinetics and rehydration characteristics of some MW-vacuum dehydrated fruits. *Drying Technol.* 17:158–174.

Pare, J., Sigomin, M., and Lapointe, J. 1991. U.S. Patent 5,002,784, filed May 7, 1990 and issued March 26, 1991.

Paré, J. R., Bélanger, J. M. R., and Stafford, S. S. 1994. Microwave-assisted process (MAP(TM)): A new tool for the analytical laboratory. *TrAC, Trends Anal. Chem.* 13:176–184.

Perkin, R. 1983. The drying of porous materials with electromagnetic energy generated at radio and microwave frequencies. Report No. ECRC/M1646, The Electricity Council Research Center, Capenhurst, England.

Perring, L., Alonso, M. I., Audrey, D., Bourqui, B., and Zbinden, P. 2001. *Fresenius J. Anal. Chem.* 370:76–81.

Plessi, M., Bertelli, D., and Miglietta, F. 2002. Effect of microwaves on volatile compounds in white and black pepper. *LWT-Food Sci. Technol.* 35:260–264.

Prabhanjan, D. G., Ramaswamy, H. S., and Raghavan, G. S. V. 1995. Microwave-assisted convective air drying of thin layer carrots. *J. Food Eng.* 25:283–293.

Prados-Rosales, R. C., Gracia, J. L. L., and Luque-de-Castro, M. D. 2003. Rapid analytical method for the determination of pesticide residues in sunflower seeds based on focused microwave-assisted Soxhlet extraction prior to gas chromatography-tandem mass spectrometry. *J. Chromatogr. A.* 993:121–129.

Prothon, F., Ahrné, L. M., Funebo, T., Kidman, S., Langton, M., and Sjöholm, I. 2001. Effects of combined osmotic and microwave dehydration of apple on texture, microstructure and rehydration characteristics. *Lebensm. Wiss. Technol.* 34:95–101.

Punt, M. M., Raghavan, G. S. V., Bélanger, J. M. R., and Paré, J. R. J. 1999. Microwave-assisted process (MAP™) for the extraction of contaminants from soil. *Soil Sediment Contam.* 8:577–592.

Raghavan, S. and Richards, M. P. 2007. Comparison of solvent and microwave extracts of cranberry press cake on the inhibition of lipid oxidation in mechanically separated turkey. *Food Chem.* 102:818–826.

Raghavan, G. S. V. and Silveira, A. M. 2001. Shrinkage characteristics of strawberries osmotically dehydrated in combination with microwave drying. *Drying Technol.* 19:405–414.

Richardson, P. Ed. 2001. *Thermal Technologies in Food Processing*. Boca Raton, FL: CRC Press.

Salek-Mery, J. 1986. Heat and mass transfer studies in fluidized beds combined with microwaves for the dehydration of food materials. PhD dissertation, University of Illinois at Urbana-Champaign, Urbana, IL.

Sanga, E. C. M., Mujumdar, A. S., and Raghavan, G. S. V. 2002. Simulation of convection-microwave drying for a shrinking material. *Chem. Eng. Process.* 41:487–499.

Schiffmann, R. F. 1995. Microwave and dielectric drying. In *Handbook of Industrial Drying*, Ed. A. S. Mujumdar, pp. 345–372. New York: Marcel Dekker.

Sham, P. W. Y., Scaman, C. H., and Durance, T. D. 2001. Texture of vacuum microwave dehydrated apple chips as affected by calcium pretreatment, vacuum level, and apple variety. *J. Food Sci.* 66:1341–1347.

Sharma, G. P. and Prasad, S. 2001. Drying of garlic (*Allium sativum*) cloves by microwave-hot air combination. *J. Food Eng.* 50:99–105.

Shivhare, U. S. 1991. Drying characteristics of corn in a microwave field with a surface-wave applicator. PhD dissertation, McGill University (Macdonald Campus), Montreal, Quebec, Canada.

Shivhare, U. S., Raghavan, G. S. V., and Bosisio, R. G. 1991. Drying of corn using variable microwave power with a surface wave applicator. *J. Microw. Power Electromagn. Energy* 26:38–44.

Shivhare, U. S., Raghavan, G. S. V., and Bosisio, R. G. 1992a. Microwave drying of corn. II. Constant power, continuous operation. *Trans. ASAE* 35:951–957.

Shivhare, U. S., Raghavan, G. S. V., and Bosisio, R. G. 1992b. Microwave drying of corn. III. Constant power, intermittent operation. *Trans. ASAE* 35:958–962.

Shivhare, U. S., Raghavan, G. S. V., Bosisio, R. G., and Giroux, M. 1993. Microwave drying of soybean at 2.45 GHz. *J. Microw. Power Electromagn. Energy* 28:11–17.

Simoneau, C., Naudin, C., Hannaert, P., and Anklam, E. 2000. Comparison of classical and alternative extraction methods for the quantitative extraction of fat from plain chocolate and the subsequent application to the detection of added foreign fats to plain chocolate formulations. *Food Res. Int.* 33:733–741.

Smith, J. S. and Hui, Y. H. Eds. 2004. *Food Processing—Principles and Applications*. Ames, IA: Blackwell Publishing.

Sparr Eskilsson, C. and Björklund, E. 2000. Analytical-scale microwave-assisted extraction. *J. Chromatogr. A* 902:227–250.

Sunjka, P. S., Rennie, T. J., Beaudry, C., and Raghavan, G. S. V. 2004. Microwave-convective and microwave-vacuum drying of cranberries: A comparative study. *Drying Technol.* 22:1217–1231.

Sutar, P. 2008. Combined osmotic and microwave vacuum dehydration of carrots. PhD dissertation, Indian Institute of Technology, Kharagpur, India.

Torringa, E., Esveld, E., Scheewe, I., van den Berg, R., and Bartels, P. 2001. Osmotic dehydration as a pre-treatment before combined microwave-hot-air drying of mushrooms. *J. Food Eng.* 49:185–191.

Tulasidas, T. 1995. Combined convective and microwave drying of grapes. *Drying Technol.* 13:1029–1031.

Tulasidas, T. N., Raghavan, G. S. V., Kudra, T., and Gariépy, Y. 1994. Microwave drying of grapes in a single mode resonant with pulsed power. Paper (Paper No. 94-6547) presented at the Annual International Winter Meeting of American Society of Agricultural Enginees—ASAE, Atlanta, GA.

Tulasidas, T. N., Raghavan, G. S. V., and Mujumdar, A. S. 1995. Microwave drying of grapes in a single mode cavity at 2450 Mhz—I: Drying kinetics. *Drying Technol.* 13:1949–1971.

Tuzen, M. and Soylak, M. 2007. Evaluation of trace element contents in canned foods marketed from Turkey. *Food Chem.* 102:1089–1095.

Venkatachalapathy, K. and Raghavan, G. S. V. 1999. Combined osmotic and microwave drying of strawberries. *Drying Technol.* 17:837–853.

Von Hippel, A. R. 1954. *Dielectric Materials and Applications.* Cambridge MA: MIT Press.

Wang, Z., Wang, L., Li, T. et al. 2006. Rapid analysis of the essential oils from dried *Illicium verum* Hook. f. and *Zingiber officinale* Rosc. by improved solvent-free microwave extraction with three types of microwave-absorption medium. *Anal. Bioanal. Chem.* 386:1863–1868.

Xu, D., Min, Z., Xinlin, L., and Mujumdar, A. S. 2008. Microwave freeze drying of sea cucumber coated with nanoscale silver. *Drying Technol.* 26:413–419.

Xu, Y., Min, Z., Xinlin, L., Mujumdar A. S., Zhou, L., and Sun, J. 2004. Studies on hot air and microwave vacuum drying of wild cabbage. *Drying Technol.* 22:2201–2209.

Yang, H. W. and Gunasekaran, S. 2001. Temperature profiles in a cylindrical model food during pulsed microwave heating. *J. Food Sci.* 66:998–1004.

Yang, H. W. and Gunasekaran, S. 2004. Comparison of temperature distribution in model food cylinders based on Maxwell's equations and Lambert's law during pulsed microwave heating. *J. Food Eng.* 64:445–453.

Yongsawatdigul, J. and Gunasekaran, S. 1996. Microwave-vacuum drying of cranberries: Part I. Energy use and efficiency. *J. Food Process. Preserv.* 20:121–143.

Young, J. C. 1995. Microwave-assisted extraction of the fungal metabolite ergosterol and total fatty acids. *J. Agric. Food Chem.* 43:2904.

Zhang, M., Tang, J., Mujumdar, A. S., and Wang, S. 2006. Trends in microwave-related drying of fruits and vegetables. *Trends Food Sci. Technol.* 17:524–534.

Zhu, M., Aviles, F. J., Conte, E. D., Miller, D. W., and Perschbacher, P. W. 1999. Microwave mediated distillation with solid-phase microextraction: Determination of off-flavors, geosmin and methylisoborneol, in catfish tissue. *J. Chromatogr. A.* 833:223–230.

13 Vacuum Frying Technology

Liu Ping Fan, Min Zhang, and Arun S. Mujumdar

CONTENTS

13.1 INTRODUCTION

Frying, along with dehydration, is an established process of food preparation and preservation worldwide. It is a simultaneous heat and mass transfer process in which moisture leaves the food in the form of vapor, while some oil is absorbed simultaneously. During the frying process, the physical, chemical, and sensory characteristics of the food are modified. Vacuum frying (VF) is a relatively new technique of frying in a suitable oil carried out under pressures well below the atmospheric pressure, preferably below 50 Torr (6.65 kPa); this lowers the boiling point of the frying oil and water making it possible to reduce the frying temperature substantially (Shyu et al. 1998a, Garayo and Moreira 2002). It is a viable option for the production of snacks from fruits and vegetables as it results in fried products with lower oil content often along with several desirable quality attributes. Moreover, the absence of air during frying inhibits some undesirable chemical reactions including lipid oxidation and enzymatic browning. Hence the natural color, flavor, and nutrients of samples can be better preserved than in normal atmospheric pressure deep-fat frying. This chapter attempts to present a short overview of recent studies on VF from a practical viewpoint.

Shyu and Hwang (2001) studied the effects of processing conditions on the quality of vacuum-fried apple. Garayo and Moreira (2002) developed a VF system to produce high quality potato chips in terms of reduced oil content, good texture, and color. Shi et al. (2001) optimized the VF process for *Colocasia esculenta* Schott's son-Taroes using an orthogonal experiment. Fan and Zhang (2004) studied VF technology for carrot chips and reported optimum conditions to obtain high quality chips. Shyu et al. (1998b, 1999) have examined the influence of VF conditions on the chemical constituents of fried carrot chips. Fan et al. (2005a) investigated the effect of frying temperature and vacuum degree on moisture, oil content, color, and texture of fried carrot chips. Many patents describe the VF method and apparatus (e.g., Imai and Kunio 1989, Yang 1989, Sakuma and KenJi 1991, Chiu and Yao-Jui 1993, Hashiguchi et al. 2000). French fries can be processed by VF. Garayo and Moreira (2002) showed that VF can produce potato chips with lower oil content but with the same texture and color characteristics as those of regular chips fried in conventional (atmospheric pressure) fryers. Yamsaengsung and Rungsee (2003) have observed that vacuum-fried potato chips and guava slices have lower oil content and more natural colorations than those fried conventionally.

Another important characteristic of VF is its safety aspect. Acrylamide, a carcinogen found to cause cancer in laboratory rats, is present in carbohydrate-rich foods cooked at high temperatures, such as fried or baked chips, bread, etc. Tests confirm that when certain amino acids such as asparagine are heated, they do react with reducing sugars to produce undesirable acrylamide, a process commonly called the Maillard reaction. This occurs at temperatures above 120°C. Recent research has revealed that lowering cooking temperature is an easy and effective way to reduce the amount of acrylamide in fried foods. Granda et al. (2004) demonstrated that VF can produce potato chips with up to 97% reduction in acrylamide content compared to the traditionally fried chips.

Various products made of vacuum-fried vegetables and fruits have been developed rapidly in recent years because of their retention of much of their original flavor, color, and nutrition. At present, there are many vacuum-fried products sold in the world market, e.g., carrot, banana, bitter melon, pumpkin, etc., as shown in Figure 13.1.

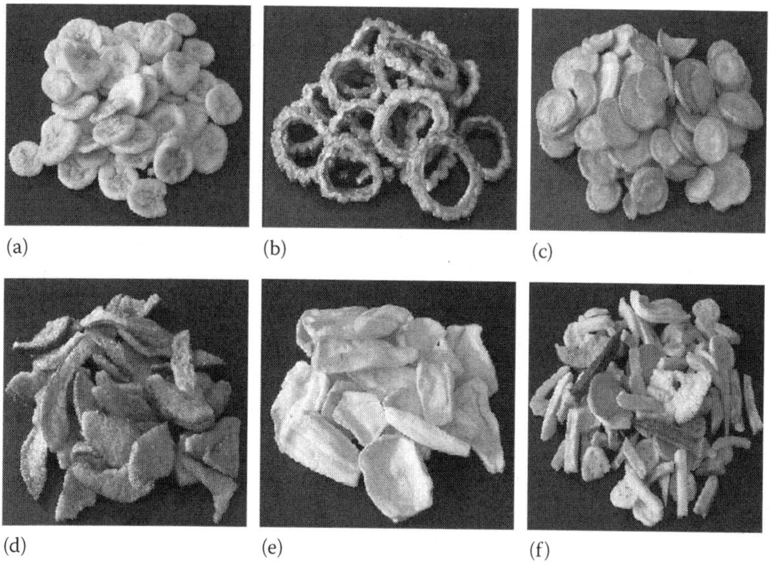

FIGURE 13.1 Vacuum-fried products: (a) banana chips; (b) bitter melon; (c) carrot; (d) pumpkin; (e) jack fruit; (f) mixed fruits.

Compared with other dehydration technologies for fruits and vegetables, VF is a viable option to obtain high quality dried products in a much shorter processing time. Conventional air drying is the most frequently used drying method in the food industry. Here the drying kinetics depend mainly on both material and air properties, such as air temperature, relative humidity, and air velocity (Islam et al. 2003). Significant color changes of dried products occur during airdrying (Krokida et al. 1998). Deep-fat frying is a method to produce dried food in which fat serves as the heat transfer medium; fat also migrates into the food, providing nutrients and flavor. To date most of the published research is related to conventional deep-fat frying.

Several models have been developed to describe moisture evaporation and oil absorption in deep-fat frying (Moreira and Bakker-Arkema 1989, Rice and Gamble 1989, Kozempel et al. 1991). Mittelman et al. (1984) reported that oil temperature and frying time are the main frying operation variables controlling mass transfer in deep-fat frying. Inevitably, at atmospheric pressure, deep-fat frying occurs at high temperatures. Surface darkening and many adverse reactions, especially the formation of acrylamide, take place because of the high temperature treatment before the food is fully cooked or dried. However, only a few works on VF are found in the literature, since research in this field is still at the initial stage.

Following a brief discussion of some of the basic aspects of VF, the effects of different operating parameters and the storage stability of vacuum-fried products are presented in subsequent sections.

13.2 THEORY

13.2.1 Basic Principles

During VF, moisture in the fruit or vegetable is rapidly removed under the reduced pressure at frying temperatures below the normal atmospheric boiling point of the oil. The VF process is portrayed in Figure 13.2. Under vacuum condition, the boiling point of water in the food immersed in the hot oil is depressed. At the same time, the surface temperature of the food rises rapidly. The free water at the food surface is lost rapidly in the form of vapor bubbles. The surrounding oil is cooled down but this is quickly compensated for by convective heat transfer. Due to evaporation, the surface dries out and improves its hydrophobicity. Fat or oil may adhere to the outer surface. When the vacuum-fried food is removed from the fryer, the vapor inside the pore gets condensed and the pressure differential between the surrounding and the pore causes the oil adhering to the surface to be absorbed into the pore space. However, if the food contains more moisture, this higher moisture content will prevent the oil from entering the pore space. Moisture evaporation also leads to shrinkage and development of surface porosity and crispiness. Water deep inside the food becomes heated and cooks the food. As frying progresses, the moisture content in the food slowly diminishes, thereby reducing the amount of steam leaving the surface.

For mass production, the relationship among the frying temperature, vacuum degree, and the frying time was introduced in the research of Yang (1997). More and more moisture is evaporated with increasing vacuum degree. The frying temperature falls as evaporation progresses due to latent heat requirements. For example, when the vacuum level increases to 93.3 kPa (70 cmHg), the frying oil temperature decreases from 110°C–115°C to 80°C–85°C. Then, the vacuum level and the frying temperature become stable with the moisture evaporating continuously. When the moisture in the food falls below the critical point, the rate of water movement from the interior to the food surface falls below the rate at which water evaporates to the surrounding fat. This results in an increase in the frying temperature while the vacuum degree remains stable until the frying process is complete.

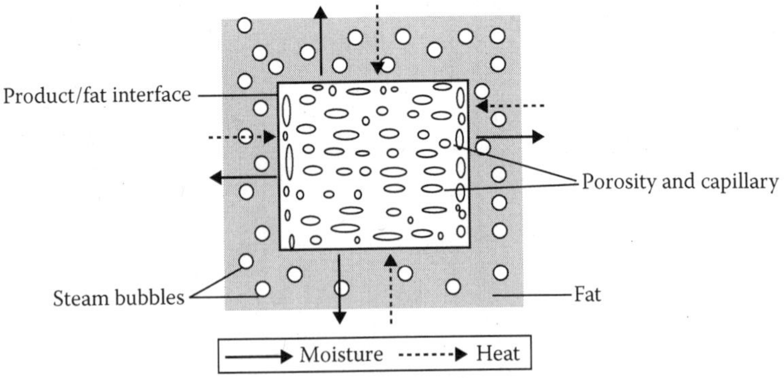

FIGURE 13.2 Schematic cross section of a piece of food during VF.

13.2.2 Process of Vacuum Frying

In general, there are two main procedures of commercial VF, which are distinguished by the method used in the freezing pretreatment, as shown in Figure 13.3.

Some detailed information is as follows:

- Material—selection of fruits and vegetables of optimum maturity.
- Rough machining—leaning, sorting, peeling, and slicing.
- Blanching—destroy enzymatic activity in vegetables and some fruits, prior to VF. The blanching method includes steam blanching and hot water blanching.
- Freezing—contribute to form a porous sponge-like structure and improve the texture of the vacuum-fried food.

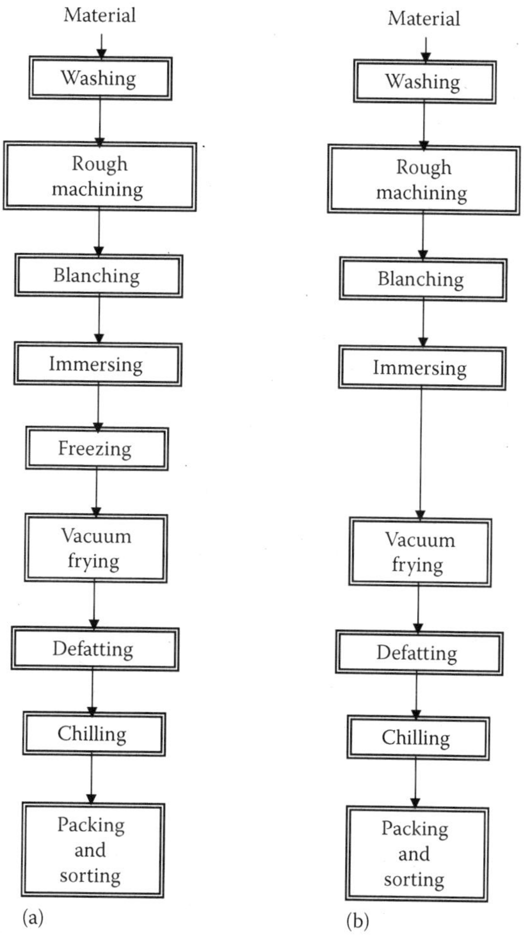

FIGURE 13.3 Procedures of commercial VF.

- VF—first, the frying fat is heated rapidly to a preset temperature. Then, the material is placed in a frying basket and the lid is closed and locked. Next, the vacuum pump is switched on to a preset vacuum degree. Finally, the frying basket is placed into cooking fat for a preset time.
- Defatting—after the vacuum-fried food is cooked, the frying basket is lifted and centrifuged (at atmospheric pressure or under vacuum) to decrease the fat content of the vacuum-fried food.

Vacuum-fried products are oxygen- and moisture-sensitive. Packaging plays a number of critical roles for fried foods. It must contain the product, protect it against moisture, oxygen, and light, as well as against shock, vibration, and mishandling during storage and shipping. Therefore, an inert gas, typically nitrogen, is used in the packaging of fried products. Very significant increases in storage life can be obtained if one of the following procedures is employed:

- Reduction in the oxygen (O_2) levels in the headspace (<2%) of the package
- Use of oxygen (O_2) barrier films as packaging material
- Incorporation of a light barrier into the packaging film

13.2.3 Equipment

A vacuum fryer is used to fry fruits or vegetables whose original color is retained with minimum browning. It consists of a fryer, frying basket, store oilcan, condenser, centrifuge, vacuum pump, and electric motor (Figure 13.4). The vacuum is created by vacuum pumps, and the whole system can be controlled by a programmable logic controller. There are three kinds of equipment including lab-scale, pilot-scale, and industry-scale according to the manufacture scale. There are two kinds of equipment including batch type and continuous type equipment according to the processing manner. These vacuum fryers have several distinctive features including the following:

1. The fryer is operated under vacuum. Typical operating pressure is <100 mmHg.
2. The frying is conducted at 100°C–120°C, and the food can be dehydrated at this temperature under vacuum.
3. The food is placed in a basket.
4. Frying begins and the oil temperature drops.
5. The oil is circulated continuously through an external heater.
6. The oil regains the temperature when the moisture content of the product reaches the predetermined value.
7. The vacuum in the fryer is broken slowly.
8. The excess oil is allowed to drain, and the product is cooled before packaging.

Batch fryers are the most commonly used models. They are employed for small-scale production or for frying very special types of products. However, the process is

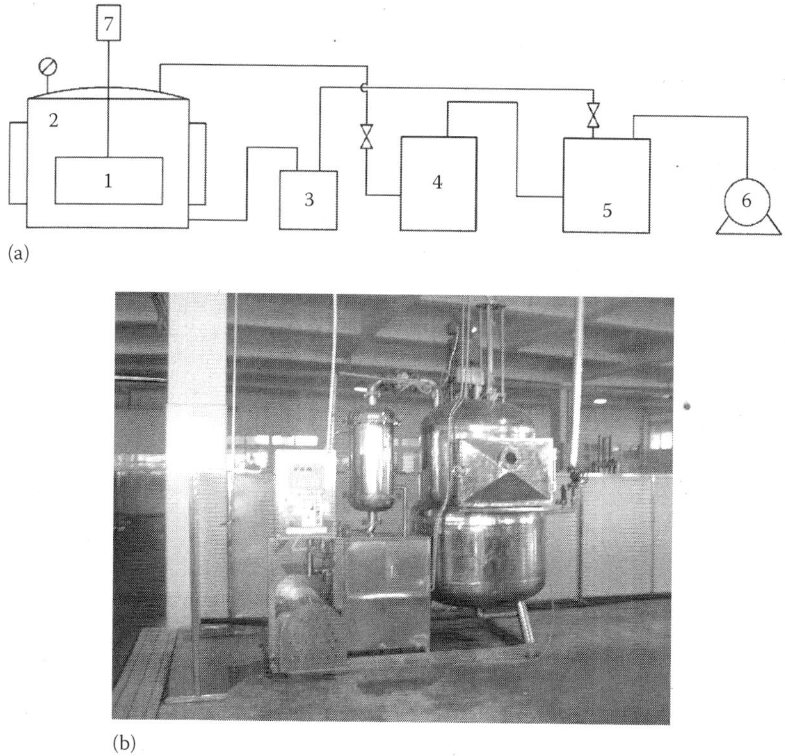

FIGURE 13.4 Schematic representation (a) and picture (b) of a VF system: 1, frying basket; 2, vacuum fryer; 3, store oilcan; 4, condenser; 5, segregator; 6, vacuum pump; 7, electric motor.

more costly. Continuous vacuum fryers are used for large-scale production of fried foods. In these systems, the VF pan is installed in a stainless steel vacuum tube. The raw material is fed through a rotary air lock. Depending on the application, the frying pan itself is designed to meet the different product specifications. A transport belt takes the finished product out of the fryer and toward the outlet system. A lock chamber at the exit of the vacuum tube prevents air from entering the vacuum zone, and a belt system takes the product from one zone to another.

13.2.4 Characteristics of Vacuum Frying

Compared with traditional deep-fat frying, the advantages of VF are

- Preservation of natural colors and flavors of the finished product
- Minimization of (health) harmful thermal reaction products, such as acrylamide
- Reduction in the fat content of the fried food
- Ability to process potatoes with higher reducing sugar levels

- High evaporation rate
- Improvement in the texture of the fried food
- Reduction of adverse effects on fat quality

On the other hand, just as in a classical drying profile, the drying process of foods is generally characterized by three distinct periods. The first one is an initial heat-up period during which the wet solid material absorbs heat from the surrounding media. This is followed by the so-called constant-rate period. The third period is known as the falling-rate period. Note that the drying rate decreases and enters the falling-rate period when the moisture level in the product is so low that its surface is no longer wetted. During this period, the drying rate is controlled by moisture diffusion mechanisms.

13.3 EFFECT OF VACUUM FRYING CONDITIONS ON FRIED FOODS

13.3.1 Effect of Pretreatment

The main purpose of VF is production of high quality fried foods with natural color, good flavor, crispy texture, and lower fat content. Some physical pretreatment procedures have been investigated in order to improve drying kinetics and product quality. The main pretreatments that control the quality of vacuum-fried food are

- Blanching
- Osmotic pretreatment
- Freezing
- Predrying

13.3.1.1 Effect on Yield and Nutritional Value of Food

The pretreatment method influences the yields of vacuum-fried foods. The yield of vacuum-fried foods can be defined as percentage of weight of fresh material before and after VF. Hot-air predrying before VF results in a lower yield (Fan et al. 2006). This may be due to a high loss of moisture and volatile components during air drying. Osmotic dehydration using salt, sugar, or other osmotic agents increases the solid content of fried food, which increases the yield (Fan et al. 2006). The freezing process destroys the cell structure, producing loss of the succi inside the cell, and that may decrease the yield.

Heat-sensitive components such as vitamin C are relatively unstable to heat, oxygen, and light, and their retention can be used as an indicator for the quality of food. Some vegetables and fruits, such as carrot and potato have been blanched to inactivate ascorbic acid oxidase and to prevent enzymatic degradation of vitamin C in the frying processes (Fan et al. 2006). On the other hand, hot air predrying could result in the oxidation of some nutritional components and decrease the food value. Under osmotic dehydration prior to VF, air is removed from the material tissue, and this should prevent the loss of some nutritional components from the food material.

13.3.1.2 Effect on Moisture and Fat Content of Food

Pretreatment affects the initial water content of material before VF, as well as the water and fat content of vacuum-fried products. Fat absorption is a complex phenomenon, which is affected by the initial water content and the pore size, as well as by the pore distribution left after water evaporation. Pretreatments (e.g., osmotic, hot air drying) that decrease the initial water content and improve the solid content can decrease the fat content after VF. Freezing, as reported by several researchers, destroys the structure and induces more disruption of the material cells, increasing fat absorption and retention during VF. In addition, the water in the frozen material can rapidly evaporate from its ice crystal state under vacuum, leaving more uniform pores and increasing overall porosity, which perhaps increases the fat absorption during VF.

Fan et al. (2006) compared the effect of different pretreatments on the quality of vacuum-dried carrot slices. Some of their results are shown in Table 13.1. It can be seen that, apart from blanching, all other pretreatments significantly decreased the initial water content of the carrot slices. As a result, for a fixed frying time, these treatments led to a smaller final water content compared to blanching alone. The pretreatment significantly affected the fat content of carrot chips (p-value < 0.05). In contrast to blanching, fat content of carrot chips treated with the other three pretreatments decreased significantly.

Similar results were obtained in the research of Shyu et al. (2005). They found that the moisture and fat contents of fried carrot chips were significantly (p-value <0.05) reduced when blanched carrot slices were pretreated by immersion in fructose

TABLE 13.1
Effects of Pretreatment on the Moisture Content, Fat Content, Water Activity, and Breaking Force of Carrot Chips

Pretreatment	Initial Water Content (kg/kg in d.b.)	Water Content (kg/kg in d.b.)	Fat Content (kg/kg in d.b.)
Blanching	8.09[a]	0.097[a]	0.43[a]
Blanching + air drying	4.31[b]	0.031[d]	0.27[c]
Blanching + osmotic	2.65[c]	0.043[b]	0.23[d]
Blanching + osmotic + freezing	2.63[c]	0.035[c]	0.32[b]

Source: From Fan, L.P. et al., *Drying Technol.*, 24, 1481, 2006. With permission.
Note: Means with different letters within a column are significantly different (p-value < 0.05) by Duncan multiple range test.

solution and freezing prior to VF. Furthermore, more uniform porosity was observed on the vertical cross section of carrot chips when examined by scanning electron microscopy.

13.3.1.3 Effects on Fat Distribution of Vacuum-Fried Products

Pretreatment varies the moisture content or texture characteristics, and that affects the fat content and its distribution. The effect of pretreatment on fat distribution in vacuum-fried carrot chips has been investigated by Fan et al. (2006). Fat was taken up in most areas of the carrot chips treated with blanching. Areas in the cortex and vascular ring had some fat blisters, whereas specific zones, usually in the central part, contained only a few small droplets. The fat distribution showed that fat was located in irregularly shaped deposits with excessive amounts of fat on the edges (Figure 13.5a). Compared to blanching alone, the volume occupied by fat in the final carrot chips treated with the other three pretreatments decreased. Predrying

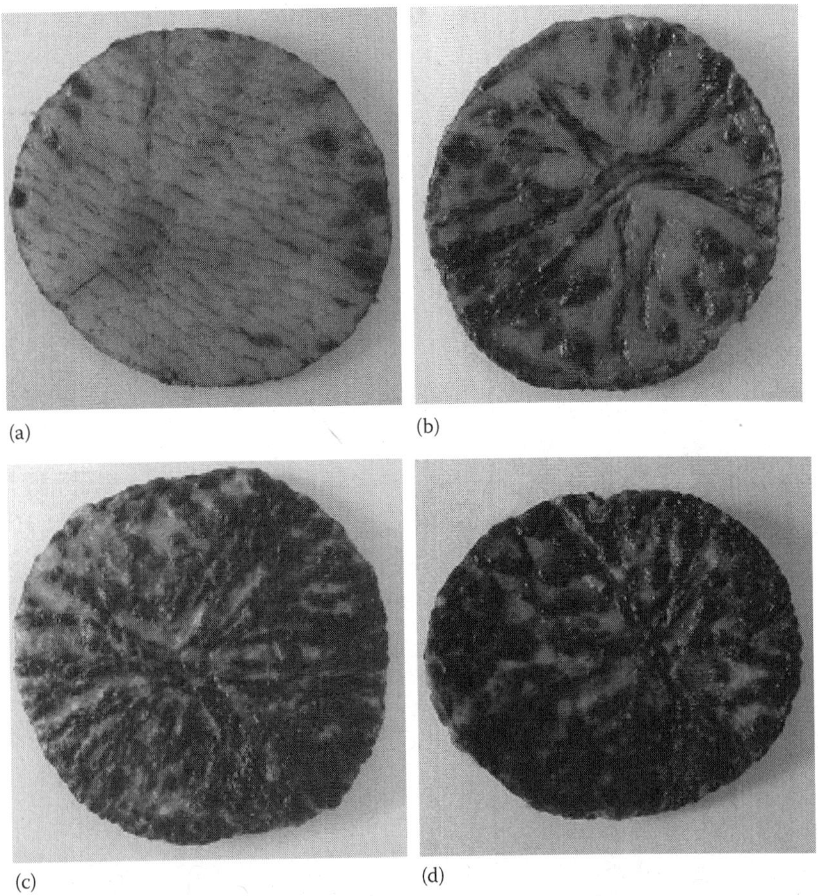

(a) (b) (c) (d)

FIGURE 13.5 Effects of frying time on the fat distribution of potato chip: (a) 5 min; (b) 10 min; (c) 20 min; (d) 30 min.

resulted in a highly heterogeneous structure with a modified water distribution. Air drying resulted in water evaporation from the edge areas prior to VF. After VF, there were areas of intense fat deposition in the central part and fat-free areas in the outer edge part.

The central fat distribution pattern showed a mottled effect with fat in numerous small droplets (Figure 13.5b). An osmotic dehydration step resulted in a fat content reduction in the final carrot chips. The degree of oil reduction varied similarly with the extent of air drying. Unlike air drying, Figure 13.5c shows a more even and larger fat distribution pattern, with fat in numerous large droplets in the central part. The osmotic and freezing slice showed a very similar distribution to the osmotic, exhibiting increased homogeneity with only small edge remaining relatively fat free (Figure 13.5d).

When the slice was fried after predrying, further water loss or fat uptake was restricted to areas that had high residual water content. Regions of the slice that were already dry did not allow fat uptake, resulting in large fat-free zones. Fat uptake was thus seen to be related to moisture loss in frying. Fat settled in areas where water was most easily lost. Therefore, the fat distribution pattern has depended on the initial water content and the structure of the material left by water evaporation.

13.3.2 Effect of Frying Temperature and Frying Time

The VF temperature is one of the key factors related to the dehydration rate, flavor, color, and nutritional value. The frying temperature is affected by the vacuum level during frying. Its target operating value depends on the difference between the initial and final water contents of the fried material and the heat sensitivity of the nutritional components. Generally, its value is controlled to around 100°C. Shyu et al. (2005) suggested that VF at moderate temperatures (90°C–100°C) for 20 min can produce carrot chips with lower moisture and oil contents, as well as good color and crispy texture. Increasing the frying temperature at the same vacuum level decreases the frying time for fried flake foods and improves the rate of drying. Garayo and Moreira (2002) found that during VF, oil temperature and vacuum pressure have a significant effect on the drying and oil absorption rates of potato chips. Frying temperature greatly affected the moisture content; the higher the temperature, the lower the moisture content for a constant frying time. This phenomenon was not in agreement with the work of Sulaeman et al. (2001). They reported that the higher the temperature, the higher the moisture content during deep-fat frying of carrot chips.

Fat absorption is a complicated process and it is affected by material properties, pretreatment, and frying condition. When the slices of material are placed in hot oil under low pressure, the free water is rapidly lost in the form of bubbles. As frying continues, the outer surface dries out, improving its hydrophobicity, and oil may adhere to the chips. When the chips are removed from the fryer, the vapor inside the pores condenses, and the difference in pressure between the surrounding tissue and the pore causes the oil to adhere to the surface. However, differences in the material to be fried should result in different effects of frying temperature on the fat content of vacuum-fried foods.

Fan et al. (2005b) studied optimization of VF to dehydrate carrot chips and found that the fat content of the product increased with increasing frying temperature and time. This result is in agreement with the work of Shyu et al. (2005). Fan et al. (2005a) also studied the effects of processing conditions on the quality of vacuum-fried carrot chips and found that, during VF, the moisture content, color, and breaking force of carrot chips decreased while the oil content increased with increasing frying temperature and time. The results for vacuum-fried apple chips also showed a decrease in the moisture content and breaking force of apple chips with increasing frying temperature (Shyu and Hwang 2001). Conversely, results for potato chips showed that in the case of VF, the final oil content of the product was not significantly affected (p-value < 0.05) by the oil temperature. This suggests that the final oil content of the potato chips is not a function of the frying temperature, but it depends mainly on the frying time (Garayo and Moreira 2002).

Frying time (t_{fry}) would affect the fat distribution of VF products in addition to effects on the fat content. Fat distributions of VF potato chips at different frying times were analyzed by Sudan black dyeing and are shown in Figure 13.5. At $t_{fry} = 5$ min, the fat adsorption mainly appeared at the outer edge of the potato chips (Figure 13.5a). This may be related to moisture evaporation from the outer edge. When $t_{fry} = 10$ min, more areas contain fat drops and their diameters enlarge (Figure 13.5b). For $t_{fry} = 20$ min, the fat drops almost appeared on the whole potato chips (Figure 13.5c). Finally, for $t_{fry} =$ 30 min, fat drops were even larger and became large black areas. Only a few areas contain no fat drops (Figure 13.5d).

The frying temperature and the frying time affect the quality of vacuum-fried foods, such as color, shrinkage, and texture. Values of lightness, L^*, for fried apple chips decreased progressively as the frying temperature and time increased. However, when apple slices were fried at 100°C for up to 20 min, both redness, a^*, and yellowness, b^*, values increased rapidly (Shyu and Hwang 2001). When carrot chips were vacuum fried, there were no significant differences in L^*, a^*, and b^* of carrot chips as a function of vacuum level and temperature. The breaking force of carrot chips decreased with increasing vacuum level during frying. There were no significant differences in the breaking force as a function of temperature; there was no apparent change in the total color difference (Hunter ΔE) with time when the frying temperature was below 100°C and the frying time was below 25 min (Shyu and Hwang 2001). Potato chips fried at lower vacuum pressure and higher temperature had less volume shrinkage. Color was not significantly affected by the oil temperature or the vacuum level. Hardness values increased with increasing oil temperature and decreasing vacuum levels (Garayo and Moreira 2002). Compared with potato chips fried at atmospheric condition (165°C), potato chips fried under vacuum (3.115 kPa and 144°C) had more volume shrinkage, were slightly softer, and lighter in color.

13.3.3 Effect of Pressure

13.3.3.1 Comparison with Atmospheric Frying

Compared with traditional atmospheric frying, VF is an efficient method of reducing the oil content in fried foods, maintaining product nutritional quality, and reducing oil deterioration. Mass transfer mechanisms in atmospheric frying can be divided into two periods: the frying period and the cooling period. During the frying period, the capillary pressure is negligible (Moreira et al. 1999), so there is no driving force for the oil to flow into the pores of the material. During the cooling period, the surface oil (adhered to the surface of the frying material) penetrates the pores. As the material cools down, the pressure inside its pores changes as a consequence of the so-called capillary pressure rise (Moreira and Barrufet 1995). This pressure difference between the surface and the pores creates a driving force for the oil and air to penetrate the pores.

The mass transfer mechanisms in VF can be understood by dividing the process into three periods (Garayo and Moreira 2002): Frying, pressurization, and cooling. At the beginning of the frying period, capillary pressure (between oil and gas) is negligible. Thus, no oil is absorbed at this stage. The second period is the pressurization from vacuum to atmospheric conditions. This step plays an important role in reducing the oil absorption during VF. At this stage, as the vessel is vented, the pressure in the pores of the potato chips rapidly increases to atmospheric levels, as air and surface oil are carried into the empty pore spaces until the pressure reaches atmospheric levels. However, because of the low pressure, gas diffuses much faster into the pore space, thus obstructing the oil passage to enter the pores. The third period starts when the product is removed from the fryer and it is known as the cooling period, when part of the adhered oil continues to penetrate the pore spaces. Since less oil adheres to the product surface during VF, less oil is absorbed during cooling.

Silva and Moreira (2008) compared the quality of sweet potato chips produced by VF and by traditional frying. Oil content of vacuum-fried products was significantly (p-value < 0.05) lower than that of traditionally fried products. Anthocyanin and total carotenoids content, on the other hand, were significantly (p-value < 0.05) higher for the VF products. Sensory panelists overwhelmingly preferred (p-value < 0.05) the vacuum-fried products for color, texture, taste, and overall quality. Most of the products retained or accentuated their original colors when fried under vacuum. The traditionally fried products showed excessive darkening and scorching.

Mariscal and Bouchon (2008) compared atmospheric and VF of apple slices in terms of oil uptake, moisture loss, and color development. The results showed VF was a promising technique that can be used to reduce oil content in fried apple slices while preserving the color of the product. In particular, drying prior to VF was shown to give the best results.

13.3.3.2 Effect of Vacuum Level

The choice of vacuum degree affects the frying temperature, T_{fry}, and therefore the quality of vacuum-fried foods. In general, the vacuum pressure, ΔP_{vac}, varies from −92.0 to −98.7 kPa. Previous research on vacuum-fried carrot chips showed that, at $T_{fry} = 80°C$, high vacuum improves the rate of the moisture evaporation (Fan et al. 2005a). VF at $\Delta P_{vac} = -95$ kPa caused the carrot slices to dry rapidly, with removal of 95% of the initial moisture content, m_0, in 15 min. During this period, maximum bath turbulence because of the release of steam bubbles could be observed. The statistical analysis showed that there were significant differences in the breaking force of the product as a function of the vacuum degree. The rate of fat absorption increased with increasing vacuum. At $T_{fry} = 80°C$, the rate of fat uptake slightly varied after frying at −60 kPa for 20 min, at −80 and −95 kPa for 15 min.

This last outcome is in accordance with the results of Garayo and Moreira (2002), who investigated the effect of vacuum level during VF of potato chips. Their results, listed in Table 13.2, indicated that P_{vac} had a significant effect on the drying rate and

TABLE 13.2
Initial Moisture Content (m_0), Moisture Content after Frying (m_f), and Oil Content ($C_{f\text{-}fat}$) for Potato Chips Fried under Different Vacuum Levels (P_{vac}) and Oil Temperatures (T_{fry})

T_{fry} (°C)	P_{vac} (kPa)	t_{fry} (s)	m_0 (wt% in w.b.)	m_f (wt% in w.b.)	$C_{f\text{-}fat}$ (wt%) (in w.b.)	(in d.b.)
118	16.661	600 (600)	78.96 ± 1.53^{a}	1.84 ± 0.19^{aa} (1.84 ± 0.19)	25.74 ± 0.36^{a}	35.54
132	16.661	480 (480)	77.92 ± 0.52^{a}	1.90 ± 0.21^{aa} (1.90 ± 0.21)	26.22 ± 0.34^{a}	36.48
144	16.661	360 (360)	73.55 ± 0.63^{a}	1.93 ± 0.02^{aa} (1.93 ± 0.02)	26.66 ± 0.26^{a}	37.33
118	9.888	600 (480)	77.98 ± 1.01^{a}	1.31 ± 0.10^{ab} (1.80 ± 0.01)	26.07 ± 0.45^{a}	35.90
132	9.888	480 (480)	75.65 ± 0.33^{a}	1.37 ± 0.04^{ab} (1.89 ± 0.19)	26.16 ± 0.20^{a}	36.09
144	9.888	360 (200)	73.45 ± 0.35^{a}	0.94 ± 0.02^{bc} (1.95 ± 0.05)	27.22 ± 0.32^{a}	37.88
118	3.115	600 (360)	79.97 ± 0.70^{a}	1.06 ± 0.19^{ac} (1.94 ± 0.08)	26.66 ± 0.31^{a}	36.89
132	3.115	480 (300)	80.11 ± 0.68^{a}	1.12 ± 0.01^{ac} (1.85 ± 0.17)	26.16 ± 0.03^{a}	35.89
144	3.115	360 (190)	75.04 ± 0.56^{a}	0.75 ± 0.00^{bd} (1.90 ± 0.06)	26.63 ± 0.02^{a}	36.67

Note: Tests were performed in duplicate. Means with the same letter are not significantly different (p-value < 0.05). Value in parenthesis in the third column is t_{fry} to obtain chips with $m_f \sim 1.90$ wt% in w.b. (actual value of m_f is given in parenthesis in the fifth column).

fat absorption rate of potato chips. However, they revealed that P_{vac} did not affect the final fat content, $C_{f\text{-}fat}$, of potato chips at any T_{fry}.

13.3.4 Vacuum Microwave Drying with Vacuum Frying

In recent years, due to consumer health concerns, much research has been concentrated on approaches to reduce oil absorption in fried products, revealing that there is a high positive correlation between the fat content and initial water content (shown in Table 13.3). Spearman correlation coefficient between the initial water content and fat content of carrot chips was 0.71 (p-value < 0.05). Fat absorption is affected by the initial water content, that is to say, combining other drying techniques with VF could reduce the fat content of vacuum-fried products.

Microwave-vacuum-drying (MWVD) is an alternative way to improve the quality of dehydrated products (Farrel et al. 2005, Zhang et al. 2006, Araszkiewicz et al. 2007). The fast mass transfer conferred by rapid energy transfer by microwave heating

TABLE 13.3
Spearman Correlation Coefficients between the Original Moisture and Fat Content of Carrot Chips ($N = 8$)

	m_0	C_{fat}
Initial moisture content—m_0	1.00000	0.71133
Prob > |R|	0.0	0.0479
Fat content—C_{fat}	0.71133	1.00000
Prob > |R|	0.0479	0.0

generates very rapid drying (Drouzas and Schurbet 1996, Duan et al. 2005, Yaghmaee and Durance 2007). Because the boiling point of water falls under vacuum conditions, MWVD can accomplish drying tasks in a shorter time at lower temperature. Moreover, the absence of air during drying inhibits oxidation, so color and nutrient content of products can be largely preserved. Because MWVD combines the advantages of both vacuum drying and microwave drying, some fruits, vegetables, and grains have been successfully dried using this technique (Kaensup et al. 2002, Xu et al. 2004). For example, Hu et al. (2006) reported that the contents of vitamin C and chlorophyll in vacuum microwave-dried edamames were largely preserved. Yongsawatdigul and Gunasekaran (1996), in turn, reported that vacuum microwave-dried cranberries had a redder color and softer texture as compared to hot air-dried cranberries.

Due to its advantages, MWVD has been used as a co-dehydration method before VF in the State Laboratory of Food Science and Safety at Jiangnan University (China). Sections 13.4.1 through 13.4.5 present some interesting information about vacuum-fried potato chips obtained in the above mentioned laboratory.

13.3.4.1 Effect on Water Loss

The moisture content of potato chips was reduced from 348 ± 16.7 wt% (d.b.) to 109 ± 4.3 wt% (d.b.) in 15 min of MWVD at 30% full power (146.5 ± 2.1 W) and $\Delta P_{vac} = -0.06$ MPa (Song et al. 2007a). As the MWVD duration increased, the moisture content decreased for the same t_{fry}. During VF, the moisture content was significantly (p-value < 0.05) affected by the MWVD time up to $t_{fry} = 30$ min. Moreover, MWVD also affected the drying rate during VF. The moisture content of potato chips untreated by MWVD was over 0.15 kg/kg (d.b.) after $t_{fry} = 20$ min, and the drying rate decreased slowly after 25 min. For the same t_{fry}, MWVD pretreatment led to equilibrium moisture content lower than 0.05 kg/kg (d.b.). This value did not change by varying the MWVD operating time (5, 10, and 15 min).

13.3.4.2 Effect on Fat Content

MWVD decreased the oil content of potato chips during VF. For the same t_{fry}, the longer the MWVD duration was, the lower the oil content. In particular, the oil content of the product fell significantly (p-value < 0.05) after $t_{fry} = 2$ min when the MWVD pretreatment was adopted (Song et al. 2007a). Moreover, the rate of fat absorption decreased with increasing MWVD duration. The oil content of potato slices untreated by MWVD was over 0.30 kg/kg (d.b.) after $t_{fry} = 20$ min, and the rate of fat absorption increased slowly after 25 min. However, the oil contents of potato chips treated by MWVD for 5, 10, and 15 min were 0.25, 0.22, and 0.17 kg/kg (d.b.), respectively, at the same t_{fry} (=20 min), and then increased only slightly. MWVD time had a significant effect (p-value < 0.05) on the equilibrium oil content of fried potato, whose values ranged from 0.19 to 0.38 kg/kg in d.b. (Song et al. 2007a). A possible reason for the reduction in oil content during VF due to MWVD is the compactness of the material matrix or an increase in the solid content.

13.3.4.3 Effect on Color

For the same t_{fry}, the longer the MWVD pretreatment, the lower is the lightness of potato chips, implying that the product gets darker, which is undesirable for potato chips. The lightness of fried potato chips processed by MWVD was significantly lower (p-value < 0.05) than the one associated with the untreated material (Song et al. 2007a).

During VF, redness (a^* values) of samples submitted to MWVD for different times (0–15 min) clearly increased compared to those of nondried samples. This indicates that browning reactions took place rapidly during the end stages of VF. It was noted that the MWVD duration had an influence on the redness, i.e., the longer the MWVD time, the more intense was the redness of the final product. The yellowness, b^*, of potato chips increased during VF, and it was also affected by MWVD. The potato chips treated by MWVD showed, during VF, an obvious increase in b^* in all cases compared to the control group (no MWVD pretreatment). This effect became more pronounced as the MWVD time increased, with b^* reaching values from 10 to 35 during VF (Song et al. 2007a), whereas, for the control groups, b^* values ranged from 5 to 10. It can therefore be concluded that the MWVD had a negative effect on the color of potato chips, increasing both redness and yellowness.

13.3.4.4 Effect on Texture

The textural properties of potato chips are measured in terms of the breaking force. A texture analyzer was used for breaking force determination. The material was placed over the end of a hollow cylinder. A stainless steel ball probe (P/0.25 s), moving at a speed of 5 mm/s over a distance of 5.0 mm, was used to break the chip. The numerical results are expressed in grams. This parameter is an indicator of the extent of crispness that has occurred during VF, with lower breaking force corresponding to higher crispiness (Fan et al. 2005a). The statistical analysis showed that the breaking force did not depend on t_{fry} (p-value < 0.05), but it was a function of the MWVD duration. For the same t_{fry}, the breaking force of potato chips decreased for low MWVD times and it then increased for longer MWVD times.

13.3.4.5 Optimal Combined Frying Conditions

In order to investigate the effects of MWVD, T_{fry}, and t_{fry}, on the quality of vacuum-fried potato chips, the well known response surface methodology (RSMt) was employed to optimize the processing conditions. A five-level, three-variable design was adopted. The three independent variables were MWVD time ($t_{pre\text{-}dry}$), T_{fry}, and t_{fry}. Five levels of each independent variable were chosen (−1.682, −1, 0, +1, and +1.682). A total of 14 level combinations were generated for the three independent variables. In this optimization process, the center point was specified by the following chosen values: $t_{pre\text{-}dry} = 6\,min$, $T_{fry} = 100°C$, $t_{fry} = 20\,min$. This point was replicated six times to evaluate the experimental error. Three response variables were selected to quantify the potato chip quality after VF: the final moisture content (m_f), the final fat content ($C_{f\text{-}fat}$), and the breaking force of the product. A summary of this experimental design is shown in Table 13.4.

RSMt results showed that m_f, $C_{f\text{-}fat}$, and the breaking force of potato chips were significantly (p-value < 0.05) affected by all the three independent variables considered. As one can see in Table 13.4, during VF, both m_f and $C_{f\text{-}fat}$ decreased as $t_{pre\text{-}dry}$ increased; m_f decreased with increasing T_{fry} and t_{fry}, while $C_{f\text{-}fat}$ increased. A good combination of frying conditions is needed to produce high quality fried products. Based on the adopted response variables for quality assessment, the optimum conditions were $t_{pre\text{-}dry} = 8$–$9\,min$, $T_{fry} = 108°C$–$110°C$, and $t_{fry} = 20$–$21\,min$ (Song et al. 2007b).

TABLE 13.4
Experimental Design for RSMt Analyses and Results

Run	Coded Variables $t_{pre\text{-}dry}$	T_{fry}	t_{fry}	m_f (wt% in w.b.)	$C_{f\text{-}fat}$ (wt% in w.b.)	Breaking Force (g)
1	−1	−1	−1	27.52	23.12	1337.2
2	−1	−1	1	9.08	28.69	974.6
3	−1	1	−1	14.55	27.82	886.7
4	−1	1	1	8.13	32.56	748.9
5	1	−1	−1	13.95	17.04	1040.3
6	1	−1	1	8.57	20.97	813.1
7	1	1	−1	3.28	20.42	719.3
8	1	1	1	2.97	24.98	656.8
9	−1.682	0	0	13.02	27.58	1007.8
10	1.682	0	0	6.87	19.01	903.3
11	0	−1.682	0	16.08	20.85	1153.4
12	0	1.682	0	6.03	24.92	488.5
13	0	0	−1.682	17.69	20.39	1308
14	0	0	1.682	5.52	24.32	517.3
15	0	0	0	7.9	22.83	759.9
16	0	0	0	8.01	23.94	682
17	0	0	0	7.73	21.27	715
18	0	0	0	7.86	21.88	735.2
19	0	0	0	8.13	22.56	636
20	0	0	0	8.05	22.98	652.7

13.3.5 Storage Stability of Vacuum-Fried Products

Vacuum-fried products have higher fat content in comparison with fat-free dried foods that may reduce their shelf life because of lipid oxidation. In order to study the stability of the vacuum-fried products during storage, it is important to know their water sorption isotherms and their glass transition temperature, T_{glass}. Material variety and processing method affect both structural and compositional characteristics of the product, which could imply modifications in the equilibrium moisture content and water activity relationships. Several mathematical models, which may be classified as theoretical, semiempirical, or empirical, have been proposed to describe the sorption behavior of various types of foods. Each of these models has had some degree of success in reproducing equilibrium moisture content values depending on the water activity range and the type of food material (Alvarez-Reyes et al. 2004, Silva and Souza 2004, Yu and Wang 2006).

In the glassy state, some reactions that depend on molecular diffusion, such as some chemical and enzymatic changes, may be prevented (Gould and Christian 1988), although this behavior does not seem to be generalized for all the chemical changes. From this point of view, state diagrams showing the relationships between water content and its physical state as a function of temperature, together with sorption isotherms are useful tools in designing storage conditions. Roos (1993) proposed the use of the Gordon–Taylor equation to model the water plasticization effect, and the GAB sorption model to predict the state of the product at a given temperature and with a given moisture content, thus allowing one to estimate its quality and stability. Nuria et al. (2004) studied the water sorption and the plasticization effect in wafers. Considerable work has been carried out to study the T_{glass} of all kinds of foods (Telis and Sobral 2002, Hua et al. 2003, Boonyai et al. 2005). In this section, based on a previous study of vacuum-fried carrot chips (Fan et al. 2005b), the relationships among water activity, a_w, equilibrium moisture content, m_{eq}, T_{glass}, the storage stability of vacuum-fried carrot chips, and the shelf life are discussed.

13.3.5.1 Sorption Isotherms

Water sorption isotherms of carrot chips at different temperatures were determined gravimetrically by exposing samples in chambers containing selected saturated salt solutions, as described by Fan et al. (2005c). The variation of m_{eq} with a_w for the vacuum-fried chips is shown in Figure 13.6. A type III isotherm was found for carrot chips at 10°C, 25°C, and 40°C, and a similar curve shape for a model fried food crust sample at 25°C was obtained by Hickey et al. (2006). The shape of the curve for fried food is typical of adsorbents with a wide range of pore size. The experimental points for vacuum-fried carrots were fitted to the GAB model:

$$m_{eq} = \frac{m_{mon} A_1 A_2 a_w}{(1 - A_2 a_w)(1 - A_2 a_w + A_1 A_2 a_w)} \tag{13.1}$$

There was good agreement between experimental and fitted values for the GAB model over the entire range of water activity and temperature analyzed experimentally (Fan et al. 2007).

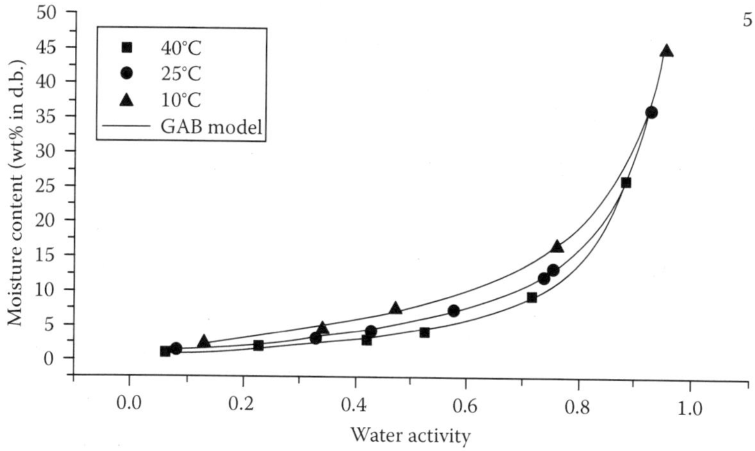

FIGURE 13.6 Water sorption isotherms of vacuum-fried carrot chips at 10°C, 25°C, and 40°C. Experimental points (dots) and fitted GAB model (line). (From Fan, L.P. et al., *Drying Technol.*, 25, 1535, 2007. With permission.)

The fitted parameters of the GAB model (monolayer moisture content m_{non}, constant A_1 and A_2) for carrot chips are reported in Table 13.5. The parameters of GAB model were fitted with nonlinear regression and the model constants were calculated by the Marquardt method. The GAB model introduces a second well differentiated sorption stage for water molecules and an additional energy constant, A_2, compared with the BET model. The m_{mon} values were lower than 10%, which is the maximum value reported for food materials (Tsami et al. 1990). The A_2 values were 0.95, 0.97, and 1.01 for 10°C, 25°C, and 40°C, respectively. The GAB constant A_1, related to the difference of magnitude between the chemical potential of the sorbate molecules in the upper layers and in the monolayer, was shown to decrease with increasing temperature, indicating a decrease in binding energy for the first adsorbed layer with increasing temperature. Such a decrease suggests an increasingly shorter residence time for the adsorbed water molecules in the first layer, with the character of the sorption process becoming less strongly localized (Calzetta et al. 1999).

TABLE 13.5
Parameters of GAB Model Fitted to Experimental Data for Carrot Chips

		Temperature(°C)		
Model	**Parameter**	**10**	**25**	**40**
GAB	m_{mon}	5.53	4.05	2.97
	A_1	3.43	2.72	2.24
	A_2	0.95	0.97	1.01
	Determination coefficient	0.9996	0.9994	0.9980

13.3.5.2 Glass Transition Temperature

Differential scanning calorimetry (DSC) was used to determine T_{glass}. A fixed amount (10 mg) of carrot chips with different moisture contents was placed in stainless steel aluminum DSC pans (model 822e, Mettler Toledo®*). The samples were scanned from −50°C to 50°C at a heating and cooling rate of 5°C/min. The midpoint of the glass transition was considered as the characteristic temperature of the transition. The constant of the Gordon–Taylor model, shown in Equation 13.2, was correlated using nonlinear regression with the Marquardt iteration method, and the value obtained was $A_3 = 9.55$. The glass transition temperature of anhydrous solid carrot chips ($T_{glass}|_{ss}$), obtained according the fitting of the Gordon–Taylor model to the data, was 40.28°C.

$$T_{glass} = \frac{T_{glass}\big|_{ss} + A_3 m\, T_{glass}\big|_{w}}{\left(1 + A_3 m\right)}, \tag{13.2}$$

As evidenced in the research of Fan et al. (2007), a pronounced drop in T_{glass} was observed as the moisture content (m) increased, which is a result of water plasticizing the amorphous structure.

The T_{glass} of carrot chips is related to m, and to the relationship between m_{eq} and a_w. This dependency can be used to obtain the critical moisture content ($m_{c\text{-glass}}$) and the critical water activity ($a_w|_{c\text{-glass}}$) at which the glass transition occurs.

At a fixed storage temperature, vacuum-fried carrot chips should lose crisp and the crystallization of amorphous compounds can take place if $m > m_{c\text{-glass}}$ or, in other words, if $a_w > a_w|_{c\text{-glass}}$. At 25°C, $m_{c\text{-glass}} = 0.99\%$ and $a_w|_{c\text{-glass}} = 0.095$, while at 10°C, $m_{c\text{-glass}} = 2.14\%$ and $a_w|_{c\text{-glass}} = 0.14$ (Fan et al. 2007). It should be noted that $m_{c\text{-glass}}$ values are higher than m_{mon} ones (see Table 13.5) at both temperatures. Thus, the carrot chip quality change depends not only on the monolayer of water, adhered to chip surface, but it is also affected by the water plasticization and the storage temperature.

13.3.5.3 Effects of the Storage Conditions

To evaluate the effect of storage conditions on the quality of the vacuum-fried carrot chips, apart from a_w and m, the following variables were measured: breaking force, fat content, C_{fat}, β-carotenoid, and ascorbic acid contents. The initial moisture content and water activity of the carrot chips were $m_0 = 2.60\%$ and $a_w = 0.27$, respectively. Its T_{glass} was about 4°C. For a storage temperature above 4°C, some reactions that depend on molecular diffusion, such as nonenzymatic browning, oxidation, and enzymatic changes, may be improved as a result of enhanced diffusion, and the system would be in a sate of instability.

Moisture content, fat content, water activity, and breaking force of the carrot chips during storage are reported in Table 13.6 (Fan et al. 2007). Results indicate that the storage temperature and time, T_{stor} and t_{stor}, and the interaction between them, $T_{stor} \times t_{stor}$, significantly affected the m and a_w values of the carrot chips. Moreover, T_{stor} and t_{stor}, significantly affected C_{fat}, but no interaction effect was observed. There were no significant differences in m and a_w as function of time at 0°C. However, these

* Registered trademark of Mettler-Toledo, Inc., Columbus, OH.

TABLE 13.6
Moisture Content, Water Activity, and Fat Content of Vacuum-Fried Carrot Chips Stored at Different Conditions

T_{stor} (°C)	t_{stor} (min)	m (wt% in w.b.)	C_{fat} (wt%)	a_w (—)	Breaking Force (g)
0	0	2.60 ± 0.02^{m}	28.32 ± 0.24^{ab}	0.27 ± 0.01^{kj}	426.9 ± 29.3^{h}
0	1	2.59 ± 0.03^{m}	28.31 ± 0.16^{ab}	0.26 ± 0.01^{k}	431.4 ± 21.8^{h}
0	2	2.60 ± 0.12^{m}	28.36 ± 0.08^{a}	0.27 ± 0.02^{kj}	439.7 ± 18.6^{h}
0	3	2.59 ± 0.06^{m}	28.35 ± 0.12^{ab}	0.26 ± 0.01^{k}	442.8 ± 20.1^{hg}
0	4	2.61 ± 0.07^{m}	28.23 ± 0.27^{abc}	0.26 ± 0.02^{k}	447.3 ± 14.3^{hg}
0	5	2.62 ± 0.03^{m}	28.29 ± 0.31^{abc}	0.28 ± 0.01^{j}	458.2 ± 17.5^{efhg}
0	6	2.63 ± 0.11^{m}	28.25 ± 0.21^{abc}	0.28 ± 0.02^{j}	460.4 ± 12.4^{efhg}
10	0	2.60 ± 0.02^{m}	28.32 ± 0.24^{ab}	0.27 ± 0.01^{kj}	426.9 ± 29.3^{h}
10	1	2.77 ± 0.05^{l}	28.26 ± 0.33^{abc}	0.28 ± 0.01^{j}	441.2 ± 15.2^{h}
10	2	3.09 ± 0.02^{k}	28.19 ± 0.19^{abc}	0.30 ± 0.03^{i}	458.7 ± 24.7^{efhg}
10	3	3.21 ± 0.13^{i}	28.11 ± 0.15^{abc}	0.31 ± 0.01^{hi}	475.5 ± 16.9^{defg}
10	4	3.37 ± 0.24^{h}	28.07 ± 0.17^{abcd}	0.33 ± 0.02^{g}	485.4 ± 22.3^{cdf}
10	5	3.62 ± 0.09^{g}	27.94 ± 0.26^{abcde}	0.35 ± 0.01^{f}	496.6 ± 21.8^{cd}
10	6	3.84 ± 0.12^{e}	27.87 ± 0.2^{bcde}	0.37 ± 0.03^{e}	504.3 ± 19.6^{cd}
25	0	2.60 ± 0.02^{m}	28.32 ± 0.24^{ab}	0.27 ± 0.01^{kj}	426.9 ± 29.3^{h}
25	1	3.15 ± 0.06^{j}	28.14 ± 0.33^{abc}	0.32 ± 0.02^{gh}	450.6 ± 20.5^{fhg}
25	2	3.77 ± 0.04^{f}	27.83 ± 0.18^{cdef}	0.37 ± 0.01^{e}	483.4 ± 24.8^{cdef}
25	3	4.16 ± 0.14^{d}	27.65 ± 0.14^{def}	0.40 ± 0.03^{d}	515.2 ± 13.7^{c}
25	4	4.85 ± 0.12^{c}	27.59 ± 0.27^{efg}	0.45 ± 0.02^{c}	549.3 ± 17.6^{b}
25	5	5.31 ± 0.08^{b}	27.41 ± 0.35^{fg}	0.48 ± 0.04^{b}	570.1 ± 28.1^{b}
25	6	5.89 ± 0.06^{a}	27.22 ± 0.19^{g}	0.52 ± 0.01^{a}	603.9 ± 26.4^{a}
p-Value					
T_{stor}		<0.0001	0.0002	<0.0001	<0.0001
t_{stor}		<0.0001	<0.0001	<0.0001	<0.0001
$T_{stor} \times t_{stor}$		<0.0001	0.0836	<0.0001	<0.0001

Note: The same letter within a column means there is no significant difference (p-value > 0.05).

parameters significantly increased with time at 10°C and 25°C. This outcome may be a consequence of the loose structure of carrot chips, after VF, with many pores absorbing water easily. In other words, the system was probably under an unstable state above the T_{glass}.

Considering that initial fat content was 28.32%, there was no significant difference in the fat content as a function of time at 0°C. However, this parameter decreased significantly with time at 10°C and 25°C. This phenomenon may be related to the increased moisture content of the carrot chips.

The value of the breaking force reflects the crispness of carrot chips. A crispy chip would be prone to puncture, and its breaking force value is therefore lower. There are significant differences in the breaking force as a function of T_{stor}, t_{stor}, and

$T_{stor} \times t_{stor}$. The breaking force increases with increasing T_{stor} and t_{stor} (Table 13.6). No significant differences were observed in breaking force after storing 6 months at 0°C, 2 months at 10°C or a month at 25°C compared with the initial product.

Fan et al. (2007) evaluated the changes of β-carotenoid and ascorbic acid contents of carrot chips as a function of the storage time at different temperatures. In both cases, a decrease in β-carotenoid and ascorbic acid contents with increasing storage temperature and time can be observed. The retentions of β-carotenoid and ascorbic acid were 90% and 82%, respectively, over 6 months at 0°C. This high retention of both components in carrot chips may be attributed to beneficial processing and storage condition. Blanching pretreatment before VF can efficiently inactivate the peroxidase and lipoxygenase, which catalyze the destruction of carotenoids and lipids during storage. In addition, during storage, the low temperature (below T_{glass}) may result in a drastic decrease in the diffusion coefficient, and some reactions that depend on molecular diffusion may be slowed. The final retentions of β-carotenoid and ascorbic acid were 81% and 63%, respectively, at a storage temperature of 10°C, values which drop to 74% and 35% at 25°C (Fan et al. 2007).

13.4 CONCLUDING REMARKS

VF is a relatively new technology that is a viable option for production of snacks from fruits and vegetables, with lower fat content and several desirable qualities. Prior to VF, some physical pretreatments, such as blanching, osmotic pretreatment, freezing and predrying etc., can be adopted that affect the drying kinetics, yield, moisture content, fat content, and fat distribution of vacuum-fried foods.

During VF, frying temperature is one of the key factors affecting the dehydration rate, flavor, color, and nutritional value. Increasing the frying temperature at the same vacuum level decreases the frying time of fried flake foods. However, differences in the material to be fried should result in different effects of frying temperature on the fat content of vacuum-fried foods. The choice of vacuum degree affects the frying temperature and therefore the quality of vacuum-fried foods, such as breaking force, the rate of fat absorption, and drying rate.

During co-dehydration by vacuum microwave and VF, vacuum microwave drying had a significant effect on moisture and oil contents, as well as color parameters and structure of potato chips. Vacuum microwave drying significantly decreased the oil and moisture contents of vacuum-fried potato chips. Vacuum microwave drying had a negative effect on color of potato chips; it decreases the L value of potato chips and increases Hunter a^* and b^* values.

REFERENCES

Alvarez-Reyes, A., Acosta-Esquijarosa, J., Sordo-Martinez, L., and Conoepcion-Martinez, F. 2004. Experimental determination and modeling of sorption isotherms of *Erythrina fusca* Lour bark. *Drying Technol.* 22:1755–1765.

Araszkiewicz, M., Koziol, A., Lupinska, A., and Lupinski, M. 2007. IR technique for studies of microwave assisted drying. *Drying Technol.* 25:569–574.

Boonyai, P., Bhandari, B., and Howes, T. 2005. Measurement of glass-rubber transition temperature of skim milk powder by static mechanical test. *Drying Technol.* 23:1499–1514.

Calzetta Resio, A., Aguerre, R.J., and Suarez, C. 1999. Analysis of the sorptional characteristics of amaranth starch. *J. Food Eng.* 42:51–57.
Chiu, and Yao-Jui. 1993. Vibrational food soaking device for a vacuum-frying machine. U.S. Patent 5239915, filed January 19, 1993, and issued August 31, 1993.
Drouzas, A.E. and Schubert, H. 1996. Microwave application in vacuum drying of fruits. *J. Food Eng.* 28:203–209.
Duan, Z.H., Zhang, M., Hu, Q.G., and Sun, J.C. 2005. Characteristics of microwave drying of bighead carp. *Drying Technol.* 23:637–643.
Fan, L.P. and Zhang M. 2004. Optimization of vacuum frying dehydration of carrot chips. *J. Wuxi Univ. of Light Indust.* 23:40–44.
Fan, L.P., Zhang M., and Mujumdar, A.S. 2005a. Vacuum frying of carrot chips. *Drying Technol.* 23:645–656.
Fan, L.P., Zhang, M., Xiao, G.N., Sun, J.C., and Tao, Q. 2005b. The optimization of vacuum frying to dehydrate carrot chips. *Int. J. Food Sci. Technol.* 40:911–919.
Fan, L.P., Zhang, M., Tao, Q., and Xiao, G.N. 2005c. Sorption isotherms of vacuum-fried carrot chips. *Drying Technol.* 23:1569–1579.
Fan, L.P., Zhang M., and Mujumdar, A.S. 2006. Effect of various pretreatments on the quality of vacuum-fried carrot chips. *Drying Technol.* 24:1481–1486.
Fan, L.P., Zhang M., and Mujumdar, A.S. 2007. Storage stability of carrot chips. *Drying Technol.* 25:1533–1539.
Farrel, G., McMinn, W.A.M., and Magee, T.R.A. 2005. Microwave-vacuum drying kinetics of pharmaceutical powders. *Drying Technol.* 23:2131–2146.
Garayo, J. and Moreira R. 2002. Vacuum frying of potato chips. *J. Food Eng.* 55:181–191.
Gould, G.W. and Christian, J.H.B. 1988. Characterization of the state of water in foods biological aspects. In *Food Preservation by Moisture Control*, ed. C. C. Seow, pp. 43–56. London, U.K.: Elsevier Applied Science.
Granda, C., Moreira, R.G., and Tichy, S.E. 2004. Reduction of acrylamide formation in potato chips by low-temperature vacuum frying. *J. Food Sci.* 69:405–411.
Hashiguchi, Myojin, T., Iwase, N., Hayashi, N., and Tetsuya. 2000, Vacuum-heat processing method. U.S. Patent 6068872, filed October 5, 1999, and issued May 30, 2000.
Hickey H., MacMillan, B., Newling, B., Ramesh, M., Eijck, P.V., and Balcom, B. 2006. Magnetic resonance relaxation measurements to determine oil and water content in fried foods. *Food REV INT.* 39:612–618.
Hu, Q.G., Zhang, M., Mujumdar, A.S., Xiao, G.N., and Sun, J.C. 2006 Drying of edamames by hot air and vacuum microwave combination. *J. Food Eng.* 77:977–982.
Hua, Z.Z., Li, B.G., and Liu, Z.J. 2003. Freeze-drying of liposomes with cryoprotectants and its effect on retention rate of encapsulated ftorafur and vitamin A. *Drying Technol.* 21:1491–1505.
Imai and Kunio, 1989, Method of and apparatus for processing vacuum fry. U.S. Patent 4828859, filed August 5, 1986, and issued May 9, 1989.
Islam, M.R., Ho, J.C., and Mujumdar, A.S. 2003. Convective drying with time-varying heat input: Simulation results. *Drying Technol.* 21:1333–1356.
Kaensup, W., Chutima, S., and Wongwises, S. 2002. Experimental study on drying of chilli in a combined microwave–vacuum-rotary drum dryer. *Drying Technol.* 20:2067–2079.
Kozempel, M.F., Tomasula, P.M., and Craig, J.C. Jr 1991. Correlation of moisture and oil concentration in French fries. *LWT-Food Sci. Technol.* 24:445–448.
Krokida, M.K., Tsami, E., and Maroulis, Z.B. 1998. Kinetics on color changes during drying of some fruits and vegetables. *Drying Technol.* 16:667–685.
Mariscal, M. and Bouchon, P. 2008. Comparison between atmospheric and vacuum frying of apple slices. *Food Chem.* 107:1561–1569.
Mittelman, N., Mizrahi, S., and Berk, Z. 1984. Heat and mass transfer in frying. In *Engineering and Food*, ed. B.M. Mckenna, pp. 109–116. London, U.K.: Elsevier Applied Science Publishers.

Moreira, R.G. and Bakker-Arkema, F.W. 1989. Moisture desorption model for non-parerl almonds. *J. Agric. Eng. Res.* 42:123–133.

Moreira, R.G. and Barrufet, M.A. 1995. Spatial distribution of oil after deep-fat frying from a stochastic model. *J. Food Eng.* 27:205–220.

Moreira, R.G., Castell-Perez, M.E., and Barrufet, M.A. 1999. *Deep Fat Frying: Fundamentals and Applications*. Gaithersburg, MD: Aspen Publishers.

Nuria, M.N., Gemma, M., Pau, T., and Amparo, C. 2004. Water sorption and the plasticization effect in wafers. *Int. J. Food Sci. Technol.* 39:555–562.

Rice, P. and Gamble, M.H. 1989. Modelling moisture loss during potato slice frying. *Int. J. Food Sci. Technol.* 24:183–187.

Roos, Y.H. 1993. Water activity and physical state effects on amorphous food stability. *J. Food Process. Preserv.* 16:433–447.

Sakuma, K. and KenJi. 1991. Process for producing fried food. U.S. Patent 4985266, filed August 30, 1989, and issued January 15, 1991.

Shi, X.Q., Deng, J.X., Liu, J.H., and Zhang Y.B. 2001. Optimal technology of vacuum frying for processing of colocasia esculenta schott's son-taroes. *J. Chin. Cereals Oil Assoc.* 16:54–59.

Shyu, S., Hau, L., and Hwang, S. 1998a. Effect of vacuum frying on the oxidative stability of oils. *J. Am. Oil Chem. Soc.* 75:1393–1398.

Shyu, S., Hau, L., and Hwang, L.S. 1998b. Influence of vacuum frying time on the chemical constituents of fried carrot chips. *Food Science-Taiwan* 25:737–747.

Shyu, S., Hau, L., and Hwang, L.S. 1999. Effect of vacuum frying temperature on the chemical components of fried carrot chips. *Food Sci. Agric. Chem.*1:61–66.

Shyu, S. and Hwang, L.S. 2001. Effects of processing condition on the quality of vacuum fried apple chips. *Food Res. Int.* 34:133–142.

Shyu, S.L., Hau, L.B., and Hwang, L.S. 2005. Effects of processing conditions on the quality of vacuum-fried carrot chips. *J. Sci. Food Agric.* 85:1903–1908.

Silva, M.A. and Souza, F.V. 2004. Drying behavior of binary mixtures of solids. *Drying Technol.* 22:165–177.

Silva, P. F. and Moreira, R.G. 2008.Vacuum frying of high-quality fruit and vegetable-based snacks, *LWT-Food Sci. Technol.*, 41:1758–1767.

Song, X.J., Zhang, M., and Mujumdar, A.S. 2007a. Effect of vacuum-microwave pre-drying on quality of vacuum-fried potato chips. *Drying Technol.* 25:2021–2026.

Song, X.J., Zhang, M., and Mujumdar, A.S. 2007b. Optimization of vacuum microwave pre-drying and vacuum frying conditions to produce fried potato chips. *Drying Technol.* 25:2027–2034.

Sulaeman, A., Keeler, L., Taylor, S.A., Giraud, D.W., and Driskell, J.A. 2001. Carotenoid content and physicochemical and sensory characteristics of carrot chips deep fried in different oils at several temperatures. *J. Food Sci.* 66:1257–1264.

Telis, V.R.N. and Sobral, P.J.A. 2002. Glass transitions for freeze-dried and air-dried tomato. *Food Res. Int.* 35:435–443.

Tsami, E., Marinos-Kouris, D., and Maroulis, Z.B. 1990. Water sorption isotherms of raisins, currants, figs, prunes and apricots. *J. Food Sci.* 55:1594–1625.

Xu, Y.Y., Min, Z., and Mujumdar, A.S. 2004. Studies on hot air and microwave vacuum drying of wild cabbage. *Drying Technol.* 22:2201–2209.

Yaghmaee, P. and Durance, T. 2007. Efficacy of vacuum microwave drying in microbial decontamination of dried vegetables. *Drying Technol.* 25:1099–1104.

Yamsaengsung, R. and Rungsee, C. 2003. Vacuum frying of fruits and vegetables. Paper presented at the 13th Annual Conference of Thai Chemical Engineering and Applied Chemistry, Thailand.

Yang, C.-S. 1989. Vacuum frying and oil separation device. U.S. patent 4873920, October 17, 1989.

Yang, R.J. 1997. Vacuum frying technology. In *Novel Technology for Modern Food Engineer*, ed. Gao, F.C., pp. 165–166. Beijing, China: Chinese Light Industry Publishers.

Yongsawatdigul, J. and Gunasekaran, S. 1996. Pulsed microwave-vacuum drying of cranberries., Part II. Quality evaluation. *J. Food Process. Preserv.* 20:145–156.

Yu, Y. and Wang, J. 2006. Modeling equilibrium moisture content of gamma-ray irradiated rough rice. *Drying Technol.* 24:671–676.

Zhang, M., Tang, J., Mujumdar, A.S., and Wang, S. 2006. Trends in microwave related drying of fruits and vegetables. *Trends Food Sci. Technol.* 17:524–534.

14 Aseptic Packaging of Food—Basic Principles and New Developments Concerning Decontamination Methods for Packaging Materials

Peter Muranyi, Joachim Wunderlich, and Oliver Franken

CONTENTS

14.1 INTRODUCTION

Packaging is a very important factor in the food and beverage industry because its function is to inhibit chemical, physical, and microbial interactions between the environment and the product. This function of packaging must guarantee high product quality and safety level for an extended period. To ensure that there is no microbial contamination of the filled product, not only the food itself but also its packaging or the packaging material has to be subjected to a decontamination process. The aim of the packaging decontamination process is to efficiently inactivate pathogens and food-spoiling microorganisms to prevent health risks and to give a high-quality product with a long shelf life. Nowadays, ever higher requirements are being put on efficient packaging decontamination due to new developments and trends in the food and packaging areas—for example, beverages with neutral pH values, innovative packaging systems made of new materials and having complex geometries, and market demands for products with extended shelf lives.

14.2 REQUIREMENTS ON THE DECONTAMINATION OF PACKAGING MATERIAL

To guarantee high-quality products with a long shelf life, the food industry puts high hygienic requirements on packaging materials and filling processes. The requirements on the efficiency of the microbial inactivation process depend on the initial microbial count of the packaging material and the aspired microbial state of the final product—sterile or just a partial reduction of the microorganisms. To select appropriate decontamination methods, various product and process-related aspects have to be considered, for example, the product properties, the microbiological state of the packaging material, the target shelf life, and the distribution channel.

14.2.1 Microbiological State of Packaging Materials

An essential point for choosing a decontamination method is knowledge about the initial microbial state of the packaging material. Materials such as glass, tin foil, and plastic films are sterile immediately after the manufacturing process. For example, polyolefins reach temperatures up to 200°C during the extrusion process, and this is high enough to inactivate most microorganisms within short times. The microbial count on the surface increases by reinfection due to handling, airborne

TABLE 14.1
Microbial Count on the Surface of Different Packaging Materials

Packaging Material	Mean Microbial Count on the Surface (CFU/dm^2)
PE, PP, PVC films	~2
Laminated films with PVC coating	0.6
Carton/Al/PE-compound	~2
Al-compound	<1
Al-film	0.4–10
Yogurt container, 180 g (PS or PP)	<1
PE bottles	8

Source: Buchner, N., *Packaging of Food* (in German), 1st edn., Springer-Verlag, Berlin, Germany, 1999. With permission.

microorganisms, electrostatic charge, and carryover effects within the filling systems (Buchner 1999, Cerny 1990). Table 14.1 gives an overview of the averaged microbial count on the surface of various packaging materials according to Buchner (1999). In addition, investigations of von Bockelmann and von Bockelmann (1986) have shown that the microbial count on the polyethylene (PE) surface of paperboard-based laminates ranged between 2 and 5 CFU/100 cm^2 and comprised 10.6% yeasts, 20.6% molds, and 68.8% bacteria.

Technical standards and guidelines for the hygienic fabrication of packaging systems exist for some foods and for the maximal allowed microbial load. Examples are the hygienic standards of the International Dairy Federation (IDF) and the Food and Drug Administration (FDA) for disposable packaging for milk and milk products (FDA 1999, IDF 1995). There is also the butter regulation, which considers the microbial state on the surface of butter wrappers according to DIN 10082.

A research group at the Fraunhofer Institute IVV and members of the Industrial Organization for Food Technology and Packaging (IVLV) have been compiling microbiological guideline values for packaging materials for some years. For example, microbiological guideline values for nonabsorbent packaging systems and materials are shown in Table 14.2 (Hennlich 2005).

The risk (RK) of contamination from the packaging material can be estimated using the following equation, which was developed by Cerf and Brissende (1981):

$$\mathrm{RK} = n_0 S \times 10^{-t_\mathrm{d}/\Delta t_\mathrm{r}} \tag{14.1}$$

where

n_0 is the initial number of microorganisms per surface area
S is the surface area
t_d is the disinfection time
Δt_r is the decimal reduction time of the most resistant organism (Holdsworth 1992)

TABLE 14.2
Microbiological Guideline Values for Nonabsorbent Packaging Systems and Materials

Microbiological Guideline Value	Mean Microbial Count on the Surface (CFU/dm²)
Packaging material (lids, films)	
Total bacterial count	<=2
Molds and yeasts	<=1
Enterobacteriaceae	Unverifiable
Packaging system (container)	Volume ≤500 mL/>500 mL
Total bacterial count	≤10/<5
Molds and yeasts	≤1/≤1
Enterobacteriaceae	Unverifiable

Source: From Hennlich, W., Guideline values for food packaging (in German), in *Handbuch Lebensmittelhygiene. Praxisleitfaden mit wissenschaftlichen Grundlagen*, Fehlhaber, K., Kleer, J., and Kley, F., Eds., Behr's Verlag, Hamburg, Germany, 2005, pp. 36–46. With permission.

The equation shows that the risk of contamination is proportional to the initial contamination level and the surface area and depends on the efficiency of the lethal treatment (von Bockelmann and von Bockelmann 1986).

14.2.2 Product-Specific Requirements on the Decontamination of Packaging Materials

The requirements for the sterilization process also depend on the specific properties of the food and its processing. Depending on the composition of the food, microbial growth can be slowed down drastically or limited. The availability of various chemical and physical inhibiting factors such as storage temperature, water activity (a_w-value), and pH value play an essential role, and the systematic application of cumulative inhibiting factors in food industry via drying, acidification, or preservatives is called a hurdle concept (Krämer 2002).

In the case of dry foods (e.g., noodles, common cereal crops), with a low a_w-value (<0.6), the microbial count of the packaging material is not an issue because microorganisms cannot propagate. According to the FDA, an a_w-value below 0.85 is the lower limit for bacterial growth. Yeasts and fungi can multiply up to a water activity of 0.6 but they can be inactivated with a mild pasteurization process. In acidic foods or in foods containing preservatives, there is only limited microbial growth. A pH value below 4.5 in foods prevents germination of bacterial endospores and restricts progeny to yeasts, fungi, and some adapted acid-forming bacterial strains, which represent only 10% of the germs on the surface (Buchner 1999). For products with a high initial microbial load, such as raw meat, the microbial load of the packaging material is negligible. Taking into account the mentioned conditions, it suffices if the inactivation process is targeted at the potential microflora that can propagate under these conditions. Distribution within the cooling chain and limited shelf lives also reduce the requirements and can make sterilization of packaging material unnecessary.

14.2.3 Process-Specific Aspects for the Decontamination of Packaging Materials

Generally, two different concepts for the decontamination of packaging material can be distinguished: complete sterilization and partial reduction of the microbial count. A partial reduction of the microbial count simply decreases the amount of the germs on the packaging surface. This leads to an extended shelf life and an improved hygienic state of the filled product. Various physical methods (e.g., infrared or ultraviolet radiation, dry or wet heat) and chemical methods (e.g., peracetic acid, hydrogen peroxide [H_2O_2]) can be used for this, and these methods are often used in combination. Partial reduction is normally used for the so-called "acid food" products with a pH value below 4.5, at which germination of bacterial endospores is not expected, and also for foods with a limited shelf life or products within the cooling chain.

The other option is complete sterilization of the material, which is achieved, for instance, by the aseptic packaging of foods. In practice, there are various interpretations of the term "sterilization." According to Wallhäußer (1995), sterility is the removal of biological entities having the ability to reproduce or transfer genetic material. This definition corresponds to the microbiological notion of sterility. In addition to this, the food industry uses the term "commercial sterility." According to the FDA, commercial sterility for a packaging material means the conditions achieved by application of heat, chemical sterilants, or other appropriate treatment that renders the equipment and containers free of microorganisms of public health significance as well as those of nonhealth significance capable of reproducing under normal nonrefrigerated conditions of storage and distribution. This is regulated by the FDA within their Code of Federal Regulations, Title 21, Part 113.3 (FDA 2007).

According to a Swiss guideline of 1987 for commercial sterile products, pathogens may not be detectable in 1 g of food and the microbial count (vegetative and spores) may not increase by two logs after incubation of the packaging for 12 and 21 days, respectively (Wallhäußer 1995).

As inactivation of microorganisms corresponds to a logarithmic order and runs asymptotically to zero, absolute sterility can only be approached, but never reached. From a statistical point of view, the appearance of a microbe-contaminated packaging unit is just a matter of enough sampling. For the food industry, it is simply the spoiling rate due to contamination that is important and this has to be defined as an acceptable maximum (Holdsworth 1992).

Both the product and its container have to be sterilized to achieve sterile conditions (von Bockelmann and von Bockelmann 1986).

The following commercial processes provide sterile conditions for packaged foods: hot filling of acid foods, subsequent heating of a filled product (autoclave procedure), and aseptic packaging processes.

The product temperature during hot filling is in the 70°C–90°C range (pasteurization), which inactivates the vegetative flora of the food and its packaging or closures respectively. To avoid thermal material damage, the hot product is chilled directly in the packaging. As this method only inactivates vegetative bacteria, yeasts, and molds, it is only suitable for acid food products where spores cannot germinate Furthermore, hot filling is only suitable for temperature-resistant materials such as glass, tin foil, carton compounds, and some plastics, for example, polypropylene (PP) (Buchner 1999).

14.3 ASEPTIC PACKAGING OF FOOD

Aseptic packaging refers to a process whereby sterile cold products are filled into presterilized packaging under sterile conditions to prevent microbial recontamination. This results in products that, even without cooling, are microbiologically stable for a long time. The aseptic packaging and filling of products is becoming an ever more popular technology, and is already established in many branches of the food industry. The main advantage of such processes is the separate sterilization of the food utilizing short heating times (heat exchanger), which improves the product quality and saves energy. One example of this is the ultrahigh temperature (UHT) heating process. In this way, heat-sensitive components of the food such as vitamins (e.g., thiamine) are retained and a high heat recovery factor is achieved. Other industrial processes such as hot filling and the subsequent heating of the filled product are mostly energetically unfavorable or cause heat damage to the product. Aseptic packaging is therefore an economical process that contributes to preservation of the sensory and nutri-physiological properties of the food. Due to the separate sterilization of the product and its packaging and the filling under sterile conditions, the product can be filled cold which allows thermosensitive materials such as low-cost plastics (e.g., polyethylene terephthalate [PET]) to be used as packaging materials. Further advantages of aseptically filled products are their stability under room temperature conditions, lower distribution and storage costs, lower process energy requirements, and absence of preservatives (Ansari and Datta 2003). The disadvantage is that aseptic packaging is a sophisticated and complex technology, and there are also restrictions for some products. Aseptic packaging is mainly used for less acidic foods with a pH value above 4.5 and a water activity greater than 0.85 (e.g., milk) and also for homogeneous products (e.g., milk, juice, or desserts) and liquid foods with small particles (e.g., rice pudding or sauces). Sometimes aseptic packaging technology is applied for the filling of nonsterile products to extend the shelf life by preventing reinfection with molds or yeasts. In this way, there is no need for additional preserving agents or heat treatment (Reuter 1986). Examples of such nonsterile foods are fermented fresh products (e.g., milk products, yoghurts), which are distributed within the cooling chain and which reduce the microbial growth due to their own microflora.

14.4 REQUIRED MICROBIAL INACTIVATION FOR THE PACKAGING STERILIZATION

The basic requirement for the aseptic packaging of foods is a sterile packaging material. The required reduction depends on various parameters such as the product properties, its shelf life, the initial microbial count, and the storage conditions.

The main product category for aseptic packaging contains sterile, neutral, or slightly sour products that can be stored at room temperature for several months. Here, Reuter (1986) recommends a microbial reduction of six logs for an appropriate test strain (e.g., *Bacillus subtilis*). For fillings where growth of *Clostridium botulinum* is possible, the packaging sterilization should be designed to realize the 12D concept, which means a reduction of this microorganism by 12 logs (Reuter 1986).

In the United States, the National Food Processors Association (NFPA) and the FDA have laid down the requirements for aseptic packaging in law (Buchner 1999). The FDA stipulates commercial sterility for products with a pH value above 4.6 (so-called low acidic food) (Ansari and Datta 2003). This is because in these foods there is theoretically a risk that pathogens such as *Salmonella* can propagate or spores of *C. botulinum* can germinate. Process authorities such as the NFPA and the National Food Laboratory (NFL) carry out the validation of filling machines. For hydrogen peroxide sterilization they stipulate a reduction of five logs of *B. subtilis* endospores for the packaging material treatment and a reduction of four logs for the machine sterilization (Wilke 1996).

Within the European community, there are no regulations laid down in law on the requirements for packaging sterilization for aseptic packaging. In Germany, the German engineering Federation (Verband deutscher Maschinen und Anlagenbau [VDMA]) has published a guideline (8742) regarding the minimum requirements and basic conditions for aseptic packaging machines for foods (VDMA 2006). According to this guideline, aseptic packaging machines have to inactivate microorganisms, including spores, effectively. For verification, there has to be a reduction of at least four logs for the packaging material and the isolator and five logs at the filler unit. The test strain has to be appropriate for the applied sterilization method. For example, *B. subtilis* SA 22 is recommended as the test strain for hydrogen peroxide technology because the endospores have high resistance to oxidation.

14.5 DECONTAMINATION METHODS FOR PACKAGING MATERIAL IN THE FOOD INDUSTRY

Nowadays, there are many different chemical and physical decontamination techniques available. Some methods such as UV radiation are only appropriate for giving a partial reduction of the microbial count. A sterilization method for packaging materials must meet a number of important criteria: it must be fast and have a reliable microbiocidal effect, it must be compatible with the packaging material and the environment, there must be no or minimal residues, it must be harmless for consumers and employees, and it must be economically efficient (Ansari and Datta 2003, Reuter 1993). Considering the earlier mentioned requirements, however, only a few methods are suitable for sterilization of packaging material in the food industry.

14.5.1 Chemical Methods

14.5.1.1 Hydrogen Peroxide

Hydrogen peroxide is one of the most common chemical agents for the sterilization of packaging material and inactivates a wide array of microorganisms. The effect is attributed to the release of atomic oxygen (nascent oxygen) and the formation of hydroxyl radicals which oxidize essential parts of the cell, such as enzymes and various proteins (Bayliss and Waites 1979, Wallhäußer 1995).

Although the microbiocidal effect of hydrogen peroxide increases with concentration and temperature (Figures 14.1 and 14.2 show the inactivation efficiency of H_2O_2 as a function of temperature and concentration), there are limitations for practical reasons. Normally, concentrations between 30% and 35% are applied in practice because of the easier handling (Kessler 2002). Practical inactivation rates in the range of seconds are reached at temperatures between 65°C and 80°C. This depends mainly on the packaging material being used. The heating of hydrogen peroxide is carried out by infrared radiation (IR) or hot air on the packaging surface and this process removes the excess solution (Cerny 1990).

According to FDA regulations (CFR, Title 21, Part 178, 1005), a maximum concentration of 35% may be used for the treatment of packaging surfaces for food contact applications and the residues in the container are not allowed to exceed 0.5 ppm H_2O_2. According to the German Ordinance on Hazardous Substances (GefStoffV) of 1986, the maximum allowable concentration (MAC) in the working place is restricted to 1 ppm in air (Reuter 1986).

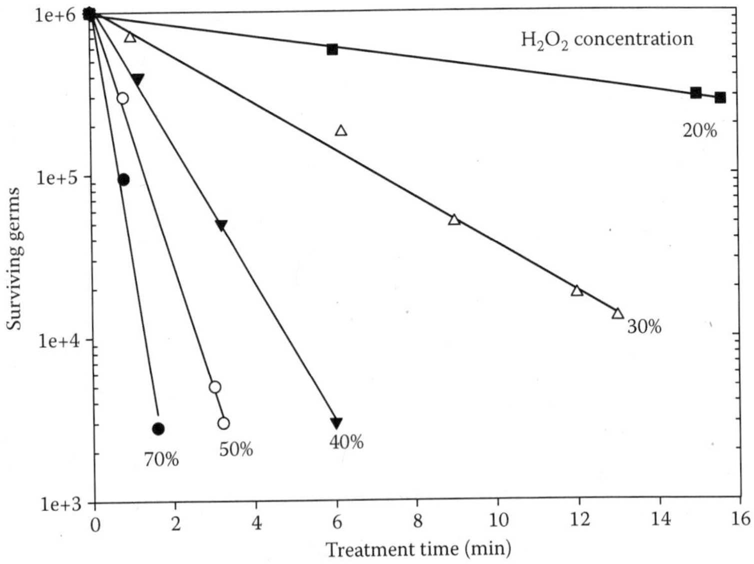

FIGURE 14.1 Influence of different hydrogen peroxide concentrations on the inactivation of *B. subtilis* endospores at a temperature of 23°C. (From Cerny, G., *Verpack.-Rundsch.*, 27(4), 29, 1976. With permission.)

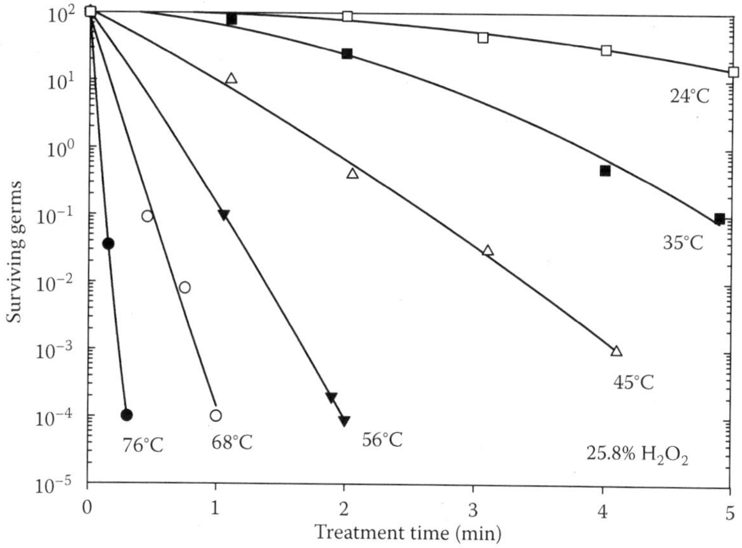

FIGURE 14.2 Influence of the temperature on the inactivation of *B. subtilis var. globigii* endospores exposed to 25.8% hydrogen peroxide. (Data from Toledo, R.T. et al., *Appl. Microbiol.*, 26(4), 595, 1973.)

Hydrogen peroxide is applied to the packaging material surface by various methods: dipping, spraying, or rinsing. The dipping method is suitable for flat packaging materials (plastic films, laminates). In this case, the film is dipped into an immersion bath of hydrogen peroxide and squeeze rolls or air jets remove the excess solution. This process results in a thin film of hydrogen peroxide on the surface, which is dried and activated by sterile, hot air. Because of the hydrophobic surface character of most plastics, wetting agents (e.g., polyoxyethylene sorbitan monooleate) are added to the immersion bath to give homogeneous wetting (Holdsworth 1992). Other methods for applying hydrogen peroxide on the surface are spraying an aerosol and condensation of vapor. Spraying is normally used for preformed containers and depending on the spraying system being used the droplets can vary between 3 and 30 μm (Ansari and Datta 2003). Because of the hydrophobic surface of plastics, the spraying method does not result in a cohesive film. The homogeneous sterilization is due to the subsequent heating process, whereby the hydrogen peroxide evaporates and acts from the gaseous phase. Another method for the treatment of preformed containers is rinsing with an aqueous hydrogen peroxide solution and subsequent drying with hot, sterile air (Ansari and Datta 2003).

14.5.1.2 Peracetic Acid

Peracetic acid is a peroxide of acetic acid. It decomposes on contact with organic materials to form the corresponding acid, oxygen, hydrogen peroxide, and water (Block 2001, Wallhäußer 1995). This agent causes oxidative damage to cell wall proteins and intracellular proteins if the undissociated acid penetrates the cell. Based on

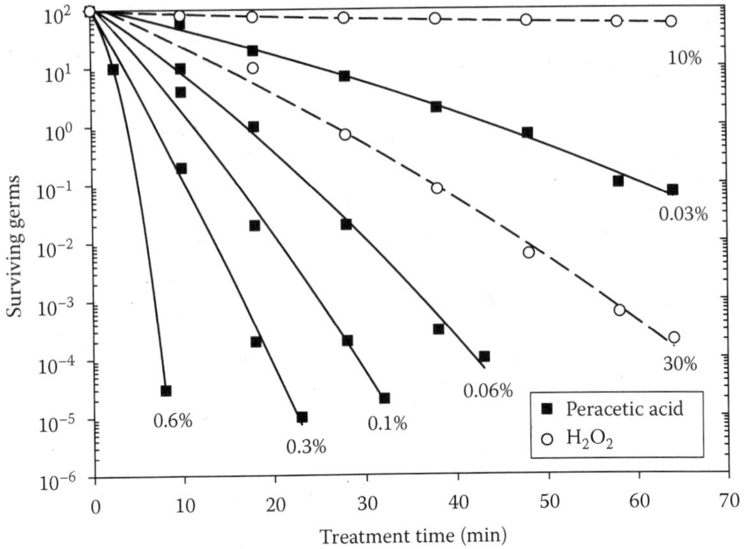

FIGURE 14.3 Inactivation efficiency of peracetic acid and hydrogen peroxide at 20°C to spores of *Clostridium sporogenes* ATCC 19404. (Adapted from Han, B.H., Destruction of microorganisms on solid surfaces (in German), PhD dissertation, Technical University Karlsruhe, Karlsruhe, Germany, 1977.)

this mechanism it inactivates a wide spectrum of microorganisms (bacteria, spores, viruses, and phages), and it is therefore a well-accepted agent for presterilizing aseptic machines and product lines (Krämer 2002). The FDA has approved peracetic acid as a sanitizer on food contact surfaces (21 CFR 178.1010). Peracetic acid is a more effective sporicide than hydrogen peroxide (Holdsworth 1992). Han (1977) has shown that peracetic acid concentrations in the 0.1%–0.6% range are more sporicidal than a 30% hydrogen peroxide solution at room temperature. Hydrogen peroxide inactivation kinetics are compared to peracetic acid inactivation kinetics in Figure 14.3. Furthermore, peracetic acid is highly effective at low temperatures (up to 0°C) and is thus used in various branches of the food industry (e.g., dairy sector, beverage industry). Unlike hydrogen peroxide, peracetic acid causes corrosion of sensitive metals such as iron, copper, and aluminum. In closed containers, higher concentrations of peracetic acid (above 60%) can lead to explosive decomposition and therefore it is normally stored at concentrations of about 40%. Residues of peracetic acid can cause off-flavors due to acetic acid.

14.5.2 Physical Methods

14.5.2.1 Thermal Processes

Thermal processes, such as dry, hot air and saturated steam, are often preferred as they do not need chemicals and have no residues. Furthermore, they are ecologically and toxicologically harmless. On the other hand, thermal processes are difficult in practice and are not suitable for thermoplastics having low conductivity or for products having heat-sensitive ingredients. Suitable materials for thermal processes are temperature-stable materials such as glass, metal, and some plastics, for example, PP and polyethylene naphthalate (PEN). The latter is only used as an additive for copolymerizates due to its high cost (Buchner 1999). In many instances, the heat that is generated in the film-manufacturing process during extrusion and calendaring is often sufficient to produce sterile films (Holdsworth 1992).

14.5.2.1.1 Saturated Steam

Saturated steam can be considered to be moist heat (Ansari and Datta 2003). The treatment of surfaces with saturated steam is a well-established method for the sterilization of various packaging materials made of metal or glass. Nowadays, this technology is also used for the treatment of plastic containers. To provide a practical inactivation rate in the range of seconds for this application, the surface of the plastic has to reach temperatures between 130°C and 140°C. This thermal stress can cause deformation of some polymers (e.g., PE and polystyrene [PS]) for prolonged treatment times. The critical factor for this technology is to apply the appropriate time–temperature settings to reach the requested microbial inactivation without causing thermal damage in a time consistent with the high production rate (Ansari and Datta 2003). Air can influence this technology negatively as it reduces the heat-transfer rate and should be removed before treatment (Reuter 1986). According to Cerny (1982) and Holdsworth (1992), deep-drawn containers made of PS and the aluminum cover foil are treated with steam at 165°C (6 bar) for 1.4 s (container) and 1.8 s (lids), respectively. To limit the temperature effect on the inner surface of the material, the outer side is chilled. This process results in an inactivation of *B. subtilis* endospores of between five and six logs (Cerny 1982). Plastics with a higher temperature resistance (like PP) and wall thickness (0.6 mm) can be treated for extended times. Cerny (1983) has also determined the efficiency of an aseptic filling system that uses saturated steam (147°C, 3.5 bar) for the sterilization of PP cups and the lid foil. He reported an inactivation rate of about seven logs for *B. subtilis* endospores in 4 s (Cerny 1983).

14.5.2.1.2 Superheated Steam and Dry, Hot Air

Superheated steam is a form of dry, hot air and is not as effective as moist heat at the same temperature (Table 14.3), meaning that longer treatment times and higher temperatures are necessary (Buchner 1999). However, in contrast to the saturated steam process, no complex pressure chambers are necessary as the process operates at atmospheric pressure. According to Cerny (1985), superheated steam should have temperatures above 200°C for a period of about 40 s to give an efficient sporicidal effect. This process is the so-called Martin–Dole system and is restricted to heat-resistant materials (e.g., cans of aluminum or tin plate) (Cerny 1985).

Dry, hot air is the preferred sterilization method for packaging materials made of paper (Ansari and Datta 2003). The temperature of the air in the hot-air process is about 300°C and the object temperature is about 180°C (Buchner 1999). According to Reuter (1986), time–temperature settings of 145°C for 3 min are used for the treatment of cardboard laminates for filling with acid products. Complete sterilization needs considerably higher temperatures and prolonged treatment times (Kessler 2002).

TABLE 14.3
***D*-Values of Various Bacterial Endospores for Dry Hot Air and Moist Heat of Different Temperatures**

Endospores on Surface	*D*-Values for Different Temperatures (Dry Hot Air)				
	120°C	140°C	150°C	160°C	180°C
B. subtilis	30 min	3–5 min	2 min	1 min	13 s
B. stearothermophilus	15 min			0.14 min	3 s
B. globigii	53 min			1.8 min	
C. sporogenes			6 min	2 min	15 s

Endospores in water	*D*-values for different temperatures (autoclave)	
	115°C	121°C
B. subtilis	2.2 min	0.4–0.7 min
B. stearothermophilus	15–24 min	1.5–4.0 min
B. megaterium	0.025 min	0.04 min
C. sporogenes	2.8–3.6 min	0.8–1.4 min

Source: From Wallhäußer, K.H., *Practice of Sterilization* (in German), 5th edn., Georg Thieme, Stuttgart, Germany, 1995. With permission.

14.5.2.2 Radiation Processes

Decontamination methods based on energetic radiation, as the commercial used UV or ionizing radiation, are reliable procedures for the treatment of heat-sensitive packaging materials. Compared to chemical processes, the radiation methods have the advantage that they work without an aqueous phase and critical residues. However, to generate the radiation, technical sophisticated systems are necessary and these are often costly and require high maintenance, especially in case of the short wavelengths. Furthermore, the high energy consumption, in conjunction with low energetic efficiency, influences profitability.

14.5.2.2.1 UV Radiation

A method for reducing the microbial count on the surface of the packaging material is the application of UV radiation. This is energetic radiation in the 380–100 nm range. This is classified in four categories depending on the wavelength: long wave UV-A radiation (380–315 nm), medium wave UV-B radiation (315–280 nm), short wave UV-C radiation (280–200 nm), and vacuum-UV radiation (200–100 nm) (DIN 5031-7 1984).

The microbiocidal effect of UV radiation is due to intracellular damage to various biomolecules, such as proteins, enzymes, and DNA. Comparison of the absorption maximum of DNA (260 nm) and the spectral inactivation behavior of microorganisms has shown that the UV radiation mainly causes DNA lesions (Schlegel 1992). Therefore, UV-C lamps (low-pressure mercury or excimer lamps) with emission maxima at 254 nm are used in practice for the disinfection of surfaces.

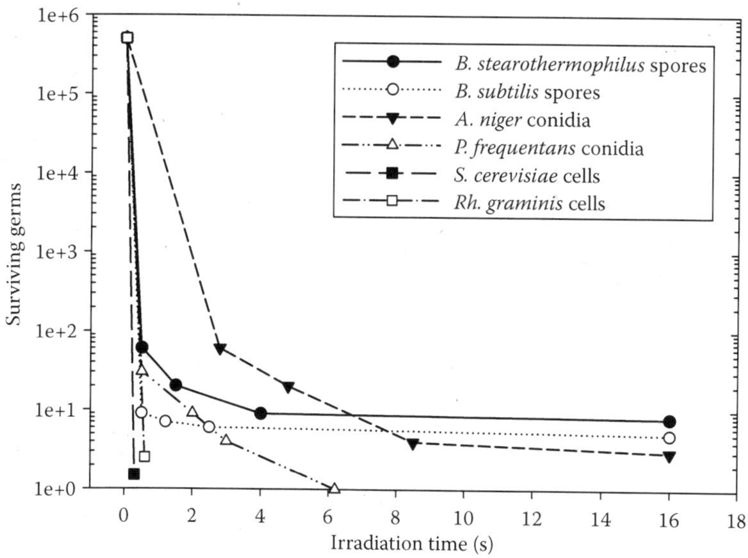

FIGURE 14.4 Inactivation of various microbial test strains by UV-C radiation (contamination level: $5 \times 10^5/36\,cm^2$, irradiance: $30\,mW/cm^2$). (From Cerny, G. et al., *Verpack.Rundsch.*, 10, 79, 1977. With permission.)

The efficiency of the UV radiation treatment depends on various parameters such as the emission spectra, the irradiance, the treatment time, the geometry of the packaging, the lamp, the relative air humidity, and the sensitivity of the microorganisms (Cerny 1977). As evidenced in Figure 14.4, UV radiation affects a wide spectrum of microorganisms and bacterial spores are more resistant than non-spore-formers (Figure 14.4 shows the antimicrobial effect of UV radiation). The lethal dose (90% inactivation) for vegetative germs is in the 2–5 mWs/cm^2 range, whereas for *B. subtilis* endospores, it is in the 8–10 mWs/cm^2 range (Krämer 2002, Reuter 1986). Most resistant are distinct spores of molds, and especially the black-pigmented conidiospores of *Aspergillus niger* (e.g., DSM 1988), which require a 20–100 times higher dose (Wallhäußer 1995). According to Cerny (1977) an irradiance of $30\,mW/cm^2$ results in a reduction of four logs for *B. subtilis* endospores after 0.3 s, whereas 1 s is required to achieve the same killing effect for fungal spores. The reason for this is the black-colored outer wall of the spore that impedes the penetration of the UV radiation into the spore (Cerny 1990). Due to its high UV resistance, this species is used as a test strain for the evaluation of the efficiency of UV-based sterilization processes.

UV radiation treatment has a limited penetration depth. The depth of penetration is about 30 cm for water, 1–2 mm for milk, and about 30 μm for synthetic films (Ansari and Datta 2003, Kessler 2002). Due to this low penetration depth, this technology is only suitable for surface applications, and the treatment of air or water in relatively thin layers (Kessler 2002). Regarding surface treatments, cell agglomerates or dust particles can cause a shadow effect, which decreases the efficiency of UV radiation. This phenomenon causes a tailing in the death kinetics and is the reason why this method is not suitable for sterilization (Krämer 2002). UV radiation is used for the disinfection of packaging for foods having a low pH value, low microbiological risk, and products with a short shelf life and is suitable for smooth surfaces such as containers or lid films. Furthermore, UV radiation is used for the disinfection of water and clear solutions in the food industry as well as for the treatment of air.

Nowadays, a combination of UV radiation and hydrogen peroxide is used for sterilization processes to improve their sporicidal activity (Ansari and Datta 2003). The effect is due to the formation of hydroxyl radicals that are produced when hydrogen peroxide is irradiated with wavelengths below 400 nm (Bayliss and Waites 1979). Bayliss and Waites (1979) have shown that the overall effect of simultaneous hydrogen peroxide and UV irradiation is much greater than the sum of the effects of the two methods alone. The advantage of this combined method is that lower concentrations of hydrogen peroxide (1%–3% (w/v)) are very effective and in this way the amount of the chemical agent can be considerably reduced (Cerny 1985). Higher concentrations of hydrogen peroxide diminish the effect. The absorption of UV radiation by hydrogen peroxide was postulated as the reason for the loss of synergism (Block 2001).

14.5.2.2.2 Infrared Radiation

IR is electromagnetic radiation of wavelength between 760 nm and 1 mm. The sterilization mechanism is due to the conversion of radiation into heat via absorption on a surface. If the area of the radiator and the packaging material are constant, the absorbed heat capacity of the packaging material is subject to the following proportionality (Engelhard 2005):

$$c_{\mathrm{R-ab}} \propto \alpha_{T,\lambda_{\mathrm{L}}}\left(T_{\mathrm{rd}}^{4} - T_{\mathrm{pk}}^{4}\right) \tag{14.2}$$

where

$c_{\mathrm{R-ab}}$ is the absorbed heat capacity
T is the temperature
λ_{L} is the wavelength
α is the absorptance
subscripts rd and pk are the radiator and the packaging material, respectively

Depending on the system being used the temperature of the IR-lamp can be in the range of 450°C–1600°C. Because of the high temperature difference between the IR lamp and the surface, the packaging material is heated to temperatures that can inactivate microorganisms within seconds. The high temperature limits the use of this method to temperature-resistant materials such as aluminum lids (Engelhard 2005). IR is comparable with UV sterilization and count reductions of the same order of magnitude, as those for UV irradiation, have been found (Ansari and Datta 2003). This means that this method is only suitable for flat materials with vertical alignment of the source.

14.5.2.2.3 Electron Beams and Gamma Rays

Electron beams (β-rays) and γ-rays are commonly used for sterilization. Both are classified as ionizing radiation, which means that they can ionize atoms by releasing electrons. γ-Radiation is highly energetic electromagnetic radiation (photons) having a wavelength below 10^{-9} cm which is generated by radioactive disintegration of a radioactive source. For sterilization, the radioisotopes cobalt 60 (Co^{60}) and cesium 137 (Cs^{137}) are mainly used. Co^{60} has a half-life of 5.3 years and emits γ-radiation of 1.3 and 1.2 MeV ($1\,eV = 1.6 \times 10^{-19}$ J) during the disintegration process (Krämer 2002). In addition, β-rays (0.318 MeV) are emitted, but the shielding of the source absorbs this radiation. Cs^{137} has a half-life of 30 years and emits γ-ray of 0.66 MeV and two β-rays of 0.51 and 1.17 MeV (Block 2001).

Electron beams, on the other hand, are generated in an accelerator or by radioisotopes and have high energies in the 0.15–10 MeV range.

The sterilization effect depends on the absorbed dose, which is given in grays (Gy) (Krämer 2002). One gray corresponds to the dose that leads to an absorption of one joule per kilogram (1 Gy = 1 J/kg). A further influencing factor is the individual resistance of the particular microorganism. The radiation dose required for inactivation of various microorganisms is given in Table 14.4. The resistance decreases in the following order: viruses, mould spores, bacterial spores, and non-spore-formers (Kessler 2002). A dose in the range of 5–10 kGy inactivates most of the vegetative germs. To achieve sterile packaging in the food industry, the radiation doses are usually between 8 and 15 kGy. The red-pigmented *Deinococcus radiodurans* has been found to be the most resistant germ to ionizing radiation, and this can survive

TABLE 14.4
Required Dose of γ-Radiation for the Inactivation of Various Microorganisms by Six Logs

Microorganisms	Dose (kGy)
Gram negative bacteria	
Pseudomonas fluorecens	<0.5–1
Enterobacteriaceae	<0.5–3
Vibrio parahaemolyticus	<0.5–1
Shigella sp.	1.5–3
Campylobacter jejuni	2
Gram positive bacteria	
Micrococcus sp.	3–5
Leuconostoc sp.	0.5–3
Lactobacillus	2–7.5
Bacillus and *Clostridium* spores	10–30
Molds	
Penicillium sp.	0.5–2
Aspergillus sp.	1.5–5
Viruses	>30

Source: From Krämer, J., *Food Microbiology* (in German), 4th edn., Eugen Ulmer, Stuttgart, Germany, 2002. With permission.

a dose of 60 kGy (Wallhäußer 1995). For implementation of radiation sterilization, knowledge about the target microorganisms, their concentration, and their resistance is necessary.

The penetration of these two types of ionizing radiation differs and depends on the thickness and density of the material. For example, 10 MeV electrons can be used to process a 3 cm thickness of unit density material by irradiating from one side, whereas electrons with energies of a few hundred kilo-electron-volts can be used to process coatings, thin films, and foamed plastics (Silverman 1977). γ-Radiation has a lower ionization density but is more penetrating. According to Mohler (1966) γ-rays of 1 MeV have a penetration depth of about 10 cm in water. Table 14.5 shows the penetration depth of different types of ionizing radiation in water.

Because of its high penetration depth, γ-radiation is used for sterilizing large packaging units. The relatively low dose rate compared to electron beams requires a treatment time of several hours. These prolonged treatment times can be compensated by the ability to irradiate large volumes in gamma facilities (Haji-Saeid et al. 2007). For highly energetic electron beams, the required dose can be applied within shorter times.

Besides sterilizing the packaging material, ionizing radiation can in some cases also modify key properties of the plastics. Depending on the applied dose, there can be degradation of polymers (chain scission), cross linking of polymer chains, a change of color, and formation of by-products such as gaseous radiolysis products, radicals, or oxidation products. Radiolysis products are oxidized decomposition products of polymers, which are responsible for the typical radiation smell. To prevent this, the applied dose has to be restricted and suitable materials have to be selected (Cerny 1976). The FDA has published a list of approved packaging materials for use with γ-radiation up to 60 kGy (21 CFR 179.45) and the American Society for Testing and Materials has prepared the standard guide F1640-03 for the selection and use of packaging materials for foods to be irradiated (Haji-Saeid et al. 2007). One advantage of this method is that the whole packaging including the product can be treated in one step, but in some countries the radiation of food is prohibited or restricted. Because of the high acquisition costs for the equipment, external companies often treat the packaging material.

TABLE 14.5
Penetration Depth of γ-Rays and Electron Beams in Water

Radiation	Energy (MeV)	Penetration (cm)
Co^{60}	1.1 and 1.3 γ	≅ 10
Electrons	1	≅ 0.5
Electrons	3	≅ 1.5

Source: Mohler, W., *Veröff des Eidg. Gesundheitsamtes*, Bern, 57, 489, 1966.

14.5.2.2.4 Pulsed-Light Technology

Pulsed or flash light is an upcoming technology and a suitable method for the decontamination of surfaces in the food packaging area. The main piece of equipment required, besides a power supply and the pulse configuration device, is an inert gas flash lamp (e.g., xenon) that emits a continuous broad spectrum of white light comprising wavelengths from 200 nm (UV) to 1000 nm (near infrared) and having a maximum at 450 nm. The emission spectrum corresponds to natural sunlight, but the intensity of the pulsed light is about 20,000 times higher. The lamp is operated in a pulsed mode (high-power pulses) with a flash duration in the range of microseconds. In this mode, the electrical energy is accumulated in an energy storage capacitor over a relatively long time (fraction of a second) and released in a very short time (Dunn et al. 1995, Elmnasser et al. 2007). This mechanism delivers high peak energy light pulses whose energy is greater than that provided by a continuous light lamp (Palmieri and Cacace 2005). To focus the light toward the particular target object, special shaped reflectors are used. The main parameters are the incident energy per cm^2 on the surface being treated (fluence) and the peak power during the duration of a single pulse (fluence rate). The required peak power depends mainly on the microorganism (size, mass, conductivity, UV-absorption), the surrounding media, and the object being sterilized (conductivity, UV-absorption) (Wekhof et al. 2001). For example, Turtoi and Nicolau (2007) have reported that the color of mould spores affects their resistance to pulsed light. Dark-colored species like *A. niger* (z = 0.81 J/cm^2) need lower z-values than the green-colored spores of *A. repens* (z = 0.927 J/cm^2) (Turtoi and Nicolau 2007). The microbiocidal efficiency is based on photochemical and photothermal mechanisms and depends on the number and power of the applied flashes. Wekhof et al. (2001) have reported that for *A. niger* and *B. subtilis* spores the photothermal effect of pulsed light is based on momentary and targeted overheating of the spores due to photon absorption and this leads to structural disintegration. The explanation for this is the short duration of the light pulse, which leads to a reduced available time for thermal conduction and this causes rapid heating of a limited surface layer up to higher temperatures. These temperatures are even higher than the steady-state temperatures that are achieved by a continuous light lamp at equivalent energy values (Dunn 1996). The lamp also emits UV-C radiation, which leads to damage of the genetic material (DNA) and this is part of the photochemical effect. Furthermore, there is an impact on proteins, membranes, and other cellular materials concomitant with the destruction of nucleic acid (Elmnasser et al. 2007).

Dunn et al. (1995) have reported that inactivation of *A. niger* spores by more than seven logs can be achieved by a few light pulses (1 J/cm^2). A dose of 4,500 μWs/cm^2 pulsed light reduces *B. subtilis* endospores by about five logs, whereas a dose of 426,00 kWs/cm^2 is required under UV light (Ansari and Datta 2003). Wekhof et al. (2001) have shown that fluencies of 1 and 5 J/cm^2 (duration 200 ns) are sufficient for a 5 log reduction of *B. subtilis* and *A. niger* spores. The pulsed-light technology could be used as a rapid, low-energy, and low thermally damaging method for the decontamination of packaging materials. The disadvantages of this technology are the low penetration depth and an efficiency reduction due to shadowing effects. It is therefore only suitable for partial reduction of the microbial count.

14.5.2.3 Plasma Technology

Although established methods are available, efforts are being made to develop new and innovative sterilization technologies for packaging materials to substitute or compete with conventional methods and overcome their disadvantages. Gas plasma treatment is an innovative technology for the nonthermal inactivation of microorganisms on packaging materials. By using the main inactivation mechanisms such as UV radiation, chemically reactive species, and charged particles, both microorganisms and biomolecules (e.g., endotoxins, prions) can be efficiently destroyed. Gas plasmas are particularly suitable for the treatment of temperature-sensitive materials such as products made of plastic. The high efficiency within treatment times of a few seconds and the versatility of the plasma technology make it suitable for various applications in the food and medical fields.

Gas plasmas have been used for many years for various surface treatments such as coating, cleaning, or structuring. Plasma sterilization is a relatively recent field of research but it was mentioned by Menashi way back in 1968 (Menashi 1968).

Gas plasma is a type of gas containing an appropriate number of free charged particles, namely electrons and ions, which are produced by energy transfer as a result of particle collision processes.

The base reactions in the plasma are dissociation and ionization of atoms and molecules. As plasma always originates in the gaseous state, gas plasmas are also called the fourth state of matter. The energy required to achieve the plasma state can be applied by thermal energy (e.g., flame), ionizing radiation (e.g., electron beams, laser), or adiabatic compression. The most commonly used method for generating low-temperature plasmas for technical applications is the application of an electrical field to a process gas. Due to natural occurrence of cosmic and radioactive radiation, there are always free electrons and ions available in a gas volume. The application of an electric or electromagnetic field results in acceleration of the charged particles and ultimately in collision processes between them. This is the basic principle for the generation of technical plasmas (Conrads and Schmidt 2000).

Compared to the heavy particles (atoms, molecules, and ions), electrons acquire very high kinetic energy in the electric field due to their very low mass. When there is a collision between electrons and heavy particles, different reactions can occur and these depend on the energy that is transferred. One such reaction is ionization due to particle collision, whereby several electronvolts of kinetic energy are transferred from the electron to the atom. This releases an electron from the outer shell of the atom and it becomes a positive charged ion. This ionization procedure causes an avalanche-like release of electrons and this keeps the process going. The dissociation of molecules leads to chemically reactive radicals, which are used, among other things, for the so-called plasma-etching processes.

A characteristic property of gas plasmas is their glow. Figure 14.5 shows the plasma glow for a low-pressure plasma system. This effect is due to radiation emitted by excited atoms and molecules when their electrons return to a lower energy state. Depending on the difference between the energy levels, electromagnetic radiation of different wavelengths is released, for example, ultraviolet, visible, or infrared light.

The large number of reactions which take place within a plasma give rise to a number of different sterilization mechanisms. These involve UV irradiation, chemical

FIGURE 14.5 Low-pressure plasma system for the sterilization of PET bottles. (From Muranyi, P. et al., *Chem. Ing. Tech.* 78, 1704, 2006a. With permission.)

reactive species, charged particle bombardment, and local dissipation heat. The penetration depth of these is limited. Therefore, plasma technology is a technique for treating the surfaces of solid materials. Each of the above-mentioned reactions has the ability to inactivate microorganisms. In plasma, all reactions take place at the same time and this enables fast and efficient inactivation by possible synergistic effects.

The main inactivation mechanism when plasma is used for surface treatment is UV radiation in combination with chemical radicals. The advantage of plasma compared to conventional UV radiators is that the gas is able to penetrate gaps. Due to the diffuse radiation, typical UV shadowing effects, caused for example, by dust particles, can be minimized. The fraction of each germicidal reaction in the plasma is controlled by external parameters, namely, the power input, the geometry of the plasma reactor, the process gas that is used, and the gas pressure. The power input affects primarily the homogeneity of the plasma, whereas varying the process gas changes the radiation spectrum and the reactive species. This in turn has an influence on the plasma chemistry and germicidal properties.

For example, air plasma shows low radiation intensity in the UV range < 300 nm but it produces highly reactive oxygen and nitrogen species such as atomic oxygen (O), ozone (O_3), hydroxyl radicals (OH), and nitrogen oxides (NO_X). In pure oxygen plasma, mainly ozone is produced whereas argon plasma only emits UV radiation (Franken et al. 2003).

There are many different systems available for plasma generation. The main difference among them is the working pressure. There are systems that work at atmospheric pressure and systems that work in a pressure range of typically 10^{-2} to 1 mbar. Atmospheric pressure plasmas are mainly driven by high voltage of few kilohertz frequencies. Low-pressure plasma can be driven by microwave excitation (2.45 GHz), DC current, or AC current, which is normally used at frequencies of 13.56 or 27.12 MHz.

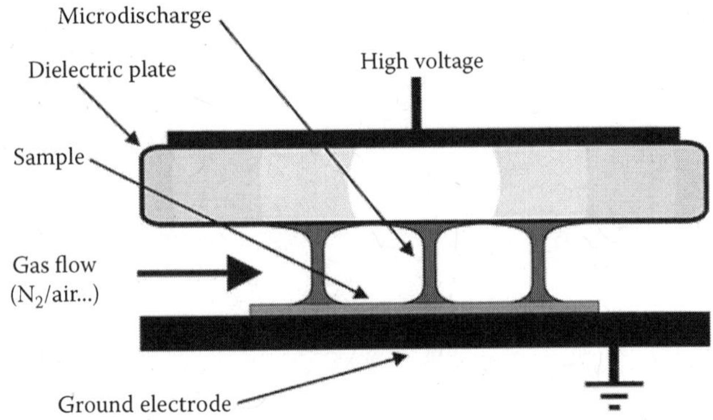

FIGURE 14.6 Setup for DBD. The high-voltage electrode is covered by a dielectric. The plasma is formed in the gap between the dielectric and the ground electrode. (From Muranyi, P. et al., *J. Appl. Microbiol.* 103, 1536, 2007. With permission.)

Capacitor plates (CCP, capacitive coupled plasma) or coils (ICP, inductive coupled plasma) can couple the electrical power. Each of these systems has its own advantages and disadvantages. Depending on the system, flat substrates (e.g., foils, films, or plates) or three-dimensional bodies (bottles, cups, and closures) can be treated.

A special type of low temperature plasma at atmospheric pressure is the dielectric barrier discharge (DBD)—also known as silent discharge. This type of plasma is suitable for the sterilization of flat substrates, for example, deep-drawn films. The setup, schematically shown in Figure 14.6, consists of a high-voltage electrode covered by a dielectric material (e.g., aluminum oxide or quartz) and a ground electrode where the material to be treated is placed. The gap between the electrodes is typically a few millimeters, and it is purged by the process gas. Application of high voltage in a typical frequency range of 10–50 kHz leads to the ignition of a nonthermal plasma that consists of many thousands of microdischarges, also called filaments. These filaments have a lifetime of several nanoseconds. Within this time, thermalization of hot electrons and cold atoms, ions, or molecules is not possible and the plasma stays cold. Inside the filaments, plasma-specific products such as UV radiation and radicals form. The microdischarges during one electrical half-wave are homogeneously distributed over the whole surface with a certain distance due to their electrical footprint. Within the following electrical half-wave, the microdischarges take place at another location due to a rest of surface charge of the previous half-wave. According to a typical frequency in the range of 20–100 kHz it can be presumed that every point of the surface will be covered by a microdischarge within a few milliseconds. The gas flow through the system causes movement of the filaments and improves homogeneity.

Studies have shown that a standard DBD system does not give sufficient sterilization for industrial applications. High doses are needed, which makes the process time and energy requirements of the process excessive. Figure 14.7 shows the inactivation of *B. subtilis* and *A. niger* spores in a standard DBD system using different process gases. The best results are achieved using Argon, for which only UV radiation is present, and humid air, in which case a large number of radicals are present and there

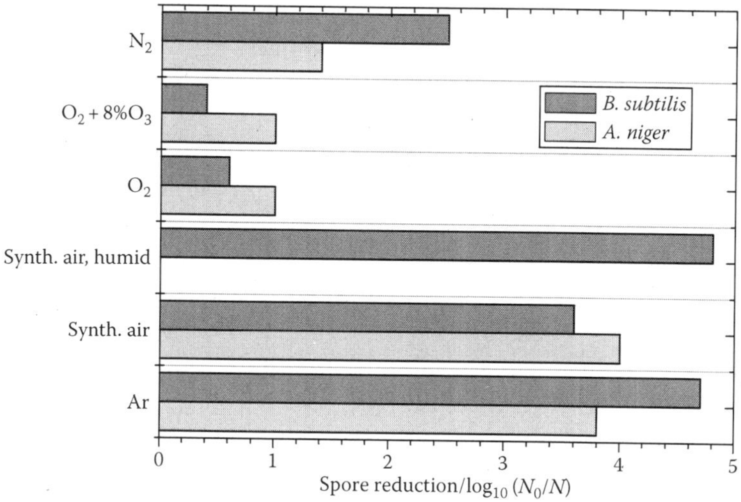

FIGURE 14.7 Inactivation of *A. niger* and *B. subtilis* spores as a function of the process gas. The treatment time was 60 s, the coupled electrical power 300 W. Argon (high UV emission) and air (high fraction of radicals) show the best inactivation efficiency. (Modified from Franken, O. et al., Research report, RWTH University, Aachen, BMBF-Project FKZ 13N7609/9, 74, 2003.)

is virtually no UV radiation. No set of process parameters was found that simultaneously produced a high UV radiation intensity and a high number of radicals.

Based on the principle of a DBD, the Fraunhofer Institute for Laser Technology (ILT) has developed a modified DBD—called a cascaded DBD (CDBD). In this system, as evidenced in Figure 14.8, a flat excimer lamp, which consists of a closed quartz system filled with a mixture of a rare gas and a halogen, replaces the dielectric.

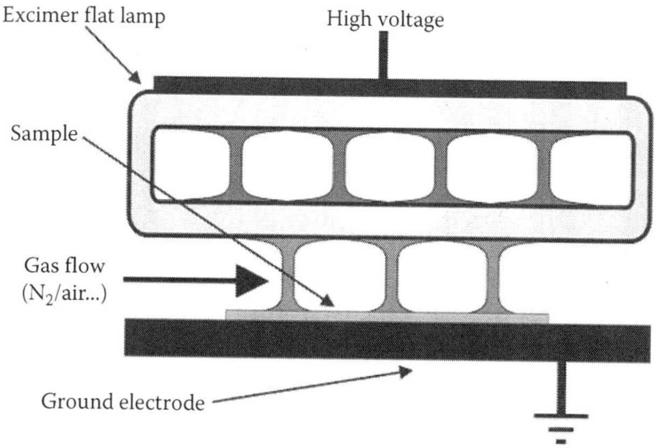

FIGURE 14.8 Setup for CDBD. The dielectric of a conventional DBD is replaced by an excimer flat lamp. High UV radiation intensity combined with chemically active radicals formed in the gap result in fast and efficient inactivation of microorganisms. (From Muranyi, P. et al., *J. Appl. Microbiol.* 103, 1537, 2007. With permission.)

On applying voltage to the system, the flat lamp starts emitting monochromatic UV light of a specific wavelength. Depending on the gas mixture, different wavelengths are produced (e.g., Xe—172 nm, KrCl—222 nm, and XeBr—282 nm). The additional UV light in the gap has several consequences. On the one hand, it is possible to use, for example, humid air in the gap to produce a large number of radicals with simultaneous high UV radiation intensity. On the other hand, the additional UV radiation increases the discharge homogeneity of the plasma inside the gap (Heise et al. 2004). This effect is known as the Joshi effect (Falkenstein 1997).

Using this modified system a wide range of microorganisms were inactivated by more than five logs within a treatment time of less than 5 s as shown in Figure 14.9 without any parameter optimization (Muranyi et al. 2006a). Spores having special resistance, for example, *A. niger*, are more resistant to UV light than many other microorganisms and can be killed by other plasma reaction products. This makes the CDBD system very efficient for inactivating a broad variety of microorganisms. Also important for industrial use is the preservation of the packaging material properties such as the permeability and sealing properties. Studies on relevant plastic films such as PET, PP, and PS at working sterilization parameters have shown no

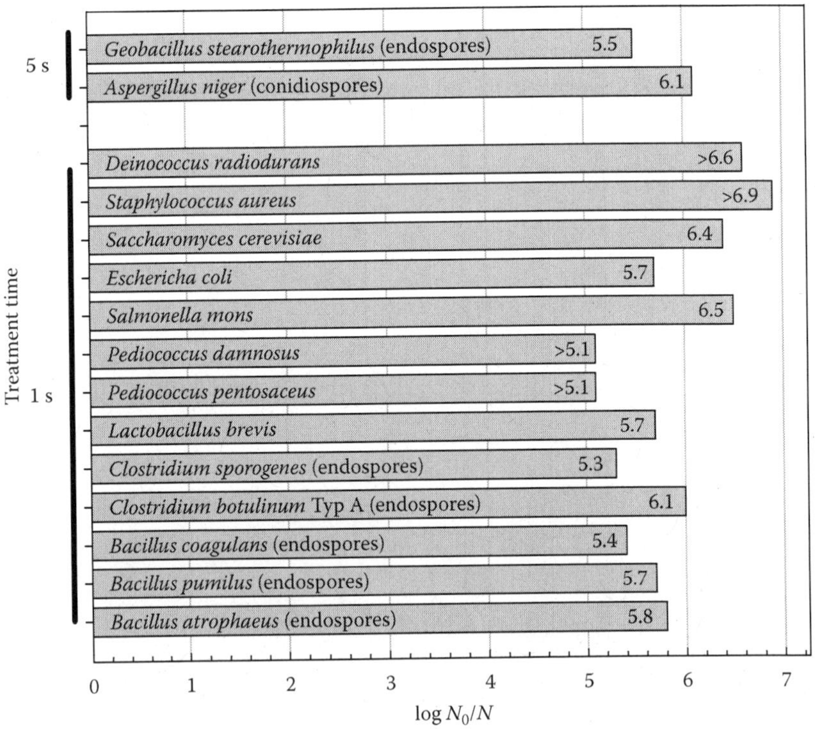

FIGURE 14.9 Inactivation efficiency of a CDBD system with a 282 nm XeBr flat lamp and using room air as the process gas. All experiments were conducted with the same parameters, no optimization has been undertaken. The power input was 130 W. (From Muranyi, P. et al., *Chem. Ing. Tech.* 78, 1703, 2006a. With permission.)

significant changes in properties. The goal is to integrate this system into existing machines to enable continuous decontamination of flat packaging materials.

However, the food packaging industry is not concerned with only flat packaging materials. For three-dimensional objects such as bottles, cups, and closures, further development of special electrodes for the CDBD system is necessary. Still, some alternative low-pressure plasma systems are available. Here, plasma excitation can be achieved by microwave radiation (2.45 GHz) or AC current (13.56 or 27.12 MHz). Although the plasma generation and the working pressure are different from the DBD system, the germicidal mechanisms inside the plasma are the same. The low pressure leads to long, free path lengths, which enables large volume plasmas. Another advantage of the low-pressure regime is the high intensity of vacuum UV radiation (V-UV, <200 nm). This radiation has a high germicidal effect, resulting in inactivation by —five to six logs within the subsecond range (Schneider et al. 2005). As with the CDBD system, the V-UV radiation in low-pressure plasmas is also responsible for fast inactivation, and this in combination with radicals and particle bombardment leads to fast and efficient decontamination within several seconds (Feichtinger et al. 2003). With the right choice of system setup it is possible to produce a plasma inside a bottle or to treat large-scale structures.

A disadvantage of this system from an industrial point of view is the low pressure itself. To create the low pressure, a vacuum system is needed. Although bottles can generally be pumped down easily, it is more difficult to treat large-scale films. In this case, it is possible to carry out the treatment batch by batch in a closed chamber or continuously in a chamber that uses vacuum sluices. At any rate, the vacuum system is much more complicated and more expensive to obtain and operate than a system at atmospheric pressure. There are plasma systems, which work at atmospheric pressure using, for example, plasma jets for treating large-scale structured films and objects. However, they are often used for surface modification because the inactivation effect is comparable to the standard DBD system and this was shown not to be efficient enough for aseptic packaging.

In a research project funded by the German ministry for education and research (BMBF, FKZ 13N7609), the suitability of all the mentioned plasma systems was demonstrated for aseptic packaging applications. Parameters for fast and efficient sterilization have been identified, meaning the technology can be implemented in industrial aseptic packaging machines.

14.6 VALIDATION OF STERILIZATION PROCESSES

For determination and quantification of the microbiological inactivation efficiency of aseptic filling machines, practical validation is carried out via challenge tests. Within these procedures, the efficiency of the packaging material and machine sterilization is tested to identify possible weak points of the filling process and to get an overview of the microbiological status of the system. For this purpose, bioindicators are used. These are carrier materials that are artificially contaminated with selected test strains. Bioindicators are measurement tools for determination of the logarithmic inactivation relative to the initial microbial count and the number of surviving microorganisms after exposure to the sterilizing agent. The need to acquire experimental data of practical microbiological validations is justified by the high complexity involved in measuring and modeling the physical and chemical processes associated with sterilization. For example, for hydrogen peroxide condensation there are process-based temporal and spatial gradients of concentration on the surface of the packaging material that are difficult to include and this can adversely influence the inactivation. The use of mathematical models to theoretically determine the inactivation rate is not recommended because of the individuality of each filling machine and their low production rate. On the other hand, statistical examination by determination of the microbial unsterility rate of a defined number of packaging units (commissioning test) is much elaborate. According to Wilke (1996), 30,000 packaging units are necessary to determinate an unsterility rate of 1:10,000 with a confidence limit of 95%. With these quantities, the method is not economically attractive, especially as further microbiological investigations could be necessary. A further disadvantage is the missing opportunity for separate consideration of packaging and machine sterilization and thus identification of weak points. Only the final product is considered. As a result, for the generation of reliable microbiological data, practical validations are essential (Muranyi et al. 2006b).

14.6.1 Test Procedures

The requirement for aseptic packaging is efficient and reliable sterilization of the packaging material and the filling machine. To validate this, specific measurement methods are used, the so-called challenge tests. The objective of a challenge test is to determine the sterilization efficiency of the packaging material and the machine as well as to highlight critical points. For this, two test methods are suitable: the count-reduction test and the end-point test (VDMA 2002).

14.6.1.1 Count-Reduction Test

The count-reduction test determines the mean inactivation rate by comparing the initial and the final microbial count of artificially contaminated bioindicators after the sterilization treatment.

$$\text{MLCR} = \log(nc_0) - \log(nc_{\mathrm{f}}) \tag{14.3}$$

where

MLCR is the mean logarithmic count reduction
nc_0 is the initial count
nc_{f} is the final count

In detail, the respective packaging material (bottles, caps, films) is contaminated with the specific test strain (e.g., *B. subtilis* endospores) and exposed to the sterilization process in the aseptic filling machine. The initial microbial count on the packaging material is normally about two logs above the actually demanded inactivation rate, for example, 10^6 CFU per sample for a requested reduction of about four logs. The number of samples comprises at least 10 samples per packaging line of the machine. Regarding the machine sterilization, this method allows determination of the inactivation rate at specific positions (isolator, packaging line) in the filling machine. In this way, weak points can be identified and optimized (Muranyi et al. 2006b).

14.6.1.2 End-Point Test

The end-point test is a practice-related test, which ends in concluding whether sterile products can be achieved. This is important for any aseptic packaging process (e.g., milk). For this purpose, a high number of samples with low initial bacterial contamination are exposed to the sterilization agent and finally filled with an appropriate product. The evaluation is qualitative with differentiation between sterile and unsterile samples. For calculation of the mean logarithmic count reduction, the following equation is used:

$$\text{MLCR} = \log(nc_{\text{pk}0}) - \log\left[\ln\left(\frac{nt_{\text{pk}}}{ns_{\text{pk}}}\right)\right] \tag{14.4}$$

where
$nc_{\text{pk}0}$ is the initial count per single package
nt_{pk} is the number of test packages
ns_{pk} is the number of sterile packages

The requirements for this kind of test are sterile conditions to avoid recontamination. Due to the large number of samples, it is more costly. The advantage of this method is that sterile products have to be produced and this increases the significance of the practical feasibility of the sterilization process. Besides the microbiological efficiency, the end-point test enables assessment of the hygienic status of the whole filling process: product sterilization, transport, filling until secure closure (Cerny 1992). Similar to what happens in the count-reduction test, the packaging materials are artificially contaminated but with graduated microbial counts (e.g., 10^2, 10^3, 10^4 CFU/object). For the end-point test, at least 100 samples of each contamination level are necessary. For calculation of the mean microbial count reduction it is essential that all germ carriers be exposed to the same sterilization conditions. Therefore, the end-point test is only suitable for packaging material sterilization, not for the evaluation of the machine sterilization, because the distribution of the sterilizing agent can differ at different machine positions. In this context, there are requirements put on certain number of sterile bioindicators when validating packaging machines. For instance, according to a VDMA guideline all samples with a contamination level of 10^4 CFU/object have to be sterile (Muranyi et al. 2006b, VDMA 2003).

14.6.2 Bioindicators

Bioindicators are carrier materials contaminated with a certain test strain. They serve as a measuring instrument for determining the sterilization efficiency of a filling machine. Bioindicators do not simulate artificial contamination because environmental safety mechanisms (e.g., for dust and oil) are not considered. Normally, packaging materials (e.g., bottles, lids, deep-drawn films) are used as carriers for validation of the packaging material sterilization. For assessing the machine sterilization, special test strips made of metal or plastic are used.

The selection of a suitable test strain depends on the applied sterilization method. Generally, the chosen microorganism should have a high and defined resistance to the sterilization agent, should be easy to detect, nonpathogenic, and of practical relevance. Normally, microorganisms in the spore state are used because of their high resistance to various physical and chemical environmental influences. Examples of such test organisms are endospores of *B. subtilis* SA 22 for hydrogen peroxide, *G. stearothermophilus* for superheated steam, and *B. pumilus* for γ-radiation (Bernard et al. 1990).

The contamination of the sample surface can be achieved by either spraying or spot-wise contamination. Spray contamination leads to a homogenous distribution of microorganisms on the surface, which reduces agglomerates and thus shading or shadow effects. The disadvantage of spray contamination is the technically sophisticated handling compared to spot-wise contamination, in which the germ suspension is just dropped onto the surface. Spot-wise contamination results in a high contamination level with agglomerates and multilayers. This can be disadvantageous for some sterilization methods such as UV radiation (Muranyi et al. 2006b).

14.6.3 Resistance of Test Strains

The result and the success of a microbiological filling machine validation depend mainly on the test strains having defined resistance. Up until a few years ago the resistance was simply controlled by selecting a certain strain of a species. However, practical experience has shown that this criterion is not sufficient. The resistance of test spores is a function of the growth condition (Cerny 1992). On determining the resistance of a spore suspension in various growth phases, remarkable fluctuations can be observed. Figure 14.10 shows the experimentally determined inactivation behavior of various *B. subtilis* SA 22 spore suspensions exposed to hydrogen peroxide. The deviations are up to two logs after 20 s. Such resistance fluctuations were observed with a frequency of 5%–10%, which means that these spore suspensions are more sensitive to the sterilization agent and can adversely influence the results of a practical validation test. For instance, a more sensitive spore suspension would give a much higher sterilization efficiency for the aseptic machine than it would be the case for another charge.

The reasons for these resistance fluctuations are numerous. Various research studies have shown that parameters such as the incubation temperature and the ingredients and properties (pH value, a_w value) of the growing medium can influence the resistance of the spores (Igura et al. 2003, Melly et al. 2002, Nicholson et al. 2000). It is remarkable that such variation also appears on standardized production of the spore suspensions. This indicates that some nonmonitorable or immeasurable parameters play a major role. The consequence of this for the practical validation is

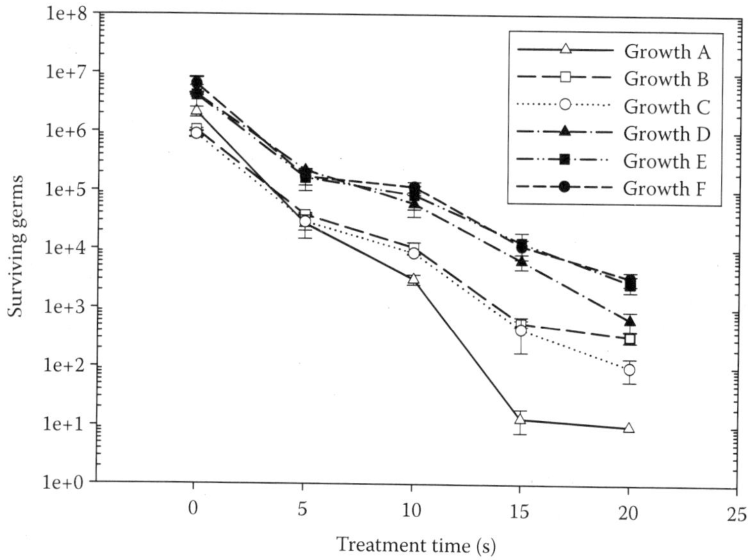

FIGURE 14.10 Variation of the measured resistance of different batches of *B. subtilis* spore suspensions to hydrogen peroxide (concentration: 35%, temperature: 55°C).

that a measured and defined resistance of the spore and germ suspensions, which are used for tests, is essential (Muranyi et al. 2006b).

14.7 CONCLUDING REMARKS

Aseptic packaging of food is a sophisticated method for a quality-preserving production of long-living products. Precondition for this process is a complete sterilization of the packaging material. However, new packaging materials made of polymers require nonthermal methods for surface sterilization. Nowadays, well-established processes work mainly with chemical agents like hydrogen peroxide or peracetic acid in combination with moderate temperatures. Nevertheless, aspects like residues in the packaging or maximal allowable concentrations are the motivation for the development of alternative decontamination methods, like the innovative plasma technology. The current state of the art shows, that plasma decontamination is suitable for a partial reduction of surfaces in any case. However, gas plasmas also have the potential for complete sterilization. Finally, the suitability of an application in the field of aseptic packaging depends on the requirements for the sterilization efficiency, which are different for various inactivation methods and have to be defined and standardized for the different plasma systems.

That applies to all new and innovative sterilization methods. For a statement on the scalability and the performance under practical conditions, the systems have to be evaluated on an industrial scale. This makes it necessary to integrate new sterilization systems in filling machines and to validate their efficiency.

REFERENCES

Ansari, I.A. and Datta, A.K. 2003. An overview of sterilization methods for packaging materials used in aseptic packaging systems. *Trans. IChemE* 81:57–65.

Bayliss, C.E. and Waites, W.M. 1979. The synergistic killing of spores of *Bacillus subtilis* by hydrogen peroxide and ultraviolet irradiation. *FEMS Microbiol. Lett.* 5:331–333.

Bernard, D.T., Gavin, A., Scott, V.N. et al. 1990. Validation of aseptic processing and packaging. *Food Technol.-Chicago* 12:119–122.

Block, S.S. 2001. *Disinfection, Sterilization and Preservation*. Philadelphia, PA: Lippincott Williams and Wilkins.

Buchner, N. 1999. *Packaging of Food*, 1st edn. Berlin, Germany: Springer-Verlag (in German).

Cerf, O. and Brissende, C.H. 1981. Aseptic packaging. New monograph on UHT milk. *Bull. Int. Dairy Fed.* 133:93–104.

Cerny, G. 1976. Sterilization of packaging materials for aseptic packaging. 1st Notification: Experiments for determination of the microbial effect of hydrogen peroxide solutions (in German). *Verpacks.-Rundsch.* 27:27–32.

Cerny, G. 1977. Sterilization of packaging materials for aseptic packaging. 2nd Notification: Experiments for determination of the microbial effect of UV-C irradiation (in German). *Verpacks.-Rundsch.* 28:77–82.

Cerny, G. 1982. Sterilization of packaging materials for aseptic packaging. 4th Notification: Decontamination of packaging surfaces with superheated steam in the thermoform-and filling-machine THM 13/37-AS of the company Hassia (in German). *Verpacks.-Rundsch.* 33:47–50.

Cerny, G. 1983. Sterilization of plastic cups by means of saturated steam in the aseptic packaging machine DOGAseptik 81 developed by GASTI (in German). *Verpacks.-Rundsch.* 34:55–58.

Cerny, G. 1985. Studies for sterilization of surfaces of packaging materials by means of a combined treatment with hydrogen peroxide and ultraviolet irradiation (in German). *Verpacks.-Rundsch.* 36:29–32.

Cerny, G. 1990: Sterilization of packaging materials for aseptic packaging (in German). *ZFL* 41:54–58.

Cerny, G. 1992. Testing of aseptic machines for efficiency of sterilization of packaging materials by means of hydrogen peroxide. *Packaging Technol. Sci.* 5:77–81.

Conrads, H. M. and Schmidt, M. 2000. Plasma generation and plasma sources. *Plasma Sources Sci. Technol.* 9:441–454.

DIN 5031-7 (Hrsg., 1984): Optical radiation physics and illumination engineering. Terms for wavebands.

Dunn, J. 1996. Pulsed light and pulsed electric field for foods and eggs. *Poult. Sci.* 75:1113–1136.

Dunn, J., Ott, T., and Clark, W. 1995. Pulsed-light treatment of food and packaging. *Food Technol.-Chicago* 9:95–98.

Elmnasser, N., Guillou, S., Leroi, F. et al. 2007. Pulsed-light system as a novel food decontamination technology: A review. *Can. J. Microbiol.* 53:813–821.

Engelhard, P. 2005. Inactivation of microorganisms on solid surfaces in atmospheres of humid air/hydrogen peroxide by infrared-treatment (in German). PhD dissertation, Technical University of Munich, Munich, Germany.

Falkenstein, Z. 1997. Influence of ultraviolet illumination on microdischarge behaviour in dry and humid N_2, O_2, air, and Ar/O_2: The JOSHI effect. *J. Appl. Phys.* 81:5975–5979.

FDA. 1999: Standards for the fabrication of single service containers and closures for milk and milk products. Washington, DC: Food and Drug Administration.

FDA. 2007. Federal Register. Code of Regulations. Title 21. Vol. 2. Part 113.3. Washington, DC: Food and Drug Administration.

Feichtinger, J., Schulz, A., Walker, M., and Schumacher U. 2003. Sterilisation with low-pressure microwave plasmas. *Surf. Coating Technol.* 174–175:564–569.

Franken, O., Pietsch, G., and Saveliev, A. et al. 2003. Fundamental investigations of plasma-based processes for the sterilization of food packaging. Research report, RWTH University, Aachen, BMBF-Project FKZ 13N7609/9.

Haji-Saeid, M., Sampa, M.H.O., and Chmielewski, A.G. 2007. Radiation treatment for sterilization of packaging materials. *Radiat. Phys. Chem.* 76:1535–1541.

Han, B.H. 1977. Destruction of microorganisms on solid surfaces (in German). PhD dissertation, Technical University Karlsruhe, Karlsruhe, Germany.

Heise, M., Lierfeld, T., Franken, O., and Neff, W. 2004. Single filament charge transfer and UV-emission properties of a cascaded dielectric barrier discharge (CDBD) set-up *Plasma Sources Sci. Technol.* 13:351–358.

Hennlich, W. 2005. Guideline values for food packaging (in German). In *Handbuch Lebensmittelhygiene. Praxisleitfaden mit wissenschaftlichen Grundlagen*, eds. K. Fehlhaber, J. Kleer, and F. Kley. Hamburg, Germany: Behr's Verlag, pp. 36–46.

Holdsworth, S.D. 1992. *Aseptic Processing and Packaging of Food Products*. Essex, U.K.: Elsevier Science Publishers LTD.

International Dairy Federation (IDF). 1995: Technical guide for the packaging of milk and milk products. Bulletin N° 300/1995 (Chapter 7: Hygiene). Brussels, Belgium: International Dairy Federation.

Igura, N., Kamimura, Y., Shahidul Islam, M. et al. 2003. Effects of minerals on resistance of *Bacillus subtilis* spores to heat and hydrostatic pressure. *Appl. Environ. Microbiol.* 69 (10):6307–6310.

Kessler, H.G. 2002. *Food and Bio Process Engineering—Dairy Technology*. München, Germany: Kessler Verlag.

Krämer, J. 2002. *Food Microbiology* (in German), 4th edn. Stuttgart, Germany: Eugen Ulmer.

Melly, E., Cowan, A.E., and Setlow P. 2002. Studies on the mechanism of killing of *Bacillus subtilis* spores by hydrogen peroxide. *J. Appl. Microbiol.* 92:1105–1115.

Menashi, W.P. 1968. Treatment of Surfaces. U.S. Patent 3 383 163, issued July 9, 1968.

Mohler, W. 1966. Informations in the range of food investigations and hygiene (in German). Veröff. des Eidg. Gesundheitsamtes Bern 57:489–505.

Muranyi, P., Langowski, H.C., and Wunderlich, J. 2006a. Plasma technology—New paths for decontamination of packaging materials (in German). *Chem. Ing. Tech.* 78:1697–1706.

Muranyi, P., Wunderlich, J., and Dobosz, M. 2006b. Sterilization of filling machines: Standardization of bio indicators, investigation methods and validation processes (in German). *Chem. Ing. Tech.* 78:1667–1673.

Muranyi, P., Wunderlich, J., and Heise, M. 2007. Sterilization efficiency of a cascaded dielectric barrier discharge. *J. Appl. Microbiol.* 103:1535–1545.

Nicholson, W.L., Munakata, N., Horneck, G. et al. 2000. Resistance of *Bacillus* endospores to extreme terrestrial and extraterrestrial environments. *J. Mol. Microbiol. Biotechnol.* 64 (3):548–572.

Palmieri, L. and D. Cacace 2005. High intensity pulsed light technology. In *Emerging Technologies for Food Processing*, ed. D.W. Sun. London, U.K.: Elsevier Academic Press, pp. 279–306.

Reuter, H. 1986. Aseptic packaging of foods—Fundamentals and state of the art. *Chem. Ing. Tech.* 58:785–793.

Reuter, H. 1993: Aseptic packaging—Processes for packaging materials sterilization and system requirements. In *Aseptic Processing of Foods*, ed. H. Reuter. Hamburg, Germany: Behr's Verlag, pp. 155–165.

Schlegel, H.G. 1992. *General Microbiology* (in German). Stuttgart, Germany: Georg Thieme Verlag.

Schneider, J., Baumgärtner, K.M., and Feichtinger, J. et al. 2005. Investigation of the practicability of low-pressure microwave plasmas in the sterilization of food packaging materials at industrial level. *Surf. Coating Technol.* 200:962–966.

Silverman, J. 1977. Basic concepts of radiation processing. *Radiat. Phys. Chem.* 9:1–15.

Toledo, R.T., Escher, F.E., and Ayres, J.C. 1973: Sporicidal properties of hydrogen peroxide against food spoilage organisms. *Appl. Microbiol.* 26:592–597.

Turtoi, M. and Nicolau, A. 2007: Intense light pulse treatment as alternative method for mould spores destruction on paper-polyethylene packaging material. *J. Food Eng.* 83:47–53.

von Bockelmann, B. and von Bockelmann, I. 1986: Aseptic packaging of liquid food products: A literature review. *J. Agric. Food Chem.* 34:384.

VDMA 2002. Code of practice—Testing the effectiveness of aseptic plants fitted with packaging sterilization devices (in German). VDMA Bulletin No. 6.

VDMA 2003. Code of practice—Testing aseptic plants sterilizing the sterile zone in a machine interior (in German). VDMA Bulletin No. 8.

VDMA 2006. Aseptic packaging machines for the food industry—Minimum requirements and basic conditions for the intended operation (in German). VDMA Guideline 8742.

Wallhäußer, K.H. 1995. *Practice of Sterilization* (in German), 5th edn. Stuttgart, Germany: Georg Thieme.

Wekhof, A., Trompeter, F.J., and Franken, O. 2001: Pulsed UV disintegration (PUVD): A new sterilization mechanism for packaging and broad medical-hospital applications. The First International Conference on Ultraviolet Technologies, June 14–16, 2001, Washington, D.C.

Wilke, B. 1996. Validation of an aseptic plant for thermoplastic packaging materials (in German). *ZFL* 1/2:24–27.

15 Controlled and Modified Atmosphere Packaging of Food Products

David O'Beirne

CONTENTS

15.1 INTRODUCTION

Modified atmosphere packaging (MAP) is used in the preservation of a wide range of food products, from uncooked fresh chilled products, through cooked perishable foods, to long-life products stable at ambient temperatures. Depending on how it is applied, this technology can substantially reduce microbial spoilage, and slow down plant physiology and a range of unwanted chemical changes. It adds to the storage-life available from chilling in fresh-cut prepared vegetables and fruits, red meats and poultry, fish, and ready meals. It is used to protect flavor in a wide range of dry and intermediate moisture foods, such as ground and instant coffee and roasted nuts.

Many applications of MAP are now mature components of the food preservation sector. Exceptions to this are applications in fresh-cut produce, where a range of technical and food safety challenges make this a difficult but dynamic sector. This chapter will focus on applications in a number of chilled foods, including the fresh-cut produce sector.

15.2 GENERAL PRINCIPLES

Although a range of gases may be used in MAP applications, most commercial practices use only the three main gases present naturally in air. As its name suggests, the technology involves packaging foods in atmospheres with compositions modified from that of air, which contains approximately 79% nitrogen (N_2), 21% oxygen (O_2), and 0.03% carbon dioxide (CO_2).

CO_2 levels ≥20% are used to inhibit microbial growth, particularly growth of gram-negative bacteria and moulds. For example, Enfors et al. (1979) have shown that, with 20% CO_2, the lag phase and the log growth rate were retarded by about 50% in pork stored at 4°C. Higher levels of CO_2, up to 40%, can delay microbial growth further.

The precise mechanisms by which CO_2 exerts its antimicrobial activity are unclear; a number of molecular effects appear to be involved. In sufficient concentration (usually ≥20%), CO_2 causes a reduction in the extracellular pH, diffusion of H_2CO_3 across the bacterial membrane, and intracellular pH changes (Wolfe 1980). Alterations in cell membrane structure at elevated CO_2 concentrations have also been observed (Castelli et al. 1969, Dixon and Kell 1989). The lowering of pH inside cells by CO_2 might cause inhibition and/or inactivation of key enzymes essential for metabolic processes, such as glycolysis, amino acid and peptide transport, active transport of ions, and proton translocation (Hutkins and Nannen 1993). In addition, the direct inhibition of bacterial enzymatic processes by CO_2 has been demonstrated (Dixon and Kell 1989, Gill and Tan 1979, King and Nagel 1975, Kritzman et al. 1977). The extent of inhibition by CO_2 varies with the microorganism, the initial bacterial population size, CO_2 concentration, temperature of incubation, and substrate (Dixon and Kell 1989, Hudson et al. 1994). Gram-negative bacteria, particularly aerobes such as *Pseudomonads*, are relatively sensitive to CO_2, whereas other bacteria, for example, lactic acid bacteria are quite resistant (Enfors and Molin 1980, Francis and O'Beirne 1998). Because the solubility of CO_2 is greater at low temperature, there are important synergistic effects between low temperature and CO_2 (Finne 1981), and in practice MAP applications in fresh perishable foods are only effective at low storage temperatures (≤4°C) (Gil and Tan 1979).

Elevated levels of N_2 and O_2 can produce useful chemical effects in some products, and MAP often delivers more than one preservation benefit, usually by combining elevated CO_2 with one of these gases. For example, MAP of fresh red meat typically involves flushing with approximately 80% O_2 and 20% CO_2, to both retain high levels of oxymyoglobin and slow microbial growth. In cured meats such as sliced bacon, CO_2 is combined with N_2, while maintaining O_2 levels $<0.05\%$. There are also some applications in dry products (e.g., ground or instant coffee) for which 100% N_2 alone is used to slow off-flavor development due to lipid oxidation. All of these applications require packaging materials with high gas barrier properties to retain the added gases.

By contrast, applications in fresh and fresh-cut produce require materials with relatively low gas barriers. The objective in this case is to balance the gas permeability of the packaging with the respiration rate of the product in such a way that a useful atmosphere modification takes place. Such modifications are typically in the range 1%–3% O_2 and 3%–10% CO_2, balance N_2. These conditions slow respiration and physiological aging, inhibit ethylene activity, and, in many products, slow enzymatic browning at cut surfaces (O'Beirne 1990).

15.3 PACKAGING TECHNOLOGY

15.3.1 Packaging Materials

The properties required of packaging materials depend on the application. For applications other than those in fresh-cut fruits and vegetables, a high gas barrier is required. In general, the packages must be strong, transparent, and resist condensation on the inside. The components must be heat sealable for package assembly. The surface may need to accept print or an adhesive label.

Packages can take a number of forms, such as sealed tray, blister pack, bag, etc. A typical package used in consumer MAP of fresh red meat might involve a sealed tray formed from a high barrier laminate containing, for example, amorphous polyethylene terephthalate (APET) and polyethylene (PE), which would be sealed using a flexible lidding film. The lidding laminate might include polyvinyledene chloride (PVdC) with polyester (protection, printable), and PE (heat sealable). There is a range of other combinations of materials for trays and lidding webs available for use in such high gas barrier applications (O'Beirne 1987).

In low gas barrier applications (respiring produce), a range of materials is also available, but oriented polypropylene (OPP) and a range of microperforated OPP films are typically used. The packages take the form of sealed bags or of rigid containers closed with the microperforated OPP.

15.3.2 Packaging Operations

In the case of rigid trays, the packages are usually made and filled by form-fill-seal machines. The essential steps are

- Thermoforming the tray
- Filling the tray
- Removing air
- Replacing with a modified atmosphere (MA)
- Sealing the package

In such tray sealing machinery, the product is placed in the tray, the lidding film is set in place above the tray, a vacuum is drawn, the package is flushed with the required MA, and the lidding film is sealed to the tray. The gas injection operation can be controlled by time or pressure. Time-controlled injection can be used to create a positive pressure (slight balloon effect) in the package. This can be useful in products that absorb CO_2 or as a visual check for leaking packs.

In the case of sealed bags or pillow packs, synchronized bag forming, product introduction, gas flushing, and sealing are required.

15.4 APPLICATIONS IN FRESH-CUT PRODUCE

Application of MAP to fresh-cut produce involves quite complex technical and safety considerations, because these products are actively respiring (consuming O_2 and producing CO_2), and because many are consumed fresh, i.e., without any thermal treatment which might destroy pathogens. The latter problem is discussed in detail under food safety issues (Section 15.6). Achieving a stable and technically useful MA for respiring products involves balancing the respiratory gas changes with appropriate package design and permeability.

Deterioration in vegetables and fruits occurs mainly through the processes of physiological aging and water loss. In fresh-cut products, enzymatic browning at cut surfaces and microbial spoilage are also important. The rate at which a product respires is a major determinant of the pace of physiological aging. The Q_{10} value for a biological process is the factor by which the rate of reaction changes per 10°C change in temperature. Although chilling slows respiration (Q_{10} 2–3), MAP can provide up to a fourfold additional reduction (Ahvenainen 1996). Useful atmospheres are generally in the range 1%–3% O_2, 3%–10% CO_2, balance N_2. The benefits of MAP arise from a general reduction in the rates of metabolic processes and the retardation of senescence. Control of enzymatic browning at cut surfaces is a key benefit. While the cut surfaces of chilled prepared vegetables in over-wrapped trays turn brown within 24h of packaging, MAP can slow this process in most products for 5–7 days (O'Beirne 1990).

The effects of reduced O_2 and elevated CO_2 in extending storage life in fruits and vegetables have been demonstrated in research and commercial application of controlled atmosphere (CA) storage (Dewey 1983, Isenberg 1979, Smock 1979). Besides reduced respiration, these benefits arise through reduced ethylene synthesis and activity. This retards ripening in climacteric fruits and toughening in some vegetables (Burg and Burg 1975). Elevated CO_2, with or without reduced O_2, has been shown to reduce chlorophyll breakdown (Singh et al. 1972). In fresh-cut products, low levels of O_2 retard enzymatic browning at cut surfaces. Although CO_2 levels are generally not high enough to have direct antimicrobial effects (Finne 1981), microbial spoilage can be indirectly slowed by the improved condition of the stored produce (Rizvi 1988).

By contrast with CA storage, tight control of gas levels is generally not possible in MAP applications. The objective is to achieve an equilibrium MA which is technically useful. This equilibrium MA is determined primarily by product respiration and gas permeability of the packaging materials used. Product weight, initial gas flushing (if used), storage temperature, exposure to light, and other factors must be taken into account (McLaughlin and O'Beirne 1999, O'Beirne 1990). In addition, the tolerance of different products for low levels of O_2 and/or high levels of CO_2 must also be considered. Specific physiological disorders, for example, brown stain in lettuce and internal browning and surface pitting of pome fruits, can be induced by unsuitable atmospheres (Rizvi 1988, Zagory and Kader 1988).

Fresh-cut products are subjected to a series of minimal processing steps. Because severe processing would destroy the fresh character of these products, the unit operations do not currently include a pasteurization step, much less a commercial sterilization step (O'Beirne and Francis 2003). Depending on the commodity, the following steps may apply: removal of unwanted parts such as outer leaves, preliminary washing to remove soil and debris, peeling and/or slicing as appropriate, antimicrobial dipping (usually in an aqueous chlorine solution), rewashing in fresh water to remove the chlorine residue, and finally packaging within flexible and/or rigid materials with suitable gas barrier properties (O'Beirne 2007). As mentioned in Section 15.3, many fresh-cut vegetables are packaged in microperforated OPP. OPP films with different numbers and sizes of microperforations are available, tailored to products with different respiration rates. When necessary, the package may be flushed with 100% nitrogen, low oxygen, or other special atmospheres. For many products, however, the respiration rate of the product combined with the gas permeability properties of the packaging are sufficient to modify the package atmosphere. Although many applications produce technically useful MAs, some products, particularly those with high respiration rates, are difficult to match with suitable gas permeability, and unintended MAs can result (Day 1994, McLaughlin and O'Beirne 1999). Suboptimal atmospheres can cause a range of biochemical and physiological changes, which affect flavor, texture, and appearance (Cliffe-Byrnes et al. 2003). These include effects of anaerobic respiration, stress response effects, polyphenol oxidase activity, lipoxygenase activity, pectolytic changes, and destruction of chlorophyll/other pigments (Varoquaux and Wiley 1994, Watada et al. 1990).

Despite such problems in a few product types, the technology has been successfully used commercially in a wide range of fresh-cut vegetable products, and in recent years it has been extended to the production of fresh-cut fruits.

15.5 APPLICATIONS IN MEAT AND FISH

15.5.1 Fresh Red Meat

Fresh red meat (e.g., beef) deteriorates through loss of its bright red color, through off-flavor development due to microbial growth and/or lipid oxidation, and through increasing exudate/drip loss. Not all aging is detrimental, however, as tenderness increases with modest age. Loss of red color is a key determinant of shelf life in retail packaging, and color chemistry considerations are also important in other MAP applications.

The changes in color are mediated through chemical changes that take place in the heme group of the myoglobin pigment of muscle. In its oxygenated form, "oxymyoglobin" is responsible for bright red color. In the absence of oxygen, its deoxygenated form "myoglobin" is dark purple in color. Either form can be oxidized to "metmyoglobin," which has an unattractive brown color.

15.5.1.1 Retail Packaging

In retail MAP of fresh red meat, the product is typically flushed with 20%–25% CO_2/75%–80% O_2. The CO_2 slows the growth of key spoilage bacteria, particularly *Pseudomonads*, and the O_2 maintains myoglobin in the red oxymyoglobin form to a depth of at least 4–5 mm at the product surface (Hood 1981, Taylor 1982). In a deep drawn sealed tray with a volume of headspace to meat of 3:1, O_2 levels remain above 60%, despite some losses due to tissue and microbial respiration (O'Beirne 1987). This is sufficient to maintain acceptable color for about 7–9 days at 4°C. For successful MAP of red meat, hygienic slaughter and tight temperature control are essential. Since exudate from the meat becomes apparent during storage, a tray with a patterned base and an absorbent pad are used. The package must be sufficiently large to prevent the meat resting on the tray sides, as this will lead to discolored patches. Condensation is minimized by avoiding temperature fluctuations, and treatment of the inner surface of the lid with an antifog coating.

This application of MAP facilitates centralized cutting and packaging, though not for distant markets. Its disadvantages are cost and the unsuitability of the packages for domestic freezing. The latter problem has been addressed by some packers through the use of an outer flexible pack (flushed bag) containing individual high-value meat portions each shrink wrapped in low-barrier film. When the outer bag is opened, unused cuts can be easily frozen in a domestic freezer.

15.5.1.2 Bulk Packaging

A simple form of short-term bulk packaging involves the use of a master pack of high gas barrier material that encloses retail packs that have permeable overwraps (e.g., plasticized PVC). The master pack is flushed with a high oxygen atmosphere (e.g., 20%–25% CO_2/75%–80% O_2) using a machine such as a CVP® A300 (CVP Systems, Inc., Downers Grove, Illinois) equipped with a snorkel-like device to evacuate and gas flush the master pack prior to heat sealing. Because of difficulties in achieving complete evacuation, the composition of such an atmosphere is not as tightly controlled as in the form–fill–seal retail packs described earlier. However, this configuration can be useful for short periods of bulk transportation or storage of over-wrapped retail packs.

The traditional bulk packaging of meat has involved boning out of carcasses to primals and vacuum packaging. This provides a chill storage life of 6–8 weeks for transportation/extended storage. The anoxic atmospheres within vacuum packages slow microbial growth, and alter the microbial ecology away from *Pseudomonads* to lactic acid bacteria (Dainty et al. 1979). The meat color is purple during storage, but the bright red color of oxymyoglobin returns when the package is opened for use. The meat is cut up for sale, or for retail MAP with a high-oxygen atmosphere as described previously. Attempts have been made to exploit the benefits of vacuum packaging directly in retail packs using vacuum skin packaging of retail cuts, but adverse consumer reaction to the dark purple color of the meat has limited its commercial application.

The possibility of simulating or improving on the benefits of vacuum packaging by using anoxic MA/CA in bulk packaging has also been explored. Although a range of potential master pack and bulk packaging applications has been investigated, commercial applications have been mostly restricted to products that cannot be successfully stored in vacuum packages, such as lamb carcasses (Gill 2003).

Both 100% N_2 and 100% CO_2 atmospheres have been used. Atmospheres of 100% N_2 result in storage lives similar to those found in vacuum packages, whereas the antimicrobial effects of the 100% CO_2 atmospheres extend product life beyond that possible with N_2. When 100% CO_2 is used, allowance must be made for substantial absorption of CO_2 by the carcass. This absorption should be equivalent to the carcass volume. In these anoxic packs, O_2 levels must be below 100 ppm, because the rate of formation of metmyoglobin is extremely high at low oxygen levels around 1%. This low O_2 level is achieved by more sophisticated levels of gas-flushing technology, involving the enclosure of the snorkel and bag-sealing elements within a hood which is evacuated during the process (Gill 2003, Penny and Bell 1993). There has also been extensive research on the use of in-pack O_2 scavengers (Doherty and Allen 1998). However, the benefits of scavenger use have been questioned, and many scavengers do not work efficiently at the low levels of O_2 and low temperatures experienced in these gas-flushed packages.

15.5.2 Cured Meats

The pigment nitrosomyoglobin is formed during curing, and is responsible for the pinkish color of cured meats. This pigment is preserved at low oxygen tension combined with low temperature, traditionally using a vacuum package. In MAP applications, an anoxic atmosphere with O_2 levels <0.1% is required. CO_2 is included to slow microbial growth, a typical atmosphere being 20% CO_2/80% N_2. Products are packaged in slim form–fill–seal packages and tested to ensure O_2 levels are <0.05%. Compared with vacuum packaging, there is no compaction of the slices and more flexibility in retail presentation, including easy-open pack features (Rizvi 1981).

15.5.3 Poultry

The shelf life of fresh poultry is mainly determined by microbial growth, particularly of *Pseudomonas* and *Acinetobacter* species (Lyijynen et al. 2003). Recommendations for optimum atmospheres suitable for fresh whole and cut poultry vary from 20% to 70% CO_2 (Goodburn and Halligan 1988). At least 20% CO_2 is used; above this level, CO_2 can lead to discoloration and off-flavor development. When 25% CO_2 is used, it is combined with 75% N_2 to limit rancidity and prevent any pack collapse resulting from CO_2 absorption. MAP can be expected to extend the chilled storage life of fresh poultry from 6 days to up to 9–12 days (O'Beirne 1987).

15.5.4 Fresh Fish

MAP has been applied to a range of fish species (Cann 1984, Fey and Regenstein 1982, Lannelonge et al. 1982). Atmospheres recommended by Torrey Research Station include 40% CO_2/30% O_2/30% N_2 (white fish), and 60% CO_2/40% N_2 (fatty fish)

(Cann 1984). Too much CO_2 can cause excessive drip; O_2 is excluded in fatty fish to minimize rancidity. Use of freshly caught fish and low (0°C) temperatures is essential.

15.6 FOOD SAFETY ISSUES

15.6.1 Extended Shelf Life and Altered Ecology

A key safety issue is whether MAs, and the production and processing systems associated with MAP of foods, increase the opportunities for pathogen contamination, survival, and growth compared with storage in air. MAP might do this by simply increasing the time available for pathogens to grow to significant numbers, and over-extension of shelf life particularly at abuse storage temperatures, is a significant risk factor.

Secondly, these atmospheres and low temperatures may inhibit aerobic spoilage organisms, and may thus facilitate the growth of pathogens without obvious signs of spoilage (Daniels et al. 1985, Farber 1991). Concerns have been raised about the possibility of *Clostridium botulinum* growth and toxin formation without obvious spoilage in MAP of fish stored at abuse temperatures (8°C) (Lindsay 1981). In fresh-cut produce, unintended anoxic atmospheres, or atmospheres high in CO_2 (>20%), develop due to poor product-package compatibility or temperature abuse. In an extreme case, anoxic conditions can enable growth and toxin production by *C. botulinum*, whereas high CO_2 atmospheres may promote the growth of pathogens, while limiting the growth of spoilage organisms (Carlin et al. 1996). For instance, CO_2 had little or no inhibitory effect on growth of *Escherichia coli* O157:H7 on shredded lettuce stored at 13°C or 22°C, and growth potential was increased in an atmosphere of $O_2/CO_2/N_2$: 5/30/65, compared with growth in air (Abdul-Raouf et al. 1993, Diaz and Hotchkiss 1996).

Francis and O'Beirne (1997), in turn, observed that when packages of shredded lettuce were flushed with inert atmospheres to minimize enzymatic browning, contaminating *Listeria monocytogenes* survived much better than in unflushed packages, at mild temperature abuse. This appeared to be due to the effects of these atmospheres and temperatures on the growth of organisms that compete with *L. monocytogenes* (Francis and O'Beirne 1998).

Data suggest that MAs in general seem safe at chill storage temperatures. However, at abuse temperatures, complex interactions between elements of the natural background microflora may have significant effects on survival and growth of pathogens.

15.6.2 Temperature

Prevention of contamination is the single most important measure for the elimination of pathogens. However, if contamination has taken place, storage temperature is a key factor affecting bacterial pathogen survival and growth. Storage at adequate refrigeration temperatures limits pathogen growth to those that are psychrotrophic, notably: *L. monocytogenes*, *Yersinia enterocolitica*, nonproteolytic *C. botulinum*, and *Aeromonas hydrophila*.

Although psychrotrophic organisms, such as *L. monocytogenes*, are capable of growth at low temperatures, reducing the storage temperature (≤4°C) will significantly reduce their rate of growth (Beuchat and Brackett 1990, Carlin et al. 1995). On packaged vegetables stored at 4°C, for example, *L. monocytogenes* populations remained constant or decreased, whereas at 8°C growth of *L. monocytogenes* was supported on all vegetables, with the exception of coleslaw mix (Francis and O'Beirne 2001). Thus, even mild temperature abuse during storage permits more rapid growth of psychrotrophic pathogens (Berrang et al. 1989, Carlin and Peck 1996, Farber et al. 1998, Conway et al. 2000, Rodriguez et al. 2000).

Mesophilic pathogens, such as *Salmonella* and *E. coli* O157:H7, are unable to grow where temperature control is good (i.e., ≤4°C). However, if temperature abuse occurs, then they may grow. Survival of *Salmonella* in produce stored for extended periods in chilled conditions may be of concern (Piagentini et al. 1997, Zhuang et al. 1995); *Salmonella* survived on a range of vegetables for more than 28 days at 2°C–4°C (ICMSF 1996). *E. coli* O157:H7 populations survived on produce stored at 4°C and proliferated rapidly when stored at 15°C (Richert et al. 2000). Reducing the storage temperature from 8°C to 4°C significantly reduced growth of *E. coli* O157:H7 on MAP vegetables; however, viable populations remained at the end of the storage period at 4°C (Francis and O'Beirne 2001). For some pathogens, such as *E. coli* O157:H7 or *L. monocytogenes*, the infectious dose can be very low for susceptible populations. In these cases any contamination can lead to illness. Temperature control cannot be the primary strategy for assuring safety from viruses and parasites.

15.6.3 Organisms Involved and Opportunities for Contamination

Pathogens of concern include *L. monocytogenes*, *E. coli* O157:H7, *Salmonella* sp., nonproteolytic *C. botulinum*, *Yersinia enterocolitica*, and *A. hydrophila*, together with viral and protozoan pathogens.

To constitute a problem, pathogens must have the opportunity to contaminate the product, generally must be able to survive and grow, and they must not be destroyed at a subsequent stage of processing. The main opportunities for contamination of meat products (e.g., with *E. coli* O157:H7 or *Salmonella* sp.) occur in the slaughter/preslaughter period, though contamination can also occur during cutting and packaging, and through cross-contamination, particularly in food preparation environments. A high percentage of poultry carry *Salmonella* contamination, and some fish species carry *C. botulinum* spores.

In fresh-cut products, the main opportunities for contamination (e.g., by *L. monocytogenes*, *E. coli* O157:H7, *Salmonella* sp.) are believed to occur during growing and harvesting of the crops, though contamination with *L. monocytogenes* has been reported during processing, and cross-contamination may occur in food preparation environments. Contamination of vegetables and fruits destined for use as fresh-cut products constitutes a very serious threats to food safety. This is because there are no subsequent effective decontamination steps in the minimal processing used, and many of these products are consumed without any further treatments such as cooking. Even rewashing prior to consumption has not been required.

The problem facing the fresh-cut sector is illustrated by the major food poisoning outbreak linked to bagged spinach in 2006 and similar outbreaks in lettuce. The outbreak occurred in September 2006 and was ultimately traced to Natural Selection Foods LLC (San Juan Bautista, CA). The investigation indicated that the outbreak was associated with bagged spinach produced under multiple labels in a single plant on a single day during a single shift (King 2006). There were 200 cases of illness in 26 states and three people died.

The U.S. Food and Drug Administration (FDA), United States Department of Agriculture (USDA), Centers for Disease Control and Prevention (CDC), and the State of California investigated the precise cause of the outbreak, particularly in terms of the source and mechanism of contamination. The on-site investigation has focused on inspections and microbial sample collections at facilities, the general environment, and water. Animal management practices, water use, and other environmental factors that could have led to contamination were reviewed and evaluated (Bracket 2006). The same *E. coli* O157:H7 strain has been found in samples from a stream, and in the feces of cattle and wild pigs from ranches implicated in the outbreak. There is also evidence that wild pigs have been in the spinach fields. The Final Report (FDA News, March 23, 2007) was unable to definitely determine the precise method by which the bacteria spread to the spinach. Potential environmental risk factors were identified as wild pigs, and the proximity of irrigation wells and surface waterways exposed to feces from cattle and wildlife. Updated guidelines on growing, harvesting, cooling, and processing have been issued by the Department of Health Services of the State of California (CHDS 2007).

15.6.4 Food Poisoning Incidents

In the early years of the development of the fresh-cut industry, there were relatively few links between these products and food poisoning. Those that have been linked include an outbreak of botulism ultimately linked to a modified atmosphere packaged dry coleslaw product (Solomon et al. 1990) and a *Salmonella* Newport outbreak linked to ready-to-eat salad vegetables (PHLS 2001). There was also an outbreak of shigellosis linked to shredded lettuce (Davis et al. 1988), though exactly how this product was packaged is unclear.

However, in the United States, there have been 18 outbreaks between 1998 and 2006 associated with spinach, lettuce, and mixed leaves. As outlined in Section 15.6.3, these have been linked to agricultural production practices (Gorny and Zagory 2000).

15.6.5 Measures to Ensure Safety

The two primary objectives in ensuring safety are prevention of contamination and storage at low temperature. These can be achieved through Good Agricultural Practice (GAP), Good Manufacturing Practice (GMP), and the application of Hazard Analysis and Critical Control Point (HACCP) principles throughout production, slaughter, processing, packaging, and storage.

Avoiding contamination of fresh-cut produce is particularly important. Table 15.1 summarizes the factors in minimizing risk, based on current (incomplete) knowledge.

TABLE 15.1
GAP, GMP, and HACCP from Preplanting to Consumption

Production and harvesting
Treatment of field as extension of factory
Land management (location, fertilization)
Water management (protection, testing programme)
Animal management (fencing, secure pack-house, birds)
Worker hygiene management (training, toilet, and washing facilities)
Equipment management (cutting knives, contact surfaces)
Fresh-cut processing
Programme for sanitizing surfaces and machines
Good preliminary decontamination and inspection
Avoid severe peeling/cutting
Eliminate/minimize human contact with processed product
Deploy effective washing/antimicrobial dipping
Avoid postdipping contamination
Packaging/distribution/retail
Careful selection of packaging material
Monitor microbial quality of packaged product
Ensure temperature is <4°C
Suitably designed vehicles
Proper vehicle loading practices
Proper chill cabinet loading
Modest shelf life labeling
Educate retailer and consumer

This is an extremely complex arena with great diversity in crop production methods, scale, environmental factors, etc. (FDA 1998), and appears to be key to addressing the current uncertainties. A high level of responsiveness to improvements in the best HACCP practices is needed. Increasing globalization of produce supplies poses serious new challenges (Tauxe 1997), and knowledge of contamination levels in imported produce is minimal (Beuchat and Ryu 1997). The only rational solution is the extension of the requirement for GAPs to wherever primary production takes place.

15.7 TECHNOLOGY TRENDS

15.7.1 Packaging Technology

MAP can benefit from rapid developments taking place in the fields of smart/active/intelligent packaging. Smart packaging refers to packaging systems that respond to the system's needs, establishing a product–package interaction that compensates for potential problems (Fonseca et al. 2000, Poças and Oliveira 1997). Active packaging generally refers to the use of agents such as gas absorbers, releasing agents, etc.

within packages, to improve shelf life, quality, and microbiological safety; for example, oxygen, carbon dioxide, or ethylene scavengers deployed in pouches or incorporated into the packaging, can control the internal atmosphere (Vermeiren et al. 1999). Intelligent packaging refers to external and internal indicators used for quality control. These include indicators of time–temperature, oxygen, leakage, and microbial growth or presence of pathogens (Ahvenained 2003). In the near future, it is expected that the intelligent package will contain complex messages that can be read at a distance through miniature radio frequency identity tags (Byrne 1997). Significant legislative developments will be needed to permit many of these developments, and they will require a positive reaction from consumers.

More reliable package atmospheres can be expected as a result of improved materials, temperature responsive packaging, and better software to define gas permeability requirements for individual products. Mathematical modeling and software development for design of MAP systems is an important developing area (Mahajan et al. 2008). It is particularly, though not exclusively, relevant to packaging design for fresh and fresh-cut produce. Modeling takes into account the complex effects of respiration rate, temperature, amount of material packed, headspace of package, surface area of product, and permeability/permeability ratio and thickness of polymer film (McLaughlin and O'Beirne 1999). Some advanced software systems store all of the data needed to specify the packaging requirements of a range of products. This includes data on gas and humidity requirements of individual respiring products, respiration rates under different atmospheres, and gas and water vapor permeability of packaging films. By specifying the product of interest and storage conditions, together with product volume, mass, etc., the physical packaging requirements can be identified and the packaging materials needed can be specified.

15.7.2 Food Safety

The principle of combining a number of different preservation techniques ("hurdle" concept, Leistner 1992) is likely to become even more important in MAP systems. The intelligent application of hurdles of different numbers and intensities can be expected (McMeekin and Ross 2002). These can be applied prior to packaging, or to packaged products.

A sanitizing step (typically dipping in a 100 ppm aqueous solution of chlorine) is generally included in minimal processing of fresh-cut produce. This reduces microbial counts, but cannot be relied upon to eliminate pathogens. Due to its limited antimicrobial effectiveness, and increasing concerns over the production of chlorinated organic compounds, a variety of other sanitizers have been evaluated as alternatives. These include acidic electrolyzed water (Park et al. 2001), peroxyacetic acid (Park and Beuchat 1999), chlorine dioxide (Zhang and Farber 1996), organic acids (Francis and O'Beirne 2002, O'Beirne and Francis 2003), and ozone (Burrows et al. 1999). Although no alternative sanitizing/antimicrobial treatment has been found to compare with chlorine for effectiveness and cost, this continues to be an area of active research and development.

The inclusion of acceptable antimicrobials, such as herbs or essential oils, on products, as coatings or as part of "smart packaging," might have the potential to prevent survival and growth of pathogens (Burt 2004). Use of buffered lactic acid treatments in combination with MAP has been effective on cooked meats and on fresh poultry (Zeitoun and Debevere 1991).

Inoculation of MAP products with organisms inhibitory to one or more pathogens may become important. For example, strains of lactic acid bacteria inhibited *A. hydrophila*, *L. monocytogenes*, *Salmonella typhimurium*, and *Staphylococcus aureus* on vegetable salads (Vescovo et al. 1996), and use of a *Lactobacillus casei* inoculum reduced growth of *A. hydrophila* on MAP vegetables such as lettuce (Vescovo et al. 1997).

Combinations of MAP and physical treatments, notably irradiation, will continue to be an attractive technical option, if consumers can find irradiation acceptable. The field has been extensively studied in relation to pasteurization of poultry, meat, fish, and produce. Some combinations of atmosphere (e.g., N_2 atmospheres) and irradiation dose are effective in improving safety without significant damage to nutrients (Diehl 1995).

15.8 CONCLUSIONS

MAP technology is poised to continue to make an important contribution in addressing consumer demand for fresh/fresh-prepared foods. Much of the technology and its applications are tried and tested, and provide safe high-quality alternatives to thermally treated and frozen foods. The one area of serious vulnerability is the fresh-cut produce system, which includes products that are eaten raw. These products can be contaminated by pathogens and are not currently subjected to any effective decontamination/pasteurization treatment prior to use. There is an urgent need to address the current gaps in our understanding of the routes by which *E. coli* O157:H7 and other pathogens can contaminate fresh-cut products. Both improved and novel processing technologies may be needed to improve safety, and perhaps to move toward pasteurization of these products.

REFERENCES

Abdul-Raouf, U.M., Beuchat, L.R., and Ammar, M.S. 1993. Survival and growth of *E. coli* O157:H7 on salad vegetables. *Appl. Environ. Microbiol.* 59:1999–2006.

Ahvenained, R. 2003. Active and intelligent packaging: An introduction. In *Novel Food Packaging Techniques*, ed. R. Ahvenainen, pp. 5–21. London, U.K.: Woodhead Publishing.

Ahvenainen, R. 1996. New approaches in improving the shelf-life of minimally processed fruit and vegetables. *Trends Food Sci. Technol.* 7:179–187.

Berrang, M.E., Brackett, R.E., and Beuchat, L.R. 1989. Growth of *Aeromonas hydrophila* on fresh vegetables stored under a controlled atmosphere. *Appl. Environ. Microbiol.* 55:2167–2171.

Beuchat, L.R. and Brackett, R.E. 1990. Survival and growth of *Listeria monocytogenes* on lettuce as influenced by shredding, chlorine treatment, modified atmosphere packaging and temperature. *J. Food Sci.* 55:755–758, 870.

Beuchat, L.R. and Ryu, J.-H. 1997. Produce handling and processing practices. *Emerg. Infect. Dis.* 3:459–465.

Bracket, R.E. 2006. On role of FDA in the outbreak of *E.coli* illness linked to fresh spinach, Statement before Committee on Health Education Labor and Pensions, United States Senate.

Burg, S.P. and Burg, E.A. 1975. Molecular requirements for the biological activity of ethylene. *Plant Physiol.* 42:144–152.

Burrows, J., Yuan, J., Novak, J., Boisrobert, C., and Hampson, B. 1999. Ozone application in food processing. *Fresh Cut*, April, 50–54.

Burt, S. 2004. Essential oils: Their antibacterial properties and potential applications in foods—A review. *Int. J. Food Microbiol.* 94:223–253.

Byrne, G. 1997. Intelligent packaging. *Prod. Image Secur.* 1:21–22.

Cann, D. 1984. Developments in modified atmosphere packaging of fish. *IFST Proc.* 17:15–16.

Carlin, F. and Peck, M.W. 1996. Growth of and toxin production by non-proteolytic *Cl. botulinum* in cooked pureed vegetables at refrigeration temperatures. *Appl. Environ. Microbiol.* 62:3069–3072.

Carlin, F., Nguyen-the, C., and Silva, A.A.1995. Factors affecting the growth of *Listeria monocytogenes* on minimally processed fresh endive. *J. Appl. Bacteriol.* 78:636–646.

Carlin, F., Nguyen-the, C., Silva, A.A., and Cochet, C. 1996. Effects of carbon dioxide on the fate of *Listeria monocytogenes*, of aerobic bacteria and on the development of spoilage in minimally processed fresh endive. *Int. J. Food Microbiol.* 32:159–172.

Castelli, A., Littaru, G.P., and Barbesl, G. 1969. Effect of pH and CO_2 concentration changes on lipids and fatty acids of *Saccharomyces cerevisiae. Archiv. Mikrobiol.* 66:34–39.

CHDS (California Health Department Services). 2007. *Recommendations in follow up to the investigation of an Escherichia coli 0157:H7 outbreak associated with Dole pre-packaged spinach.* http://www.dhs.ca.gov. (March, 2007).

Cliffe-Byrnes, V., McLaughlin, C.P., and O'Beirne, D. 2003. Effects of packaging film and storage temperature on the quality of modified atmosphere packaged dry coleslaw mix. *Int. J. Food Sci. Technol.* 38:187–199.

Conway, W.S., Leverentz, B., Saftner, R.A., Janisiewicz, W.J., Sams, C.E., and Leblanc, E. 2000. Survival and growth of *Listeria monocytogenes* on fresh-cut apple slices and its interaction with *Glomerella cingulata* and *Penicillium expansum. Plant Dis.* 84:177–181.

Dainty, R.M., Shaw, B.G., Harding, C.O., and Michanie, S. 1979. The spoilage of vacuum packaged beef by cold tolerant bacteria. In *Cold Tolerant Microbes in Spoilage and the Environment*, eds. A.D. Russell and R. Fuller, pp. 83–100. London, U.K.: Academic Press.

Daniels, J.A., Krishnamurthi, R., and Rizvi, S.S.H. 1985. A review of the effect of carbon dioxide on microbial growth and food quality. *J. Food Prot.* 48:532–537.

Davis, H., Taylor, J.P., Perdue, J.N. et al. 1988. A shigellosis outbreak traced to commercially distributed shredded lettuce. *Am. J. Epidemiol.* 128:1312–1321.

Day, B.P.F. 1994. Modified atmosphere packaging and active packaging of fruits and vegetables. In *Minimal Processing of Foods, VTT Symposium Series* 142, eds. R. Ahvenainen, T. Mattila-Sandholm, and T. Ohlsson, pp. 173–209. Espoo Finland: VTT.

Dewey, D.H. 1983. Controlled atmosphere storage of fruits and vegetables. In *Developments in Food Preservation – 2*, ed. S. Thorne, pp. 1–24. London, U.K.: Applied Science.

Diaz, C. and Hotchkiss, J.H. 1996. Comparative growth of *E. coli* O157:H7, spoilage organisms and shelf life of shredded iceberg lettuce stored under modified atmospheres. *J. Sci. Food Agric.* 70:433–438.

Diehl, J.F. 1995. *Safety of Irradiated Foods*, 2nd revised edition. New York: Marcel Dekker Inc.

Dixon, N.M. and Kell, D.B. 1989. A review—The inhibition by CO_2 of the growth and metabolism of micro-organisms. *J. Appl. Bacteriol.* 67:109–136.

Doherty, A.M. and Allen, P. 1998. The effect of oxygen scavengers on the color stability and shelf-life of CO_2 master packaged pork. *J. Muscle Foods* 9:351–363.

Enfors, S.O. and Molin, G.1980. Effect of high concentrations of carbon dioxide on growth rate of *Pseudomonas fragi, Bacillus cereus* and *Streptococcus cremoris. J. Appl. Bacteriol.* 48:409–416.

Enfors, S.O., Molin, G., and Ternstrom, A.1979. Effect of packaging under carbon dioxide, nitrogen, or air on the microbial flora of pork stored at 4°C. *J. Appl. Bacteriol.* 14:197.

Farber, J.M.1991. Microbiological aspects of modified atmosphere packaging technology: A review. *J. Food Prot.* 54:58–70.

Farber, J.M., Wang, S.L., Cai, Y., and Zhang, S.1998. Changes in populations of *Listeria monocytogenes* inoculated on packaged fresh-cut vegetables. *J. Food Prot.* 61:192–195.

FDA.1998. *Guide to minimize microbial food safety hazards for fresh fruits and vegetables*, Washington, DC, 36 p.

Fey, M. and Regenstein J.M.1982. Extending shelf-life in fresh water red hake and salmon using CO_2-O_2 modified atmosphere and potassium sorbate ice at 1°C. *J. Food Sci.* 47:1048–1054.

Finne, G. 1981. The effect of carbon dioxide on the bacteriological and chemical properties of foods: A review. In *Proceedings of the First National Conference on Modified and Controlled Atmosphere Packing of Seafood Products*, ed. Martin, R.E., pp. 2–25. San Antonio, TX: National Marine Fisheries Service.

Fonseca, S.C., Oliveira, F.A.R., Brecht J.K., and Chau, K.V. 2000. Modeling O_2 and CO_2 exchange for development of perforation-mediated atmosphere packaging. *J. Food Eng.* 43:9–15.

Francis, G. and O'Beirne, D.1997. Effects of gas atmosphere, antimicrobial dip, and temperature on the fate of *Listeria innocua* and *Listeria monocytogenes* on minimally processed lettuce. *Int. J. Food Sci. Technol.* 32:141–152.

Francis, G.A. and O'Beirne, D.1998. Effects of storage atmosphere on *Listeria monocytogenes* and competing microflora using a surface model system. *Int. J. Food Sci. Technol.* 33:465–476.

Francis, G.A. and O'Beirne, D.2001. Effects of vegetable type, package atmosphere and storage temperature on growth and survival of *Escherichia coli* O157:H7 and *Listeria monocytogenes*. *J. Ind. Microbiol. Biotechnol.* 27:111–116.

Francis, G.A. and O'Beirne, D.2002. Effects of vegetable type and antimicrobial dipping on survival and growth of *Listeria innocua* and *E. coli*. *Int. J. Food Sci. Technol.* 37:711–718.

Gill, C.O. 2003. Active packaging in practice: Meat. In *Novel Food Packaging Techniques*, ed. R. Ahvenainen, pp. 365–383. London, U.K.: Woodhead Publishing.

Gill, C.O. and Tan, K.H. 1979. Effect of carbon dioxide on growth of *Pseudomonas fluorescens*. *Appl. Environ. Microbiol.* 38:237–240.

Goodburn, K.E. and Halligan, A.C. 1988. *Modified Atmosphere Packaging: A Technology Guide*. Leatherhead, U.K.: BFMIRA. 44 pp.

Gorny, J.R. and Zagory, D. 2000. Produce food safety. In *The Commercial Storage of Fruits Vegetables and Florist and Nursery Stocks* (*USDA Handbook 66*), eds. Gross, K.C., Saltveit, M.E., and Wang, C.Y., 35 p. Washington, DC: USDA.

Hood, D.E. 1981. Technology back-up for beef processors. *Farm Food Res.* 12:69–72.

Hudson, J.A., Mott, S.J., and Penney, N. 1994. Growth of *Listeria monocytogenes*, *Aeromonas hydrophila* and *Yersinia enterocolitica* on vacuum and saturated carbon dioxide controlled atmosphere-packaged sliced roast beef. *J. Food Prot.* 57:204–208.

Hutkins, R.W. and Nannen, N.L.1993. pH homeostasis in lactic acid bacteria. *J. Dairy Sci.* 76:2354–2365.

ICMSF (International Commission on Microbiological Specifications for Foods). 1996. *Microorganisms in Foods. 5. Microbiological Specifications of Food Pathogens*. London, U.K.: Blackie Academic and Professional.

Isenberg, F.M.R. 1979. Controlled atmosphere storage of vegetables. *Hort. Rev.* 1:301–336.

King, L.J. 2006. *Testimony on CDC food safety activities and the recent E. coli spinach outbreak, before Committee on Health, Education, Labour and Pensions*, United States Senate. CDC, 2006.

King, A.D. Jr. and Nagel, C.W. 1975. Influence of carbon dioxide upon the metabolism of *Pseudomonas aeruginosa*. *J. Food Sci.* 40:362–366.

Kritzman, G., Chel, L., and Henis, Y. 1977. Effect of carbon dioxide on growth and carbohydrate metabolism in *Sclerotium rolfsii*. *J. Gen. Microbiol.* 100:167–175.

Lannelonge, M., Hanna, M.O., Finne, G., Nickelson, R., and Vanderzant, C. 1982. Storage characteristics of finfish fillets (*Archosarqus probatocephalus*) packed in modified gas atmospheres containing carbon dioxide. *J. Food Prot.* 45: 440–444.

Leistner, L. 1992. Food preservation by combined methods. *Food Res. Int.* 25:151–158.

Lindsay, R. 1981. Modified atmosphere packaging systems for refrigerated fresh fish providing itself life extension and safety from *Clostridium botulinum* toxigenesis. In: *Proceedings of the First National Conference on Modified and Controlled Atmosphere Packaging of Seafood Products*, Ed. R.E. Martin, pp. 30–51. National Fisheries Institute, Washington, DC.

Lyijynen, T., Hurme, E., and Ahvenainen, R. 2003. Optimizing packaging. In *Novel Food Packaging Techniques*, ed. R. Ahvenainen, pp. 441–458. London, U.K.: Woodhead Publishing.

Mahajan, P.V., Oliveira, F.A.R., Montanez, J.C., and Iqbal, T. 2008. Packaging design for fresh produce: An engineering approach. *New Food*, issue 1: 35–36.

McLaughlin, C.P. and O'Beirne, D. 1999. The effect of storage temperature and respiratory gas concentration on the respiration rate of dry coleslaw mix. *J. Food Sci.* 64:116–119.

McMeekin, T.A. and Ross, T. 2002. Predictive microbiology: Providing a knowledge-based framework for change management. *Int. J. Food Microbiol.* 78:133–53.

O'Beirne, D. 1987. Concept and benefits of modified atmosphere packaging. *Food Ireland*, May/June, 43–44.

O'Beirne, D. 1990. Modified atmosphere packaging of fruits and vegetables (Review). In *Chilled Foods: The Revolution in Freshness*, ed. R. Gormley, pp. 133–199. London, U.K.: Elsevier Applied Science.

O'Beirne, D. 2007. Microbial safety of fresh-cut vegetables. *Proceedings of the International Conference on Quality Management of Fresh-Cut Produce*, eds. S. Kanlayanarat, P.M.A. Toivonen, and K.C. Cross, pp. 159–172. Bangkok Thialand, August 6–8: Acta Horticulturae 746.

O'Beirne, D. and Francis, G.A. 2003. Analysing pathogen survival in modified atmosphere packaged (MAP) produce. In: *Novel Food Packaging Techniques*, ed. R Ahvenainen, pp. 231–275. London, U.K.: Woodhead Publishing.

Park, C.-M. and Beuchat, L.R.1999. Evaluation of sanitizers for killing *Escherichia coli* O157:H7, *Salmonella* and naturally occurring microorganisms on cantaloupes, honeydew melons and asparagus. *Dairy Food Environ. Sanit.* 19:842–847.

Park, C.-M., Hung, Y.-C., Doyle, M.P., Ezeike, G.O.I., and Kim, C. 2001. Pathogen reduction and quality of lettuce treated with electrolysed oxidizing and acidified chlorinated water. *J. Food Sci.* 66:1368–1372.

Penny, N. and Bell, R.G. 1993. Effect of residual oxygen on the color, odor, and taste of carbon dioxide–packaged beef, lamb, and pork, during short-term storage at chill temperatures. *Meat Sci.* 33: 245–252.

PHLS (Public Health Laboratory System) 2001. *Salmonella* Newport infection in England associated with the consumption of ready to eat salad, *Eurosurveillance Weekly*, June, 26.

Piagentini, A.M., Pirovani, M.E., Güemes, D.R., Di Pentima, J.H., and Tessi, M.A.1997. Survival and growth of *Salmonella hadar* on minimally processed cabbage as influenced by storage abuse conditions. *J. Food Sci.* 62:616–618, 631.

Poças, M.F. and Oliveira, F.A.R. 1997. *Food packaging: A guide for package users*. FIPA, Portugal.

Richert, K.J., Albrecht, J.A., Bullerman, L.B., and Sumner, S.S. 2000. Survival and growth of *Escherichia coli* O157:H7 on broccoli, cucumber and green pepper. *Dairy Food Environ. Sanit.* 20:24–28.

Rizvi, S.S. 1988. Controlled atmosphere packaging of fruits and vegetables. *N.Y. Food Life Sci.* 18:19–23.

Rizvi, S.S. 1981. Requirements for foods packaged in polymeric films. *Crit. Rev. Food Sci. Nutr.* 14:111–134.

Rodriguez, A.M.C., Alcala, E.B., Gimeno, R.M.G., and Cosano, G.Z. 2000. Growth modeling of *Listeria monocytogenes* in packaged fresh green asparagus. *Food Microbiol.* 17:421–427.

Singh, B., Wang, D.J., Selunkhe, D.K., and Rahman, A.R. 1972. Controlled atmospheric storage of lettuce. 2. Effects on biochemical composition of the leaves. *J. Food Sci.* 37:52–55.

Smock, R.M. 1979. Recent advances in controlled atmosphere storage of fruits. *Hort. Rev.* 1:301–336.

Solomon, H.M., Kautter, D.A., Lilly, T., and Rhodehamel, E.J. 1990. Outgrowth of *Cl. botulinum* in shredded cabbage at room temperature under modified atmosphere. *J. Food Prot.* 53:831–833.

Tauxe, R.V.1997. Emerging foodborne diseases: An evolving public health challenge. *Dairy Food Environ. Sanit.* 17:788–795.

Taylor, A.A. 1982. Retail packaging systems for fresh meat. *Proceedings on the International Symposium "Meat Source and Technology"*, Lincoln, NE, pp. 352–366.

Varoquaux, P. and Wiley, R.C. 1994. Biological and biochemical changes in minimally processed refrigerated fruits and vegetables. In *Minimally Processed Refrigerated Fruits and Vegetables*, ed. R.C. Wiley, pp. 226–267. New York: Springer.

Vermeiren, L., Devlieghere, F., van Beest, M., de Kruijf, N., and Debevere, J. 1999. Developments in the active packaging of foods. *Trends Food Sci. Technol.* 10:77–86.

Vescovo, M., Scolari, G., Orsi, C., Sinigaglia, M., and Torriani, S. 1997. Combined effects of *Lactobacillus casei* inoculum, modified atmosphere packaging and storage temperature in controlling *Aeromonas hydrophila* in ready-to-use vegetables. *Int. J. Food Sci. Technol.* 32: 411–419.

Vescovo, M., Torriani, S., Orsi, C., Macchiarolo, F., and Scolari, G. 1996. Application of antimicrobial-producing lactic acid bacteria to control pathogens in ready-to-use vegetables. *J. Appl. Bacteriol.* 81:113–119.

Watada, A.E., Kazuhiro, A., and Yamuuchi, N. 1990. Physiological activities of partially processed fruits and vegetables. *Food Technol.* 44:116.

Wolfe, S.K. 1980. Use of CO- and CO_2-enriched atmospheres for meats, fish and produce. *Food Technol.* 34:55–58.

Zagory, D. and Kader, A.A. 1988. Modified atmosphere packaging of fresh produce. *Food Technol.* 42:70–77.

Zeitoun, A.A.M. and Debevere, J.M. 1991. The effect of treatment with buffered lactic acid on microbial decontaminating and the shelf-life of poultry. *Int. J. Food Microbiol.* 11:305–312.

Zhang, S. and Farber, J.M. 1996. The effects of various disinfectants against *Listeria monocytogenes* on fresh-cut vegetables. *Food Microbiol.* 13:311–321.

Zhuang, R.-Y., Beuchat, L.R., and Angulo, F.J. 1995. Fate of *Salmonella montevideo* on and in raw tomatoes as affected by temperature and treatment with chlorine. *Appl. Environ. Microbiol.* 61:2127–2131.

16 Latest Developments and Future Trends in Food Packaging and Biopackaging

Jose M. Lagaron and Amparo López-Rubio

CONTENTS

16.1 PLASTICS AND BIOPLASTICS IN PACKAGING

In general, the selection of the most adequate preservation technology is a major consideration when designing food products, particularly if these are going to be packaged and distributed. Thus, by means of a correct selection of materials and packaging technologies, it is possible to keep the product quality and freshness during the needed period for its commercialization and consumption. Plastic packaging offers excellent advantages, multifunctionality, and versatility for this purpose.

Nowadays, plastic packaging is the largest application for plastics (ca. 37% in Europe), and within the packaging niche, food packaging stands out as the largest plastic-demanding application. Plastics bring in enormous advantages, such as thermoweldability, flexibility in thermal and mechanical properties, lightness, integrated projects (integrating forming, filling, and sealing), and low price (Arvanitoyannis et al. 1997, Haugaard et al. 2001, Petersen et al. 1999). Nevertheless, plastics also have some limitations when compared with more traditional materials like metals, alloys, or ceramics. One of the major disadvantages of plastics is the possibility of transport of low molecular weight components by means of permeation, migration (of polymer residues and/or additives), or scalping (sorption of aroma compounds). Transport or barrier properties are determined by permeability ($\wp = \zeta \times D_m$), diffusion (D_m), solubility (ζ), and partition coefficients. Other limitations of plastics are a comparatively low thermal resistance and a strong interdependence among thermal, mechanical, and barrier properties. In spite of this, plastic materials continue to expand and replace the conventional use of paperboard, tinplate cans, and glass, due to their previously mentioned positive characteristics and also on account of the development of multilayer systems, which can include metalized layers for enhanced barrier properties and UV protection. This has resulted in the complex, multicomponent structures widely used today. However, there is still great interest in simplifying such multilayer structures and/or increasing the barrier performance of the composing layers due to cost saving, ease of recycling, reduction of the amount of materials used (ecopackaging), and/or substitution by renewable biomass-derived plastics. As a result, strong efforts in developing new materials and blends have been made over the last decades to reduce complexity and enhance performance in plastic food packaging.

In this context, there is also a current trend to substitute petroleum-based materials by renewable bio-based derived plastics, which will reduce the oil dependence, facilitate the after-life of the packaging (by composting, for instance), and reduce the carbon footprint of the food-packaging industry. Regarding biodegradable (renewable and nonrenewable) materials, three families are usually considered. The first one includes polymers directly extracted from biomass, such as the polysaccharides chitosan, starch and cellulose, and proteins, such as gluten and zein. A second family makes use of either oil-based or biomass-derived monomers and applies classical chemical synthetic routes to obtain the final biodegradable polymer; this is the case, for instance, of polycaprolactones (PCL), polyvinyl-alcohol (PVOH), ethylene-vinyl alcohol copolymers (EVOH), and sustainable monomers of polylactic acid (PLA) (Arvanitoyannis et al. 1997, Haugaard et al. 2001, Petersen et al. 1999). The third family consists of polymers produced by natural or genetically modified microorganisms such as polyhydroxyalcanoates (PHA) and polypeptides (Reguera et al. 2003).

The most commercially viable materials at the moment are some biodegradable polyesters, which can be processed by conventional equipment. In fact, these materials are already used in a number of monolayer and also multilayer applications, particularly in the food-packaging and biomedical fields. Among the most widely researched thermoplastic, sustainable biopolymers for monolayer packaging applications are starch, PHA, and PLA. Specifically, starch and PLA biopolymers are, undoubtedly, the most interesting families of biodegradable materials, as they present an interesting balance of properties and have become commercially available (for instance, by the companies Novamont and Natureworks, respectively), being produced in a large industrial scale. Of particular interest in food packaging is the case of PLA due to its excellent transparency and relatively good water resistance.

Nevertheless, these materials still present shortcomings, such as inferior barrier and thermal properties when compared, for instance, to polyethylene terephthalate (PET), and, therefore, it is of great industrial interest to enhance the barrier properties of these materials while maintaining their inherently good properties like transparency and biodegradability (Bastiolo et al. 1992, Chen et al. 2003, Jacobsen and Fritz 1996, Jacobsen et al. 1999, Koening and Huang 1995, Park et al. 2002, Tsuji and Yamada 2003).

There are also other materials extracted from biomass resources, such as proteins (e.g., zein), polysaccharides (e.g., chitosan), and lipids (e.g., waxes), with excellent potential as gas and aroma barriers, and as carriers of active and bioactive compounds. The main drawbacks of these families of materials are their inherently high rigidity, difficulty of processing in conventional equipment, and, for proteins and polysaccharides, the very strong water sensitivity arising from their hydrophilic character, which leads to a strong plasticization, deteriorating the excellent oxygen barrier characteristic (in dry state) as relative humidity and water sorption in the material increase. This low water resistance of proteins and polysaccharides strongly handicap their use in food-packaging applications. Nevertheless, chitosan and zein biopolymers exhibit two very interesting characteristics; on one hand, chitosan displays antimicrobial properties (Dong and Manjeet 2004, Möller et al. 2004) and on the other, zein shows an unusually high water resistance compared to similar biomaterials (Wang and Padua 2004). Furthermore, zein in a resin form can also be heat processed. In spite of that, and from an application point of view, it is of great relevance to diminish the water sensitivity of proteins and polysaccharides, and to enhance the gas barrier properties and overall functionalities of thermoplastic biopolyesters to make them more adequate for food-packaging applications.

In another line, traditionally, food package has been defined as a passive barrier to delay the adverse effect of the environment over the contained product. Nevertheless, the current tendencies include the development of packaging materials that interact with the environment and with the food, playing an active role in its preservation. Moreover, the packaging can also be designed to impact the consumer health by integrating functional ingredients in the packaging structure. These new food-packaging systems have been developed as a response to trends in consumer preferences toward mildly preserved, fresh, tasty, healthier, and convenient food products with prolonged shelf life. These novel packaging technologies can also be used to compensate for shortcomings in the packaging design, for instance, in order to control the oxygen, water, or carbon dioxide levels in the package headspace. In addition, changes in retail practices, such as a globalization of markets resulting in longer distribution distances, present major challenges to the food-packaging industry, which finally act as driving forces for the development of new and improved packaging concepts that extend shelf life while maintaining the safety, quality, and health aspects of the packaged foods. Novel active and bioactive packaging technologies, combined with biopackaging and nanotechnology can best help to do so (see Figure 16.1). Therefore, proper combination of these three technological cornerstones will provide innovation in the food-packaging sector over the next few years.

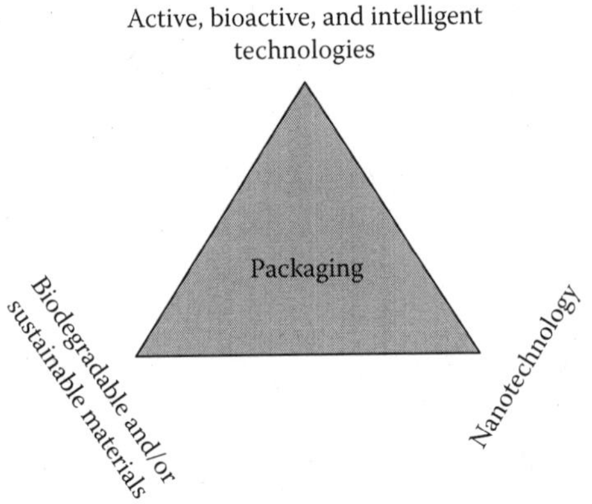

FIGURE 16.1 Innovation in food packaging.

16.2 NANOCOMPOSITES IN FOOD PACKAGING

Nanotechnology is by definition the fabrication and utilization of structures with at least one dimension in the nanometer length scale that creates novel properties and phenomena otherwise not displayed by either isolated molecules or bulk materials. Nanoscale structures display a high surface-to-volume ratio, which is ideal for applications that involve composite materials, chemical reactions, drug delivery, controlled release of substances in active and functional packaging technologies, and energy storage in intelligent packaging. Among the various options involving nanotechnology, inorganic layered nanocomposites have attracted most of the attention as far as food packaging is concerned (Lange and Wyser 2003, Markarian 2005). Since Toyota researchers in the late 1980s found that mechanical, thermal, and barrier properties of nylon-clay composite materials improved dramatically by reinforcing with less than 5 wt% of clay (Okada et al. 1987), extensive research work has been performed in the study of nanocomposites for packaging applications (see Figure 16.2). These advantages in properties are not nullified by significant reductions in other relevant properties including impact resistance and, above a critical loading level, transparency. In terms of processing, nanocomposites can be mainly obtained by three routes: (1) solution mixing followed by casting, (2) *in situ* polymerization, and (3) melt blending, which is the most interesting one for its easy implementation in most plastic processing methods (Pinnavaia and Beall 2000).

To date, the most interesting packaging technology based on blending to improve the barrier properties of packaging materials has been the so-called oxygen scavengers (see Section 3.1). This technology is known to lead to relatively low levels of oxygen in contact with the food because it traps permeated oxygen from both the headspace and the environment surrounding the package.

When a high barrier to carbon dioxide is also needed, as in the case of carbonated beverages, the combination of nanotechnology with oxygen scavengers in multilayer materials can provide the proper characteristics to preserve and guarantee the quality of the food during an extended shelf life (in comparison to PET, for instance). These systems are claimed, by Colormatrix, United Kingdom, for example, to typically

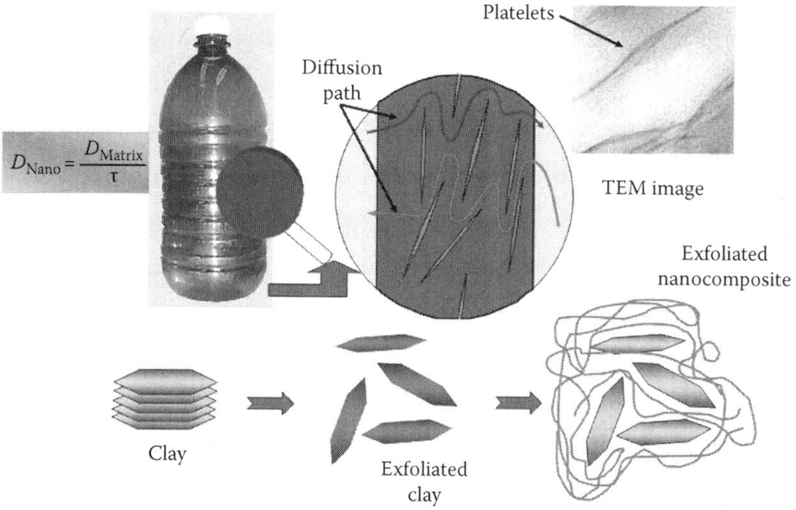

FIGURE 16.2 Schematic showing how nanoclays get dispersed in polymers during processing and how they lead to enhanced barrier properties by decreasing the diffusion coefficient.

generate up to 6 months shelf life in carbonated drinks made of PET (Moloney 2006). Therefore, the nanocomposite technology can be used alone or in combination with scavengers, active and passive barrier, to enhance the barrier performance of multilayer systems.

Some food companies are already producing packaging materials based on nanotechnology that are extending the life of foods and drinks and improving food safety. Several large beer makers, including South Korea's Hite Brewery and Miller Brewing Company, are already using this technology. However, this nanotechnology for packaging has so far been applied to a limited number of polymers, for example, polyamides (PA), thermosetting epoxies, and, to some extent, PET (Carlblom and Seiner 1998, Strupinsky and Brody 1998).

Physical models predict that nanocomposites containing less than 5 wt% nanolayered particulate loadings can reduce permeability of the matrix by ca. 10-fold. In most real cases, however, the barrier improvements to gases and vapors achieved by direct melt blending are between two- and sixfold, depending on the system (Lagaron et al. 2005). Therefore, there is still some room to improve current nanocomposite formulations.

It should be highlighted that most of the nanocomposite formulations in the market (first-generation nanocomposites), which are devised to enhance the properties of engineering polymers in structural applications, are currently making use of organophilic chemical modifiers. However, for food-packaging applications, only food-contact approved materials and additives should be used, and always below their corresponding threshold migration levels. In this respect, NanoBioMatters S.L. and its second-generation nanocomposite products, traded as NanoBioTer® (NanoBioMatters S.L., Spain), claim to have developed proprietary products that comply with the current European food contact legislation. Second-generation nanocomposites are, thus, referred to as nanocomposite formulations that are specifically designed to comply with current regulations and, at the same time, are cost effective and formulated to target specific materials (including biopolymers), their properties, or production technologies. In essence, second-generation nanocomposites are materials with targeted specifications rather than wide-spectrum generic formulations (Lagaron et al. 2006b).

Despite the fact that multilayer solutions are currently needed in many food-packaging applications including the previously mentioned carbonated beverages, a monolayer solution would be of great interest for many reasons, including recyclability, technology, and material costs.

Monolayer composite materials with proper characteristics for specific applications can be produced by simple melt or solution blending routes. The reason why, in PET bottles, nanocomposites are still devised to be used in multilayer systems (such as Imperm® nanocomposites of PA from Nanocor, United States) is because PET nanocomposites are extremely difficult to obtain as a monolayer due to the high temperature needed to process the polymer, which causes degradation of the organophilic chemical modification of the layered clay particles. Therefore, some sort of modification of the inorganic layered particles is needed in nanocomposites to facilitate exfoliation of these into nanolayers during processing, and also to enhance compatibility with the polymer. In fact, there exist new formulations reported in the literature capable of obtaining high-barrier PET nanocomposites by direct blending; therefore rendering monolayer formulations possible for PET (see Table 16.1).

Another example for which ultrahigh barrier properties are needed is high-barrier food-packaging applications, such as aromatic products, retortable pouches, dehydrated products, vacuum packaging, and some modified atmosphere packaging applications. In these and other similar uses, aluminum, EVOH, or other high-barrier materials and technologies are currently applied. However, all of the existing technologies are associated with drawbacks of some kind. For example, it would be highly desirable to enhance the barrier properties of EVOH by 10 times over the whole humidity range, rendering the material virtually impermeable at dry conditions and capable of substituting typical impermeable materials or technologies in many applications for which transparency or more simple packaging structures provide added value.

In this context, ultrahigh barrier properties to oxygen via nanocomposites have also been achieved by specific food-contact-complying nanoclay formulations in EVOH polymers over the whole humidity range (Lagaron et al. 2005). This makes the nanocomposites technology unique in the sense that, unlike other blending technologies, it can improve the oxygen barrier properties of EVOH by more than three times over the whole humidity range with a nanoclay addition of 5 wt%, retaining transparency and flexibility.

Biodegradable materials can also strongly benefit from nanocomposites. By means of novel nanobiocomposites, disadvantages in the original biodegradable materials, that is, relative humidity dependency, low thermal and barrier properties, or inadequate mechanical properties, can be overcome.

In this growing market within the food-packaging industry, specific formulations have to be developed that can be compatible with the biodegradable matrices and renewable concepts associated with these new packaging materials. Table 16.1

TABLE 16.1
Reductions (%) in Oxygen and Water Vapor Permeability Reported for Some Nanocomposites of Thermoplastic Polymers and Biopolymers of Use in Food Packaging

Matrix	Type of Clay	Nanoclay (wt%)	$\wp O_2$ (% Red)	$\wp H_2O$ (% Red)	Reference
PA	Modified-MMT	2	—	54	Yano et al. (1993)
PET	Organically modified-MMT	5	55	—	Harrison et al. (1996)
		1	—	43	
PU	Organically modified-MMT	3	30	50	Osman et al. (2003)
EVOH	Food-contact-clay	5	75	—	Lagaron et al. (2005)
PLA	Modified-MMT	5	48	50	Gu et al. (2007)
PET	MMT	5	94	—	Choi et al. (2006)
PLA	Food-contact-MMT	5	32	54	Cava et al. (2006)
PET	Food-contact-MMT	1	25	—	Sanchez-Garcia et al. (2007a)
PHBV	Food-contact-MMT	5	28	52	Sanchez-Garcia et al. (2007b)
PCL	Food-contact-MMT	5	—	63	
PVC	SiO_2	3	40	45	Zhu et al. (2008)

Note: MMT stands for montmorillonite clay.

compiles some examples of nanocomposite and nanobiocomposite developments for packaging materials and typical barrier improvements, that is, permeability reductions, achieved for oxygen and water vapor.

Many nanocomposite technologies are under investigation at the moment by a number of food-packaging and material manufacturers, and products will soon begin to make their way into the market.

16.3 ACTIVE PACKAGING

The concept of active packaging for food applications has been exploited for about two decades, obtaining the package an active role in the preservation of the products. Packages may be termed active when they perform some desired role in food preservation other than providing an inert barrier to external conditions (Rooney 1995).

The opportunity of modifying the inner atmosphere of the package or even the product by simply incorporating certain substances in the package wall has made this group of technologies very attractive, representing an increasingly productive research area. Even though the first active packaging developments and most of the commercialized technologies consist of sachet technologies, which make use of a small permeable pouch (sachet) containing the active compound that is inserted inside the package, current trends tend toward the incorporation of active ingredients directly into the package wall. This strategy is associated with a number of advantages, such as reduction in package size, higher effectiveness of the active principles (that are now completely surrounding the product), and, in many cases, higher throughput in packaging production, since the additional step of incorporating the sachet is eliminated (López-Rubio et al. 2008). Polymers and in particular biomass-derived polymers are the preferred materials for active packaging because of their intrinsic properties, constituting an ideal carrier for active principles, with the advantage of being tuneable in terms of controlled release and the possibility of combining several polymers through blending or multilayer extrusion to tailor the application.

Active packaging has been used with many products and is under investigation for numerous others. The combinations of polymers and active substances that can be studied for potential use as active packages are in principle unlimited, and it is forecasted that the number of applications will increase in the near future. In fact, the global value of the total active packaging market will be worth \$2.649 billion by 2010 as predicted by Pira International Ltd. (Anon 2005), a significant increase in comparison with the not inconsiderable \$1.558 billion in 2005.

It is not the aim of this section to review all existing technologies, and the reader is directed to specific books and reviews on the subject for an exhaustive description of both basic aspects and applications (Brody et al. 2001, López-Rubio et al. 2004, Ozdemir and Floros 2004, Rooney 1995, Vermeiren et al. 1999). Instead, in this chapter, the latest active packaging developments in which the active substances are part of the package wall will be focused.

Among the existing active packaging technologies, oxygen scavengers and antimicrobial packaging stand out over the other developments. Both technologies were initially based on the sachet concept, using reducing and inhibitory substances, respectively. Lately, the growth in both areas has been enormous, especially in the case of antimicrobials. Other active packaging applications include systems capable of absorbing carbon dioxide, moisture, ethylene, and/or flavor/odor taints; releasing carbon dioxide and/or flavor/odor.

While in the United States and Japan active packaging technologies are already present in the market, in Europe the legislation regarding active packaging was only developed very recently, starting by the fall of 2008 to guide commercial implementation of such technology within the European Union (EU) area.

16.3.1 Oxygen Scavengers

Oxygen is ubiquitous and triggers numerous spoilage reactions in food products. Thus, the presence of oxygen inside the package of certain products needs to be limited, reduced, or even avoided. Oxygen scavenging packages represent an option to modified atmosphere packaging and a complement or alternative to vacuum packaging in certain products. It is normally used in a complementary way, because this technology is able to absorb the residual oxygen normally left inside the package after vacuum packaging, and it will also absorb the oxygen permeating through the packaging walls during storage and commercialization, allowing for a longer shelf life, and promoting quality and freshness maintenance of the products.

Most of the existing oxygen scavenging systems are iron based, such as the thermoformed Oxyguard® (Toyo Seikan Kaisha, Ltd., Tokyo, Japan), commercially used for cooked rice (Day 2008). The iron-based chemicals often react with water supplied by the food product or the process (retorting involves pressurized water vapor) to produce a reactive, hydrated metallic reducing agent that scavenges oxygen within the food package and irreversibly converts it into a stable oxide.

Nonmetallic oxygen scavengers have also been developed using organic reducing agents such as ascorbic acid, ascorbate salts, or cathecol. Immobilization of enzymes like glucose oxidase or ethanol oxidase onto packaging film surfaces is yet another way of developing oxygen scavenging systems (Day 2003). An example of this technology is $ZerO_2$® (Food Science Australia, North Ryde, New South Wales, Australia) in which the active ingredient is ascorbic acid. Prototypes with this material have been observed to remove oxygen from orange juice packages in less than 3 days and to halve the vitamin C loss over a storage period of 1 year (Scully and Horsham 2005).

To initiate the scavenging process, some kind of activation mechanism is normally required, and, thus, the reducing substances can be blended with the polymer and subjected to conventional extrusion processes. Apart from the layer containing the scavenger, an external high barrier layer like EVOH must be included, using a laminated material as package wall, to avoid extensive entrance of oxygen from the exterior and, thus, early exhaustion of the scavenging capacity.

Finally, the use of oxygen scavengers for beer, wine, and other beverages is a potentially huge market that has only recently begun to be exploited. Either blended with the polymer (usually polyester or PET) or incorporated into the bottle closures, crowns, and caps, the nonmetallic reagents react with the oxygen from the bottle headspace, and with any entering oxygen. Some examples of commercially available oxygen scavenging systems for beverages are the PureSeal® bottle crowns (W.R. Grace Co. Inc., United States), oxygen scavenging PET beer bottles (Continental PET Technologies, United States), OS2000® cobalt-catalyzed oxygen scavenger films (Cryovac Sealed Air Corporation, United States), and the light-activated $ZerO_2$ materials.

16.3.2 Ethylene Scavengers

Ethylene is a plant growth-simulating hormone that accelerates ripening and senescence by increasing the respiratory rate of fresh produce, such as fruits and vegetables, thereby decreasing its shelf life. Ethylene also accelerates the rate of chlorophyll degradation in leafy vegetables and fruits (Knee 1990).

In recent years, numerous packaging films and bags for produce have appeared in the market that are based on the ability of certain finely ground minerals to scavenge ethylene (Day 2008). Some commercial examples are Evert-Fresh Green Bags® (Evert-Fresh Inc., United States) and Peakfresh® (Peakfresh Products, Australia) (Knee 1990). These materials contain several minerals, such as zeolites, Japanese oya, clays, or nanoclays, blended with the polymer (usually polyethylene) as finely dispersed powder. However, little direct evidence for their ethylene-absorbing capacity has been published in peer-reviewed scientific journals, and it is not completely clear whether the minerals perform as an active ingredient or they just change the permeability of the package wall, increasing the diffusion rate of the internal gases (including ethylene). In any case, these packages have proven to increase the shelf life of strawberries, lettuce, broccoli, and other ethylene-sensitive products (Tregunno and Tewary 2000), making them very attractive for the commercialization of fresh produce.

16.3.3 CO_2 Controllers

In order to increase the shelf life of fresh produce and maintain its quality and freshness, ethylene not only needs to be removed, but an optimal gas composition is also to be kept within the package, which will depend on the type of product. Fresh fruits and vegetables, after harvesting, keep on consuming oxygen and releasing CO_2, and, thus, the active package has to be designed taking into account the respiration rate of the product and the permeability of the packaging material used. Addition of finely dispersed zeolites to polymeric materials, apart from providing ethylene absorbance, also serves to control the CO_2 concentration in the package. These substances are said to absorb the gas until equilibrium is achieved, and when the CO_2 concentration in the headspace decreases below a given value, the gas starts to flow from the absorber into the internal atmosphere. Therefore, this technology is referred to as CO_2 controllers (Lee et al. 2001).

Lee et al. (2001) studied the suitability of zeolites and active carbon, in sachet form, to control the CO_2 concentration inside kimchi packages. Zeolite was more effective than active carbon in terms of CO_2 absorption capacity, and could also maintain CO_2 adsorption equilibrium as long as it was protected from water vapor. Although research in this area is still in its early stages, it is foreseen that the development of CO_2 controllers is likely to involve a change from sachet form to direct incorporation into the polymeric package wall.

On the other hand, high levels of CO_2 usually play a beneficial role in retarding microbial growth on meat and poultry surfaces and, consequently, CO_2 emitters could also be incorporated in meat packages to control bacterial spoilage. To the best of our knowledge, the only commercial developments of CO_2 emitters are based on the sachet technology, such as the Verifrais® package that has been manufactured by SARL Codimer (Paris, France) and used for extending the shelf life of fresh meat and fish. This innovative package consists of a standard modified atmosphere package tray that has a perforated false bottom under which a porous sachet containing sodium bicarbonate/ascorbate is positioned. When juice exudate from the packed meat or fish drips onto the sachet, carbon dioxide is emitted and this antimicrobial gas can replace the carbon dioxide already absorbed by the fresh food, so avoiding pack collapse (Rooney 1995).

16.3.4 Moisture Absorbers

Excess moisture is a major cause of food spoilage, favoring the growth of pathogenic microorganisms. In the case of dry products, limiting the moisture content is essential to maintain their textural properties and suppress mould growth. Soaking up moisture with absorbers or desiccants is very effective at maintaining food quality and extending shelf life by inhibiting microbial growth and moisture-related degradation of texture and flavor (Day 2008).

However, addition of desiccants to the package wall could alter the optical and mechanical properties of the films and, as a result, some integrated active packaging technologies with water absorption capacity have been developed for high water activity foods such as meat, fish, fruits, and vegetables. These technologies are in the form of absorbent pads, sheets, and blankets that basically consist of two layers of a microporous plastic film, such as polyethylene or polypropylene, between which a superabsorbent polymer with a strong affinity for water is placed. Polyacrylate salts, carboxymethyl cellulose, and starch copolymers constitute examples of superabsorbent materials used. These systems are capable of absorbing water up to 500 times their weight (Reynolds 2007), and they are used as drip-absorbing pads placed under meat or fish cuts (Suppakul et al. 2003). Commercial moisture absorber sheets, blankets, and trays include Toppan Sheet® (Toppan Printing Co. Ltd., Japan), Thermarite® (Thermarite Pty Ltd., Australia), Luquasorb® (BASF, Germany), and Fresh-R-Pax® (Maxwell Chase Inc., Douglasville, Georgia, United States). Some absorbing pads used to soak up the exudates in fish and meat trays also incorporate organic acids and surfactants in order to prevent microbial growth, because food exudates are rich in nutrients (Hansen et al. 1988).

Instead of superabsorbent polymers, dessicants can also be used for intercepting humidity by placing them between two layers of a plastic material highly permeable to water vapor. A commercial example of this active packaging concept is Pitchit® (Showa Denko, Japan), which consists of a layer of propyleneglycol sandwiched between two layers of PVOH (Labuza and Breene 1989). These materials are marketed for home use in a roll or single sheet form for wrapping fresh meat or fish products, reducing their surface water activity (Kruijf et al. 2002).

16.3.5 Flavor/Odor Absorbers and Releasers

The undesirable adsorption of food flavor components by polymeric packaging materials, also known as scalping, has long been recognized. However, this negative attribute of plastics can be used, through a proper design of the packaging structure, to selectively remove undesirable compounds produced during the commercialization of the food products. A similar approach to ethylene and CO_2 scavenging can be employed, that is, incorporating certain zeolites or activated carbon into the polymer blend. These materials can act as molecular sieves, removing malodorous substances, such as aldehydes (hexanal and heptanal), from the package headspace.

Unpleasant smelling amines, formed from the break down of fish muscle proteins, are alkaline and can hence be neutralized by acidic compounds. The Japanese company Anico has marketed Anico® (Anico Co. Ltd., Japan) bags that make use of such concept. These bags are made from films containing a ferrous salt and an organic acid, such as citrate or ascorbate, and they are claimed to oxidize amines as they are absorbed by the polymer film (Rooney 1995).

However, the commercial use of the previously mentioned technologies is controversial due to concerns arising from their ability to mask natural spoilage reactions, misleading consumers about the condition of the packaged food. For this reason these active packaging technologies are banned in Europe and the United States (Anon 2006). Nevertheless, flavor-releasing films, commercially available in Japan, have a number of interesting applications, such as making ready-to-eat food for the army more appetizing. There are a number of commercial developments of flavor-releasing films that aim to create "in-store awareness" of the product (Markarian 2006) or to stimulate the olfactory bulb so that the flavor vapors are interpreted as taste by the brain (Todd 2003). An example of this type of package is ScentSational® (ScentSantional Technologies, LLC, Jenkintown, United States). ScentSational utilizes its patented and proprietary Encapsulated Aroma Release® technology to incorporate FDA-approved food grade FEMA-GRAS (generally recognized as safe) flavors directly into food and beverage packaging components and fragrances into consumer products packaging. This technology has been licensed to a company that commercializes Aroma Water® (Aroma Water, LLC, United States), adding either lemon-lime or mandarin orange scent to bottled water. The preferred materials for flavor release packages are polyolefins, such as polypropylene or polyethylene, and there is a wide range of flavors like chocolate, banana, caramel, and strawberry that have been used in these innovative packages (Mohan 2006).

16.3.6 Antimicrobial Packaging

The main objective of the antimicrobial packaging technology is to control undesirable microorganisms on foods by incorporating or depositing antimicrobial substances into or onto packaging materials (Cooksey 2001, Han 2000). The principle of action of antimicrobial packaging is based on either intended migration (controlled release, see Figure 16.3) of biocide species to the liquid phase and/or to the headspace or by direct contact (nonintended migration).

Different systems have already been developed and the interested reader is referred to the review of Appendini and Hotchkiss (2002) for details. Ag-substituted zeolite is commercialized in Japan and other countries. Strong and broad antimicrobial attributes are claimed (Vermeiren et al. 1999), but the real efficacy of this system has not been fully evaluated or understood. The requisite migration from polymers is minimal and silver ions' antimicrobial effects are weakened by sulfur-containing amino acids in many food products (Brody 2001).

The most practical application of this technology seems to be for low-nutrient beverages, such as tea or mineral water. Its use in Europe is not regulated for food applications. Commercial examples of Ag-zeolites are Zeomic® (Shinanen New Ceramics Co. Ltd., Japan), AgIon® (AgIon Technologies Inc., United States), and Apacider® (Sangi Group America, United States).

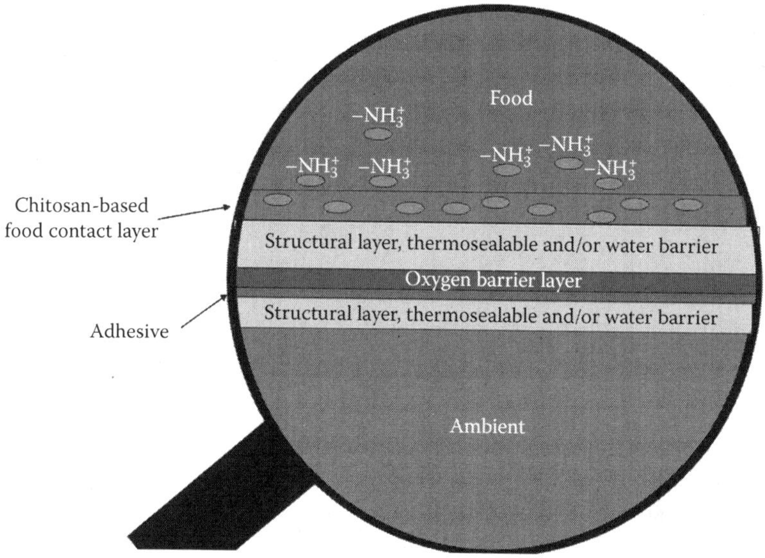

FIGURE 16.3 Schematic showing a feasible design of a packaging multilayer system, where the antimicrobial layer is in direct contact with the food and allows controlled release of protonated glucosamine biocide groups.

Another option is the use, within plastics and ceramic substrates, of EU and Food and Drugs Administration, FDA, food-contact approved nanotechnology systems based on biocide, metals, natural extracts, and other principles. This is commercially available by a nanotech company under the general trade name of NanoBioTer® (NanoBioMatters S.L., Spain) (Lagaron 2006, 2007a, Sanchez-Garcia et al. 2008). Nanodispersion of these systems in plastics and bioplastics leads to synergies with other properties (enhanced barrier, etc.), while retaining transparency and other good properties of the polymeric matrix.

Volatile substances, like SO_2, chlorine dioxide, or allyl isothiocyanate have also been studied to be incorporated into package systems (Appendini and Hotchkiss 2002). Chlorine dioxide received FDA acceptance as an antimicrobial agent for package materials. It is an antimicrobial gas released from a basic, chlorine-containing chemical upon exposure to moisture. Its main advantage is that once moisture permeates through the package, the chlorine dioxide gas is released functioning at a distance and, thus, it is one of the few packaging antimicrobials that do not require direct contact with the food. The antimicrobial properties of dermaseptin S4 derivative have also been discussed for food-packaging applications (Miltz et al. 2006).

Other possible antimicrobial substances are food preservatives, such as sorbates, benzoates, propionates, and parabens, all of them covered by FDA regulations (Floros et al. 1997). Sorbate-releasing plastic film for bologna and cheese packaging is a good example of successful research and development on antimicrobial packaging. The plastic resin, polyvinylidene chloride, and the antimicrobial agents were mixed, extruded, and pelletized to produce masterbatch resins. Films containing 1.5% and 3.0% (w/v) sorbic acid prevented growth of *L. monocytogenes* on bologna slices with populations as much as 7.1 logs lower after 28 days of storage at 4 °C compared with the sorbic acid–free controls. With use of the sorbic acid–containing films, common spoilage organisms were also inhibited on both cheese and bologna products (Limjaroen et al. 2005). Films containing sodium propionate, in turn, have been proven useful in prolonging the shelf life of bread by retarding microbial growth (Soares et al. 2002). An interesting commercial development is the marketing of food-contact approved Microban® (Microban Products Co., United States) kitchen products such as chopping boards, dish cloths, etc., which contain triclosan, an antimicrobial chlorinated aromatic compound, which is also used in soaps, shampoos, etc. (Chung et al. 2003). The use of triclosan for food-contact applications was allowed in EU countries by the Scientific Committee for Food, SCF, in the 10th additional list of monomers and additives for food contact materials (SCF, 2000), with a quantitative restriction on migration of 5 mg/kg of food.

In the literature, there are numerous examples of different systems being used for antimicrobial purposes and incorporated into plastics. The most relevant and with more potential are those based on natural plant extracts and food components, bacteriocins, and natural biopolymers, such as chitosan derivatives. Examples of these systems are those based on the following principles: thymol (Del Nobile et al. 2008) and thymol in nanocomposites (Sanchez-Garcia et al. 2008); chitosan and chitosan derivatives (Belalia et al. 2008, Fernandez-Saiz et al. 2008, 2009); nisin (Nguyen et al. 2008); 2-nonanona (Almenar et al. 2007); natamycin (Oliveira et al. 2007); cinnamon, rosemary, garlic essential oils, oregano, and thyme (Becerril et al. 2007, López et al. 2007, Seydim and Sarikus 2006); clove oils (Matan et al. 2006); carvacrol (Charlier et al. 2007); lysozyme (Buonocore et al. 2005), among many others (Kenawy et al. 2007, López-Rubio et al. 2004, Rhim 2007, Rodriguez et al. 2007). The EU intended to launch in late 2008 the directive that will regulate active products to come into contact with foods, and the cited systems will more likely be contemplated as feasible commercial alternatives to be used in the EU area.

The incorporation of plant extracts and essential oils in different packaging formulations, like for example adhesive layers, is without doubt one of the most interesting areas of current research (Rodriguez et al. 2007). Some of these compounds also have antioxidant properties, a fact that renders them very suitable for designing packaging materials with greater stability for the contained foods without stringent requirements for oxygen barrier.

Also of particular interest is the bio-based polymer chitosan. This system has been widely researched, but it was only recently that the phenomenology of the polysaccharide and its biocide properties were more clearly understood and adequate characterization methods put in place in order to optimize biocide capacity of this polymer in food-packaging and coating applications (Fernandez-Saiz et al. 2008, 2009, Lagaron et al. 2007) (see Figure 16.3).

Recent developments and optimization in nanofabrication by electrospinning of this biopolymer indicate that nanostructured fiber mats of this material, such as

FIGURE 16.4 Nanostructured fiber mats of electrospun chitosan with antimicrobial performance as determined by SEM. (From Torres-Giner, S. et al. *Eng. Life Sci.* 8, 303–314, 2008b. With permission.)

the one shown in Figure 16.4, have very strong biocide properties (Torres-Giner et al. 2008b). The electrospinning conditions for obtaining the fibers were fixed at 15 kV of power voltage, 10 cm of tip-to-collector distance, and 0.15 mL/h of the volumetric flow rate. Antimicrobial activity tests were performed by adding different samples containing 100, 75, 50, and 25 mg of the electrospun mats in 10 mL of Mueller Hinton Broth (MHB) of pH 6.2 containing 10^5 colony forming units/mL of an early mid-log phase culture of *Staphylococcus aureus* incubated at 37°C for 24 h. After the incubation period, appropriate serial dilutions were carried out, and 0.1 mL of each MHB sample was plated on TSA (Tryptone Soy Agar) plates. Finally, after overnight incubation at 37°C, bacterial colonies were counted (Torres-Giner et al. 2008b).

16.4 BIOACTIVE PACKAGING

Bioactive packaging is a new concept in food technology that seeks to transform foods into functional foods upon packaging (López-Rubio et al. 2006). The idea is to use similar concepts as in active packaging to deliver functional or bioactive substances like vitamins, pre-, and probiotics, overcoming thereby some of the existing drawbacks in the fabrication of functional foods. The extensive knowledge generated in the biomedical and pharmaceutical areas can then be exploited for the development of bioactive packages, aimed at incorporating the desired bioactive substances in the food product by controlled release during storage, either just before consumption or even during digestion, taking into account the specific product/functional substance requirements needed in each particular case.

The development of this new concept can be carried out by

1. Integration and controlled release of bioactive ingredients from biodegradable and/or sustainable packaging systems
2. Micro- and nanoencapsulation of these bioactive substances in the packaging
3. Packaging provided with enzymatic activity exerting a health-promoting benefit through transformation of specific food-borne components

Integration and controlled release of bioactive ingredients from the package wall can be attained through smart blending, generating the so-called controlled release packaging (CRP). CRP is one of the most innovative and challenging technologies for releasing bioactive compounds at controlled rates suitable for enhancing the quality and safety of a wide range of foods (Lacoste et al. 2005). The idea is to create different morphologies by blending various polymers under specific conditions, so as to achieve controlled release of the incorporated substances. From medical and pharmaceutical research, the procedures for achieving release under various conditions are well established, but this knowledge has been of limited use in the food-packaging area. Several polymeric systems have been developed for the controlled release of antimicrobials (Iconomopoulou and Voyiatzis 2005), antioxidants (Heirlings et al. 2004), and even enzymes (Buonocore et al. 2003).

Although bioactive packaging is a rather novel concept at an early stage of development, micro- and nanoencapsulation technologies are drawing a lot of attention, and there are a number of interesting systems with potential commercial application. Cyclodextrins containing bioactives, for instance, have been incorporated as additives for controlled release into polylactide-co-polycaprolactone films (Plackett et al. 2007). Specifically, α-cyclodextrins were found suitable for long-term controlled release of an antimicrobial substance from polymer films.

Polymeric sustainable and/or biodegradable biomaterials, such as chitosan, polycaprolactone, and zein, are the preferred vehicles for the delivery of the bioactive substances. However, there are several commercial developments for canned beverages. One example is FreshCan® Wedge technology, a patented delivery system, jointly developed by Ball Packaging Europe and Degussa FreshTech Beverages, which enables dry sensitive ingredients, such as vitamins, to be dispensed into a canned beverage only when the can is opened (Mohan 2006). But even in this type of technology, the active ingredient is contained within a polymeric container, the so-called wedge, which is a polypropylene device containing 10 mL of the dry active ingredient. This plastic container is air- and watertight, and when the can is opened, the internal pressure decreases, causing the wedge end to spring open. Thus, sensitive substances are not dissolved in the beverage until it is consumed (O'Sullivan and Kerry 2008).

Various nanotechnologies, such as the use of nanoclays and electrospun or electrosprayed active or bioactive nanostructured materials, are offering invisibility within the package (i.e., transparency and required mechanical properties) and will also become excellent carrying systems in which the active principle can be better dispersed and, therefore, be more effective (Lagaron et al. 2006a). These novel nanotechnologies will also enhance the plastic or bioplastic properties and will act as functional barriers against unintended migration or as scavengers of potential toxic by-products, increasing, therefore, the quality and safety of food products during commercialization.

In this section, some examples of these novel technologies are presented, focusing on two main areas: nutrient delivery and enzymatically active packages.

16.4.1 Nutrient Release Packaging

Some packages have been developed that add a specific nutrient that cannot be preserved in the food product. Among the ingredients or nutrients that are sensitive to be lost during storage or cannot be added directly to the food due to matrix incompatibility, vitamins, phytochemicals, marine oils, and flavonoids are among the most interesting ones.

For instance, the precursors of vitamin A, that is, carotenoids, have been associated with lower incidence of several types of cancer, including lung, stomach, and skin (Mayne et al. 1994). An increasing trend in the food industry is toward replacing synthetic additives with natural products. However, creating suitable water dispersible forms of carotenoids is difficult because of the limited solubility of pure carotene crystals. Encapsulation is a potential approach to transform liquids into stable and free-flowing powders that are easy to handle and incorporate into dry food systems. Encapsulation technologies targeting incorporation into foods are thoroughly discussed in Chapter 7. However, one particular encapsulation methodology, namely, electrospinning, used in the biomedical area for tissue engineering but with promising applications in food technology, is briefly explained here. In electrospinning, polymer nanofibers may be obtained by the application of a strong electrical field between a grounded target and a polymer solution that is pumped from a container through a needle. Recently, encapsulation and stabilization of light-sensitive β-carotene inside ultrafine fibers of edible zein prolamine has been achieved by means of electrospinning (Lagaron 2007b) (see Figure 16.5). These zein prolamine nanostructures obtained by electrospinning exhibit a range of unique features and properties that make them ideal carriers for the protection and stabilization of light-sensitive added value food components (Torres-Giner et al. 2008a).

16.4.2 Enzymatic Packaging

Incorporation of enzymes to the package wall can facilitate in-package processing (Brody and Budny 1995). Unlike the current situation in which most foods deteriorate in quality during storage, products exposed to enzymes bound to packaging materials might improve during storage (Soares and Hotchkiss 1998). Many enzymes are currently being used in several food transformation processes; however, more recently there have been a number of trials reported in which these enzymes are also immobilized in packaging materials. The choice of the immobilization method (adsorption, ionic binding, covalent attachment, crosslinking, or entrapment/encapsulation) and of the biomaterial support for the manufacturing of enzymatic packages depends on the nature of the biocatalyst (e.g., whole cells or purified enzymes, from fungal or bacterial origin, either native or genetically modified),

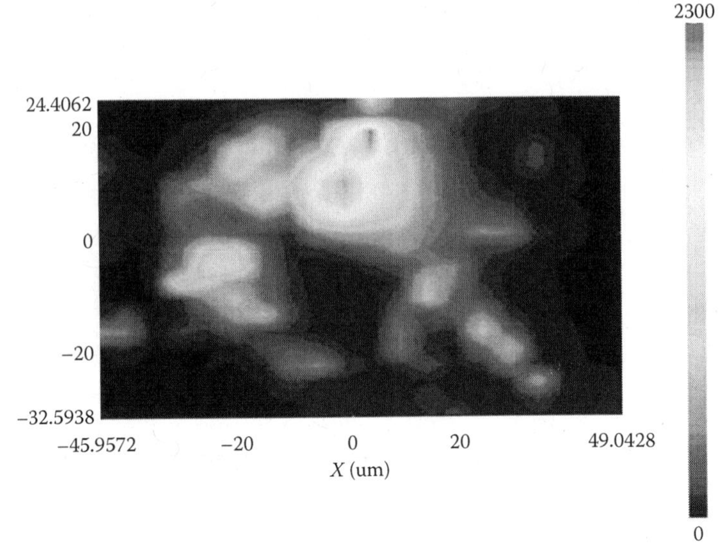

FIGURE 16.5 Raman imaging of ultrathin encapsulates of β-carotene in zein prolamine obtained by electrospinning. The shining areas indicate the location of rich domains of the stabilized bioactive in the capsules. (From Fernandez, A. et al. *Food Hydrocoll.* 23, 1427–1432, 2009. With permission.)

the envisaged storage conditions, the type of food to be packed, and the specific application of the biocatalyst (López-Rubio et al. 2006).

Enzymatic packaging is mainly designed for liquid foods, as the packaging materials, where the enzymes are immobilized, need to be in direct contact with the food product. Two promising concepts deal with milk. The first one is based on the ability of the enzyme cholesterol reductase to transform the cholesterol in coprosterol, which is not absorbed in the intestine. However, to the best of our knowledge, the previous enzyme has not been successfully immobilized to date, but using such a technology, untreated milk could be packaged and in the time taken to transport the package to the consumer, it conceivably could become free of cholesterol (Brody and Budny 1995). The second concept is in-package production of lactose-free milk through the attachment of the enzyme lactase in order to produce milk for lactose-intolerant people. Goddard et al. (2007) used a yeast-derived β-galactosidase that was covalently attached to a surface-modified polyethylene film, and sustained enzyme activity over a range of temperature and pH similar to that of free lactase enzyme. However, the method of surface modification for covalent attachment involves the use of toxic chemicals, such as glutaraldehyde, which is usually unsuited for food applications.

In order to avoid the use of toxic chemicals involved not only in covalent attachment but also in the crosslinking of the materials (methods used for enzyme immobilization), mineral supports could be added to the biopolymer materials. Addition of tiny mineral particles, usually referred to as nanoclays, apart from improving the mechanical and barrier properties of the package walls, have proved to be very efficient in enzyme binding. Within this area there are also some ventures to develop novel technologies based on the use of specific inorganic nanocarriers for the production of bioactive enzymatic (β-galactosidases) films (Lagaron 2005a,b). Several research studies have been carried out concerning the attachment of enzymes in diverse clay mineral supports. There are several mechanisms by which enzymes can be immobilized on clay minerals. What remains essential in all of these novel technologies is to maintain the enzyme functionality and here is where current research efforts are being focused.

16.5 CONCLUDING REMARKS

The use of polymeric packages for food applications has increased considerably over the last decades. Apart from the intrinsic benefits associated with polymers, significant improvements in their physicochemical characteristics, specifically regarding barrier, mechanical, and thermal properties, have been attained as a consequence of extensive research work. In this way, generation of nanocomposites, that is, addition of small amounts of exfoliated nanoclays into polymer matrices, has proven to be a great breakthrough, with barrier improvements to gases and vapors between two- and sixfold, which can be potentially increased with new formulations.

Furthermore, due to the shortage of oil resources and waste-management issues, research focus is shifting from synthetic oil-based plastics to biomass-derived biodegradable and environmentally friendly polymers. The drawbacks initially showed by these biopolymers in terms of poor barrier properties and high instability have, in turn, resulted in novel applications, making polymers an ideal partner for active and bioactive packaging, in which the package is not a passive barrier anymore, but actively contributes to the preservation of food. Biopolymers are, thus, the ideal matrix for the incorporation and controlled release of a number of substances to be added to foods. Probably the area that is more quickly evolving is antimicrobial packaging, but it is foreseen that biopackages will also serve as reservoirs for vitamins, antioxidants, pre-, and probiotics.

REFERENCES

Almenar, E., Del-Valle, V., Hernández-Muñoz, P., Lagarón, J.M., Catalaá, R., and Gavara, R. 2007. Equilibrium modified atmosphere packaging of wild strawberries. *J. Sci. Food Agric.* 87:1931–1939.

Anon. 2005. Up and active. Pira's latest market report plots a healthy future for active packaging. *Active Intell. Pack. News* 3:5.

Anon. 2006. UK Food Standards Agency. *Active Intell. Pack News* 4:5.

Appendini, P. and Hotchkiss, J.H. 2002. Review of antimicrobial food packaging. *Innov. Food Sci. Emerg. Technol.* 3:113–126.

Arvanitoyannis, I., Psomiadou, E., Biliaderis, C.G., Ogawa, H., Kawasaki, H., and Nakayama, O. 1997. Biodegradable films made from low density polyethylene (LDPE), ethylene acrylic acid (EAA), polycaprolactone (PLC) and wheat starch for food packaging applications: Part 3. *Starch-Starke* 49:306–322.

Bastiolo, C., Bellotti, V., Del Tredici, G.F., Lombi, R., Montino, A., and Ponti R. 1992. Biodegradable polymeric compositions based on starch and thermoplastic polymers. World Patent WO 9219680, filed May 4, 1992, and issued Nov. 12, 1992.

Becerril, R., Gómez-Lus, R., Goni, P., López, P., and Nerin, C. 2007. Combination of analytical and microbiological techniques to study the antimicrobial activity of a new active food packaging containing cinnamon or oregano against *E-coli* and *S-aureus*. *Anal. Bioanal. Chem.* 388:1003–1011.

Belalia, R., Grelier, S., Benaissa, M., and Coma, V. 2008. New bioactive biomaterials based on quartenized chitosan. *J. Agric. Food Chem.* 56:1582–1588.

Brody, A.L. 2001. What's active in active packaging? *Food Technol.* 55:104–106.

Brody, A.L. and Budny, J.A. 1995. Enzymes as active packaging agents. In *Active Food Packaging*, ed. M. L. Rooney. Glasgow, U.K.: Blackie Academic Professional, pp. 174–192.

Brody, A.L., Strupinsky, E.R., and Kline, L.R. 2001. *Active Packaging for Food Applications*. London, U.K.: CRC Press.

Buonocore, G.G., Del Nobile, M.A., Panizza, A., Bove, S., Battaglia, G., and Nicolais, L. 2003. Modeling the lysozyme release kinetics from antimicrobial films intended for food packaging applications. *J. Food Sci.* 68:1365–1370.

Buonocore, G.G., Conte, A., Corbo, M.R., Sinigaglia, M., and Del Nobile, M.A. 2005. Mono- and multilayer active films containing lysozyme as antimicrobial agent. *Innov. Food Sci. Emerg. Technol.* 6:459–464.

Carlblom, L.H. and Seiner, J.A. 1998. Gas barrier coating compositions containing platelet-type fillers. World Patent WO 9824839, filed Oct. 7, 1997, and issued Jun. 11, 1998.

Cava, D., Gimenez, E., Gavara, R., and Lagaron, J.M. 2006. Comparative performance and barrier properties of biodegradable thermoplastics and nanobiocomposites versus PET for food packaging applications. *J. Plast. Film Sheet.* 22:265–274.

Charlier, P., Ben Arfa, A., Preziosi-Belloy, L., and Gontard N. 2007. Carvacrol losses from soy protein coated papers as a function of drying conditions. *J. Appl. Polym. Sci.* 106:611–620.

Chen, C.C., Chueh, J.Y., Tseng, H., Huang, H.M., and Lee, S.Y. 2003. Preparation and characterization of biodegradable PLA polymeric blends. *Biomaterials* 24:1167–1173.

Choi, W.J., Kim, H.-J., Yoon, K.H., Kwon, O.H., and Hwang, C.I. 2006. Preparation and barrier property of poly(ethylene terephthalate)/clay nanocomposite using clay-supported catalyst. *J. Appl. Polym. Sci.* 100:4875–4879.

Chung, D., Papadakis, S.E., and Yam K.L. 2003. Evaluation of a polymer coating containing triclosan as the antimicrobial layer for packaging materials. *Int. J. Food Sci. Technol.* 38:165–169.

Cooksey, K. 2001. Antimicrobial food packaging materials. *Addit. Polym.* 8:6–10.

Day, B.P.F. 2003. Active packaging. In *Food Packaging Technologies*, eds. R. Coles, D. McDowell, and M. Kirwan. Boca Raton, FL: CRC Press, pp. 282–302.

Day, B.P.F. 2008. Active packaging of food. In *Smart Packaging Technologies for Fast Moving Consumer Goods*, eds. J. Kerry and P. Butler. Chichester, U.K.: John Wiley & Sons, pp. 1–18.

Del Nobile, M.A., Conte, A., Incoronato, A.L., and Panza, O. 2008. Antimicrobial efficacy and release kinetics of thymol form zein films. *J. Food Eng.* 89:57–63.

Dong, S.C. and Manjeet, S.C. 2004. Biopolymer-based antimicrobial packaging: A review. *Crit. Rev. Food Sci. Nutr.* 44:223–237.

Fernandez, A., Torres-Giner, S., and Lagaron, J.M. 2008. Novel route to stabilization of bioactive antioxidants by encapsulation in electrospun fibers of zein prolamine. *Food Hydrocoll.* 23:1427–1432.

Fernandez-Saiz, P., Lagarón, J.M., Hernández-Muñoz, P., and Ocio, M.J. 2008. Characterization of antimicrobial properties on the growth of *S-aureus* of novel renewable blends of gliandins and chitosan of interest in food packaging and coating applications. *Int. J. Food Microbiol.* 124:13–20.

Fernandez-Saiz, P., Lagaron, J.M., and Ocio, M.J. 2009. Optimization of the biocide properties of chitosan for its application in the design of active films of interest in the food area. *Food Hydrocoll.* 23:913–921.

Floros, J.D., Dock, L.L., and Han, J.H. 1997. Active packaging technologies and applications. *Food Cosmetics Drug Packag.* 20:10–17.

Goddard, J.M., Talbert, J.N., and Hotchkiss, J.H. 2007. Covalent attachment of lactase to low-density polyethylene films. *J. Food Sci.* 72:36–41.

Gu, S.-Y., Ren, J., and Dong, B. 2007. Melt rheology of polylactide/montmorillonite nanocomposites, *J. Polym. Sci, Part B: Polym. Phys.* 45:3189–3196.

Han, J.H. 2000. Antimicrobial food packaging. *Food Technol.* 54:56–65.

Hansen, R., Rippl, C., Miidkiff, D., and Neuwirth, J. 1988. Antimicrobial absorbent food pad. U.S. Patent 4,865,855, filed Jan. 11, 1988, and issued Sep. 12, 1989.

Harrison, A.G., Meredith, W.N.E., and Higgins, D.E. 1996. Polymeric packaging film coated with a composition comprising a layer mineral and a crosslinked resin. U.S. Patent 5,571,614, filed Feb. 4, 1994, and issued Nov. 5, 1996.

Haugaard, V.K., Udsen, A.M., Mortensen, G., Hoegh, L., Petersen, K., and Monahan, F. 2001. Food biopackaging. In *Biobased Packaging Materials for the Food Industry—Status and Perspectives*, ed. C.J. Weber. A report from the EU concerted action project: Production and application of biobased packaging materials for the food industry (Food Biopack), funded by DG12 under the contract PL98 4046. http://www.biomatnet.org/publications/f4046fin.pdf

Heirlings, L., Siró, I., Devlieghere, F. et al. 2004. Influence of polymer matrix and adsorption onto silica materials on the migration of α-tocopherol into 95% ethanol from active packaging. *Food Addit. Contam.* 21:1125–1136.

Iconomopoulou, S.M. and Voyiatzis, G.A. 2005. The effect of the molecular orientation on the release of antimicrobial substances from uniaxially drawn polymer matrixes. *J. Control. Release* 103:451–464.

Jacobsen, S. and Fritz, H.G. 1996. Filling of poly(lactic acid) with native starch. *Polym. Eng. Sci.* 36:2799–2804.

Jacobsen, S., Degée, P.H., Fritz, H.G., Dubois, P.H., and Jérome, R. 1999. Polylactide (PLA)—a new way of production. *Polym. Eng. Sci.* 39:1311–3119.

Kenawy, E.R., Worley, S.D.M., and Broughton, R. 2007. The chemistry and applications of antimicrobial polymers: A state-of-the-art-review. *Biomacromolecules* 8:1359–1384.

Knee, M. 1990. Ethylene effects in controlled atmosphere storage of horticultural crops. In *Food Preservation by Modified Atmospheres*, eds. M. Calderon and R. Barkai-Golan. Boca Raton, FL: CRC Press, pp. 225–236.

Koening, M.F. and Huang, S.J. 1995. Biodegradable blends and composites of polycaprolactone and starch derivatives. *Polymer* 36:1877–1882.

Kruijf, N.D., Beest, M.V., Rijk, R., Sipilainen, M.T., Paseiro, L.P., and Meulenaer, B.D. 2002. Active and intelligent packaging: Applications and regulatory aspects. *Food Addit. Contam.* 19:144–162.

Labuza, T.P. and Breene, W.M. 1989. Applications of "active packaging" for improvement of shelf-life and nutritional quality of fresh and extended shelf-life of foods. *J. Food Proc. Preserv.* 13:1–69.

Lacoste, A., Schaich, K.M., Zumbrunnen, D., and Yam, K.L. 2005. Advancing controlled release packaging through smart blending. *Packaging Technol. Sci.* 18:77–87.

Lagaron, J.M. 2005a. Bioactive packaging: A novel route to generate healthier foods. Paper presented at the Second Conference in Food Packaging Interactions, CAMPDEM (CCFRA), Chipping Campden, U.K., July 14–15.

Lagaron, J.M. 2005b. Biodegradable and sustainable plastics as essential elements in novel bioactive packaging technologies. Paper presented at the First Conference on Biodegradable Polymers for Packaging Applications, PIRA International, Leatherhead, U.K., July 5–6.

Lagaron, J.M. 2006. Novel antimicrobial and bioactive packaging technologies based on biomaterials and nanotechnology. Paper presented at the Pira International Conference: "Future Markets for Active and Antimicrobial Packaging," London, U.K.

Lagaron, J.M. 2007a. Application of nanotechnology to active and bioactive compounds. Paper presented at the 5th FunctionalFoodNet (FFNet) Congress, Valencia, Spain.

Lagaron, J.M. 2007b. Natural products for antimicrobial applications. Paper presented at the First Conference on Antimicrobials, PIRA, Prague, Czech Republic.

Lagarón, J.M., Cabedo, L., Cava, D., Feijoo, J. L., Gavara, R., and Gimenez, E. 2005. Improving packaged food quality and safety. Part 2: Nanocomposites. *Food Addit. Contam.* 22:994–998.

Lagaron, J.M., Giménez-Torres, E., and Cabedo, L. 2006a. Method for producing nanocomposite materials for multi-sectorial applications, World Patent WO2007074184, filed Dec. 29, 2005, and issued Jul. 5, 2006.

Lagaron, J.M., Gimenez, E., Sánchez-García, M.D., Ocio, M.J., and Fendler, A. 2006b. Second generation nanocomposites: A must in passive and active packaging and biopackaging applications. Technical session. "The 15th IAPRI World Conference on Packaging" Tokyo, Japan: International Association of Packaging Research Institutes Japan Packaging Institute, May 2006.

Lagaron, J.M., Fernandez-Saiz, P., and Ocio, M.J. 2007. Using ATR-FTIR spectroscopy to design active antimicrobial food packaging structures based on high molecular weight chitosan polysaccharide. *J. Agric. Food Chem.* 55:2554–2562.

Lange, J. and Wyser, Y. 2003. Recent innovations in barrier technologies for plastic packaging: A review. *Packaging Technol. Sci.* 16:149–158.

Lee, D.S., Shin, D.H., Lee, D.U., Kim, J.C., and Cheigh, H.S. 2001. The use of carbon dioxide absorbents to control pressure build-up and volume expansion of Kimchi packages. *J. Food Eng.* 48:183–188.

Limjaroen, P., Ryser, E., Lockhart, H., and Harte, B.2005. Inactivation of *Listeria monocytogenes* on beef bologna and cheddar cheese using polyvinylidene chloride films containing sorbic acid. *J. Food Sci.* 70:M276–M271.

López, P., Sánchez, C., Batlle, R., and Nerin, C. 2007. Vapor-phase activities of cinnamon, thyme, and oregano essential oils and key constituents against foodborne microorganisms. *J. Agric. Food Chem.* 55:4348–4356.

López-Rubio, A., Almenar, E., Hernández-Muñoz, P., Lagaron, J.M., Catalá, R., and Gavara, R. 2004. Overview of active polymer-based packaging technologies for food applications. *Food Rev. Int.* 20:357–386.

López-Rubio, A., Gavara, R., and Lagaron, J.M. 2006. Bioactive packaging: Turning foods into healthier foods through biomaterials. *Trends Food Sci. Technol.* 17:567–575.

López-Rubio, A., Lagaron, J.M., and Ocio, M.J. 2008. Active polymer packaging of non-meat food products. In *Smart Packaging Technologies for Fast Moving Consumer Goods*, eds. J. Kerry and P. Butler. Chichester, U.K.: John Wiley & Sons, pp. 19–32.

Markarian, J. 2005. Automotive and packaging offer growth opportunities for nanocomposites. *Plast. Addit. Compound.* 7:18–21.

Markarian, J. 2006. Compounders smell success in packaging. *Plast. Addit. Compound.* 8:24–27.

Matan, N., Rimkeeree, H., Mawson, A.J., Chompreeda, P., Haruthainthanasan, V., and Parker, M. 2006. Antimicrobial activity of cinnamon and clove oils under modified atmosphere conditions. *Int. J. Food Microbiol.* 107:180–185.

Mayne, S.T., Janerich, D.T., Greenwald, P. et al. 1994. Dietary β-carotene and lung cancer risk in U.S. nonsmokers. *J. Natl. Cancer Inst.* 86:33–38.

Miltz, J., Rydlo, T., Mor, A., and Polyakov, V. 2006. Potential evaluation of a dermaseptin S4 derivative for antimicrobial food packaging applications. *Packag. Technol. Sci.* 19:345–354.

Mohan, A.M. 2006. Smart solutions pour forth for beverage delivery. *Packag. Dig.* 43:22–26.

Möller, H., Grelier, S., Pardon, P., and Coma, V. 2004. Antimicrobial and physicochemical properties of chitosan-HPMC-Based Films. *J. Agric. Food Chem.* 52:6585–6591.

Moloney, S. 2006. Integrating active and passive barrier technologies. Paper presented at the International Packaging Exhibition, Hispack2006 Conference in Packaging Developments, Barcelona, Spain.

Nguyen, V.T., Gidley, M.J., and Dykes, G.A. 2008. Potential of a nisin-containing bacterial cellulose film to inhibit *Listeria monocytogenes* on processed meats. *Food Microbiol.* 25:471–478.

O'Sullivan, M.G. and Kerry, J.P. 2008. Smart packaging technologies for beverage products. In *Smart Packaging Technologies for Fast Moving Consumer Goods*, eds. J. Kerry and P. Butler. Chichester, U.K.: John Wiley & Sons, pp. 211–232.

Okada, A., Kawasumi, M., Kurauchi, T., and Kamigaito, O. 1987. Synthesis of nylon 6-clay hybrid. *Polym Preprints* 28:447–448.

Oliveira, T.M., Soares, N.D.F., Pereira, R.M., and Fraga, K.D. 2007. Development and evaluation of antimicrobial natamycin-incorporated film in Gorgonzola cheese conservation. *Packag. Technol. Sci.* 20:147–153.

Osman, M.A., Mittal, V., Morbidelli, M., and Suter, U.W. 2003. Polyurethane adhesive nanocomposites as gas permeation barrier. *Macromolecules* 36:9851–9858.

Ozdemir, M. and Floros, J.D. 2004. Active food packaging technologies. *Crit. Rev. Food Sci.* 44:185–193.

Park, E.S., Kim, M.N., and Yoon, J.S. 2002. Grafting of polycaprolactone onto poly(ethylene-co-vinyl alcohol) and application to polyethylene-based biodegradable blends. *J. Polym. Sci. Part B: Polym. Phys.* 40:2561–2569.

Petersen, K., Nielsen, P.K., Bertelsen, G. et al. 1999. Potential of biobased materials for food packaging. *Trends Food Sci. Technol.* 10:52–68.

Pinnavaia, T.J. and Beall, G.W. 2000. *Polymer–Clay Nanocomposites*. Chichester, U.K.: John Wiley & Sons Ltd.

Plackett, D., Ghanbari-Siahkali, A., and Szente, L. 2007. Behavior of α- and β-cyclodextrin-encapsulated allyl isothiocyanate as slow-release additives in polylactide-co-polycaprolactone films. *J. Appl. Polym. Sci.* 105:2850–2857.

Reguera, J., Lagaron, J.M., Alonso, M., Reboto, V., Calvo, B., and Rodriguez-Cabello, J.C. 2003. Thermal behavior and kinetic analysis of the chain unfolding and refolding and of the concomitant nonpolar solvation and desolvation of two elastin-like polymers. *Macromolecules* 36:8470–8476.

Reynolds, G. 2007. Superabsorbent soaks up packaging problems. *Food Production Daily USA*, January 22, 2007. http://www.foodproductiondaily-usa.com

Rhim, J.W. 2007. Potential use of biopolymer-based nanocomposite films in food packaging applications. *Food Sci. Biotechnol.* 16:691–709.

Rodríguez, A., Batlle, R., and Nerin, C. 2007. The use of natural essential oils as antimicrobial solutions in paper packaging. Part II. *Prog. Org. Coatings* 60:33–38.

Rooney, M.L. 1995. *Active Food Packaging*. Glasgow, U.K.: Blackie Academic Professional.

Sanchez-Garcia, M.D., Gimenez, E., and Lagaron, J.M. 2007a. Comparative barrier performance of novel pet nanocomposites with biopolyester nanocomposites of interest in packaging food applications. *J. Plast. Film Sheet.* 23:133–148.

Sanchez-Garcia, M.D., Gimenez, E., and Lagaron, J.M. 2007b. Novel PET Nanocomposites with enhanced barrier performance of interest in food packaging applications. Paper presented at the Plastics Encounter ANTEC Conference, Cincinnati, OH.

Sanchez-Garcia, M.D., Gimenez, E., Ocio, M.J., and Lagaron, J.M. 2008. Novel nanobiocomposites with antimicrobial and barrier properties of interest in active packaging applications. Paper presented at the ANTEC Conference 2008, Milwaukee, WI.

SCF (Scientific Committee for Food). 2000. http://ec.europa.eu/food/fs/sc/scf/reports_en.html

Scully, A. and Horsham, M. 2005. Emerging packaging technologies for enhanced food preservation. *Food Sci. Technol.* 20:16–19.

Seydim, A.C. and Sarikus, G. 2006. Antimicrobial activity of whey protein based edible films incorporated with oregano, rosemary and garlic essential oils. *Food Res. Int.* 39:639–644.

Soares, N.F.F. and Hotchkiss, J.H. 1998. Naringinase immobilization in packaging films for reducing naringin concentration in grapefruit juice. *J. Food Sci.* 63:61–65.

Soares, N.F.F., Rutishauser, D.M., Melo, N., Cruz, R.S., and Andrade, N.J. 2002. Inhibition of microbial growth in bread through active packaging. *Packag. Technol. Sci.* 15:129–132.

Strupinsky, G. and Brody, A. 1998. A twenty-year retrospective on plastics: Oxygen barrier packaging materials. Paper presented at TAPPI Polymers, Laminations and Coatings Conference, San Francisco, CA.

Suppakul, P., Miltz, J., Sonneveld, K., and Bigger, S.W. 2003. Active packaging technologies with an emphasis on antimicrobial packaging and its applications. *J. Food Sci.* 68:408–420.

Todd, H. 2003. Top it off–value-added closures give consumers safety, convenience and a little fun. *Beverage World*, August 15, 52.

Torres-Giner, S., Gimenez, E., and Lagaron, J.M. 2008a. Characterization of the morphology and thermal properties of zein prolamine nanostructures obtained by electrospinning. *Food Hydrocoll.* 22:601–614.

Torres-Giner, S., Ocio, M.J., and Lagaron, J.M. 2008b. Development of active antimicrobial fiber based chitosan polysaccharide nanostructures using electrospinning. *Eng. Life Sci.* 8:303–314.

Tregunno, N. and Tewary, G. 2000. Innovative packaging solutions add value to Canadian produce. *Fruit Veg.* May/June.

Tsuji, H. and Yamada, T. 2003. Blends of aliphatic polyesters. VIII. Effects of poly(L-lactide-co-ε-caprolactone) on enzymatic hydrolysis of poly(L-lactide), poly(ε-caprolactone), and their blend films. *J. Appl. Polym. Sci.* 87:412.

Vermeiren, L., Devlieghere, F., van Beest, M., de Kruijf, N., and Debevere, J. 1999. Developments in the active packaging of foods. *Trends Food Sci. Technol.* 10:77–86.

Wang, Y. and Padua, G.W. 2004. Water sorption properties of extruded zein films. *J. Agric. Food Chem.* 52:3100–3105.

Yano, K., Usuki, A., Okada, A., Kurauchi, T., and Kamigaito, O. 1993. Synthesis and properties of polymide-clay hybrid. *J. Polym. Sci. Part A: Polym. Chem.* 31:2493–2498.

Zhu, A., Cai, A., Zhang, J., Jia, H., and Wang, J. 2008. PMMA-grafted-silica/PVC nanocomposites: Mechanical performance and barrier properties. *J. Appl. Polym. Sci.* 108:2189–2196.

Part II

New Materials, Products, and Additives

17 Biodegradable Films Based on Biopolymers for Food Industries

Ana Cristina de Souza, Cynthia Ditchfield, and Carmen Cecilia Tadini

CONTENTS

17.1 INTRODUCTION

Packages have become an essential element in current developed societies. In particular, food packaging has experienced an extraordinary expansion, because most commercialized foodstuffs, including fresh fruits and vegetables, are being sold in packages. Over the last few decades, the use of polymers, as food packaging materials, has increased enormously due to their advantages over other traditional materials such as glass or tin plate. A great advantage of plastics is the large variety of materials and compositions available, which makes it possible to adopt the most convenient packaging design to the very specific needs of each product. Relevant characteristics of plastics are, for example, low cost, low weight, good thermosealability, ease of printing, and the fact that they are microwaveable. They can also be conformed into an unlimited variety of sizes and shapes, and converters can easily modify them. The optical properties (brightness and transparency) can also be adapted to the specific requirements of the product and allows the consumer to see the packaged product, providing it with a nice appeal (López-Rubio et al. 2004).

However, it is widely accepted that the use of long-lasting polymers for short-lived applications (packaging, catering, surgery, hygiene) is not entirely adequate. This is not justified when increased concern exists about the preservation of ecological systems. Most of today's synthetic polymers are produced from petrochemicals and are not biodegradable. These persistent polymers are a significant source of environmental pollution, harming wildlife when they are dispersed in nature. Plastic materials play a large part in waste management, and the collectivities (municipalities, regional, or national organizations) are becoming aware of the significant savings that the collection of compostable wastes would provide (Avérous 2004). An enormous amount of garbage is generated daily, and food packaging represents a considerable part of this. It is estimated that the United States generates 607,000 ton/day of garbage, Brazil generates 240,000 ton/day, and Germany generates 85,000 ton/day (Vilpoux and Avérous 2004). This waste is composed of many different types of material, some of which are not biodegradable and will remain without decomposing for hundreds, sometimes thousands of years. The disposal of this waste material requires considerable expenses as well as adequate sanitary conditions. Inadequately disposed garbage can cause contamination of soil, water, and water tables, as well as spread diseases.

Consumer demands for safer, environmentally friendly, and higher quality products have increased, and thus technical solutions for the garbage disposal problem are sought. Historically, most of this waste material was destined to landfills, but increasing cost of land and diminishing availability of land for this use has limited further use of this method for dealing with garbage. Incineration has also been frequently used despite issues related to air pollution control. More recently, recycling was introduced and developed allied to selective waste collection programs and public incentive policies such as "reduce, reuse, and recycle." Full implementation of these policies worldwide is still deficient, and other issues like population growth and increasing industrialization have been the driving force for research on other technical solutions. In this context, the development of biodegradable films for packaging materials that can be used as a substitute for petrochemical polymers is an interesting perspective, since it provides an alternative to nondegradable products, and increases income in the agricultural sector.

17.2 BIODEGRADABLE POLYMERS

The first manufactured polymers were derived from biomass resources (animal bones, horns, and hooves, often modified). However, they were increasingly replaced by petrochemical polymers parallel to the growth of the petrochemical industry since the 1930s. Since the 1980s and especially in the 1990s, however, a comeback of bio-based polymers has been observed in certain application areas. One of the main drivers for this development was the goal to provide the market with polymers that are biodegradable. In principle, biodegradable polymers can also be manufactured entirely from petrochemical raw materials. But bio-based polymers, defined here as polymers that are fully or partially produced from renewable raw materials, have so far played a more important role in the domain of biodegradable polymers. These developments have also been a stimulus for research on bio-based polymers that are not biodegradable.

For many decades, cellulose-based polymers played a key role in a wide range of applications, for example, apparel, food (e.g., for sausages), and nonplastics (e.g., varnishes). In the meantime, these bio-based polymers have lost important markets mainly to polyolefins. Since the 1980s, more and more types of starch polymers have been introduced. To date, starch polymers are one of the most important groups of commercially available bio-based materials. At the outset, simple products such as pure thermoplastic starch and starch/polyolefin blends were introduced. Due to the incomplete biodegradability of starch/polyolefin blends, these products had a negative impact on the market toward biodegradable polymers, which achieved largely development of copolymers consisting of thermoplastic starch and biodegradable petrochemical copolymers (Patel et al. 2003).

Major progress has been made concerning the production of other types of bio-based polymers at industrial scale, based on biotechnology, contributing to the establishment of a sustainable bio-based economy. Another technological driver is the progress of nanotechnology, which also offers new possibilities for bio-based polymers.

With regard to the type of bio-based and/or biodegradable polymers, it is important to describe the biodegradability of these materials. The American Society for Testing of Materials (ASTM 1998) and the International Standards Organization (ISO 2005, 2007) define degradable plastics as those that undergo a significant change in chemical structure under specific environmental conditions. These changes result in a loss of physical and mechanical properties, as measured by standard methods. Bio-based polymers can be, but not necessarily are, biodegradable polymers. For example, starch polymers are generally biodegradable, while crystalline polylactic acid (PLA) is virtually nonbiodegradable.

There are three primary classes of polymeric materials on which scientists are currently focusing (Kolybaba et al. 2003, Sorrentino et al. 2007, Wolf et al. 2005). The first class comprises conventional plastics, which are resistant to biodegradation, as the surfaces in contact with the soil in which they are disposed are characteristically smooth. Microorganisms within the soil are unable to consume a portion of the plastic, which in turn, cause a more rapid breakdown of the supporting matrix. This group of materials usually has an impenetrable petroleum-based matrix, which is reinforced with carbon or glass fibers.

The second class under consideration is composed of polymeric materials that are partially degradable. They are designed with the goal of more rapid degradation than that of conventional synthetic plastics. Production of this class of materials typically includes surrounding naturally produced fibers with a conventional (petroleum-based) matrix. When disposed of, microorganisms are able to consume the natural macromolecules within the plastic matrix. This leaves a weakened material, with rough, open edges. Further degradation may then occur.

The third and final class of polymeric materials is currently attracting a great deal of attention from researchers and industry. These plastics are designed to be completely biodegradable. The polymeric matrix is derived from natural sources (such as starch or microbiologically grown polymers), and the fiber reinforcements are produced from common crops such as flax or hemp. Under appropriate conditions of moisture, temperature, and oxygen availability, biodegradation leads to fragmentation or disintegration of the plastics with no toxic or environmentally harmful residue, and microorganisms are able to consume these materials in their entirety, eventually leaving carbon dioxide and water as by-products.

It should be highlighted that materials must meet specific criteria set out by the ASTM and ISO to be classified as biodegradable.

17.2.1 Classification

According to their source, biodegradable polymers can be classified into four main classes (Guilbert 2000, Vilpoux and Avérous 2004):

- Agro-polymers directly extracted or removed from biomass (i.e., polysaccharides, proteins, polypeptides, polynucleotides). They are compostable and renewable polymers and can be processed directly, either plasticized, as fillers or modified by chemical reactions.
- Polymers produced by classical chemical synthesis using renewable bio-based monomers or mixed sources of biomass and petroleum, such as PLA or biopolyester.
- Polymers produced by genetically modified bacteria or by microorganisms through fermentation of agricultural products used as substrate. As examples, one may cite polyhydroxy alkanoates (PHA), bacterial cellulose, xanthan, curdian, and pullan.
- Polymers conventionally obtained from the petrochemical industry by chemical synthesis, such as polycaprolactone (PCL) and polyesteramide (PEA).

Figure 17.1 shows the main types of biodegradable polymer and examples of materials in each category. Out of these, only the petrochemical polymers are from nonrenewable sources.

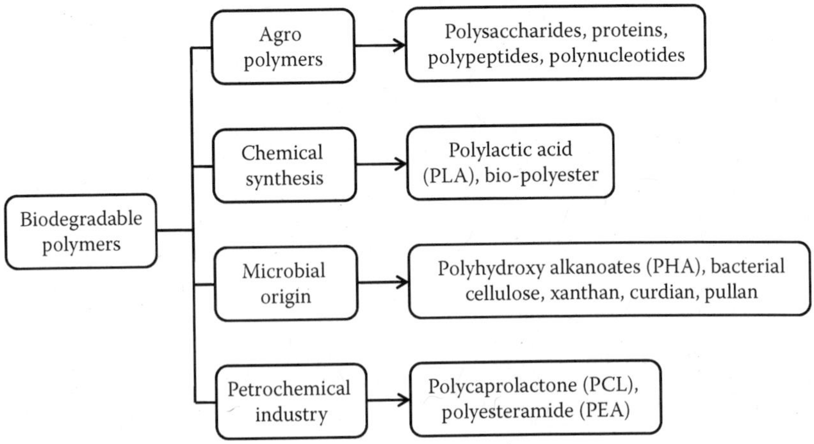

FIGURE 17.1 Types of biodegradable polymers and examples of materials in each category.

Although many types of biodegradable polymers are currently produced at industrial scale (PLA, PHA, PCL, PEA, and others), polymers from agricultural sources are the most studied ones by researchers, especially polysaccharides. This family is represented by different products such as starch and cellulose, both based on glucose units linked in macromolecular chains.

Among the films made from polysaccharides, those obtained from starch are the most important (Alves et al. 2007, Bertuzzi et al. 2007, Famá et al. 2006, 2007, Garcia et al. 2006, Hayashi et al. 2006, Mali et al. 2006, Parra et al. 2004, Talja et al. 2008, Veiga-Santos et al. 2008), because starch is one of the most commonly used agricultural raw materials, since it is a renewable source, inexpensive (even cheaper than polyethylene), widely available, and relatively easy to handle (Rodríguez et al. 2006). Besides, it is found in several forms according to the origin of its raw material.

Nowadays, starch-based polymers are a large part of the bioplastics market, representing from 85% to 90% of the total market. In 2002, about 30,000 metric tons per year were produced, and the market share of these products was about 75%–80% of the bio-based polymers global market (Degli Innocenti and Bastioli 2002). Moreover, 75% of starch polymers are used for packaging applications including soluble films, films for bags and sacks, and loose fill.

Since 1999, in many supermarkets in Scandinavia and in the Mediterranean coast, biodegradable shopping bags for the separate collection of organic waste already exist. Starch has been used as native or slightly modified starch, as well as singly or blended to other materials, which can be conventional synthetic materials or other natural polymers resulting in biodegradable polymers.

17.2.2 Starch-Based Plastics

Starch is a major storage carbohydrate (polysaccharide) in higher plants and is available in an abundance surpassed only by cellulose as a naturally occurring organic compound. It can be found in fruits, seeds, rhizomes, and tubers. It is composed of a mixture of two polymers: amylose, an essentially linear polysaccharide, and amylopectin, a highly branched polysaccharide. As can be observed in Figure 17.2, amylose is a linear polymer of anhydroglucose units linked by many units of glucose units bonded by α (1–4) bonds with a molecular weight of (5×10^4 to 2×10^5) kg/mol, whereas amylopectin (Figure 17.3) is a highly multiple-branched polymer with high molecular weight of ($>10^5$) kg/mol, with α (1–4) bonds and branching occurring at α (1–6) bonds every 24–70 glucose units. Simplified three-dimensional models for these two molecules, evidencing the formation of the helical structure, can be found online (Steane 2008a,b).

Industrially, starch is used, as ingredient, in the manufacture of paper, textiles, adhesives, and foods. The relative amount of amylose and amylopectin depends upon the starch source, such as amylomaize (rich-amylose starch) or waxy maize (rich-amylopectin starch), and film properties are a function of the amount of each macromolecule present within the starch granule (Tharanathan 2003). There are many possible sources from which starch can be obtained to make bioplastics, and thus each place can use the most abundant starch source available locally. The main starch

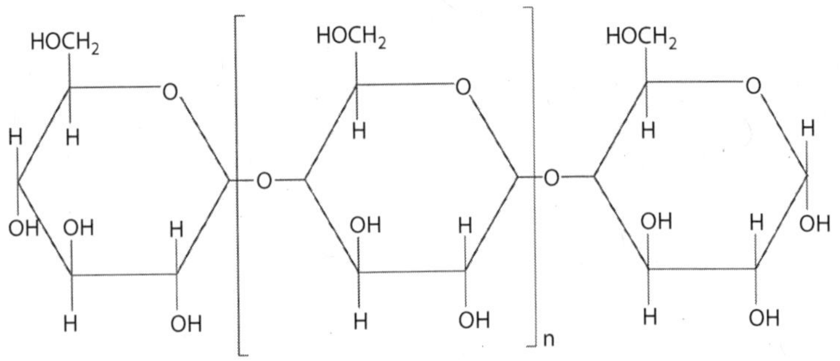

FIGURE 17.2 Structural representation of amylose, a linear polymer of glucose molecules existing in starch.

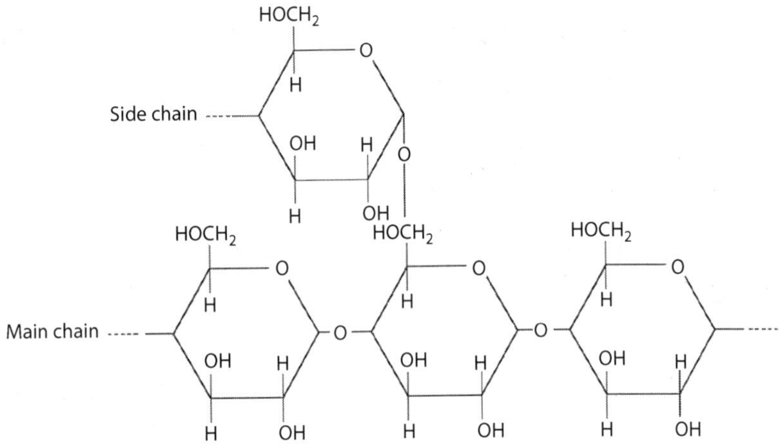

FIGURE 17.3 Structural representation of amylopectin, a highly multiple-branched polymer of glucose molecules existing in starch.

sources produced worldwide are corn (64%), sweet potato (13%), and cassava (11%) (FAO 2004). Many other sources are available like wheat, rice, yam, green banana, potato, peas, beans, sorghum, arrowroot, and lentils, among others. The amylose content in relation to total starch content of some sources is presented in Table 17.1.

Currently, the predominant raw material for the production of starch polymers is corn. Other sources of starch are also being used where price and availability permit. Examples include the use of potato starch in Germany and the Netherlands, and cassava (tapioca) in some parts of the world, like Brazil, which is the second largest cassava-producing country (Mali et al. 2006), after Nigeria, whose production reached 26.6 million metric tons (Mt) in 2005, representing approximately 13% of the world production (FAO 2005). Cassava starch production in Brazil is very significant and 5,465,000 tons were produced in 2005, out of which 11,000 tons were exported as native starch and 27,000 tons were exported as modified starches. Cassava starch is appreciated for its paste clarity, low gelatinization temperature, and good gel stability (Mali et al. 2006).

TABLE 17.1
Amylose Content in Relation to Total Starch Content according to Botanical Origin

Source	Amylose Content (g/100 g)
Maize	30 (Franco 2002)
Rice	8–30 (Franco 2002)
Potato	18–20 (Franco 2002)
Waxy rice	0 (Ribeiro and Seravalli 2004)
Waxy maize	0 (Ribeiro and Seravalli 2004)
Wheat	24 (Ribeiro and Seravalli 2004)
Cassava	20–25 (Franco 2002)

Sources: Franco, G.V., *Table of Chemical Composition of Foods*, Atheneu, Rio de Janeiro, Brazil, 2002; Ribeiro, E.P. and Seravalli, E.A.G., Carbohydrates, in *Food Chemistry*, Edgard Blüncher—Instituto Mariá de Teenologia, Sao Paulo, Brazil, 2004.

Starch is unique among carbohydrates because it occurs naturally as discrete granules with dimensions from 0.5 to 175 μm. This is because the short-branched amylopectin chains are able to form helical structures that crystallize. Starch granules exhibit hydrophilic properties and strong intermolecular association via hydrogen bonding due to the hydroxyl groups on the granule surface. The melting point (T_{melt}) of native starch is higher than the thermal decomposition temperature (T_{td}); hence the poor thermal processability of native starch and the need for conversion to a starch polymer that has a much improved property profile (Mali et al. 2006, Wolf et al. 2005).

Starch biopolymers are transparent, easy to process, provide moderate to good barriers against oxygen and carbon dioxide, are readily biodegradable, and are compatible with most materials, allowing blending to be easily performed. Their main disadvantages are higher water vapor permeability (WVP), solubility in water, and lower mechanical resistance when compared to conventional plastics, as well as high manufacturing cost.

17.2.2.1 Production of Starch Polymers

Starch is not a real thermoplastic, but, in the presence of a plasticizer (water, glycerin, sorbitol, etc.), shearing, and high temperatures (90°C–180°C), it melts and fluidizes, enabling its use in injection, extrusion, and blowing equipment, such as those for synthetic plastics (Vilpoux and Avérous 2004). With this combination, (starch, water, and heat) starch gelatinization can be achieved, which is the disruption of the granule organization. The granules swell, forming a viscous paste, with destruction of most of intermacromolecule hydrogen links. Figure 17.4 illustrates the main technologies applied to manufacture starch polymers.

The production begins with the extraction of starch, whose process depends on botanical origin, followed by milling, fiber separation, slurry washing, and drying, thus obtaining pure starch. According to the desired performance of starch polymers, either before or after the drying step, the pure starch can be chemically modified and converted to a thermoplastic material, which can be accomplished by extrusion only, by sequential steps of extrusion and blending, or by a combined extrusion/blending step.

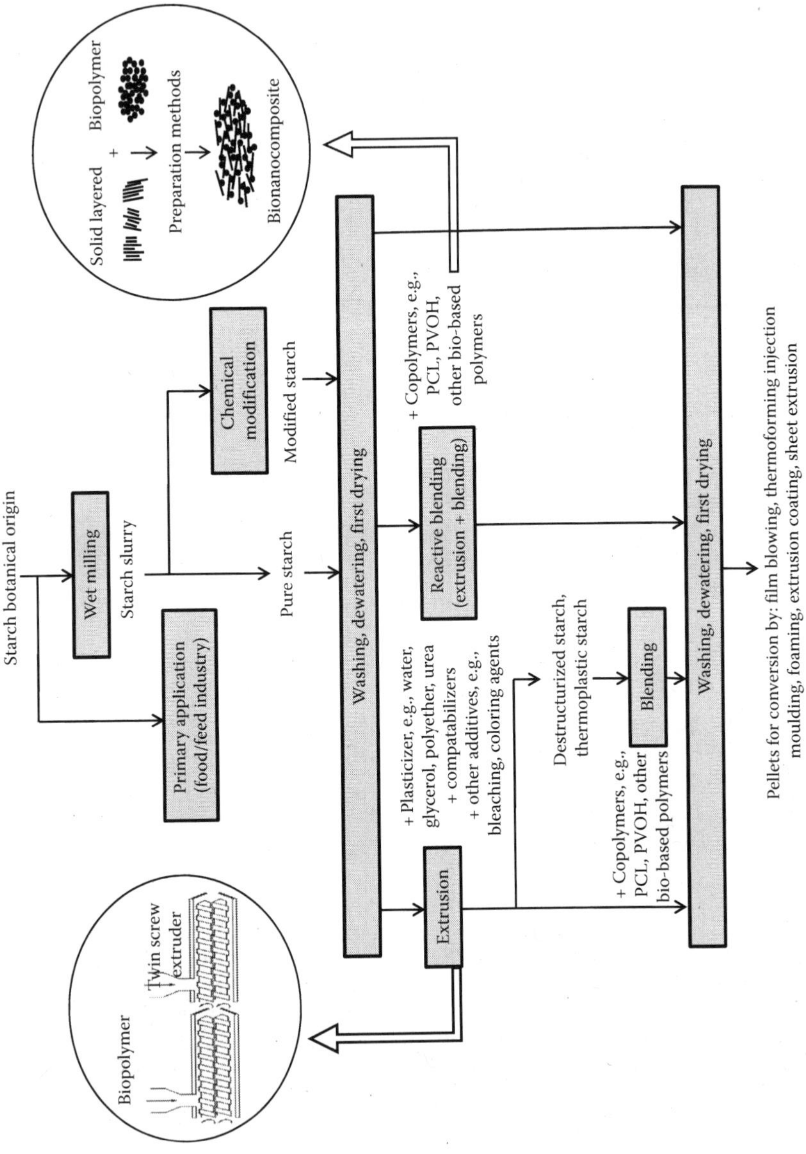

FIGURE 17.4 Main technologies applied for manufacturing starch polymers.

In the past, the primary production method for starch films was solvent casting. In this method, the polymer is dissolved in a suitable solvent to make a viscous solution sufficiently fluid for spreading out quickly on the casting surface. Once this solution is cast, it is dried to give the final film. This technique, used by researchers, has several limitations that include the small quantities of films produced and lengthy production time. On an industrial scale, the polymer solution is fed continuously through a slit die onto a large rotating drum or onto a moving metal belt. A hood assembly may be used to remove organic solvents from the work area.

Another process used since the 1930s is compression molding (wafer technology). Polymers that are thermally stable above their melting or softening temperatures can be fabricated into films by a combination of heat and pressure. This technique is more often used in the laboratory than in the manufacturing plant because large films are difficult to prepare by this method, and the process is batch rather than continuous. The apparatus used consists of two electrically heated platens, one forced against the other by means of a hydraulic unit (usually handpumped). The polymer is placed between two sheets of aluminum or copper foil, and this sandwich is placed between the two heated platens. Pressure is then applied (13.8–34.5 MPa for about 30 s), whereupon the sandwich is removed and cooled, and the film is separated from the foil. Metal shims or gaskets can be used in the sandwich to define the thickness of the film (Allcock et al. 2003).

For economic viability of starch films, a continuous and therefore cost-effective process, such as extrusion, needs to be used: the combination of thermal and mechanical inputs can be obtained by extrusion, a common plastic processing technique.

An alternative method for melt extrusion of films is a blowing film technique, which involves a simple extrusion of the molten polymer, and shape formation by forcing into an annular die, together with a low-pressure air supply, to create a balloon. The bubble is flattened by rollers, slit length-ways to form a continuous film, and then wound into a roll. Film made in this way has a high degree of biaxial orientation. The final film properties are dependent on the efficiency of the machine, blowing conditions, and formulation.

Films can also be produced by calendaring, which consists of squeezing molten polymer into a thin sheet between heated rollers. This is a method normally used for the manufacture of thick films (0.05 mm to more than 0.25 mm thick).

As already mentioned, except for its use as filler in reinforced plastics, native starch presents poor thermoprocessability and must be modified for applications as bioplastics, by granule destructuring. Moreover, blending with other polymers and/or plasticizers is required for improving its mechanical and barrier properties. The main destructuring agent is water, which plays two roles, acting as an agent that promotes starch gelatinization (i.e., starch swells, forming a viscous paste, with destruction of most of intermacromolecule hydrogen links), and as a plasticizer. But another plasticizer is necessary apart from water to decrease T_{melt}. For pure dried starch, T_{melt} varies from 220°C to 240°C, a range that includes the temperature for the onset of starch decomposition (T_{td} = 220°C). If a nonvolatile plasticizer, such as a polyol, is added, T_{melt} decreases, and under high temperature and shear, starch can be processed into a moldable thermoplastic, known as thermoplastic starch (TPS). In addition, plasticizers also limit the microbial growth by lowering film water activity (Avérous 2004, Weber 2000).

During the thermoplastic process, water contained in starch and the added plasticizers play an indispensable role because they can form hydrogen bonds with the starch, replacing the strong interactions between the hydroxyl groups of the starch molecules, and thus transforming it into a thermoplastic (Tang et al. 2008b).

17.2.3 Thermoplastic Starches

The main factors that influence starch destructuring are water content and temperature. TPS is obtained under conditions of low water contents and high levels of destructuring, with added plasticizers. As can be observed in Figure 17.5, the process comprises one or two stages. In a one-stage process, the extruder, which is usually a twin-screw extruder, is fed with native starch, while water and liquid plasticizer are successively introduced along the barrel. On the other hand, in a two-stage process, a dry blend is initially prepared in a turbomixer by slowly adding, under high speed, the plasticizer into the native starch until a homogeneous dispersion is obtained. The mixture is then put in a vent oven, allowing the plasticizer to diffuse into the granule. After cooling, the right amount of water is added to the mixture using a turbomixer, and this blend is finally introduced into an extruder.

The screw is the heart of the extrusion process, and its design and speed of rotation greatly influence the extrusion operation. The molten polymer is then extruded through the die slit. The flat sheet of molten polymer is collected by a rotating drum, which cools the film down below its softening temperature. Subsequent rollers complete the cooling and orientation processes. Typically, the sheet of molten polymer emerging from the die is 10–40 times thicker than the final film because the speed of the rotating drum exceeds the speed at which the polymer is extruded from the die. The final film may be 0.01–0.1 mm thick, and is oriented uniaxially, in the direction of the extrusion.

Plasticizing agents commonly used for TPS production include glycerol (Alves et al. 2007, Famá et al. 2006, 2007, Jangehud and Chinnan 1999, Mali et al. 2006, Parra et al. 2004), polyethylene glycol (Parra et al. 2004), and other polyols, such as sorbitol, mannitol, and sugars (Veiga-Santos et al. 2008), resulting in materials with better flexibility and extensibility.

Sorbitol, also known as glucitol, is a sugar alcohol that occurs naturally in many stone fruits and berries and can be obtained by reduction of glucose, changing the aldehyde group to an additional hydroxyl group. Polyethylene glycol is a

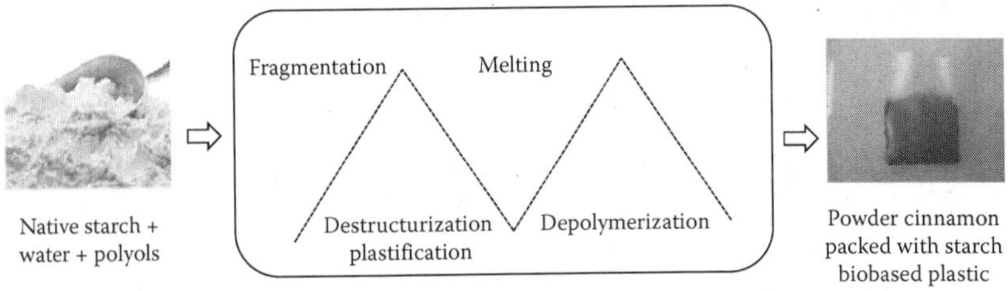

FIGURE 17.5 Thermoplastic starch obtained by extrusion process.

nontoxic compound that can be used in pharmaceuticals and as a food additive. It is produced by the interaction of a calculated amount of ethylene oxide with water, ethylene glycol, or ethylene glycol oligomers. Glycerol, in turn, is a polyalcohol found naturally in a combined form, as glycerides, in animal and vegetable fats and oil. It can be obtained by several chemical processes (oil saponification, sucrose hydrogenation, and others), but, nowadays, glycerol is largely obtained as a by-product of biodiesel production, which contributes to lowering its price. The hydroxyl groups present in sorbitol, polyethylene glycol, and glycerol are responsible for inter- and intramolecular interactions (hydrogen bonds) in starch-based polymeric chains, providing films with a more flexible structure and adjusting them to the packaging fabrication process.

In plasticized starches, the efficiency of plasticization with the same amount of plasticizer is dependent on the amylose/amylopectin ratio. This effect was also verified in edible films based on blends of sodium caseinate and starches. Percent elongation at break (%) of cornstarch/sorbitol blend was slightly greater compared to the one presented by corn starch/glycerol blend, and showed a behavior similar to that of corn starch/xylose blend. A possible explanation might be the higher hydroxyl number of sorbitol (four free –OH) than glycerol (three free –OH) (Arvanitoyannis et al. 1996). However, in the case of sucrose, which is a nontoxic, edible, low-cost biodegradable raw material, and contains seven free –OH, the plasticizing efficiency decays compared to sorbitol due to crystallinity development during storage. Such phenomenon can be attributed to sucrose crystallization, which starts with a crystalline nucleus. In fact, evidence of sucrose crystallization in cassava starch-based films blended with sucrose 0–2 g/100 g, propylene glycol 0–1 g/100 g, sodium phosphate 0–0.2 g/100 g, and soybean oil 0–0.06 g/100 g in aqueous solution, was observed during storage (Veiga-Santos et al. 2008). The same authors reported the possibility of substituting sucrose with inverted sugar to retard crystallization, due to the lower tendency to crystallize when compared to sucrose, increasing film forming suspension viscosity, making it more difficult for crystals to form (Veiga-Santos et al. 2008).

Owing to some structural modifications within the starch network, plasticizers also promote other alterations. By decreasing intermolecular attractions between adjacent polymeric chains, plasticizers lead to a less dense film matrix that facilitates the movement of polymer chains under stress, resulting in an increase in elongation and a decrease in tensile strength with increasing plasticizer content. Besides, the decrease in the intermolecular attraction and the hydrophilic character of the plasticizers facilitate the migration of water vapor molecules, and therefore, WVP values are always significantly higher for plasticized films. Moreover, plasticizer addition decreases film glass transition temperature (T_{glass}) due to weakened strength of macromolecular interactions, and this behavior is also responsible for the increase observed in film permeability (Talja et al. 2007).

Plasticizers are usually added at the ratio of 10:60 g/100 g of starch, but it is difficult to determine an ideal quantity of each plasticizer to obtain a film with good mechanical and barrier properties because the plasticizer content depends upon the quantity and type of starch that is used in the formulation. It is known that films with high plasticizer content can exhibit phase separation, and these compositions become more sensitive to ambient humidity as the plasticizer content increases (da Róz et al. 2006). Furthermore, depending on the concentration used, plasticizers can cause an effect known as antiplasticizing: instead of increasing flexibility and hydrophilicity, they actually cause a contrary effect. Usually this fact occurs when low concentrations of plasticizers are used (below 20 g/100 g of starch), at which the interaction of the plasticizer with the polymeric matrix is not enough to increase molecular mobility.

Other additives are used during extrusion operation to overcome high permeability caused by the plasticizer and to improve mechanical properties of the films. In this area, the production of bionanocomposites has proven to be a promising option.

Bionanocomposites formed by combinations of different clays, used as fillers, and starch have been studied and interesting results were obtained. Montmorillonite/starch composites have been the most frequently studied demonstrating a potential for improvement of tensile strength, Young's modulus, water resistance, and WVP of starches from many different sources (Avella et al. 2005, Chiou et al. 2007, Cyras et al. 2008, Kampeerapappun et al. 2007, McGlashan and Halley 2003, Tang et al. 2008a). Increases in tensile strength of as much as 92% and decrease in WVP of up to 70% in relationship to pure starch (corn, wheat, potato, waxy corn, high amylose cornstarch) films were reported (Tang et al. 2008a). An increase in Young's modulus of 500% has also been observed for potato starch/montmorillonite composites (Cyras et al. 2008). Other clays have been studied like hectorite (Wilhelm et al. 2003) and kaolin (Carvalho et al. 2001) with similar results. TPS reinforced by clay (or TPS–clay nanocomposites) have been investigated because of the undisputed potential of clays in improving film mechanical and barrier properties. Mineral clays are technologically important and are mainly composed of hydrated aluminosilicate with neutral or negative charged layers (Wilhelm et al. 2003). Clay is a potential filler; itself a naturally abundant mineral that is toxin free and can be used as one of the components for food, medical, cosmetic, and health care products (Chen and Evans 2005).

During processing, polymer chains penetrate into the clay interlayer forming melt intercalation or exfoliation. For real nanocomposites, the clay layers must be uniformly dispersed in the polymer matrix (intercalated or exfoliated), as opposed to being aggregated as tactoids (Figure 17.6). Since plasticizers play an indispensable role in the TPS process due to the interactions between starch and plasticizers, it has been hypothesized that they might also participate in the interactions between starch and clay and therefore could greatly affect the formation of the nanostructure and further influence mechanical and water vapor barrier properties of starch–clay nanocomposite films.

This matter was addressed by Tang et al. (2008b), who worked with biodegradable nanocomposites fabricated from corn starch and montmorillonite (MMT) nanoclays by melt extrusion processing and the influence of glycerol content and different plasticizers (glycerol, urea, and formamide) on the formation of nanostrucuture and properties of the starch–clay nanocomposite films was studied. The nanocomposite was prepared from dried starch, plasticizers, clay (6 g/100 g), and water (19%) mixtures, using a lab-scale twin-screw extruder. Nanocomposite films with 5% of glycerol exhibit the lowest WVP, indicating that the occurrence of

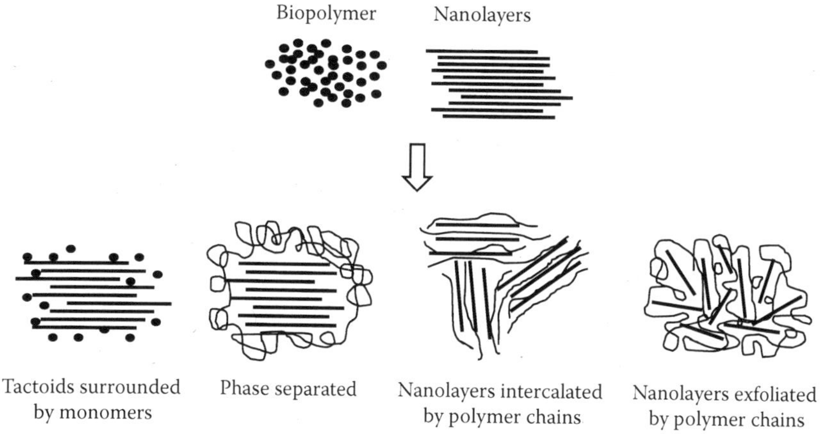

FIGURE 17.6 Representation of the production of thermoplastic starch reinforced by clay (or TPS–clay nanocomposites).

exfoliation was very helpful for improving the barrier properties of the films. Small amounts of glycerol may facilitate interactions between starch chains and silicate layers of clay, allowing glycerol and starch to diffuse together inside the layers of silicates (glycerol forms hydrogen bonds with starch, replacing the strong interactions between the hydroxyl groups of the starch molecules). When sufficient glycerol content is incorporated into the starch–nanoclay samples, intercalation process is inhibited to a certain extent because an increase in glycerol–starch interactions might compete with interactions between glycerol, starch, and the clay surface. Nevertheless, corn starch-based nanocomposite films with 5 g/100 g of MMT using different plasticizers at the same level (15 g/100 g), exhibited different WVP: formamide plasticized film presented the lowest WVP, followed by urea and glycerol plasticized ones.

Recently, composite materials based on cellulose and/or its derivatives have become a topic of intense research (Avérous and Boquillon 2004, Borges et al. 2001, Chiellini et al. 2001, Lu et al. 2005, Seavey et al. 2001). This large interest is justified by the fact that cellulose fibers are (1) renewable and biodegradable; (2) abundant and inexpensive; (3) available in a large variety of different morphologies, geometries, and surface properties depending on the source and/or separation process; and (4) comparable in specific strength to commodity fibers, including glass fibers. In addition, it has been demonstrated that cellulose fibers show good compatibility with polysaccharide matrices (López-Rubio et al. 2007). The chemical similarity between fibers and polysaccharides is one of the reasons for the notable improvement in mechanical and barrier properties of films. The material is referred to as micro fibrillated cellulose (MFC) that consists of long nanoscale bundles of micro fibrils forming entangled networks. MFC is normally prepared by the delamination of delignified wood fibers in high-pressure homogenizers. The high specific strength of MFC enables strong composite materials to be manufactured (López-Rubio et al. 2007).

17.2.4 Properties of Starch Thermoplastics

The majority of starch polymers are produced via extrusion and blending of pure or modified starch, as shown in Figure 17.5. Chemical, mechanical, and thermal properties of these materials are briefly summarized below:

- Plasticized starch polymers are less crystalline compared to native starch. Crystallinity induced by processing is influenced by parameters such as extrusion residence time, screw speed, and temperature. It is mainly caused by recrystallization (also known as starch retrogradation) of amylose into the single-helical structure.
- Density of starch polymers is higher than most conventional thermoplastics and higher than most bio-based polymers.
- TPS and starch blend films have reasonable transparency.
- The mechanical properties of starch polymers are in general inferior to those of petrochemical polymers.
- Starch polymers are reasonably easy to process but are vulnerable to degradation. Different authors have shown that after processing, plasticized starch undergoes aging, with a strong evolution of mechanical properties such as the tensile modulus, which increases during several weeks (Avérous et al. 2000, van Soest and Knooren 1997). If aging occurs at $T<T_{glass}$, the TPS will present a physical aging, with material densification. If $T>T_{glass}$, a retrogradation phenomenon with evolution of crystallinity will occur, and plasticizer molecules will rearrange into the material during storage. The retrogradation kinetics depends upon mobility of the macromolecules, plasticizer type, and content.
- TPS shows different levels of permeability to moisture and oxygen. Although water permeability is high due to the polar character of TPS, oxygen permeability is low compared to those of most polyesters. Therefore, possible applications for starch polymers are restricted by their sensitivity to moisture and water contact and high vapor permeability.
- Starch polymers are biodegradable, although too high copolymer content can adversely affect biodegradability due to the complex interaction between starch and the copolymer molecules.

17.3 APPLICATION OF BIOPOLYMERS IN FOOD PACKAGING

Packaging is used as a mechanical protection and as a food preservation technology, retarding food product deterioration, extending shelf life, and maintaining the quality and safety of the packaged foods. One of the main functions of food packaging is to protect the food from microbial and chemical contamination, delaying the adverse effect of the environment on the contained product.

The use of bio-based packaging materials for food depends on availability, quantities, prices, and properties of the materials. To date, considerable resources have been allocated to research and development on this subject, including pilot scale studies, but usage of bio-based packaging materials in the food industry is still limited (Helén et al. 2000). Recent applications of bio-based packaging are summarized in

Table 17.2. The majority of these investigations have been undertaken in academic environments but a few have apparently found commercial application.

Starch polymers are used in applications including biodegradable film for lawn and leaf collection compost bags. Its relatively high water vapor permeability is useful in applications such as fog-free packaging of warm foodstuffs (see Table 17.2). Applications in the agricultural sector include starch polymer blends for agricultural mulch films, planters, and planning pots. Further novel applications include materials for encapsulation and slow release of active agents such as agrochemicals (Degli Innocenti and Bastioli 2002).

Starch polymers have been widely used in blends with other polymers, a powerful route to obtain materials with improved property/cost performances. This approach is cheaper than the development of new polymers. Many patents were published on this topic, some related to association of starch polymers with agro-polymers, such as proteins or pectins. But, according to Avérous (2004), most of this research is focused on the blending of TPS starch with biodegradable polyesters: PCL, PEA,

TABLE 17.2
Recent Applications of Bio-Based Packaging for Foodstuffs and Their Limitations

Food Category	Packaging	Product Example	Limitations and/or Value-Added Function	Reference
Dairy products	Biodegradable copolymers films	Cheese	Moisture uptake and reduced polymer molecular weight	Plackett et al. (2006)
	Starch laminate	Cheese	Formation of vacuum	van Tuil et al. (2000)
Beverage	Biodegradable polylactic acid polymer	Orange juice	—	Jin and Zhang (2008)
	Paper cups coated with bio-based degradable plastics	Beverages in general	Poor moisture barrier	Sobol (1995)
Fruits and vegetables	Starch laminate	Cut vegetables	Antifogging	van Tuil et al. (2000)
Dry products	Starch and starch laminates	In general	Antifogging	van Tuil et al. (2000), Bastioli (2001)
	Paper bags coated with bio-based plastics	Bread	Poor moisture barrier	Helén et al. (2000)
Others	Biodegradable foams based on cassava starch, sunflower proteins and cellulose fibers	—	Reduced water absorption	Salgado et al. (2008)

PLA, polyhydroxybutyrate-*co*-hydroxyvalerate, polybutylene succinate-adipate, poly(butylene adipate-*co*-terephtalate), and poly(hydroxyl ester ether). These commercially available polyesters show some interesting and reproducible properties such as a more hydrophobic character, lower water permeability, and some improved mechanical properties.

17.3.1 Active Packaging

To keep product quality and freshness, it is necessary to select correct materials and packaging technologies. In this way, the current tendencies include the development of packaging materials that interact with the food, playing an active role in preservation.

Trends in consumer preferences toward fresh and tasty foods that are also practical and exhibit a prolonged shelf life, together with the globalization of markets resulting in longer distribution distances, are acting as driving forces for the development of active packages for the replacement of food preservation treatments (heat treatments, salting, acidification, drying, and others).

Packaging is referred as active when it performs some desired role in food preservation other than providing an inert barrier to external conditions. Active packaging has been defined as packaging, which "changes the condition of the packed food to extend shelf life or to improve safety or sensory properties, while maintaining the quality of packaged food" (Kerry at al. 2006). Many types of active packaging have been developed, mostly concerned with substances that absorb oxygen, ethylene, moisture, carbon dioxide, flavors, odors; those that release carbon dioxide, antimicrobial agents, antioxidants, flavors, and those that indicate temperature/time–temperature exposure or pH (Vermeiren et al. 1999).

The first designs in active packaging made use of a small pouch (sachet) containing the active ingredient inserted inside the permeable package. This technology yields some attractive characteristics, especially a high activity rate and lack of complex equipment or modification of packaging procedures. However, there are many disadvantages related to the use of sachets, the most important one being the presence inside the package of substances that are often toxic and could be accidentally eaten, could migrate into the food or may cause consumer rejection (López-Rubio et al. 2004).

The alternative, which is being extensively studied, is the incorporation of the active substance within the package material wall. Plastics are convenient materials for this sort of technology, not only as vehicles of the active substance, but also as participating active parts of the active principle. Hence, an important objective here is to design functional plastic materials that include the active agent in their structure, acting, or releasing adequate substances in a controlled manner. The additional advantages of incorporating this active agent into the polymeric structure (package wall) over their use in sachets are, for example, package size reduction, sometimes higher efficiency of the active substance (which is completely surrounding the product), and higher output in packaging production (as the incorporation of the sachet means an additional step, generally manual) (López-Rubio et al. 2004).

Some precautions and considerations have to be taken into account when applying these active plastics. The active agent may change the plastic properties, desorption kinetics are variable and dependent upon plastic permeability, the active capacity may get shortened by an early reaction if there is no effective triggering mechanism, and there is a potential for undesired migration of active substances or low molecular weight reaction products into the food. It is important to highlight that each type of food has a specific spoilage mechanism that must be studied and understood before designing and applying an active technology (López-Rubio et al. 2004).

The development and application of these technologies is limited due to two main factors. First, there is a lack of knowledge about the effectiveness of most systems, consumer resistance, and economic impact of this technology. Second, there are no specific regulations for active packaging up to now. As a result, a more exhaustive study of the chemical, microbiological, and physiological effects of the applied technologies must be carried out. A careful environmental impact study must also be performed before commercial implementation of these packages (López-Rubio et al. 2004).

Examples of active packaging applications for use within the food industry are briefly considered in Table 17.3. The first broad class of such applications relates to absorbers and scavengers. Many food deterioration reactions, such as off-flavor

TABLE 17.3
Examples of Active Packaging Applications for Use within the Food Industry

Absorbing/scavenging properties	Oxygen	Sachets inserted into the package or as adhesive bonded to the inner wall of the package
	Carbon dioxide	Zeolite and other finely divided minerals can be employed as CO_2 absorbers
	Moisture	Sachets containing desiccants (CaO, silica gel, natural clay) have normally been used
	Ethylene, flavors, taints, UV light	Ethylene-removing films incorporated into their structure, with a finely dispersed powdered material (minerals), such as zeolites, clays, or Japanese oya
Releasing/emitting properties	Ethanol, carbon dioxide, antioxidants, preservatives, sulphur dioxide, flavors, pesticides	—
Removing properties	Catalyzing food component removal: lactose, cholesterol	—
Temperature control	Insulating materials, self-heating and self-cooling packaging, microwave susceptors and modifiers, temperature-sensitivity packaging	—
Microbial and quality control	UV and surface-treated packaging materials	Many antimicrobial agents have been used in food packaging, such as potassium sorbate, nisin, imazalil, and triclosan

development, nutrient losses, color changes, and microbial growth, are caused by the presence of oxygen, which may derive from O_2 entrance through the packaging material, air enclosed in the food and packaging material, small leakages due to poor sealing and inadequate evacuation, and/or gas flushing of the headspace. However, deterioration and/or respiration reactions can increase CO_2 concentration inside the package of some foods, which may damage the package or may affect the product (anaerobic metabolism, pH reduction, and color and flavor changes). Although these problems can be partially overcome using high barrier materials with vacuum or modified atmosphere packaging, or high carbon dioxide permeability, recently oxygen scavenger and CO_2 systems have been used for minimizing food quality changes and extending product shelf life. Moisture absorbents are used to eliminate remaining trapped water during the packaging process or released during the storage and transportation, to avoid microbial growth or consumer rejection by the appearance of condensate water within the package. Ethylene gas, which acts as a plant hormone, is produced by fruits and vegetables during ripening and accelerates respiration, leading to maturity, softening product tissues and, therefore, accelerating senescence. On the other hand, its accumulation can cause yellowing of green vegetables and may be responsible for a number of undesirable reactions, such as the development of bitter flavors and loss of chlorophyll. Thus, to increase shelf life and maintain acceptable visual and sensorial qualities of packaged fresh fruits and vegetables, accumulation of ethylene inside the package should be avoided.

Antimicrobial packaging constitutes the second broad class of active packaging applications. It is gaining interest from researchers and industry due to its potential to provide quality and safety benefits. This packaging technology could play a role in extending shelf life of foods and reducing the risk of foodborne illnesses by minimizing, inhibiting, and retarding the growth of microorganisms that may be present within the packed food or packaging material itself (Appendini and Hotchkiss 2002). There are two basic categories of antimicrobial films: one involves the direct incorporation of the antimicrobial additive into the packaging film, while the second type of film is coated with a material, which acts as a carrier for the additive. In either case, direct contact between the packaging material and the foodstuff is necessary for the system to be effective. However, one emerging concept is the use of vapor-active antimicrobial additives. Direct contact with the food product would not be required for these types of systems. It is important that the films and coatings be formulated to allow the controlled release of the antimicrobial additives into foodstuffs. Slow diffusion of an antimicrobial agent is significant because it allows for longer and more effective antimicrobial action and therefore longer shelf life (Cooksey 2001).

For further details on active packaging and the latest developments on this particular area, the interested reader is referred to Chapter 16.

17.4 APPLICATIONS OF STARCH POLYMERS AS FOOD PACKAGING

The main use of starch-based films in this market segment is as bags for fruits and vegetables and for bread. The dimension of this market is still very limited because of the price difference in comparison with traditional plastics. An increase in its use may come from the aggregated value perceived by consumers of an environmentally friendly brand. The main advantage of starch-based materials over the traditional ones stays is their breathability, which permits better storage conditions.

Starch-based compostable bags for the separate collection of organic waste are an example of successful products in the sector of biodegradable materials. Bags of different sizes are already used by millions of European citizens for the separate collection of the organic fraction of municipal solid waste to be composted (Bastioli 2001).

Another potential application is the substitution of expanded polystyrene (EPS) food trays, due to their environmental impact related to the management of solid waste disposal. Possible replacements for EPS trays are paper trays, laminated or extrusion coated with biodegradable polymers from renewable resources. Today cups, plates, and other containers laminated or extrusion coated with starch-based and PLA-based films are available in the market at industrial level for hot and cold liquids (Bastioli 2001).

17.4.1 Application of Bio-Based Polymers as Active Packaging

Postprocessing protection using antimicrobial agents incorporated into polyethylene or other edible polymers has been proposed as an innovative approach that can be applied to some food products against various microorganisms, including *Lactobacillus plantarum*, *Listeria monocytogenes*, *Salmonella* spp., and *Escherichia coli* O157:H7, during storage. A variety of polymer films have been used to deliver nisin, including sodium caseinate films (Kristo et al. 2008), glucomana–gellan gum blends (Xu et al. 2007), alginate films (Natrajan and Sheldon 2000, Millete et al. 2007), corn zein films (Hoffman et al. 2001), whey protein, soy protein, egg albumin, and wheat gluten films (Ko et al. 2001).

Biodegradable PLA was evaluated by Jin and Zhang (2008) as a material for antimicrobial food packaging. PLA films prepared by solvent casting technique were incorporated with nisin, calculated to be 6.25 mg/g of film or 0.04 mg/cm^2 of surface area of film, against pathogens growth in liquid foods (orange juice and liquid egg white), using an agar diffusion method and a liquid incubation method. This combination of a biopolymer and a natural bacteriocin demonstrated its antimicrobial effectiveness against *E. coli* O157:H7 in orange juice. After 24 h, PLA/nisin reduced the population to 5.2 log CFU/mL (colony forming unit per milliliter), at 48 h and to 3.5 log CFU/mL at 72 h, whereas the control had declined to about 6 log CFU/mL.

Plackett et al. (2006) evaluated the impact of modified and unmodified L-polyactide and L-polyactide–PCL copolymer films as materials for cheese packaging. The materials were designed to provide slow release of cyclodextrin-encapsulated antimicrobials to control the mold growth on packaged cheeses. The overall and specific migration properties of various PLAs were assessed, with no indication of significant problems in terms of migration of substances, including the allyl isothiocyanate (AITC) antimicrobial component, into either cheese or selected food simulants. According to the authors, moisture uptake by films and a decrease in polymer molecular weight with time of exposure to high humidity were identified as areas of concern.

Although the AITC can be extracted from natural sources, and in some countries, its use as flavoring substance is permitted; there is a concern about the potential contamination with traces of allyl chloride, which is applied in the manufacturing of synthetic AITC.

Efforts have been focused on the incorporation of natural and edible compounds derived from natural sources (plants, spices, etc.) like cinnamon, clove, orange, coffee, honey, propolis, and pepper, as antimicrobial agents into edible films and coatings (Jin and Zhang 2008). It is important to know that the antimicrobial agents must be able to withstand the temperatures that are required to melt and form the polymers. When antimicrobial agents are added, it is presumed that they do not bind to the polymeric structure, but are probably entrapped within the spaces between the polymer chains (Cooksey 2001).

An edible antimicrobial film based on yam starch and chitosan was developed by Durango et al. (2006) and evaluated against *Salmonella enteritidis*. The chitosan-treated films caused a reduction of one to two log cycles in number of microorganisms, whereas pure chitosan presented a reduction of four to six log cycles compared to control.

Seydim and Sarikus (2006) tested the antimicrobial activity of whey-protein-based edible films incorporated with oregano, rosemary, and garlic essential oils against *E. coli* O157:H7, *Staphylococcus aureus*, *L. monocytogenes*, *S. enteritidis*, and *Lactobacillus plantarium*. The authors reported that the film containing oregano essential oil was the most effective against these bacteria at 2 g/100 mL of essential oil in film solution than those with garlic and rosemary extracts.

The antimicrobial activity of nisin, varied from 2000 to 5000 IU/mL, supported in edible films prepared, by casting technique, with suspensions of tapioca starch (5.0%) containing glycerol (2.5%) and water (92.5%) was evaluated by Sanjurjo et al. (2006) against *Listeria innocua*. The antimicrobial effect determined by agar diffusion technique was observed in tapioca starch films with a concentration of nisin higher than 2200 IU/cm^2 of film, indicating a useful barrier to further product contamination.

Biodegradable films based on cassava starch, incorporated with cinnamon and clove as natural antimicrobial agents, prepared by the casting technique, were evaluated by Tadini et al. (2007) against mold in pan bread slices as shown in Figure 17.7. Moisture uptake by films with time promoted mold growth after 7 days of ambient storage, whereas the addition of antimicrobial agents reduced the WVP in the films compared to the control. The extent of this influence was more significant when clove was incorporated into the biofilm matrix, at the same level of cinnamon, probably because of different particle sizes, as clove has larger particles than cinnamon.

17.4.2 Application of Bio-Based Polymers as Intelligent Packaging

Intelligent packaging (also more loosely described as smart packaging) is the one which in some way reacts to changes in some properties of the food it encloses, or the environment in which it is kept, and it is thereby able to inform the manufacturer, retainer, or consumer of the real state of these properties (Kerry et al. 2006). Although distinctly different from the concept of active packaging, features

FIGURE 17.7 Pan bread slice in contact with a cassava starch film incorporated with antimicrobial agents. (From Kechichian, V., Incorporation of antimicrobial ingredients into biodegradable films of cassava starch, MSc dissertation, University of Sao Paulo, Sao Paulo, Brazil, 2007.)

of intelligent packaging can be used to check the effectiveness and integrity of active packaging systems.

Relatively little information is available in scientific literature about intelligent packaging. Many ideas have been proposed, numerous patent applications filled, and much research in a wide range of disciplines undertaken, but very little commercial application has resulted. Examples include time–temperature indicators (TTIs), ripeness indicators, biosensors, and radio-frequency identification (Brody et al. 2008).

Many intelligent packaging concepts involve the use of sensors and indicators. A sensor is defined as a device used to detect, locate, or quantify energy or matter, giving a signal for the detection or measurement of a physical or chemical property to which the device responds. These smart devices may be incorporated into packaging materials or attached to the inside or outside of a packaging (Kerry et al. 2006). Mills (2005) presents oxygen indicators and intelligent inks for packaging food, especially optical sensors applied in modified food packaging. Kerry et al. (2006) mentioned the use of thin film coatings for the sensing material that results in low diffusion barrier properties and very fast response to changes in oxygen concentration. This feature is important for real-time, on-line quality control of large volume throughput of packages.

An indicator may be defined as a substance that signals the presence or absence of another substance or the degree of reaction between two or more substances by means of a characteristic change, especially in color (Kerry et al. 2006). For example, many visual oxygen indicators consisting of redox dyes have been patented, and a visual carbon dioxide indicator system consisting of calcium hydroxide and a redox indicator dye incorporated in polypropylene resin was described by Hong and Park (2000).

A pH-indicator packaging reports the correlation between the packed product and its pH along the storage period (Hong and Park 2000). It is based on pH-sensitive sensors that can be physically or chemically attached to the packaging (Ensafi and Kazaemzadeh 1999, Hazneci et al. 2004). Product pH is usually related to microbial spoilage. For food products packed in low oxygen environments, for example, a pH above 4.6 can represent a growth risk for *Clostridium botulinum*, which produces a toxin that can cause consumer illness and even death (Evangelista 2002).

Since it would be difficult for consumers to detect pH variation in a product, it is advantageous to have a sensor that could indicate these changes at the store (Chen et al. 2001). The pH-indicator packaging allows the consumer to determine if a product is safe without the need to open the packaging. The use of pH indicators presents an extra security for manufacturers and consumers, indicating product spoilage that can result from tampering, inadequate temperature control, and other factors that could result in illnesses from products still within the expiration date. It can also be used to minimize losses because it can indicate if a product is still intact even if the expiration date has been reached. This kind of packaging is therefore considered intelligent, because it only presents a change in response to alterations in conditions within the packaging.

The pH-indicators studied so far are mainly based on three different principles:

1. Detection of pH variation due to microbial growth (Chen et al. 2001, Freadman and Beach 2000, Hui et al. 2002)
2. Materials that acquire or lose fluorescence with pH variation (Tsien et al. 2000)
3. Chemical substances that change color with pH variation of the packed product (Iguchi et al. 1998, Park 2003)

Synthetic pH indicators like bromocresol green, phenolphtalein, and methyl red among others can be used. Natural compounds such as carotenoids (Bamore et al. 1998) and anthocyanins can also be considered (Rossi 2002). Food contact and migration of these compounds into the product are issues particularly for the synthetic indicators, thus safe incorporation of these into packaging is under investigation.

Tadini et al. (2007) have described biodegradable films based on cassava starch and incorporated with red cabbage and grape pomace extracts as sources of anthocyanins (a natural pH indicator). These extracts have an ability to change color when in contact with buffer solutions at different pHs (2, 4, 7, 10, and 12). A color change in the extract-incorporated films, at different pHs, was visible to the naked eye, and increasing extract concentration provided a more readily identifiable color change. Further research should be conducted to identify possible food applications of these films in which a given pH change can be directly related to spoilage.

17.5 CONCLUSIONS

The development of bio-based polymers is just beginning, and the use of such materials, until now, does not represent an expressive share of the plastics market. The main field regards the use of films packaging for food products, loose film for transport packaging, service packaging like carry bags, cups, plates, and bio-waste bags, among others. The production costs are still high in comparison to those of conventional plastics, rendering unfeasible an economic large-scale supply.

Biopolymers fulfill all requirements associated with current environment concerns but they show some limitations in terms of performance like thermal resistance, barrier and mechanical properties, apart from their higher costs. Then, this kind of packaging materials needs more research, more added values like the introduction of "smart and intelligent" molecules able to give information about the properties of the food inside the packaging (quality, shelf life, microbiological safety) and its nutritional values.

Since nanomaterials are abundant in nature, and numerous techniques are available for producing various nanomaterials, nanotechnology has the potential to transform food packaging in the future. Such nanoscale innovation could potentially improve mechanical and barrier properties, detection of pathogens, and smart and active packaging with food safety and quality benefits. Moreover, future research in the area of microbial active packaging should focus on naturally derived antimicrobial agents, biopreservatives, and biodegradable packaging technologies.

REFERENCES

Allcock, H. R., Lampe, F. W., and Mark, J. E. 2003. *Contemporary Polymer Chemistry*, 3rd edn. Upper Saddle River, NJ: Pearson Education Inc.

Alves, V. D., Mali, S., Beléia, A., and Grossmann, M. V. E. 2007. Effect of glycerol and amylase enrichment on cassava starch film properties. *J. Food Eng.* 78: 941–946.

Appendini, P. and Hotchkiss, J. H. 2002. Review of antimicrobial food packaging. *Innov. Food Sci. Emerg. Technol.* 3: 113–126.

Arvanitoyannis, I., Psomiadou, E., and Nakayama, A. 1996. Edible films made from sodium caseinate, starches, sugars or glycerol. Part 1. *Carbohydr. Polym.* 31: 179–192.

ASTM Standard D 5338-98, 1998 (2003). Test Method for Determining Aerobic Biodegradation of Plastic Materials under Controlled Composting Conditions, ASTM International, West Conshohocken, PA, 2005, DOI: 10.1520/D5338-98R03, www.astm.org/Standards.

Avella, M., de Vlieger, J. J., Errico, M. E., Fischer, S., Vacca, P., and Volpe, M. G. 2005. Biodegradable starch/clay nanocomposite films for food packaging applications. *Food Chem.* 93: 467–474.

Avérous, L. 2004. Biodegradable multiphase systems based on plasticized starch: A review. *J. Macromol. Sci. – Part C, Polym. Rev.* C44: 231–274.

Avérous, L. and Boquillon, N. 2004. Biocomposites based on plasticized starch: Thermal and mechanical behaviours. *Carbohydr. Polym.* 56: 111–122.

Avérous, L., Moro, L., Dole, P., and Fringant, C. 2000. Properties of thermoplastics blends: Starch–polycaprolactone. *Polymer* 41: 4157–4167.

Bamore, C. R., Luthra, N. P., Mueller, W. B., Pressley, W. W., and Beckwith, S. W. 1998. Additive transfer film suitable for cook-in end use. U.S. Patent 6,667,082, filed January 20, 1998, and issued December 23, 2003.

Bastioli, C. 2001. Global status of the production of biobased packaging materials. *Starch* 53: 351–355.

Bertuzzi, M. A., Castro Vidaurre, E. F., Armanda, M., and Gottifredi, J. C. 2007. Water vapor permeability of edible starch based films. *J. Food Eng.* 80: 972–978.

Borges, J. P., Godinho, M. H., Martins, A. F., Trindade, A. C., and Belgacem, M. N. 2001. Cellulose-based composite films. *Mech. Compos. Mater.* 37: 257–264.

Brody, A. L., Bugusu, B., Han, J. H., Sand, C. K., and McHugh, T. 2008. Innovative food packaging solutions. *J. Food Sci.* 73: 107–116.

Carvalho, A. J. F., Curvelo, A. A. S., and Agnelli, J. A. M. 2001. A first insight on composites of thermoplastic starch and kaolin. *Carbohydr. Polym.* 45: 189–194.

Chen, B. and Evans, J. R. G. 2005. Thermoplastic starch–clay nanocomposites and their characteristics. *Carbohydr. Polym.* 61: 455–463.

Chen, N., Chen, N., and Chen, N. 2001. Food freshness indicator. U.S. Patent 6,723,285, filed April 11, 2001, and issued April 20, 2004.

Chiellini, E., Cinelli, P., Imam, S. H., and Mao, L. 2001. Composite films based on biorelated agro-industrial waste and poly(vinyl alcohol). Preparation and mechanical properties characterization. *Biomacromolecules* 2: 1029–1037.

Chiou, B. -S., Wood, D., Yee, E., Imam, S. H., Glenn, G. M., and Orts, W. J. 2007. Extruded starch-nanoclay nanocomposites: Effects of glycerol and nanoclay concentration. *Polym. Eng. Sci.* 47: 1898–1904.

Cooksey, K. 2001. Antimicrobial food packaging materials. *Addit. Polym.* 8: 6–10.

Cyras, V. P., Manfredi, L. P., Ton-That, M. T., and Vázquez, A. 2008. Physical and mechanical properties of thermoplastic starch/montmorillonite nanocomposite films. *Carbohydr. Polym.* 73: 55–63.

da Róz, A. L., Carvalho, A. J. F., Gandini, A., and Curvelo, A. A. S. 2006. The effect of plasticizers on thermoplastic starch compositions obtained by melt processing. *Carbohydr. Polym.* 63: 417–424.

Degli Innocenti, F. and Bastioli, C. 2002. Starch-based biodegradable Polymeric materials and plastics-History of a decade activity. United Nations Industrial Development Organization-UNIDO. http://www.ics.trieste.it/chemistry/plastics/egm-edp2002.htm (accessed December 18, 2008).

Durango, A. M., Soares, N. F. F., Benevides, S. et al. 2006. Development and evaluation of an edible antimicrobial film based on yam starch and chitosan. *Packag. Technol. Sci.* 19: 55–59.

Ensafi, A. A. and Kazaemzadeh, A. 1999. Optical pH sensor based on chemical modification of polymer film. *Microchem. J.* 63: 381–388.

Evangelista, J. 2002. Food contamination. In *Foods: A Comprehensive study*, ed. J. Evangelista, pp. 176–231. Rio de Janeiro, Brazil: Atheneu (in Portuguese).

Famá, L., Flores, S. K., Gerschenson, L., and Goyanes, S. 2006. Physical characterization of cassava starch biofilms with special reference to dynamic mechanical properties at low temperatures. *Carbohydr. Polym.* 66: 8–15.

Famá, L., Goyanes, S., and Gerschenson, L. 2007. Influence of storage time at room temperature on the physicochemical properties of cassava starch films. *Carbohydr. Polym.* 70: 265–273.

FAO Food and Agriculture Organization of the United Nations. 2004. Global cassava market study. In *Proceedings of the Validation Forum on the Global Cassava Development Strategy*. http://www.fao.org/docrep/007/y5287eb/y5287e00.htm (accessed December 12, 2008).

FAO Food and Agriculture Organization of the United Nations. 2005. Statistical Databases. FAOSTAT Agricultural data. http://faostat.fao.org/faostat/ (accessed October 30, 2006).

Franco, Guilherme V. 2002. *Table of Chemical Composition of Foods*. Rio de Janeiro, Brazil: Atheneu (in Portuguese).

Freadman, M. and Beach, H. C. 2000. Method and apparatus for detecting bacteria. U.S. Patent 6,589,761, filed September 14, 2000, and issued July 08, 2003.

Garcia, M. A., Pinotti, A., and Zaritzky, N. E. 2006. Physicochemical, water vapor barrier and mechanical properties of corn starch and chitosan composite films. *Starch* 58: 453–463.

Guilbert, S. 2000. Potential of the protein based biomaterials for the industry. Paper presented at The Food Biopack Conference, Copenhagen, Denmark.

Hazneci, C., Ertekin, K., Yenigul, B., and Cetinkaya, E. 2004. Optical pH sensor based on spectral response of newly synthesized Schiff bases. *Dyes Pigments* 62: 35–41.

Hayashi, A., Veiga-Santos, P., Ditchfield, C., and Tadini, C. C. 2006. Investigation of antioxidant activity of cassava starch bio-based materials. Paper presented at the 2nd CIGR Section VI International Symposium on Future of Food Engineering, Warsaw, Poland.

Helén, H., Kantola, M., and Kotilainen, E. 2000. A finish study of biobased packaging materials for food applications. Paper presented at The Food Biopack Conference, Copenhagen, Denmark.

Hoffman, K. L., Han, I. Y., and Dawson, P. L. 2001. Antimicrobial effects of corn zein films impregnated with nisin, lauric acid, and EDTA. *J. Food Prot.* 64: 885–889.

Hong, S. -I. and Park, W. -S. 2000. Use of color indicators as an active packaging system for evaluating kimchi fermentation. *J. Food Eng.* 46: 67–72.

Hui, H. K., Feldman, L. A., and Gorham, R. A. 2002. Biological indicator for sterilization processes with double buffer system. U.S. Patent 6,458,554, filed February 27, 2002, and issued October 1, 2002.

Iguchi, S., Shinomiya, S., Hamamoto, R., Abe, A., Inai, M., and Kawakami, K. 1998. Package for container of liquid medicine containing bicarbonate and pH indicator. U.S. Patent 6,232,128, filed December 16, 1998, and issued May 15, 2001.

ISO Standard 14855-1, 2005. Determination of the ultimate aerobic biodegradability of plastic materials under controlled composting conditions—Method by analysis of evolved carbon dioxide—Part 1: General method, ISO International, http://www.iso.org (accessed December 19, 2008).

ISO Standard 14855-2, 2007. Determination of the ultimate aerobic biodegradability of plastic materials under controlled composting conditions—Method by analysis of evolved carbon dioxide—Part 2: Gravimetric measurement of carbon dioxide evolved in a laboratory-scale test, ISO International, http://www.iso.org (accessed December 19, 2008).

Jangehud, A. and Chinnan, M. S. 1999. Properties of peanut protein film: Sorption isotherm and plasticizer effect. *Lebens-Wiss. Technol.* 32: 89–94.

Jin, T. and Zhang, H. 2008. Biodegradable polylactic acid polymer with nisin for use in antimicrobial food packaging. *J. Food Sci.* 73: 127–134.

Kampeerapappun, P., Aht-ong, D., Pentrakoon, D., and Srikulkit, K. 2007. Preparation of cassava starch/montmorillonite composite film. *Carhobyr. Polym.* 67: 155–163.

Kechichian, V. 2007. Incorporation of antimicrobial ingredients into biodegradable films of cassava starch. MSc dissertation, University of São Paulo (in Portuguese), Brazil.

Kerry, J. P., O'Grady, M. N., and Hogan, S. A. 2006. Past, current and potential utilization of active and intelligent packaging systems for meat and muscle-based products: A review. *Meat Sci.* 74: 113–130.

Ko, S., Janes, M. E., Hettiarachchy, N. S., and Johnson, M. G. 2001. Physical and chemical properties of edible films containing nisin and their action against *Listeria monocytogenes*. *J. Food Sci.* 66: 1006–1011.

Kolybaba, M., Tabil, L. G., Panigrahi, S., Crerar, W. J., Powell, T., and Wang, B. 2003. Biodegradable polymers: Past, present and future. Paper presented at the CSAE/ASAE Annual Intersectional Meeting, Fargo, North Dakota.

Kristo, E., Koutsoumanis, K. P., and Biliaderis, C. G. 2008. Thermal, mechanical and water vapor barrier properties of sodium caseinate films containing antimicrobials and their inhibitory action on *Listeria monocytogenes*. *Food Hydrocoll.* 22: 373–386.

López-Rubio, A., Almenar, E., Hernadez-Muñoz, P., Lagarón, J. M., Catalá, R., and Gavara, R. 2004. Overview of active polymer-based packaging: Technologies for food applications. *Food Rev. Int.* 20: 357–387.

López-Rubio, A., Lagaron, J. M., Ankerfors, M., et al. 2007. Enhanced film forming and film properties of amylopectin using micro-fibrillated cellulose. *Carbohydr. Polym.* 68: 718–727.

Lu, Y., Weng, L., and Cao, X. 2005. Biocomposites of plasticized starch reinforced with cellulose crystallites from cottonseed linter. *Macromol. Biosci.* 5: 1101–1107.

Mali, S., Grosmann, M. V. E., Garcia, M. A., Martino, M. N., and Zaritsky, N. E. 2006. Effects of controlled storage on thermal, mechanical and barrier properties of plasticized films from different starch sources. *J. Food Eng.* 75: 453–460.

McGlashan, S. A. and Halley, P. J. 2003. Preparation and characterization of biodegradable starch-based nanocomposite materials. *Polym. Int.* 52: 1767–1773.

Millete, M., Le Tien, C., Smoragiewicz, W., and Lacroix, M. 2007. Inhibition of *Staphylococcus aureus* on beef by nisin-containing modified alginate films and beads. *Food Control* 18: 878–884.

Mills, A. 2005. Oxygen indicators and intelligent inks for packaging food. *Chem. Soc. Rev.* 34: 1003–1011.

Natrajan, N. and Sheldon, B. W. 2000. Efficacy of nisin-coated polymer films to inactivate *Salmonella typhimurium* on fresh broiler skin. *J. Food Prot.* 63: 1189–1196.

Park, S. -K. 2003. A freshness indicator of foodstuffs. WO Patent 03,106, 995, filed June 14, 2003, and issued December 24, 2003.

Parra, D. F., Tadini, C. C., Ponce, P., and Lugão, A. B. 2004. Mechanical properties and water vapor transmission in some blends of cassava starch edible films. *Carbohydr. Polym.* 58: 475–481.

Patel, M., Bastioli, C., Marini, L., and Würdinger, E. 2003. Life-cycle assessment of bio-based polymers and natural fibres. In *Encyclopedia Biopolymers*, vol. 10, ed. A. Steinbüchel, pp. 409–452. Weinheim: Wiley-VCH.

Plackett, D. V., Holm V. K., Johansen, P. et al. 2006. Characterization of L-polylactide and L-polylactide-polycaprolactone co-polymer films for use in cheese-packaging applications. *Packag. Technol. Sci.* 19: 1–24.

Ribeiro, E. P. and Seravalli, E. A. G. 2004. Carbohydrates. In *Food Chemistry*. São Paulo: Edgard Blüncher—Instituto Mauá de Tecnologia (in Portuguese).

Rodríguez, M., Osés, J., Ziani, K., and Maté, J. I. 2006. Combined effect of plasticizers and surfactants on the physical properties of starch based edible films. *Food Res. Int.* 39: 840–846.

Rossi, A. V. 2002. Universal pH indicator paper using qualitative filter paper impregnated with alcoholic extracts from fruits containing anthocyanins. MU Patent 8,201,475, filed July 30, 2002, and issued May 4, 2004 (in Portuguese).

Salgado, P. R., Schmidt, V. C., Ortiz, S. E. M., Mauri, A. N., and Laurindo J. B. 2008. Biodegradable foams based on cassava starch, sunflower proteins and cellulose fibers obtained by a baking process. *J. Food Eng.* 85: 435–443.

Sanjurjo, K., Flores, S., Gerschenson, L., and Jagus, R. 2006. Study of the performance of nisin supported in edible films. *Food Res. Int.* 39: 749–754.

Seavey, K. C., Ghosh, I., Davis, R. M., and Glasser, W. G. 2001. Continuous cellulose fibre-reinforced cellulose ester composites I: Manufacturing options. *Cellulose* 8: 149–159.

Seydim, A. C. and Sarikus, G. 2006. Antimicrobial activity of whey protein based edible films incorporated with oregano, rosemary and garlic essential oils. *Food Res. Int.* 39: 639–644.

Steane, R. G. 2008a. *The amylose molecule*. http://www.biotopics.co.uk/JmolApplet/amylose40jdisplay.html (accessed January 20, 2009).

Steane, R. G. 2008b. *The amylopectin molecule*. http://www.biotopics.co.uk/JmolApplet/amylopectinjdisplay.html (accessed January 20, 2009).

Sobol, R. E. 1995. Biodegradable thermally insulated beverage cup. U.S. Patent 5,542,599, filed August 07, 1995, and issued August 06, 1996.

Sorrentino, A., Gorrasi, G., and Vittoria, V. 2007. Potential perspectives of bio-nanocomposites for food packaging applications. *Trends Food Sci. Technol.* 18: 84–95.

Tadini, C. C., Veiga-Santos, P., Ditchifield, C., and Kechichian, V. Biodegradable film based on cassava starch containing natural antimicrobial ingredients and its uses. BR Patent 0.704.589-1, filed April 13, 2007 (in Portuguese).

Talja, R. A., Helén, H., Roos, Y. H., and Jouppila, K. 2007. Effect of various polyols and polyol contents on physical and mechanical properties of potato starch-based films. *Carbohydr. Polym.* 67: 288–295.

Talja, R. A., Helén, H., Roos, Y. H., and Jouppila, K. 2008. Effect of type and content of binary polyol mixtures on physical and mechanical properties of starch-based edible films. *Carbohydr. Polym.* 71: 269–276.

Tang, S., Zou, P., Xiong, H., and Tang, H. 2008a. Effect of nano-SiO_2 on the performance of starch/polyvinyl alcohol blend films. *Carbohydr. Polym.* 72: 521–526.

Tang, X., Alavi, S., and Herald, T. J. 2008b. Effects of plasticizers on the structure and properties of starch–clay nanocomposite films. *Carbohydr. Polym.* 74: 552–558.

Tharanathan, R. N. 2003. Biodegradable films and composite coatings: Past, present and future. *Trends Food Sci. Technol.* 14: 71–78.

Tsien, R. Y., Miyawaki, A., and Llopis, J. 2000. Fluorescent protein sensors for measuring the pH of a biological sample. U.S. Patent 6,627,449, filed October 31, 2000, and issued September 30, 2003.

van Soest, J. J. G. and Knooren, N. 1997. Influence of glycerol and water content on the structure and properties of extrudated starch plastic sheets during aging. *J. Appl. Polym. Sci.* 64: 1411–1422.

van Tuil, R., Fowler, P., Lawther, M., and Weber, C. J. 2000. Properties of bio-based packaging materials. In *Bio-Based Packaging Materials for the Food Industry: Status and Perspectives*, ed. C. J. Weber, pp. 13–44. Denmark: CPL Scientific Publishing Services.

Veiga-Santos, P., Suzuki, C. K., Nery, K. F., Cereda, M. P., and Scamparini, A. R. P. 2008. Evaluation of optical microscopy efficacy in evaluating cassava starch biofilms microstructure. *LWT – Food Sci. Tecnol.* 41: 1506–1513.

Vermeiren, L., Devlieghere, F., Beest, M. V., Kruijf, N., and Debevere, J. 1999. Developments in the active packaging of foods. *Trends Food Sci. Technol.* 10: 77–86.

Vilpoux, O. and Avérous, L. 2004. Starch-based plastics. In *Technology, Use and Potentialities of Latin American Starchy Tubers*, eds. M. P. Cereda, and O. Vilpoux, pp. 521–553. São Paulo: NGO Raízes and Cargill Foundation.

Xu, X., Li, B., Kennedy, J. F., Xie, B. J., and Huang, M. 2007. Characterization of konjac glucomannan-gellan gum blend films and their suitability for release of nisin incorporated therein. *Carbohydr. Polym.* 70: 192–197.

Weber, C. 2000. *Bio-Based Packaging Materials for the Food Industry: Status and Perspectives.* Denmark: CPL Scientific Publishing Services.

Wilhelm, H. M., Sierakowski, M. R., Souza, G. P., and Wypych, F. 2003. Starch films reinforced with mineral clay. *Carbohydr. Polym.* 52: 101–110.

Wolf, O., Crank, M., Patel, M. et al., eds. 2005. *Techno-Economic Feasibility of Large-scale Production of Bio-Based Polymers in Europe.* Sevilla: JRC http://ftp.jrc.es/EURdoc/eur22103en.pdf

18 Goat Milk Powder Production in Small Agro-Cooperatives

Uliana K. L. Medeiros, Maria de Fátima D. Medeiros, and Maria Laura Passos

CONTENTS

18.1 INTRODUCTION

This chapter is aimed at developing a strategic methodology to identify a nonconventional process route to produce, at low cost, goat milk powder. This route, being technically and economically feasible, will be implemented in some agro-cooperatives located in the Northeast of Brazil. Since goat milk powder is a new product-process investment for these small enterprises, a strategic plan should be created to develop the most adequate route for them. Therefore, Figure 18.1 shows schematically the main steps of the proposed plan to meet such a challenge of a new product-process route.

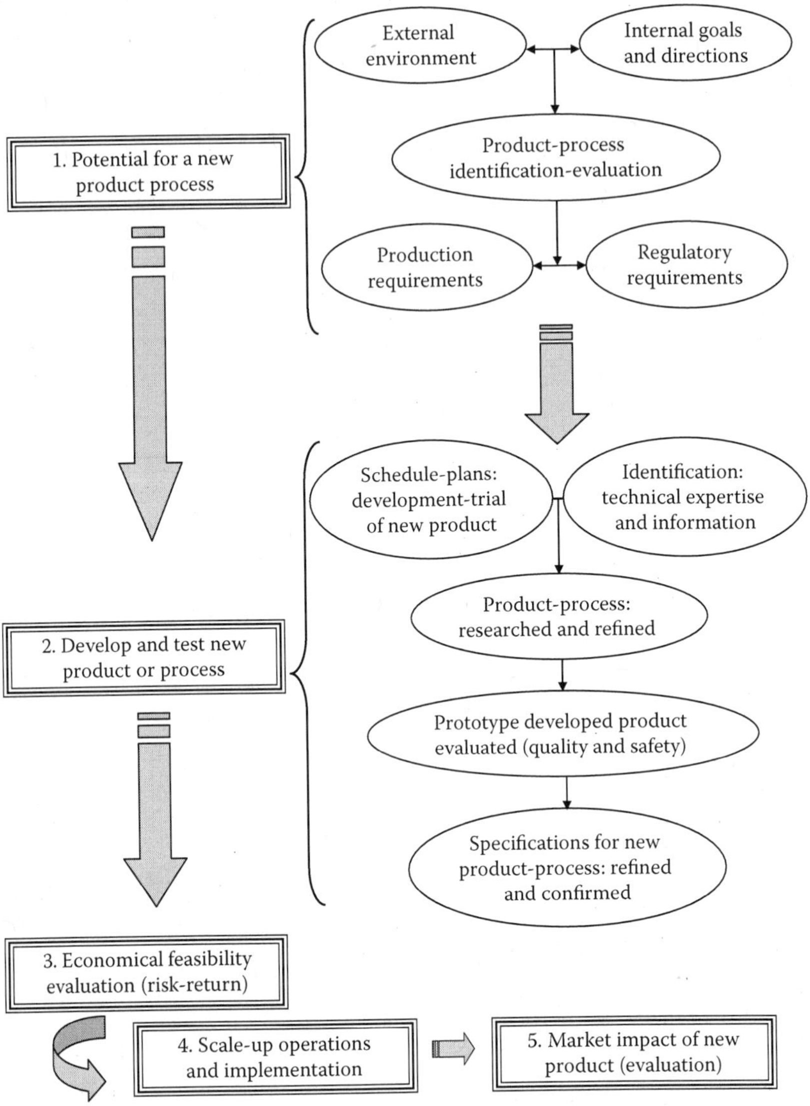

FIGURE 18.1 Scheme of the basic steps to follow in developing a new product process.

The first step identifies the potential for developing the new product process. This comprises organized research for acquiring appropriate knowledge of

- The external environment and the internal enterprise organization to identify emerging opportunities and to clarify the cooperative goals for the new product process
- The product itself and the available routes for its production, to evaluate them and identify the possible processing route(s) that can attend the cooperative goals

- Definitions of the general production and regulatory requirements (i.e., production scale, main market, food quality and safety according to legislation, etc.)

The second step must determine the technical feasibility of the new product process. Since this step involves experimental tests, trials, analyses, and work, it ought to be well designed and organized, following a rigid schedule to provide, within a short and precise time period, the answer regarding the technical feasibility of the processing route identified as potential. In addition to this well-structured schedule, scientific knowledge and technical information are necessary to refine and build, at laboratory scale, the processing route to produce goat milk powder. Following the experimental work, the prototype of the new product should be obtained and evaluated in accordance with the requirements proposed earlier. Based on the results, the product-process route is confirmed, and then it can be refined.

The third step comprises the economical feasibility evaluation, in which investment risks, as well as its return, are calculated with regard to the industrial implementation of the processing route. If this economical evaluation is feasible, the fourth step starts, consisting of scaling-up the process and implementing the industrial plant in the agro-cooperatives. The final step is to introduce the new product to the market and monitor its performance by evaluating its impact and proposing strategies to enlarge its competitiveness and market.

Although these last three steps are under development, they are not within the scope of this chapter, whose objective is to develop a nonconventional process route, technically feasible, to produce high-quality goat milk powder at low cost.

18.2 POTENTIAL FOR GOAT MILK POWDER PRODUCTION

18.2.1 Economic Scenario of Goat Activity

Nowadays, goat milk production and processing constitutes an economic activity of increasing importance since this milk is of high nutritional interest, providing high-quality protein, fat, carbohydrates, vitamins, and several minerals, such as iron, calcium, and phosphorus, as detailed later. Parallel to milk production, another relevant branch of this emergent economy is the production of goat meat, mohair, and cashmere fibers.

Goats seem to be the earliest domesticated animal and, due to their inherent nature, they have been raised usually in small herds maintained by individuals either as a source of income or as a hobby. Nevertheless, as pointed out in the literature (Haenlein 1996), goats have provided important sustenance, self-sufficiency, and survival for people and countries during economically difficult times. Although the goat nickname "cow of the poor man" still persists in some countries, there is a new consensus in the world that dairy goats provide foods not only for starving and poor people but also for prosperous periods and affluent people (Haenlein 1996). Several works published recently confirm the economic feasibility of investing in the goat dairy business in both developed and developing countries (Harris and Springer 1992, Rubino and Haenlein 1997, Perosa et al. 1999, Stoney and Francis 2001, Holanda et al. 2006).

The increase, not only in research, but also in goat livestock and milk production during the last years, as shown in Figure 18.2, corroborates changes in status and interest in the goat activity.

According to statistical data (FAO 2006), although China has the biggest goat stock in the world (199 millions of heads in 2006, i.e., 24% of all goat stock in the world), India is the first producer of goat whole fresh milk (3.8 million metric tons in 2006, representing 27.6% of the world production). In this context, Brazil has the largest goat livestock and goat milk production in South America. In 2006, Brazil

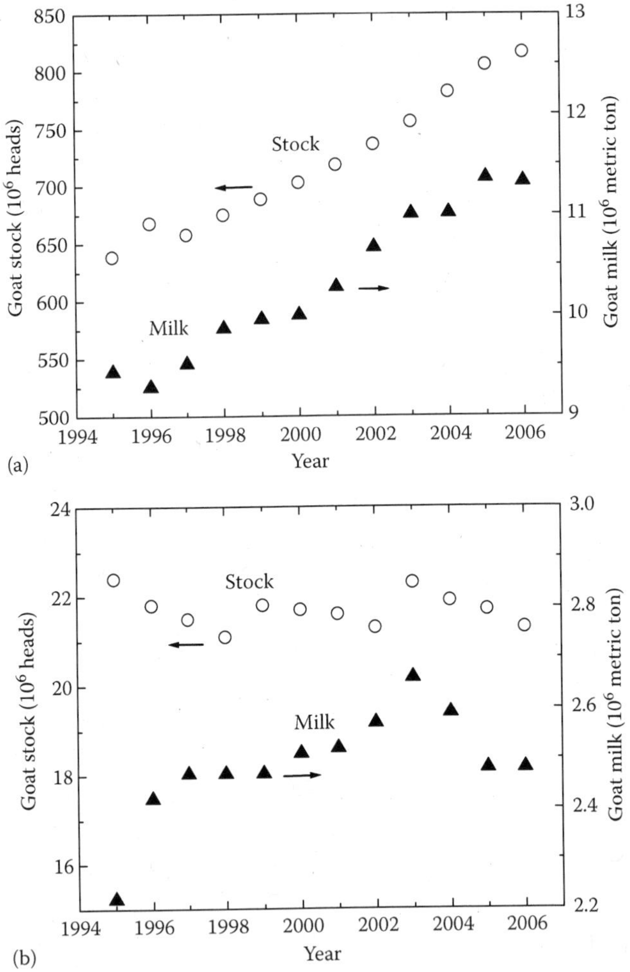

FIGURE 18.2 Evolution of the goat livestock and the goat milk production in the world from 1995 to 2006: (a) developing countries and (b) developed countries. (Based on data from Food and Agricultural Organization of the United Nations, FAOSTAT Statistics Division, 2006. http://faostat.fao.org/site/573 or 569/pageID=573 or 569)

was the 15th country in the list of world goat herds (10.4 millions of heads), and 18th where world goat milk production was concerned (135,000 metric tons).

Indeed, goat production systems are moving toward quality and/or quantity. Developed countries, which hold 3% of goat livestock and 19.5% of milk production (see Figure 18.2b), invest greatly in quality, developing a well-organized goat milk production chain, because their consumers demand more sophisticated products. France, whose dairy goat sector has been developed since the early 1950s and is well organized into cooperatives and industries, remains the leader of the goat cheese production. In the United States and Canada, goat milk is a niche market that started growing in the early 1980s. Another developed country with a well-organized goat sector, although small, is New Zealand. Dairy Goat Co-operative New Zealand Ltd., which is based on the farmer–owner cooperative system, focuses on nutritional milk products for babies and young children, ultrahigh temperature (UHT) milk and whole milk powder. Processing goat milk basically from August to March, this dairy is able to export its products to other countries, like Taiwan.

Although developing countries tend to move toward quantity to assure the maximum whole fresh goat milk for population needs, some countries have already improved technology and animal genetics for increasing not only the amount of milk production but also the investment in the dairy goat sector. Note, for instance, that Mexico increased its goat milk production by 17% from 2001 to 2006, surpassing the Brazilian production, without any additional increase in its goat livestock (FAO 2006).

Atypically, Brazil presents two different systems for raising dairy goat: the intensive farming under confinement, representative of developed countries, and the extensive farming used in developing countries. The former is characterized by

1. Smaller and restricted areas, commonly near urban centers or metropolitan regions
2. Crossbred and purebred animals (Alpina, Saanen, Toggenburg, Anglo-Nubian, Bhuj, and Boer breeds)
3. Use of supplementation with protein sources and cultivated grounds with leftovers from bean and corn cultures for feeding goats
4. Sanitary measures, such as vaccination, parasite control, and other facilities

This system is found in the Southeast, which retains 2.5% of the Brazilian goat stock (IBGE 2006) and where a dairy industry is structured, processing the collected goat milk by UHT and providing a regular offer of goat milk products to Brazilian markets. The biggest goat dairy enterprise in this region is Celles Cordeiro Agroindustrial Ltda, Rio de Janeiro, RJ, as shown in Table 18.1. Although private, it is associated with universities, research centers, and cooperatives. Since 1995 when it was founded, this company has imported, basically from the Netherlands, 60–70 tons of goat milk powder per year to continuously commercialize its products. More recently, in 2008, this company started reducing its milk import by contracting small dairy Brazilian factories to dry and powder its own raw goat milk (Villar 2008).

TABLE 18.1
Information about Dairy Goat Industries and Cooperatives in 2007

Enterprise	Country	Processing Volume ($\times 10^6$ L/year)	Products	Price Ratio (Goat to Cow Fresh Milk)
SOIGNON® (Eurial Poitouraine, Nantes, Pays de la Loire)	France	140	Different brands of cheese	1.4–1.5
Amalthea van Dijk	Netherlands	25	Different brands of cheese; yogurt	
ANGULO® (Angulo General Quesera, SL, Burgos)	Spain	18	Different brands of cheese	1.4–1.7
St. Helens Farm	United Kingdom	6	Fresh milk; probiotic yogurt; cheeses; butter; cream	
MEYENBERG® (Jackson-Mitchell, Turlock, California)	United States	3.6	Fresh whole, lowfat, evaporated, powder milks; butter; cheeses	
CAPRILAT® (Celles Cordeiro Agroindustrial, Ltda, CCA, Nova Friburgo, Rio de Janeiro)	Brazil	1.6	UHT whole and lowfat milks; instantaneous and regular milk powders	2.0–2.6
Paraiba State Government	Brazil	5.3	Pasteurized milk for institutional program	
Cooperatives associated with RN State Government	Brazil	3.6	Pasteurized milk for institutional program	

Sources: Modified from Cordeiro, P.R.C., Benefits and strategies of integrated commercialization, Keynotes presented at II Goats and Ovine Symposium, Belo Horizonte, MG, Brazil, May 18–20, 2007; Cordeiro, P.R.C. and Cordeiro, A.G.P.C., Goat milk business in Brazil and its chain of production, Paper presented at XII Northeast Cattle Raising Seminary, Fortaleza, CE, Brazil, June 23–26, 2008.

The extensive farming system, on the other hand, is located in the Northeast, where 92.4% of the Brazilian livestock are found (IBGE 2006). The basic characteristics of this system are

1. Larger areas, without delimiting fences and where animals can be easily lost
2. Native breeds (Moxotó, Canindé, Marota, Repartida) or animals without specific breed

3. Use of native feeding grounds (caatinga—a region covered with semiarid scrub forest)
4. Lack of feeding supplementation except during periods of critical drought
5. Little sanitary care and facilities

Governmental and State Social Programs have been supporting goat producers in the Northeast as a form of improving family life and helping them to survive. Since 1995, State Governments in the Northeast have strengthened their social programs to improve milk production for child feeding needs and for malnourished handicapped, pregnant women, and indigent families. Inasmuch as this goat milk production is already domestic and comes from small farms, producers have created rural cooperatives and associations to organize themselves. This type of organization first occurred in the state of "Rio Grande do Norte" (RN), in 1995. Since then, specialists, researchers, and consultants from federal universities, research centers, and public institutions have been engaged in projects for guiding these pioneer producers toward actual techniques and methods to raise goats healthfully and sanitarily, preserve, pasteurize, and process their milk production for industrial use. Therefore, the RN State has structured its agribusiness by investing in the goat-production chain, producing, in 2006, about 2.3 million L of fresh whole goat milk (IBGE 2006). This amount might rise to 2.4 million L if the informal production was added. From the amount produced in 2006, 2.1 million L, belonging to 46 municipalities that participated in the institutional goat milk program for feeding children, were pasteurized in five dairy factories of this program. From Figure 18.3, it can be inferred that the goat agribusiness has grown successfully under the government institutional support. From 1995 to 2006, while the herd of goats increased by 26% in the RN State, its production of whole fresh

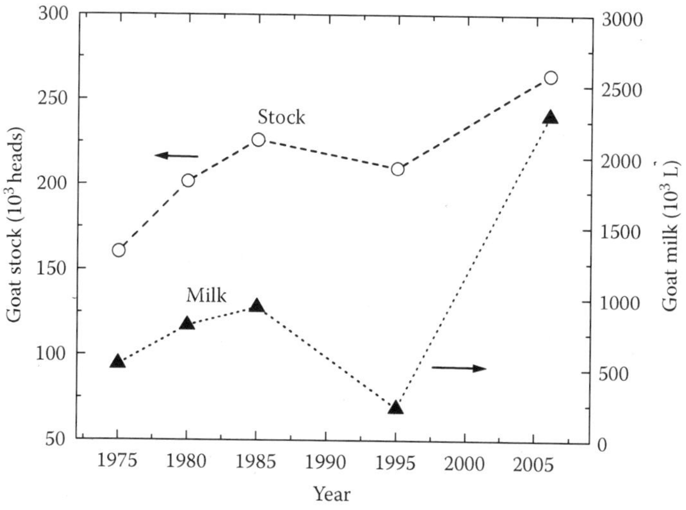

FIGURE 18.3 Evolution of the goat livestock and the goat milk production in the "Rio Grande do Norte" State, Brazil, from 1975 to 2006. (Based on data from IBGE, Brazilian Institute of Geography and Statistic, 2006, http://www.ibge.gov.br/home/estatistica/ppm/2000–2006/ppm200.pdf to ppm2006.pdf)

goat milk expanded by about 900%. In 2007, as shown in Table 18.1, the goat milk production continued to increase (70% when compared to that of 2006). Nowadays, other viable optional applications for the exceeding of goat milk production urge to be developed to avoid losing this socially oriented agribusiness, since the quota system for selling goat milk to the RN government is limited threatening the growth of goat milk production, other viable optional applications for the exceeding of goat milk production urge to be developed to avoid losing this socially orientated agribusiness.

Following this successful experience, the goat producers of "Paraíba," another State in the Northeast located near RN, started organizing themselves into cooperatives under the government support, surpassing, in the last three years, the goat milk production of the RN State, as confirmed in Table 18.1.

This global and local state of the art in goat activity provides good perspectives for investing in a dairy goat production, being relevant to specify clearly the basic needs and goals of the RN goat milk producers.

18.2.2 Local Needs and Goals

To identify the economical aspect of this goat activity in the Northeast of Brazil, basically in the RN State, budget analyses of goat milk production units have already been made during these years of prosperity. Table 18.2 shows a summary of these analyses based on technical efficiency coefficients:

- Case 1—developed in 2003 (Pereira 2003) based on data collected during 5 months in a farm located in the RN State
- Case 2—developed in 2006 (Holanda et al. 2006) based on data collected by the government agencies in the RN State during 1 year
- Case 3—developed in 2007 (Santos et al. 2008) based on data collected from 19 farms in the Southeast of Brazil (Rio de Janeiro [RJ] and Minas Gerais States). Santos et al. (2008) have divided their data into five groups according to milk production. The values presented in Table 18.2 are means related to the group whose production ranges from 30 to 50 L/day.

In these analyses, the herd numbers were maintained constant. Consequently, only the herd maintenance cost has been considered in these calculations. Revenue cost included those obtained by selling milk and by-products (fertilizer, animals, etc.) during the period analyzed. The effective operational cost represented direct expenses covered by farmers (wages, expenses with grazing and animal feeds, vaccination and medicines, pasture maintenance, buildings and equipment maintenance, energy, water, fuel and transportation, technical and administration assistance). In the RN State, these expenses are free from taxes and income according to the government regulations, and they have not been included in the calculations for cases 1 and 2. In the Southeast, expenses with artificial inseminations exist and they have been considered in the calculations for case 3. The total operational cost was determined by adding the effective operational cost to the family labor opportunity cost for daily tasks and/or the depreciation of buildings, equipment, and animals. The analysis performed in 2003, case 1, excluded the familiar labor cost; however, in cases 2 and 3, this cost was added. The total cost comprised the total operating cost as well as the

TABLE 18.2
Unitary Goat Milk Production Cost in Farms Located in Two Brazilian States (RN and RJ), with Values Expressed in the Brazilian Currency (R$/U.S.$ ≅ 0.421)

Specification	Case 1 Reported in 2003 (Pereira 2003)	Case 2 Reported in 2006 (Holanda et al. 2006)	Case 3 Reported in 2008 (Santos et al. 2008)
Production for selling (L/day)	51.71	14.11	36
Budget items	R$/L	R$/L	R$/L
Revenues	0.76	1.23	1.64
Milk	0.73	1.00	1.32
Others	0.03	0.23	0.32
Effective operational cost	0.70	0.45	1.15
Total operational cost	0.92	0.99	1.93
Total cost	1.18	1.18	2.48
Gross margin (=1–2)	0.06	0.78	0.49
Net income (=1–3)	–0.16	0.24	–0.29
Profit (=1–4)	–0.42	0.05	–0.46
Additional information			
Period analyzed	February–June 2003	Year 2006	Year 2008
Herd characteristics	Nonstable	Stable	Stable
Heads	261	33	44
Bucks (head)	4	1	2
Does (heads)	185	26	25
Lactating does (heads)	85	16	20
Stable capital/(amount of milk produced per year) (R$/L)	5.25	4.06	9.60

Sources: Pereira, G.F., Study about the rentability of the goat milk production system in the RN State, Master's thesis, Federal University of Rio Grande do Norte, Rio Grande do Norte, Brazil, 2003, Table 4.4 (in Portuguese); Holanda E.V. Jr., et al., Economic performance of the goat milk production by individuals in Rio Grande do Norte State, Technical Communication no. 74, 4, ISSN 1676-7675, Sobral Emprapa Caprinos, 2006, http://www.cnpc.embrapa.br/cot74.pdf, accessed April 2, 2008 (in Portuguese); Santos E. Jr., et al., *Rev. Bras. Zootec.*, 37, 780, 2008.

interest in the stable and circulating capitals, calculated based on the savings account interest.

For easy comparison, all costs have been divided by the amount of milk sold in the period analyzed in each case to estimate the unitary goat milk production cost (Brazilian currency per liter = R$/L, with R$/U.S.$ ≅ 0.421).

The gross margin, which reflects short-term results and by which producers evaluate the business success, is positive for all three cases but higher for the smallest producer (case 2). The net income reflects long-term results and shows, when it is negative as in cases 1 and 3, that the business earning is in jeopardy particularly for those producers whose income generation depends only on this farm activity. These producers, after a long period, might become incapable of maintaining the investment needs, unless they can increase the price of their product, for example, by adding value to it. In case 1, if the price of the final product increases by 45%, the net income becomes positive (=1.103). Even in the Southeast, where this goat activity is consolidated, the net income result is negative (only two out of the six farms analyzed presented a positive net income). The profit shows the economic stability of the activity. When it is negative but with the gross margin positive (as in cases 1 and 3), the farmer can produce goat milk during a short period with a crescent decapitalization. Not only for cases 1 and 3 but also for case 2, the goat milk production activity leads to null or insignificant profitability in the long term.

These results corroborate the urgent necessity to seek new options for improving the profit of this goat milk production activity, which is already well structured in the RN State. In addition, it indicates that adding value to whole goat milk can be a possible option for avoiding the extinction of this agro-cooperative business. Focusing on dairy goat products, one should stress that any informal subproduct, such as domestic cheeses, yogurts, ice creams, or cosmetics, deals with fluctuations in their selling market, which can be detrimental to a long-term business. Conversely, an intermediate product with a need-value in the food industry provides a long-term stable market.

Goat milk powder attends the requisite of an industrial need product, emphasizing its high nutritional interest and its good opportunities in the institutional, industrial, and commercial, markets. First, due to the goat milk seasonality, the government, the food industry, and the local market are likely to buy its powder during, at least, the 2 months of the doe's dry period. By knowing the relevance of this dry period, in which does should be bred to freshen, repair, and regenerate their mammary system for the next lactation, it can be assured that, once a year, there is a seasonal demand for goat milk powder in the market. Moreover, among other industrial applications, goat milk powder can also be used as an ingredient for baby foods due to its nutritional value (to be discussed in Section 18.3.1). This should probably be a market to explore. Another option is to transport this powdery product, after packaging, to industrial centers in the Southeast of Brazil that demand this type of product. As previously discussed, the goat dairy Celles Cordeiro Agroindustrial Ltda enterprise imports powdery milk to maintain its industrial line of products. Since there is an intention of this enterprise to use raw materials and subproducts from Brazil, it will be another probable market for goat milk powder. Based on data presented by Pereira (2003), the importation of goat milk powder from this enterprise reached 1.4×10^6 L in 2002. This is more than 20% of the current goat milk production in the RN cooperatives, supposing roughly that the governmental milk program can buy 80% of this production.

Based on these considerations, there is a potential market with capacity to absorb, as a processed powdery product, the exceeding of goat milk production of the RN cooperatives. According to Dubeuf (2007), after identifying the market and its capacity for a new goat-milk-based product, the two other indicators to be taken into account before deciding to investigate the corresponding process are the local price of cow milk and the acceptance for goat milk. Table 18.1 shows that, in Brazil, prices of fresh goat milk are twice (or more) as much as those of fresh cow milk. Furthermore, a brief search on the Web has shown that goat milk powder can reach prices up to 3.5 times that of cow milk powder in Brazilian supermarkets.

In Brazil, there are no religious or political restrictions about goats and their products. An inquiry online developed by Agripoint Consultoria Ltd. (Danes 2007), from November to December 2006, showed that cultural habits limited consumers to buy goat or ovine food products (30%) as well as the lack of advertisements about these products (29%), followed by prices (14%), quality (13%), and prejudices (6%).

As a final conclusion, one can state that the actual local scenario favors the search for a processing route to produce goat milk powder in agro-cooperatives or associations of the RN State. From Table 18.2, it can be inferred that such route must present lower investment and maintenance costs than those of the conventional process.

18.3 GOAT MILK: PROCESS AND PRODUCTS

Haenlein (2004) has stressed the threefold significance of goat milk and its products in human nutrition: feeding more starving and malnourished people than cow milk; treating people afflicted with cow milk allergies and gastrointestinal disorders, and filling the gastronomic needs of connoisseur consumers. It is essential to briefly characterize raw goat milk, comparing it with other types of milk, not only to understand its advantages over cow milk, but also to differentiate these two types of milk in relation to the manufacture process.

18.3.1 Milk Characterization

18.3.1.1 Chemical Composition

Milk can be described as a dilute emulsion consisting of an oil/fat dispersed phase and an aqueous continuous phase (serum), or as a colloidal suspension of proteins, or even as a solution of lactose, soluble protein, minerals, vitamins, and other components. Walstra et al. (1999) have linked these definitions to the different milk configurations seen under microscope lens. At a low magnification, a uniform, but turbid, liquid is visualized, characterizing a dilute emulsion. Increasing the magnification by 1000 times, spherical fat droplets (globules) appear floating in a turbid liquid (plasma), characterizing fat globules as the dispersed emulsion phase. Finally, at magnifications over 10,000 times more, one can see clearly the plasma, which is composed of protein particles (casein micelles) dispersed in an opalescent liquid (serum). This serum is a colloid, containing globular and lipoproteins in suspension.

Although the milk composition depends on several factors such as breeding, state of lactation, feeding, individual animal differences, and climatic and geographic conditions, compiled data from the literature were used to obtain mean values of the milk chemical composition from three different mammals, as shown in Table 18.3. These values are averaged over 15 data sets for raw goat milk (Lythgoe 1940,

TABLE 18.3
Mean Chemical Composition of Milk from Three Mammals

Constituents (%w/w)	Goat	Cow	Human
Water	87.1 (±1.7)	87.5 (±0.3)	87.6 (±0.4)
Protein	3.5 (±0.3)	3.1 (±0.1)	1.3 (±0.6)
Casein	2.3	2.6	0.4
α_{s1}-Casein	0.3	1.2	N/A
α_{s2}-Casein	0.5	0.1	N/A
β-Casein	1.1	0.9	0.06
κ-Casein	0.3	0.4	N/A
Whey protein	0.7	0.5	0.7
Lipids (wt% total lipids)	3.9 (±0.8)	3.7 (±0.2)	3.8 (±0.1)
Saturates	66.0	60.8	46.3
Monounsaturates	27.4	33.5	38.7
Polyunsaturates	4.5	4.0	10.8
Carbohydrate (lactose)	4.5 (±0.3)	4.8 (±0.1)	7.0 (±0.03)
Ash	0.7 (±0.1)	0.7 (±0.2)	0.21 (±0.0)

Sources: Lythgoe, H.C., *J. Dairy Sci.*, m123, 1103, 1940; Webb, B.H. and Johnson, A.H., *Fundamentals of Dairy Chemistry*, Avi Publishing Co., Inc., Westport, CT, 1965; Jenness, R., *J. Dairy Sci.*, 63, 1606, 1980; Sawaya, W.N. et al., *J. Dairy Sci.*, 67, 1657, 1984; Walstra, P. et al., *Dairy Technology: Principles of Milk, Properties and Processes*, Marcel Dekker, New York, 1999; Guo, M.R. et al., *J. Dairy Sci.*, 84E, E80, 2001; Haenlein, G., Lipids and proteins in milk, particularly goat milk, Technical Report, 2–3, University of Delaware, Newark, DE, 2002, http://ag.udel.edu/extension/information/goatmgt/gm-08.htm, accessed March 12, 2008; FSA, Food Standards Agency, *McCance and Widdowson's the Composition of Foods*, Sixth summary edition, Royal Society of Chemistry, Cambridge, U.K., 2002, Table Milk; Maree, H.P., Goat milk and its use as a hypo-allergenic infant food, Article 152, 2003, Table 1, http://www.goatconnection/articles/publish/article_152.shtml, accessed March 10, 2008; St-Gelais, D. et al., Composition of goat's milk and processing suitability, Technical Report, 4–5, Agriculture and Agri-Food Canada: Food Research and Development Center, Saint-Hyacinthe, Quebec, Canada, 2003; Pereira, R.A.G. et al., *Rev. Inst. Adolfo Luz*, 64, 208, 2005; Oliveira, M.A. et al., *Rev. Inst. Adolfo Luz*, 64, 106, 2005; Costa, R.G. et al., *Rev. Inst. Adolfo Luz*, 66, 138, 2007; Eddleman, H., Composition of human, cow, and goat milk, Article B120A, 2007, Table 1, http://www.goatworld.com/articles/goatmilk/colostrum.shtml, accessed May 20, 2008; Thompkinson, D.K. and Kharb, S., *CRFSFS*., 6, 80, 2007; OGMPA, Ontario Goat Milk Producers' Association, 2008, Table, http://www.ontariogoatmilk.org/frames.htm, accessed March 10, 2008.

Webb and Johnson 1965, Jenness 1980, Sawaya et al. 1984, Guo et al. 2001, FSA 2002, Haenlein 2002, Maree 2003, St-Gelais et al. 2003, Oliveira et al. 2005, Pereira et al. 2005, Costa et al. 2007, Eddleman 2007, OGMPA 2008), 7 data sets for raw cow milk (Webb and Johnson 1965, Walstra et al. 1999, FSA 2002, Haenlein 2002, Maree 2003, Eddleman 2007, OGMPA 2008), and 5 data sets for mature human milk (Webb and Johnson 1965, Haenlein 2002, Maree 2003, Eddleman 2007, Thompkinson and Kharb 2007). For each mean value, 95% confidence bands were computed, and these are given in brackets as the error of each value.

As seen in Table 18.3, the lipid content is statistically the same for the three milks analyzed, being the data scatter (quantified by the error value) higher for goat milk probably due to its seasonality associated with its domestic production.

Although this result seems contradictory in relation to the goat milk advantages already confirmed by specialists, there must be differences in the composition of these lipids, since they play an important role in dairy products. They contribute to flavor and to establish the price of milk on the basis of its fat content. Note that *fat* refers correctly to lipid because at ambient temperature milk lipids are solid (in opposition to *oil*, which is liquid). Furthermore, the main milk lipids belong to the triglyceride class (about 98% of milk fat), which is characterized by a glycerol backbone bound to three different fatty acids. A carboxyl group and a hydrocarbon chain compose these fatty acids, whose classification is based on the number of carbon atoms in their chain. A short-chain fatty acid is composed by 4 (C4 = butyric acid), 6 (C6 = caproic acid), 8 (C8 = caprylic acid), or 10 (C10 = capric acid) carbon atoms. A medium-chain fatty acid (MCFA) or a medium-chain triglyceride (MCT) has 6, 8, 10, or 12 carbon atoms (C12 = lauric acid) in its hydrocarbon chain. Following the discussion presented in Chapter 22, a fatty acid chain of 12 carbon atoms is considered as the border between MCT and a long-chain triglyceride (LCT). The major LCTs found in milk are formed by one or more of these long-chain acids: myristic (C14), palmitic (C16), stearic (C18), arachidic (C20), or unsaturated one (C14:1 = myristoleic, C16:1 = palmitoleic, C18:1 = oleic, C18:2 = linoleic, C18:3 = linolenic, or C20:4 = arachidonic).

LCTs require a large amount of bile acids and several digestive steps to be emulsified and hydrolyzed to fatty acids and monoglycerides, which are dissolved in micelles and able to be absorbed by intestinal epithelial cells. In these cells, these compounds are transformed into chylomicrons, released into the bloodstream via the lymph vessels, and then transported to peripheral tissues (adipose tissue and muscle). Conversely, since MCTs are smaller molecules and more soluble in water, they are completely hydrolyzed to MCFAs and glycerol, and rapidly absorbed by intestinal cells. Once in these cells, these MCFAs are mostly bound with albumin and transferred to the bloodstream via portal vein. Subsequently, MCFAs are transported straight to the liver, becoming an available source of energy, since they can be rapidly degraded by oxidation inside the mitochondria. This briefly explains why LCTs contribute to fat storage and MCTs to a healthy energy source in human body. Details of these mechanisms are well presented by Odle (1997).

Therefore, the quantification of these fatty acids is essential to identify the nutritional value of milk. In Figure 18.4, these fatty acids are quantified by their content per total fat content (wt%) for goat, cow, and human milk. As corroborated by these data, goat milk presents the highest percentage of MCTs (17.6 wt% of total fat), confirming its metabolic ability to provide energy instead of storing fats in adipose tissues. Additionally, studies under development have identified the beneficial effect of goat milk on the metabolism of iron and copper in animals with malabsorption syndrome (Barrionuevo et al. 2002). This also means that goat milk has a better digestibility and potential use as medical treatment for an array of clinical disorders such as malabsorption syndrome, premature infant feeding, and infant malnutrition, among others.

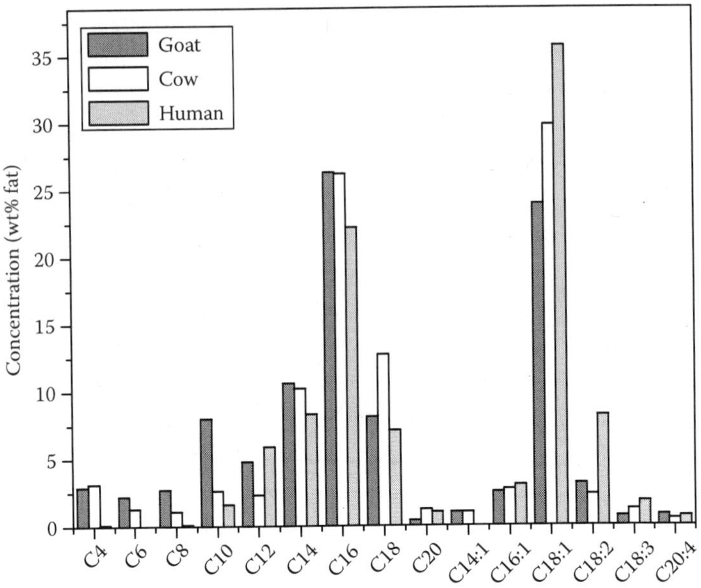

FIGURE 18.4 The principal fatty acids and their concentrations in three types of milk: goat, cow, and human. (Based on data from Haenlein, G., *Small Ruminant Res.*, 51, 160, 2004; Haenlein, G., Lipids and proteins in milk, particularly goat milk, Technical Report, University of Delaware, Newark, DE, 2002, Table 2, http://ag.udel.edu/extension/information/goatmgt/gm-08.htm, accessed March 12, 2008; Maree, H. P., Goat milk and its use as a hypo-allergenic infant food, Article 152, 2003, Table 3, http://www.goatconnection/articles/publish/article_152.shtml, accessed March 10, 2008; FSA, Food Standards Agency, *McCance and Widdowson's the Composition of Foods*, 6th Summary Edition, Royal Society of Chemistry, Cambridge, U.K., 2002.)

Moreover, since most of milk fat is in a globular form, the sizes of these globules, as well as their structure, should affect milk properties and nutritional value. As demonstrated in the literature (Walstra et al. 1999, Vignolles et al. 2007), these globules are covered by a thin membrane, composed of several materials such as bipolar materials, proteins, phospholipids (other lipid class in milk), cholesterol, and enzymes. This membrane protects fat against lipolytic enzymes and avoids their aggregation and coalescence. During milk homogenization, this membrane is recombined, absorbing layers containing casein micelles and whey proteins to lower surface tension. Bearing in mind that the data analyzed here do not include those concerning homogenized fat globules, one can see in Table 18.4 the mean Sauter diameter (d_{32}) and size range of these fat globules for raw goat, cow, and human milks. These data show that the smallest mean size of fat globules is found in raw goat milk, which means that goat milk has a larger proportion of smaller fat globules than the other two types of milk. This explains the apparent absence of agglutination in cold goat milk and its slower rate to cream. Smaller fat globules are better dispersed resulting, at high proportion, in a more uniform emulsion. It may contribute to enhance the goat milk digestibility, though there is some controversy over this matter

TABLE 18.4
Mean Sauter Diameter and Size Range of Fat Globules for Three Types of Milk

Milk Type	d_{32} (μm)	Size Range (μm)
Goat (Attaie and Richter 2000)	2.76 ± 0.07	0.73–8.58
Cow (Attaie and Richter 2000)	3.51 ± 0.08	0.92–15.75
Human (Michalsk et al. 2005)	3.5 ± 0.1	N/A
Mature (>45 days postpartum)	1.5 ± 0.1	
Transitional (2–5 days postpartum)	4.3 ± 0.9	
Colostrum (≤48 h postpartum)		

Sources: Attaie, R. and Richter, R.L., *J. Dairy Sci.*, 83, 942, 2000; Michalsk, M.C. et al., *J. Dairy Sci.*, 88, 1937, 2005.

because of the scarce knowledge on this subject. As fats provide lubrication, which is corroborated by a creamy mouth taste, their content, composition, and globule size must also influence milk properties, specially its viscosity.

Another important difference between goat and cow milk is related to their protein content. As detailed in the literature (Walstra et al. 1999), proteins are chains of amino acid molecules connected by peptide bonds. The structure of the protein molecules can vary from a single straight shape to a complex, tightly compacted globule form. There are two major categories of milk protein, the casein family and the whey protein. The former contains phosphorus and precipitates at pH 4.6 and 20°C, a fact that actually constitutes the basis of cheese production. The high level of phosphorus allows casein to associate with calcium, maintaining a high concentration of this mineral in milk. The whey (serum) proteins do not contain phosphorus, remaining in milk solution even at pH 4.6.

Returning to Table 18.3, it is seen that the casein family includes α_{s1}-casein, α_{s2}-casein, β-casein, and κ-casein, which exist in milk mostly as casein micelles. Their levels in goat and cow milk are quite different; the α_{s1}-casein content is higher in cow milk, whereas the α_{s2}-casein and β-casein contents are higher in goat milk. The κ-casein content is nearly the same in both cases. Jenness (1980) has analyzed the basic differences between casein micelles in goat and cow milk, concluding that the goat-milk casein micelles present a lower sedimentation rate, due to their smaller sizes, low heat stability, higher β-casein solubilization, less solvation, and more calcium and phosphorus contents. This author has attributed these differences to lower levels of α_{s1}-casein in goat milk and, also, to a softer, more friable curd obtained from goat milk upon acidification. His conclusions indicate that the softer and more friable goat milk curd probably facilitates the digestion of this milk. Data from St-Gelais et al. (2003) corroborate these properties of goat milk curd associated with the protein content in goat milk. Experiments developed in animals show that goat milk with null α_{s1}-casein content is less allergenic than the one that contains this type of casein (Haenlein 2004). Consequently, goat milk can be indicated as a possible alternative for a person who produces antibodies to cow milk proteins (keeping in mind that goat milk must be tested by a person before becoming a real alternative).

In the context of human nutrition, one of the most important milk contributions is the source of calcium and phosphorus. The calcium and phosphorus contents in goat milk are somewhat higher than those in cow milk (134 mg of Ca and 110 mg of P in 100 mL of goat milk; 121 mg of Ca and 70 mg of P in 100 mL of cow milk).

Based on the chemical composition, it is corroborated that goat milk is of high nutritional interest, providing high-quality protein, fat, carbohydrates, vitamins, and minerals. For human nutrition, goat milk offers superior digestibility than cow milk does, due to its MCFA content, its reduced level of α_{s1}-casein, and, probably, its higher concentration of small-size fat globules. In addition, goat milk is a suitable alternative for the production of baby food formulae since many bioactive compounds presented in this milk (nucleotides, free amino acids, and polyamines) are at similar levels to those in human milk (Martinez-Ferez et al. 2006).

18.3.1.2 Thermal and Physical Properties

As goat milk property data are scarce in the literature, goat milk obtained from the RN farms was characterized according to its main thermal and physical properties (Table 18.5). Milk samples were collected, poured into 1 L containers, frozen (−18°C), and stored in a freezer.

Before each analysis, samples were equilibrated at ambient temperature (T_{amb} = 25°C) and gently homogenized by hand. Analyses of pH and acidity (% lactic acid per 100 g of sample) resulted in values within the expected range, i.e., 6.3 and 0.166, respectively, at T_{amb} (standard values for cow milk—pH = 6.6 ± 0.2 at 25°C and acidity = 0.16).

Table 18.5 lists the properties measured as function of temperature, the equipment, and procedure adopted in each case, the empirical data correlation, and, when it was possible, a comparison of data at 20°C with those of cow milk. Measurements are made in triplicate (as seen in Figures 18.5 and 18.6) to evaluate the experimental error, which is also presented in Table 18.5. The adopted empirical equations are similar to those used for water properties (Perry et al. 1999), and these equations were fitted to data by the least-squares technique. Statistical analyses of residues were performed to obtain the best fit (random residues with a normal distribution over the mean), whose correlation coefficient was always higher than 0.90, with the mean standard deviation from experimental data within the allowable range of acceptance (F-distribution test, 95% of confidence).

Figure 18.5 shows the goat milk density as function of temperature, and compares these data with those of cow milk density reported in the literature (Rutz et al. 1955, Walstra et al. 1999). Values of milk density are similar for both types of milk, since the fat content and the liquid–solid ratio are in the same range. From the literature (Ueda 1999, Walstra et al. 1999) it is known that fat content and physical sate (liquid–solid fat ratio) mainly affect the milk density. A different behavior for the goat milk density could erroneously be inferred from Figure 18.4, as goat milk has a greater percentage of MCTs, and it would probably present a larger amount of liquefied fat at low temperatures. However, fat solidification or melting in milk is a complex phenomenon due to several fat constituents and its interactions, emphasizing that higher melting triglycerides dissolve in the liquid fat. Consequently, solidification or melting cannot be associated

TABLE 18.5
Main Properties of Whole Goat Milk (Raw)

Properties	Measurement (Equipment—Methods)	Empirical Equation or Correlation	Values at 20°C	
			Goat	Cow
Moisture content m_{milk} (w.b.)	Drying: oven at 90°C until equilibrium	Drying curves	0.88	0.87
Density ρ_{milk} (kg m^{-3}) (±1 kg m^{-3})	Pyicnometry: previously calibrated 50 mL standard pycnometers and thermostatic tanks	$\frac{\rho_{milk}(T)}{1037.8} = 1 - 1.97\times10^{-4}T - 2.15\times10^{-6}T^2 - 1.58\times10^{-8}T^3$ (18.1) for $10°C \le T \le 75°C$ ($T \pm 2°C$)	1032 ($C_{fat} \approx 4.2$ wt%)	1030 ($C_{fat} = 3$ wt%)
Viscosity μ_{milk} (mPa s) (±0.4 mPa s)	Shear stress vs. shear rate curves: RheoStress HAAKE (Thermo Fisher Scientific Inc., Karlsruhe, Germany) model RS-150, with a well-controlled thermostatic sample tank	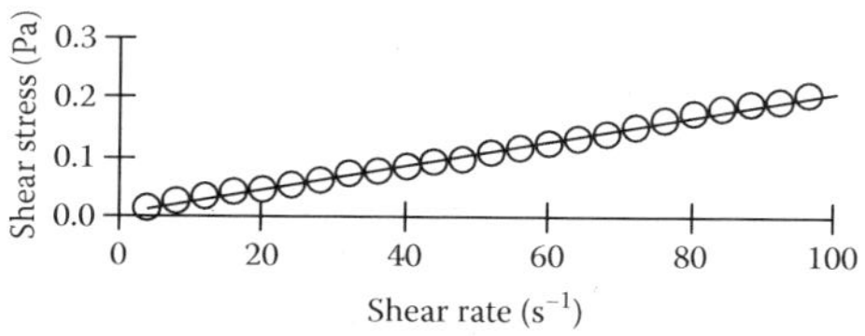 Newtonian behavior for $25°C < T < 45°C$	5.2	2.13
Surface tension β_{milk} (mN m^{-1}) (±0.9 mNm^{-1})	Ring method—Du NOUY (CSC Scientific Co., Inc., Fairfax, Virginia)	Decreases abruptly as temperature rises from 50°C to 60°C (without a good equation for prediction)	72	52
Specific heat capacity $c_{p\text{-}milk}$ (kJ kg^{-1} K^{-1}) (±0.2 kJ kg^{-1}K^{-1})	Determining thermal diffusivity: KD2 Decagon	$c_{p\text{-}milk} = 2.97 + 1.09\times10^{-1}T - 1.3\times10^{-3}T^2$ (18.2)	4.5	3.93
Thermal conductivity λ_{milk} (W m^{-1} K^{-1}) (±0.04 W m^{-1} K^{-1})	Calibrated thermal analyzer	$\lambda_{milk} = 0.19 + 2.53\times10^{-2}T - 4.74\times10^{-4}T^2$ (18.3) for $5°C \le T \le 25°C$	0.5	0.56

Source: Cow milk data from Fox, P.F. and McSweeney, P.L.H., *Dairy Chemistry and Biochemistry*, Plenum Publishers, New York, 1998 437–458.

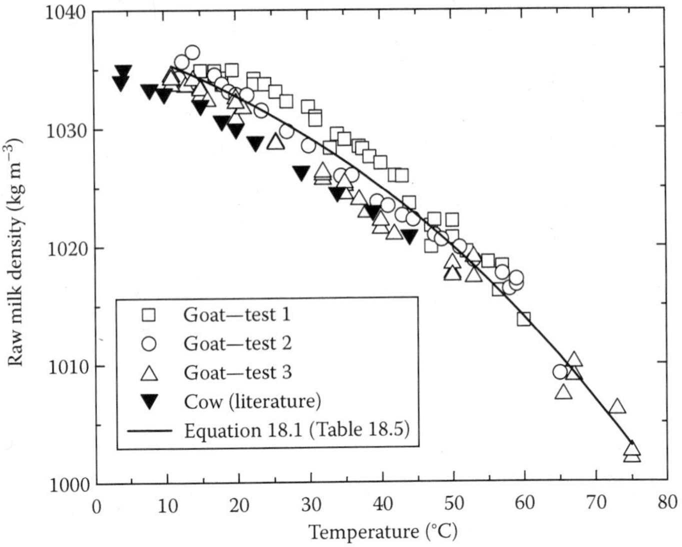

FIGURE 18.5 Milk density as function of temperature for the raw goat milk analyzed in this work and its comparison with data for raw cow milk reported in the literature (Cow milk fat content = 4 wt% and solid content = 12 wt%). (Based on data from Rutz, W.D. et al., *J. Dairy Sci.*, 38, 1315, 1955; Walstra, P. et al., *Dairy Technology: Principles of Milk, Properties and Processes*, Marcel Dekker, New York, 1999.)

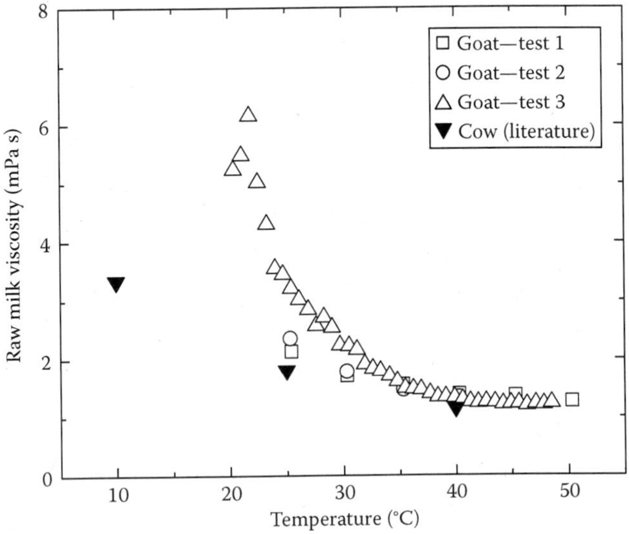

FIGURE 18.6 Milk viscosity as a function of temperature from 10°C to 50°C for raw goat milk analyzed in this work and its comparison with data for processed cow milk reported in the literature (cow milk fat content = 3.5 wt%, pasteurized and homogenized milk). (From Kristensen, D. et al., *J. Dairy Sci.*, 80, 2285, 1997.)

with the individual melting temperature of each triglyceride presented in milk, but only with the final milk fat melting temperature of about 37°C (Walstra et al. 1999).

From Equation 18.1 (Table 18.5), the thermal volumetric expansion coefficient $\left(-\frac{1}{\rho_{\mathrm{milk}(0)}}\frac{d\rho}{dT}\right)$ can be obtained to estimate an approximate fat content value for this goat milk (Ueda 1999), which is evaluated at 4.2 wt% as shown in Table 18.5.

The goat milk analyzed behaves as a Newtonian fluid for 25°C < T < 45°C as seen in Table 18.5. However, at low temperatures, viscosity increases significantly as well shown in Figure 18.6, due mainly to the increased voluminosity of casein micelles acting against shear flow. Note that casein micelles are spherical and their sizes are distributed from 20 to 270 nm. As temperature decreases, the hydration shell of the casein micelle augments in size, and the volume occupied by the micelle increases markedly. Although complex interactions exist among micelle components, the increase in micelle volume should occur due to the β-casein dissociation from micelles and, subsequently, the dissolution of colloidal calcium phosphate (which acts by limiting submicelle aggregation) (Jenness 1980, Kristensen et al. 1997, Walstra et al. 1999). A comparison between viscosity data of cow and goat milks is also provided in Figure 18.6. Since the β-casein content is a little higher in goat milk, a larger extent of increased voluminosity of casein micelles (higher degree of micelle hydration) is expected. Furthermore, given that goat milk casein micelles have a higher mean size diameter because of their lower α_{s1}-casein content (Pierre et al. 1995), the micelle voluminosity may be greater than the one of cow milk, interfering more effectively in increasing milk viscosity at low temperature.

On the other hand, at higher temperatures (≥60°C), milk viscosity increases over a small range of temperature, as shown in Figure 18.7, due the denaturation of whey

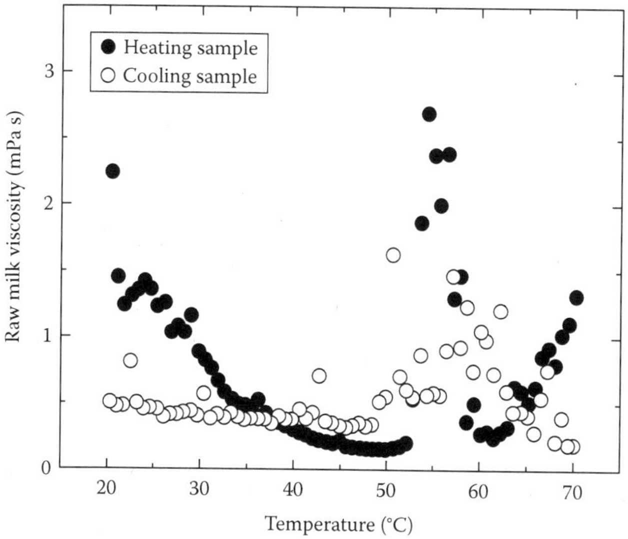

FIGURE 18.7 Milk viscosity as a function of temperature for concentrated goat milk (m_{milk} = 0.807 w.b.) analyzed in this work, during successive steps of heating and cooling.

proteins (breakage of some bonds inside the molecules with new associations resulting in less stability of remaining bonds). Since most of the bond dissociation and formation is irreversible, cooling down the milk does not restore the previous viscosity values (Walstra et al. 1999), as corroborated by the data in Figure 18.7.

As shown in Figure 18.8, goat milk surface tension is almost constant from 10°C to 50°C, although there is a slight decreasing trend as temperature increases. An increase in temperature to values above 50°C causes goat milk surface tension to decrease abruptly, reaching a new quasiconstant value. This corroborates the temperature range of whey protein denaturation, with the migration of β-casein out of the micelles, lowering the milk–air and plasma–fat surface energies (β-casein is a very amphiphilic protein and acts like a detergent, as pointed out by Walstra et al. 1999). Compared with cow milk, goat milk surface tension is higher up to the point of whey denaturation. This can be attributed to the higher surface area of fat globules (lower size fat globules) that favors more adsorption of casein and whey protein from plasma.

As expected, goat milk heat capacity and thermal conductivity increase as temperature rises from 5°C to 25°C. These two thermal properties are fundamental to describe heat transfer in the dryer. Equations for predicting their values as a function of temperature are of the polynomial type. Milk heat capacity is strongly influenced by the milk fat content, composition, and its physical state. Goat milk presents, at 20°C, a heat capacity value somewhat higher than cow milk, probably due to their different fat contents (see Table 18.5). Thermal conductivity, in turn, decreases as the fat content or the solid content are raised, explaining the small difference between the experimental values of goat and cow milk at 20°C in Table 18.5.

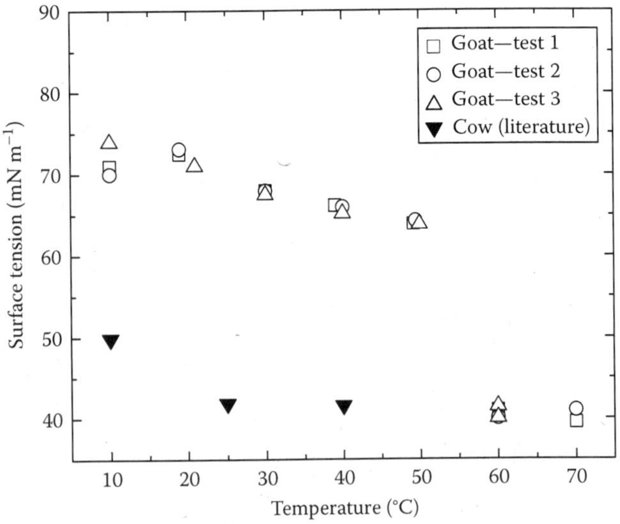

FIGURE 18.8 Milk surface tension as a function of temperature from 10°C to 70°C for raw goat milk analyzed in this work and its comparison with data for processed cow milk reported in the literature. (Cow milk fat content = 3.5 wt%, pasteurized and homogenized milk). (Based on data from Kristensen, D. et al., *J. Dairy Sci.*, 80, 2285, 1997.)

With the preliminary characterization of raw goat milk and knowing more about its behavior, the subsequent step includes a brief review of the traditional process for drying and powdering milk, followed by a summarized description of the nonconventional process studied here.

18.3.2 Milk-Powder Production

18.3.2.1 Traditional Process

The traditional process for producing milk powder, commonly used by the industrial dairy sector, consists of the following consecutive steps:

1. Reception—in which whole fresh milk coming from farms is inspected with regard to chemical composition, physical properties, sensorial attributes, and bacteriological content, as specified by the law regulation.
2. Pretreatment—in which the milk approved by the initial control (step 1) is clarified in centrifugal separators or filters, cooled in plate heat exchangers, and stored into tanks at 4°C. Next, it is standardized by adjusting the milk fat to total solids ratio, according to the final product type.
3. Heat treatment—which includes thermization, pasteurization, sterilization, UHT heating for killing all pathogenic microorganisms and improving the milk resistance to storage.
4. Concentration—consisting of two operations: water evaporation (or milk concentration) and milk homogenization. Milk is concentrated in a multiple-effect evaporator system (plate or falling film) under vacuum conditions, from 12 to 50 wt% of total solids. This assures low temperatures (<70°C), less cost for water removal, and concentrated milk emulsion with sensorial attributes similar to those of raw milk with less occluded air. Homogenization is usually applied to decrease the fat content and/or to stabilize the concentrated milk emulsion, breaking up free fat globules (1–15 μm) into small ones (1–2 μm).
5. Drying—involving milk drying and powdering, air separation, and product cooling, as shown in Figure 18.9 and detailed in the subsequent paragraph.
6. Packaging—which is the final step involving the packing of milk powder product in suitable containers, its storage, if necessary, at specific conditions to allow its safe conservation, and its shipping to markets or consumers.

Drying is the most important industrial step in defining the quality of the powder product. In spite of its high energy requirements, spray drying (SD) is the most used technology by the industrial food sector due to the reduced heat effect on powder products. This technique is responsible for more than 95% of dairy powder products in the world (Pisecky 1995).

As discussed in Chapter 9, SD transforms liquid containing dissolved solids into droplets and gently removes water from these droplets. Concentrated milk is atomized into fine and uniform droplets (50–100 μm) by a centrifugal bowl-type rotary wheel with curved vanes or a high-pressure spray nozzle, located at the top of the spray chamber. These droplets fall into the spray chamber in a concurrent flow with

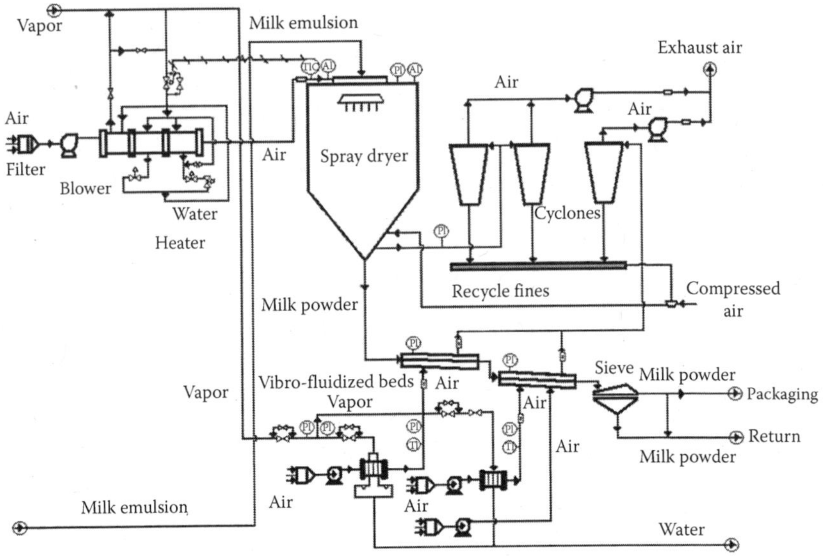

FIGURE 18.9 Schematic flowchart of an industrial two-stage drying milk plant. (Modified from Ribeiro C.P., Jr., Modeling of plate heat-transfer systems and simulation of a Brazilian milk powder plant, MSc thesis, Federal University of Minas Gerais, Belo Horizonte, Brazil, 2000 [in Portuguese].)

hot, filtered air, reducing their moisture content. The inlet air temperature must be in the range of 140°C–180°C to avoid any thermal damage to the product (Birchal et al. 2005). In the first drying period, liquid droplets shrink as water is evaporated from their surface. At the end of this first period, droplets lose their liquid constitution and become particles with a thinner solid crust formed at their surface. In the second drying period, the solid crust thickens, increasing its resistance to the vapor diffusion (Birchal and Passos 2005). As a consequence, the drying rate decreases and the spray dryer becomes less efficient in water removal. For a moisture content below 7 wt%, milk powder should be dried in another type of dryer that can well adjust heat transfer and water evaporation rates to minimize energy consumption, to avoid excessive increase in the particle temperature, and to improve the instant powder quality.

In a single-stage drying operation, composed of a spray dryer with a pneumatic conveying system, the removal of the last moisture portion from milk powder occurs slowly and expensively, and the final product is fine nonagglomerated particles (30–50 μm) that are difficult to disperse into water.

The two-stage drying operation, shown in Figure 18.9 by a schematic flowchart of a dairy industry, uses the spray dryer followed by a vibro-fluidized bed (VFB) system, which consists of two rectangular vibrated chambers with a slightly inclined perforated plate base for air distribution.

In the three-stage drying operation, as pointed out in Chapter 9, besides the VFB system, a static fluid bed, inserted into the conical base of the spray dryer, works as a dryer–agglomerator (Pisecky 1995).

In the VFB system, the powder moisture content is reduced from about 7–6 to 3 wt% (whole milk) or to 4 wt% (skim milk). Hot air, generally at 80°C (up to 120°C), injected gradually throughout the shallow moving bed of particles, fluidizes this bed and keeps the heat transfer rate close to that required for water evaporation. Although particles are moving and drying, their temperature is falling. At the end section of this system, air, generally at ambient conditions, is supplied for faster cooling of particle agglomerates. Drying and cooling operations can be well controlled in this system because the particle residence time is much longer than in the spray dryer. Moreover, exit air carries out fine particles, which are not agglomerated, into cyclones (see Figure 18.9) or sintered bag filters (see Chapter 9), where they are separated from air and recycled to the spray chamber. This reduces dustiness and improves powder agglomeration, producing milk powder with a mean size of 150–200 μm, which is within the desirable range for fast or instant reconstituting (Pisecky 1995).

18.3.2.2 Nonconventional Process

The spouted bed (SB) dryer, which was accidentally discovered in 1955 by Mathur and Gisher (Mathur and Epstein 1974), has been one of the few drying techniques most investigated in Brazil. As summarized by Costa et al. (2006), works developed in this area have focused on drying of grains and, more recently, on drying of pasty materials. Although great results have been obtained during the last 10 years, only a few examples translated into the pursuance of actual applications in rural cooperatives or the industrial sector due to the complexity of scaling-up SB dryers and their limited volumetric operational capacity.

However, for small-scale production, SBs of inert particles continue to be a potential alternative for drying high viscous and thermal sensitive suspensions efficiently and at a low cost. Recently, as shown in Chapter 8, the SB technology has begun to be used industrially for granulating pharmaceutical products, with commercial units already being supplied by Glatt® (Glatt Ingenieur Technik—GmbH, Weimar, Germany).

As schematized in Figure 18.10, this technique, when used to dry and powder suspensions, consists of injecting the suspension into a bed of inert particles spouted

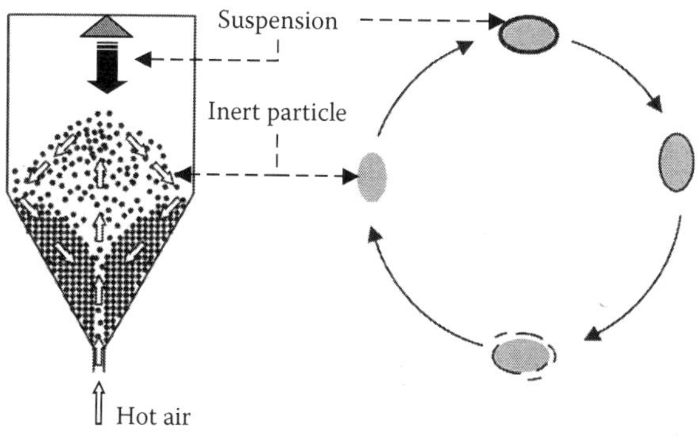

FIGURE 18.10 Scheme of the spouted bed technique applied for drying pasty materials. (Modified from Trindade, A.L.G., New technique to concentrate black liquor for producing powdery fuel, Master's thesis, Federal University of Minas Gerais, Belo Horizonte, Brazil, 2004 [in Portuguese].)

by hot air. The suspension covers these particles forming a thin film on their surface, which is dried and powdered during the circulation of the inert particles inside the dryer. The powder formed is then carried by the exhaust air and separated from it in a filter or cyclone system. Note that there are three distinct regions in the spouted bed: the spout or dilute solid region, generally in the center of the column, where inert particles are accelerated by airflow, moving upward and colliding among themselves. The fountain, above the bed of particles, is the region where particles change their velocity direction, falling onto the surface of the annulus. This annular region is a moving bed, where air flows upward and particles move downward and inward, sliding one against others, until reaching the spout region once again. During this cyclic particle path, the suspension layer covering the particle must be dried and detached from the particle surface by the intensive particle–particle attrition. This leads particles to be free of suspension when they become wet again at the top of dryer, reducing particle agglomeration. Furthermore, powder formed during each particle cycle must be carried out by air to a cyclone or a filter, avoiding any destabilization of the SB operation due to an accumulation of a large amount of powder inside the bed column.

Most pasty food materials behave as a pseudoplastic fluid as their solid content increases. For these pasty materials, the SB technique can overcome problems related to increasing suspension viscosity during drying. Note that the SB causes intense collisions between particles, which lead to elevated shear rates in the thin suspension film adhered to the particle surface, reducing its apparent viscosity and rendering the drying operation feasible for a small-scale powder production (Passos et al. 2004).

Freire (1992) has experimentally confirmed that the physicochemical and biological properties of pasty food materials (such as blood, egg emulsion, vegetable extracts) can be well preserved in the SB powder product. However, the dryer thermal efficiency is still low for industrial applications, requiring modifications in the path of the exhaust air to recycle it and recover its sensible heat. Therefore, modeling and simulation of this drying operation are required to direct dryer modifications toward the improvement of its performance, and subsequently to optimize the operational variables.

Based on this specific need, Costa et al. (2001) developed a simulator for predicting the drying of diluted Newtonian suspension in conical spout-fluid beds. Although the effect of interparticle forces on the fluid dynamic has been neglected, the structure of this simulator is flexible to changes. This program comprises three interactive modules associated with the numerical solution of three models: the gas flowing through the bed of inert particles; the particle circulation inside the spout, annulus, and fountain regions; and the drying of suspensions in these three regions. In addition, this program generates its own database for air properties as a function of air humidity and temperature.

Trindade et al. (2004) proposed a methodology to introduce into this simulator the effect of the cohesive forces between particles on the gas flow characteristics. Based on the semiempirical formulations developed by Passos and Mujumdar (2000), this effect of interaction forces between inert particles coated with the suspension should be quantified by the following experimental curves: (Q_{ms}/Q_{ms0}) vs. $(l_{fl}/\varepsilon_{ag})(1 - \varepsilon_{ag})$ and $(\Delta P_{ms}/\Delta P_{ms0})$ vs. $(l_{fl}/\varepsilon_{ag})(1 - \varepsilon_{ag})(\phi/d_p)$. By varying the type and the amount of suspension injected into the bed of inerts, the minimum spouting gas flow rate (Q_{ms})

greatly or slightly increases or decreases, in relation to the one obtained in the bed of particles without suspension (Q_{ms0}). This leads to a reduction or an increase in the minimum spouting pressure drop (ΔP_{ms}), compared to the one evaluated in the bed of particles without suspension (ΔP_{ms0}). The type of interaction forces must be directly related to the suspension properties, but the strength of these forces should depend on the thickness of the suspension layer that covers the particle surface, l_{fl}, as well as the bed failure mechanism during the spout formation. This bed failure mechanism is expected to be related to $(l_{fl}/\varepsilon_{ag})(1 - \varepsilon_{ag})$, with ε_{ag} representing free porosity of the annulus region (actual pore volume through which air flows per total bed volume) and the ratio (ϕ/d_p), which represents the inert particle sphericity per its mean diameter. Therefore, values, form, and shape of these two experimental curves differ from one suspension to another as well as from one bed failure mechanism to another.

Using this methodology proposed by Trindade et al. (2004), Costa et al. (2006) have introduced the effect of adhesion and cohesion forces into the fluid-particle dynamic model and simulated fluid and solids flow motion. Work is currently under development to include these modifications into the drying model of the simulator in order to describe all the drying operation of pasty materials, considering the actual effect of particle agglomeration and particle-to-wall adhesion.

It ought to be emphasized that particle agglomeration and particle-to-wall adhesion destabilize the SB operation, causing an abrupt drop in powder production efficiency, as well as the sudden interruption of the drying operation. Passos et al. (2004) corroborate that this efficiency can be enhanced overcoming the SB instability, if the suspension is injected intermittently into the bed of inert particles, with a precise on-line control.

On basis of this brief review, there are several research works in the literature that corroborate the potential use of the SB technique to dry pasty materials, such as milk emulsion. Moreover, this technique is already in the industrial use for coating pharmaceutical products, corroborating its industrial feasibility.

Having proven the potential use of SB rather than SD for small-scale production, the subsequent step in the search for a new process-product route is to add experimental information to confirm the technical feasibility of the SB processing route for the RN agro-cooperatives interested in producing goat milk powder.

18.4 TECHNICAL FEASIBILITY OF THE SPOUTED-BED PROCESSING ROUTE

18.4.1 Spouted-Bed Processing Route

For the RN agro-cooperatives interested in producing goat milk powder, due to space requirements and the initial investment cost, the step of milk concentration (in evaporators) must be withdrawn. Since the SB technique is used for drying diluted suspensions, this milk concentration step does not exist in the SB processing route. Moreover, the produced powder from these cooperatives is to supply local industries or government agencies with milk ingredient for infant or child dairy products. Because of this principal use, a particle agglomerator, like a VFB system, is not necessary to be attached to the SB dryer-cyclone. Therefore, the basic steps proposed for this SB processing route are

1. Reception (milk acceptance as described in Section 18.3.2.1)
2. Pretreatment (filtration, cooling, storage, and standardization as described in Section 18.3.2.1)
3. Heat treatment (thermization, pasteurization, sterilization, and UHT treatment as required by the Brazilian regulation for goat milk production)
4. Drying and powdering (comprising a unit similar to the one presented in Figure 18.11)
5. Packaging and shipping

A pilot scale drying unit, similar to that displayed in Figure 18.11, was built in the laboratory to perform the experimental tests. In this pilot unit, the SB column includes a conical base (60° included angle, 0.13 m in height with a 0.03 m nozzle diameter) attached to a cylindrical vessel (0.18 m in diameter and 0.72 m in height). The Lapple cyclone, attached to the top of the SB column, has a cylindrical diameter of 0.10 m. A digital anemometer, Lutron Electronic Enterprise Co. model YK-80AM with an accuracy of 0.1 m/s, measures the airflow rate. The inlet air temperature can be maintained constant by a device controller OMRON® (Omron Corporation,

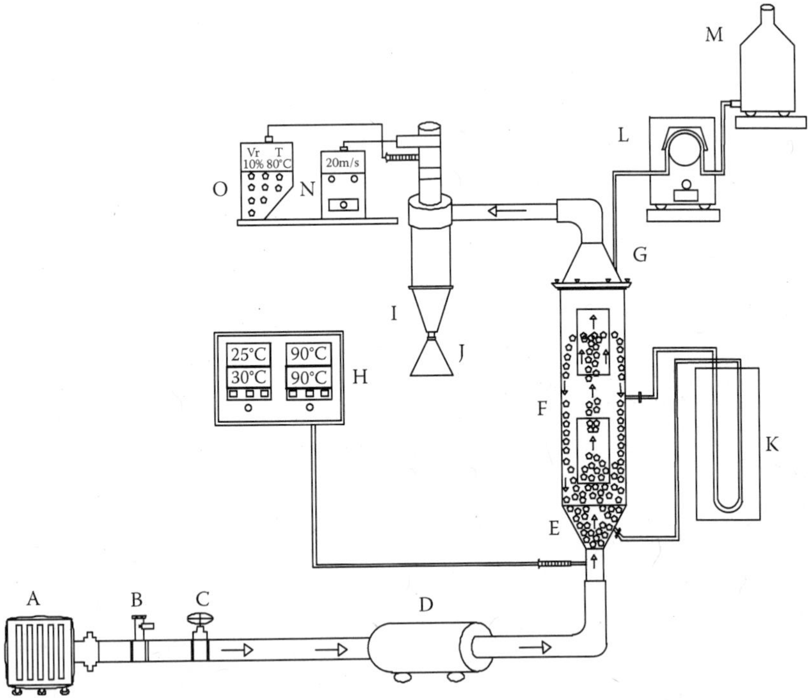

FIGURE 18.11 Schematic unit for drying goat milk, comprising: A, filter and blower; B, by-pass valve; C, ball-valve (adjust control); D, electrical heater; E, inert particles; F, Plexiglas door-visor; G, stainless steel spouted bed; H, control panel; I, Lapple cyclone; J, powder collector container; K, U-water-manometer; L, peristaltic pump; M, milk storage tank; N, anemometer; O, thermo-hydrometer.

Kyoto, Honshu) model E5AW. A digital thermo-hygrometer, Cole-Parmer® (Cole-Parmer Instrument Company, Vernon Hills, IL) model 4085, measures the relative humidity and temperature of the air leaving the cyclone.

To generate the useful information required for quantifying the technical feasibility of this drying unit, the following substeps have been designed and performed:

1. Preliminary tests, aiming at defining the basic parameters to operate the SB unit. This included the following main decisions or selections: type(s) of inert particles, type(s) of milk injector, operational values for the inlet air temperature, mass of inert particles, air and milk emulsion flow rates, and finally, the drying operation methodology to overcome problems with particle agglomeration.
2. Drying tests, aiming at quantifying the technical feasibility of the SB unit to produce high-quality milk powder. This included the following criteria: the most appropriate SB configuration (type of injector and inert particle, mode of milk feeding, processing time), operation efficiency (powder production, accumulation, and collection), and powder quality parameters.
3. Specifications of data needs for using the simulator to scale-up this pilot unit to industrial size.

18.4.2 Experimental Results

18.4.2.1 Preliminary Tests

Based on the experimental knowledge from previous works developed in this area, four types of particles were preselected: glass beads, polystyrene pellets, high-density polyethylene (PE) particles, and polypropylene (PP) particles.

Since glass beads require not only a pretreatment to avoid their breakage during the spouting regime (high attrition at moderate temperature), but also a high pressure drop and airflow rate to attain a stable spouting regime, the use of these inert particles was discouraged. Because of the low sphericity of polystyrene pellets (<0.7), the milk film formed on their surface was not well distributed under a stable spouting regime, and, hence, this type of particle was also eliminated from the selected list.

Both PE and PP particles maintain a stable spouting regime and a good circulation without and with the presence of goat milk emulsion, which improves the milk distribution over the entire bed and leads to a homogeneous milk film formation on the particle surface. Restrictions on the use of both PE and PP particles are related to the drying temperature, which must be below 130°C (melting temperature of PE is about 133°C and PP has a temperature range for melting that starts at 140°C). Such restriction is not critical, since the drying temperature must be lower than 130°C to avoid thermal damage of the product. Therefore, these PE and PP particles were chosen as the inert materials for the experimental runs.

The densities of the PE and PP particles used in this work are, respectively, (991 ± 10) kg/m^3 and (867 ± 5) kg/m^3. Their mean diameter (diameter of sphere with the same particle volume) and sphericity ($\phi = \pi d_p^2/S_p$) are $d_p = (3.52 \pm 0.02)$ mm and $\phi = 0.941$ for PE particles, and $d_p = (4.02 \pm 0.02)$ mm and $\phi = 0.902$ for PP particles.

Two types of milk injectors available in the laboratory were selected. The first one, a two-fluid nozzle atomizer, needs to be attached to an air compressor and to a peristaltic pump for spraying the milk emulsion into the center-top of the inert particle bed fountain. The second type, a drop-injector, consists of a graduated pipette that needs to be manually controlled to inject the milk emulsion, drop by drop, at the center-top of the inert particle bed fountain. Based on results presented by Passos et al. (2004), the intermittent and continuous modes of injecting the milk emulsion into the SB dryer were investigated.

Taking into account some preliminary tests performed in the SB pilot unit (Andrade et al. 2005, Marciano et al. 2007, Nascimento et al. 2007), the following decisions were made:

- The mass of inert particles, M_{inert}, was fixed at 2.5 kg, based on the required condition to better operate the blower (7 HP), which limits M_{inert} to a maximum of 3 kg. Working at $M_{inert} = 3.0$ kg and $M'_{inert} = 2.5$ kg did not affect the powder production.
- The airflow rate, Q, could be fixed at the minimum condition for maintaining the stable spouting ($Q/Q_{ms0} = 1.2$) because this condition supplied enough air to attend all requirements for the milk-drying operation.
- The elimination of the milk concentration step in the SB processing route was confirmed by the no-significant effect of the initial milk concentration on the amount of power produced and on the powder moisture content for same operating conditions.
- The most appropriate flow rate for injecting milk emulsion into bed of inert particles (2.5 kg) was at 10 mL/min (continuous feeding).
- The value of the air inlet temperature, T_{glin}, was fixed at 90°C, since the amount of powder produced increased as the inlet air temperature was raised (from 70°C to 90°C). However, the quality of this milk powder product must be quantified to reinforce or redefine this value.
- The type of inert particles could affect the milk film adherence to the particle surface. Therefore, this hypothesis was checked in the drying test to choose the more appropriate type of particles for avoiding instabilities in the spouting regime due to increased powder retention on the particle surface.

In addition, the best procedure established to operate and control the SB dryer under intermittent feed of milk emulsion consisted of

1. Injecting air at T_{glin} and Q to spout M_{inert} allowing enough time to reach the stable regime and the thermal equilibrium (T_{glout} = constant)
2. Injecting the aliquot of milk emulsion into the bed during a controlled time interval, short enough for assuring only a small variation in the spouting conditions (quantified by the following two measured variables: the fountain height and T_{glout})
3. Discontinuing the injection of milk during another time interval, long enough to restore the fountain height and T_{glout} initial values
4. Repeating steps 2 and 3 until the total volume of milk was processed

Note that, at industrial scale, changes in T_{glout} can be monitored by a computer software for controlling the extension of each time interval. By setting this control in the industrial unit, powder will be continuously produced at replicable results, as confirmed by Passos et al. (2004).

Similarly, the best procedure established to operate and control the SB dryer under continuous feed of milk emulsion consisted of

1. Injecting air at T_{glin} and Q to spout M_{inert} allowing enough time to reach the stable regime and the thermal equilibrium (T_{glout} = constant).
2. Continuously injecting the milk emulsion at 10 mL/min.
3. Controlling the stability of SB regime by measuring the fountain height and T_{glout} during the whole test. If necessary, adjust the Q value to maintain the stable regime.

18.4.2.2 Drying Tests

In this substep, to better investigate and quantify the technical feasibility of continuously producing goat milk powder of high quality in the SB pilot unit, an experimental methodology was developed. This comprised the analysis of two groups of variables, one related to the powder production and the other associated with the goat milk powder quality. In the first group, the powder production efficiency was defined, calculated, and analyzed using data obtained from a two-level full factorial design. In the second group, the quality of the goat milk powder produced in the SB pilot unit was inferred through the comparison of its main physicochemical properties with those of goat milk powder produced industrially in a spray dryer.

As described in Section 18.4.2.1, the most adequate operating conditions to perform the drying tests were fixed at $M_{\text{inert}} = 2.5\,\text{kg}$, $Q/Q_{\text{ms}} = 1.2$, $T_{\text{glin}} = 90°\text{C}$, $m_{\text{milklin}} =$ 0.87 (w.b.), milk volume to be processed = 500 mL. With continuous feed, the milk flow rate was fixed at 10 mL/min resulting in tests with 50 min of processing. On the other hand, with intermittent feed, the processing time was set to 100 min (twice that of continuous feed) with five repetitions of the sequence: 10 min to inject 100 mL of milk (10 mL/min) followed by 10 min without injection.

Drying experiments were performed following a 2^3 full factorial design with replicates, as presented in Table 18.6. The response variables for these experiments were the powder production efficiency, η_{pd}, the powder retention on the surface of all particles in the bed, $\chi_{\text{pd-ret}}$, and the powder lost by adhesion to equipment walls, χ_{wall}. These variables were defined as follows:

$$\eta_{\text{pd}} = \frac{M_{\text{pd}}\left(1 - m_{\text{pd}}\right)}{M_{\text{milk}}\left(1 - m_{\text{milklin}}\right)} \tag{18.4}$$

$$\chi_{\text{pd-ret}} = \frac{M_{\text{fl}}\left(1 - m_{\text{fl}}\right)}{M_{\text{milk}}\left(1 - m_{\text{milklin}}\right)} \tag{18.5}$$

$$\chi_{\text{wall}} = 1 - \eta_{\text{pd}} - \chi_{\text{pd-ret}} \tag{18.6}$$

TABLE 18.6
Experimental Matrix of Experiments and Results

Matrix Cell = (Inert, Injector, Feed)	**Inert Type: PP = +1 and PE = −1; Injector Type: Drop-Injector = +1 and Atomizer = −1; Mode of Feed Emulsion: Intermittent = +1 and Continuous = −1**							
Design	(−1, −1, −1)	(+1, −1, −1)	(−1, +1, −1)	(+1, +1, −1)	(−1, −1, +1)	(+1, −1, +1)	(−1, +1, +1)	(+1, +1, +1)
Experiment	1	2	3	4	5	6	7	8
(Replication)	1R	2R	3R	4R	5R	6R	7R	8R
η_{pd} (%)	19.95	32.15	41.65	64.48	3.53	44.30	21.81	54.31
	22.05	27.26	32.96	49.41	2.38	26.56	19.07	47.35
$\chi_{pd\text{-}ret}$ (%)	33.16	17.86	32.92	16.99	26.32	28.33	30.91	22.51
	32.48	18.79	32.11	22.45	34.86	24.78	36.55	22.29
χ_{wall} (%)	46.89	49.99	25.43	18.53	70.15	27.37	47.28	23.18
	45.47	53.95	34.93	28.14	62.76	48.66	44.38	30.36
m_{pd} (% w.b.)	3.42	2.75	2.59	3.66	3.01	2.33	1.84	2.70
	2.03	3.32	2.12	2.43	3.22	3.25	1.87	3.63

where

M_{milk} and $m_{milklin}$ represent, respectively, the total mass and the moisture content (w.b.) of the raw goat milk emulsion injected into the bed

M_{pd} and m_{pd} are, respectively, the total mass and the moisture content (w.b.) of powder collected in the cyclone

M_{fl} and m_{fl} are, respectively, the total mass and the moisture content (w.b.) of the milk film covering the particle surface

These last two variables were measured by randomly sampling wetted inert particles from center to wall and from top to bottom of the SB column after each test. These samples were weighted, dried in the oven, weighted again, washed well, dried, and weighted again. The first weight measurement provides the mass of inert particles covered by the milk film, the second one provides the mass of dried samples (particles and solids milk powder), and the third one provides the mass of inert particle samples, whose value must match the mass of nonprocessed inert particles.

Table 18.6 also presents the experimental values obtained for m_{pd} and the three response variables: η_{pd}, $\chi_{pd\text{-}ret}$, and χ_{wall}. The software STATISTICA® (Statsoft, Inc., Tulsa, Oklahoma) was used to analyze the effect of the independent variables and their interactions on the response parameters. From this analysis, m_{pd} does not vary significantly with the type of inert particles, injector or mode of feeding emulsion, and its mean value remains equal to (2.8 ± 0.3)% w.b. This result was expected since the milk drying operation occurred under the same Q and T_{glin} conditions, corroborating the simulated results obtained by Costa et al. (2001). Consequently, in all experiments performed, η_{pd} is straightly related to the powder production mechanisms (i.e., breakage of the milk film that covers the particle, its detachment from the particle surface, and its powdering by particle attrition and collision) and to the powder adherence to dryer walls. As shown in Table 18.7, η_{pd} depends significantly (p-value of coefficient <0.05 to be representative in the linear model) on the type of inert particles, injector, and feeding, as well as on the linear interaction between the type of inert and the mode of feeding milk. Operating the SB column with PP particles and using the drop-injector to feed milk into the SB column continuously or intermittently increases η_{pd}. Passos et al. (2004) have reported obstructions in the nozzle orifice by the bamboo black liquor when it was atomized into the SB of inert particles. However, this seems unlikely to occur in the case of milk atomization, unless the inlet milk emulsion has previously undergone intensive mechanical treatments, releasing fat due to breakage of the globule membrane. Otherwise, the free fat content decreases with atomization pressure, inhibiting potential lactose crystallization (Vignolles et al. 2007). Thus, the milk emulsion injection drop by drop into the SB fountain may enhance the emulsion distribution among particles, leading to better operation.

The data from Table 18.7 demonstrate that $\chi_{pd\text{-}ret}$ depends only on the type of particles, decreasing by using PP particles as inert. PP particles must favor breakage of the milk film, releasing it from the particle surface and/or powdering this milk film. This is an interesting result since physical properties of these particles are close (PE/PP density ratio = 1.14, size diameter ratio 0.88, and sphericity ratio = 1.04), reinforcing that PP and PE materials are similar in many aspects. Because of their hydrophobic nature (high contact angle and low surface energy), both materials are recommended for use as inert materials in SB drying of suspensions (Rocha and Taranto 2008), even though PP is more resistant to "environmental stress cracking" and abrasion (Roylance 1993). This may explain the higher $\chi_{pd\text{-}ret}$ for PE particles,

TABLE 18.7

Statistical Analysis of the Factorial Experiment Detailed in Table 18.6, with Coefficients of a Linear Model Including Main Factors and Their Second-Order Interactions

Correlation	Variable η_{pd}		Variable $\chi_{pd\text{-}ret}$		Variable χ_{wall}	
Factors	Coeff.	p-Value	Coeff.	p-Value	Coeff.	p-Value
Intersect	31.83	0.0000	27.08	0.0000	41.09	0.0000
1. Inert type	11.40	0.0001	−5.33	0.0001	−6.07	0.0080
2. Injector type	9.55	0.0004	0.01	0.9905	−9.56	0.0006
3. Feed mode	−4.41	0.0287	1.23	0.1438	3.17	0.1041
1 by 2	1.11	0.5234	−0.70	0.3864	−0.41	0.8203
1 by 3	4.31	0.0315	1.49	0.0866	−5.80	0.0108
2 by 3	−1.33	0.445	−0.26	0.7392	1.59	0.3841
R-sq	0.91		0.85		0.85	
Lack of fit (F-test)	Non p-value = 0.35		Non p-value = 0.15		Non p-value = 0.18	

considering their reuse, after washing and cleaning, in consecutive tests. Although χ_{wall} is dependent on η_{pd} and $\chi_{pd\text{-}ret}$, the data in Table 18.7 indicate that the mode of feeding emulsion significantly affects χ_{wall} through its linear interaction with the type of inert particles. Consequently, to minimize χ_{wall} and $\chi_{pd\text{-}ret}$, and, at the same time, maximize η_{pd}, the SB pilot unit must operate with PP inert particles, under intermittent feed of milk emulsion using the drop-injector to insert this emulsion drop by drop into the top of the SB fountain.

Considering this condition as the most appropriate one to operate the pilot SB dryer within the experimental range of variables analyzed, goat milk powder produced under this condition was selected for quality analysis. This powder, stored in a closed glass vessel under ambient conditions, was sampled twice for characterization (1 month and 6 months of storage). Goat milk powder, industrially produced by SD (two-stage drying), was acquired in the local market. This commercial product is well known and is well accepted by consumers. Based on dates stamped on its package, this SD powder had undergone 1 month of storage, which matches the storage time of the first SB sample.

Based on the previous review (Section 18.3.1) and on the literature (Pisecky 1995, Hardy et al. 2002, Birchal et al. 2005, Ilari and Mekkaoui 2005), the quality of SB goat milk powder can be probed and evaluated by comparing its relevant physicochemical properties with those of industrial SD goat milk powder. Powder properties selected for this comparison are shown in Table 18.8, as well as methods and procedures used to measure them. To complement these data, some properties of the reconstituted milk emulsion obtained from SB and SD powders were also measured and compared with those of raw goat milk (presented earlier). All measurements were performed in triplicate.

The results of this analysis are shown in Tables 18.9 and 18.10. In fact, globally, for 1 month of storage, the properties of SB goat milk powder (SB-1) compare well with those of SD powder (SD-1), corroborating the quality of SB goat milk powder.

The moisture content of commercialized goat milk powder varies from 2.5 to 4.0 wt% (w.b.), showing that the m_{pd} value of SB-1 powder is within the characteristic value of the industrialized products (even after 1 month of storage). Returning to Table 18.6, one can see that the m_{pd} values obtained for almost all powders produced are in this range, the only exception being test 7 powder (PE particles are used as inert). However, after 6 months of storage, m_{pd} increases, reaching a high value that is out of the commercial range. Note that this SB-6 powder was stored under ambient conditions in a closed (but not hermetically sealed) container.

Water activity, a_w, refers to water availability in food products, and values lower than 0.60 assure no microbial proliferation. At $a_w = 0.60$, few molds and osmophilic yeasts start developing inside food products (Beuchat 1981). The recommended value for food products with 3–5 wt% (w.b.) moisture content is 0.30, which is attended by SD-1 and SB-1 powders. This confirms that milk powder produced in the SB pilot presents the required microbiological quality, with an a_w value compatible with the commercialized product. However, a_w for SB-6, even though lower than 0.6, is high enough to consider that some event took place in the powder during storage. The increase in a_w (>0.4) for milk powder storage at inappropriate conditions is a strong indicative of lactose crystallization. As pointed out by Hardy et al. (2002), during the

TABLE 18.8
Methods and Analyses for Characterizing Goat Milk Powder

Powder Properties	Methods and Procedures
Moisture content m_{pd} (w.b.)	Drying of samples in oven at controlled temperature (90°C) until constant mass
Water activity a_w (—)	Determining by model 4TE, AquaLab® (Decagon Devices, Inc., Pullman, WA)
Particle size distribution	Cumulative percentage undersize (in volume) distribution curve of random samples laser particle size analyzer, model 1180-CILAS® (Compagnie Industrielle de Lasers, La Source, Orleans)
Mean particle diameter d_p (μm)	Integration of the particle size distribution curve—mean volumetric volume
Dispersion index span I_d (—)	$$I_d = \frac{d_{p90\%} - d_{p10\%}}{d_{p50\%}} \quad (18.7)$$
Apparent powder density ρ_{ap} (kg/m³)	Volume of 20 g of powder poured gently into a graduated cylinder without tapping (Birchal et al. 2005)
Tapped powder density $\rho_{ap\text{-}max}$ (kg/m³)	Volume of 20 g of powder poured in a closed graduated cylinder and tapped (drop it from a horizontal surface of 20 cm height, 40 times) (Birchal et al. 2005)
Hausner cohesion number, HR (—)	$$HR = \frac{\rho_{ap\text{-}max}}{\rho_{ap}} \quad (18.8)$$
Angle of repose θ (°)	Internal angle between the powder pile and the horizontal surfaces, after allowing powder to flow freely through a funnel onto the horizontal surface forming a stable pile
Solubility (in water) ζ (—)	Mass of soluble solids per g of powder (after 5 min of magnetic stirring, 5 min of centrifugation at 2600 rpm)
Sorption curve at 25°C m_{eq} vs. a_w	Consecutive measurements of water activity and powder equilibrium moisture content, starting at 92% of air relative humidity, model 4TE-AquaLab® (Decagon Devices, Inc., Pullman, WA)

first cycle of water adsorption in milk powder at a_w about 0.4, there is an apparent disappearance of water, indicating that lactose changes from amorphous to crystalline. Jouppila et al. (1997) have suggested that the glass transition of lactose mainly controls its crystallization. Fontana (2008) has found, based on the water sorption isotherm for SD milk powder at 25°C, the lactose glass transition at a_w= 0.437, the onset of lactose crystallization at a_w= 0.548, and the onset of lactose dissolution

TABLE 18.9
Goat Milk Powder Properties for the Three Samples Analyzed at 23°C–25°C

Properties	SD-1 (SD Powder 1 Month Storage)	SB-1 (SB Powder 1 Month Storage)	SB-6 (SD Powder 6 Months Storage)
m_{pd} (%w.b. = wt%)	4.94 (±0.09)	3.44 (±0.10)	5.84 (±0.17)
a_w (—)	0.30 (±0.04)	0.27 (±0.01)	0.57 (±0.03)
d_p (μm)	265	219	221
I_d (—)	0.94	1.4	1.2
ρ_{ap} (kg m^{-3})	421 (±14)	357 (±12)	358 (±16)
$\rho_{ap\text{-}max}$ (kg m^{-3})	667 (±14)	556 (±N/A)	538 (±20)
HR (—)	1.62 (±0.06)	1.56 (±0.05)	1.51 (±0.09)
θ (°)	55 (±1)	49 (±6)	47 (±4)
ζ (—)	1.00 (±0.01)	0.89 (±0.10)	0.76 (±0.02)
Sorption curve—data fitted by the GAB model (25°C):			N/A
$m_{eq} = \dfrac{c_1 c_2 c_3 a_w}{(1 - c_2 a_w)(1 - c_2 a_w + c_2 c_3 a_w)}$ (18.9)			
GAB parameters (the least-squares fitting)	$c_1 = 1.481$ $c_2 = 0.989$ $c_3 = 0.099$ R-sq = 0.977	$c_1 = 0.073$ $c_2 = 0.882$ $c_3 = 31.9$ R-sq = 0.996	N/A

TABLE 18.10
Properties of Reconstituted Milk from SD-1 and SB-1 Powders (23°C–25°C)

Properties	SD-1 Reconstituted Milk	SB-1 Reconstituted Milk
a_w(—)	0.987 (±0.003)	0.987 (±0.003)
m_{milk} (%w.b.)	88.2	88.0
ρ_{milk} (kg m^{-3})	1035	1038
μ_{milk} (Pa s)	3.0	3.0
β_{milk} (mN m^{-1})	37.5 (±0.2)	39.3 (±0.1)
c_{pmilk} (kJ kg^{-1} K^{-1})	4.3 (±0.4)	4.6 (±0.7)
λ_{milk} (W m^{-1} K^{-1})	0.46 (±0.05)	0.56 (±0.05)

at 0.731. Basically, for goat milk powder, more research needs to be developed for identifying these changes. Although Table 18.9 presents data for the sorption curve of SB and SD goat milk powder, these are still scarce to analyze in detail the evolution of lactose crystallization. The decrease in the milk powder solubility, by comparing the solubility of SB-1 with the one of SB-6, is also associated with lactose

crystallization and its effects, destabilizing the structure of proteins, which reduces the milk solubility.

With regard to the mean particle size, the value measured for the SD-1 powder is higher than the range of 150–200 μm, previously mentioned for powder produced in the two-stage SD process, but it is within the range of instant powder. Although the d_p value of SB-1 is 17% lower, it is still in the range of instant powder. However, the dispersion index span, I_d, indicates a higher amount of fines in the SB-1 powder, which can reduce its capacity of being instant. This result was already expected due to the high attrition between inert particles. However, if it is necessary to produce instant powder, a VBF system can be attached to the SB dryer. The initial idea is to produce goat milk powder to be used as an ingredient in child food. The d_p value of SB-6 suggests an increase in d_p and a decrease in Id as storage proceeds. The particle size distribution data shown in Figure 18.12a indicate that changes in d_p occur at $d_p \leq 300\,\mu m$, with a clear increase in particle size at $d_p \leq 200\,\mu m$. This behavior could probably be related to the lactose glass transition and crystallization. Figure 18.12b presents the powder size distribution curves of SD-1 powder and the SB powder obtained in the pilot unit using the atomizer injector and stored during 6 months. Comparing the latter curve with the one of SB-6 powder obtained using the drop-injection (Figure 18.12a), a quite different behavior of particle rearrangement during storage can be seen for these two powders, basically in the range $d_p \leq 200\,\mu m$. For skim cow milk powder, Ilari and Mekkaoui (2005) have identified the particle size range of 100–112 μm as the one with the better properties for handling and processing. Contrary to this result reported for skim cow milk, it seems that, in the case of goat milk, this 100–112 μm size range is responsible for the changes in powder characteristics. Once again this is an indicative that, in the further steps of this new product-process development, more data concerning the behavior of goat milk powder and its properties during storage will have to be collected.

Although the stick point temperature has been used more recently to quantify particle cohesion and wall adhesion (Ozemen and Langrish 2002, Jaya and Das 2009), the Hausner cohesion number, HR, defined by Equation 18.8 in Table 18.8, was chosen in this work to quantify the strength of the cohesive force between particles. As HR deals with properties relevant to dynamic motion (Geldart 1986) rather than static parameters, it may better evaluate the flowability of the three powders analyzed. Additionally, the repose angle, θ, of these powders was also measured. All the three powders analyzed presented HR > 1.4, which means powders with very poor flowability. These values are similar to those of whole cow milk powder obtained by Birchal et al. (2005). The poor flowability of these powders is corroborated by their values of θ higher than 40°.

The apparent density, ρ_{ap}, and the tapped density, $\rho_{ap\text{-}max}$, of the SD-1 powder are about 18% higher than those of the SB-1. This difference is associated with the solid composition and concentration (density of solids), the amount of air entrapped into individual particles (mainly in SD), the amount of interstitial air (i.e., the packed bed porosity), and/or the type of cohesive forces. Since the SB-1 powder has lower d_p, its bulk or apparent density is lower due to the higher packed bed porosity (Yu et al. 2003). Therefore, an increase in d_p by enhancing agglomeration can improve the ρ_{ap} value, if necessary (requirement for instant powder).

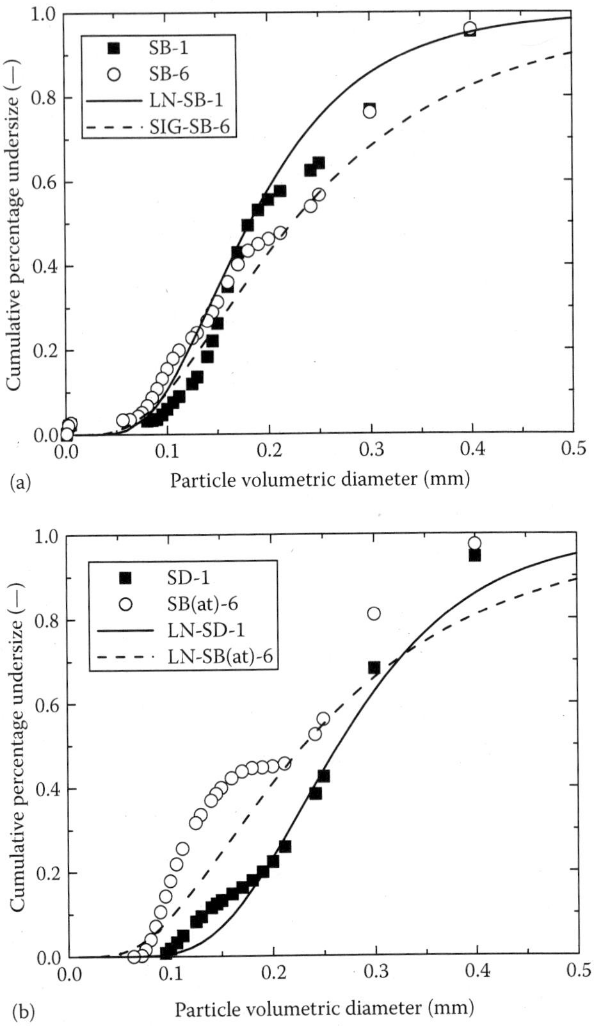

FIGURE 18.12 Cumulative percentage undersize (in volume) distribution curve for random samples: (a) samples of goat milk powder produced in the SB pilot unit using a drop-injector, SB-1 for 1 month storage and SB-6 for 6 months storage; (b) samples of milk powder produced by different techniques (spray drying [SD] and spouted bed [SB]) using atomizers, SD-1 for 1 month storage and SB(at)-6 for 6 months storage. LN, lognormal model prediction and SIG, sigmoid model prediction.

Finally, in Table 18.10, the properties measured for the reconstituted goat milk from SD-1 and SB-1 powders are presented. Values of these properties are similar between these two reconstituted milks. Returning to Table 18.5, it is seen that the measured properties of SB-1 reconstituted milk, except the surface tension β_{milk}, are similar to those of raw goat milk. This confirms, as discussed by Fox and McSweeny (1998), the similar chemical composition regarding to relevant constituents of raw goat milk. Concerning β_{milk}, there is a decrease in value of this property in the reconstituted milk in comparison to raw milk. This decrease was already expected, since the reconstituted milk is processed milk and β_{milk} is sensitive to homogenization and thermal changes occurred in proteins and fatty acids within the milk structure. However, due to their complexity, these changes cannot be quantified. Otherwise, the β_{milk} values of reconstituted cow milk reported in the literature are from 40 to 60 mN m^{-1} at 20°C–25°C (Fox and McSweeney 1998). This evidence and the β_{milk} value for the SD-1 reconstituted milk are used to accept the β_{milk} value for the SB-1 reconstituted milk as a correct one regarding the powder quality.

This demonstrates the high quality of the powder produced in the SB pilot unit. However, the storage of this SB powder needs attention due to the occurrence of lactose crystallization, which will affect the quality of the stored powder.

Based on the results from these tests, it can be concluded that the technical feasibility of producing goat milk powder using the low-cost SB technology is well confirmed by the high product quality. Within the experimental range of the operational variables analyzed, the most adequate conditions to operate the SB pilot unit are PP particles as inert and drop-injector for feeding milk emulsion intermittently. Packaging of this product is an important operation to be investigated in the further step of this product-process development. In addition, tests must be designed to quantify the minimum spouting conditions as function of the milk emulsion added to the bed of inert particles.

18.5 CONCLUDING REMARKS

This chapter has developed, analyzed, and discussed the first two steps to develop a new product process for the production of goat milk powder in small agro-cooperatives. The third and the fourth steps shown in Figure 18.1 are under development, combining experimental and simulation data to scale up the pilot unit to industrial size and to optimize this product process with regard to costs and powder quality. This is without any doubt the great challenge nowadays for the food industry.

REFERENCES

Andrade, W.M., Jr., Xavier, C.H., Araújo, A.A.L. et al. 2005. Drying milk in spouted beds. In *Proceedings of VI Brazilian Congress on Chemical Engineering Undergraduate Research*, ed. Silva, M.G.C. et al. Campinas: UNICAMP, CD-5pp (in Portuguese).

Attaie, R. and Richter, R.L. 2000. Size distribution of fat globules in goat milk. *J. Dairy Sci.* 83:940–944.

Barrionuevo, M., Alferez, M.J.M., Lopez-Aliaga, I., Sanz-Sampelayo, M.R., and Campos, M.S. 2002. Beneficial effect of goal milk on nutritive utilization of iron and copper in malabsorption syndrome. *J. Dairy Sci.* 85:657–664.

Beuchat, L. 1981. Microbial stability as affected by water activity. *Cereal Foods World* 26:345–351.

Birchal, V.S. and Passos, M.L. 2005. Modeling and simulation of milk emulsion drying in spray dryers. *Braz. J. Chem. Eng.* 22:293–302.

Birchal, V.S., Passos, M.L., Wildhagen, G.R.S., and Mujumdar, A.S. 2005. Effect of spray-dryer operating variables on the whole milk powder quality. *Drying Technol.* 23:611–636.

Cordeiro, P.R.C. 2007. Benefits and strategies of integrated commercialization. Keynotes presented at II Goat and Ovine Symposium, Belo Horizonte, Brazil (in Portuguese).

Cordeiro, P.R.C. and Cordeiro, A.G.P.C. 2008. Goat milk business in Brazil and its chain of production. Paper presented at XII Northeast Cattle Raising Seminary, Fortaleza, Brazil (in Portuguese).

Costa, E.F., Jr., Cardoso, M., and Passos, M.L. 2001. Simulation of drying suspension in spout-fluid beds of inert particles. *Drying Technol.* 19:1975–2001.

Costa, E.F., Jr., Freire, F.B, Freire, J.T., and Passos, M.L. 2006. Spouted beds of inert particles for drying suspension. *Drying Technol.* 24:315–325.

Costa, R.G., Beltrão Filho, E.M., Queiroga, R.C.R.E., Medeiros, A.N., Oliveira, C.J.B., and Guerra, I.C.D. 2007. Physicochemical characteristics of goat milk marketed in the Paraíba State, Brazil. *Rev. Inst. Adolfo Luz* 66:136–141 (in Portuguese).

Danes, M.A.C. 2007. Inquiry: The cultural habits and the lack of advertisements limit the consuming. http://www.farmpoint.com.br/?actA=7&areaID=1&secaolD=10&mesN=1&anoN=2007&paginaN=2 (accessed May 20, 2008).

Dubeuf, J.-P. 2007. Characteristics and diversity of the dairy goat production systems and industry around the world. Paper presented at the III International Goat and Ovine Symposium, João Pessoa, Brazil.

Eddleman, H. 2007. Composition of human, cow, and goat milk. Article B120A. http://www.goatworld.com/articles/goatmilk/colostrum.shtml (accessed May 20, 2008).

FAO. 2006. Food and Agricultural Organization of the United Nations. FAOSTAT Statistics Division. http//:faostat.fao.org/site/573 or 569/pageID=573 or 569 (accessed May 30, 2008).

Fontana, A.J. Jr. 2008. Investigating the moisture sorption behavior of foods using a dynamic dew point isotherm method. Paper presented at 5th Conference on Water in Food—Euro Food's Water-2008, Germany.

Fox, P.F. and McSweeney, P.L.H. 1998. *Dairy Chemistry and Biochemistry*. New York: Plenum Publishers.

Freire, J. T. 1992. Drying of pastes in spouted beds. In *Special Topics in Drying*, ed. J.T. Freire. São Carlos SP, Brazil: UFSCar, pp. 41–86 (in Portuguese).

FSA (Food Standards Agency). 2002. *McCance and Widdowson's the Composition of Foods*. Sixth summary edition. Cambridge, U.K.: Royal Society of Chemistry.

Geldart, D. 1986. *Gas Fluidization Technology*. New York: John Wiley & Sons.

Guo, M.R., Dixon, P.H., Park, Y.W., Gilmore, J.A., and Kindstedt, P.S. 2001. Seasonal changes in the chemical composition of commingled goat milk. *J. Dairy Sci.* 84(E. Suppl.):E79–E83.

Haenlein, G.F. 1996. Status and prospects of dairy goat industry in the United States. *J. Anim. Sci.* 74:1173–1181.

Haenlein, G. 2002. Lipids and proteins in milk, particularly goat milk. Technical Report. University of Delaware, Newark, DE. http://ag.udel.edu/extension/information/goatmgt/gm-08.htm (accessed March 12, 2008).

Haenlein, G. 2004. Goat milk in human nutrition. *Small Ruminant Res.* 51:155–163.

Hardy, J., Scher, J., and Banon, S. 2002. Water activity and hydration of dairy powders. *Lait* 8:441–452.

Harris, B. and Springer, F. 1992. *Dairy Goat Production Guide*. Animal Science Department Series CIR452, Florida Cooperative Extension Service, Institute of Food and Agricultural Sciences, University of Florida, Gainesville, FL. http://edis.ifas.ufl.edud/dsi3400.pdf (accessed May 30, 2008).

Holanda, E.V., Jr., Franca, F.C.M.C., and Lobo, R.N.B. 2006. Economic performance of the goat milk production by individuals in Rio Grande do Norte State. Technical Communication no. 74 ISSN 1676-7675 Sobral Emprapa Caprinos. http://www.cnpc.embrapa.br/cot74.pdf (accessed April 2, 2008) (in Portuguese).

IBGE. 2006. Brazilian Institute of Geography and Statistic. http://www.ibge.gov.br/home/estatistica/ppm/2000–2006/ppm200.pdf to ppm2006.pdf (accessed May 30, 2008) (in Portuguese).

Ilari, J.-L. and Mekkaoui, L. 2005. Physical properties of constitutive size classes of spray-dried skim milk powder and their mixtures. *Lait* 85:279–294.

Jaya, S. and Das, H. 2009. Glass transition and sticky point temperatures and stability/mobility diagram of fruit powders. *Food Bioprocess Technol.* 2:89–95

Jenness, R. 1980. Composition and characteristics of goat milk: Review 1968–1979. *J. Dairy Sci.* 63:1605–1630.

Jouppila, K., Kansikas, J., and Roos, Y.H. 1997. Glass transition, water plasticization, and lactose crystallization in skim milk powder. *J. Dairy Sci.* 80:3152–3160.

Kristensen, D., Jensen, P.Y., Madsen, F., and Birdi, K.S. 1997. Rheology and surface tension of selected processed dairy fluids: Influence of temperature. *J. Dairy Sci.* 80:2282–2290.

Lythgoe, H.C. 1940. Composition of goat milk of known purity. *J. Dairy Sci.* 23:1097–1108.

Marciano, L.A., Medeiros, U.K.L., Medeiros, M.F.D., and Passos, M.L. (2007). Powder retention into the spouted bed during the goat milk drying. In *Proceedings of VII Brazilian Congress on Chemical Engineering Undergraduate Research*, eds. J.T. Freire et al. São Carlos, Brazil: UFSCar, CD-6pp (in Portuguese).

Maree, H.P. 2003. Goat milk and its use as a hypo-allergenic infant food. Article 152. http://www.goatconnection/articles/publish/article_152.shtml (accessed March 10, 2008).

Martinez-Ferez, A., Rudloff, S., Guadix, A. et al. 2006. Goats' milk as a natural source of lactose-derived oligosaccharides: Isolation by membrane technology. *Int. Dairy J.* 16:173–181.

Mathur, K.B. and Epstein, N. 1974. *Spouted Beds*. New York: Academic Press.

Michalsk, M.C., Briard, V., Michel, F., Tasson, F., and Poulain, P. 2005. Size distribution of fat globules in human colostrum, breast milk and infant formula. *J. Dairy Sci.* 88:1927–1940.

Nascimento, R.J.A., Medeiros, U.K.L., Passos, M.L., and Medeiros, M.F.D. 2007. Effect of solid concentration, temperature and emulsion-feed injector on the drying of milk in spouted beds. In *Proceedings of VII Brazilian Congress on Chemical Engineering Undergraduate Research*, eds. J.T. Freire et al. São Carlos, Brazil: UFSCar, CD-6pp (in Portuguese).

Odle, J. 1997. New insights into the utilization of medium-chain triglycerides by the neonate: Observations from a piglet model. *J. Nutr.* 127:1061–1067.

OGMPA (Ontario Goat Milk Producers' Association). 2008. http://www.ontariogoatmilk.org/ANALYSIS.HTM (accessed March 10, 2008).

Oliveira, M.A., Favaro, R.M.D., Okada, M.M., Abe, L.T., and Iha, M.H. 2005. Physicochemical and microbiological quality of frozen pasteurized goat whole milk and of ultra high temperature milk commercialized in the region of Ribeirão Preto—SP. *Rev. Inst. Adolfo Luz* 64:104–109 (in Portuguese).

Ozemen, L. and Langrish, T.A.G. 2002. Comparison of glass transition temperature and sticky point temperature for skim milk powder. *Drying Technol.* 20:1177–1192.

Passos, M.L. and Mujumdar, A.S. 2000. Effect of cohesive forces on fluidized and spouted beds of wet particles. *Powder Technol.* 110:222–238.

Passos, M.L., Trindade, A.L.G., d'Angelo, J.V.H., and Cardoso, M. 2004. Drying of black liquor in spouted bed of inert particles. *Drying Technol.* 22:1041–1067.

Pereira, G.F. 2003. Study about the rentability of the goat milk production system in the RN State. Master's thesis, Federal University of Rio Grande do Norte, Rio Grande do Norte, Brazil (in Portuguese).

Pereira, R.A.G., Queiroga, R.C.R.E., Vianna, R.P.T., and Oliveira, M. L.G. 2005. Chemical and physical quality of goat milk distributed in the social program "New Pact Cariri" of the state of Paraiba, Brazil. *Rev. Inst. Adolfo Luz* 64:205–211 (in Portuguese).

Perosa, J.M.Y., Goncalves, H.C., Noronha, C.C., Andrighetto, C., and Yokoi, C.H. 1999. Economic indexes for milk production in small goatherds. *Inf. Econ. São Paulo* 29:7–14 (in Portuguese).

Perry, R.H., Green, D.W., and Maloney, J.O. 1999. *Perry's Chemical Engineers' Handbook*. New York: The McGraw-Hill Companies, Inc.

Pierre, A., Michel, F., and Le-Graet, Y. 1995. Variation in size of goat milk casein micelles related to casein genotype. *Lait* 75:489–502.

Pisecky, J. 1995. Evaporation and spray drying in the dairy industry. In *Handbook of Industrial Drying*, ed. A.S. Mujumdar, 2nd ed., vol. 1. New York: Marcel Dekker Inc, pp. 715–742.

Ribeiro Jr., C.P. 2000. Modeling of plate heat-transfer systems and simulation of a Brazilian milk powder plant. MSc thesis, Federal University of Minas Gerais, Belo Horizonte, Brazil (in Portuguese).

Rocha, S.C.S. and Taranto, O.P. 2008. Advances in spouted bed drying of foods. In *Advances in Food Dehydration*, ed. C. Ratti. Boca Raton, FL: CRC Press, pp. 153–186.

Roylance, D. 1993. Assessment of olefin-based tail strings. *J. Appl. Biomater*. 4:289–301.

Rubino, R. and Haenlein, G.F.W. 1997. Goat milk production systems: Sub-systems and differentiation factors. *Cahiers Options Mediterraneennes* 25:9–16. http:// ressources.ciheam.org/om/pdf/c25/97605950.pdf

Rutz, W.D., Whitnah, C.H., and Baetz, G.D. 1955. Some physical properties of milk. I. Density. *J. Dairy Sci*. 38:1312–1318.

Santos, E., Jr., Vieira, R.A.M., Henrique, D.S., and Fernandes, A.M. 2008. Characteristics of the dairy goat primary sector in Rio de Janeiro State, Brazil. *Rev. Bras. Zootec*. 37:773–781.

Sawaya, W.N., Safi, W.J., Al-Shalhat, F., and Al-Mohammad, M.M. 1984. Chemical composition and nutritive value of goat milk. *J. Dairy Sci*. 67:1655–1659.

St-Gelais, D., Ali, O.B., and Turcot, S. 2003. Composition of goat's milk and processing suitability. Technical Report. Agriculture and Agri-Food Canada: Food Research and Development Center, Saint-Hyacinthe, Quebec, Canada.

Stoney, K. and Francis, J. 2001. Dairy goat products developing new markets. RIRDC Publication no. 01/055. Rural Industries Research and Development Corporation Australia. http://www.rirdc.gov.au/reports/index.htm (accessed March 8, 2008).

Thompkinson, D.K. and Kharb, S. 2007. Aspects of infant food formulation. *CRFSFS* 6:79–102.

Trindade, A.L.G. 2004. New technique to concentrate black liquor for producing powdery fuel. Master's thesis, Federal University of Minas Gerais, Belo Horizonte, Brazil (in Portuguese).

Trindade, A.L.G., Passos, M. L., Costa, E.F., Jr., and Biscaia Jr., E.C. 2004. The effect of interparticle cohesive forces on the simulation of fluid flow in spout-fluid beds. *Braz. J. Chem. Eng*. 21:113–125.

Ueda, A. 1999. Relationship among milk density, composition and temperature. Masther's thesis, University of Guelph, Guelph, Ontario, Canada.

Vignolles, M.-L., Jeantet, R., Lopez, C., and Schuck, P. 2007. Free fat, surface fat and dairy powders: Interactions between process and product. A review. *Lait* 87:187–236.

Villar, L. 2008. Enterprises will expand dairy products derived from goat milk. *DCI* Commerce Industry Newspaper (Diário de Comércio Indústria), January 10, Agribusiness Section (in Portuguese) http://www.dci.com.br/noticias.asp?id_editoria=7&id_noticia=209218&editoria=(accessed August 8, 2008).

Walstra, P., Geurts, T.J., Noomen, A., Jellema, A., and van Boekel, M.A.J.S. 1999. *Dairy Technology: Principles of Milk, Properties and Processes*. New York: Marcel Dekker.

Webb, B.H. and Johnson, A.H. 1965. *Fundamentals of Dairy Chemistry*. Westport, CT: Avi. Publishing Co., Inc.

Yu, A.B., Fenz, C.L., Zou, R.P., and Yang, R.Y. 2003. On relationship between porosity and interparticle force. *Powder Technol*. 130:70–76.

19 Meat Products as Functional Foods

Juana Fernández López and
José Angel Pérez Alvarez

CONTENTS

19.1 INTRODUCTION

The meat industry is one of the most important industries in the world, and research into new products is continuous, whether as a result of consumer demand or because of the ferocious competition in the industry. However, such research and the launch of new products are directed at providing healthy alternatives to what has frequently been accused of causing a variety of pathologies. This negative image derives mainly from the content of fat, saturated fatty acids, and cholesterol, and their association with cardiovascular diseases, some types of cancer, obesity, etc. (Ovensen 2004, Valsta et al. 2005). Also, intake of sodium chloride, which is added to most meat products during processing, has been linked to hypertension (Ruusunen and Poulanne 2005).

New demands for healthier meat products have brought about the emergence of so-called functional foods on the market. Functional foods can be defined in a one way as those foods that contain a component (nutrient or nonnutrient) with a selective activity related to one or several functions of the organism and having an added physiological effect over and above its nutritional value, and whose positive action justifies the denomination as functional (physiological) or even healthy (Pascal and Collet-Ribbing 1998). Nevertheless, scientifically proved substantiation about the health effect is required and some legal authorizations are needed to probe the "healthy effect" when manufacturers develop specific, functional meat products (Fernández-López et al. 2006).

Meat and meat products can be modified by including ingredients considered beneficial or by eliminating/reducing those components considered harmful to health (Fernández-Ginés et al. 2005). The meat itself can be modified by manipulating animal feed or by *postmortem* manipulation of the carcass. As regards meat products, efforts are principally directed toward modifying the salt and lipid content (fatty acids) or incorporating a series of functional ingredients (fruit or cereal fiber, vegetal proteins, monounsaturated or polyunsaturated fatty, vitamins, calcium, inulin, etc.) (Jiménez-Colmenero et al. 2001, Mendoza et al. 2001, Arihara 2006, Fernández-Ginés et al. 2005).

19.2 PROCESSING STRATEGIES IN THE DEVELOPMENT OF FUNCTIONAL MEAT PRODUCTS

Insofar as meat is concerned, the modifications to which it may be subjected to confer functional properties on it are based on modifications to the feed an animal receives or postmortem manipulation of the carcass. In the first case, the lipid, fatty acid, and vitamin E content can be modified, whereas in the second, fat can be removed by mechanical processes. As regards meat products, efforts are mainly directed toward their reformulation by modifying the lipid and salt content, and/or adding a series of functional ingredients (fiber, vegetal proteins, mono- or polyunsaturated fatty acids, vitamins, calcium, phytochemicals, etc.) (Jimenez-Colmenero et al. 2001, Fernández-Ginés et al. 2005). However, one of the inconveniences associated with the incorporation of a new ingredient in any food, and meat products are not an exception, is the effect that this might have on the corresponding technological, nutritional, and sensorial properties. All these aspects will be developed in the following sections.

19.2.1 Partial Replacement of Nonhealthy Ingredients

Food composition, and in particular the presence or absence of some ingredients, is specially taken into account and it is reflected in consumption tendencies. Food calorie input, its fat content, and the proportion of incorporated salt are some of the factors determining consumer acceptance (Demos et al. 1994). The public has become aware of the effect of animal fat on serum cholesterol and the relationship between salt content and hypertension (Appel et al. 2006). There is evidence that fat-rich diets, apart from causing obesity, are also directly related to the risk of colon cancer. Fat and cholesterol are also associated with cardiovascular diseases. Consequently, fats, cholesterol, and salt are considered as nonhealthy ingredients.

19.2.1.1 Fats

The limitations about the fat intake refer not only to the amount of fat but also to the fatty acid composition and the cholesterol levels in meat products. Meat fat content can vary widely depending on various factors such as species, feeding, cut, etc. The average fresh meat cuts, e.g., in Europe contain about 10% fat but within a wide range (from 6% in turkey meat to 10.5% in pork meat). Meat products however are more fatty and contain on average 25% fat (from 3% to 50%) in Europe (Bauer and Honikel 2007). Meat lipids usually contain less than 50% saturated fatty acids and up to 70% unsaturated fatty acids (Romans et al. 1994). The amount of cholesterol in meat and meat products depends on numerous factors, but in general it is less than 75 mg/100 g, except in the case of some edible offal, in which the concentrations are much higher (Chizzolini et al. 1999). However, animal fat in meat products is essential for many of their properties and characteristics, which no other component can give. Fat fulfils basic functions that to a great extent contribute to the sensory properties (palatability, juiciness, aroma, and toughness), which characterize many meat products. It also has functional properties (consistency, binding, and humidity), and it is vital in rheological and structural properties (Keeton 1994). The characteristics of pork fat (unsaturated, melting point, and the presence of connective tissue, etc.) are ideal for producing meat products; it is easier to handle and its acid fat composition contributes greatly to the formation of aroma and flavor. The most commonly used pork fat is back fat, which is rich in saturated acid fats, has a greater melting point and is, therefore, firmer.

The production of low-fat products involves a series of technological problems related to texture, flavor, and mouthfeel (Keeton 1994), and with the acceptance of these products (Giese 1996), the basic concern is always to reduce fat but retain traditional flavor and texture.

Reduction of fat in meat products processing is achieved by reformulation. This consists of combining preselected meat raw materials (leaner) with appropriate amounts of water, fat (animal or vegetable), flavorings, and other ingredients (fat replacements or substitutes), which coupled with technological processes give the product certain desirable characteristics (Keeton 1994, Jiménez-Colmenero et al. 2001). Fatty acid composition can be altered (simultaneously with fat reduction or otherwise) by genetic and feeding strategies (at animal production level) to improve the degree of lipid unsaturation, and by replacing part of the animal fat normally present in the product with another more suited to human needs, i.e., with less saturated fatty acids and more monounsaturated (oleic) or polyunsaturated acids, and with no cholesterol. Although this substitution does not mean a decrease in the calorie content, it does favor an improvement in the product's nutritive qualities. With this in mind, fish oils (rich in n-3) have been used in products like hamburgers and sausages (Marquez et al. 1989, Park et al. 1989, Paneras et al. 1998). The reduction of cholesterol could be achieved by substituting raw meat materials, fat, and proteins for others from vegetables, which do not contain cholesterol. Several meat products have been reformulated (sausages, hamburgers, etc.), by modifying their traditional composition by reducing and/or partially replacing animal fat with vegetable oils (peanut, sunflower, olive, etc.) and incorporating vegetable proteins (soy, oats, etc.). Marquez et al. (1989) reduced the level of cholesterol in frankfurter sausages by 35% upon substituting animal fat for peanut oil. Other authors (Paneras et al. 1998) used another type of oil (olive, cotton, and soy) in substitution for animal fat and obtained frankfurter sausages with cholesterol levels lower than 59% of what they usually contain.

By incorporating olive oil in the formulation of meat products, one can expect an improvement in the nutritive value of the fatty fraction: an increase in the percentage of monounsaturated fatty acids (MUFA) and polyunsaturated fatty acids (PUFA), as well as a decrease in the saturated fraction (SFA). Additionally, it could also affect oxidative phenomena, since the incorporation of tocopherols and polyphenols, common in oil, could increase antioxidant strength (Bloukas and Paneras 1994, Muguerza et al. 2001, 2002, 2003). There are several studies based on this approach, in which olive oil is used as a total or partial substitute for pork fat, in cooked meat products, such as frankfurters (Bloukas and Paneras 1993, Bloukas et al. 1997a, Pappa et al. 2000, Lurueña-Martínez et al. 2004), as well as in traditionally dried or fermented products (Bloukas et al. 1997b, Muguerza et al. 2001, 2002, 2003, Kayaardi and Gök 2003, Severino et al. 2003, Ansorena and Astiasarán 2004). The incorporation of olive oil in the formulation of meat products can be carried out in a liquid form or pre-emulsified with sodium caseinate or with isolate soy proteins. Products with a better appearance and better technological behavior are obtained when the olive oil is incorporated pre-emulsified (Bloukas et al. 1997a).

Different substances (proteins, micro particles, fat substitutes, hydrocolloids, etc.) have been tested for solving the problems involved in the reduction of fat in meat products (lower in processing yield and firmer and less juicy), and, so far, hydrocolloids have been the most effective (starch and derivates, cellulose and derivates, gums and others), especially in low fat frankfurters (Bloukas and Paneras 1994). Gums are the most numerous: carrageen, alginate, pectin, xanthan gum, and galactomanan (guar gum and locust bean gum) all belong to this group (Lurueña-Martínez et al. 2004).

In dry-fermented meat products, the important role of fat in appearance, texture, and color significantly limits the degree of substitution with more unsaturated fats. Attempts to substitute pork fats with vegetable oil have proven to be difficult, but the results are very favorable, and the conclusion is that cold meat can be made with olive oil without the sensory and technological attributes being adversely modified. The studies have focused on typical traditional dry-fermented products (chorizo, salami, soudjouk, etc.) from Mediterranean countries (Spain, Greece, Italy, and Turkey), where olive oil has an important role in their culture (Wu and Brewer 1994, Muguerza et al. 2001, Bloukas et al. 1997b).

In conclusion, the use of olive oil in the formulation of meat-based products is at present a viable action for substituting pork fat, as shown in the studies carried out in low-fat sausages and in traditional Mediterranean fermented products. Both these products reflect two of the most used technologies in the production of meat-based products; one involves thermal treatment and the other a fermentation process (Sayas 2008).

19.2.1.2 Salt

There exist evidences about a relationship between salt consumption and high blood pressure (Desmond 2006). Sodium consumption in a diet should be restricted to at least less than 6 g of salt per day (2400 mg sodium/day), considering the consumption coming from all food sources (WHO 2003).

Salt-content reduction in elaborated food represents a significant contribution to sodium amount drop in a diet. In the last 70 years, salt content in meat products has been decreasing (Hutton 2002), and in the last 10–15 years, the meat industry is making serious efforts to elaborate this type of product (Collins 1997). In general, the consumer has a positive attitude with meat products with low salt levels, a fact that is more evident in women than in men (Guardia et al. 2006). Any action being done by the society to reduce the amount of salt consumed will probably improve people's health and diminish social costs (Selmer et al. 2000).

Products with low salt content will contribute to satisfy the demand from an important sector in the market (Demos et al. 1994) looking for diet improvement as part of a life style. Nonetheless, from the technological point of view, there exist some limitations to reduce salt content due to the functional properties salt possesses, many of which render elaborate processes giving specific characteristics to the finished product easier.

Salt is one of the most used ingredients in processed meats for various reasons: it has a flavoring and enhancing effect, it also contributes to texture (improving the muscular protein extraction, enhancing water-binding capacity, and binding components and increasing batter viscosity), and it helps to increase shelf life and safety aspects as well (bacteriostatic effect) (Collins 1997, Hutton 2002, Ruusunen and Puolanne 2005, Desmond 2006).

There are different alternatives to reduce the salt content in meat products, and these have been thoroughly analyzed for each meat product mainly because not all of them are applicable to all the different products (Rosmini et al. 2008).

In some types of meat products (minced, restructured, cured, and cooked products), the reduction in the sodium content does not present major problems from the technological point of view, so, in this case, it has been associated with the application of new technologies (high-pressure application) to contribute to extend the product shelf life (MacFarlane et al. 1984, Rubio et al. 2006). On the other hand, different aspects must be taken into account in the case of raw-cured products. In this specific case, salt has important actions in the curing process (texture, safety, color, and aroma) that render it an irreplaceable ingredient. For example, in dry-fermented sausage manufacturing, a salt content up to 2.5% would be indicated as the limit to obtain a good quality product, but when the salt content drops to 2.25%, sausages are less firm, and the typical aroma is weaker (Ruusunen and Puolanne 2005).

In other cases, the reduction in salt can be made by modifying the elaboration process, incorporating others additives that partially or totally replace the salt content, and other factors of conservation that guarantee the product safety. Different studies have been carried out to evaluate diverse ingredients (soy proteins, potato starch, caseinates, gums, phosphates, etc.) as partial substitutes for sodium chloride (Kuraishi et al. 1997, Tsai et al. 1998, Xiong et al. 1999, Ruusunen et al. 2002, Ruusunen and Puolanne 2005, Desmond 2006). The effects of these substitutes depend on the concentration and the meat product, but in any case all of them provided binding effects.

Another alternative to reduce the sodium content is to use equivalent additives without sodium (Terrel et al. 1983). In meat products like emulsions, different salt mixtures have already been studied, including formulas with potassium chloride, which seems to compensate the problems associated with texture and salt reduction, and magnesium chloride, which seems to be adequate when phosphates are not used (Lin et al. 1991, Ruusunen et al. 2001, Guardia et al. 2006, Rosmini et al. 2008), even though further investigation is required.

19.2.2 Addition of Functional Ingredients

The object of including functional ingredients in the case of meat products is not only concerned with providing it with certain desirable properties, but it is also an attempt to change their image in these health-conscious days. Such ingredients include dietary fiber (DF) from different origins (fruits, vegetable, or cereals), calcium, nuts, unsaturated fatty acids, probiotic bacteria, etc., as shown in Table 19.1.

TABLE 19.1
Studies Conducted about the Addition of Several Functional Ingredients to Different Types of Meat Products

Functional Ingredient	Type of Meat Product	References
Dietary fiber		
Fruit fiber	Restructured	Desmond et al. (1998); Pralongpan et al. (2002); California Prune Boards's European Office (2000)
	Cooked	Griguelmo-Miguel and Martín-Belloso (1999); Sanchez-Escalante et al. (2000); Fernández-Ginés et al. (2003, 2004); García et al. (2004); Fernández-López et al. (2005)
	Dry-fermented	García et al. (2002); Aleson-Carbonell et al. (2003, 2004); Fernández-López et al. (2008)
Vegetable fiber	Restructured	Chevalier and Galvin (1995); Jamora and Rhee (2002)
	Cooked	Cofrades et al. (2000)
Cereal fiber	Restructured	Keeton (1994); Bollinger (1996); Dawkins et al. (1999); Kim et al. (2000)
	Cooked	Cofrades et al. (2000); Pacheco-Delaye and Vivas-Lovera (2003)
	Dry-fermented	García et al. (2002); Tyburcy et al. (2003).
Nuts	Restructured	Jiménez-Colmenero et al. (2003); Cofrades et al. (2004); Ruíz-Capillas et al. (2004); Ayo et al. (2005); Serrano et al. (2005); Turhan et al. (2005)
	Cooked	Jiménez-Colmenero et al. (2005); Ayo et al. (2008)
Calcium	Cooked	Boyle et al. (1994); Korstanje and Hoek (2001); Cáceres et al. (2006)
	Dry-fermented	Gimeno et al. (1998, 2001); Flores et al. (2005); Salazar et al. (2005)
Meat protein-derived bioactive peptides	Dry-fermented	Arihara (2006)
Probiotic bacteria	Dry-fermented	Andersen (1998); Bunte et al. (2000); Jahreis et al. (2002); Pennachia et al. (2006)

19.2.2.1 Dietary Fiber

The generally accepted definition of DF includes polysaccharides and lignin that are neither digested nor absorbed in the human small intestine (Asp 1987). Total DF is divided into two fractions, one that is soluble in water at 100°C and pH 6–7, and another that is insoluble, and both of them have different beneficial effects on human health.

The insoluble DF fraction is mainly related to intestinal motility. Cummings (2000) reported that the mechanism by which fiber may be protective against gastrointestinal diseases is related to the decreased transit time, increased stool weight, and decreased intracolonic pressure. The soluble DF fraction, in turn, is chiefly involved in lowering effects on blood cholesterol and glucose intestinal absorption; both considered a risk factor for coronary diseases (Hallfrisch et al. 1995). Finally, positive benefits of fiber with respect to colon cancer have been reported (Fuchs et al. 1999).

Due to the chemical structure and inherent fiber-specific properties (capillary effect, network, etc.), DFs show functional effects that add important benefits to food and food production. The fiber addition to meat products can be justified by three principal reasons. First, for their nutritional properties: products with fibers or fibers enrichment. Second, because of their technological properties: texture improvement, prevention of loss due to drying out, prevention of drying out due to juice loss during cooking, prevention of gel separation, rehydration capacity improvement, and freezing–thawing stability improvement. Third, for their characteristics as an ingredient without calories, which allow their use in low-fat meat product.

19.2.2.1.1 Dietary Fiber from Fruits

The main advantage of DF from fruits, compared with alternative sources of fiber such as cereals, is its higher proportion of soluble fraction: about 33% in fruits and only 7% in wheat bran (Griguelmo-Miguel and Martín-Belloso 1999, Gorinstein et al. 2001). This is an important point to remember since DF intake must be balanced; i.e., the water-soluble fraction should represent between 30% and 50% of the total DF (Eastwood 1987). Components in fruits thought to be associated with the reduction of "lifestyle diseases" include soluble and insoluble DF, antioxidant nutrients (vitamins C, E, selenium, β-carotene), and phytonutrients (bioactive plant compounds that impart color, flavor, and other functional properties to foods), including polyphenols, flavonoides, anthocyanins, and carotenoids.

Fruit fiber is suitable for meat products and has been applied not only for their nutritional properties but also for their functional and technological properties (Thebaudin et al. 1997). In this regard, fiber has been successful in improving cooking yield, in reducing formulation costs, and in enhancing texture (Iyengar and Gross 1991, Jiménez-Colmenero 1996). Various types of fruit fiber have been studied alone or in combination with other ingredients for formulations of different types of meat products.

Peach fiber has been successfully used in conventional and reduced-fat cooked meat sausages (15–30 g/kg in bologna) (García et al. 2004), as well as in normal and reduced-fat dry fermented sausages (1.5%–3%) (García et al. 2002). Pineapple cores, a high-fiber part of pineapple fruit, have been added into beef burgers, improving their yield and texture, the water interaction of fibers, especially the water retention capacity, having a marked influence on their functional properties (Prakongpan et al. 2002). Plum fiber has been used as fat replacer, moisture binder, and flavor enhancer in frozen precooked meats (such as precooked hamburgers), beef–pork sausages, turkey sausages, and pepperoni with excellent results (California Prune Board's European Office 2000). Conventional and reduced-fat cooked meat sausages (bologna type) with added apple fiber (15–30 g/kg) were successfully manufactured by Sánchez-Escalante et al. (2000) and García et al. (2004). Different types of citrus fiber have been added to different meat products with excellent results. Lemon albedo was added at different concentrations (up to 10%) to cooked sausages (bologna type) (Fernández-Ginés et al. 2004) and dry-cured sausages (Aleson-Carbonell et al. 2003, 2004). Orange fiber powder (from citrus juice by-products) has been added at different concentrations (0%–2%) to cooked sausages (bologna type) (Fernández-Ginés et al. 2003) and to dry-cured sausages (Fernández-López et al. 2008). The most important finding observed when these citrus fibers have been added to cured products (raw or cooked) is the reduction in the residual nitrite level. This reduction has health-related effects because of the reduced possibility of nitrosamine formation, which is a risk associated with the consumption of meat products with nitrite in their formulation. Orange and lemon fiber powders have been used in cooked Swedish-style meatballs formulation (Fernández-López et al. 2005) with important findings related to lipid oxidation and microbial deterioration.

19.2.2.1.2 Dietary Fiber from Vegetables

Vegetables are usually used as a source of plant protein in the human diet. However, DF is also present as one of the main constituent of vegetables (Zia-ur-Rehman et al. 2003).

The use of different pea products as functional ingredients in meat products have been studied by several authors. Noort (2001) studied the use of pea fiber as a water-binding agent in a range of sausages and other meat products. Chevalier and Galvin (1995) used a natural ingredient extracted from smooth yellow peas with 60% DF as a ground beef extender for hamburgers with excellent results, and they recommended its use as a substitute for other fibrous extenders (textured soy protein concentrate, potato, and oat fibers). Parrheim Foods Group (Parrish and Heimbecker Co., Canada) has developed a functional pea fiber from Golden Canadian field peas, which contains >50% total DF (in dry matter) and is successfully used in the meat industry. It improves freeze–thaw stability and can be applied in different meat products (Anon 2002).

Two different carrot products (carrot powder and freeze-dried carrot extract) have been used in extruded meat products and both have shown antioxidant properties, which means that the intrinsic antioxidant in such ingredients (carotenoids and certain phenolic acids) may survive the thermoplastic extrusion process to a sufficient degree to minimize lipid oxidation in the extrudate (Jamora and Rhee 2002).

Soy fiber (82% DF, out of which 85% is insoluble fiber) has been used in bologna sausages (Cofrades et al. 2000). Owing to their water-binding ability and swelling properties, insoluble fibers can influence food texture. Insoluble fiber can increase the consistency of meat products through the formation of an insoluble three-dimensional network (Backers and Noll 2001) capable of modifying rheological properties of the continuous phase of emulsions.

Potato fiber is a highly functional ingredient that is stable under a wide range of meat processing conditions (heat processing, freezing–thawing conditions, etc.), and that can improve many meat products. The fiber porous structure facilitates the retention of large amounts of water (9 g of cold water or 14 g of hot water per gram of fiber). Even in frozen storage, this product has excellent water holding capacities, making it useful in frozen raw burger and sausage production. This type of fiber has been also successfully used in meat analog emulsions, such as for ground-beef type products, resulting in a firmer and more formable mixture (Dolata et al. 2002).

19.2.2.1.3 Dietary Fiber from Cereals

Vitacel® is a wheat fiber composition developed by Rettenmaier & Sons (Rosenberg, Germany), which consists of 98% DF (cellulose, hemicellulose), and it is therefore of great nutritional importance. With an addition of only 1% Vitacel wheat fiber, significant functional advantages can be achieved in meatballs and burgers. In both products, the addition of 2% Vitacel wheat fiber has shown technological advantages, particularly a reduced loss (14%) during heating (Bollinger 1996). García et al. (2002) added wheat fiber (1.5%–3%) to normal (25% fat) and low-fat (6% fat) dry-fermented sausages. The effect of wheat fiber on the physicochemical properties of salami type sausages has been investigated by Tyburcy et al. (2003).

Oat bran and oat fiber have been used in beef burgers at a concentration of 3% (Keeton 1994) or mixed with flavorings and seasonings (Giese 1992) with favorable results. Cofrades et al. (2000) studied the effects of oat fiber (2%) on the texture of frankfurters formulated with low (5%) and high fat (30%). García et al. (2002) used oat fiber and oat bran as nonmeat ingredients in ground beef and pork sausages, and concluded that both of them provide acceptable flavor and texture. Dawkins et al. (1999) added different concentrations of oat bran (15%–50%) to chevon (goat) meat-based patties. The effect of rice bran on quality characteristics of frozen sausages was studied by Pacheco-Delahaye and Vivas-Lovera (2003). Rice fiber (RF) and rice bran oil (RBO) were added to restructured beef roast at levels of 3% RF or (2% RBO)/(3% RF) and examined for their effects on oxidative stability and consumer acceptability (Kim et al. 2000).

19.2.2.2 Nuts

Unlike other types of fruit, nuts are high in calories and fat. However, on average, 85% of the fat is unsaturated. Walnuts, in particular, are unique in their composition relative to alpha-linolenic acid (ALA), containing a ratio of linoleic acid to ALA of 4:1, a value that can significantly decrease mortality due to coronary diseases. In addition, nuts are a good source of many minerals including magnesium, copper, zinc, selenium, phosphorus, and potassium. Nuts are also a good source of fiber (5%–10%), as well as vitamins E, B_6, folic acid, thiamin, and niacin. Because of this composition, epidemiological studies show that frequent consumption of nuts in general correlates inversely with myocardial infarction or death by vascular ischemic disease, regardless of other risk factors such as age, sex, smoking, hypertension, weight, and exercise (Fraser et al. 1992, Sabaté 1993). Due to these properties, nuts have been incorporated into meat products to confer potential heart-healthy benefits.

Most studies on the incorporation of nuts into meat products have been carried out with walnuts because of their special healthy attributes. Walnuts have been incorporated into different restructured meat products, namely, restructured beef steaks (Jiménez-Colmenero et al. 2003, Cofrades et al. 2004, Ruíz-Capillas et al. 2004, Serrano et al. 2005), meat batters (Ayo et al. 2005), frankfurters (Jiménez-Colmenero et al. 2005, Ayo et al. 2008), and beef burgers (Turhan et al. 2005). These authors have reported that the meat products to which walnuts were added could be considered functional foods. Further work is necessary to analyze the effect of nuts on the stability of meat products. Their high level of lipids and the fatty acid composition (unsaturated) would favor lipid oxidation; meanwhile, antioxidants contained in nuts could avoid such oxidation. In some cases, walnuts are added to promote high levels of components associated with health benefits related to cardiovascular diseases.

19.2.2.3 Polyunsaturated Fatty Acids

Studies suggest that long-chain n-3 PUFAs support good cardiovascular health, may have a beneficial effect in several forms of cancer and in diseases with an immunoinflammatory component, and they also play a role in the brain (Connor 2000, Ruxton et al. 2005). PUFAs may be introduced into meat products by enrichment with PUFA-rich ingredients such as fish oil or algae oil, and by manipulation of feed for animals for production of omega-3-rich meat (Augustin and Sanguansri 2003, Ward and Singh 2005, Valencia et al. 2007). The use of fish oil, although efficient from the nutritional point of view, has given rise to some sensory problems (Elmore et al. 2005). Algae oil shows some benefits over fish oils as a long-chain PUFA supplier, including low taste intensity and off-odor problems. In particular, the oil from the microalgae *Schizochytrium sp.*, supplying at least 32% of docosahexahenoic acid (DHA), has been allowed as a novel food ingredient in the European Union (2003/427/CE).

A specific challenge to increase the concentration of n-3 PUFA is the expected enhancement in the susceptibility of fortified meat products to lipid oxidation. Inclusion of n-3 PUFAs to meat has affected flavor (O'Keefe et al. 1995), and routine meat processing procedures (e.g., grinding, cooking) can exacerbate oxidation. In lipids with high oxidative susceptibility (e.g., n-3 oils), it is necessary to use multiple antioxidants to control rancidity development. The use of antioxidants has been an effective means to maintain lipid stability in meat products without alteration of the fatty acid profile (Rhee et al. 1997, Wilkinson et al. 2001).

19.2.2.4 Calcium

Calcium intake is necessary for the proper maintenance of physiological systems. Calcium deficiency has serious consequences during growth since it prevents the development of bones. The problems of chronic calcium deficiency generally involve osteomalacia, osteoporosis, and a deficient calcification of teeth (Fairé and Frasquet 1999). Given the well-established relationship between calcium intake and the reduction in the risk of several diseases, the intake of calcium by a large part of the population should be increased (Selgas and García 2008).

One way to increase calcium intake without causing great changes to eating habits is to add it to habitually eaten foods. Indeed, calcium-enriched commercial foods—including enriched meat products—have begun to appear (Boyle et al. 1994).

Two main methods of enrichment have been developed for both cooked and ripened products: the direct addition of calcium salts or their incorporation as a substitute for sodium chloride, either alone or in combination with other mono and bivalent cations (Devatkal and Mendiratta 2001). The majority of research has in fact been aimed at the production of meat products for low-sodium diets, given the link between high sodium chloride levels and high blood pressure. Generally, the final products contain 8 mg/100 g more calcium than the raw material. Thus, a lower sodium level and increased nutritional value is simultaneously obtained. Therefore, calcium plays an important technological role via its interactions with meat product components and can contribute to their improvement.

Several authors have applied different sources of calcium in cooked meat products. Boyle et al. (1994) manufactured reduced-fat frankfurters fortified with calcium carbonate or calcium citrate–malate. These were added in sufficient quantities to provide 25%–100% of the adult recommended dietary allowance (RDA) in one 45 g frankfurter. These authors also described an increase in the pH of the product after the addition of calcium salts from 5.99 to 7.15. Cáceres et al. (2006) manufactured bologna-type cooked meat sausages with different fat levels and added calcium lactate, calcium gluconate, or calcium citrate, three high-bioavailability salts (Korstanje and Hoek 2001). The authors have concluded that it was possible to manufacture cooked sausages with added calcium salts up to a maximum level of 25% of the RDA (350 mg/100 g product).

In contrast to what is observed for cooked meats, little work has been performed on the enrichment of ripened meat products with calcium. Whenever calcium has been added, this has been done either as a partial substitute for NaCl (for low sodium products) or as an additional ingredient. Gimeno et al. (1998) studied the effect of a mixture of sodium, potassium, magnesium, and calcium chlorides as a partial replacement for NaCl in traditional Spanish dry fermented sausages. The mixture contained 4.64 g/kg $CaCl_2$, which increased the calcium level from 82.6 mg/100 g in the initial batter to 182.9 mg/100 g after ripening. No effects were found on lactic acid bacteria, but *Micrococcaceae* showed significantly lower counts during ripening—a consequence of the greater acidification observed when calcium salts were incorporated. Gimeno et al. (2001) studied the effect of calcium ascorbate on the sensorial characteristics and hygienic safety of Spanish dry fermented sausages. The calcium salt was added mainly as a partial substitute for NaCl to produce a calcium-enriched product with reduced sodium content. Calcium ascorbate was added at different concentrations to provide from 16.25% of the RDA in the control batch, to 50% in the most enriched batch. The calcium ascorbate reduced the redox potential of meat product, and, together with nitrite, enhanced the inhibition of *Clostridium botulinum* and accelerated the curing reactions in the cured meat product. Changes in the textural properties involved mainly hardness and gumminess.

In combination with starter cultures, Flores et al. (2005) used calcium chloride at final concentrations of 0.05% and 0.5% in the manufacture of dry fermented sausages and studied the effect on their volatile compound profile and sensorial acceptance. The addition of 0.05% calcium chloride had and important effect on the volatile pattern by inhibiting lipid oxidation and reducing the concentration of linear aldehydes, products than can produce a high impact on sausage flavor due to their low detection thresholds. The authors also observed that carbohydrate fermentation was greater in the 0.05% batch, and that the addition of this salt increased the generation of ethyl esters, leading to significantly higher levels than the controls. Sensorially, the 0.5% batch was the least acceptable, due to their greatest oxidation. Salazar et al. (2005), in turn, manufactured *salchichón*, a traditional, Spanish, dry fermented sausage, enriched with several calcium salts (calcium lactate, calcium gluconate, and calcium citrate) selected for their high calcium bioavailability. These were independently added to the original batter and in sufficient amounts to give a concentration close to 25% of the RDA in the ripened product (275 mg/100 g approx.). The salts were incorporated into a mixture with the spices to facilitate their distribution. The results showed the salts not to negatively influence the pH, water activity, color, growth of microbiota, or sensorial acceptance.

19.2.2.5 Meat Protein-Derived Bioactive Peptides

Bioactive peptides are a group of promising functional components of meat (Arhiara 2006). Although the activities of these peptides in the sequences of proteins are latent, they are released by proteolytic enzymes (i.e., muscle, microbial, and digestive proteinases). Therefore, meat proteins have possible bioactivities beyond a nutritional source of amino acids alone. Although information on bioactive peptides generated from meat proteins is still limited, there is a possibility of using such components for developing novel functional meat products. Most proteins contain bioactive sequences, but those sequences are inactive within the parent proteins. Active peptide fragments are released from native proteins only via proteolytic digestion. Once such peptides are liberated, they can act as regulatory compounds. During gastrointestinal proteolysis, bioactive peptides would be generated from food proteins (Pihlanto and Korhonen 2003).

The most extensively studied bioactive peptides generated from food proteins are angiotensin I-converting enzyme (ACE) inhibitory peptides (Vermeirssen et al. 2004, Mine and Shahidi 2005). ACE inhibitory peptides have attracted much attention because of their ability to prevent hypertension. Among the bioactive peptides derived from meat proteins, ACE inhibitory peptides have been studied most extensively (Arihara 2006). Fujita et al. (2000) isolated ACE inhibitory peptides generated from chicken muscle proteins by thermolysin treatment. Arihara et al. (2001) reported ACE inhibitory peptides in enzymatic hydrolyzates of porcine skeletal muscle proteins. Saiga et al. (2003a) reported antihypertensive activity of *Aspergillus* protease-treated chicken muscle extracts.

Information on meat protein-derived bioactive peptides other than ACE inhibitory peptides is still limited (Arihara 2006). Carnosine and anserine are endogenous antioxidative dipeptides found in skeletal muscle. Apart from endogenous nonprotein peptides, several antioxidative peptides have been generated from meat proteins by enzymatic digestion, as reported in the literature (Saiga et al. 2003b, Arihara et al. 2005). Opioid peptides are defined as peptides that have an affinity for an opiate receptor as well as opiate-like effects (Pihlanto and Korhonen 2003). Basically, opioid peptides have effects on the nerve system. Although there has been no report on the generation of opioid sequences thought to be present in the sequences of muscle proteins, it should be possible to find opioid peptides in meat proteins by proteolytic treatment (Arhiara 2006). In addition to these bioactive peptides, several bioactive peptides (i.e., immunomodulating, antimicrobial, prebiotic, mineral binding, antithrombotic, and hypocholesterolemic peptides) have been found from enzymatic hydrolyzates of various food proteins (Mine and Shahidi 2005). It is expected that interest will be shown in research aimed at finding such meat protein-derived peptides.

Generation of bioactive peptides in meat products, such as fermented meat products, is a possible direction for introducing physiological functions. Bioactive peptides would be generated in fermented meat products since meat proteins are hydrolyzed by proteolytic enzymes during fermentation and storage. Developing functional fermented meat products could be a good strategy in the meat industry. On the other hand, rediscovery of traditional fermented meats as functional foods is also an interesting direction. Many traditional fermented foods, such as fermented dairy products, have been rediscovered as functional foods (Farnworth 2003). Numerous physiologically active components, including bioactive peptides, have been discovered in these traditional fermented foods. For these reasons, traditional fermented meats are attractive targets for finding new functional meat products (Arihara 2006).

19.2.2.6 Probiotic Bacteria

The use of probiotic lines is another attractive approach for designing functional meat products. Probiotic therapy is the ingestion of viable microorganisms, which have documented health benefits for the consumers by maintaining or improving their intestinal microbiota balance (Stanton et al. 2003, Arihara et al. 2006). So, using probiotic bacteria, potential health benefits can be introduced into meat products. Target products with probiotic bacteria are mainly dry-cured sausages because in their elaboration process a fermentation stage with microorganism (starter cultures) takes place. Microorganisms, usually used as starter cultures, are lactic acid bacteria (LAB), enterococcus, moulds, and yeast. Many species are used as probiotics, mainly LAB but also *Bacillus* spp., and fungi as *Saccharomyces* spp., and *Aspergillus* spp. The most common probiotics belong to the genera *Lactobacillus* and *Bifidobacterium* (Sendra 2008). The use of probiotic starter cultures for dry-cured sausages production is one of the actions technically possible to produce probiotic meat products (Erkkilä and Petäjä 2000). The first probiotic meat products were commercialized in Germany, in 1998, where a company introduced a salami containing *L. casei*, *L. acidophilus*, and *Bifidobacterium* spp.; and in Japan, in 1998, a company marketed a new range of meat spread product fermented with probiotic LAB (*L. rhamnosus* FERM-P-15120) (Arihara 2006). Several authors have studied the use of different probiotic LAB as starter cultures for fermented sausages (Bunte et al. 2000, Jahreis et al. 2002, Pennachia et al. 2006).

19.2.2.7 Other Health Promoting Nutraceuticals

Nutraceuticals are a group formed by different phytochemicals (resveratrol, citofalvan-3-ol, etc.) and plant extracts (aromatic herb extracts, grape seed, bearberry, olive leaf extracts, etc.) which provide medicinal or health benefits, including the prevention and treatment of diseases. Nutraceuticals are believed to modulate the etiology of many chronic diseases such as cancer (Gao et al. 2003), coronary heart disease (Donaldson 2004), diabetes (Maghrani et al. 2004), hypertension, and osteoporosis (ADA 2004). The present popularity of this type of compounds has prompted a surge of in vitro studies examining the protective properties of physiologically active components against oxygen-induced damage (Carpenter et al. 2006, 2007). Some of these compounds have been assayed in different meat and meat products, in which some of them have demonstrated antioxidant and antibacterial activities (Fernández-López et al. 2003, Fernández-López et al. 2005, Carpenter et al. 2007).

19.3 CONCLUDING REMARKS

Different studies related to functional foods highlight a strong interest in this new category of food within certain consumer sectors, as much for the beneficial effects on health as for other attributes related to convenience. Consumers are increasingly more familiar with these products and are convinced a healthy diet can lessen the risk of having an illness or lessen the symptoms. The diet–health relationship is acquiring great importance within the food market, which has been realized by agro-food suppliers and is generating opportunities for a large majority of agro-food companies (Sánchez and Barrena 2004). For this reason, the acceptance of functional food is increasingly greater and demand is growing for new products, which besides satisfying nutritional needs also provide benefits to health.

In industrialized countries, and especially in European ones, the market for such products has exhibited high rates of growth (Verbeke 2005). This reflects in part a broader trend toward the increased use of dietary supplements (Kwak and Jukes 2001), but also increasing scientific evidence of the links between certain functional ingredients (e.g., omega-3 fatty acids, probiotics, and docosahexaenoic acid) and enhanced general well-being and/or reductions in the risk of specific diseases (Diplock et al. 1999). The food-processing sector has responded with high rates of new-product development that present consumers with increasing choices regarding the vehicle through which functional ingredients are delivered, including pills and a wide variety of food products (Verbeke 2005, 2006, Herath et al. 2007, Henson et al. 2008).

A sizeable empirical research literature on consumer attitudes toward functional foods and nutraceuticals is becoming established (Cox et al. 2004, Lucknow and Delehunty 2004, Sorenson and Bogue 2005, Verbeke 2006). Some studies suggest that belief in the efficacy of functional ingredients play a critical role in determining the propensity to consume. However, there is also increasing evidence that, apart from perceived efficacy, the mode through which functional ingredients are delivered mitigates consumer acceptability (Delong et al. 2003, Wennstrom and Mellentin 2003), in part because of variation in perceptions of the ability to use such products effectively (Cox et al. 2004). Most of these studies have also shown that consumer acceptance of functional foods is far from being unconditional, with one of the main conditions for acceptance pertaining to taste, besides trustworthiness of health claims. Taste expectations and experiences have been reported as extremely critical factors when selecting foods from the functional foods category (Gilbert 2000, Verbeke 2005). These authors have stressed that consumers will hardly be willing to compromise on the taste of functional foods for eventual health benefits.

As previously mentioned, the increase in concern about the risks to health associated with meat products with a high fat and cholesterol level has led the industry to develop new products or to improve traditional products to make them healthier (Aleson-Carbonell et al. 2004, Fernández-López et al. 2004), and further studies are required to demonstrate the clear benefits of meat and meat components for human health. Along with the accumulation of scientific data, there is an urgent need to inform consumers of the exact physiological value of meat and meat products including novel functional meat products. Since food safety is another critical aspect of food quality, efforts should also be directed to ensure that new functional meat products are safe. Without proof of product safety, most consumers would hesitate to adopt new foods into their diet. From the consumer point of view, it is also very important for functional meat products to show almost the same sensorial characteristics as similar or traditional meat products. In this case, technological and sensorial characteristics of functional meat products must be clearly identified and understood. More information on the role of these compounds in cognition and any psychological effects of these functional components would be useful and helpful.

REFERENCES

ADA. 2004. Position of the American Dietetic Association: Functional foods. *J. Am. Diet. Assoc.* 104: 814–826.

Aleson-Carbonell, L., Fernández-López, J., Sayas-Barberá, M.E., Sendra, E., and Pérez-Alvarez, J.A. 2003. Utilization of lemon albedo in dry-cured sausages. *J. Food Sci.* 68: 1826–1830.

Aleson-Carbonell, L., Fernández-López, J., Sendra, E., Sayas-Barberá, E., and Pérez-Alvarez, J.A. 2004. Quality characteristics of a non-fermented dry-cured sausage formulated with lemon albedo. *J. Sci. Food Agric.* 84: 2077–2084.

Andersen, L. June 1998. Pre- and probiotics—Sausage science. *Functional Foods. June,* 26–29.

Anon. 2002. Uptake 80 functional pea fiber. *Technical Information.* Canada: Parris and Heimbecker, Limited.

Ansorena, D. and Astiasarán, I. 2004. Effect to storage and packing on fatty acid composition and oxidation in dry fermented sausages made with added olive oil and antioxidants. *Meat Sci.* 67: 237–244.

Appel, L.J., Brands, M.W., Daniels, S.R., Naranja, N., Elmer, P., and Sacks, F.M. 2006. Dietary approaches to prevent and treat hypertension: A scientific statement from the American Heart Association. *Hypertension* 47: 296–308.

Arihara, K. 2006. Strategies for designing novel functional meat products. *Meat Sci.* 74: 219–229.

Arihara, K., Nakashima, Y., Mukai, T., Ishikawa, S., and Itoh, M. 2001. Peptide inhibitors for angiotensin I-converting enzyme from enzymatic hydrolysates of porcine skeletal muscle proteins. *Meat Sci.* 57: 319–324.

Arihara, K., Tomita, K., Ishikawa, S., Itoh, M., Akimoto, M., and Sameshima, T. 2005. Anti-fatigue peptides derived from edible meat proteins. Japan Patent 2007-045794, filed August 12, 2005, and issued February 22, 2007.

Arihara, K., Ishikawa, S., and Ito, M. 2006. Bifidobacteria proliferation-promoting peptide. Japan Paten 2007-189914, filed January 17, 2006, and issued August 02, 2007.

Asp, N.G. 1987. Dietary fibre. Definition, chemistry and analytical determination. *Mol. Aspects Medic.* 9: 17–29.

Augustin, M.A. and Sanguansri, L. 2003. Polyunsaturated fatty acids: Delivery, innovations and incorporation into foods. *Food Aust.* 55: 294–296.

Ayo, J., Carballo, J., Solas, M.T., and Jiménez-Colmenero, F. 2005. High pressure processing of meat batters with added walnuts. *Int. J. Food Sci. Technol.* 40: 47–51.

Ayo, J., Carballo, J., Solas, M.T., and Jiménez-Colmenero, F. 2008. Physicochemical and sensory properties of healthier frankfurters as affected by walnut and fat content. *Food Chem.* 107: 1547–1552.

Backers, T. and Noll, B. 2001. Safe plant based ingredients for meat processing: Dietary fibers + lupine protein. *Int. Food Market. Technol.* 15: 12–15.

Bauer, F. and Honikel, K.O. 2007. Meat—A food of high nutrient density. *Fleischw. Int.* 22: 39–42.

Bloukas, J.G. and Paneras, E.D. 1993. Substituting olive oil for pork back fat affects quality of low-fat frankfurters. *J. Food Sci.* 58: 705–709.

Bloukas, J.G. and Paneras, E.D. 1994. Vegetable oils replace pork backfat for low-fat frankfurters. *J. Food Sci.* 59: 725–733.

Bloukas, J.G., Paneras, E.D., and Fournitzis, G.C. 1997a. Sodium lactate and protective culture effects on quality characteristics and shelf-life of low-fat frankfurters produced with olive oil. *Meat Sci.* 45: 223–228.

Bloukas, J.G., Paneras, E.D., and Fournitzis, G.C. 1997b. Effect of replacing pork backfat with olive oil on processing and quality characteristics of fermented sausages. *Meat Sci.* 45: 133–144.

Boyle, E.A.E., Addis, P.B., and Epley, R.J. 1994. Calcium fortified, reduced fat beef emulsion product. *J. Food Sci.* 59: 928–932.

Bollinger, H. 1996. Wheat fibre-A new generation of dietary fibres. *Food Tech. Eur.* 3: 35–38.

Bunte, C., Hertel, C., and Hammes, W.P. 2000. Monitoring and survival of *Lactobacillus paracasei* LTH 2579 in food and the human intestinal tract. *Syst. Appl. Microbiol.* 23: 260–266.

Cáceres, E., García, M.L., and Selgas, M.D. 2006. Design of a new cooked meat sausage enriched with calcium. *Meat Sci.* 73: 368–377.

California Prune Board's European Office. 2000. Not a prune, more a dried plum. *Int. Food Ing.* 6: 37–38.

Carpenter, R., O'Callaghan, Y.C., O'Grady, M.N., Kerry, J.P., and O'Brien, N.M. 2006. Modulatory effects of resveratrol, citroflavan-3-ol and plant derived extracts on oxidative stress in U937 cells. *J. Med. Foods* 9: 187–195.

Carpenter, R., O'Grady, M.N., O'Callaghan, Y.C., O'Brien, N.M., and Kerry, J.P. 2007. Evaluation of the antioxidant potential of grape seed and bearberry extracts in raw and cooked pork. *Meat Sci.* 76: 604–610.

Chevalier, O. and Galvin, P.A. 1995. Use of inner pea fiber in frozen hamburger. Paper presented at the Annual Meeting of the Institute of Food Technologists. Chicago IL.

Chizzolini, R., Zanardi, E., Dorigoni, V., and Ghidini, S. 1999. Calorific value and cholesterol content of normal and low-fat meat and meat products. *Trends Food Sci. Technol.* 10: 119–128. Cofrades, S., Guerra, M.A., Carballo, J., Fernández-Martín, F., and Jimenez-Colmenero F. 2000. Plasma protein and soy fiber content effect on bologna sausage properties as influenced by fat level. *J. Food Sci.* 65: 281–287.

Cofrades, S., Serrano, A., Ayo, J., Solas, M.T., Carballo, J., and Jiménez-Colmenero, F. 2004. Restructured beef with different proportions of walnut as affected by meat particle size. *Eur. Food Res. Technol.* 18: 230–238.

Collins, J.E. 1997. Advances in meat research. In *Production and Processing of Healthy Meat, Poultry and Fish Products*, eds. A.M. Pearson and T.R. Dutson, pp. 283–301. London, U.K.: Blackie Academic & Professional.

Connor, W.E. 2000. Importance of n-3 fatty acids in health and disease. *Am. J. Clin. Nutr.* 71: 171S-175S.

Cox, D.N., Koster, A., and Russell, C.G. 2004. Predicting intentions to consume functional foods and supplements to offset memory loss using an adaptation of protection motivation theory. *Appetite* 33: 55–64.

Cummings, J.H. 2000. Nutritional management of diseases of the gut. In *Human Nutrition and Dietetics*, eds. J.S. Garrow, W.P.T. James, and A. Ralph, pp. 547–573. Edinburgh, U.K.: Churchill Livingston.

Dawkins, N.L., Phelps, O., McMillin, K.W., and Forrester, I.T. 1999. Composition and physicochemical properties of Chevon patties containing oat bran. *J. Food Sci.* 64: 597–600.

Delong, N., Ocke, M.C., Branderhorst, H.A.C., and Friele, R. 2003. Demographic and lifestyle characteristics of functional food consumers and dietary supplement users. *Brit. J. Nutr.* 89: 55–64.

Demos, B.P., Forrest, J.C., Grant, A.L., Judge, M.D., and Chen, L.F. 1994. Chen, low-fat, no added salt in restructured beef steaks with various binders, *J. Muscle Foods* 5: 407–418.

Desmond, E. 2006. Reducing salt: A challenge for the meat industry. *Meat Sci.* 74: 188–196.

Desmond, E., Troy, D.J., and Bucley, J. 1998. Comparative studies on non-meat ingredients used in the manufacture of low-fat burgers. *J. Muscle Foods* 9: 221–224.

Devatkal, S. and Mendiratta, S.K. 2001. Use of calcium lactate with salt-phosphate and alginate calcium gels in restructured pork rolls. *Meat Sci.* 58: 371–379.

Diplock, A.T., Aggett, P.J., Ashwell, M., Bornet, F., Fern, E.B., and Roberfroid, M.B. 1999. Scientific concepts of functional foods in Europe: Consensus document. *Brit. J. Nutr.* 81: S1–S27.

Dolata, W., Piotrowska, E., Makała, H., Krzywdzińska-Bartkowiak, M., and Olkiewicz, M. 2002. The effect of partial replacement of fat with the potato fiber preparation on the quality of model batters and finely comminuted meat products. *Acta Sci. Pol., Technol. Aliment.* 1: 5–12.

Donaldson, M.S. 2004. Nutrition and cancer: A review of the evidence for an anti-cancer diet. *Nutrition* 3: 19.

Eastwood, M.A. 1987. Dietary fibre and risk of cancer. *Nutr. Rev.* 7: 193.

Elmore, J.S., Cooper, S.L., Enser, M., Mottram, D.S., Sinclair, L.A., and Wildinsion, R.G. 2005. Dietary manipulation of fatty acid composition in lamb meat and its effect on the volatile aroma compounds of grilled lamb. *Meat Sci.* 69: 233–242.

Erkkilä, S. and Petäjä, E. 2000. Screening of commercial meat starter cultures at low pH and in the presence of bile salts for potential probiotic use. *Meat Sci.* 55: 297–300.

Fairé R. and Frasquet, C. 1999. Calcium, phosphorous and magnesium. In *Nutrition Handbook*, eds. M. Hernández and A. Sastre, pp. 217–229. Madrid, Spain: Díaz de Santos (in Spanish).

Farnworth, E.R. 2003. *Handbook of Fermented Functional Foods*. Boca Raton, FL: CRC Press.

Fernández-Ginés, J.M., Fernández-López, J., Sayas-Barberá M.E., Sendra, E., and Pérez-Alvarez, J.A. 2003. Effects of storage conditions on quality characteristics of bologna sausages made with citrus fiber. *J. Food Sci.* 68: 710–715.

Fernández-Ginés, J.M., Fernández-López, J., Sayas-Barberá M.E., Sendra, E., and Pérez-Alvarez, J.A. 2004. Lemon albedo as a new source of dietary fiber: Application to bologna sausages. *Meat Sci.* 67: 7–13.

Fernández-Ginés, J.M., Fernández-López, J., Sayas-Barberá, E., and Pérez-Alvarez, J.A. 2005. Meat products as functional foods: A review. *J. Food Sci.* 70: R37–R43.

Fernández-López, J., Sevilla, L., Sayas, E., Navarro, C., Marín, F., and Pérez-Alvarez, J.A. 2003. Evaluation of the antioxidant potential of hyssop (*Hyssopus officinalis,* L.) and rosemary (*Rosmarinus officinalis* L.) extracts in cooked pork meat. *J. Food Sci.* 68: 660–664.

Fernández-López, J., Fernández-Ginés, J.M., Aleson-Carbonell, L., Sendra, E., Sayas, E., and Pérez-Alvarez, J.A. 2004. Application of functional citrus by-products to meat products. *Trends Food Sci. Technol.* 15: 176–185.

Fernández-López, J., Zhi, N., Aleson-Carbonell, L., Pérez-Alvarez, J.A., and Kuri, V. 2005. Antioxidant and antibacterial activities of natural extracts: Application in beef meatballs. *Meat Sci.* 69: 371–380.

Fernández-López, J., Sendra, E., Sayas, E., Navarro, C., and Pérez-Alvarez, J.A. 2006. Use of citric fiber in the development of functional meat products. *Alimentación, Equiposy Tecnología.* 25: 63–69 (in Spanish).

Fernández-López, J., Sendra, E., Sayas, E., Navarro, C., and Pérez-Alvarez, J.A. 2008. Physicochemical and microbiological profile of "salchichón" (Spanish dry-fermented sausage) enriched with orange fiber, *Meat Sci.* 80: 410–417.

Flores, M., Nieto, P., Ferrer, J.N., and Flores, J. 2005. Effect of calcium chloride on the volatile pattern and sensory acceptance of dry-fermented sausages. *Eur. Food Res. Technol.* 221: 624–630.

Fraser, G.E., Sabaté, J., Beeson, W.L., and Strahan, T.M. 1992. A possible protective effect of nut consumption on risk of coronary heart disease. *Arch. Intern. Med.* 152: 1416–1424.

Fuchs, C.S., Giovannucci, E.L., Colditz, G.A., et al. 1999. Dietary fiber and the risk of colorectal cancer and adenoma in women. *Engl. J. Med.* 340: 169–176.

Fujita, H., Yokoyama, K., and Yoshikawa, M. 2000. Classification and antihypertensive activity of angiotensin I-converting enzyme inhibitory peptides derived from food proteins. *J. Food Sci.* 65: 564–569.

Gao, X., Deeb, D., Media, J., et al. 2003. Immunomodulatory activity of Resveratrol: Discrepant in vitro and in vivo immunological effects. *Biochem. Pharmacol.* 66: 2427–2435.

García, M.L., Domínguez, R., Galvez, M.D., Casas, C., and Selgas, M.D. 2002. Utilization of cereal and fruit fibres in low fat dry fermented sausages. *Meat Sci.* 60: 227–236.

García, M.L., Cáceres, E., and Selgas, M.D. 2004. Utilisation of fruit fibres in conventional and reduced fat cooked meat sausages. *Meat Sci.* 90: 121–129.

Giese, J. 1992. Developing low-fat meat products. *Food Technol.* 46: 100–108.

Giese, J. 1996. Fats, oils and fat replacers. *Food Technol.* 50: 78–83.

Gilbert, L. 2000. The functional food trend: What's next and what Americans think about eggs. *J. Am. College Nutr.* 19: 507S–512S.

Gimeno, O., Astiasarán, I., and Bello, J. 1998. A mixture of potassium, magnesium and calcium chlorides as a partial replacement of sodium chloride in dry-fermented sausages *J. Agric. Food Chem.* 46: 4372–4375.

Gimeno, O., Astiasarán, I., and Bello, J. 2001.Calcium ascorbate as potential parcial substitute for sodium chloride in dry-fermented sausages: Effect on colour, texture and hygienic quality at different concentrations. *Meat Sci.* 57: 23–29.

Gorinstein, S., Martín-Belloso, O., Park, Y.S., et al. 2001. Comparison of some biochemical characteristics of different citrus fruits. *Food Chem.* 74, 309–315.

Griguelmo-Miguel, N. and Martín-Belloso, O. 1999. Comparison of dietary fibre from by-products of processing fruits and greens and from cereals. *Lebens.- Wissens. Technol.* 32, 503–508.

Guardia, M.D., Guerrero, L., Gelabert, J., Gou, P., and Arnau, J. 2006. Consumer attitude towards sodium reduction in meat products and acceptability of fermented sausages with reduced sodium content. *Meat Sci.* 73: 484–490.

Hallfrisch, J., Scholfield, D.J., and Behall, K.M. 1995. Diets containing soluble oat extracts improve glucose and insulin responses of moderately hypercholesterolemic men and women. *Am. J. Clin. Nutr.* 61: 379–384.

Henson, S., Masakure, O., amd Cranfield, J. 2008. The propensity for consumers to offset health risks through the use of functional foods and nutraceuticals: The case of lycopene. *Food Qual. Pref.* 19: 395–406.

Herath, D., Cranfield, J.L., and Henson, S. 2007. Health claims as credence attributes: Evidence from functional foods and nutraceuticals sector in Canada. In *International Food Economy Group*, Department of Food, Agricultural and Resource Economics, University of Guelph, Guelph, Ontario, Canada.

Hutton, T. 2002. Sodium technological functions of salt in the manufacturing of food and drink products. *Brit. Food J.* 104: 126–152.

Iyengar, R. and Gross, A. 1991. Fat substitutes. In *Biotechnology and Food Ingredients*, eds. I. Goldberg and R. Williams, pp. 287–313. New York: Van Nostrand Reinhold.

Jahreis, G., Vogelsang, H., Kiessling, G., Schubert, R., Bunte, C., and Hammes, W.P. 2002. Influence of probiotic sausage (*Lactobacillus paracasei*) on blood lipids and inmunological parameters of healthy volunteers. *Food Res. Int.* 35: 133–138.

Jamora, J.J. and Rhee, K.S. 2002. Storage stability of extruded products from blends of meat and nonmeat ingredients: Evaluation methods and antioxidative effects of onion, carrot and oat ingredients. *J. Food Sci.* 67: 1654–1659.

Jiménez-Colmenero, F. 1996. Technologies for developing low-fat meat products. *Trends Foods Sci. Technol.* 7: 41–48.

Jiménez-Colmenero, F., Carballo, J., and Cofrades, S. 2001. Healthier meat and meat products: Their role as functional foods. *Meat Sci.* 59: 5–13.

Jiménez-Colmenero, F., Serrano, A., Ayo, J. Solas, M.T., Cofrades, S., and Carballo, J. 2003. Physicochemical and sensory characteristics of restructured beef steak with added walnuts. *Meat Sci.* 65: 1391–1397.

Jiménez-Colmenero, F., Ayo, J., and Carballo, J. 2005. Physicochemical properties of low sodium frankfurter with added walnut: Effect of transglutaminase combined with caseinate, KCl and dietary fibre as salt replacers. *Meat Sci.* 69: 781–789.

Kayaardi, S. and Gök, V. 2003. Effect of replacing beef fat with olive oil on quality characteristics of Turkish soudjouk (sucuk). *Meat Sci.* 66: 249–257.

Keeton, J.T. 1994. Low-fat meat products-technological problems with processing. *Meat Sci.* 36: 261–276.

Kim, J.S., Godber, J.S., and Prinaywiwatkul, W. 2000. Restructured beef roasts containing rice bran oil and fiber influences cholesterol oxidation and nutritional profile. *J. Muscle Food* 11: 111–127.

Korstanje, R. and Hoek, M. 2001. Calcium and others mineral. In *Guide to Functional Foods Ingredients*, ed. J. Young, pp. 197–210. London, U.K.: LFRA Ltd.

Kuraishi, Ch., Sakamoto, J., Yamazaki, K., Susa, Y., Kuhara, Ch., and Soeda, T. 1997. Production of restructured meat using microbial transglutaminase without salt or cooking. *J. Food Sci.* 62: 488.

Kwak, N.S. and Jukes, D.J. 2001. Functional foods part 1: The development of a regulatory concept. *Food Control* 12: 99–107.

Lin, G.C., Mittal, G.S., and Barbut, S. 1991. Optimization of tumbling and KCl substitution in low sodium restructured hams. *J. Muscle Foods* 2: 71–91.

Lucknow, T. and Delehunty, C. 2004. Consumer acceptance of orange juice containing functional ingredients. *Food Res. Int.* 37: 805–814.

Lurueña-Martinez, M.A., Vivar-Quintana, A.M., and Revilla, I. 2004. Effect of locust bean/xanthan gum addition and replacement of pork fat with olive oil on the quality characteristics of low-fat frankfurters. *Meat Sci.* 68: 383–389.

Maghrani, M., Lemhadri, A., Zeggwagh, N.A., et al. 2004. Effects of an aqueous extract of Triticum repens on lipid metabolism in normal and recent-onset diabetic rats. *J. Ethnopharma.* 90: 331–337.

Marquez, E.J., Ahmed, E.M., West, R.L., and Johnson, D.D. 1989. Emulsion stability and sensory quality of beef frankfurters produced at different fat and peanut oil levels. *J. Food Sci.* 54: 867–870,873.

Mendoza, E., García, M.L., Casas, C., and Selgas, M.D. 2001. Inulin as fat substitute in low fat, dry fermented sausages. *Meat Sci.* 57: 387–393.

Mine, Y. and Shahidi, F. 2005. *Nutraceutica Proteins and Peptides in Health and Disease.* Boca Raton, FL: CRC Press.

Muguerza, E., Gimeno, O., Ansorena, D., Bloukas, J.G., and Astiasarán, I. 2001. Effect of replacing pork backfat with pre-emulsified olive oil on lipid fraction and sensory quality of Chorizo de Pamplona—A traditional Spanish fermented sausage. *Meat Sci.* 59: 251–258.

Muguerza, E., Fista, G., Ansorena, D., Astiasarán, I., and Bloukas, J.G. 2002. Effect of fat level and partial replacement of pork backfat with olive oil on processing and quality characteristics of fermented sausages. *Meat Sci.* 61: 397–404.

Muguerza, E., Ansorena, D., Bloukas, J.G., and Astiasarán, I. 2003. Effect of fat level and partial replacement of pork backfat with olive oil on the lipid oxidation and volatile compounds of Greek dry fermented sausages. *J. Food Sci.* 68: 1531–1536.

Noort, M. 2001. Peas as functional ingredients in meat products. *Voedingsmiddelentechnologie* 34: 23–27.

O'Keefe, S.F., Proudfoot, F.G., and Ackman, R.G. 1995. Lipid oxidation in meats of omega-3 fatty acid-enriched broiler chickens. *Food Res. Int.* 28: 417–424.

Ovensen, L. 2004. Cardiovascular and obesity health concerns. In *Encyclopaedia of Meat Science*, eds. W. K. Jensen, C. Devine, and M. Dikeman, pp. 623–628. Oxford, U.K.: Elsevier.

Pacheco-Delahaye, E. and Vivas-Lovera, N. 2003. Effect of defatted corn germ flour and rice bran on some chemical, physical and sensory properties of sausages. *Acta Cient. Venez.* 54: 274–283.

Paneras, E.D., Bloukas, J.G., and Filis, D.G. 1998. Production of low-fat frankfurters with vegetables oils following the dietary guidelines for fatty acids. *J. Muscle Foods* 9: 111–126.

Pappa, I.C., Bloukas, J.G., and Arvanitoyannis, I.S. 2000. Optimization of salt, olive oil and pectin level for low-fat frankfurters produced by replacing pork backfat with olive oil. *Meat Sci.* 56: 81–88.

Park, J., Rhee, K.S., Keeton, J.T., and Rhee, K.C. 1989. Properties of low–fat frankfurters containing monounsaturated and omega-3 polyunsaturated oils. *J. Food Sci.* 54: 500–504.

Pascal, G. and Collet-Ribbing, C. 1998. European perspectives on functional foods. *IPTS Rep.* 24: 1–7. (In Spanish).

Pennachia, C., Vaughan, E.E., and Villani, F. 2006. Potential probiotic *Lactobacillus* strains from fermented sausages: Further investigation on their probiotic properties. *Meat Sci.* 73: 90–101.

Pihlanto, A. and Korhonen, H. 2003. Bioactive peptides and proteins. *Adv. Food Nutr. Res.* 47: 175–276.

Prakongpan, T., Nitithamyong, A., and Luangpituksa, P. 2002. Extraction and application of dietary fiber and cellulose from pineapple cores. *J. Food Sci.* 67: 1308–1313.

Rhee, K.S., Krahl, L.M., Lucía, L.M., and Acuff, G.R. 1997. Antioxidative/antimicrobial effects and TBARS in aerobically refrigerated beef as related to microbial growth. *J. Food Sci.* 62: 1205–1210.

Romans, J.R., Costello, W. J., Carlson, C.W., Greaser, M.L., and Jones, K.W. 1994. *The Meat We Eat.* Danville, IL: Interstate Publisher, Inc.

Rosmini, M.R., Frizzo, L., and Zogbi, A. 2008. Meat products with low sodium content: Processing and properties. In *Technological Strategies for Functional Meat Products Development*, ed. J. Fernández-López, pp. 87–108. Kerala, India: Transworld Research Network.

Ruíz-Capillas, C., Cofrades, S., Serrano, A., and Jiménez-Colmenero, F. 2004. Biogenic amines in restructured beef steaks as affected by added walnuts and cold storage. *J. Food Prot.* 67: 607–612.

Ruusunen, M., Tirkkonem, M.S., and Puolanne, E. 2001. Saltiness of coarsely ground cooked ham with reduced salt content. *Agr. Food Sci. Finland* 10: 27–32.

Ruusunen, M., Niemistö, M., and Puolanne, E. 2002. Sodium reduction in cooked meat products by using commercial potassium phosphate mixture. *Agr. Food Sci. Finland* 11: 199–207.

Ruusunen, M. and Puolanne, E. 2005. Reducing sodium intake from meat products. *Meat Sci.* 70: 531–541.

Ruxton, C.H.S., Calder, P.C., Reed, S.C., and Simpson, M.J.A. 2005. The impact of long-chain n-3 polyunsaturated fatty acids on human health. *Nutr. Res. Rev.* 18: 113–129.

Sabaté, J. 1993. Does nut consumption protect against ischaemic heart disease? *Eur. J. Clin. Nutr.* 47: S71–S75.

Saiga, A., Okumura, T., Makihara, T., et al. 2003a. Angiotensin I-converting enzyme inhibitory peptides in a hydrolyzed chicken breast muscle extract. *J. Agric. Food Chem.* 51: 1741–1745.

Saiga, A., Tanabe, S., and Nishimura, T. 2003b. Antioxidant activity of peptides obtained from porcine myofibrillar proteins by protease treatment. *J. Agric. Food Chem.* 51: 3661–3667.

Salazar, M.P., García, M.L., and Selgas, M.D. 2005. Meat products enriched with calcium. Paper presented at the *3rd National congress of food science and technology*, Burgos, Spain (in Spanish).

Sánchez, M. and Barrena, R. 2004. The consumer and the novel foods: Functional foods and transgenic foods. *Revista española de estudios agrosociales y agropesqueros* 204: 25–127 (in Spanish).

Sánchez-Escalante, A., Torrescano, G., Camou, J.P., Ballesteros, M.N., and González-Mendez, N.F. 2000. Utilization of applesauce in a low-fat bologna-type product. *Food Sci. Technol. Int.* 6: 379–386.

Sayas, E. 2008. The use of olive oil in processing meat products. In *Technological Strategies for Functional Meat Products Development*, ed. J. Fernández-López, pp. 139–162. Kerala, India: Transworld Research Network.

Selgas, M.D. and García, M.L. 2008. Meat products with calcium. In *Technological Strategies for Functional Meat Products Development*, ed. J. Fernández-López, pp. 57–85. Kerala, India: Transworld Research Network.

Selmer, R., Kristiansen, I.S., Haglerod, A., et al. 2000. Cost and health consequences of reducing the population intake of salt. *J. Epidemiol. Commun. Health* 54: 697–702.

Sendra, E. 2008. Application of prebiotics and probiotics in meat products. In *Technological Strategies for Functional Meat Products Development*, ed. J. Fernández-López, pp. 117–137. Kerala, India: Transworld Research Network.

Serrano, A., Cofrades, S., Ruiz-Capillas, C., Olmedilla-Alonso, B., Herrero-Barbudo, C., and Jiménez-Colmenero, F. 2005. Nutritional profile of restructured beef steak with added walnuts, *Meat Sci.* 70: 647–654.

Severino, C., De Pelli, T., and Baiano, A. 2003. Partial substitution of pork backfat with extra-virgin olive oil in "salami" products: Effects on chemical, physical and sensorial quality. *Meat Sci.* 64: 323–331.

Sorenson, D. and Bogue, J. 2005. Market-oriented new product design of functional orange juice beverages: A qualitative approach. *J. Food Products Mark.* 11: 57–73.

Stanton, C., Desmond, C., Coackley, M., Collins, J.K., Fitzgerald, G., and Ross, P. 2003. Challenges facing development of probiotic-containing functional foods. In *Handbook of Fermented Functional Foods*. ed. E.R. Farnworth, pp. 27–58. Boca Raton, FL: CRC Press.

Terrel, R.N., Quintanilla, M., Vanderzant, C., and Gardner, F.A. 1983. Effect of reduction or replacement of sodium chloride on growth of *Micrococcus, Moraxella* and *Lactobacillus* inoculated ground pork. *J. Food Sci.* 48: 122–124.

Thebaudin, J.Y., Lefebvre, A.C., Harrington, M., and Bourgeois, C.M. 1997. Dietary fibers: Nutritional and technological interest. *Trends Food Sci. Technol.* 8: 41–48.

Tsai, S.J., Unklesbay, N., Unklesbay, K., and Clarke, A. 1998. Thermal properties of restructured beef products at different isotherm temperatures. *J. Food Sci.* 63: 481–484.

Turhan, S., Sager, I., and Sule-Ustun, N. 2005. Utilization of hazelnut pellicle in low-fat beef burgers. *Meat Sci.* 71: 312–319.

Tyburcy, A., Cwiek, J., and Adamczak, L. 2003. The effect of wheat fibre, transglutaminase, and a mincing method of raw meat material on properties of salami-imitating sausages. *Zywnosc* 10: 72–80.

Valencia, I., Ansorena, D., and Astiasarán, I. 2007. Development of dry fermented sausages rich in docosahexaenoic acid with oil from the microalgae *Schixochytrium* sp.: Influence on nutritional properties, sensorial quality and oxidation stability. *Food Chem.* 104: 1087–1096.

Valsta, L.M., Tapanainen, H., and Männistö, S. 2005. Meat fats in nutrition. *Meat Sci.* 70: 525–530.

Verbeke, W. 2005. Consumer acceptance of functional foods: Sociodemographic, cognitive and attitudinal determinants. *Food Qual. Prefer.* 16: 45–47.

Verbeke, W. 2006. Functional foods: Consumer willingness to compromise on taste for health? *Food Qual. Prefer.* 17: 126–131.

Vermeirssen, V., Camp, J.V., and Verstraete, W. 2004. Bioavailability of angiotensin I converting enzyme inhibitory peptide. *Br. J. Nutr.* 92: 357–366.

Ward, O.P. and Singh, A. 2005. Omega-3/6 fatty acids: alternative sources of production. *Process Biochem.* 40: 3267–3652.

WHO. 2003. Diet, nutrition and the prevention of chronic diseases. In *Technical Report Series* 916, pp. 10–36. Geneva, Switzerland: World Health Organization.

Wilkinson, A.L., Sun, Q., Senecal, A., and Faustman, C. 2001. Antioxidants effects on TBARS and fluorescende measurements in freeze-dried meats. *J. Food Sci.* 66: 20–24.

Wennstrom, P. and Mellentin, J. 2003. *The Food and Health Marketing Handbook*. London, U.K.: New Nutrition Business.

Wu, S.Y. and Brewer, M.S. 1994. Soy protein isolate antioxidant effect on lipid peroxidation of ground beef and microsomal lipids. *J. Food Sci.* 59: 702–706.

Xiong, Y.L., Noel, D.C., and Moody, W.G. 1999. Textural and sensory properties of low-fat beef sausages with added water and polysaccharides as affected by pH and salt. *J. Food Sci.* 64: 550–554.

Zia-ur-Rehman, Islam M., and Shah W.H. 2003. Effect of microwave and conventional cooking on insoluble dietary fibre components of vegetables. *Food Chem.* 80: 237–240.

20 Probiotics and Prebiotics in Fermented Dairy Products

Gabriel Vinderola, Clara González de los Reyes-Gavilán, and Jorge Reinheimer

CONTENTS

20.1 INTRODUCTION TO PROBIOTICS: DEFINITION AND MICROORGANISMS USED

The microbiota of the large intestine are thought to account for 95% of the total cells in the human body, to have a population of approximately 10^{11}–10^{12} CFU/g of intestinal contents, and to represent approximately 10^{12} cells/g dry weight feces. A lot of metabolic activity is carried out by the intestinal microbiota to support the constant elimination through feces. The intestine of the human baby is sterile at birth; the composition of the intestinal microbiota is relatively simple in infants but becomes more complex in adults. The intestinal microbiota is thought to derive from the microbiota of the mother (vagina, skin, and breast milk) and from the environment. With regard to bacterial species and strains, there is a high degree of variability among human subjects, and this variability depends on age, diet, immune status, stress factors, and many other factors not yet completely known. Indeed, the normal intestinal microbiota is as yet an unexplored organ of host defense (Isolauri et al. 2004). The main genera or species found within the intestinal microbiota include *Bacteroides*, *Eubacterium*, *Ruminococcus*, *Clostridium*, and *Bifidobacterium*, and, as subdominant microbiota, *Escherichia coli*, *Veillonella*, *Staphylococcus*, *Proteus*, *Streptococcus*, and *Lactobacillus* (Tannock 2003).

The metabolic activities of the intestinal microbiota are very diverse: for example, the formation of short-chain fatty acids that act as carbon source for colonocytes, the synthesis of vitamins, and the activation or inactivation of bioactive food components. Its function is crucial for the development of the host gut mucosal immune system, for the anatomy of the intestine (Moreau and Gaboriau-Routhiau 2000), and it also has many implications in health and disease (Blaut and Clavel 2007). In general, it is possible to categorize the gut microbiota components on the basis of whether they exert potentially pathogenic activities, a mixture of potentially pathogenic and health-promoting activities, or exclusively potentially health-promoting aspects (Gibson et al. 2003). Species from the genera *Bifidobacterium* and *Lactobacillus*, commonly found among the human intestinal microbiota in healthy subjects, have a long tradition of being considered health-promoting bacteria, and it is from these genera that the most studied probiotic bacteria come from.

The term probiotic was first introduced by Dr. Roy Fuller in 1989, to refer to "a live microbial feed supplement which beneficially affects the host animal by improving its intestinal microbial balance" (Fuller 1989). The large amount of animal and human studies carried out in the last two decades concerning the health-promoting capacities of probiotic bacteria, beyond their capacity to improve the intestinal microbial balance, has led to a reformulation of this definition. One of the most widely used definitions of probiotic bacteria is the one adopted by the joint FAO/WHO working group (FAO/WHO 2002), which states that probiotics are "live microorganisms which when administered in adequate amounts confer a health benefit on the host." This joint commission also released guidelines for the evaluation of probiotics in foods.

Species from which strains have been isolated, characterized and proposed as probiotics are *Lactobacillus casei*, *L. paracasei*, *L. rhamnosus*, *L. acidophilus*, *L. gasseri*, *L. johnsonii*, *L. plantarum*, *L. reuteri*, *L. crispatus*, *L. fermentum*, *Bifidobacterium bifidum*, *B. adolescentis*, *B. lactis*, *B. breve*, *B. infantis*, *B. longum*, *Saccharomyces boulardii*, *S. cerevisiae*, *Enterococcus faecium*, and the safe strain *E. coli* Nissle. The role of enterococci in disease still raises questions on their safety for use in foods or as probiotics (Franz et al. 2003, Kayser 2003, Rinkinen et al. 2003). The definition of "probiotics" requires that the term only be applied to live microbes having a substantiated beneficial effect. However, microbes administered alive are considered probiotics regardless of their ability to survive intestinal transit. As such, strains or species that do not survive intestinal transit, such as *Streptococcus thermophilus*, *L. delbrueckii* subsp. *bulgaricus*, or *Lactococcus lactis* (*Lc. lactis*) might be considered probiotics (Reid et al. 2003a, Guarner et al. 2005). It is important to state that no new isolate of a strain, belonging to one of the species listed earlier, should be considered a "probiotic" until animal studies and double-blind, randomized, placebo-controlled (DBPC) phase 2 human trials or other appropriate designs have been carried out to determine if the strain is safe and efficacious.

The facts that species from the genera *Bifidobacterium* and *Lactobacillus* can be found as members of the human intestinal microbiota, that some strains have a satisfactory tolerance to gastrointestinal conditions, and that only health-promoting effects have been ascribed to them, make these genera the ones of choice for the isolation, characterization, and commercialization of probiotic strains. The strategy would be then to reinforce the intestinal microbiota, at least as a transient intestinal colonization, with a health-promoting candidate. Beyond the human intestinal microbiota, potential strains for probiotic use, mainly from the genus *Lactobacillus*, can be isolated from fermented foods such as traditionally fermented milks, cheeses, sausages, or cereal-based fermented foods (Farnworth 2008), since bifidobacteria poorly tolerate oxygen and do not naturally occur in those food products. One might think that probiotic candidates isolated from the intestinal microbiota of the host in which the probiotic will be used would perform better on this host because it is ecologically better adapted to the intestinal environment; this is known as the species-specificity criterion. However, doubts arise about the capacity of those intestinal isolates to stand the harsh conditions frequently imposed by the food that is used as carrier for the probiotic strain (acidity in fermented milks, dissolved oxygen, metabolites derived from the growth of other species in the food, chemicals, etc.). It has been shown that many commercial probiotic strains of intestinal origin perform well in acidic conditions found in fermented milks (Lourens-Hattingh and Viljoen 2001), and tolerate the chemical environment found in these products (Vinderola et al. 2002a). The same can be said about probiotic strains isolated from fermented products: they are naturally adapted to the food matrix from where they were isolated, but their performance in the gastrointestinal tract remains unknown. Haller et al. (2001) determined that metabolic and functional properties of intestinal lactobacilli were also found in certain bacteria of fermented food origin. Dogi and Perdigón (2006) reported that commensal and noncommensal strains of *L. fermentum* and *L. acidophilus* were both able to activate the gut mucosal immune response in mice. It seems then that the technological and in vivo performances of probiotic strains remain an intrinsic property of each isolate and results are strongly strain dependent.

20.2 HEALTH BENEFITS OF FERMENTED MILKS CONTAINING PROBIOTIC BACTERIA

Some of the health benefits of the consumption of probiotic bacteria or fermented dairy products containing them are the beneficial modulation of the gut microbiota activity by the reduction of the risk associated with mutagenicity and carcinogenicity, alleviation of lactose intolerance, reinforcement of gut mucosal immunity, acceleration of intestinal mobility, hypocholesterolemic effect, prevention or shortening of the duration of diarrheas, prevention of inflammatory bowel disease (IBD) and pouchitis, prevention of colon cancer, inhibition of *Helicobacter pylori* and intestinal pathogens, and treatment and prevention of allergy, among many others. A detailed discussion of these health benefits, the strains involved, and the mechanisms of action is far beyond the scope of this chapter in the context of this book. However, the interested reader is strongly advised to read some reviews dealing in depth with these aspects of probiotic bacteria (Reid et al. 2003b, Adolfsson et al. 2004, Saxelin et al. 2005, Senok et al. 2005, Shah 2007, de Vrese and Schrezenmeir 2007, Vasiljevic and Shah 2008) or books (Fuller and Perdigón 2003, Salminen et al. 2004, Goktepe et al. 2007, Farnworth 2008). We will limit ourselves to highlight that is unlikely that any probiotic strain might accomplish all the different health benefits attributed to probiotics, and that a particular effect must be associated with a defined strain.

The mechanisms of action for the majority of health benefits ascribed to probiotic bacteria are related to an adequate activation of the gut-associated immune response. For this, the specific strain, the food matrix used as a vehicle, the dose, and the period and way of administration (continuous or in a cyclical way) seem to be crucial for the achievement of the desired health benefit. It has been recently reported (de Moreno de LeBlanc et al. 2008) that a continuous long-term (98 days) administration of a fermented milk to mice, which contained a probiotic strain, induced to mice induced a peak in the gut immune response after 7 days of administration of the product. This response was evaluated as the number of IgA + cells in the small intestine lamina propria. During the subsequent consumption, the response diminished almost to control values, showing the importance of considering a continuous or cyclical administration for the achievement of a desired immune state, able to be effective in the prevention or therapeutic treatment of a pathological condition or an enteric infection.

20.3 WHY ADD PROBIOTIC BACTERIA TO TRADITIONAL YOGURTS?

The association between man and fermented milks stretches back thousands of years before the Christian era. The main geographical areas in which the traditional use of fermented milk dates back to ancient times include North and North-Eastern Europe, South-Eastern Europe including the Caucasus, Transcaucasia, and North-East Asia in the Soviet Union, the Near East and the Indian subcontinent (Kanbe 1992). The natural starter microbiota of these traditional fermented milks include *S. thermophilus*, *Lactococcus*, *L. delbrueckii* subsp. *bulgaricus*, *L. helveticus*, *L. acidophilus*, *L. kefir*, *L. kefiranofaciens*, and species from the *L. casei* group, among others.

There are no precise records available regarding the origin of yogurt. It is believed that the ancient Turkish people in Asia, where they lived as nomads, first made yogurt. One of the first industrial productions of yogurt in Europe was undertaken by Danone in 1922 in Madrid, Spain (Rasic and Kurmann 1978). The microorganisms involved from then on for commercial yogurt production were *S. thermophilus* and *L. delbrueckii* subsp. *bulgaricus*. After World War II, the technology of yogurt production, its consumption, and the understanding of its properties have increased rapidly (Prajapati and Nair 2003). However, the massive use of probiotic bacteria, belonging to the genera *Lactobacillus* (strains of the *L. casei* and *L. acidophilus* groups) and *Bifidobacterium*, and the development of the great variety of fermented milks with added probiotic bacteria, available nowadays in the market, took place from the early 1980s (Stanton et al. 2003).

Yogurt has long been recognized as a product with many desirable health-promoting effects for consumers (Adolfsson et al. 2004). What are the facts then that justify the addition of probiotic cultures to traditional yogurts, for which health benefits are already recognized?

Microorganisms used for yogurt production, *S. thermophilus* and *L. delbrueckii* subsp. *bulgaricus*, are not of intestinal origin and hardly survive the harsh conditions found during the gastrointestinal transit (Reid et al. 2003a). However, *Lactobacillus* and bifidobacteria used for probiotic purposes are members of the intestinal microbiota and show only positive effects on the host (Mitsuoka 1992). Certain probiotic strains present the capacity to withstand the low pH of the stomach and the inhibitory compounds found during the intestinal transit (bile salts, lysozyme, lactoferrin, defensins) (Russell et al. 2005, Lehrer et al. 2005). This confers on them the possibility to have desirable metabolic activity during their passage thorough the gut (Bron et al. 2004) and to interact with immune cells from the gut associated immune system (Vinderola et al. 2005a), launching desirable immune responses. Finally, numerous studies have demonstrated the importance of cell viability for the accomplishment of many health-promoting effects (Ouwehand and Salminen 1998, Galdeano and Perdigón 2004).

Although there is still debate about whether starter lactic acid bacteria (LAB) should be regarded as probiotics (Senok et al. 2005), many authors consider that these cultures clearly fulfill the current concept of probiotics (Guarner et al. 2005). It is fair then to acknowledge that traditional yogurts containing *S. thermophilus* and *L. delbrueckii* subsp. *bulgaricus* do exert many health-promoting effects (Adolfsson et al. 2004, Guarner et al. 2005) due to their content of lactase (released from LAB cells sensitive to the detergent action of bile salts, for example), the release of bioactive peptides from milk proteins (Vinderola et al. 2008), the production of soluble exopolysaccharides (Vinderola et al. 2006a) or antimicrobial components (Ouwehand and Vesterlund, 2004), and related to the release of cell-wall rests (Morata de Ambrosini et al. 1996, Tejada-Simon and Pestka 1999) when whole cells are disrupted by bile salts and lysozyme during gastrointestinal transit. However, the presence of probiotic bacteria able to survive the gastrointestinal transit will also assure many other health benefits for which cell viability and cell integrity in the intestinal lumen is required.

20.4 ENUMERATION OF PROBIOTICS IN FERMENTED PRODUCTS: CHALLENGES, HURDLES, AND ACHIEVEMENTS

The enumeration of probiotic bacteria in fermented food products is a challenging task in the microbiology laboratory. The close phylogenetic proximity between probiotic and starter LAB becomes one of the greatest hurdles at the moment of differentiating colonies of probiotic strains from those specifically taking part of the starter culture. However, the sustained efforts in this field in the last two decades have allowed the formulation of a great variety of culture media for enumerating probiotic bacteria, although the final use of a given medium will greatly depend on the particular combination of starter and probiotic strains and on the food to be analyzed.

The control of the level of viable cells of probiotic bacteria in pure cultures or in the final product is important for several reasons; it allows: (1) the quality control of frozen or lyophilized concentrated cultures intended for direct inoculation, (2) to monitor the viability of added probiotic bacteria during the different steps of production and shelf life of the food used as carrier, and finally, (3) the determination of a functional dose for a specific health-promoting effect.

When formulating or choosing a selective, elective, or differential plate count technique for enumerating probiotic bacteria some considerations must be taken into account:

- Probiotic bacteria (*L. casei*, *L. acidophilus*, bifidobacteria, etc.) are closely related to starter LAB (*S. thermophilus*, *L. delbrueckii* subsp. *bulgaricus*, *Lc. lactis*) from a nutritional point of view. It is then difficult to favor the growth of the former while inhibiting the growth of the latter on agar media.
- The probiotic strains are mainly used as adjunct cultures due to their poor growth in milk; then they are generally added to the final desired level (10^7–10^8 CFU/mL) whereas starter LAB may reach, after milk fermentation or during cheese ripening, levels ranging from 10^7 to 10^8 CFU/mL. If the growth of starter LAB is not adequately inhibited, their proliferation on agar plates might interfere with or even eclipse the growth of probiotic bacteria colonies.
- Probiotic bacteria do not constitute a physiological and taxonomical homogeneous group. It is then very difficult to formulate a unique culture medium suitable for all the species and strains already in the market is, at the same time, suitable to inhibit the great variety of starter LAB used for the manufacture of the different fermented dairy products all around the world.
- Difficulties in the enumeration will arise in those food matrixes in which more than one probiotic strain is intended to be added, especially if the strains belong to the same genera. Figure 20.1 shows the selective and differential capacity of MRS-bile (Man Rogosa Sharpe agar added of 0.15% [w/v] bovine bile salts) and MRS-LP (MRS agar added of 0.20% [w/v] lithium chloride and 0.30% [w/v] sodium propionate), two culture media used for the simultaneous enumeration of three probiotic strains (bifidobacteria, *L. acidophilus* and *L. paracasei*) in a cheese matrix containing *S. thermophilus* and *Lc. lactis* as starter LAB (Vinderola and Reinheimer 2000). The enumeration of probiotic bacteria would certainly be very difficult if mixtures of probiotic strains of the same species were intended to be used, an attractive idea for designing novel probiotic products if those strains possess different health-promoting effects. However, until now no multiprobiotic products—containing different strains of the same species—have been released into the market.

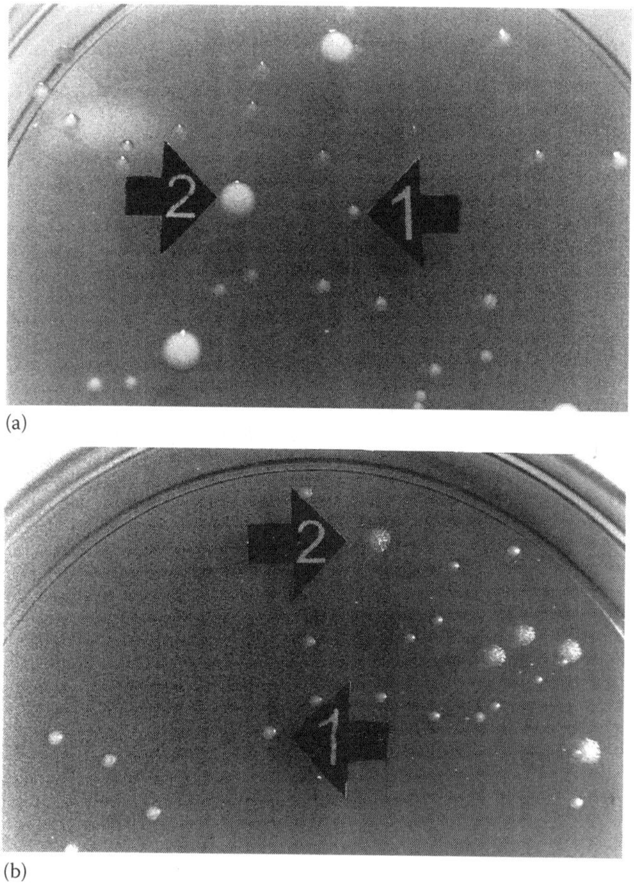

FIGURE 20.1 Morphology of colonies of *B. bifidum* (1) and *L. paracasei* (2) on MRS-LP agar (a) incubated for 4d at 37°C in anaerobiosis. Morphology of colonies of *L. paracasei* (1) and *L. acidophilus* (2) on MRS-Bile agar (b) incubated for 3d at 37°C in aerobiosis. (Photographs taken by Ruben Pantanalli, Facultad de Ingeniería Química, Santa Fe, Argentina.)

A suitable culture medium for the control of probiotic bacteria should then be able to inhibit the growth of the accompanying starter LAB, or at least offer a suitable differentiating capacity between probiotic and starter LAB. It must offer, at the same time, an adequate recovery rate of probiotic bacteria. It must also be selective, elective, or differential if two or more probiotic species are used. Finally, if it will be used for routine quality control in dairy plants, it should be easy to prepare and stable. In this way, the use of expensive or unstable antibiotics or selective agents would render the culture medium of little utility for the routine microbiology laboratory.

Table 20.1 lists some of the selective, elective, or differential compounds that have been proposed in the last two decades for the enumeration of probiotic bacteria in culture media. Attention was initially focused on the enumeration of *L. acidophilus* and bifidobacteria. However, because of viability problems associated with certain strains of *L. acidophilus* and *Bifidobacterium* spp. during cold storage, the trend

TABLE 20.1
Strategies for the Formulation of Culture Media for the Enumeration of Probiotic Bacteria in Dairy Products

Year	Proposed for	Selective/Elective/Differential Agent Used	Reference
1984	*L. acidophilus*	Maltose, salicin, raffinose, melibiose	Hull and Roberts (1984)
1990	*Bifidobacterium*	Propionic acid—pH 5	Beerens (1990)
1992	*Bifidobacterium*	Lithium chloride and sodium propionate	Lapierre et al. (1992)
1993	*L. acidophilus* and *B. bifidum*	Prussian blue	Onggo and Fleet (1993)
1993	*L. acidophilus*	X-Glu (chromogenic reagent)	Kneifel and Pacher (1993)
1995	*Bifidobacterium*	Lithium chloride, sodium propionate and iodoacetate	Arroyo et al. (1995)
1995	*Bifidobacterium*	Bile salts and gentamicin	Lim et al. (1995)
1996	*L. acidophilus* and *Bifidobacterium*	Prussian blue—pH 5 or maltose	Rybka and Kailasapathy (1996)
1996	*L. acidophilus*	Salicin	Lankaputhra and Shah (1996)
1996	*Bifidobacterium*	Neomycin, paramomycin, nalidixic acid, lithium chloride, sodium propionate	Ghoddusi and Robinson (1996)
1996	*Bifidobacterium*	Aztreonam, nalidixic acid, netilmycin, paramomycin	Pacher and Kneifel (1996)
1996	*L. acidophilus* and *Bifidobacterium*	Maltose, salicin, sorbitol, neomycin, nalidixic acid, lithium chloride, paramomycin	Dave and Shah (1996)
1997	*L. casei*	Medium acidified to pH 5.4 and incubated for 14 days at 15°C	Champagne et al. (1997)
1998	*L. casei*	Bromocresol green, ribose	Ravula and Shah (1998a)
1999	*L. acidophilus* and *Bifidobacterium*	Bile salts, lithium chloride, and sodium propionate	Vinderola and Reinheimer (1999)
1999	*Bifidobacterium*	Lactulose, methylene blue, lithium chloride, and sodium propionate	Nebra and Blanch (1999)
1999	*L. acidophilus* and *Bifidobacterium*	Sodium acetate, acetic acid, iodoacetic acid	Ingham (1999)
1999	*Bifidobacterium*	Polymixin B, kanamycin, iodoacetate, triphenyltetrazolium	Payne et al. (1999)
2000	*L. acidophilus*, *Bifidobacterium*, and *L. casei*	Bile salts, lithium chloride, and sodium propionate	Vinderola and Reinheimer (2000)
2000	*Bifidobacterium*	Mupirocin	Rada and Koc (2000)
2001	*Bifidobacterium*	Dicloxacillin, propionic acid	Bonaparte et al. (2001)
2003	*L. acidophilus*, *Bifidobacterium*, and *L. casei*	19 media assessed: sugar, antibiotic, organic, and inorganic salt based	Tharmaraj and Shah (2003)

TABLE 20.1 (continued)
Strategies for the Formulation of Culture Media for the Enumeration of Probiotic Bacteria in Dairy Products

Year	Proposed for	Selective/Elective/Differential Agent Used	Reference
2004	*L. acidophilus*, *Bifidobacterium*, and *L. casei*	Ribose + incubation at 27°C	Talwalkar and Kailasapathy (2004)
2006	*L. acidophilus*, *L. rhamnosus*, *L. paracasei*, *Bifidobacterium*	Clindamycin, lithium chloride, sodium propionate, Neomycin, paramomycin, nalidixic acid	Van de Casteele et al. (2006)
2006	*L. acidophilus*, *Bifidobacterium*	Bromocresol green, clindamycin, aniline blue, dicloxacillin	Darukaradhya et al. (2006)
2007	*L. acidophilus*, *L. paracasei*, *B. lactis*	Maltose, raffinose, lithium chloride	Tabasco et al. (2007)

was, later on, to add strains of the *L. casei* group to fermented milks (Shah 2000). When an increasing number of fermented milk manufacturers started incorporating *L. casei* in their products, the culture media optimized for *L. acidophilus* and bifidobacteria were then tested for the recovery, selection, and differentiation of *L. casei*, or alternatively, new culture media were formulated for its selective enumeration (Champagne et al. 1997, Ravula and Shah 1998a, Vinderola and Reinheimer 2000). The great variety of basal culture media used and the diversity of selective, elective, or differential agents added (antibiotics, salts, sugars, chromogenic compounds) give an idea of the complexity of this task. It seems that for each new combination of probiotic culture and starter LAB in a particular food matrix, an adequate screening of selective/elective/differential culture media must be carried out before determining the most appropriate one for testing viability of probiotic bacteria.

No official protocol or defined standard method has been released yet about the control of cell viability of probiotics in food products. In 1990, the International Dairy Federation (IDF) released a bulletin (IDF 252/1990) proposing culture media for the detection and enumeration of bifidobacteria in stools and in fermented milk products. In this bulletin, four culture media were suggested for detecting bifidobacteria in stools, whereas 14 culture media were suggested for their detection and enumeration in fermented milks, starter cultures, and other materials. In 1995, another IDF bulletin (IDF 306/1995) proposed 16 different culture media for the colony count of *L. acidophilus* combined with yogurt bacteria or combined with both yogurt bacteria and bifidobacteria. At present, no bulletin has been released by this international organization concerning the enumeration of strains of the *L. casei* group such as *L. casei*, *L. paracasei*, *L. rhamnosus*, which are perhaps the most used species in commercial fermented dairy products nowadays.

The continuous efforts by the IDF and ISO (International Organization for Standardization) to design a unique and reliable method for the accurate enumeration of *L. acidophilus* in milk products led to the publication, in 2006, of a technique for the enumeration of presumptive *L. acidophilus* on a selective medium (ISO 20128/IDF 192/2006). This technique relies on the use of the antibiotics clindamycin and ciproflaxin, both able to inhibit the growth of the most common microorganisms used in fermented milks, nonfermented milks, and infant formulae, such as *L. delbrueckii* subsp. *bulgaricus*, *L. delbrueckii* subsp. *lactis*, *S. thermophilus*, bifidobacteria, lactococci, *L. casei*, *L. paracasei*, *L. rhamnosus*, *L. reuteri*, and *Leuconostoc* species. However, the user is advised that the method cannot distinguish between *L. acidophilus*, *L. johnsonii*, *L. gasseri*, and *L. crispatus*, four closely related species from the *L. acidophilus* group (Klein et al. 1998). Additionally, the introduction to the technique also warns the reader that because of the large variety of dairy products, the method may not be appropriate in every detail for certain products. This could be the case, states the standard, when the number of presumptive *L. acidophilus* is very much lower than the number of other lactobacilli such as *L. rhamnosus*, *L. reuteri*, *L. plantarum*, *L. helveticus*, or yeasts (ISO 20128/IDF 192/2006).

To develop a standard method for the selective enumeration of bifidobacteria in dairy products, the IDF launched in 2003 a highly organized and controlled trial, comprising 20 qualified laboratories from Europe, Japan, and New Zealand (IDF 411/2007). These laboratories received seven blind samples of dairy products containing bifidobacteria alone or in combination with *L. acidophilus*, *L. gasseri*, *S. thermophilus*, and *L. delbrueckii* subsp. *bulgaricus*. They were asked to use a culture medium containing the antibiotic mupirocin (MUP) and the substrate TOS (transgalactosylated oligosaccharide) for the elective enumeration of bifidobacteria in those samples. Results of the trails have shown that the antibiotic does not influence the growth of bifidobacteria while inhibiting the growth of commonly used LAB. Additionally, a faster individual growth, bigger bifidobacterial colonies, and higher colony count have resulted from using TOS–MUP medium (IDF 411/2007). Based on these results of the performed trials, a technique for the enumeration of presumptive bifidobacteria on milk products is expected to be released jointly by the ISO and the IDF in the near future.

Until now, most of the culture media and enumeration techniques proposed by scientists present a certain degree of limitation for the enumeration of probiotic bacteria. Simultaneously, no definite culture medium has been officially proposed yet for the enumeration of probiotic bacteria occurring alone or in combination (several species). Payne et al. (1999) have concluded that colony differentiation is extremely difficult when mixed cultures of different *Bifidobacterium* species are plated together. Therefore, when designing a novel food containing probiotics, it would be advisable to assess first the response (growth, recovery, colony morphology, and color) of each strain as pure culture to a group of culture media, so as to identify the best option for a reliable assessment of the viability in the final product. However, screening for an optimal growth medium is a laborious task, and it is not often feasible to test all potential growth media. In the *Instituto de Lactología Industrial* (INLAIN, UNL-CONICET) in Argentina, after a screening study of 14 different general, differential, or selective media incubated in aerobiosis or anaerobiosis (Vinderola and Reinheimer 1999, 2000), two culture media have been chosen to enumerate *L. acidophilus*, *L. paracasei*, and *B. bifidum* occurring together with *S. thermophilus* and *Lc. lactis* in Argentinian fresh cheese (Vinderola et al. 2000a). The aspects of these colonies can be seen in Figure 20.1.

Currently, procedures for enumerating probiotic bacteria fermented milks rely solely on plate counts. Plating methodologies are time consuming, tedious, and susceptible to false counts due to autoaggregation of some strains (Talwalkar and Kailasapathy 2004). Moreover, bacterial stress during the shelf life of the product could lead to the inability of some sublethally injured cells ("dormant" bacteria) to grow on agar plates or even form colonies, which can make the enumeration even more complicated and confusing. The development of rapid, convenient, reliable methodologies or new agar media is still essential. Fortunately, new techniques such as the use of quantitative real-time polymerase chain reaction (RT-PCR), fluorescent in situ hybridization (FISH), or already available commercial viability assays (Lahtinen et al. 2006) are being developed, and the results seem promising.

20.5 VIABILITY VERSUS FUNCTIONALITY

From the first time the term "probiotic bacteria" was defined, it has always been stated that probiotics are "live microorganisms." In this way, the assessment of probiotic viability during their industrial production as pure cultures, after their addition to food, and during the shelf life of the products containing them has been the only parameter monitored to support the probiotic capacity of the strain or the product containing it. Although the plate count approach is often used as the gold standard method to measure bacterial viability, it actually only indicates how many cells can replicate under the conditions provided for growth (Ben Amor et al. 2007). "Viable cells" is, for the moment, a synonym for "functional cells." Nonetheless, recent evidence suggests that cell viability would no longer be the only parameter to indicate the real probiotic capacity or functionality of a microorganism. Adhesion of probiotic to intestinal epithelial cells is important for at least a "transient colonization" of the gut, and because it has been related to several beneficial effects such as shortening the duration of diarrhea, immunogenic effects, and competitive exclusion, among others (Isolauri et al. 1991, Saavedra et al. 1994, Salminen et al. 1996, Malin et al. 1997). Tuomola et al. (2001) have reported a significant diminution in the adhesion capacity of a commercial strain of *L. acidophilus* isolated at different points of the same production line without changes in cell counts. If adhesion is modified during industrial processes, other probiotic traits may also be altered, changing their functionality even without alterations of cell viability.

Just like any other probiotic characteristic, adhesion, and its stability during industrial processes, is also a strain-dependent feature. Elo et al. (1991) tested the stability of *Lactobacillus* GG adhesion from different production lots and products by comparing the original strain with cultures used for a longer period in industrial processes, but only slight variations in adhesion properties were observed. Later on, Saarela et al. (2003) demonstrated that resistance to gastric acidity, a desirable trait for probiotic bacteria, depended on the carbon source used as substrate for production of probiotic cultures. Matto et al. (2006) showed that for a particular strain of *B. animalis*, acid tolerance was influenced by the processing and storage conditions, whereas the level of viable cells in the product was not affected during storage. In the *Instituto de Lactología Industrial* (INLAIN, UNL-CONICET), it was observed (Figure 20.2) that the resistance of a strain of *L. acidophilus* to simulated gastric digestion (the pH dropped from 4.5 to 2 in 90 min) diminished during the

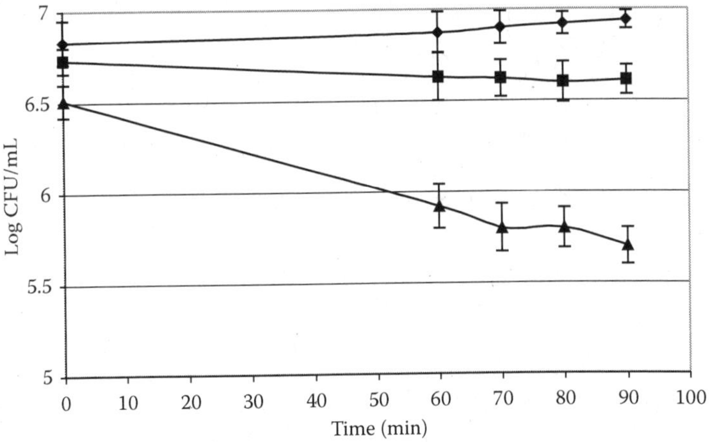

FIGURE 20.2 Cell counts of *L. acidophilus* DRU during the simulated gastric digestion (pH dropped gradually from 4.5 to 2 in 90 min) at different times (♦: 0 days, ■: 15 days and ▲: 30 days) during the refrigerated (5°C) storage in milk acidified to pH 4.5 with lactic acid.

refrigerated storage of the strain in acidified milk, although the level of viable cells remained almost unchanged during this period.

Additionally, there might be other biological factors beyond gastric acidity that can also change the functionality of probiotic strains during the gastrointestinal transit. Thus, bile-salt resistant derivatives from lactobacilli were recently isolated (Burns et al. 2008a). These isolates were adapted to grow under physiological concentrations of bile salts (up to 0.5% [w/v]). Bile-salt resistant derivatives grown in a culture medium with 0.5% (w/v) bile salts were less hydrophobic and less autoaggregative than parent strains (Figure 20.3). Autoaggregation has been related to the capacity of the strains to adhere to the intestinal epithelium (Collado et al. 2007, Voltan et al. 2007),

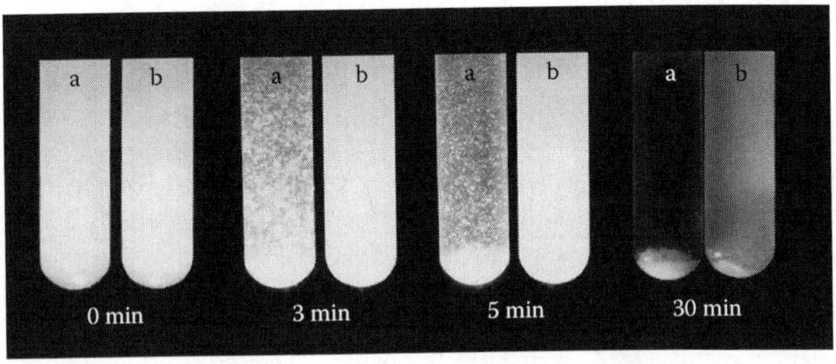

FIGURE 20.3 Time course of the autoaggregation, in phosphate buffer, of cells of *L. delbrueckii* subsp. *lactis* 200 (a) (grown in MRS broth) and its bile-resistant derivative (b) (grown in MRS broth with of 0.5% [w/v] bile-salts). (Photographs taken by David Pagura, Santa Fe, Argentina.)

whereas a high hydrophobicity would allow probiotic cells to better interact with immune cells such as macrophages and dendritic cells (Ofek and Doyle 1994), as those found in the gut mucosa. In this way, bile exposure during gastrointestinal transit might also alter the functionality of the strain without altering cell viability.

Up to now, most studies developed to evaluate the influence of technological variables on functionality relied on the determination of acid or bile sensitivity. Since acid and bile resistances are desirable characteristics for a probiotic strain but not an indicator of probiotic activity, there is yet a need to develop new tools or assays to evaluate the real functionality beyond cell viability. For example, the evaluation of the capacity for activating the gut immune response in animal trials appears as an attractive strategy, as well as the use of physiological fluorescent probes (Ben Amor et al. 2007) to discriminate among viable, stressed, or injured cells.

20.6 PROBIOTICS IN FERMENTED DAIRY PRODUCTS

20.6.1 Yogurt and Fermented Milks

Probiotic cultures have had a long association with dairy food products. Several reasons have made fermented milks, such as yogurt, one of the first choices for the incorporation of probiotic bacteria into commercial dairy products. First, certain probiotics share the same ecological niche in the gut as some lactobacilli, and it is generally thought that "closely related bacteria would perform well together," although, as will be discussed later, this argument does not apply to some strain combinations. Additionally, yogurt has long been recognized as a product with many desirable attributes for consumers, making it a suitable choice as a carrier for probiotics since it is largely incorporated in the daily diet of many consumers.

For the manufacture of yogurt containing probiotic bacteria, conventional yogurt processing procedures can be applied, with the probiotic bacteria being added prior to fermentation, simultaneously with yogurt cultures, or after fermentation to the cooled product before packaging. The methods used to manufacture stirred and drinking yogurt, in particular, are well suited to the addition of probiotics after fermentation (Stanton et al. 2003). There are many reports about the successful addition of probiotic bacteria and their survival during refrigerated storage in fermented dairy products (Shah 2000, Vinderola et al. 2000b,c, Kailasapathy 2006). However, as soon as commercial product development took place, other studies warned about the loss of cell viability for certain strains during the refrigerated period of the product and under certain conditions (Gilliland and Speck 1977, Dave and Shah 1998, Donkor et al. 2006). This could be due to the inhibitory activity of certain ingredients or process steps used during the manufacture of dairy products (Heller 2001, Vinderola et al. 2002a), as well as to negative interactions between lactic acid starter and probiotic bacteria (Joseph et al. 1998, Vinderola et al. 2002b).

Thus, probiotic viability depends on factors such as the production method and the strain, the point in time at which the probiotic bacteria are added during the manufacturing process, interactions between the strains used, temperature of fermentation, chemical composition of the fermentation medium, final acidity, accumulation of organic acids, milk solids content, availability of nutrients, dissolved oxygen (mainly in the case of bifidobacteria), or storage temperature (Stanton et al. 2003). This list is not exhaustive, and possible interaction effects between different parameters have not been fully investigated yet. In spite of that, low pH in fermented milks and the postacidification phenomena during cold storage have been pointed out as the main factors responsible for the loss of viability of probiotic bacteria.

20.6.2 Cheese

Cheese has also been a food target for carrying probiotic bacteria. Incorporation of these microorganisms into cheese appears to be an encouraging alternative to the problem of their survival because of the higher pH values in this kind of product, the closed matrix, and the high fat content, which might act as a protective factor for these organisms both during food storage and the passage through the gastrointestinal tract (Stanton et al. 1998). However, the manufacturing process of certain cheeses may have to be modified and adapted to the requirements of probiotics since some particular characteristics of certain cheeses (low a_w, cooking of curds, long ripening times) may render them inadequate for probiotic bacteria. In this way, fresh cheese appears to be ideally suited to act as a carrier for probiotic bacteria. Fresh cheese is an unripened product, and thus storage occurs at refrigeration temperatures and shelf life is rather limited (Heller et al. 2003).

One of the main hurdles to overcome is to determine when and how to add probiotic bacteria to avoid their loss from the curd during whey drainage. A solution must be found for each particular cheese: for example, the addition of probiotic bacteria together with cream and salt for cottage cheese (Blanchette et al. 1995, 1996) or the use of an ultrafiltration process to concentrate milk solids as in the manufacture of Argentine fresh cheese (Vinderola et al. 2000a). Successful incorporation into cheeses of probiotic bacteria has been reported for Cheddar (Furtado et al. 1993, Dinakar and Mistry 1994, Shaw and White 1994, Daigle et al. 1997, Gardiner et al. 1998, 1999, McBrearty et al. 2001, Phillips et al. 2006, Ong et al. 2007), Cottage (Blanchette et al. 1995, 1996, Riordan et al. 1998), Gouda (Gomes et al. 1995), Crescenza (Ghodussi and Robinson 1996, Gobbetti et al. 1998, Burns et al. 2008b), Egyptian Kariesh (Murad et al. 1998), semihard goat's (Gomes and Malcata 1998), Argentine fresh (Vinderola et al. 2000a), Canestrato Pugliese hard (Corbo et al. 2001), Egyptian Tallaga (El-Zayat and Osman 2001), Estonian "Pikantne" (Songisepp et al. 2004), white brined (Yilmaztekin et al. 2004), white (Kasimoglu et al. 2004), Minas fresh (Buriti et al. 2005), ewe's (Kourkoutas et al. 2006), semihard (Bergamini et al. 2005), and petit-suisse (Cardarelli et al. 2008) cheeses.

20.6.3 Frozen Products

Frozen dairy products have also been targeted carriers for the development of probiotic foods. Products such as ice cream, frozen yogurt, frozen desserts, and sour milk have been proposed as effective carriers for probiotic lactobacilli (Holocomb et al. 1991, Hekmat and McMahon 1992, Ravula and Shah 1998b). It is argued that frozen temperatures would be the main advantage over the storage of dairy products at 5°C–10°C, even though a considerably longer shelf life compared to that of traditional yogurts could also act against probiotic viability. Additionally, the low water activities and high osmotic pressures due to the presence of a large amount of sugars in oversweetened products, as is the case of frozen deserts, could be an additional cause for losing cell viability. Alterations of cell membrane permeability, and intracellular dehydration caused by ice crystal formation that may break cells, are also likely causes of microbial inactivation during freezing. On the other hand, the presence of various natural cryoprotectants (casein, sucrose, fat) in certain products could be attractive for the design of probiotic frozen products. Anyway, the ability of probiotic bacteria to withstand frozen storage is still strain dependent. An example of a successful development is an ice cream named Biogarde®,* released in Germany in the 1980s, containing *B. bifidum* and *L. acidophilus* as probiotic cultures.

20.7 MICROENCAPSULATION OF PROBIOTICS

Microencapsulation is a technology for packaging solids, liquids, or gaseous materials in miniature, sealed capsules, as detailed in Chapter 7. The controlled release of the encapsulated ingredient at the appropriate place and at the right time is a useful property for many applications in the food industry. This strategy can be used to deliver different ingredients (flavors, peptides, oils, enzymes, acidulants, colors, sweeteners) in food formulations. The microcapsule can be opened by different means as pH changes, solvation, temperature, enzymatic activity, time, etc. (Gouin 2004, Anal and Singh 2007).

To exert beneficial effects on the host, probiotics must maintain their viability or functionality, remaining in the food at high numbers until the end of the shelf life and, then, being capable of surviving the passage through the gastrointestinal tract (Holzapfel 2006). In fermented dairy products, factors such as low pH, organic acids, oxygen, bacterial competition, temperature, and time of storage are frequent causes of probiotic viability losses. On the other hand, the very low pH of the stomach and the presence of bile salts in the small intestine are the main biological barriers that cause loss in viable cell numbers after food ingestion. Microencapsulation of living probiotic cells is a very interesting approach to increase the resistance of sensitive strains against adverse conditions (Gismondo et al. 1999).

Microencapsulation of probiotics in a biodegradable polymer matrix is a very advantageous strategy. In this way, the cells entrapped in microcapsules are easy to handle and quantify. The survival of cells during processing and storage can be improved by the addition of cryo- and osmoprotective components. Biopolymer microcapsules are easy to prepare on a lab-scale but the scaling-up of the method is difficult and their costs are high. Spherical beads with diameters from 0.3 to 3.0 mm can be produced by extrusion or emulsification techniques (Anal and Singh 2007). For this purpose, diverse food-grade polymers can be used to make gel beads: κ-carrageenan or κ-carrageenan and locust bean gum, alginate (alone or combined with fructooligosaccahrides, starch, hydroxipropyl cellulose, or poly-L-lysine), cellulose acetate phthalate (CAP), proteins and polysaccharide mixtures, chitosan, and starch (Doleyres and Lacroix 2005, Ross et al. 2005, Anal and Singh 2007,

* Registered trademark of Sanofi Bio-Industries GmbH, Biogarde Labor Hamm, Hamm (Germany).

Ding and Shah 2007, Mortazavian et al. 2008). After drying the beads, a further surface coating can be applied. As coating compounds, gelatin and whey proteins (Reid et al. 2007) have been reported to offer protection during processing and storage, as well as against stomach and upper intestine environments. Acylation of alginate and chitosan produced "functionalized beads" with an improved resistance to enzymatic and acidic attacks during transit through the stomach. Recently, Cellbiotech (Cellbiotech International CO., Ltd., Korea) used soy peptides as a coating material to make protein-coated bacteria protected against moisture attack in dry products for extending shelf life. These can be additionally coated with cellulose and gum to make a double-coated product. This technology can be applied to all strains of bacteria, increasing their shelf life at room temperature and their survival into the intestine.

Spray drying (SD) has been widely used to encapsulate food ingredients (flavors, lipids, etc.), and it is the most common technology used in the food industry due to the relative low cost and available multiscale equipment. As detailed in Chapter 9, with this technology, a liquid product (a solution, an emulsion, or a suspension) is atomized in a hot gas current to produce a powder instantaneously (Gharsallaoui et al. 2007). Microencapsulation by SD is a valuable technique to encapsulate probiotic bacteria that produce small, uniformly coated microspheres containing viable cells. This method can produce large amounts of material. However, high cell mortality, resulting from simultaneous dehydration and thermal inactivation of bacteria, has been reported often (Ross et al. 2005, Doleyres and Lacroix 2005, Anal and Singh 2007). Typical materials to encapsulate bacteria cells are skimmed milk, starch, modified waxy maize starch, whey proteins, gelatin, acacia gum, cellulose acetate phthalate, etc. (Fávaro-Trindade and Grosso 2002, Doleyres and Lacroix 2005, Gharsallaoui t al. 2007). To improve bacterial survival, the addition of protectants to the media before drying is a strategy used by researchers. The incorporation of thermoprotectants such as trehalose, nonfat milk solids, adonitol, and granular starch have been reported as effective to improve cell viability during drying and storage (Ross et al. 2005, Anal and Singh 2007).

In any case, the selection of the encapsulation method to be used depends on the strain used. The scale of the process is also important since costs linked to encapsulation in gel beads are higher than those related, for example, with SD, as is well pointed out in Chapter 7.

20.8 TRANSLATION OF FUNCTIONAL EFFECTS TO OTHER FOOD MATRICES: DRIED CELL-FREE FRACTIONS OF FERMENTED MILKS, POSSIBILITIES FOR THE DEVELOPMENT OF FUNCTIONAL ADDITIVES

During milk fermentation, LAB produce a range of secondary metabolites, some of which have been associated with health-promoting properties. One of the most noticeable is B vitamins and bioactive peptides released from milk proteins. Some of the beneficial effects obtained by the oral administration of fermented milks containing viable bacteria were also reported for the same product containing nonviable cells, although in this last case the effect was generally more moderate. For example, the oral administration of pasteurized kefir to mice increased the number of IgA producing cells in the lamina propria of the gut as occurs with kefir containing viable cells, but while pasteurized kefir was administered diluted 1:10 in drinking water, viable kefir was diluted 1:100 to achieve the same effect (Vinderola et al. 2005b).

Some effects observed after the oral administration of fermented milks containing live probiotic bacteria are the production of the regulatory/proinflammatory cytokine IL-6 by intestinal epithelial cells, the increase in the number of IgA producing cells in the gut as well as the polyvalent secretory IgA in the luminal content, the production of proinflammatory and regulatory cytokines by peritoneal macrophages, the enhancement of their phagocytic activity (Perdigón et al. 2001, Maldonado Galdeano et al. 2007), and the prevention of enteric infections by *Salmonella enteritidis* serovar Typhimurium (Perdigón et al. 2001) or enteroinvasive *E. coli* (Medici et al. 2005). However, other studies (Vinderola et al. 2006b,c, 2007a,b) in which only the cell-free fraction of these fermented milks was administered to mice demonstrated the induction of the same effects described previously for fermented milks containing viable bacteria.

Several factors in fermented milks can affect the expression of genes in the intestinal epithelium, particularly those that are associated with enterocyte differentiation (Sanderson 1998). These factors seem to be also present in the cell-free fraction of milk fermented by *L. helveticus* R389 since its oral administration to mice induced the proliferation of goblet cells and mast cells in the lamina propria of the small intestine of mice and enhanced the expression of TRPV6 calcium channels in the brush borders of enterocytes (Vinderola et al. 2007c).

Commercial drying processes such as SD are routinely used to dry protein meals such as blood meal, soy isolates, and milk proteins (Fellows 2000). This process could be a low-cost accessible tool to produce dried powders from the cell-free fraction of fermented milks, for which functional properties were described earlier. The application of SD to the cell-free fraction of fermented milks allows nutritional and sensorial qualities to be retained, together with an extreme reduction in weight, higher solubility, longer shelf life at moderate temperatures, and the possibility to perform rehydration at any desired level or time, as well as the possibility to add the cell-free fraction directly to almost any food matrix, especially those with low water activity or high acidity, in which the probiotic viability would be seriously threatened. A functional powder from the cell-free fraction of fermented milks would create a novel characteristic on the food without necessarily changing the sensory quality of the product. However, and depending on the functional effect desired, the dose of powder used could influence the organoleptic characteristics of the product (Friedrich and Acree 1998), thus transforming a traditional food into a new product with novel organoleptic characteristics, which contributes to the expansion of the market of functional products.

The use of a dried cell-free fraction of fermented milks opens the possibility of bringing some of the functional effects observed for fermented milks to nondairy foods. This can broaden the market of functional foods to new matrices whose physicochemical composition (extreme pH values, high salt or sugar concentration, presence of inhibitory additives, low water activities) or manufacturing technological processes (heating, high pressure, mechanical stress, pasteurization, exposure to radiation) are not suitable for maintaining a high level of viable cells during shelf life until consumption. Products with an unfavorable water activity are, for example, cereals, chocolate, marmalade, honey, and toffees. These products are "too dry" for applying live bacteria and "too wet" for the application of freeze-dried bacteria (Schmid et al. 2006). Other vehicles not suitable for viable probiotic bacteria but representing potential food matrices for dried cell-free fractions would be soups, sweets, muffins, snacks, tea, chewing-gums, high acidic beverages, infant foods, dressings, energy drinks, biscuits, jams, cakes, sweeteners, and sports drinks.

There is an increasing array of functional foods available that are being designed to confer health benefits. Fermented dairy products are foods largely incorporated into the dietary habits of the population and their safety has been largely demonstrated in time. The relative technological simplicity for incorporating the same biologically active metabolites, presented in fermented milks, but as a dried cell-free fraction, into other food matrices opens the field for the development of a great variety of new functional foods using matrices not currently suitable nowadays for viable probiotic bacteria (Vinderola 2008).

20.9 THE PREBIOTIC CONCEPT

A "prebiotic" can be defined as a "nondigestible food ingredient that beneficially affects the host by selectively stimulating the growth and/or activity of one or a limited number of bacteria in the colon and thus improves host health" (Gibson and Roberfroid 1995). To accomplish these effects, a prebiotic has to reach the colon undigested and preferably persist through the large intestine, as benefits are more apparent distally (Gibson et al. 2004). A prebiotic must stimulate the growth and/or metabolic activity of a limited number of bacteria leading to changes in the overall microbial balance of the human colon toward the increase of beneficial bacteria (mainly *Lactobacillus* and *Bifidobacterium*), and the desirable decrease of the potentially harmful microorganisms (enterobacteria and others).

Although any dietary component reaching the colon undigested could be a potential prebiotic, in practice, most prebiotic substrates are carbohydrates. The original straight concept of prebiotic has recently been brought into question by several results indicating that the effect of some prebiotic substrates could not be confined to such a reduced number of beneficial bacteria as initially thought, also affecting other intestinal populations (Klessen et al. 2001, Duncan et al. 2003) whose hypothetical beneficial effects had not been explored yet. In view of that, a WHO Expert Group has recently recommended revising the definition of prebiotic by considering not only the effect on bifidobacteria and lactobacilli but also broadening it to other ecological interactions among members of the human microbiota (WHO/FAO 2007).

20.10 CHEMICAL COMPOSITION, CHAIN LENGTH, AND SOURCES OF PREBIOTICS

Many substrates of dietary origin, mainly from vegetables, or endogenously produced by the host, are available for fermentation by the colonic microbiota. Through the normal diet, resistant starch is the most quantitatively important prebiotic (Cummings and Macfarlane 1991). Other plant polysaccharides such as gums, inulin, and hemicelluloses (xylan, arabinan, glucan, galactans, and mannans) come next in quantitative importance. Sugars and oligosaccharides from plants such as raffinose, stachyose, and fructooligosaccharides (FOS) are also fermented by colonic bacteria. Disaccharides derived from lactose such as lactulose and lactitol have recognized bifidogenic effects. Human milk contains a huge variety of bifidogenic factors that include aminosugars (*N*-acetyl-glucosamine, *N*-acetyl-galactosamine, *N*-acetyl-manosamine), galactooligosaccharides (GOS), tryptic and chemotryptic digested caseins, and whey proteins, as well as sugars and peptidic components of κ-casein. Among the endogenous substrates, mucin glycoproteins produced by goblet cells in the colonic epithelium are important substances fermented in the colon. Related mucopolysaccharides such as chondroitin sulfate, or even exopolysaccharides produced by the endogenous intestinal microbiota or by bacteria ingested with foods can act as prebiotics (Manning et al. 2004). Finally, and to a lesser extent, proteins and peptides from the diet, originating by pancreatic secretions, or being produced by bacteria, are also available in the colon.

Oligosaccharides are short-chain carbohydrates consisting approximately of between 2 and 20 residues. Apart from those naturally occurring in the normal diet, others can be extracted from vegetables or are produced by the industry. FOS, GOS, isomaltooligosaccharides (IMO), glucooligosacccharides, and xylooligosaccharides (XOS) are oligosaccharides with recognized prebiotic effect (Manning et al. 2004).

The prebiotic properties of carbohydrates are likely to be influenced by the monosaccharide composition, glycosidic linkages between the monosaccharide residues, and the molecular weight. Thus, most recognized prebiotics are composed of glucose, galactose, xylose, and fructose. With respect to glycosidic linkages, for example, both maltose and IMO are composed of α-glucosyl linkages; however, the 1–4 linkages of maltose are quickly degraded by glycosidases of the small intestine, whereas the 1–6 linkages of IMO confers resistance to digestion, being selectively fermented in the colon. In general, the microbial selectivity of a substrate, as prebiotic, increases as the degree of polymerization decreases. It is also generally assumed that the longer the oligosaccharide the slower the fermentation, and hence the effect of prebiotic would penetrate further into the colon. This is the case of the polysaccharide inulin that can exert the effect in more distal regions of the colon than the corresponding oligosaccharides FOS, which are more quickly fermented and exert their action in the proximal part.

20.11 PRODUCTION OF OLIGOSACCHARIDES AND TECHNOLOGICAL PROPERTIES OF OLIGO- AND POLYSACCHARIDES

Currently, more than 20 different oligosaccharides are commercially available. They can be extracted from natural sources (e.g., soybean oligosaccharides), obtained by the enzymatic hydrolysis of polysaccharides (e.g., XOS, IMO, and FOS), by chemical isomerization (e.g., lactose isomerized to lactulose), or produced by enzymatic transglycosylation (e.g., GOS and FOS) of an initial disaccharide molecule using glycosidases from bacteria or fungi (Reuter et al. 1999, Mayer et al. 2004, Fernández-Arrojo et al. 2007). In transglycosylation reactions the starting material for FOS is sucrose and for GOS is lactose. Some oligosaccharides are produced from two disaccharides, for instance, lactosucrose, which is produced from lactose and sucrose, and glycosylsucrose, which is produced from sucrose and liquid starch. In general, oligosaccharides obtained enzymatically are shorter than those obtained from polysaccharides by hydrolysis, and both the final components and the efficiency of the reaction depend on the enzymes and the conditions used. The most abundantly used prebiotic oligosaccharides are GOS and FOS (Sako et al. 1999). The polysaccharide inulin is also widely used as a food ingredient.

Some prebiotics, mainly long-chain polysaccharides, can be used to improve rheological and textural properties of food products. This is the case for the polysaccharides called fat replacers, which are similar to fat in physicochemical behavior and also possess prebiotic properties (Warrand 2006). Inulin is the prebiotic most widely assayed for improving physicochemical properties of dairy products. This polysaccharide can reduce syneresis and can improve the body and texture of fat-free yogurts (Aryana and McGrew 2007, Aryana et al. 2007) and ice-cream (El-Nagar et al. 2002). The effect of prebiotic carbohydrates on the rheology of dairy products is strongly influenced by the chain length, with the longest chains generally having the strongest effect.

Intensive research efforts have been focused on the protection of the viability of certain probiotic cultures both during food technological processes and during the gastric transit. A significant increase of probiotics viability via encapsulation has been obtained with a variety of carriers, some of which are complex prebiotic carbohydrates (Ross et al. 2005). These combined products can be also regarded as synbiotics (see Section 20.12).

Some authors suggest that the resistance of probiotics to specific gastrointestinal or environmental stressing conditions can improve in the presence of certain sugars (Perrin et al. 2000, Noriega et al. 2006). Moreover, Saarela et al. (2003) indicated that the prebiotic lactulose might improve the bile tolerance and cold-storage stability of *Lactobacillus salivarius*.

20.12 HEALTH BENEFITS ATTRIBUTED TO PREBIOTICS

In humans, the proximal colon is a saccharolytic environment and in this part mostly carbohydrate fermentation occurs, whereas in the distal colon, proteins and amino acids become a more dominant metabolic energy source through fermentation (Macfarlane et al. 1992). Carbohydrates in the colon are fermented to short chain fatty acids (SCFA)—acetate, propionate, and butyrate—and several other compounds such as lactate, pyruvate, ethanol, succinate, and the gases H_2, CO_2, CH_4, and H_2S.

Acetate is mainly metabolized in the muscles, kidney, heart, and brain, whereas propionate enters the liver and is a gluconeogenic precursor that suppresses cholesterol and fatty acids synthesis. Butyrate is metabolized by the colonic epithelium, serving as a regulator of cell growth and differentiation. Proteins that reach the colon are fermented to branched-chain fatty acids such as isobutyrate, isovalerate, and several nitrogenous compounds. In contrast to the case of carbohydrate fermentation, some of these protein-derived compounds may be toxic. Excessive protein fermentation, especially in the distal colon, has therefore been related to diseases such as colon cancer, bowel cancer, and ulcerative colitis. Therefore, bacterial activity in the human colon is involved in beneficial actions or in a number of acute (diarrheal infections) or chronic diseases (inflammatory bowel diseases, colon cancer, and pseudomembranous colitis) that have been linked, to varying extents, to microbiota composition and activities.

Fermentation of prebiotics in the colon contributes toward a decrease in pH and the production of SCFA, resulting in reduced numbers of pathogenic bacteria and other beneficial actions. Prebiotics are also being studied for their potential in improving, alleviating, or preventing acute or chronic disorders (bacterial gastroenteritis, rotavirus infections, irritable and inflammatory bowel disease, atopic eczema, gastric and peptic ulcer, bowel cancer) based on the competitive exclusion of enteropathogens and their immonomodulating capacity (Manning et al. 2004). Additionally, there is interest in the possibility of increasing mineral (particularly calcium) absorption and modulating blood lipids such as cholesterol and triglycerides by the use of prebiotics. Although several mechanisms have been postulated for these two last effects, the way by which they occur is not clearly understood at the present time.

It has been shown that the presence of FOS, GOS, XOS, IMO, and lactulose in culture fermentations with human fecal bacteria alter the microbiota, increasing the level of bifidobacteria and/or lactobacilli, and in some cases causing clostridia and bacteroides declination (Rycroft et al. 2001). In general, the health benefits of prebiotic intake can be similar to those described for targeted probiotic microorganisms, if the selectively stimulated bacteria possess beneficial properties. In addition, there are benefits apart from the effects attributable to a direct increase in the number of LAB. Some of them are due to a switch in the metabolism of microorganisms themselves or to more or less complex metabolic interactions among the gut microbiota. One of these beneficial effects can be ascribed to the recently named "metabolic cross-feeding" phenomenon, in which metabolic products produced from dietary prebiotics by one bacterial species may then provide substrates to support the growth of other populations. Therefore, cross-feeding can result in metabolic consequences that would not be predicted simply from the substrate preferences of target bacteria considered as separate from their intestinal environment. Recently, two routes of metabolic cross-feeding have been described between bifidobacteria and butyrate producers from the human gut, one due to the consumption of fermentation end products (lactate and acetate), and another due to cross-feeding of partial breakdown products from complex substrates (Belenguer et al. 2006, Duncan et al. 2004, Rossi et al. 2005). A simple and schematic representation of interaction between prebiotic substrates and probiotics in the colon is provided in Figure 20.4.

20.13 SYNBIOTICS: THEIR USE IN DAIRY PRODUCTS

"Synbiotics" are the combination of prebiotics and probiotics and they can show additive or synergistic effects with respect to each of the components considered separately. This enhancement of effects can be attributed to the increased survival of the bacteria due to a protective effect of the prebiotic or to its use as a selective substrate by the probiotic; the prebiotics can also act on beneficial microbiota (Roberfroid 1998), or even a synbiotic can be designed to act in two different regions of the

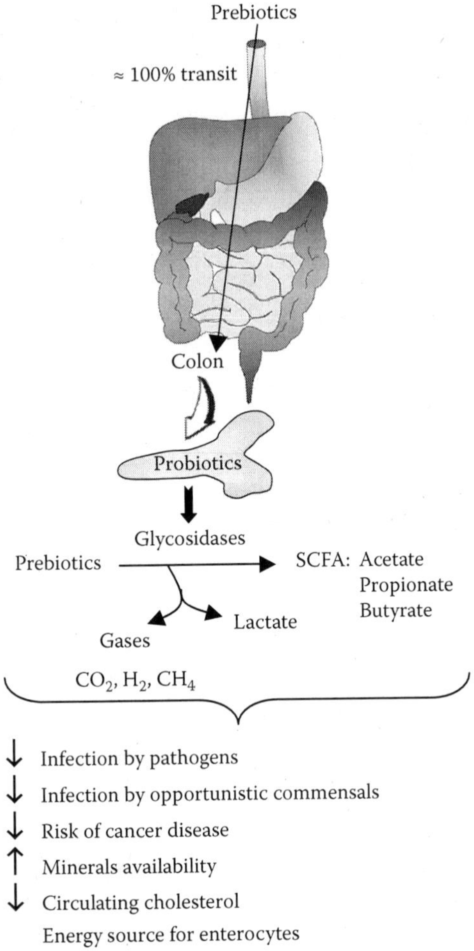

FIGURE 20.4 Interaction in the host between prebiotic substrates and probiotics. Prebiotics reach the colon undigested and they are hydrolyzed by microbial glycosidases. Fermentation of released oligo- or monosaccharides give rise to the production of short chain fatty acids that can achieve different beneficial functions.

gastrointestinal tract (Holzapfel and Schillinger 2002). In dairy products, synbiotics have been studied for cheese, fermented milks, and other milk-based preparations (Shin et al. 2000, Buriti et al. 2007, Cardarelli et al. 2008), and not only health benefits but also improvement or at least maintenance of sensory and rheological properties of the product have been pursued (Cardarelli et al. 2008). The most commonly used substrates in synbiotics are inulin, GOS, and FOS. A promising new synbiotic product combines *Bifidobacterium animalis* subsp. *lactis* (Lafti®B94*) selected for interacting with the resistant starch Hi-Maize®† (Crittenden et al. 2001). Finally, it

* Registered trademark of Koninklijke DSM N.V., Heerlen (the Netherlands).
† Registered trademark of Penford Australia Ltd., Lane Cove, New South Wales (Australia).

is worth mentioning the increasing use of synbiotics in infant formulas, in which GOS are frequently the substrate chosen (Pérez-Conesa et al. 2006 and 2007, Puccio et al. 2007).

20.14 CONCLUDING REMARKS

The inclusion of probiotic bacteria and prebiotic substrates into dairy foods has experienced an enormous expansion, leading to the introduction of a great variety of new food products into the market. Some well-designed, double-blind, placebo-controlled trials have been carried out in humans which focused on the study of functional properties of certain prebiotic substrates and probiotic strains used as pure cultures or added into specific food matrices. However, many problems arise at the industry level when probiotic strains are incorporated into different food matrices or in combination with other strains. These difficulties are related to the selective enumeration of probiotics when they are included in combination with lactic acid starter or other probiotic bacteria, changes in their functionality induced by certain food components or by the processing conditions of the food product, and loss or declination of the probiotic properties during the shelf life of the product acting as a vehicle for delivery. There are many probiotic strains with demonstrated and encouraging beneficial characteristics. However, sometimes the great efforts and economic resources put into the study of functional properties might be lost in the way that brings probiotic bacteria from laboratory studies to the consumers: this long way moves from their production as frozen or dried cultures, through their inclusion into the food matrix and the manufacture of the food product, to the shelf life until the final consumption.

There is a huge variety of known natural and industrially synthesized prebiotics. Their combination in some cases with probiotic microorganisms in synbiotic mixtures will contribute to reinforce or at least to maintain the beneficial properties of probiotic strains in food products. Fortunately, there is also a great array of molecular, biochemical, and microbiological tools for the study of the fate of the probiotic property in this long way from laboratory to the market. Certainly in the future, scientific studies addressing these issues will give us a clearer picture of the mechanisms of beneficial effects induced by probiotic bacteria and prebiotic substrates, as well as the technological, microbiological, and biological factors affecting them.

REFERENCES

Adolfsson, O., Meydani, S.N., and Russell, R.M. 2004. Yogurt and gut function. *Am. J. Clin. Nutr.* 80:245–256.

Anal, A.K. and Singh, H. 2007. Recent advances in microencapsulation of probiotics for industrial applications and targeted delivery. *Trends Food Sci. Technol.* 18:240–251.

Arroyo, L., Cotton, L.N., and Martin, J.H. 1995. AMC agar—A composite medium for selective enumeration of *Bifidobacterium longum*. *Cult. Dairy Prod. J.* 30:12–15.

Aryana, K.J. and McGrew, P. 2007. Quality attributes of yogurt with *Lactobacillus casei* and various prebiotics. *LWT-Food Sci. Technol.* 40:1808–1814.

Aryana, K.J., Plauche, S., Rao, R.M., McGrew, P., and Shah, N.P. 2007. Fat-free plain yogurt manufactured with inulins of various chain lengths and *Lactobacillus acidophilus*. *J. Food Sci.* 72:M79–M84.

Beerens, H. 1990. An elective and selective isolation medium for *Bifidobacterium* spp. *Lett. Appl. Microbiol.* 11:155–157.

Belenguer, A., Duncan, S.H., Calder, A.G., et al. 2006. Two routes of metabolic cross-feeding between *Bifidobacterium adolescentis* and butyrate-producing anaerobes from the human gut. *Appl. Environ. Microbiol.* 72:3593–3599.

Ben Amor, K., Vaughan, E.E., and de Vos, W.M. 2007. Advanced molecular tools for the identification of lactic acid bacteria. *J. Nutr.* 137:741S–747S.

Bergamini, C.V., Hynes, E.R., Quiberoni, A., Suárez, V.B., and Zalazar, C.A. 2005. Probiotic bacteria as adjunct starters: Influence of the addition methodology on their survival in a semi-hard Argentinean cheese. *Food Res. Int.* 38:597–604.

Blanchette, L., Roy, D., and Gauthier, S.F. 1995. Production of cultured cottage cheese dressing by bifidobacteria. *J. Dairy Sci.* 78:1421–1429.

Blanchette, L., Roy, D., Bélanger, G., and Gauthier, S.F. 1996. Production of cottage cheese using dressing fermented by bifidobacteria. *J. Dairy Sci.* 79:8–15.

Blaut, M. and Clavel, T. 2007. Metabolic diversity of the intestinal microbiota: Implications for health and disease. *J. Nutr.* 137:751S–755S.

Bonaparte, C., Klein, G., Kneifel, W., and Reuter, G. 2001. Development of a selective culture medium for the enumeration of bifidobacteria in fermented milks. *Lait* 81:227–235 (in French).

Bron, P.A., Grangette, C., Mercenier, A., de Vos, W.M., and Kleerebezem, M. 2004. Identification of *Lactobacillus plantarum* genes that are induced in the gastrointestinal tract of mice. *J. Bacteriol.* 186:5721–5729.

Buriti, F.C.A., da Rocha, J.S., and Saad, S.M.I. 2005. Incorporation of *Lactobacillus acidophilus* in Minas fresh cheese and its implications for textural and sensorial properties during storage. *Int. Dairy J.* 15:1279–1288.

Buriti, F.C.A., Cardarelli, H.R., Filisetti, T.M.C.C., and Saad, S.M.I. 2007. Synbiotic potential of fresh cream cheese supplemented with inulin and *Lactobacillus paracasei* in co-culture with *Streptococcus thermophilus*. *Food Chem.* 104:1605–1610.

Burns, P., Vinderola, G., Binetti, A., Quiberoni, A., de los Reyes-Gavilán, C., and Reinheimer J. 2008a. Bile-resistant derivatives obtained from non-intestinal dairy lactobacilli. *Int. Dairy J.* 18:377–385.

Burns, P., Patrignani, F., Serrazanetti, D., et al. 2008b. Probiotic Crescenza cheese containing *Lactobacillus casei* and *Lactobacillus acidophilus* manufactured with high pressure-homogenized milk. *J. Dairy Sci.* 91:500–512.

Cardarelli, H.R., Buriti, F.C.A., Castro, I.A., and Saad, S.M.I. 2008. Inulin and oligofructose improve sensory quality and increase the probiotic viable count in potentially synbiotic petit-suisse cheese. *LWT-Food Sci. Technol.* 41:1037–1046.

Champagne, C.P., Roy, D., and Lafond, A. 1997. Selective enumeration of *Lactobacillus casei* in yoghurt-type fermented milks based on a 15°C incubation temperature. *Biotechnol. Tech.* 11:567–569.

Collado, M.C., Surono, I., Meriluoto, J., and Salminen, S. 2007. Indigenous dadih lactic acid bacteria: Cell-surface properties and interactions with pathogens. *J. Food Sci.* 72:89–93.

Corbo, M.R., Albenzio, M., De Angelis, M., Sevi, A., and Gobbetti, M. 2001. Microbiological and biochemical properties of Canestrato Pugliese hard cheese supplemented with bifidobacteria. *J. Dairy Sci.* 84:551–561.

Crittenden, R.G., Morris, L.F., Harvey, M.L., Tran, L.T., Mitchell, H.L., and Playne, M.J. 2001. Selection of a *Bifidobacterium* strain to complement resistant starch in a synbiotic yoghurt. *J. Appl. Microbiol.* 90:268–278.

Cummings, J.H. and Macfarlane, G.T. 1991. The control and consequences of bacterial fermentation in the human colon. *J. Appl. Bacteriol.* 70:443–459.

Daigle, A., Roy, D., Bélanger, G., and Vuillemard, J.C. 1997. Production of hard pressed cheese (Cheddar cheese-like) using cream fermented by *Bifidobacterium infantis* ATCC 27920 G. Abstract D25. *J. Dairy Sci.* 80 (Suppl. 1) 103.

Darukaradhya, J., Phillips, M., and Kailasapathy, K. 2006. Selective enumeration of *Lactobacillus acidophilus*, *Bifidobacterium* spp., starter lactic acid bacteria and non-starter lactic acid bacteria from Cheddar cheese. *Int. Dairy J.* 16:439–445.

Dave, R.I. and Shah, N.P. 1996. Evaluation of media for selective enumeration of *Streptococcus thermophilus*, *Lactobacillus delbrueckii* ssp. *bulgaricus*, *Lactobacillus acidophilus*, and bifidobacteria. *J. Dairy Sci.* 79:1529–1536.

Dave, R.I. and Shah, N.P. 1998. Ingredient supplementation effects on viability of probiotic bacteria in yogurt. *J. Dairy Sci.* 81:2804–2816.

de Moreno de LeBlanc, A., Chaves, S., Carmuega, E., Weill, R., Antóine, J., and Perdigón, G. 2008. Effect of long-term continuous consumption of fermented milk containing probiotic bacteria on mucosal immunity and the activity of peritoneal macrophages. *Immunobiology* 213:97–108.

de Vrese, M. and Schrezenmeir, J. 2007. Supplement: Effects of probiotics and prebiotics. *J. Nutr.* 137:739S–853S.

Dinakar, P. and Mistry, V.V. 1994. Growth and viability of *Bifidobacterium bifidum* in Cheddar cheese. *J. Dairy Sci.* 77:2854–2864.

Ding, W.K. and Shah, N.P. 2007. Acid, bile and heat tolerance of free and microencapsulated probiotic bacteria. *J. Food Sci.* 72:446–450.

Dogi, C. and Perdigón, G. 2006. Importance of the host specificity in the selection of probiotic bacteria. *J. Dairy Res.* 73:357–366.

Doleyres, Y. and Lacroix, C. 2005. Technologies with free and immobilised cells for probiotic bifidobacteria production and protection. *Int. Dairy J.* 15:973–988.

Donkor, O.N., Henriksson, A., Vasiljevic, T., and Shah, N.P. 2006. Effect of acidification on the activity of probiotics in yoghurt during cold storage. *Int. Dairy J.* 16:1181–1189.

Duncan, S.H., Scott, K.P., Ramsay, A.G., et al. 2003. Effects of alternative dietary substrates on competition between human colonic bacteria in an anaerobic fermentor system. *Appl. Environ. Microbiol.* 69:1136–1142.

Duncan, S.H., Holtrop, G., Lobley, G.E., Calder, G., Stewart, C.S., and Flint, H.J. 2004. Contribution of acetate to butyrate formation by human faecal bacteria. *Br. J. Nutr.* 91:915–923.

El-Nagar, G., Clowes, G., Tudorica, C.M., Kuri, V., and Brennan, C.S. 2002. Rheological quality and stability of yog-ice cream with added inulin. *Int. J. Dairy Technol.* 55:89–93.

Elo, S., Saxelin, M., and Salminen, S. 1991. Attachment of *Lactobacillus casei* strain GG to human colon carcinoma cell line Caco-2: Comparison with other strains of dairy starter bacteria. *Lett. Appl. Microbiol.* 13:154–156.

El-Zayat, A.I. and Osman, M.M. 2001. The use of probiotics in Tallaga cheese. *Egypt. J. Dairy Sci.* 29:99–106.

FAO/WHO 2002. Guidelines for the evaluation of probiotics in food. Food and Agriculture Organization of the United Nations and World Health Organization Working Group Report. http://www.who.int/foodsafety/fs_management/en/probiotic_guidelines.pdf (accessed November 5, 2008).

Farnworth, Edward. 2008. *Handbook of Fermented Functional Foods*, 2nd edition. Boca Raton, FL: CRC Press.

Fávaro-Trindade, C.S. and Grosso, C.R.F. 2002. Microencapsulation of *L. acidophilus* (La-05) and *B. lactis* (Bb-12) and evaluation of their survival at the pH values of the stomach and in bile. *J. Microencapsul.* 19:485–494.

Fellows, P.J. 2000. *Dehydration, Food Processing Technology*, 2nd edition. Cambridge, U.K.: Woodhead Publishing Limited.

Fernández-Arrojo, L., Marin, D., De Segura, A.G., et al. 2007. Transformation of maltose into prebiotic isomaltooligosaccharides by a novel alpha-glucosidase from *Xantophyllomyces dendrorhous*. *Process Biochem*. 42:1530–1536.

Franz, C.M., Stiles, M.E., Schleifer, K.H., and Holzapfel, W.H. 2003. Enterococci in foods—A conundrum for food safety. *Int. J. Food Microbiol.* 88:105–122.

Friedrich, J.E. and Acree, T.E. 1998. Gas chromatography olfactometry (GC/O) of dairy products. *Int. Dairy J.* 8:235–241.

Fuller, R. 1989. A review: Probiotics in man and animals. *J. Appl. Bacteriol.* 66:365–378.

Fuller, R. and Perdigón, G. 2003. *Gut Flora, Nutrition, Immunity and Health.* Oxford, U.K.: Blackwell Publishing Ltd.

Furtado, M.M., Partridge, J.A., and Ustunol, Z. 1993. Ripening characteristics of reduced fat cheddar cheese using heat-shocked and live *Lactobacillus casei* as adjunct culture. Abstract D28. *J. Dairy Sci.* 76 (Suppl. 1), 101.

Galdeano, C.M. and Perdigón, G. 2004. Role of viability of probiotic strains in their persistence in the gut and in mucosal immune stimulation. *J. Appl. Microbiol.* 97: 673–681.

Gardiner, G., Ross, R.P., Collins, J.K., Fitzgerald, G., and Stanton, C. 1998. Development of a probiotic Cheddar cheese containing human-derived *Lactobacillus paracasei* strains. *Appl. Environ. Microbiol.* 64:2192–2199.

Gardiner, G., Stanton, C., Lynch, P.B., Collins, J.K., Fitzgerald, G., and Ross, R.P. 1999. Evaluation of Cheddar cheese as a food carrier for delivery of a probiotic strain to the gastrointestinal tract. *J. Dairy Sci.* 82:1379–1387.

Gharsallaoui, A., Roudaut, G., Chambin, O., Voilley, A., and Saurel, R. 2007. Applications of spray-drying in microencapsulation of food ingredients: An overview. *Food Res. Int.* 40:1107–1121.

Ghoddusi, H.B. and Robinson, R.K. 1996. The test of time. *Dairy Ind. Int.* 61:25–28.

Gibson, G.R. and Roberfroid, M. B. 1995. Dietary modulation of the human colonic microbiota-Introducing the concept of prebiotics. *J. Nutr.* 125:1401–1412.

Gibson, G.R., Rastall, R.A., and Fuller, R. 2003. The health benefits of probiotics and prebiotics. In *Gut Flora, Nutrition, Immunity and Health*, eds. R. Fuller and G. Perdigón, pp. 52–76. Oxford, U.K.: Blackwell Publishing.

Gibson, G.R., Probert, H.M., Van Loo, J., Rastall, R.A., and Roberfroid, M.B. 2004. Dietary modulation of the human colonic microbiota: Updating the concept of prebiotics. *Nutr. Res. Rev.* 17:259–275.

Gilliland, S.E. and Speck, M.L. 1977. Instability of *Lactobacillus acidophilus* in yogurt. *J. Dairy Sci.* 60:1394–1398.

Gismondo, M.R., Drago, L., and Lombardi, A.1999. Review of probiotics available to modify gastrointestinal flora. *Int. J. Antimicrob. Agents* 12:287–292.

Gobbetti, M., Corsetti, A., Smacchi, E., Zocchetti, A., and De Angelis, M. 1998. Production of Crescenza cheese by incorporation of bifidobacteria. *J. Dairy Sci.* 81:37–47.

Goktepe, I., Juneja, V.K., and Ahmedna, M. 2007. *Probiotics in Food Safety and Human Health.* Boca Raton, FL: CRC Press.

Gomes A.M.P. and Malcata, F.X. 1998. Development of probiotic cheese manufactured from goat milk: Response surface analysis via technological manipulation. *J. Dairy Sci.* 81:1492–1507.

Gomes, A.M.P., Malcata, F.X., Klaver, F.A.M, and Grande, H.J. 1995. Incorporation and survival of *Bifidobacterium* sp. strain Bo and *Lactobacillus acidophilus* strain Ki in a cheese product. *Neth. Milk Dairy J.* 49:71–95.

Gouin, S. 2004. Microencapsulation: Industrial appraisal of existing technologies and trends. *Trends Food Sci. Technol.* 15:330–347.

Guarner, F., Perdigón, G., Corthier, G., Salminen, S., Koletzko, B., and Morelli, L. 2005. Should yoghurt cultures be considered probiotics? *Brit. J. Nutr.* 93:783–786.

Haller, D., Colbus, H., Ganzle, M.G., Scherenbacher, P., Bode, C., and Hammes W.P. 2001. Metabolic and functional properties of lactic acid bacteria in the gastro-intestinal ecosystem: A comparative in vitro study between bacteria of intestinal and fermented food origin. *Syst. Appl. Microbiol.* 24:218–226.

Hekmat, S. and McMahon, D.J. 1992. Survival of *Lactobacillus acidophilus* and *Bifidobacterium bifidum* in ice cream for use as a probiotic food. *J. Dairy Sci.* 75:1415–1422.

Heller, K.J. 2001. Probiotic bacteria in fermented foods: Product characteristics and starter organisms. *Am. J. Clin. Nutr.* 73:374S–379S.

Heller, K.J., Bockelmann, W., Schrezenmeir, J., and de Vrese, M. 2003. Cheese and its potential as a probiotic food. In *Handbook of Fermented Functional Foods*, ed. E.R. Farnworth, pp. 203–226. Boca Raton, FL: CRC Press.

Holocomb, J.E., Frank, J.F., and McGregor, J.U. 1991. Viability of *Lactobacillus acidophilus* and *Bifidobacterium bifidum* in soft-serve frozen yogurt. *Cult. Dairy Prod. J.* 26:4–5.

Holzapfel, W.H. 2006. Introduction to prebiotics and probiotics. In *Probiotics in Food Safety and Human Health*, Eds. I. Goktepe, V.K. Juneja, and M. Ahmedna, pp. 1–33. Boca Raton, FL: CRC Press.

Holzapfel, W.H. and Schillinger, U. 2002. Introduction to pre and probiotics. *Food Res. Int.* 35:109–116.

Hull, R.R. and Roberts, A.V. 1984. Differential enumeration of *Lactobacillus acidophilus* in yogurt. *Aust. J. Dairy Technol.* 39:160–163.

IDF. 1990. Culture media for detection and enumeration of bifidobacteria in fermented milk products. Bulletin No. 252, International Dairy Federation, Brussels, Belgium.

IDF. 1995. Detection and enumeration of *Lactobacillus acidophilus*. Bulletin No. 306, International Dairy Federation, Brussels, Belgium.

IDF. 2006. Milk products—Enumeration of presumptive *Lactobacillus acidophilus* on a selective medium—Colony-count technique at 37°C. International Standard ISO 21128, IDF 192.

IDF. 2007. Selective enumeration of bifidobacteria in dairy products: Development of a standard method. Bulletin No. 411, International Dairy Federation, Brussels, Belgium.

Ingham, S.C. 1999. Use of modified *Lactobacillus* selective medium and *Bifidobacterium* iodoacetate medium for differential enumeration of *Lactobacillus acidophilus* and *Bifidobacterium* spp. in powered nutritional products. *J. Food Prot.* 62:77–80.

Isolauri, E., Juntunen, M., Rautanen, T., Sillanaukee, P., and Koivula, T. 1991. A human *Lactobacillus* strain (*Lactobacillus* GG) promotes recovery from acute diarrhea in children. *Pediatrics* 88:90–97.

Isolauri, E., Salminen, S., and Ouwehand, A.C. 2004. Probiotics. *Best Pract. Res. Clin. Gastroenterol.* 18:299–313.

Joseph, P.J., Dave, R.I., and Shah, N.P. 1998. Antagonism between yogurt bacteria and probiotic bacteria isolated from commercial starter cultures, commercial yogurts, and a probiotic capsule. *Food Aust.* 50:20–23.

Kailasapathy, K. 2006. Survival of free and encapsulated probiotic bacteria and their effect on the sensory properties of yoghurt. *LWT—Food Sci. Technol.* 39:1221–1227.

Kanbe, M. 1992. Traditional fermented milks of the world. In *Functions of Fermented Milk*, eds. Y. Nakazawa and A. Hosono, pp. 41–60. London, U.K.: Elsevier Science Publishers Ltd.

Kasimoglu, A., Goncuozlu, M., and Akgun, S. 2004. Probiotic white cheese with *Lactobacillus acidophilus*. *Int. Dairy J.* 14:1067–1073.

Kayser, F.H. 2003. Safety aspects of enterococci from the medical point of view. *Int. J. Food Microbiol.* 88:255–262.

Kleessen, B., Hartmann, L., and Blaut, M. 2001. Oligofructose and long-chain inulin: Influence on the gut microbial ecology of rats associated with a human faecal flora. *Br. J. Nutr.* 86:291–300.

Klein, G., Pack, A., Bonaparte, C., and Reuter, G. 1998. Taxonomy and physiology of probiotic lactic acid bacteria. *Int. J. Food Microbiol.* 41:103–125.

Kneifel, W. and Pacher, B. 1993. An X-glu based agar medium for the selective enumeration of *Lactobacillus acidophilus* in yogurt-related milk products. *Int. Dairy J.* 3:277–291.

Kourkoutas, Y., Bosnea, L., Taboukos, S., Baras, C., Lambrou, D., and Kanellaki, M. 2006. Probiotic cheese production using *Lactobacillus casei* cells immobilized on fruit pieces. *J. Dairy Sci.* 89:1439–1451.

Lahtinen, S.J., Gueimonde, M., Ouwehand, A.C., Reinikainen, J.P., and Salminen, S.J. 2006. Comparison of four methods to enumerate probiotic bifidobacteria in a fermented food product. *Food Microbiol.* 23:571–577.

Lankaputhra, W.E.V. and Shah, N.P. 1996. A simple method for selective enumeration of *Lactobacillus acidophilus* in yogurt supplemented with *L. acidophilus* and *Bifidobacterium* spp. *Milchwissenschaft* 51:446–451.

Lapierre, L., Undeland, P., and Cox, L.J. 1992. Lithium chloride-sodium propionate agar for the enumeration of bifidobacteria in fermented dairy products. *J. Dairy Sci.* 75:1192–1196.

Lehrer, R.I., Bevins, C.L., and Ganz, T. 2005. Defensins and other antimicrobial peptides and proteins. In *Mucosal Immunology*, 3rd edition, eds. J. Mestecky, M.E. Lamm, W. Strober, J. Bienenstock, J.R. McGhee, and L. Mayer, pp. 94–110. San Diego, CA: Academic Press.

Lim, K.S., Huh, C.S., and Baek, Y.J. 1995. A selective enumeration medium for bifidobacteria in fermented dairy products. *J. Dairy Sci.* 78:2108–2112.

Lourens-Hattingh, A., and Viljoen, B.C. 2001. Yogurt as probiotic carrier food. *Int. Dairy J.* 11:1–17.

Macfarlane, G.T., Gibson, G.R., and Cummings, J.H. 1992. Comparison of fermentation reactions in different regions of the human colon. *J. Appl. Bacteriol.* 72:57–64.

Maldonado Galdeano, C., de Moreno de Leblanc, A., Vinderola, G., Bibas Bonet, M.E., and Perdigón, G. 2007. A proposal model: Mechanisms of immunomodulation induced by probiotic bacteria. *Clin. Vac. Immunol.* 14:485–492.

Malin, M., Verronen, P., Korhonen, H., et al. 1997. Dietary therapy with *Lactobacillus* GG, bovine colostrum or bovine immune colostrum in patients with juvenile chronic arthritis: Evaluation of effect on gut defence mechanisms. *Inflammopharmacol.* 5:219–236.

Manning, T.S., Rastall, R., and Gibson, G. 2004. Prebiotics and lactic acid bacteria. In *Lactic Acid Bacteria, Microbiological and Functional Aspects*, eds. S. Salminen, A. von Wright, and A. Ouwehand, pp. 407–418. New York: Marcel Dekker AG.

Matto, J., Alakomi, H.L., Vaari, A., Virkajarvi, I., and Saarela, M. 2006. Influence of processing conditions on *Bifidobacterium animalis* subsp. *lactis* functionality with a special focus on acid tolerance and factors affecting it. *Int. Dairy J.* 16:1029–1037.

Mayer, J., Conrad, J., Klaiber, I., Lutz-Wahl, S., Beifuss, U., and Fischer, L. 2004. Enzymatic production and complete nuclear magnetic resonance assignment of the sugar lactulose. *J. Agric. Food Chem.* 23:6983–6990.

McBrearty, S., Ross, R.P, Fitzgerald, G.F., Collins, J.K., Wallace, J.M., and Stanton, C. 2001. Influence of two commercially available bifidobacteria cultures on Cheddar cheese quality. *Int. Dairy J.* 11:599–610.

Medici, M., Vinderola, C.G., Weill, R., and Perdigón, G. 2005. Effect of fermented milk containing probiotic bacteria in the prevention of an enteroinvasive *Escherichia coli* infection in mice. *J. Dairy Res.* 72:243–249.

Mitsuoka, T. 1992. Intestinal flora and aging. *Nutr. Rev.* 50:438–446.

Morata de Ambrosini, V., Gonzalez, S., Perdigón, G., Pesce de Ruiz Holgado, A., and Oliver, G. 1996. Chemical composition of the cell wall of lactic acid bacteria and related species. *Chem. Pharm. Bull.* 44:2263–2267.

Moreau, M.C. and Gaboriau-Routhiau, V. 2000. Influence of resident intestinal microflora on the development and functions of the intestinal-associated lymphoid tissue. In *Probiotics Immunomodulation by the Gut Flora and Probiotics*, eds. R. Fuller and G. Perdigón, pp. 69–114. Dordrecht, the Netherlands: Kluwer Academic Publishers.

Mortazavian, A.M., Ehsani M.R., Azizi A., et al. 2008. Viability of calcium-alginate-microencapsulated probiotic bacteria in Iranian yogurt drink (Doogh) during refrigerated storage and under simulated gastrointestinal conditions. *Aust. J. Dairy Technol.* 63:24–29.

Murad, H.A., Sadek, Z.I., and Fathy, F.A. 1998. Production of bifidus kariesh cheese. *Deutsche Lebensmittel-Rundschau.* 94:409–412.

Nebra, Y. and Blanch, A.R. 1999. A new selective medium for *Bifidobacterium* spp. *Appl. Environm. Microbiol.* 65:5173–5176.

Noriega, L., Cuevas, I., Margolles, A., and de los Reyes-Gavilán, C.G. 2006. Deconjugation and bile salts hydrolase activity by *Bifidobacterium* strains with acquired resistance to bile. *Int. Dairy J.* 16:850–855.

Ofek, I. and Doyle, J.D. 1994. Interaction of bacteria with phagocytic cells. In *Bacterial Adhesion to Cells and Tissues*, eds. I. Ofek and J.D. Doyle, pp. 171–194. New York: Chapman & Hall.

Ong, L., Henriksson A., and Shah, N.P. 2007. Chemical analysis and sensory evaluation of Cheddar cheese produced with *Lactobacillus acidophilus*, *Lb. casei*, *Lb. paracasei* or *Bifidobacterium* sp. *Int. Dairy J.* 17:937–945.

Onggo, I. and Fleet, G.H. 1993. Media for the isolation and enumeration of lactic acid bacteria from yoghurts. *Aust. J. Dairy Technol.* 48:89–92.

Ouwehand, A.C. and Salminen, S.J. 1998. The health effects of cultured milk products with viable and non-viable bacteria. *Int. Dairy J.* 8:749–758.

Ouwehand, A.C. and Vesterlund, S. 2004. Antimicrobial components from lactic acid bacteria. In *Lactic Acid Bacteria. Microbiological and Functional Aspects* (3rd edition), revised and expanded, eds. S. Salminen, A. von Wright, A. Ouwehand, pp. 375–396. New York: Marcel Dekker, Inc.

Pacher, B. and Kneifel, W. 1996. Development of a culture medium for the detection and enumeration of bifidobacteria in fermented milk products. *Int. Dairy J.* 6:43–64.

Payne, J.F., Morris, A.E.J., and Beers, P. 1999. Note: Evaluation of selective media for the enumeration of *Bifidobacterium* sp. in milk. *J. Appl. Microbiol.* 86:353–358.

Perdigón G, Fuller R, and Raya R. 2001. Lactic acid bacteria and their effect on the immune system. *Curr. Issues Intest. Microbiol.* 2:27–42.

Pérez-Conesa, D., López, G., Abellán, P., and Ros, G. 2006. Bioavailability of calcium, magnesium and phosphorus in rats fed probiotic, prebiotic and synbiotic powder follow-up infant formulas and their effect on physiological and nutritional parameters. *J. Sci. Food Agric.* 86:2327–2336.

Pérez-Conesa, D., López, G., and Ros, G. 2007. Effects of probiotic, prebiotic and synbiotic follow-up infant formulas on large intestine morphology and bone mineralisation in rats. *J. Sci. Food Agric.* 87:1059–1068.

Perrin, S., Grill, J.P., and Schneider, F. 2000. Effects of fructooligosaccharides and their monomeric components on bile salt resistance on three species of bifidobacteria. *J. Appl. Microbiol.* 88:968–974.

Phillips, M., Kailasapathy, K., and Tran, L. 2006. Viability of commercial probiotic cultures (*L. acidophilus*, *Bifidobacterium* sp., *L. casei*, *L. paracasei* and *L. rhamnosus*) in cheddar cheese. *Int. J. Food Microbiol.* 108:276–280.

Prajapati, J.B. and Nair, B.M. 2003. The history of fermented foods. In *Handbook of Fermented Functional Foods*, ed. E.R. Farnworth, pp. 1–26. Boca Raton, FL: CRC Press.

Puccio, G., Cajozzo, C., Meli, F., Rochat, F., Grathwohl, D., and Steenhout, P. 2007. Clinical evaluation of a new starter formula for infants containing live *Bifidobacterium longum* BL999 and prebiotics. *Nutrition* 23:1–8.

Rada, V. and Koc, J. 2000. The use of mupirocin for selective enumeration of bifidobacteria in fermented milk products. *Milchwissenschaft* 55:65–67.

Rasic, J.L.J. and Kurmann, J.A. 1978. *Yoghurt-Scientific Grounds, Technology, Manufacture and Preparations*. Copenahen, Denmark: Technical Dairy Pub. House.

Ravula, R. R. and Shah, N. P. 1998a. Selective enumeration of *Lactobacillus casei* from yogurts and fermented milks. *Biotechnol. Tech.* 12:819–822.

Ravula., R.R. and Shah, N.P. 1998b. Viability of probiotic bacteria in fermented frozen dairy desserts. *Food. Aust.* 50:136–139.

Reid, A.A., Champagne, C.P., Gardner, N., Fustier P., and Vuillemard, J.C. 2007. Survival in food systems of *Lactobacillus rhamnosus* R011 microentrapped in whey protein gel particles. *J. Food Sci.* 72:31–37.

Reid, G., Sanders, M.E., Gaskins, H.R., Gibson, G.R., et al. 2003a. New scientific paradigms for probiotics and prebiotics. *Clin. Gastroenterol.* 37:105–118.

Reid, G., Jass, J., Sebulsky, M.T., and McCormick, J.K. 2003b. Potential uses of probiotics in clinical practice. *Clin. Microbiol. Rev.* 16:658–672.

Reuter, S., Nygaard, A.R., and Zimmermann, W. 1999. Beta-galactooligosaccharide synthesis with beta-galactosidases from *Sulfolobus solfataricus*, *Aspergillus oryzae*, and *Escherichia coli*. *Enzyme Microbiol. Technol.* 25:509–516.

Rinkinen, M., Jalava, K., Westermarck, E., Salminen, S., and Ouwehand, A.C. 2003. Interaction between probiotic lactic acid bacteria and canine enteric pathogens: A risk factor for intestinal *Enterococcus faecium* colonization? *Vet. Microbiol.* 92:111–119.

Riordan, K.O. and Fitzgerald, G.F. 1998. Evaluation of bifidobacteria for the production of antimicrobial compounds and assessment of performance in cottage cheese at refrigeration temperature. *J. Appl. Microbiol.* 85:103–114.

Roberfroid, M.B. 1998. Prebiotics and synbiotic concepts and nutritional properties. *Br. J. Nutr.* 80:S197–S202.

Ross R.P., Desmond C., Fitzgerald G.F., and Stanton C. 2005. Overcoming the technological hurdles in the development of probiotic foods. *J. Appl. Microbiol.* 98:1410–1417.

Rossi, M., Corradini, C., Amaretti, A., et al. 2005. Fermentation of fructooligosaccharides and inulin by bifidobacteria: A comparative study of pure and fecal cultures. *Appl. Environ. Microbiol.* 71:6150–6158.

Russell, M.W., Bobek, L.A., Brock, J.H., Hajishengallis, G. and Tenuovo, J. 2005. Innate humoral defense factors. In *Mucosal Immunology* (3rd edition), eds. J. Mestecky, M.E. Lamm, W. Strober, J. Bienenstock, J.R. McGhee and L. Mayer, pp. 73–93. San Diego, CA: Academic Press.

Rybka, S. and Kailasapathy, K. 1996. Media for the enumeration of yoghurt bacteria. *Int. Dairy J.* 6:839–850.

Rycroft, C.E, Jones M.R., Gibson G.R., and Rastall, R.A. 2001. A comparative in vitro evaluation of the fermentation properties of prebiotic oligosaccharides. *J. Appl. Microbiol.* 91:878–887.

Saarela, M., Hallamaa, K., Mattila-Sandholm, T., and Matto, J. 2003. The effect of lactose derivatives lactulose, lactitol and lactobionic acid on the functional and technological properties of potentially probiotic *Lactobacillus* strains. *Int. Dairy J.* 13:291–302.

Saavedra, J.M., Bauman, N., Oung, I., Perman, J., and Yolken, R. 1994. Feeding of *Bifidobacterium bifidum* and *Streptococcus thermophilus* to infants in hospital for prevention of diarrhea and shedding of rotavirus. *Lancet* 344:1046–1049.

Sako, T., Matsumoto, K., and Tanaka, R. 1999. Recent progress on research and applications of non-digestible galacto-oligosaccharides. *Int. Dairy J.* 9:69–80.

Salminen, S., Isolauri, E., and Salminen, E. 1996. Clinical uses of probiotics for stabilizing the gut mucosal barrier: Successful strains and future challenges. *Antonie Van Leeuwenhoek* 70:347–58.

Salminen, S., von Wright, A., and Ouwehand, A. 2004. *Lactic Acid Bacteria. Microbiology and Functional Aspects* (3rd edition revised and expanded). New York: Marcel Dekker, Inc.

Sanderson, I.R. 1998. Dietary regulation of genes expressed in the developing intestinal epithelium. *Am. J. Clin. Nutr.* 68:999–1005.

Saxelin, M., Tynkkynen, S., Mattila-Sandholm, T., and Vos, W.M. 2005. Probiotic and other functional microbes: From markets to mechanisms. *Curr. Opin. Biotechnol.* 16:204–211.

Schmid, K., Schlothauer, R.C., Friedrich, U., Staudt, C., Apajalahti, J., and Hansen, E.B. 2006. Development of probiotic food ingredients. In *Probiotics in Food Safety and Human Health*, eds. I. Goktepe, V.K. Juneja and M. Ahmedna, pp. 35–66. Boca Raton, FL: Taylor & Francis Group.

Senok, A.C., Ismaeel, A.Y., and Botta, G.A. 2005. Probiotics: Facts and myths. *Clin. Microbiol. Infect.* 11:958–966.

Shah, N.P. 2000. Probiotic bacteria: Selective enumeration and survival in dairy foods. *J. Dairy Sci.* 83:894–907.

Shah, N.P. 2007. Functional cultures and health benefits. *Int. Dairy J.* 17:1262–1277.

Shaw, S.K. and White, C.H. 1994. Survival of bifidobacteria in reduced fat Cheddar cheese. *J. Dairy Sci.* 77 (Suppl. 1), 4.

Shin, H.S., Lee, J.H., Pestka, J.J., and Ustunol, Z. 2000. Growth and viability of commercial *Bifidobacterium* spp. in skim milk containing oligosaccharides and inulin. *J. Food Sci.* 65:884–887.

Songisepp, E., Kullisaar, T., Hütt, P., et al. 2004. A new probiotic cheese with antioxidative and antimicrobial activity. *J. Dairy Sci.* 87:2017–2023.

Stanton, C., Desmond, C., Coakley, M., Collins, J.K., Fitzgerald, G., and Ross, R.P. 2003. Challenges facing development of probiotic-containing functional foods. *In Handbook of Fermented Functional Foods*, ed. E.R. Farnworth, pp. 27–58. Boca Raton, FL: CRC Press.

Stanton, C., Gardiner, G., Lynch, P.B., Collins, J.K., Fitzgerald, G., and Ross, R.P. 1998. Probiotic cheese. *Int. Dairy J.* 8: 491–496.

Tabasco, R., Paarup, T., Janer, C., Peláez, C., and Requena, T. 2007. Selective enumeration and identification of mixed cultures of *Streptococcus thermophilus, Lactobacillus delbrueckii* subsp. *bulgaricus*, *L. acidophilus, L. paracasei* subsp. *paracasei* and *Bifidobacterium lactis* in fermented milk. *Int. Dairy J.* 17:1107–1114.

Talwalkar, A. and Kailasapathy, K. 2004. Comparison of selective and differential media for the accurate enumeration of strains of *Lactobacillus acidophilus, Bifidobacterium* spp. and *Lactobacillus casei* complex from commercial yoghurts. *Int. Dairy J.* 14:143–149.

Tannock, G.W. 2003. The intestinal microflora. In *Gut flora, Nutrition, Immunity and Health*, eds. R. Fuller and G. Perdigón, pp. 1–23. Oxford, U.K.: Balckwell Publishing.

Tejada-Simon, M.V. and Pestka, J.J. 1999. Proinflammatory cytokine and nitric oxide induction in murine macrophages by cell wall and cytoplasmic extracts of lactic acid bacteria. *J. Food Prot.* 62:1435–1444.

Tharmaraj, N. and Shah, N.P. 2003. Selective enumeration of *Lactobacillus delbrueckii* ssp. *bulgaricus*, *Streptococcus thermophilus, Lactobacillus acidophilus*, Bifidobacteria, *Lactobacillus casei, Lactobacillus rhamnosus*, and *Propionibacteria*. *J. Dairy Sci.* 86:2288–2296.

Tuomola, E., Crittenden, R., Playne, M., Isolauri, E., and Salminen, S. 2001. Quality assurance criteria for probiotic bacteria. *Am. J. Clin. Nutr.* 73(Suppl. 2):393S–398S.

Van de Casteele, S., Vanheuverzwijn, T., Ruyssen, T., Van Asschea, P., Swings, J., and Huys, G. 2006. Evaluation of culture media for selective enumeration of probiotic strains of lactobacilli and bifidobacteria in combination with yoghurt or cheese starters. *Int. Dairy J.* 16:1470–1476.

Vasiljevic, T. and Shah, N.P. 2008. Probiotics from Metchnikoff to bioactives. *Int. Dairy J.* 18:714–728.

Vinderola, G. 2008. Dried cell-free fraction of fermented milks: New functional additives for the food industry. *Trends Food Sci. Technol.* 19:40–46.

Vinderola, C.G. and Reinheimer, J.A. 1999. Culture media for the enumeration of *Bifidobacterium bifidum* and *Lactobacillus acidophilus* in the presence of yoghurt bacteria. *Int. Dairy J.* 9:497–505.

Vinderola, C.G. and Reinheimer, J.A. 2000. Enumeration of *Lactobacillus casei* in the presence of *L. acidophilus*, bifidobacteria and lactic starter bacteria in fermented dairy products. *Int. Dairy J.* 10: 271–275.

Vinderola, C.G., Prosello W., Ghiberto D., and Reinheimer, J.A. 2000a. Viability of probiotic (*Bifidobacterium*, *Lactobacillus acidophilus* and *Lactobacillus casei*) and nonprobiotic microflora in Argentinian fresco cheese. *J. Dairy Sci.* 83:1905–1911.

Vinderola, C.G., Bailo, N., and Reinheimer, J.A. 2000b. Survival of probiotic microflora in Argentinian yoghurts during refrigerated storage. *Food Res. Int.* 33:97–102.

Vinderola, C.G., Gueimonde, M., Delgado, T., Reinheimer, J.A., and de los Reyes-Gavilan, C.G. 2000c. Characteristics of carbonated fermented milk and survival of probiotic bacteria. *Int. Dairy J.* 10:213–220.

Vinderola, C.G., Costa, G.A., Regenhardt, S., and Reinheimer, J.A. 2002a. Influence of compounds associated with fermented dairy products on the growth of lactic acid starter and probiotic bacteria. *Int. Dairy J.* 12:579–589.

Vinderola, C.G., Mocchiutti, P., and Reinheimer, J.A. 2002b. Interactions among lactic acid starter and probiotic bacteria used for fermented dairy products. *J. Dairy Sci.* 85:721–729.

Vinderola, C.G., Matar, C., and Perdigón, G. 2005a. Role of intestinal epithelial cells in the immune effects mediated by Gram-positive probiotic bacteria. Involvement of Toll-like receptors. *Clin. Diag. Lab. Immunol.* 12:1075–1084.

Vinderola, C.G., Duarte, J., Thangavel, D., Perdigón, G., Farnworth, E., and Matar, C. 2005b. Interactions among lactic acid starter and probiotic bacteria used for fermented dairy products. *J. Dairy Sci.* 85:721–729.

Vinderola, C.G., Perdigón, G., Duarte, J., Farnworth, E., and Matar, C. 2006a. Modulation of the gut immune response by the exopolysaccharide produced by *Lactobacillus kefiranofaciens*. *Cytokine* 36:254–260.

Vinderola, C.G., Perdigón, G., Duarte, J., Farnworth, E., and Matar, C. 2006b. Effects of the oral administration of the products derived from milk fermentation by kefir microflora on the immune stimulation. *J. Dairy Res.* 73:472–479.

Vinderola, C.G., Perdigón, G., Duarte, J. Farnworth, E. and Matar, C. 2006c. Effects of kefir fractions on innate immunity. *Immunobiology* 211:149–156.

Vinderola, C.G., Matar, C., Palacios, J., and Perdigón, G. 2007a. Mucosal immunomodulation by the non-bacterial fraction of milk fermented by *Lactobacillus helveticus* R389. *Int. J. Food Microbiol.* 115:180–186.

Vinderola, C.G., Matar, C., and Perdigón, G. 2007b. Milk fermented by *Lactobacillus helveticus* R389 and its non-bacterial fraction confer enhanced protection against *Salmonella enteritidis* serovar Typhimurium infection in mice. *Immunobiology* 212:107–118.

Vinderola, C.G., Matar, C., and Perdigón, G. 2007c. The non-bacterial fraction of milk fermented by *L. helveticus* R389 activates calcineurin as a signal to stimulate the gut mucosal immunity. *BMC Gastroenterol.* 8:19.

Vinderola, C.G., de Moreno de LeBlanc, A., Perdigón, G., and Matar, C. 2008. Biologically active peptides released in fermented milks: Role and functions. In *Handbook of Fermented Functional Foods* (2nd edition), ed. E.R. Farnworth, pp. 177–202. Boca Raton, FL: CRC Press.

Voltan, S., Castagliuolo, I., Elli, M., et al. 2007. Aggregating phenotype in *Lactobacillus crispatus* determines intestinal colonization and TLR2 and TLR4 modulation in murine colonic mucosa. *Clin. Vac. Immunol.* 14:1138–1148.

Warrand, J. 2006. Healthy polysaccharides-The next chapter in food products. *Food Technol. Biotechnol.* 44:355–370.

WHO/FAO. 2007. FAO Technical meeting on prebiotics. http://www.fao.org/ag/agn/agns/files/Prebiotics_Tech_Meeting_Report.pdf (accessed *November 5th*, 2008).

Yilmaztekin, M., Ozer, B.H., Atasoy F. 2004. Survival of *Lactobacillus acidophilus* LA-5 and *Bifidobacterium bifidum* BB-02 in white-brined cheese. *Int. J. Food Sci. Nutr.* 55:53–60.

21 Uses of Whole Cereals and Cereal Components for the Development of Functional Foods

Dimitris Charalampopoulos, Severino S. Pandiella, and Colin Webb

CONTENTS

21.1 INTRODUCTION

The interest in developing functional foods is thriving, driven largely by the market potential for foods that can improve the health and well being of consumers. The concept of functional foods includes foods or food ingredients that improve health and well being (Katan and De Roos 2004). Established types of functional food products include, among others, products designed to reduce high blood pressure, cholesterol, risk of heart disease, and relief from constipation (Katan and De Roos 2004). Over the last 15 years the functional food research has moved progressively toward the development of novel functional food ingredients. Notable examples include omega 3 fatty acids, plant sterols, polyphenols, probiotics, and prebiotics (Katan and De Roos 2004, Kern 2006, Jones and Jew 2007).

Probiotics are defined as "live microorganisms, which when administered in adequate amounts confer a health benefit on the host" (Pineiro and Stanton 2007). Common microorganisms used in probiotic preparations are predominantly *Lactobacillus* species, such as *L. acidophilus*, *L. johnsonii*, *L. gasseri*, *L. casei*, *L. reuteri*, *L. rhamnosus*, *L. brevis*, and *L. plantarum*, and *Bifidobacterium* species, such as *B. longum*, *B. infantis*, and *B. lactis* (Reid et al. 2006). The incorporation of probiotic strains in traditional food products has been established in the dairy industry, leading to the production of novel types of fermented milks, bio-yoghurts, and cheeses, as detailed in Chapter 20. Probiotics are also available as dietary supplements in dried form.

A prebiotic is a selectively fermented ingredient that allows specific changes, both in the composition and/or activity in the gastrointestinal microbiota that confers benefits upon host well being and health (Gibson et al. 2004). The selective properties of prebiotics are mainly related to the growth of bifidobacteria and lactobacilli at the expense of other bacterial groups such as bacteroides, clostridia, and enterobacteria (Macfarlane et al. 2006). The established prebiotics include fructooligosaccharides (FOS), galactooligosaccharides (GOS), and inulin, although prebiotic potential has been shown for xylo-oligosaccharides, soy-oligosaccharides, gluco-oligosaccharides, isomalto-oligosaccharides, and gentio-oligosaccherides. The products of prebiotic fermentation include short chain fatty acid (SCFA). They are very important for the host health as they affect the proliferation of enterocytes, inflammation, colocteral carcinogenesis, colonization by pathogens, and the production of nitrogenous metabolites (Ouwehand et al. 2005, Tuohy et al. 2005).

The development of products consisting of probiotics and prebiotics is a challenge for the food industry in its effort to use the abundant natural resources by producing high-quality functional products and meet the consumer's demands for healthier foods. In this respect, cereals have been investigated in recent years regarding their potential uses. Cereals contribute over 60% of the world food production providing dietary fiber, proteins, energy, minerals, and vitamins required for human health. Possible applications in functional food formulations could be

- As fermentable substrates for the growth of probiotic microorganisms
- As potential prebiotics due to their content in nondigestible carbohydrates
- As dietary fiber promoting several beneficial physiological effects

21.2 CEREALS AS SUBSTRATES FOR PROBIOTICS

Lactic acid fermentation of cereals is a long established processing method and is being used in Asia and Africa for the production of foods in various forms, such as beverages, gruels, and porridges. In the Western countries, cereals, like wheat and rye, are used for sourdough production, which is traditionally prepared by adding a prefermented sourdough of good quality to the dough. Similar to such processes, the production of a conceptual probiotic fermented cereal product involves the incorporation of a well-defined probiotic strain into a fermentable cereal substrate. The fermentation is carried out under controlled conditions to produce a fermented food with defined and consistent organoleptic characteristics, as well as functional properties. However, in designing such products several technological aspects have to be considered. These include the composition and processing of the cereal grains, the growth capability and productivity of the probiotic culture, the stability of the probiotic during storage, and the organoleptic properties and nutritional value of the final product. These will be looked at in more detail in the following sections.

21.2.1 Effect of Cereal Composition on Growth of Probiotics

For successful delivery in foods probiotics must be able to remain viable during processing and storage. A concentration of at least 10^7 viable cells/mL at the time of consumption is recommended (Gomes and Malcata 1999, Ross et al. 2005). Therefore, the ability of the strain to use the available nutrients in the medium is a very important criterion in the selection of a suitable strain, especially taking into account that lactobacilli and bifidobacteria have complex nutritional requirements, which vary a lot from species to species (Severson 1998).

The main carbohydrate components of cereal grains are starch, water soluble and insoluble components of dietary fiber, and several free sugars including glucose, glycerol, stachyose, xylose, fructose, maltose, sucrose, and arabinose. The contents of these components depend on the variety (Becker and Hanners 1991), the processing, and the amount of water added to the cereal flour. Table 21.1, which is compiled from various published works, presents the composition of different varieties of

TABLE 21.1
Composition of Foods Expressed as 100 g of Edible Portion

Parameter	Malt	Rice	Corn	Wheat	Oats	Sorghum	Millet	Milk (Liquid)
Water (%)	8	12	13.8	12	13.1	11	11.8	87.4
Protein (g)	13.1	7.5	8.9	13.3	9.3	11	9.9	3.5
Fat (g)	1.9	1.9	3.9	2.0	5.9	3.3	2.9	3.5
Carbohydrates (g)	77.4	77.4	72.2	71.0	62.9	73.0	72.9	4.9
Fiber (g)	5.7	0.9	2.0	2.3	2.3	1.7	3.2	N.D.
Ash (g)	2.4	1.2	1.2	1.7	2.3	1.7	2.5	0.7
Ca (mg)	40	32	22	41	110	28	20	118
P (mg)	330	221	268	372	380	287	311	93
Fe (mg)	4.0	1.6	2.1	3.3	11	4.4	68	Trace
K (mg)	400	214	284	370	170	350	430	144
Thiamin (mg)	0.49	0.34	0.37	0.55	0.60	0.38	0.73	0.03
Riboflavin (mg)	0.31	0.05	0.12	0.12	0.14	0.15	0.38	0.17
Niacin (mg)	900	1.7	2.2	4.3	1.3	3.9	2.3	0.1
Mg (mg)	140	88	147	113	130	ND	162	13

Sources: Compiled from Morgan, D.E., *J. Sci. Food Agric.*, 19, 394, 1968; Chavan, J.K. and Kadam, S.S., *Crit. Rev. Food Sci. Nutr.*, 28, 349, 1989; Peterson, D.M. et al., *J. Agric. Food Chem.*, 23, 9, 1975; Severson, D.K., *Nutritional Requirements of Commercially Important Microorganisms*, Nagodawithana, W. and Reed, G., Eds., Esteekay Associates Inc., Milwaukee, WI, 1998, 258–297.

cereals compared to that of milk (Morgan 1968, Chavan and Kadam 1989, Peterson et al. 1975, Severson 1998). Cereals have higher content in some of the essential vitamins than milk, higher content of dietary fiber, and increased amount of minerals, but lower amount of fermentable carbohydrates.

Most of the published work has studied the growth of probiotics in oat, wheat, and barley based media. Marklinder and Lonner (1992) investigated the potential of a fermented oatmeal soup (18.5%) containing intestinal *Lactobacillus* strains as the base for a nutritive solution for enteral feeding. After testing several strains, it was concluded that oats are in general a suitable substrate for the growth of lactobacilli, regardless of the differences between species and strains. Among the strains tested, *L. acidophilus* exhibited the slowest rates of pH reduction, and the lowest levels of viable cells in the final product, probably due to its high requirements for several nutrients (Morishita et al. 1981). The highest viable cell counts, 3×10^9 CFU/mL and 1×10^9 CFU/mL, were achieved using *L. plantarum* and *L. reuteri*, respectively. Addition of malted barley flour, proteases, and amino acids increased the rate of pH decrease and the total amount of lactobacilli in the final product (Marklinder and Lonner 1994). In similar studies, human derived *L. acidophilus*, *L. brevis*, and *L. plantarum* strains were successfully cultivated in oat media, reaching concentrations higher than 10^8 CFU/mL (Bekers et al. 2001, Martensson et al. 2003, Patel et al. 2004a, Angelov et al. 2006, Kedia et al. 2008). Oat media (unsupplemented and supplemented with soy protein) were also shown to support the growth of certain bifidobacteria isolated from the "feces" of human elderly subjects, albeit to low levels (Laine et al. 2003).

In another study, human derived *L. reuteri*, *L. plantarum*, and *L. acidophilus* strains were cultured in malt, barley, and wheat extracts formulated without the addition of any supplements (Charalampopoulos et al. 2002). The malt medium supported better the cell growth than barley and wheat probably due to the increased amounts of fermentable sugars (i.e., maltose, sucrose, glucose, and fructose) and free amino nitrogen. Each strain demonstrated a specific preference for one or more sugars, which has been reported for lactobacilli isolated from fermented cereal products (Gobbetti and Corsetti 1997). *L. plantarum* exhibited the highest cell population probably due to its unique ability to tolerate low pH values by maintaining a proton (pH) and charge gradient between the inside and outside of the cells even in the presence of high amounts of lactate and protons (Giraud et al. 1998). *L. acidophilus* exhibited the poorest growth probably because of substrate deficiency in specific nutrients. In a recent study by Rozada-Sanchez et al. (2008), a malt medium supplemented with yeast extract was used for the growth of several *Bifidobacterium* species including *B. breve*, *B. adolescentis*, *B. longum*, and *B. infantis*. All bacteria grew well in the medium, reaching cell concentrations higher than 10^8 CFU/mL. However, interestingly the main sugars of the medium (fructose, glucose, maltose, and maltotriose) and free amino nitrogen were consumed in very low quantities during the fermentation.

In the studies of Helland et al. (2004a,b), several commercial and well-established research probiotic strains were used to develop porridges and puddings based on maize. It was observed that in the case of the porridges the maximum cell count was obtained after 12 h fermentation (between 10^7 and 10^8 CFU/mL), with a pH below 4.0.

Similar viable cell counts were obtained in the case of the puddings; the pH in this case ranged between 3.4 and 4.4, depending on the strain used.

21.2.2 Effect of Cereals on Probiotic Survival

Besides the growth characteristics of the probiotic strain, a key factor in the selection of a suitable probiotic starter is its ability to survive the acidic environment of the fermented product during storage and the adverse conditions of the gastrointestinal tract, such as the low pH of the stomach and the high concentration of bile acids in the small intestine.

There are several factors influencing the survival of a probiotic microorganism during storage, including the pH, the levels of organic acids, the presence of hydrogen peroxide, the presence or absence of oxygen, the interactions with other microorganisms, etc. Among these, pH is the most important one, especially from the perspective of a fermented food product (Gomes and Malcata 1999, Ross et al. 2005). Although differences exist between species and specific strains, lactobacilli are generally considered to be intrinsically resistant to low pH values, e.g., between 3 and 4 (Kashket 1987, Hood and Zottola 1988, Jin et al. 1998, Corcoran et al. 2008, Kotsou et al. 2008). The above suggest that in the case of a fermented cereal product, the final pH, the concentrations of the organic acids in the fermented cereal product, and the intrinsic properties of the probiotic strain used are important in order to maximize probiotic viability during storage.

There is a very limited number of studies that have investigated the survival of probiotics in nondairy products, such as fermented cereals (Kabeir et al. 2005, Charalampopoulos and Pandiella 2008) or fermented vegetables, such as olives and cabbage (Lavermicocca et al. 2005, Yoon et al. 2006). Although there was a considerable dependence on the strains used, the studies with cereals showed that probiotic lactobacilli and bifidobacteria survived well during refrigerated storage at the low pH of the fermented cereals, ranging between 3.4 and 4. More specifically, in the study of Kabeir et al. (2005), the concentration of *B. longum* decreased less than 1 log after 14 days of storage, whereas in the study of Charalampopoulos and Pandiella (2008) the concentration of a human-derived *L. plantarum* strain decreased less than 2 log after 25 days of storage in malt, barley, and wheat media extracts. Overall, malt extract offered significantly better protection compared to the other extracts. Based on supporting experiments using model solutions containing varying amounts of glucose and lactic acid similar to those found in the fermented cereal extracts, it was deduced that the concentrations of fermentable sugars, and to a lesser extent lactic acid, were most likely influencing the survivability of *L. plantarum* in the cereal extracts (Charalampopoulos and Pandiella 2008). This could be due to the fact that, at low pH values, lactobacilli regulate their cytoplasmic pH mainly by the translocation of protons from the cytoplasm to the environment by an ATPase at the expense of ATP (Sanders et al. 1999, van de Guchte et al. 2002). Therefore, the presence of fermentable sugars in the product, which are metabolized slowly during storage at 4°C through the glycolytic system, probably resulted in the accumulation of additional ATP, permitting optimal H^+ extrusion, and consequently leading to increased viability during storage. The increased survival of probiotic lactobacilli in the presence of glucose in very acidic environments has also been reported by Corcoran et al. (2005).

The survival of the probiotic strains in the upper parts of the gastrointestinal part, i.e., the stomach and small intestine, is also influenced by the physicochemical properties of the food carrier used for delivery. The buffering capacity and the pH of the carrier medium are significant factors, since food formulations with pH ranging from 3.5 to 4.5 and high buffering capacity would increase the pH of the gastric tract and thus enhance the stability of the probiotic strain (Kailasapathy and Chin 2000, Zarate et al. 2000). In relation to cereals, it has been shown based on experiments using *in vitro* model systems mimicking the acidic conditions of the stomach that cereal components, such as whole cereal extracts and cereal fiber, protected *L. plantarum* and *L. acidophilus* (Charalampopoulos et al. 2003, Michida et al. 2006). Similarly, cereal extracts were also shown to exert a protective effect on various intestinal lactobacilli in model solutions containing high concentrations of bile acids, as pointed out by Patel et al. (2004b). These authors related the protective effect with the presence of fermentable sugars in the cereals extracts.

Specific cereal fractions have also been shown to physically protect probiotic bacteria from adverse conditions. Wang et al. (1999a) have shown that *Bifidobacterium* strains were able to adhere to high amylose maize starch granules and thus be protected protect in conditions simulating the stomach and small intestine. In addition, according to *in vivo* experiments using mice, a sixfold better recovery from mice feces was observed for the amylase adhered cells compared to the control. The mechanisms of adhesion of probiotic bifidobacteria to starch granules from various sources including maize potato, oat, and barley have been investigated in detail by Crittenden et al. (2001a) and O'Riordan et al. (2001).

21.2.3 Nutritional and Organoleptic Properties of Fermented Cereal Products

Lactic acid fermentation usually improves the nutritional value and digestibility of cereals. Cereals are limited in essential amino acids such as threonine, lysine, and tryptophan, thus making their protein quality poorer compared to meat and milk (Chavan and Kadam 1989, Kohajdova and Karovicova 2007). Their protein digestibility is also lower than that of meat, partially due to the presence of phytic acid, tannins, and polyphenols, which bind to proteins, thus making them indigestible (Oyewole 1997). Lactic acid fermentation of different cereals, such as maize, sorghum, and finger millet, has been found to reduce the amounts of phytic acid and tannins, and improve protein availability (Lorri and Svanberg 1993, Tou et al. 2007, Palacios et al. 2008a). Moreover, in a recent publication by Palacios et al. (2008b), 23 bifidobacteria strains were screened for their availability to degrade phytic acid, and the one showing the highest degrading ability was used for whole-wheat bread making. To a large extent related to phytate degradation, the fermentation of cereals, such as maize, sorghum, and millet, by lactic acid bacteria (LAB) and yeasts has been shown to increase the bioavailability of minerals including iron and zinc (Towo et al. 2006, Hemalatha et al. 2007, Proulx and Reddy 2007). However, in a recent study by

Bering et al. (2007), it was reported that a fermented oat product containing a probiotic *L. plantarum* strain did not increase iron absorption and no absorption seemed to occur in the distal part of the colon in healthy women subjects. Increased amounts of riboflavin, thiamine, niacin, and lysine due to the action of LAB in fermented blends of cereals were also reported (Hamad and Fields 1979, Sanni et al. 1999). In addition, increased amounts of folates and total phenolic compounds have been reported in cereal fermentation due to the action of folate-producing LAB and yeasts (Jagerstad et al. 2005, Katina et al. 2005, Kariluoto et al. 2006, Katina et al. 2007).

The grains of corn, sorghum, millet, barley, rye, and oats contain an appreciable amount of crude fiber and lack the gluten-like proteins of wheat. The traditional foods made from these grains usually lack flavor and aroma (Chavan and Kadam 1989). Lactic acid fermentation improves the sensorial value of the final product, which is very much dependent on the amounts of lactic acid, acetic acid, and several aromatic volatiles, such as higher alcohols and aldehydes, ethyl acetate, and diacetyl. The compounds are mainly produced via the homofermentative or heterofermentative metabolic pathways, depending on the strain used (Salim et al. 2006, Kohajdova and Karovicova 2007). The interaction of LAB with yeast, which are costarting cultures in most types of cereal products, such as sourdough, contributes significantly to the flavor characteristics of the final product (Annan et al. 2003, Salim et al. 2006). The above suggest that from a product development perspective, it is important to select the appropriate strain to efficiently control the distribution of the metabolic end products for both natural LAB starters (Damiani et al. 1996, Hansen and Schieberle 2005, Van der Meulen et al. 2007) and probiotic starters (Helland et al. 2004a,b, Escamilla-Hurtado et al. 2005, Gerez et al. 2008). Moreover, the end product distribution and the overall sensory profile of the product depend also on the composition of the substrate and the environmental conditions (pH, temperature, aerobiosis/anaerobiosis), whose control would allow specific fermentation routes to be channeled toward a more desirable product (Gobbetti et al. 1995, Katina et al. 2004, Hansen and Schieberle 2005, Katina et al. 2006).

Although there is not a lot of information specifically on the flavor profile of nondairy probiotic food products, the general consensus is that probiotic products obtained using a single probiotic strain as starter are hardly acceptable to consumers, as they lack sensory appeal due to a rather sour and acidic taste. For this reason, in the case of dairy-based products, the probiotic strains are often mixed with other bacteria, such as *Streptococcus thermophilus*, *Lactococcus lactis*, *L. delbrueckii* subsp. *bulgaricus* (Saarela et al. 2000, Martin-Diana et al. 2003, Xu et al. 2006, Savoie et al. 2007). Similarly, for cereal-based probiotic products, it may be necessary to include supporting strains that are able to bring out the preferred flavor to the product. As in the case of the dairy products, it is important that the supporting strains grow in the substrate and do not act antagonistically toward the probiotic strain. As mentioned previously, the most common microorganisms used in conjunction with LAB in fermented cereal products, such as sourdough, are yeasts, predominantly *Candida* and *Saccharomyces* species. The interaction between the yeast and LAB is considered to play a vital role toward the flavor of the sourdough (Annan et al. 2003, Hansen and Schieberle 2005, Guerzoni et al. 2007).

21.3 DIETARY FIBER FROM CEREAL GRAINS AND ITS POTENTIAL PREBIOTIC EFFECTS

21.3.1 Definition

Dietary fiber is the edible parts of plants or analogous carbohydrates, which resist the hydrolysis by alimentary tract enzymes. It includes nonstarch polysaccharides (NSP) as well as oligosaccharides. Dietary fiber can be divided into two categories according to their water solubility, each providing different physiological effects. Water-soluble fiber consists mainly of nonstarchy polysaccharides, mainly β-glucan and arabinoxylan. Water-insoluble fiber includes lignin, cellulose, and a part of hemicellulose, such as insoluble water-unextractable arabinoxylans (Asp 1996, Oscarsson et al. 1996, Gray 2008). The water-soluble fiber is readily fermented by colonic bacteria, leading to the production of SCFA, predominately acetate, propionate, and butyrate, whereas the water-insoluble fiber to a large extent is not degraded.

The content of dietary fiber varies between cereal grains and varieties (Table 21.2) (Herrera et al.1998, Malkki and Virtanen 2001, Nelson 2001, Cyran and Lapinski 2006, Holtekjolen et al. 2006). In cereal botanical components, the majority of soluble dietary fiber generally occurs in the cell wall of the endosperm and the aleurone layer. The procedures for the isolation of dietary fiber are based on the physical fractionation of grains using various combinations of pearling, grinding, sieving, and air classification (Knuckles and Chiu 1995, Yoon et al. 1995, Doehlert and Moore 1997, Izydorczyk et al. 2003, Mousia et al. 2004, Dornez et al. 2006, Hemery et al. 2007, Wang et al. 2007, Izydorczyk et al. 2008). The combination of such technologies gives the opportunity to produce processing streams of specified botanical components, such as the outer pericarp, the inner pericarp, the seed coat, the aleurone cells,

TABLE 21.2
Comparison of Total Dietary Fiber Content in Cereal Grains

Cereals	Total Dietary Fiber (%, db)
Legumes	13.6–28.9
Rye	15.5
Corn	15
Triticale	14.5
Oats	14
Wheat	12
Sorghum	10.7
Barley	10
Finger millet	6.2–7.2
Rice	3.9–0.2

Sources: Compiled from Herrera, I.M. et al., *Arch. Latinoam. Nutr.*, 48, 181, 1998; Nelson, A.L., *Cereal Foods World*, 46, 96, 2001.

the embryo, and the starchy endosperm, which are enriched in different components, e.g., specific types of dietary fiber. If purer fractions of dietary fiber are required then further extraction is required, which is usually performed with hot water, alkali, or ethanol (Bhatty et al. 1993, Westerlund et al. 1993, Brennan and Cleary 2005, Mandalari et al. 2005, Papageorgiou et al. 2005). According to Brennan and Cleary (2005) the cost is the main factor prohibiting the industrialization of these extraction methodologies. As a result, pure preparations of dietary fiber have often been ignored as functional ingredients for food products at the expense of more crude fractions.

21.3.2 The Physiological Effects of β-Glucan and Arabinoxylan

β-Glucans are the predominant components of the cell walls of cereal grains. They are unbranched polysaccharides composed of (1→4) and (1→3) linked β-D-glucopyranosyl units in varying proportions. On average, two or three (1→4)-linked units are separated by a single (1→3)-linkage (Holtekjolen et al. 2006). High molecular weight β-glucans, up to 3 million Da, are viscous due to labile cooperative associations (Wood et al. 1991). When exposed to physical forces and chemical or enzymatic hydrolysis, the molecular size of β-glucan reduces to achieve molecular weights of 0.4–2 million Da in typical food preparations (Beer et al. 1997). Among all the cereal grains, barley and oats contain the highest level of β-glucan, covering the ranges of 3%–11% and 3%–7% on a dry basis, respectively. It is usually concentrated in the inner aleurone cell walls and subaleurone endosperm cell walls of barley and oats (Wood et al. 1991, Brennan and Cleary 2005). Wheat is not recognized as a source of β-glucan because of its low content, which is usually below 1% on a dry basis.

Arabinoxylans are composed of (1→4) linked β-D-xylopyranosyl units, to which α-L-arabinofuranosyl units are linked as side chains. The degree and distribution of side chains are important factors for their physicochemical properties (Hopkins et al. 2003). Arabinoxylans account for 30% of the dry weight of most cereals. They occur in cell walls of the starchy endosperm cells and the aleurone layer in most cereals, constituting 60%–70% w/w of the total carbohydrates (Grootaert et al. 2007). Nonendospermic tissues of wheat, particularly the pericarp, also contain a very high concentration of arabinoxylan (Selvendran and Du Pont 1980).

Besides their contribution toward the nutritional and functional characteristics of foods, β-glucans and arabinoxylans have been shown to be beneficial for human health. β-glucans have been shown to lower blood cholesterol levels and thus reducing the risk of heart disease and controlling blood glucose levels (Delaney et al. 2003, Kerckhoffs et al. 2003, Li et al. 2003a,b, Brennan and Cleary 2005, Topping 2007).

Regarding the beneficial effects of arabinoxylans, which overall have been investigated to a lesser extent than β-glucans, it has been reported that arabinoxylan supplementation in rats improved fecal bulking and reduced the epithelial proliferation indices (Lu et al. 2000). In addition, there are indications that arabinoxylans have an effect against type II diabetes (Lu et al. 2004), and decrease postprandial glucose levels (Lu et al. 2000) and insulin response (Weickert et al. 2005).

21.3.3 Cereal Fractions as Potential Prebiotics

There is growing interest in the use of prebiotic oligosaccharides as functional food ingredients. A prebiotic is "a selectively fermented ingredient that allows specific changes both in the composition and/or activity in the gastrointestinal microbiota that confers benefits upon host well being and health" (Gibson et al. 2004). This is currently accepted as including increases in the populations of *Bifidobacterium* spp. and *Lactobacillus* spp., as both genera are viewed as positive for the host health and have a long history as probiotics (Hughes et al. 2007). Current prebiotics include fructans, i.e., FOS and inulin, as well as GOS. These are either produced by hydrolysis of plant polysaccharides, e.g., inulin, or in the case of GOS by enzymatic synthesis. Recently, there is a significant interest in investigating the potential prebiotic effect of cereal components, such as β-glucans, arabinoxylans, native oligosaccharides, and resistant starch.

21.3.3.1 Arabinoxylans

Crittenden et al. (2002) showed that out of 18 *Bifidobacterium* and 17 *Lactobacillus* strains tested in total, only four strains were able to grow in a medium supplemented with rye arabinoxylan. These included one *B. adolescentis* strain and three *B. longum* strains. According to Crittenden et al. (2002), it is possible that enzymatic hydrolysis of arabinosyl groups from the arabinoxylan occurs through the action of arabinofuranohydrolase, which is able to remove arabinosyl moieties from double-substituted xylose units in the polymer. This reaction takes place extracellularly, and the released arabinose is then imported into the cell where it can be used as carbon source. *B. adolescentis* has been shown to produce arabinofuranohydrolase (Van Laere et al. 1999). In the study of Crittenden et al. (2002) the arabinoxylan was fermented by several *Bacteroides* spp including *B. fragilis*, *B. thetaiotaomicron*, and *B. vulgatus*. The ability of *Bacteroides* to break down arabinoxylan has also been suggested by other researchers and was attributed to a range of depolymerizing enzymes (Berg et al. 1978, Salyers et al. 1981, Macfarlane and Gibson 1991, Hopkins et al. 2003, Chassard et al. 2007, Grootaert et al. 2007).

In the study of Hopkins et al. (2003), which investigated the degradation of cross-linked and non-cross-linked arabinoxylan by fecal bacteria from children, it was observed that arabinoxylan digestion was principally associated with increased viable counts of *B. fragilis*, which was supported by increases in the *Bacteroides–Porphyromonas–Prevotella* group, measured by fluorescent *in situ* hybridization (FISH). However, arabinoxylan did not support the growth of bifidobacteria. The breakdown of arabinoxylan resulted in increased propionate formation. In the *in vitro* study by Hughes et al. (2007), which investigated the fermentation of various arabinoxylan fractions by fecal bacteria, a structure–function relationship was observed between the molecular weight of arabinoxylan fractions and their fermentation properties, whereby the selectivity of arabinoxylan for bifidobacteria and lactobacilli increased as the molecular weight decreased. In addition, eubacteria increased significantly, whereas small increases in clostridia and bacteroides were observed. In similar *in vitro* studies using arabinoxylan fractions of different properties, i.e., a water unextractable fraction, a xylanase pretreated fraction, and a water-soluble extractable fraction, a stimulatory effect upon bifidobacteria and lactobacilli was observed for all fractions, with the greatest being in the case of the xylanase treated arabinoxylan (Vardakou et al. 2007, 2008). Overall, further studies on the potential prebiotic effects of arabinoxylan and arabinoxylan hydrolysates are needed, investigating their breakdown by intestinal bacteria. Ideally, these *in vitro* studies should be complemented by *in vivo* studies with animals and humans.

21.3.3.2 β-Glucans

Studies with β-glucans are less conclusive that the studies with arabinoxylans. In the study of Snart et al. (2006), it was shown that β-glucans and β-glucan hydrolysates effectively stimulated the growth of lactobacilli in a rat model. In addition, curdlan, which is similar in structure to β-glucan but comprises only 1→4 D-glucopyranosyl units, was previously shown to selectively increase the bifidobacteria population in the caecum of curdlan-fed rats (Shimizu et al. 2001). Moreover, based on pure culture experiments, oat β-glucan hydrolyzates were found to enhance the growth of certain bifidobacteria and lactobacilli (Jaskari et al. 1998), whereas in experiments conducted using a simulator of the human intestinal microbial ecosystem (SHIME), oat bran feeding favored the growth of bifidobacteria and resulted in an increase in the concentrations of acetate, propionate, and butyrate (Kontula 1999).

On the other hand, in the study of Crittenden et al. (2002), which used a complex growth medium supplemented with barley β-glucan, it was shown that β-glucan was not fermented by the strains of bifidobacteria (18 strains) and lactobacilli (17 strains) tested, but was fermented by *Bacteroides* and *Clostridium* spp. Similar observations were also reported by Hughes et al. (2008), this time using a growth medium supplemented with oat and barley β-glucan fractions of various molecular weights (from 137 to 327 kDa), inoculated with a fecal inoculum. The fermentation experiments were carried out for 24 h and the microbiota changes were monitored by FISH. The *Lactobacillus* and *Bifidobacterium* group did not increase, whereas the *Clostridium histolyticum* subgroup, and in some cases the *Bacteroides–Prevotella* groups, increased. In addition, there was a significant increase in propionate concentration compared to the control (no supplementation), which according to the authors can be attributed to the increases in the *C. histolyticum* subgroup, the *Bacteroides–Prevotella* group, and the Clostridia cluster IX. The significant increase in butyrate concentration could be attributed to the increases in *Ruminococcus–Eubacterium–Clostridium* cluster (Monsma et al. 2000, Hughes et al. 2008). Hughes et al. (2008) comment that although β-glucan was not "prebiotic" in the classical sense of increasing lactobacilli and bifidobacteria, it significantly modulated the microbial communities and the SCFA. These changes could potentially have hypocholesterolemic effects (Malkki et al. 1992, Drzikova et al. 2005, Furgal-Dierzuk 2006, Topping 2007) and should therefore be further investigated by conducting *in vivo* studies.

21.3.3.3 Resistant Starch

Resistant starch is defined as the "total amount of starch and products of starch degradation that resist starch digestion in the small intestine of healthy people" (Asp and Bjorck 1992). Depending on the various reasons for enzyme resistance resistant starch might be classified into four categories, i.e., physically inaccessible starch (RS1), resistant starch granules (RS2), retrograded starch (RS3), and chemically modified starches (RS4) (Sharma et al. 2008). Native high amylose starch is known to be high in the RS2 type, which is starch in its native granular form. This, after cooking and cooling gives the RS3 type (Berry 1986, Sievert and Pomeranz 1989). The manufacture of resistant starch usually involves partial acid hydrolysis and hydrothermal treatments (Brumovsky and Thompson 2001) or chemical modification (Wolf et al. 1999, Sharma et al. 2008). Similar to NSP, resistant starch escapes digestion in the small intestine and is fermented in the colon to SCFA. Resistant starch is thought to be the greatest contributor of SCFA in the large intestine (Bird et al. 2000). Also, compared to dietary fiber, e.g., oat bran, wheat bran, and cellulose, resistant starch produces a higher proportion of butyrate as well as higher ratio of butyrate to SCFA, which is believed to be consistent with good colonic health (Phillips et al. 1995, Martin et al. 1998). Resistant starch has been associated with various health benefits including reduction in the risk of colon cancer, increased fecal bulking, modulation of blood glucose level and blood cholesterol level, as well as potential prebiotic effects (Sharma et al. 2008).

The prebiotic potential of resistant starch has not been investigated extensively. Crittenden et al. (2001b), while examining 40 *Bifidobacterium* strains in monoculture experiments, identified only one strain (*B. lactis*) that was able to hydrolyze Hi-maize resistant starch. Moreover, it was reported that out of 38 types of human colonic bacteria including *Bifidobacterium*, *Bacteroides*, *Fusobacterium*, *Clostridium*, and *Eubacterium* species, a wide range of bacterial strains, were able to degrade gelatinized maize starch, whereas only a few strains of *Bifidobacterium* and *Clostridium* strains were capable of using high-amylose maize starch granules (Wang et al. 1999b). In a recent human study with 200 healthy volunteers, it was found that resistant starch (RS3 type) stimulated the growth of bifidobacteria, as measured by selective viable counting (Bouhnik et al. 2004). All these results suggest that resistant starch is likely to have a prebiotic effect. However, further research is needed to evaluate the fermentability both *in vitro* and *in vivo*, and investigate the effects of the type of resistant starch on its fermentability by the intestinal microbiota.

21.3.3.4 Oligosaccharides

There are mainly two types of oligosaccharides in cereal grains. They are galactosyl derivatives of sucrose (stachyose and raffinose) and fructosyl derivatives of sucrose (FOS) (Henry and Saini 1989, Sampath et al. 2008). The exact distributions of these polymers within the cereal grains have not been fully established. For example, in the case of wheat, reported values suggest their distributions in all milling products, including bran (Yamada et al. 1993) and germ (Pomeranz 1988). The wheat germ is particularly rich in the raffinose family of oligosaccharides. The extraction of oligosaccharides from natural resources and in particular cereals has not been fully developed due to the complexity of these substances and their connections with other macromolecules, particularly proteins. Depending on the location of the oligosaccharides in the grain, it could be carried out based on physical or enzymatic methods (Henry and Saini 1989). Most likely due to the difficulties in isolating them, there is not a lot of information regarding the potential prebiotic effects of cereal oligosaccharides. Arrigoni et al. (2002) reported that the *in vitro* fermentation of a commercial wheat germ preparation with high content in raffinose by human fecal microbiota resulted in an increase in the concentration of bifidobacteria, as measured by FISH, followed by an increase in propionate concentration. Matteuzzi et al. (2004) performed a double-blind, placebo-controlled human study using the same product and reported reduced fecal pH values, whereas the number of bifidobacteria and lactobacilli, as measured by FISH, increased significantly only in subjects with low basal levels.

21.4 CONCLUSIONS

The multiple beneficial effects of cereals can be exploited in different ways leading to the design of novel cereal foods or cereal ingredients that can target specific populations. For example, cereals can be used as fermentable substrates for the growth of probiotic microorganisms. The main parameters that have to be considered are the composition and processing of the cereal grains, the growth of the probiotic strain in the cereal based medium, the stability of the probiotic strain during storage, the organoleptic properties, and the nutritional value of the final product. Additionally, cereals can be used as sources of nondigestible carbohydrates that besides promoting several beneficial physiological effects can also selectively stimulate the growth of the bifidobacteria and lactobacilli present in the colon and act as prebiotics. Promising results have been reported so far for arabinoxylan and resistant starch indicating their potential prebiotic activities. Other cereal polysaccharides that could potentially have a prebiotic effect include β-glucan and oligosaccharides, such as GOS and FOS, although for the former the results so far are contradictory. Overall, further *in vitro* and *in vivo* research is needed to elucidate the prebiotic effect of cereal dietary fiber and understand its metabolism in the colon. Separation of fiber fractions from different cereal varieties or cereal by-products can be achieved through a variety of processing technologies, such as milling, sieving, debranning, and pearling in combination with extraction methods. Finally, it could be concluded that functional foods based on cereals is a challenging perspective, however, the implementation of new technologies for cereal processing enhancing their health potential, and the acceptability of cereal-based food products are of primary importance.

REFERENCES

Angelov, A., Gotcheva, V., Kuncheva, R., and Hristozova, T. 2006. Development of a new oat-based probiotic drink. *Int. J. Food Microbiol.* 112:75–80.

Annan, N.T., Poll, L., Sefa-Dedeh, S., Plahar, W.A., and Jakobsen, M. 2003. Influence of starter culture combinations of *Lactobacillus fermentum*, *Saccharomyces cerevisiae* and *Candida krusei* on aroma in Ghanaian maize dough fermentation. *Eur. Food Res. Technol.* 216:377–384.

Arrigoni, E., Jorger, F., Kolloffel, B., et al. 2002. In vitro fermentability of a commercial wheat germ preparation and its impact on the growth of bifidobacteria. *Food Res. Int.* 35:475–481.

Asp, N.G. 1996. Dietary carbohydrates: Classification by chemistry and physiology. *Food Chem.* 57:9–14.

Asp, N.G. and Bjorck, I. 1992. Resistant starch. *Trends Food Sci. Technol.* 3:111–114.

Becker, R. and Hanners, G.D. 1991. Carbohydrate composition of cereal grains. In *Handbook of Cereal Science and Technology*, eds. K.I. Lorenz and K. Kulp, pp. 469–496. New York: Marcel Dekker.

Beer, M.U., Wood, P.J., Weisz, J., and Fillion, N. 1997. Effect of cooking and storage on the amount and molecular weight of (1->3)(1->4)-beta-D-glucan extracted from oat products by an *in vitro* digestion system. *Cereal Chem.* 74:705–709.

Bekers, M., Marauska, M., Laukevics, J., et al. 2001. Oats and fat-free milk based functional food product. *Food Biotechnol.* 15:1–12.

Berg, J.O., Nord, C.E., and Wadstrom, T. 1978. Formation of glycosidases in batch and continuous culture of *Bacteroides fragilis*. *Appl. Environ. Microbiol.* 35:269–273.

Bering, S., Sjoltov, L., Wrisberg, S.S., et al. 2007. Viable, lyophilized lactobacilli do not increase iron absorption from a lactic acid-fermented meal in healthy young women, and no iron absorption occurs in the distal intestine. *Br. J. Nutr.* 98:991–997.

Berry, C.S. 1986. Resistant starch—Formation and measurement of starch that survives exhaustive digestion with amylolytic enzymes during the determination of dietary fiber. *J. Cereal Sci.* 4:301–314.

Bhatty, R.S. 1993. Extraction and enrichment of (1->3),(1->4)-beta-D-glucan from barley and oat brans. *Cereal Chem.* 70:73–77.

Bird, A.R., Hayakawa, T., Marsono, Y., et al. 2000. Coarse brown rice increases fecal and large bowel short-chain fatty acids and starch but lowers calcium in the large bowel of pigs. *J. Nutr.* 130:1780–1787.

Bouhnik, Y., Raskine, L., Simoneau, G., et al. 2004. The capacity of nondigestible carbohydrates to stimulate fecal bifidobacteria in healthy humans: A double-blind, randomized, placebo-controlled, parallel-group, dose-response relation study. *Am. J. Clin. Nutr.* 80:1658–1664.

Brennan, C.S. and Cleary, L.J. 2005. The potential use of cereal (1->3,1->4)-beta-D-glucans as functional food ingredients. *J. Cereal Sci.* 42:1–13.

Brumovsky, J.O. and Thompson, D.B. 2001. Production of boiling-stable granular resistant starch by partial acid hydrolysis and hydrothermal treatments of high-amylose maize starch. *Cereal Chem.* 78:680–689.

Charalampopoulos, D. and Pandiella, S.S. 2008. Survival of human derived *Lactobacillus plantarum* in fermented cereal extracts during refrigerated storage. *LWT-Food Sci. Technol.* (submitted).

Charalampopoulos, D., Pandiella, S.S., and Webb, C. 2002. Growth studies of potentially probiotic lactic acid bacteria in cereal-based substrates. *J. Appl. Microbiol.* 92:851–859.

Charalampopoulos, D., Pandiella, S.S., and Webb, C. 2003. Evaluation of the effect of malt, wheat and barley extracts on the viability of potentially probiotic lactic acid bacteria under acidic conditions. *Int. J. Food Microbiol.* 82:133–141.

Chassard, C., Goumy, V., Leclerc, M., Del'homme, C., and Bernalier-Donadille, A. 2007. Characterization of the xylan-degrading microbial community from human faeces. *FEMS Microbiol. Ecol.* 61:121–131.

Chavan, J.K. and Kadam, S.S. 1989. Nutritional improvement of cereals by fermentation. *Crit. Rev. Food Sci. Nutr.* 28:349–400.

Corcoran, B.M., Stanton, C., Fitzgerald, G.F., and Ross, R.P. 2005. Survival of probiotic lactobacilli in acidic environments is enhanced in the presence of metabolizable sugars. *Appl. Environ. Microbiol.* 71:3060–3067.

Corcoran, B.M., Stanton, C., Fitzgerald, G., and Ross, R.P. 2008. Life under stress: The probiotic stress response and how it may be manipulated. *Curr. Pharm. Des.* 14:1382–1399.

Crittenden, R., Karppinen, S., Ojanen, S., et al. 2002. *In vitro* fermentation of cereal dietary fibre carbohydrates by probiotic and intestinal bacteria. *J. Sci. Food Agric.* 82:781–789.

Crittenden, R., Laitila, A., Forssell, P., et al. 2001a. Adhesion of bifidobacteria to granular starch and its implications in probiotic technologies. *Appl. Environ. Microbiol.* 67:3469–3475.

Crittenden, R.G., Morris, L.F., Harvey, M.L., Tran, L.T., Mitchell, H.L., and Playne, M.J. 2001b. Selection of a *Bifidobacterium* strain to complement resistant starch in a symbiotic yoghurt. *J. Appl. Microbiol.* 90:268–278.

Cyran, M. and Lapinski, B. 2006. Physico-chemical characteristics of dietary fibre fractions in the grains of tetraploid and hexaploid triticales: A comparison with wheat and rye. *Plant Breed. Seed Sci.* 54:77–84.

Damiani, P., Gobbetti, M., Cossignani, L., Corsetti, A., Simonetti, M.S., and Rossi, J. 1996. The sourdough microflora. Characterization of hetero- and homofermentative lactic acid bacteria, yeasts and their interactions on the basis of the volatile compounds produced. *LWT-Food Sci. Technol.* 29:63–70.

Delaney, B., Nicolosi, R.J., Wilson, T.A., et al. 2003. Beta-glucan fractions from barley and oats are similarly antiatherogenic in hypercholesterolemic Syrian golden hamsters. *J. Nutr.* 133:468–475.

Doehlert, D.C. and Moore, W.R. 1997. Composition of oat bran and flour prepared by three different mechanisms of dry milling. *Cereal Chem.* 74:403–406.

Dornez, E., Gebruers, K., Wiame, S., Delcour, J.A., and Courtin, C.M. 2006. Insight into the distribution of arabinoxylans, endoxylanases, and endoxylanase inhibitors in industrial wheat roller mill streams. *J. Agr. Food Chem.* 54:8521–8529.

Drzikova, B., Dongowski, G., and Gebhardt, E. 2005. Dietary fibre-rich oat-based products affect serum lipids, microbiota, formation of short-chain fatty acids and steroids in rats. *Br. J. Nutr.* 94:1012–1025.

Escamilla-Hurtado, M.L., Valdes-Martinez, S.E., Soriano-Santos, J., et al. 2005. Effect of culture conditions on production of butter flavor compounds by *Pediococcus pentosaceus* and *Lactobacillus acidophilus* in semisolid maize-based cultures. *Int. J. Food Microbiol.* 105:305–316.

Furgal-Dierzuk, I. 2006. The effect of cereal by-product dietary fibre on the serum lipid profile and short-chain fatty acid concentration in caecal digesta of rats. *J. Anim. Feed Sci.* 15: 57–60.

Gerez, C.L., Cuezzo, S., Rollan, G., and de Valdez, G.F. 2008. *Lactobacillus reuteri* CRL 1100 as starter culture for wheat dough fermentation. *Food Microbiol.* 25:253–259.

Gibson, G.R., Probert, H.M., Van Loo, J., Rastall, R.A., and Roberfroid, M.B. 2004. Dietary modulation of the human colonic microbiota: Updating the concept of prebiotics. *Nutr. Res. Rev.* 17:259–275.

Gobbetti, M. and Corsetti, A. 1997. *Lactobacillus sanfrancisco* a key sourdough lactic acid bacterium: A review. *Food Microbiol.* 14:175–187.

Gobbetti, M., Simonetti, M.S., Corsetti, A., Santinelli, F., Rossi, J., and Damiani, P. 1995. Volatile compound and organic acid productions by mixed wheat sour dough starters: Influence of fermentation parameters and dynamics during baking. *Food Microbiol.* 12:497–507.

Gomes, A.M.P. and Malcata, F.X. 1999. *Bifidobacterium* spp. and *Lactobacillus acidophilus*: Biological, biochemical, technological and therapeutical properties relevant for use as probiotics. *Trends Food Sci. Technol.* 10:139–157.

Gray, J. 2008. Dietary fibre—Definition, analysis, physiology, health. *Agro Food Ind. Hi-Tech.* 19:4–5.

Grootaert, C., Delcour, J.A., Courtin, C.M., Broekaert, W.F., Verstraete, W., and Van de Wiele, T. 2007. Microbial metabolism and prebiotic potency of arabinoxylan oligosaccharides in the human intestine. *Trends Food Sci. Technol.* 18:64–71.

Guerzoni, M.E., Vernocchi, P., Ndagijimana, M., Gianotti, A., and Lanciotti, R. 2007. Generation of aroma compounds in sourdough: Effects of stress exposure and lactobacilli-yeasts interactions. *Food Microbiol.* 24:139–148.

Hamad, A.M. and Fields, M.L. 1979. Evaluation of the protein quality and available lysine of germinated and fermented cereals. *J. Food Sci.* 44:456–459.

Hansen, A. and Schieberle, P. 2005. Generation of aroma compounds during sourdough fermentation: Applied and fundamental aspects. *Trends Food Sci. Technol.* 16:85–94.

Helland, M.H., Wicklund, T., and Narvhus, J.A. 2004a. Growth and metabolism of selected strains of probiotic bacteria in milk- and water-based cereal puddings. *Int. Dairy J.* 14:957–965.

Helland, M.H., Wicklund, T., and Narvhus, J.A. 2004b. Growth and metabolism of selected strains of probiotic bacteria, in maize porridge with added malted barley. *Int. J. Food Microbiol.* 91:305–313.

Hemalatha, S., Platel, K., and Srinivasan, K. 2007. Influence of germination and fermentation on bioaccessibility of zinc and iron from food grains. *Eur. J. Clin. Nutr.* 61:342–348.

Hemery, Y., Rouau, X., Lullien-Pellerin, V., Barron, C., and Abecassis, J. 2007. Dry processes to develop wheat fractions and products with enhanced nutritional quality. *J. Cereal Sci.* 46:327–347.

Henry, R.J. and Saini, H.S. 1989. Characterization of cereal sugars and oligosaccharides. *Cereal Chem.* 66:362–365.

Herrera, I.M., Gonzalez, E.P., and Romero, J.G. 1998. Soluble, insoluble and total dietary fiber in raw and cooked legumes. *Arch. Latinoam. Nutr.* 48:179–182.

Holtekjolen, A.K., Uhlen, A.K., Brathen, E., Sahlstrom, S., and Knutsen, S.H. 2006. Contents of starch and non-starch polysaccharides in barley varieties of different origin. *Food Chem.* 94:348–358.

Hood, S.K. and Zottola, E.A. 1988. Effect of low ph on the ability of *Lactobacillus acidophilus* to survive and adhere to human intestinal-cells. *J. Food Sci.* 53:1514–1516.

Hopkins, M.J., Englyst, H.N., Macfarlane, S., Furrie, E., Macfarlane, G.T., and McBain, A.J. 2003. Degradation of cross-linked and non-cross-linked arabinoxylans by the intestinal microbiota in children. *Appl. Environ. Microbiol.* 69:6354–6360.

Hughes, S.A., Shewry, P.R., Li, L., Gibson, G.R., Sanz, M.L., and Rastall, R.A. 2007. In vitro fermentation by human fecal microflora of wheat arabinoxylans. *J. Agr. Food Chem.* 55:4589–4595.

Hughes, S.A., Shewry, P.R., Gibson, G.R., McCleary, B.V., and Rastall, R.A. 2008. In vitro fermentation of oat and barley derived beta-glucans by human faecal microbiota. *FEMS Microbiol. Ecol.* 64:482–493.

Izydorczyk, M.S., Dexter, J.E., Desjardins, R.G., Rossnagel, B.G., Lagasse, S.L., and Hatcher, D.W. 2003. Roller milling of Canadian hull-less barley: Optimization of roller milling conditions and composition of mill streams. *Cereal Chem.* 80:637–644.

Izydorczyk, M.S., Chornick, T.L., Paulley, F.G., Edwards, N.M., and Dexter, J.E. 2008. Physicochemical properties of hull-less barley fibre-rich fractions varying in particle size and their potential as functional ingredients in two-layer flat bread. *Food Chem.* 108:561–570.

Jagerstad, M., Piironen, V., Walker, C., et al. 2005. Increasing natural-food folates through bioprocessing and biotechnology. *Trends Food Sci. Technol.* 16:298–306.

Jaskari, J., Kontula, P., Siitonen, A., Jousimies-Somer, H., Mattila-Sandholm, T., and Poutanen, K. 1998. Oat beta-glucan and xylan hydrolysates as selective substrates for *Bifidobacterium* and *Lactobacillus* strains. *Appl. Microbiol. Biotechnol.* 49:175–181.

Jin, L.Z., Ho, Y.W., Abdullah, N., and Jalaludin, S. 1998. Acid and bile tolerance of *Lactobacillus* isolated from chicken intestine. *Lett. Appl. Microbiol.* 27:183–185.

Jones, P.J. and Jew, S. 2007. Functional food development: Concept to reality. *Trends Food Sci. Technol.* 18:387–390.

Kabeir, B.M., Abd-Aziz, S., Muhammad, K., Shuhaimi, M., and Yazid, A.M. 2005. Growth of *Bifidobacterium longum* BB536 in medida (fermented cereal porridge) and their survival during refrigerated storage. *Lett. Appl. Microbiol.* 41:125–131.

Kailasapathy, K. and Chin, J. 2000. Survival and therapeutic potential of probiotic organisms with reference to *Lactobacillus acidophilus* and *Bifidobacterium* spp. *Immunol. Cell Biol.* 78: 80–88.

Kariluoto, S., Aittamaa, M., Korhola, M., Salovaara, H., Vahteristo, L., and Piironen, V. 2006. Effects of yeasts and bacteria on the levels of folates in rye sourdoughs. *Int. J. Food Microbiol.* 106:137–143.

Kashket, E.R. 1987. Bioenergetics of lactic-acid bacteria—Cytoplasmic ph and osmotolerance. *FEMS Microbiol. Rev.* 46:233–244.

Katan, M.B. and De Roos, N.M. 2004. Promises and problems of functional foods. *Crit. Rev. Food Sci. Nutr.* 44:369–377.

Katina, K., Kaisa, P., and Karin, A. 2004. Influence and interactions of processing conditions and starter culture on formation of acids, volatile compounds, and amino acids in wheat sourdoughs. *Cereal Chem.* 81:598–610.

Katina, K., Arendt, E., Liukkonen, K.H., Autio, K., Flander, L., and Poutanen, K. 2005. Potential of sourdough for healthier cereal products. *Trends Food Sci. Technol.* 16:104–112.

Katina, K., Heinio, R.L., Autio, K., and Poutanen, K. 2006. Optimization of sourdough process for improved sensory profile and texture of wheat bread. *LWT-Food Sci. Technol.* 9:1189–1202.

Katina, K., Liukkonen, K.H., Kaukovirta-Norja, A., et al. 2007. Fermentation-induced changes in the nutritional value of native or germinated rye. *J. Cereal Sci.* 46:348–355.

Kedia, G., Vazquez, J.A., and Pandiella, S. 2008. Fermentability of whole oat flour, PeriTec flour and bran by *Lactobacillus plantarum*. *J. Food Eng.* 89:246–249.

Kerckhoffs, D., Hornstra, G., and Mensink, R.P. 2003. Cholesterol-lowering effect of beta-glucan from oat bran in mildly hypercholesterolemic subjects may decrease when beta-glucan is incorporated into bread and cookies. *Am. J. Clin. Nutr.* 78:221–227.

Kern, M. 2006. 2025: Global trends to improve human health—From basic food via functional food, pharma-food to pharma-farming and pharmaceuticals—Part 1: Basic food, functional food. *Agro Food Ind. Hi-Tech.*17:39–42.

Knuckles, B.E. and Chiu, M.C.M. 1995. Beta-glucan enrichment of barley fractions by air classification and sieving. *J. Food Sci.* 60:1070–1074.

Kohajdova, Z. and Karovicova, J. 2007. Fermentation of cereals for specific purpose. *J. Food Nutr. Res.* 46:51–57.

Kontula, P. 1999. In vitro and in vivo characterization of potential probiotic lactic acid bacteria and prebiotic carbohydrates. *Finnish J. Dairy Sci.* 54:1–2.

Kotsou, M.G., Mitsou, E.K., Oikonomou, I.G., and Kyriacou, A.A. 2008. In vitro assessment of probiotic properties of *Lactobacillus* strains from infant gut microflora. *Food Biotechnol.* 22:1–17.

Laine, R., Salminen, S., Benno, Y., and Ouwehand, A.C. 2003. Performance of bifidobacteria in oat-based media. *Int. J. Food Microbiol.* 83:105–109.

Lavermicocca, P., Valerio, F., Lonigro, S.L., et al. 2005. Study of adhesion and survival of lactobacilli and bifidobacteria on table olives with the aim of formulating a new probiotic food. *Appl. Environ. Microbiol.* 71:4233–4240.

Li, J., Kaneko, T., Qin, L.Q., Wang, J., and Wang, Y. 2003a. Effects of barley intake on glucose tolerance, lipid metabolism, and bowel function in women. *Nutrition* 19:26–929.

Li, J., Kaneko, T., Qin, L.Q., Wang, J., Wang, Y., and Sato, A. 2003b. Long-term effects of high dietary fiber intake on glucose tolerance and lipid metabolism in GK rats: Comparison among barley, rice, and cornstarch. *Metabolism* 52:1206–1210.

Lorri, W. and Svanberg, U. 1993. Lactic-fermented cereal gruels with improved in vitro protein digestibility. *Int. J. Food Sci. Nutr.* 44:29–36.

Lu, Z.X., Walker, K.Z., Muir, J.G., Mascara, T., and O'Dea, K. 2000. Arabinoxylan fiber, a byproduct of wheat flour processing, reduces the postprandial glucose response in normoglycemic subjects. *Am. J. Clin. Nutr.* 71:1123–1128.

Lu, Z.X., Walker, K.Z., Muir, J.G., and O'Dea, K. 2004. Arabinoxylan fibre improves metabolic control in people with Type II diabetes. *Eur. J. Clin. Nutr.* 58:621–628.

Macfarlane, G.T. and Gibson, G.R. 1991. Co-utilization of polymerized carbon-sources by *Bacteroides ovatus* grown in a 2-stage continuous culture system. *Appl. Environ. Microbiol.* 57:1–6.

Macfarlane, S., Macfarlane, G.T., and Cummings, J.H. 2006. Review article: Prebiotics in the gastrointestinal tract. *Aliment. Pharmacol. Ther.* 24: 701–714.

Malkki, Y. and Virtanen, E. 2001. Gastrointestinal effects of oat bran and oat gum: A review. *LWT-Food Sci. Technol.* 34:337–347.

Malkki, Y., Autio, K., Hanninen, O., et al. 1992. Oat bran concentrates: Physical-properties of beta-glucan and hypocholesterolemic effects in rats. *Cereal Chem.* 69:647–653.

Mandalari, G., Faulds, C.B., Sancho, A.I., et al. 2005. Fractionation and characterisation of arabinoxylans from brewers' spent grain and wheat bran. *J. Cereal Sci.* 42:205–212.

Marklinder, I. and Lonner, C. 1992. Fermentation properties of intestinal strains of *Lactobacillus*, of a sour dough and of a yogurt starter culture in an oat-based nutritive solution. *Food Microbiol.* 9:197–205.

Marklinder, I. and C. Lonner. 1994. Fermented oatmeal soup—Influence of additives on the properties of a nutrient solution for enteral feeding. *Food Microbiol.* 11:505–513.

Martensson, O., Duenas-Chasco, M., Irastorza, A., Oste, R., and Holst, O. 2003. Comparison of growth characteristics and exopolysaccharide formation of two lactic acid bacteria strains, *Pediococcus damnosus* 2.6 and *Lactobacillus brevis* G-77, in an oat-based, non-dairy medium. *LWT-Food Sci. Technol.* 36:353–357.

Martin, L.J.M., Dumon, H.J.W., and Champ, M.M.J. 1998. Production of short-chain fatty acids from resistant starch in a pig model. *J. Sci. Food Agric.* 77:71–80.

Martin-Diana, A.B., Janer, C., Pelaez, C., and Requena, T. 2003. Development of a fermented goat's milk containing probiotic bacteria. *Int. Dairy J.* 13:827–833.

Matteuzzi, D., Swennen, E., Rossi, M., Hartman, T., and Lebet, V. 2004. Prebiotic effects of a wheat germ preparation in human healthy subjects. *Food Microbiol.* 21:119–124.

Michida, H., Tamalampudi, S., Pandiella, S.S., Webb, C., Fukuda, H., and Kondo, A. 2006. Effect of cereal extracts and cereal fiber on viability of *Lactobacillus plantarum* under gastrointestinal tract conditions. *Biochem. Eng. J.* 28:73–78.

Monsma, D.J., Thorsen, P.T., Vollendorf, N.W., Crenshaw, T.D., and Marlett, J.A. 2000. *In vitro* fermentation of swine ileal digesta containing oat bran dietary fiber by rat cecal inocula adapted to the test fiber increases propionate production but fermentation of wheat bran ileal digesta does not produce more butyrate. *J. Nutr.* 130:585–593.

Morgan, D.E. 1968. Note on variations in the mineral composition of oat and barley grain grown in Wales. *J. Sci. Food Agric.* 19:393–395.

Morishita, T., Deguchi, Y., Yajima, M., Sakurai, T., and Yura, T. 1981. Multiple nutritional-requirements of lactobacilli—Genetic lesions affecting amino-acid biosynthetic pathways. *J. Bacteriol.* 148:64–71.

Mousia, Z., Edherly, S., Pandiella, S.S., and Webb, C. 2004. Effect of wheat pearling on flour quality. *Food Res. Int.* 37:449–459.

Nelson, A.L. 2001. Properties of high-fiber ingredients. *Cereal Foods World* 46:93–97.

O'Riordan, K., Andrews, D., Buckle, K., and Conway, P. 2001. Evaluation of microencapsulation of a *Bifidobacterium* strain with starch as an approach to prolonging viability during storage. *J. Appl. Microbiol.* 91:1059–1066.

Oscarsson, M., Andersson, R., Salomonsson, A.C., and Aman, P. 1996. Chemical composition of barley samples focusing on dietary fibre components. *J. Cereal Sci.* 24:161–170.

Ouwehand, A.C., Derrien, M., de Vos, W., Tiihonen, K., and Rautonen, N. 2005. Prebiotics and other microbial substrates for gut functionality. *Curr. Opin. Biotechol.* 16:212–217.

Oyewole, O.B. 1997. Lactic fermented foods in Africa and their benefits. *Food Control* 8:289–297.

Palacios, M.C., Haros, M., Sanz, Y., and Rosell, C.M. 2008a. Phytate degradation by *Bifidobacterium* on whole wheat fermentation. *Eur. Food Res. Technol.* 226:825–831.

Palacios, M.C., Haros, M., Rosell, C.M., and Sanz, Y. 2008b. Selection of phytate-degrading human bifidobacteria and application in whole wheat dough fermentation. *Food Microbiol.* 25:169–176.

Papageorgiou, M., Lakhdara, N., Lazaridou, A., Biliaderis, C.G., and Izydorczyk, M.S. 2005. Water extractable (1->3,1->4)-beta-D-glucans from barley and oats: An intervarietal study on their structural features and rheological behaviour. *J. Cereal Sci.* 42:213–224.

Patel, H.M., Wang, R.H., Chandrashekar, O., Pandiella, S.S., and Webb, C. 2004a. Proliferation of *Lactobacillus plantarum* in solid-state fermentation of oats. *Biotechnol. Prog.* 20:110–116.

Patel, H.M., Pandiella, S.S., Wang, R.H., and Webb, C. 2004b. Influence of malt, wheat, and barley extracts on the bile tolerance of selected strains of lactobacilli. *Food Microbiol.* 21:83–89.

Peterson, D.M., Senturia, J., Youngs, V.L., and Schrader, L.E., 1975. Elemental composition of oats groats. *J. Agr. Food. Chem.* 23:9–13.

Phillips, J., Muir, J.G., Birkett, A., et al. 1995. Effect of resistant starch on fecal bulk and fermentation-dependent events in humans. *Am. J. Clin. Nutr.* 62:121–130.

Pineiro, M. and Stanton, C. 2007. Probiotic bacteria: Legislative framework—Requirements to evidence basis. *J. Nutr.* 137: 850–853.

Pomeranz, Y. 1988. Chemical composition of kernel structures. In *Wheat: Chemistry and Technology*, vol. I, ed. Y. Pomeranz, pp. 91–158. St. Paul, MN: AACC.

Proulx, A.K. and Reddy, M.B. 2007. Fermentation and lactic acid addition enhance iron bioavailability of maize. *J. Agric. Food Chem.* 55:2749–2754.

Reid, G., Kim, S.O., and Kohler, G.A. 2006. Selecting, testing and understanding probiotic microorganisms. *FEMS Immunol. Med. Microbiol.* 46:149–157.

Ross, R.P., Desmond, C., Fitzgerald, G.F., and Stanton, C. 2005. Overcoming the technological hurdles in the development of probiotic foods. *J. Appl. Microbiol.* 98:1410–1417.

Rozada-Sánchez, R., Sattur, A.P., Thomas, K., and Pandiella, S.S. 2008. Evaluation of *Bifidobacterium* spp. for the production of a potentially probiotic malt-based beverage. *Process Biochem.* 43:848–854.

Saarela, M., Mogensen, G., Fonden, R., Matto, J., and Mattila-Sandholm, T. 2000. Probiotic bacteria: Safety, functional and technological properties. *J. Biotechnol.* 84:197–215.

Salim, R., Paterson, A., and Piggott, J.R. 2006. Flavour in sourdough breads: A review. *Trends Food Sci. Technol.* 17:557–566.

Salyers, A.A., Gherardini, F., and Obrien, M. 1981. Utilization of xylan by 2 species of human colonic bacteroides. *Appl. Environ. Microbiol.* 41:1065–1068.

Sampath, S., T. M. Rao, K. K. Reddy, K. Arun, and P. V. M. Reddy. 2008. Effect of germination on oligosaccharides in cereals and pulses. *J. Food Sci. Technol.-Mysore* 45:196–198.

Sanders, J.W., Venema, G., and Kok, J. 1999. Environmental stress responses in *Lactococcus lactis*. *FEMS Microbiol. Rev.* 23:483–501.

Sanni, A.I., Onilude, A.A., and Ibidapo, O.T. 1999. Biochemical composition of infant weaning food fabricated from fermented blends of cereal and soybean. *Food Chem.* 65:35–39.

Savoie, S., Champagne, C.P., Chiasson, S., and Audet, P. 2007. Media and process parameters affecting the growth, strain ratios and specific acidifying activities of a mixed lactic starter containing aroma-producing and probiotic strains. *J. Appl. Microbiol.* 103:163–174.

Selvendran, R.R. and Du Pont, M.S. 1980. Simplified methods for the preparation and analysis of dietary fiber. *J. Sci. Food and Agric.* 31:1173–1182.

Severson, D. K. 1998. Lactic acid fermentations. In *Nutritional Requirements of Commercially Important Microorganisms*, eds. W. Nagodawithana, and G. Reed, pp. 258–297. Milwaukee, WI: Esteekay Associates Inc.

Sharma, A., Yadav, B.S., and Ritika 2008. Resistant starch: Physiological roles and food applications. *Food Rev. Int.* 24:193–234.

Shimizu, J., Tsuchihashi, N., Kudoh, K., Wada, M., Takita, T., and Innami, S. 2001. Dietary curdlan increases proliferation of bifidobacteria in the cecum of rats. *Biosci. Biotechnol. Biochem*. 65:466–469.

Sievert, D. and Pomeranz, Y. 1989. Enzyme-resistant starch.1. Characterization and evaluation by enzymatic, thermoanalytical, and microscopic methods. *Cereal Chem*. 66:342–347.

Snart, J., Bibiloni, R., Grayson, T., et al. 2006. Supplementation of the diet with high-viscosity beta-glucan results in enrichment for lactobacilli in the rat cecum. *Appl. Environ. Microbiol*. 72:1925–1931.

Topping, D. 2007. Cereal complex carbohydrates and their contribution to human health. *J. Cereal Sci*. 46:220–229.

Tou, E.H., Mouquet-Rivier, C., Picq, C., Traore, A.S., Treche, S., and Guyot, J.P. 2007. Improving the nutritional quality of ben-saalga, a traditional fermented millet-based gruel, by co-fermenting millet with groundnut and modifying the processing method. *LWT-Food Sci. Technol*. 40:1561–1569.

Towo, E., Matuschek, E., and Svanberg, U. 2006. Fermentation and enzyme treatment of tannin sorghum gruels: Effects on phenolic compounds, phytate and *in vitro* accessible iron. *Food Chem*. 94:369–376.

Tuohy, K.M., Rouzaud, G.C.M., Bruck, W.M., and Gibson, G.R. 2005. Modulation of the human gut microflora towards improved health using prebiotics—Assessment of efficacy. *Curr. Pharm. Des*. 11:75–90.

van de Guchte, M., Serror, P., Chervaux, C., Smokvina, T., Ehrlich, S.D., and Maguin, E. 2002. Stress responses in lactic acid bacteria. *Anton. Leeuw. Int. J. G*. 82:187–216.

Van der Meulen, R., Scheirlinck, I., Van Schoor, A., et al. 2007. Population dynamics and metabolite target analysis of lactic acid bacteria during laboratory fermentations of wheat and spelt sourdoughs. *Appl. Environ. Microbiol*. 73:4741–4750.

Van Laere, K.M.J., Voragen, C.H.L., Kroef, T., Van den Broek, L.A.M., Beldman, G., and Voragen, A.G.J. 1999. Purification and mode of action of two different arabinoxylan arabinofuranohydrolases from *Bifidobacterium adolescentis* DSM 20083. *Appl. Microbiol. Biotechnol*. 51:606–613.

Vardakou, M., Palop, C.N., Christakopoulos, P., Faulds, C.B., Gasson, M.A., and Narbad, A. 2008. Evaluation of the prebiotic properties of wheat arabinoxylan fractions and induction of hydrolase activity in gut microflora. *Int. J. Food Microbiol*. 123:166–170.

Vardakou, M., Palop, C.N., Gasson, M., Narbad, A., and Christakopoulos, P. 2007. *In vitro* three-stage continuous fermentation of wheat arabinoxylan fractions and induction of hydrolase activity by the gut microflora. *Int. J. Biol. Macromol*. 41:584–589.

Wang, X., Brown, I.L., Evans, A.J., and Conway, P.L. 1999a. The protective effects of high amylose maize (amylomaize) starch granules on the survival of *Bifidobacterium* spp. in the mouse intestinal tract. *J. Appl. Microbiol*. 87:631–639.

Wang, X., Conway, P.L., Brown, I.L., and Evans, A.J. 1999b. *In vitro* utilization of amylopectin and high-amylose maize (amylomaize) starch granules by human colonic bacteria. *Appl. Environ. Microbiol*. 65:4848–4854.

Wang, R.H., Koutinas, A.A., and Campbell, G.M. 2007. Effect of pearling on dry processing of oats. *J. Food Eng*. 82:369–376.

Weickert, M.O., Mohlig, M., Koebnick, C., et al. 2005. Impact of cereal fibre on glucose-regulating factors. *Diabetologia* 48:2343–2353.

Westerlund, E., Andersson, R., and Aman, P. 1993. Isolation and chemical characterization of water-soluble mixed-linked beta-glucans and arabinoxylans in oat milling fractions. *Carbohyd. Polym*. 20:115–123.

Wolf, B.W., Bauer, L.L., and Fahey, G.C. 1999. Effects of chemical modification on *in vitro* rate and extent of food starch digestion: An attempt to discover a slowly digested starch. *J. Agric. Food Chem*. 47:4178–4183.

Wood, P.J., Weisz, J., and Mahn, W. 1991. Molecular characterization of cereal beta-glucans. 2. Size-exclusion chromatography for comparison of molecular-weight. *Cereal Chem.* 68:530–536.

Xu, S., Boylston, T.D., and Glatz, B.A. 2006. Effect of inoculation level of *Lactobacillus rhamnosus* and yogurt cultures on conjugated linoleic acid content and quality attributes of fermented milk products. *J. Food Sci.* 71:275–280.

Yamada, H., Itoh, K., Morishita, Y., and Taniguchi, H. 1993. Structure and properties of oligosaccharides from wheat bran. *Cereal Foods World* 38:490–492.

Yoon, S.H., Berglund, P.T., and Fastnaught, C.E. 1995. Evaluation of selected barley cultivars and their fractions for beta-glucan enrichment and viscosity. *Cereal Chem.* 72:187–190.

Yoon, K.Y., Woodams, E.E., and Hang, Y.D. 2006. Production of probiotic cabbage juice by lactic acid bacteria. *Bioresource Technol.* 97:1427–1430.

Zarate, G., Chaia, A.P., Gonzalez, S., and Oliver, G. 2000. Viability and beta-galactosidase activity of dairy propionibacteria subjected to digestion by artificial gastric and intestinal fluids. *J. Food Prot.* 63:1214–1221.

22 Advances in Development of Fat Replacers and Low-Fat Products

James R. Daniel

CONTENTS

22.1 INTRODUCTION

Fats and oils in foods have long been appreciated for their contribution to taste and texture, among other properties. Fats and oils are chemically triglycerides, and whether the particular triglyceride is a fat (solid at room temperature) or an oil (liquid at room temperature) depends on several factors, among them fatty acid chain length and degree of hydrogenation. These triglycerides have numerous uses in food, for example, as tenderizers in baked goods, leaveners (for plastic fats) in baked goods such as cakes, and as a component of emulsions (typically this is a function of liquid oils but it can be fulfilled by solid fats). Dietary triglycerides also provide essential fatty acids and carry fat-soluble vitamins to where they are needed. Fat replacers usually do only one (or perhaps two) of the jobs that are done by conventional fats and oils. Their main function is to provide a fat-like texture to the foods in which they are found. Because fats and oils make foods taste good, there is a concern regarding the overconsumption of these ingredients in the diet. The American Heart Association (AHA) recommends a diet in which lipids provide no more than 30% of calories and in which saturated fat (typically an animal product) provides not more than 10% of the calories. Data from the NHANES III study (McDowell et al. 1994) indicate that Americans are typically consuming 34% of the calories in their diets as fat. Although this is still above AHA recommendations, it is not nearly the 40%–42%

they were consuming in the late 1970s. The main concern regarding dietary fat is in regard to its contribution to obesity, which is epidemic in the United States at present. Moreover, obesity has been identified as a risk factor for coronary heart disease (CHD) and some types of cancer. As unsaturated fats are synthesized in nature, their double bonds are of the cis type. However, the hydrogenation process used in making plastic fats or vegetable margarines causes the cis double bonds to be isomerized to the trans form. The trans fats are of concern because they raise levels of low-density lipoprotein (LDL), which in turn could lead to an increased risk of CHD.

To address these concerns, food technologists and chemists have been examining the production and use of fat replacers, sometimes referred to as fat mimetics or fat substitutes (see Table 22.1). As indicated previously, these materials are meant to replace one (or possibly two) of the properties of fats and oils. Usually, the property of interest relates to the texture and mouthfeel provided by the natural fat or oil. Fats replacers are normally classified as being carbohydrate based, protein based, or fat like (or fat based). Among the carbohydrate-based fat replacers are maltodextrins (from corn, tapioca, and potato starch), microcrystalline cellulose, polydextrose, methylcellulose gums, pectin, inulin, galactomannans, glucomannans, xanthan gum, and β-glucans. These typically just provide the fatty mouthfeel that is so desirable in fat-containing systems. The genesis of their ability to do this is their ability to control the mobility and structure of water. In the case of the protein-based fat replacers there are whey proteins, such as Simplesse® (Nutrasweet Company, Deerfield, IL) and Dairy-Lo® (Pfizer, Inc., New York). Their correct functioning depends on partial denaturation or a microparticulation process producing particles with the correct diameter. In the case of the fat-like or fat-based replacers, these may be caloric or not depending on their metabolism and absorption. They include medium-chain triglycerides (MCTs), caprenin, Salatrim® (Vogeler, Inc., Philadelphia, PA), and Olestra® (Procter and Gamble Company, Cincinnati, OH). Depending on their level of use, these may contribute other characteristics of fats and oils more than just texture and mouthfeel. These would include tenderization in baked goods and vitamin transport. There are also ingredients that reduce fat absorption in deep fat frying.

The number of foods produced with reduced levels of triglycerides is truly staggering (more than 5000 as of 1998), with more than 1000 reduced or low-fat products being introduced per year during the 1990s. Meat, poultry, and fish provide 30% of the fat in the diet, baked goods and other grain products provide 25%, and oils and dairy products provide 11% and 18%, respectively. When it comes to fat-modified products, the most popular categories are cheese and dairy products, salad dressing, sauces, mayonnaise, and fat-free or low-fat milk products. Because of the plethora of products that have a variety of different levels of fat replaced, a new labeling lexicon was required. Products that contain <0.5 g of fat per serving may be labeled *fat free*, while those with ≤3 g per serving may be labeled *low fat*. If a product contains 25% less fat than the full-fat product it can be described as *reduced* or *less fat*. If a modified product contains one-third fewer calories or half the fat of the regular product it may bear the descriptor light. If more than half of the calories of a given food come from triglycerides, then the term *reduced fat* may only be used if the fat content in the modified version is reduced by 50%. *Calorie free* and *low calorie* are terms that may be appropriate if each serving of the food contributes <5 or ≤40 calories per serving,

TABLE 22.1
Examples of Fat Replacers, Their Sources, and Potential Uses

Source	Example	Potential Uses
Carbohydrate based		
Corn maltodextrins	Maltrin® (Grain Processing Corporation, Muscatine, IA), Sta-Slim® (Tate and Lyle Ingredients, Decatur, IL), Stellar	Baked goods, candy, dairy products, dressings, margarine, frozen desserts, meat, fish, poultry
Resistant starch	Crystalean® (Opta Food Ingredients, Bedford, MA)	Biscuits, crackers, cookies
Modified starch	FirmTex, Fantesk	Baked goods, frozen desserts
Tapioca dextrins	N-Oil	Candy, frozen desserts, dressings, yogurt, puddings
Potato dextrins	Paselli	Dressings, sauces, frozen desserts, cakes, fillings
β-Glucans	Oatrim	Baked goods, beverages, dairy products, dressings
Microcrystalline cellulose	Avicel	Baked goods, sauces, dressings, dairy products, frozen desserts, meat products
Inulin (fructooligosaccharides)	Raftiline, Raftilose	Frozen desserts, dressings, meat products, baked goods
Polydextrose	Litesse	Candy, dressings, bakery fillings, toppings
Carrageenan	Genu carrageenan	Meat products
Alginate	Sodium alginate	Meat products
Glucomannan	Konjac flour	Baked goods, sausages, processed meat, frozen desserts
Guar gum (mixtures)	N-Flate, Novagel	Baked goods, whipped products
Locust bean gum	Locust bean gum	Meat products
Xanthan gum	Xanthan gum	Dressings, sauces, frozen desserts, baked goods, spreads
Fat based		
Structured lipids	Caprenin, Salatrim (Benefat)	Candy, dressings, baked goods, cheese
SPEs	Olestra	
Synthetic fats	DDM, TATCA	Mayonnaise, margarine, frying
Protein based		
Whey, partially denatured	Dairy-Lo	Frozen desserts, yogurt, spreads, sour cream
Whey, microparticulated	Simplesse	Frozen desserts, baked goods, dressings, frostings

respectively. If the modified version of the food contains ≤25% of the calories in the regular (full fat) food it may be described as *reduced calorie* or *fewer calories*.

The question still remains about the effectiveness of these fat substitutes in reducing obesity and risk factors for other weight- and diet-related diseases. While they may be part of the solution, they are by no means a "magic bullet." In fact, they will probably be useless for weight control unless combined with other well-known, effective strategies such as overall reduced dietary intake of calories and an increase in physical activity. The American Dietetic Association has issued a position paper on the use of fat replacers in the diet (Mattes 1998).

A number of excellent reviews and monographs on the subject of fat replacers and their uses have appeared in recent years. Roller and Jones (1996) have edited a comprehensive monograph on fat replacers, while Mela (1997), Callaway (1998), Lichenstein et al. (1998), Porter et al. (1998), and Roberts et al. (1998) have all addressed the physiological and metabolic implications of fat in the diet and the use of fat replacers. The triglyceride-based fat replacer Salatrim has been reviewed by Dreher et al. (1998), and Olestra has been examined with regard to its formulation and effect in the gastrointestinal tract (Jandacek et al. 1999, Bimal and Guonong 2006). Regiospecific synthesis of structured lipids has been reviewed by Xu (2000) and Gupta et al. (2003). Lermer and Mattes (1999) have reviewed the dietary perception of fat and its implications, both metabolic and ingestive, and Sajilata and Singhal (2005) have extensively reviewed the use of specialty starches (maltodextrins) in low-fat foods such as butter/margarine spreads, pourable and spoonable salad dressings, no-fat or low-fat ice cream, and baked low-fat snacks.

22.2 CARBOHYDRATE-BASED FAT REPLACERS

Many carbohydrate-based fat replacers are based on starch. They can be starch maltodextrins (starch degradation products), resistant starches, or starch–lipid complexes. McMahon et al. (1996) compared the fat replacing ability of Stellar® (a starch maltodextrin—A. E. Staley Mfg. Company, Decatur, IL), Novagel® (a mixture of microcrystalline cellulose and guar gum—FMC Corporation, Philadelphia, PA), Simplesse, and Dairy-Lo (two protein-based fat replacers) in low-fat mozzarella cheeses. The apparent viscosity measured at 80°C was not affected by the fat replacers but the Stellar cheese and the Simplesse cheese had greater overall meltability than the Dairy-Lo or control cheeses. Interestingly, the Novagel cheese had the largest amount of water but melted the least. Novagel was incorporated as large particles that led to increased openness in the cheese and large serum channels being formed. None of the other fat replacers increased openness in the cheese structure. Perhaps part of the reason for these differences were the different levels of fat replacer actually retained in the cheeses (Novagel 1.7 wt%, Dairy-Lo 0.6 wt%, Stellar 0.4 wt%, and Simplesse 0.2 wt%). Increased moisture content of the low-fat cheeses compared to the control suggests that at least one process at work is the greater retention of water by the hydrophilic fat replacers in the experimental cheeses. Stellar containing cheeses had the most melt, suggesting that this maltodextrin provides lubrication in the cheese, allowing the protein strands to move to a greater degree. Modified starches have also been used, in the form of a fat replacer called N-Flate in low-fat cakes (Khalil 1998). N-Flate is a blend of emulsifiers, modified food starch, guar gum, and nonfat dried milk. Cakes made with this material were compared with those made using a potato maltodextrin (Paselli® MD 10—Cooperatie Avebe U.A., Veendam, the Netherlands) and a synthetic carbohydrate known as polydextrose (Litesse®, Danisco Cultor America, Ardsley, New York). These replacers were used to replace 0%, 25%, 50%, or 75% of the original fat in the cake. The cake was a Madeira-type cake with a significant amount of fat (24 wt%) in the control version. It was suggested that the carbohydrate-based replacers functioned by binding water and providing a desirable mouthfeel and texture. The replacers were used alone and in the presence of mono- and diglyceride emulsifiers, and a number of physical attributes, as well as sensory attributes, were measured. The specific gravity of the cake batters was affected by fat replacers and emulsifier levels but the type of fat replacer was not important to this effect. It was noted that this reduction in specific gravity was indicative of greater incorporation of air into the batter, which will affect the baked cake texture. It is possible that substitution of carbohydrate-based fat replacers for fat may also contribute to an increase in both gluten formation and toughness of the cake. That was generally not seen in this set of experiments. The pH was affected by the type and level of fat replacers, but incorporating emulsifiers with the fat replacers did not have an effect on pH. All of the batters were in the pH range of 6.5–7.7. Viscosity of the batters increased as the level of fat replacer increased, perhaps due to increased water binding by the replacer. Addition of emulsifiers did not affect the viscosity. Cake volumes and standing heights generally increased as the level of fat replacer increased. An exception was the N-Flate cake at 75% replacement, for which the standing height decreased. Compressibility of the cakes increased with increasing fat replacer content, except for the 75% N-Flate level, which correlated with the decrease in cake standing height. Increased fat replacer contents were correlated with an increase in browning, probably due to a greater degree of Maillard browning in the fat replaced cakes. Subjective crust color, crumb color, flavor, softness, and eating quality were all affected. Panelists did not see an improvement, relative to the control, in crust or crumb color at any level of fat replacer but did see an increase in flavor, softness, and eating quality at the 25% and 50% levels of fat replacers. Addition of emulsifiers improved only softness and eating quality. Paselli MD 10 cakes showed the highest sensory scores, followed by N-Flate, and then Litesse.

Mansour et al. (2003) investigated the use of N-Flate and N-Oil® (a tapioca dextrin—National Starch and Chemical, Bridgewater, NJ) in combination with FirmTex® (a modified corn starch—Henry Enrich and Company, New York) in low-fat cookies and their nutritional and metabolic effects in rats. A polydextrose and two potato starch maltodextrins were also used in initial screening experiments but it was determined that they produced inferior cookies compared to the replacers chosen for further sensory and rat studies. The replacement of butter oil in the cookie composition was generally 25% and 50% of the butter oil on a weight basis. Cookies made with N-Oil, N-Flate, FirmTex, and N-Flate combined with FirmTex had sensory scores similar to the control. Panelists rated the appearance, flavor, and mouthfeel of the experimental cookies higher than the control, whereas color, crispness, chewiness, and overall acceptability were rated similar to those of the control. These cookies were incorporated in a rat diet, and total serum cholesterol, LDL-cholesterol,

HDL-cholesterol, and serum triglycerides all decreased compared to the control. The type of fat replacer in the cookies added to the rat diet had no effect on food intake, body weight, or weight gain of the rats. Organ weights were generally not significantly affected by fat replacer type, except when the replacers were at 50% of the butter oil amount, in which case liver and spleen weights were increased. This may have been due to the fatty acid composition of the diet, which elevated the utilization of the fatty acids by the rats. In the test groups (cookie containing diet), the percentages of palmitic, stearic, myristic, and palmitoleic acids were increased in liver and kidney lipids compared to the control, whereas the percentages of oleic, linoleic, and linolenic acids were decreased in the test groups compared to the control.

The use of maltodextrins with or without xanthan gum in muffins has been reported (Khouryieh et al. 2005). Following a proposal advanced by Glicksman (1991), the authors used xanthan gum to provide thickening, lubricity and flow control, the maltodextrin being ($DE \leq 3$) a bulking agent. The sweetening agent used in this case was trichlorogalactosucrose (Sucralose®—McNeil Specialty Products Company, New Brunswick, NJ). The low-fat muffins were formulated to achieve approximately a 2/3 reduction in fat. Nutritional analysis showed a reduction from 5 to 2 g of fat per muffin. The muffins were formulated as either no-sugar-added or no-sugar-added/low-fat muffins with or without added xanthan gum (0.1 wt%). All treatment groups had significantly higher moisture content, less moisture loss after baking, and less volume compared to the control muffin (either on day 1 or day 4). As expected, added xanthan gum had no effect on water activity due to its low level of use and its large molecular weight. Objective texture profile analysis showed that the test muffins were significantly firmer and chewier than the control in spite of their higher moisture content. This may be due to an over formation of gluten caused by the reduction in the fat tenderizer. The no-sugar-added/low-fat muffin was significantly preferred for taste liking over a commercial no-sugar-added/low-fat muffin. However, the commercial muffin was less chewy and thus preferred for that characteristic compared to the test no-sugar-added/low-fat muffin.

The ice cream/frozen dessert field has been a fertile area for research into the use of carbohydrate-based fat replacers. One such endeavor was reported by Byars (2002), who used a starch-lipid cojet cooked material called FanTesk as the fat replacer in frozen desserts. This material was previously reported to be used as a fat replacer in cookies, cheese, and meat products (Brandt 2000). When FanTesk was used at levels to provide from 0.5 to 1.1 wt% fat (compared to 3.5 wt% fat in the control) the overrun, melting behavior, and rheological properties of the reduced-fat ice creams were quite similar to those of the commercial product. Except for the lowest level of FanTesk (3%) at low shear rates, all the reduced-fat frozen dessert mixes showed shear thinning behavior. The higher was the level of FanTesk the greater the mix viscosity. The FanTesk frozen dessert samples also showed greater elasticity than the commercial ice cream. It would be interesting to have a sensory analysis of these products to see if correlations can be made between rheological parameters and sensory texture parameters.

Carrageenan, a food gum derived from seaweed, has been widely investigated in the production of reduced fat sausages and other reformed meat products. Xiong et al. (1999) examined the effect of a few types of carrageenan and several other food gums (sodium alginate, locust bean gum, and a combination of locust bean gum and xanthan gum) in a fat-replacing role in low-fat beef sausages. Gums were used in this case in an attempt to restore the juiciness and mouthfeel that was lost when the fat was removed. The low-fat sausages had 23 wt% added water and were formulated to 1 or 2.5 wt% sodium chloride. Gums were added at 0.5 wt%. ι- and κ-carrageenan significantly increased cooking yield, hardness, and bind strength when the NaCl content was 1% but had little effect on the 2.5 wt% NaCl sausages. When the other gums were used in a similar way, the sausages were more soft, deformable, crumbly, and slippery. Thermal analysis by differential scanning calorimetry showed that the carrageenans slightly increased the thermal stability of the meat protein myosin. Both sensory texture analysis and texture profile analysis by Instron indicated that the carrageenans increased the textural attributes of the sausages. The conclusion reached in this study was that ι- and κ-carrageenan were the only acceptable gums for low- or high-salt beef sausages.

Totosaus et al. (2004) investigated the reduction of fat and sodium chloride in sausages and their replacement with κ-carrageenan. Fat content was reduced from 15 to 10 wt%, and sodium chloride was reduced from 2.5 to 1.5 and 1 wt%. Potassium chloride and calcium chloride were also added at levels well below 1 wt%, along with κ-carrageenan at 0.5 wt%. Potassium ion is required for gelation interaction between κ-carrageenan molecules, and calcium ion would likely reduce repulsive interactions between negatively charged protein molecules and negatively charged carrageenan molecules. The incorporation of κ-carrageenan improved the yield and textural and sensory attributes of the low-fat sausages. It was suggested that the potassium and calcium ions stabilized the carrageenan structures, allowing them to trap water and contributing to a fat-like texture. Another report on the use of carrageenan in reformed meat products was contributed by Koutsopoulos et al. (2008), who examined the partial replacement of fat in sausage with olive oil. Olive oil is the most monounsaturated vegetable oil, as well as a good source of antioxidants, and it is widely believed that these monounsaturates contribute to a resistance to certain types of cancer and cardiovascular disease (Avellone et al 2003). Control sausage (30 wt% fat) was compared to a variety of low-fat (10 wt% fat; 8% pork fat and 2% olive oil) sausages containing ι- or κ-carrageenan (0, 1, 2, or 3 wt%). Although the addition of carrageenan significantly improved the quality characteristics of sausages made by partially replacing the fat with olive oil, they were not equal to the full-fat control sausage. ι-Carrageenan had a better effect than κ-carrageenan in this investigation on qualities like pH, weight loss, and TBA (thiobarbituric acid) values (a measure of fat oxidation). Carrageenan levels above 2 wt% had negative effects on the texture of the sausages. When ι-carrageenan was used at levels up to 2 wt% in sausages that were ripened for 4 weeks (with vacuum packaging for the last 2 weeks of ripening), there was a positive effect on the physicochemical (pH, weight loss, TBA value, and Hunter L^*, a^*, b^* color values) and microbiological (lactic acid bacteria and *Enterobacteriaceae*) characteristics of the reduced-fat, partial olive oil replacement sausages, but again the sensory attributes were lower than those of the full-fat control commercial sausage. Again, food gums, ι-carrageenan in this case, were found somewhat useful in replacing the flavor, texture, juiciness, and lubricity contributed by fat in reformed meat products such as sausages.

κ-Carrageenan was compared against a number of potential fat replacers (isolated soy protein, modified waxy maize starch, wheat gluten, CarraFat® [Zaragoza, Vicente R., Santa Ana, California], a dairy-based gelled food product, isolated muscle protein, and konjac flour) in low-fat frankfurters (Yang et al. 2001). κ-Carrageenan was not the optimal fat replacer (50 wt% fat reduction in the frankfurters) in this study. Modified waxy maize starch, isolated soy protein, and isolated muscle protein produced frankfurters most similar to the high fat controls. Those containing κ-carrageenan and wheat gluten were considerably different from the controls. CarraFat, the dairy-based gel, and konjac flour fat-replaced frankfurters also received poor palatability scores as they were overly soft and excessively juicy.

Vegetable fibers have been investigated as fat replacers, principally due to their water-holding ability. These products are frequently β-glucans or mixtures of cellulose and other plant cell wall materials, and, depending on their source, are referred to as Oatrim® (ConAgra, Omaha, Nebraska), Nutrim® (VDF FutureCeuticals Inc., Momence, Illinois), and Z-trim (Brandt 2000, Inglett 2001). Oatrim is a mixture of amylodextrin (a branched starch molecule degradation product) and a β-glucan (see Figure 22.1). Nutrim contains a β-glucan as well, and Z-trim is derived from the hulls of such vegetable materials as corn, rice, soybean, peas, or from corn and wheat bran, and likely contains cellulose, hemicelluloses, and lignin.

Two types of oat fiber (OF), namely, bleached and high absorption, have been used in preparation of light bologna and fat-free frankfurters (Steenblock et al. 2001). For bologna, addition of bleached oat fiber (BOF) resulted in slightly higher moisture levels than did high absorption oat fiber (HAOF), and in keeping with this observation, addition of the highest levels of OF resulted in the highest processing yields. In general, HAOF showed greater processing yield for frankfurters than did BOF. Purge loss (loss of water from the reformed meat product) was inversely related to the level of OF. Regarding color, generally, OF addition increased lightness (L^*) and yellowness (b^*) but decreased redness (a^*). Adding OF had a hardening effect (according to texture profile analysis by Instron) on bologna compared to the control, however, increasing the OF amount from 1 to 3 wt% (based on control formula weight) caused a decrease in hardness in the test bologna. In bologna, increasing OF levels increased springiness and chewiness. Sensory hardness was also increased by OF addition

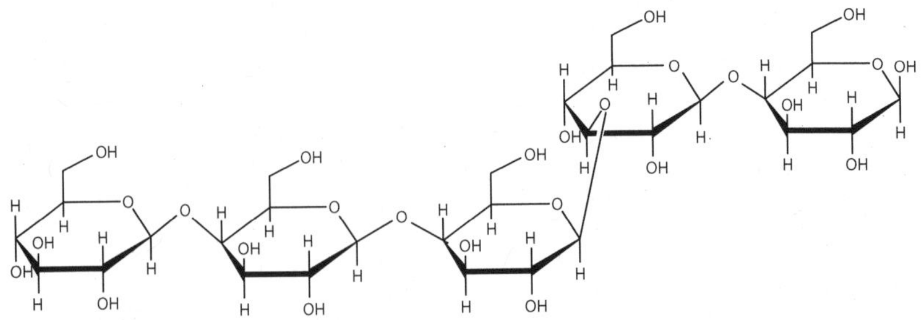

FIGURE 22.1 Representative structure of a section of a β-glucan. These glucans have a mixture of β-1,3 and β-1,4 glycosidic linkages.

relative to the control, and this increase was directly related to the added OF amount. Denseness and moistness values decreased as the added OF amount increased, which would be in keeping with the ability of vegetable fibers to bind water. Sensory toughness was greater for all BOF levels than the full-fat control for frankfurters as well as all HAOF levels (except for 1% HAOF). Sensory hardness in frankfurters was also greater for all BOF products than the control, which is consistent with the Instron results. Thus, the plethora of effects on physicochemical and sensory effects may be positive or negative, depending on the meat product under consideration, and must be used to provide optimal sensory qualities to the consumer.

One of the reasons why the use of OF or OF components is deemed desirable is the effect that oat products have on serum cholesterol. A study was done by Davy et al. (2002) in males comparing the effects of high-fiber oat cereals with wheat cereals. Males were fed oat or wheat cereal to provide 14 g of fiber/day for 12 weeks. It was observed that two large servings of oat cereal (containing β-glucan) compared to wheat cereal (containing much less β-glucan) resulted in lower concentrations of small, dense LDL particles, as well as lower LDL particle numbers. These desirable changes in the oat group were not accompanied by an increase in plasma triacylglycerols (TAGs) or a decrease in HDL-cholesterol, but a trend toward such undesirable changes was observed in the wheat group. The authors suggested that the difference between these two cereal types might be related to its predominant fiber type (soluble vs. insoluble).

An application of Oatrim (as well as avocado fruit puree) was reported by Wekwete and Navder (2008). They replaced 50% of the fat (butter) in oatmeal cookies and assessed the physical, textural, and sensory properties of the resulting cookies. A preliminary study by these authors (Wekwete and Navder 2005) had shown that cookies made using fat replacers were acceptable at 25% and 50% butter replacement levels but became unacceptable at 75% replacement. The fat-replaced products had stickier dough that was harder to work with, and the control cookies decreased in height on baking whereas the fat-replaced cookies increased in height. The spread ratio (ratio of diameter/height) was also significantly decreased for the Oatrim and avocado puree reduced-fat cookies. Moisture loss was lower in the control group, which also had the lowest water activity, followed by the Oatrim cookie and then the avocado puree cookie. The control cookies had a higher initial stiffness, hardness, and brittleness as measured by texture analyzer. Toughness, as measured by total work under the texture analyzer curve, increased in both the Oatrim cookie and the avocado puree cookie. With regard to the sensory analysis of the cookies, there were no significant differences in color, tenderness, sweetness, flavor, aftertaste, or overall acceptability between the control cookie and either of the fat-replaced products. The only significant difference in the sensory data was in appearance. Because the 50% replacement of butter with Oatrim or avocado puree reduced the fat by 39% and 35%, respectively, these cookies would be termed reduced-fat cookies, and according to the sensory scores, they would not be significantly different from the full-fat cookies except in appearance.

Use of a β-glucan (comparing its effect to N-Flate and polydextrose) in peanut butter cookies has been reported by Swanson (1998). The fat replacers replaced about 35% of the fat in these cookies. All of the reduced-fat cookies were less acceptable from a sensory point of view than the full-fat cookie. Among the fat replacers tested, polydextrose led to a slightly more preferred cookie than the β-glucan or N-Flate. Not surprisingly, there was a significant relationship between how well school lunch managers (the taste panel in this work) liked the reduced-fat cookie and their intent to serve it in their school. The managers indicated they would serve the reduced-fat cookies more infrequently than the full-fat cookies.

Sahan et al. (2008) examined the use of β-glucans in low-fat cheeses and compared the β-glucan cheese to those made using Avicel® (a mixture of microcrystalline cellulose and carboxymethylcellulose [CbMC]) or Simplesse Dry 100. Reduction of fat was accompanied by an increase in moisture and protein content, as would be expected. In addition, the use of fat replacers increased the yield to a small extent. Fat-replaced cheeses received lower sensory scores for appearance, probably due to a slight greenish tint as opposed to the normal yellow color. Fat replacers improved texture, and the Simplesse and Avicel containing cheeses had texture scores close to those of the full-fat cheese. Flavor scores of the fat-replacer containing cheeses were not different from the full-fat cheese, except for the cheese containing the β-glucan. Meltability was reduced in fat-replaced cheeses. In general, reduction of fat in cheese increased the texture profile analysis hardness value, the springiness value, the gumminess value, and the cohesiveness value. The overall conclusion was that for this low-fat Kashar cheese, the best option for fat replacement was Avicel rather than β-glucan or Simplesse.

Another carbohydrate-based fat replacer that has been investigated is polydextrose (see Figure 22.2). Polydextrose is obtained by melting and condensation of various ingredients, which are about 90 parts by weight D-glucose, 10 parts by weight sorbitol, and up to 1 part by weight citric acid or 0.1 part by weight phosphoric acid. Zoulias et al. (2000) used polydextrose to replace 35% of the fat in a low-fat cookie recipe. Although the main thrust of this paper is on the use of alternative sweetening agents in the context of a low-fat cookie, it was shown that maltitol, lactitol, and sorbitol can replace sucrose and produce an acceptable product (if lower sweetness). Addition of acesulfame-K (an intense, synthetic sweetener) increased the sweetness of these cookies and improved flavor perception and overall acceptability. In a related study from the same group (Zoulias et al. 2002a), polydextrose was compared to four other types of fat replacers in cookies: a low DE maltodextrin, a β-glucan rich product, an oligofructose (inulin), and a microparticulated whey protein, Simplesse Dry 100. The cookies were analyzed by a texture analyzer, and from each test a maximum stress value was obtained, as well as the ratio of maximum stress to maximum strain. The former is related to hardness, whereas the latter is associated with brittleness. Just removing the fat from the cookies resulted in an increase of hardness and brittleness, whereas adding any one of the fat replacers resulted in some cookies that were harder and some that were softer. The polydextrose and β-glucan cookies were harder than the control while the maltodextrin, protein, and inulin-containing cookies were not. The brittleness of all cookies increased with the addition of fat mimetics but only moderate increases were observed in the maltodextrin, protein, and inulin products. In this study, polydextrose was found not to be an acceptable replacement for 50% of the fat in cookies.

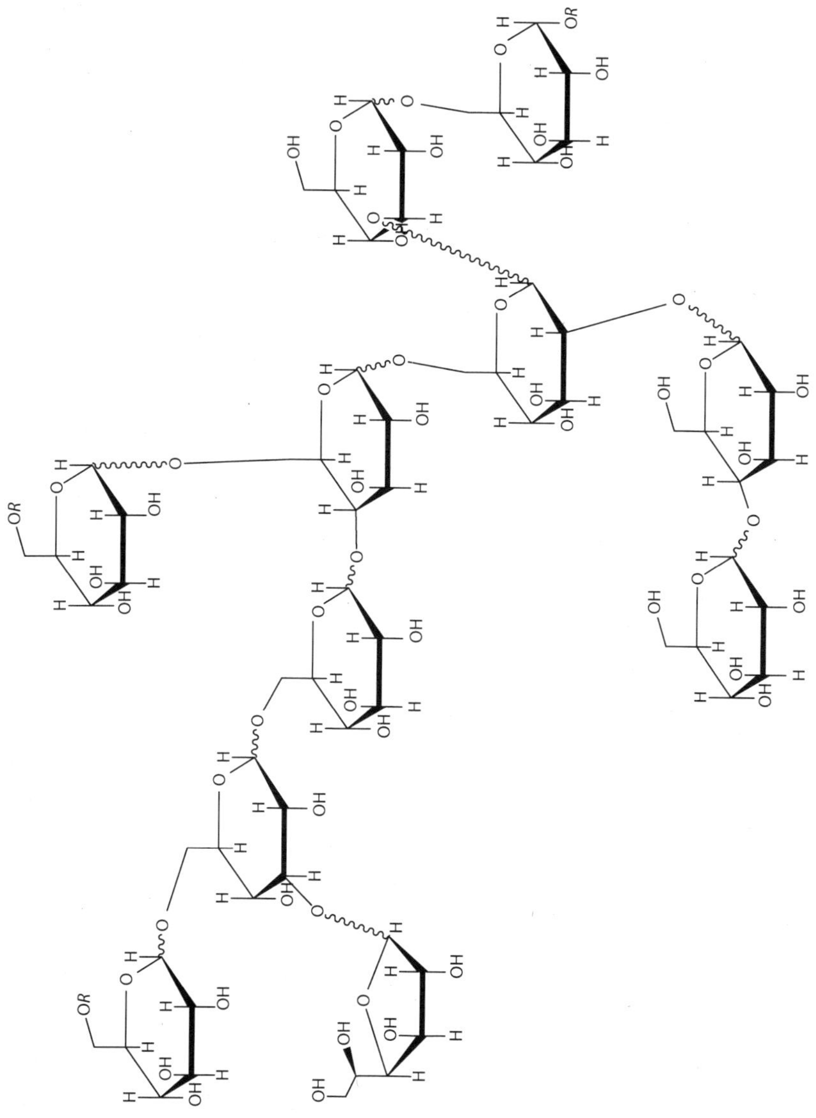

FIGURE 22.2 Representative structure of polydextrose. This oligosaccharide is prepared by a thermal reaction between glucose, sorbitol, and citric acid under vacuum. R can be D-glucose, sorbitol, citric acid, or polydextrose moieties.

Polydextrose was also used as a fat replacer in a totally fat-free frozen wild blueberry–tofu–soymilk dessert product (Camire et al. 2006). Low-fat products in which the blueberry source was blueberry juice were less well liked overall than both the high-fat product and the low-fat product in which the blueberry source was blueberry puree. The effect of fat mimetics (polydextrose, Simplesse, and a mixture of poydextrose and Simplesse) on the flavor perception of strawberry-flavored chemicals in ice cream has been reported (Liou and Grun 2007). The full-fat ice cream had 10 wt% fat and the reduced-fat test ice cream had 4 wt% fat. The choice of fat mimetic had a significant effect on color and rate of melting (the Simplesse containing ice cream melted at a greater rate). Full-fat ice creams were smoother, creamier, and had a greater mouthcoating than 4% fat ice creams containing fat mimetics. Ice creams containing no fat mimetics were the lowest in these textural characteristics. There appears to be a synergistic effect between Simplesse and polydextrose with regards to smoothness, creaminess, and mouthcoating. Ice creams using a mixture of these fat mimetics scored higher than ice creams using either alone. The 4% fat ice creams containing either one or both fat mimetics were slightly sweeter (and had a sweeter aftertaste) than either the 10% full-fat ice cream or the 4% fat ice cream with no fat mimetics. Full-fat ice cream and 4% fat ice cream with fat mimetics had slightly stronger strawberry flavor than the 4% fat ice cream with no fat mimetic. In general, the 4% fat ice cream with Simplesse had a more similar flavor profile to that of the 10% full-fat ice cream than did the 4% fat ice cream containing polydextrose. Discriminant analysis of the ice cream showed that the product with Simplesse had properties more similar to those of the full-fat ice cream, while the ice cream containing polydextrose had properties more similar to the low-fat (4 wt%) ice cream without fat mimetics. Thus, in this application, the carbohydrate-based polydextrose proved to be inferior to the aggregated-protein fat replacer.

Other vegetable gums or hydrocolloids are sometimes used as fat replacers, among them guar gum, a seed gum from *Cyamopsis tetragonolobus*, and xanthan gum, which is a bacterial gum from *Xanthomonas campestris* (see Figure 22.3). Use of these gums in low-fat cakes has been reported by Zambrano et al. (2004). They observed that with cakes containing up to 50 wt% fat substitution no difference was noted with respect to firmness and elasticity for up to 15 days, compared to the full-fat cake. Cakes substituted with guar gum or xanthan gum were not different from each other in moisture content or dough specific density but were different from the control. The specific volume of the cakes did not differ up to 75 wt% fat substitution. When internal characteristics were considered, only the 25% fat-reduced cake was equal to the control. At high levels (75 wt%) of fat substitution by guar and xanthan gums, the softness scores of the cakes were significantly reduced relative to the control, however, no difference in crumb color was noted with either gum at any level of fat substitution. The fat-reduced cakes tended not to show any difference in water activity at day zero but did tend to go to lower water activities more rapidly than the full-fat products. Sensory analysis of cakes replacing 50 wt% of the fat with either guar or xanthan gum showed significant differences for overall acceptance, texture, and flavor. The xanthan gum cake was favored overall and for texture, while there were no differences in flavor. The strong sensory point of the guar gum cake was its flavor, while the xanthan gum cake was appreciated more because of its texture. The weak points for

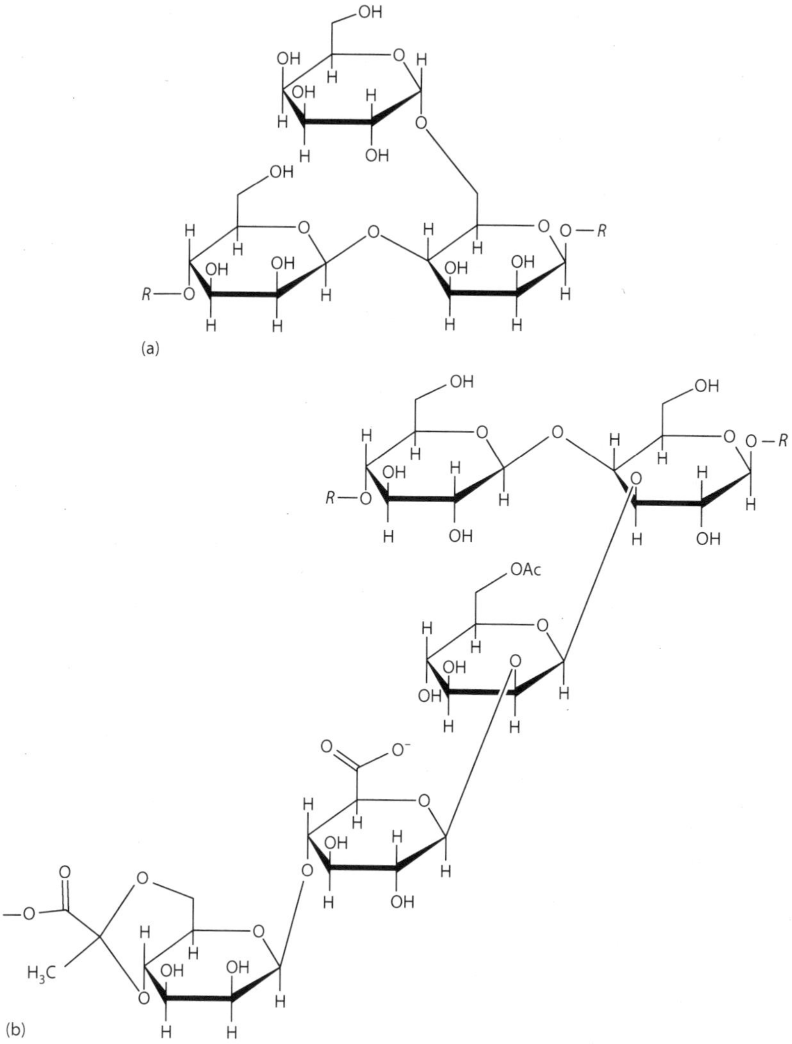

FIGURE 22.3 Some vegetable gums used as fat replacers. (a) Representative structure of guar repeating unit. Although guar is not actually precisely regular in its substitution pattern, this structure provides a good functional idea of what the polysaccharide will be like based on its high degree of substitution. (b) Repeating unit of xanthan gum. It may be helpful to think of xanthan gum as a highly substituted cellulose derivative.

the guar cake were its texture and a dry, heavy crumb, while the xanthan cake received few negative comments. Overall, in this set of experiments, xanthan gum replacing 50 wt% of the fat in a cake formulation was the preferred cake.

Guar gum (in combination with microcrystalline cellulose in a product called Novagel) has been compared to the protein-based fat replacer Dairy-Lo (Adapa et al. 2000) in the formulation of low-fat ice creams. Fat levels in the ice creams ranged from 6 to 12 wt%, and the viscoelastic properties of the mixes and frozen desserts were measured by rheometer. The overall conclusion of this study was that addition of either the carbohydrate-based or protein-based fat replacer had no effect on the elastic properties of the ice creams but did affect their viscous properties. This is important because creating and stabilizing the correct structure (a combination of foam, an emulsion, and a colloidal dispersion) is critical to achieving the desired ice cream texture. Addition of the fat replacers did not promote good overrun in the frozen ice cream dessert, probably because these materials provide increased viscous properties. Such a viscosity change decreases whipping ability (the ability to incorporate air) but would increase the stability of the frozen foam (i.e., stabilize the air once it has been incorporated). The ice cream containing Novagel had the least fat destabilization. The overall conclusion here is that in such complex systems a proper balance of milk fat, protein, and carbohydrate may be critical to produce the correct structure in the frozen dessert, rather than just relying on the idea of replacing milk fat by a single type of fat replacer. This notion of interactions and synergisms among fat-replacer molecules is not unlike the concept of multiple intense sweetener systems in use in many beverages and other food systems currently.

Another alternative is the use of protein–polysaccharide blends. The interaction of protein with polysaccharides has been reviewed (Schmitt et al. 1997, Turgeon et al. 2003), and the use of such blends has a number of perceived advantages. First of all, protein-based replacers are more "natural" than fat-based ones. Secondly, polysaccharides can bind large amounts of water and provide lubricity and melting qualities, apart from being able to protect the protein from over aggregation during thermal treatment. Furthermore, any included polysaccharide will contribute fiber and may have a beneficial effect on cholesterol levels.

Laneuville et al. (2005), for instance, studied the use of xanthan gum in combination with whey protein isolate as a fat replacer in cake frostings and sandwich cookie fillings. Brookfield apparent viscosities were measured along with texture profile analysis, melting profiles, and water activity. In the case of 75% fat-reduced cake frosting, several samples were identified by principal component analysis to be soft and have relatively high cohesiveness, while having intermediate adhesiveness and viscosity at the same time. Response surface methodology analysis of the frostings showed that hardness and cohesiveness of the samples were influenced mainly by the proportion of whey protein isolate–xanthan gum fat replacer present. Adhesiveness and apparent viscosity, on the other hand, were primarily influenced by the water–fat replacer interaction, which is related to the water:fat replacer ratio. For cake frostings, the ideal value of the water:fat replacer ratio is about 8:1, while for cookie fillings it should be somewhat lower, about 5:1. The water activity level of one of the fat-replaced samples was very similar to that of commercial cake frostings, but for cookie fillings the variability of water activity was greater. If using the whey protein isolate–xanthan gum as a fat replacer in cookie filling, it would also be important to use a humectant to lower water activity and to prevent migration of water into the cookie itself, thus producing sogginess.

A similar strategy, using protein and polysaccharides, was used by Andres et al. (2006) in an investigation of the production of low-fat chicken sausages. These authors used a mixture of whey protein concentrate and a guar/xanthan gum mixture (ratio 3:7) to replace various amounts of added fat. The reason for the guar:xanthan mixture is that xanthan provides good viscosity, salt tolerance, and thermal stability, while guar produces high-viscosity dispersions, alone or in synergistic interaction with xanthan gum. Weight losses were not affected by the addition of gums. Increasing gum content did increase the lightness of the sausages, but this could be expected, as the gums themselves are light colored to white. Sausages were tested for hardness, adhesiveness, cohesiveness, and chewiness, and significant differences between the formulations were found. The main effect of added gums seemed to be a decrease in hardness, chewiness, and resilience, perhaps due to an increase in retained water in the sausage. As the level of hydrocolloid increased, scanning electron micrographs showed a more continuous and less granular meat emulsion, essentially a gel-type structure rather than a meat emulsion. Sensory analysis of the low-fat chicken sausages (0.22–6.09 wt% fat) showed that they had generally acceptable scores. Two levels of gum mixture were used (0.13 and 0.32 wt%) and sausages containing the higher level of gum were preferred by the sensory panel.

22.3 PROTEIN-BASED FAT REPLACERS

Two protein-based fat replacers have dominated investigation recently: Simplesse, a microparticulated whey protein produced by a process involving high shear and heat, and Dairy-Lo, a whey protein produced by ultrafiltering sweet whey followed by partial heat denaturation. Most of their uses have been in dairy products of one kind or another but other applications have also been examined. Singh et al. (2000) studied the use of Simplesse and microcrystalline cellulose (incorrectly identified by the authors as a starch-based fat replacer) in the production of >50 wt% fat-reduced peanut butter. They looked at each replacer individually and also in combination to see if an optimal product could be obtained. What they found was that the test peanut pastes produced had a non-Newtonian (pseudoplastic) flow behavior that was time dependent. The test pastes were lighter in color, easier to spread, less firm, and less grainy when compared to a commercial 30% reduced-fat product. The textural attributes of the test products were deemed to be similar to those of a full-fat peanut butter by a sensory panel. The reduced-fat pastes were similar in adhesiveness and mouthcoating (both among themselves and in comparison to the control) but were oilier than the control. Two of the several treatments, containing both, Simplesse and microcrystalline cellulose, were suggested as formulations that would allow 50 wt% of fat reduction with acceptable rheological and textural properties.

Zoulias et al. (2002b) examined the effects of Simplesse, Raftiline® (an inulin, Raffienerie Tirlemontoise S.A., Bruxelles, Belgium), a maltodextrin, and polydextrose as fat replacers in low-fat, sugar-free cookies. At 35 wt% fat replacement (using a 20 wt% gel of the fat replacer), Simplesse, the inulin, and the maltodextrin, in combination with lactitol or sorbitol, produced cookies of comparable hardness and brittleness to the control. The polydextrose product, on the other hand, gave very hard and brittle cookies. Simplesse and polydextrose produced cookies with a diameter increase comparable to the control cookies on baking. When fat replacement was increased to 50 wt%, Simplesse, the inulin, and the maltodextrin, when combined with a mixture of lactitol and sorbitol, produced cookies that were hard, brittle, and did not expand properly on baking. The Simplesse cookies showed the greatest expansion but the least acceptable flavor. The maltodextrin cookies were the most acceptable in flavor, followed by the inulin cookies. The problem of hardness in the cookie at the 50 wt% fat replacement level was solved in one of two ways: (1) decreasing the amount of lactitol/sorbitol mix or (2) increasing the fat replacer (only in the case of the maltodextrin fat replacer). All of the fat-replaced cookies had increased water activity compared to the control, but not to the extent that it would influence shelf life.

Simplesse was also used in combination with resistant starch, inulin, and dairy protein (caseinate) in biscuits (cookies) as reported by Gallagher et al. (2003). They used response surface methodology to optimize the mixture of fat replacers to be used in the biscuits. Characteristics measured included dough hardness, breaking strength, moisture content, water activity, surface color, and cookie dimensions. Dough hardness was strongly correlated with the presence and amount of sodium caseinate. All trials were thicker than the control, and there was a positive correlation between cookie thickness and breaking strength. Surface brownness (L^* values) was strongly correlated with the amount of protein-based fat replacers present. It is likely that this is related to their contribution to Maillard browning as a result of the reaction of basic amino acid side chains with sugars at the elevated temperatures of baking. The optimum levels of replacers (based on flour addition) in the multi-ingredient fat replacer were determined to be 14 wt% (based on flour) resistant starch (Novelose 330®, National Starch and Chemical Company, Bridgewater, NJ), 14.5 wt% (based on flour) sodium caseinate, 25 wt% (based on flour) inulin (Raftilose®, Raffienerie Tirlemontoise S.A., Bruxelles, Belgium), and 25 wt% (based on flour) Simplesse Dry 100. The texture and appearance of the cookies using the optimized fat-replacer mix were reported to be of a quite high standard.

Simplesse has also been compared to inulin (in gel or powdered form) in bread formulations as reported by O'Brien et al. (2003). In this work, inulin in gel form at a 2.5 wt% level was found to be a more effective fat replacer than powdered inulin or Simplesse. As inulin level increased, water absorption increased as well. Complex modulus for doughs with fat or inulin was lower than for control doughs without fat, and addition of Simplesse to the dough increased the complex modulus significantly. Loaves with fat or inulin had higher volumes than other test loaves. Loaf volume is an important predictor of shelf life and staling rate, so a high loaf volume is desirable. Fat containing loaves had the softest crumb but inulin gel containing breads were only slightly harder. Overall, the incorporation of 2.5 wt% inulin in gel form was found to be the most effective fat replacer in bread, whereas breads containing Simplesse were not acceptable. It was postulated that Simplesse weakens the gluten network, resulting in a more porous texture which would be unable to retain leavening gases. As a result, the loaves produced using this replacer would exhibit low loaf volume and high hardness.

In a quite different application, Yildiz-Turp and Serdaroglu (2008) have recently reported the use of hazelnut oil, pre-emulsified with Simplesse, to replace 15, 30, or 50 wt% of the beef fat in a fermented Turkish sausage called sucuk. Substitution of the beef fat with the fat replacer/hazelnut mixture caused no changed in TBA value at any level of fat replacement after 12 days of fermentation and ripening. Higher levels of substitution caused a statistically significant increase in softness noted both by a penetrometer (at 50 wt%) and by a sensory panel (at 30 and 50 wt%). On the positive side, as hazelnut/Simplesse substitution increased, cholesterol levels decreased. Generally, the use of the fat replacer in this application caused significant decreases in slice appearance, texture, and taste scores. Interestingly, there was no difference in the overall acceptability scores, except when the fat replacer replaced 15 wt% of the beef fat, in which case the highest overall acceptability score was observed.

Simplesse and Dairy-Lo have been widely investigated in a number of dairy applications, which may be only fitting as they are milk protein based and should hence be compatible with a wide variety of dairy products. McMahon et al. (1996) have reported on the use of Simplesse and Dairy-Lo as fat replacers in mozzarella cheese (fat content 4–5 wt%), comparing cheeses made with these protein-based replacers to those made with the carbohydrate-based replacers Stellar (corn maltodextrin) and Novagel (a mixture of microcrystalline cellulose and guar gum). There was no significant difference in apparent viscosity at 80°C for any of the test cheeses, but the cheeses made with Stellar and Simplesse had significantly greater meltability (important in its use on pizza) than did the cheeses made with Dairy-Lo or Novagel. Interestingly, the cheese with the most moisture (the Novagel cheese) melted to the least extent, at least up to 14 days postpreparation. The Novagel cheese had the most open structure as determined by scanning electron microscopy (SEM), probably because it had the largest particle size of any of the replacers. The other fat replacers were of sufficiently small particle size that they did not influence the openness of the cheese curd structure.

A very similar study was reported by Aryana and Haque (2001) but using cheddar cheese as the vehicle in this case. The same fat replacers were used and scanning and transmission microscopy was performed on the cheeses so obtained. These authors have concluded that fat replacers do indeed alter the microstructure of the cheeses, albeit by different mechanisms. Simplesse and Novagel soften the low-fat cheeses (compared to the low-fat, no-fat-replacer cheese) by imparting discontinuities to the casein matrix and weakening it. Such discontinuities were not seen in the case of the Dairy-Lo or Stellar cheeses, but it was suggested that they soften the low-fat cheddar by imparting the fewest layers to the fat–protein interface during aging.

Kavas et al. (2004) reported on the use of Simplesse and Dairy-Lo, as well as two starch-based fat replacers, in the production of low-fat white pickled cheese. Interestingly, in this case, the use of fat replacers caused an increase in hardness, as opposed to what was seen in the case of mozzarella and cheddar cheeses. The fat-replaced products were also gummier and chewier after 90 days of aging compared to the full-fat cheese. Fat reduction also caused an increase in moisture and total nitrogen for the aged cheeses. Sensory evaluation showed that the fat-replaced cheeses were acceptable, and no off-flavor or bitterness was observed. Simplesse was the preferred fat replacer in this study, as it produced a cheese very similar to the full-fat one. Use of the other replacers produced a cheese that was different from the full-fat product but still quite acceptable.

Theophilou and Wilbey (2007) have recently reported on the effect of Simplesse in halloumi cheese, the traditional cheese of Cyprus. Even though this was not a typical fat-replacer-type study, Simplesse was included in the cheeses, and it was noted that those test cheeses with the protein-based fat replacer gained higher sensory preference scores than would be expected based on their fat in dry matter (FDM) values. It was suggested that this indicates that the microparticulated whey protein was exhibiting some fat mimetic function in these cheeses. As such, it would seem to be a way to improve the textural and acceptability properties of reduced-fat halloumi cheese.

Simplesse has been examined as a fat mimetic to be added to skim milk to approximate the sensory qualities of higher fat milks (1 wt%, 2 wt%). This work by Phillips and Barbano (1997) compared the effects of Simplesse, polydextrose, whey protein concentrate, Dairy-Lo, and titanium dioxide added to skim milk. The function of the titanium dioxide is to whiten the milk, an important sensory quality in fluid dairy products. When polydextrose or sodium caseinate were added to skim milk there was no difference in the L^* value (lightness) compared to skim milk. Dairy-Lo, whey protein concentrate, and Simplesse all provided L^* values that were greater than that of skim milk but less than that of 1% fat milk. Sodium caseinate plus titanium dioxide produced milk that had a higher L^* value than 1% fat milk. Addition of Simplesse, sodium caseinate or a blend of sodium caseinate and titanium dioxide produced milks with increased viscosities compared to skim milk. Initial screening led to the selection of these three fat replacers for additional sensory screening. Samples containing Simplesse had a cooked aroma similar to that of 2% fat milk but there was no difference in oxidized aroma. Simplesse also slightly improved the creamy aroma of skim milk and produced a product whose creamy aroma was not different from that of 1% fat milk. Simplesse-containing samples were both more astringent and sweeter than other milk samples. Addition of Simplesse increased the viscosity of the fluid milk, and, as a consequence, it had more mouthcoating and residual mouthcoating than skim milk. Addition of sodium caseinate by itself had little effect on the sensory properties of the skim milk, but when accompanied by titanium dioxide it did positively affect the sensory response to the milk. Whatever fat substitute is used to emulate the properties of higher fat fluid milk products it is very important that it impart the necessary opacity and make the milk less blue and green.

A report by Prindiville et al. (2000) examined low-fat formulation of chocolate ice creams (low-milk-fat or low-fat cocoa butter) versus nonfat ice creams containing Simplesse or Dairy-Lo. Use of the whey-based fat replacers versus low levels of fat did not affect consumer acceptance of the ice creams. The fat-replacer ice creams had a greater intensity of cocoa flavor but they developed an icy texture at a greater rate than the low-fat ice creams. In freshly produced ice cream, Simplesse behaved more similar to milk fat than did Dairy-Lo as indicated by brown color, cocoa flavor, and cocoa character. However, Dairy-Lo was more similar to milk fat in terms of thickness and mouthcoating. In comparing storage stability of the fat-replaced ice creams, Simplesse was better able to slow the development of iciness over 12 weeks of frozen storage than was Dairy-Lo. There was no difference in hardness in any of the ice creams and no significant difference in viscosity of the mixes (although the Dairy-Lo product tended to be slightly more viscous). Melting rates did differ, with the cocoa butter ice cream melting the fastest, followed by the Dairy-Lo product, then Simplesse and finally the low-fat ice cream containing milk fat (2.5 wt%).

A subsequent work from the same laboratory (Welty et al. 2001) investigated the effect of Simplesse, Dairy-Lo, and Oatrim on flavor volatiles in chocolate ice cream. This involved the use of gas chromatography-mass spectrometry and headspace analysis. The level of milk fat, but not the level or type of fat replacers, influenced the levels of benzaldehyde and phenylacetaldehyde in the headspace. Simplesse affected the release of 2-methyl-5-propyl pyrazine, as well as the headspace concentration of other pyrazines. Volatiles released into the headspace increased when fat was replaced by a fat replacer, indicating that the fat replacers are less hydrophobic than fat, which makes sense from what is known about their structures (they are denatured or aggregated whey proteins and thus quite hydrophilic).

Yazici and Akgun (2004) have reported on the use of Simplesse and Dairy-Lo as a fat replacer in strained yogurt, a widely consumed dairy product in Turkey. The fat-replaced products had similar total solids, hardness, L* and b* color values, and similar texture values. Panelists preferred the Dairy-Lo yogurt to the Simplesse yogurt for flavor, appearance, and color. The reduced-fat yogurts had higher viscosity than the yogurts containing fat, and the Dairy-Lo samples were more viscous than the Simplesse ones. Higher fat yogurts were preferred, indicating that the fat replacers could not totally make up for the lower fat level in this product.

Mattes (2001) compared Simplesse, Paselli, and Olestra to butter and analyzed their effects on TAG concentration in blood. He concluded that only exposure to real fat, i.e., butter, caused an increase in postprandial TAG, and, in addition to replacing fat and fat calories, these fat replacers may have a favorable effect on lipid metabolism.

22.4 FAT-LIKE OR FAT-BASED REPLACERS

As previously noted, lipids contribute numerous qualities to foods, among them flavor, mouthfeel, palatability, lubricity, and creaminess. In the case of the fat-based replacers, the basic approach is to manipulate the fatty acids bonded to the central glycerol backbone or to change the glycerol backbone to another moiety (e.g., sucrose), to reduce the ester linkage to an ether linkage, or to reverse the direction of the ester linkage linking the fatty acid to the glycerol. This type of chemical modification produces low-calorie fats and oils. The chemical reactions used to produce these types of fat substitutes are typically random interesterifications, and the products are referred to as structured lipids. These substitutes resemble conventional fats to a much greater degree than the other types of replacers (carbohydrate or protein based). In addition, they are lipophilic and generally stable to the elevated temperatures associated with baking and deep fat frying. The basis of the use of these chemicals as calorie-control agents is that due to their altered structures they are hydrolyzed and absorbed to a much lower degree than their normal triglyceride counterparts, thus resulting in decreased caloric contribution to the diet (Hamm 1984).

Structured lipids are obtained via hydrolysis and interesterification of MCTs and long-chain triglycerides (LCTs). In general, the dividing line between MCTs and LCTs is 12 carbon atoms. However, some structured lipids may have a mixture of short-chain fatty acids such as acetic, propionic, and butyric, and of long-chain fatty acids such as stearic. Such lipids provide many of the physical properties of normal dietary fats and oils but contribute only about one-half of the normal calories. Short-chain fatty acid esters may be hydrolyzed but the short fatty acids contribute considerably fewer calories, whereas long-chain fatty acid esters are poorly hydrolyzed and absorbed (Finley et al. 1994). Pivk et al. (2008) have recently reported on a novel method for analyzing the deposition of MCTs on the tongue surface and have established a correlation between thickness of the deposition and the sensory properties of "fatty film" and "lubricating film." The MCT oils were introduced at various volumes into the subjects' mouths, moved about the mouth, and then spat out. Determination of the lipid thickness was by fluorescence and was judged both on the tongue and palate. More oil was distributed onto the tongue (nonuniformly) that on the palate.

Caprenin was an early entry into the structured lipids area. It is a mixed MCT and LCT containing caprylic (C8:0), capric (C10:0), and behenic acids (C22:0). A representative structure is shown in Figure 22.4. It was developed by Procter and Gamble® (Procter and Gamble Company, Cincinnati, Ohio), as a cocoa butter substitute for use in candies and confectionery coatings (Sandrou and Arvanitoyannis 2000). The C8 and C10 fatty acids in this compound are derived from coconut and palm kernel oil, while the behenic acid comes from hydrogenated canola (rapeseed) oil. A study using caprenin as a fat source in muffins fed to rats showed that it contained about 4.3 cal/g (Ranhotra et al. 1994). Caprenin provides 5 cal/g in humans, and it was initially proposed for use in Milky Way II and Milky Way Lite (Mars, Inc., McLean, VA) candy bars. However, difficulties in actual use in production and slight undesirable effects on serum cholesterol levels resulted in its later voluntary withdrawal from the market in the 1990s.

Salatrim (this acronym stands for short-chain triglyceride and LCTmolecules) is another structured lipid that has been investigated. It is composed of a mixture of short-chain (acetic, propionic, and/or butyric) and long-chain (mostly stearic) fatty acids. By varying the ratio of short-chain to long-chain fatty acids in this structured lipid (see Figure 22.5), it is possible to achieve flexibility in both physical and functional properties. This product was developed by the Nabisco (Kraft Foods Holding, Inc., Northfield, IL) Foods Group (Smith et al 1994, Softly et al. 1994). Potential uses for Salatrim include snacks and baked products, molded chocolates, ice cream, and other dairy products. One of its more well-known brand names is Benefat® (Danisco A/S Corporation, Copenhagen, Denmark). Hershey has developed

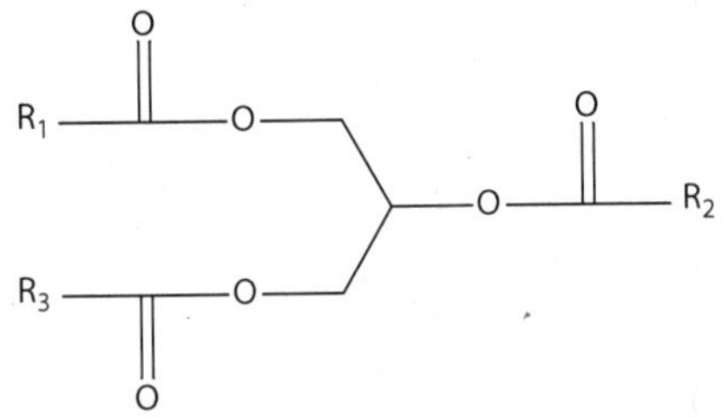

FIGURE 22.4 Representative structure of caprenin. R_1, R_2, and R_3 are, variously, caprylic acid (C8:0), capric acid (C10:0), and behenic acid (C22:0).

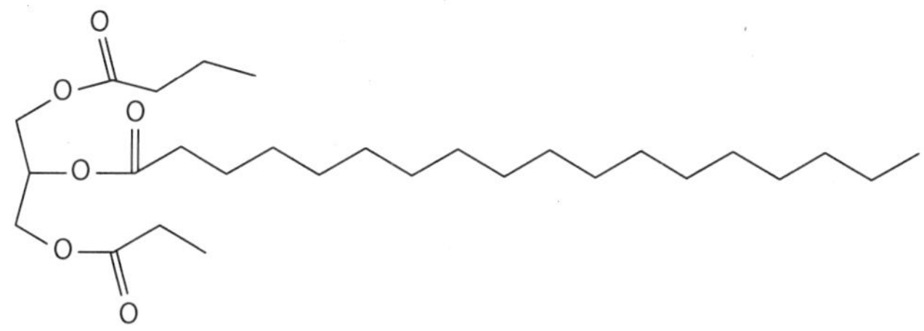

FIGURE 22.5 A representative structure of Salatrim. The propionyl, butyryl, and stearyl groups may be distributed on the glycerol backbone differently than represented in this figure. In fact, Salatrim is a randomized TAG.

a reduced-fat chocolate chip for baking purposes using this fat replacer. Other uses include chocolate-flavored coatings, caramel, fillings, spreads, dips, sauces, and some dairy products (Sandrou and Arvanitoyannis 2000). Taste tests on these products showed that they were equally preferred to the full-fat food. One of the benefits of Salatrim is that it crystallizes in its stable crystal form and further tempering (as in chocolates) is not required (Kosmark 1996). This reduces the cost of producing Salatrim-containing foods compared to their full-fat counterparts. The caloric availability has been studied (Finley et al. 1994) and was found to be 4.5–6.0 cal/g in rats and 4.7–5.1 cal/g in humans.

Olestra is the other major class of fat-like replacers. Development of these fat substitutes is described in detail by Akoh and Swanson (1994). Olestra is defined as a mixture of octa-, hepta-, and hexaesters of sucrose (see Figure 22.6). This material, sometimes referred to as sucrose polyester (SPE), is produced by the reaction of sucrose and long-chain fatty acids isolated from vegetable oils under a variety of alkaline conditions. SPE has many physical and functional properties in common with the fats and oils (lipophilic but nonabsorbable and nondigestible) they are meant to replace except that they contribute no calories. Its principle use is in the frying of savory foods such as Pringles® (Procter and Gamble Company, Cincinnati, OH) and Frito-Lay® (Frito-Lay North America, Plano, TX) brand chips and snacks. Its synthesis is primarily by transesterification, acylation (with either fatty acid anhydrides

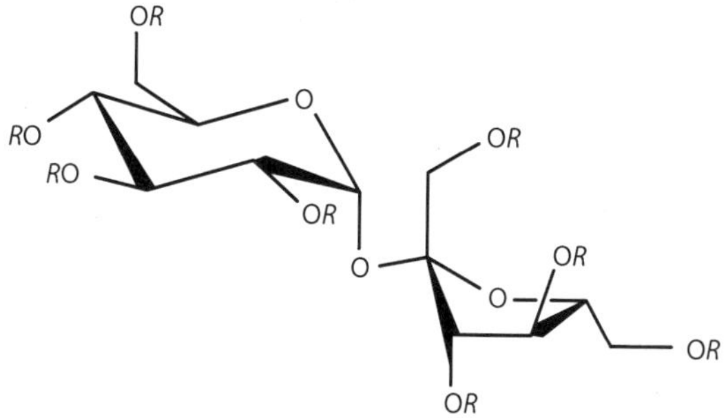

FIGURE 22.6 Representative structure of Olestra in which 6, 7, or 8 of the *R* groups would be long-chain fatty acyl groups and 2, 1, or 0 would be H atoms. Olestra is a mixture of these SPEs.

or fatty acid acyl chlorides), or direct esterification with fatty acids (McCoy et al. 1989). SPE is similar in color, consistency, density, and refractive index to natural vegetable oils. Olestra has been used in ice cream (Wei 1984), sausage (Linares 1995), and cheddar cheese (Drake et al 1994b). Somewhat more recently Crites et al. (1997) have reported on the effect of SPE on the microstructure of low-fat cheddar cheese. SPE was substituted for 10%, 25%, 50%, or 75% of the total fat content in milk, and cheddar cheese was manufactured and aged for 8 months prior to experimental examination. SEM was used to examine the microstructure of the cheese. Other analyses showed no significant differences between regular and low-fat cheddar cheeses for moisture, pH, or whey titratable acidity. The function of fat in regular cheese is to interrupt the protein matrix and provide a smooth creamy consistency. The substitution of SPE for milk fat seemed to provide similar textural qualities, although it was noted that the SPE-containing cheese had a texture more like processed cheese rather than natural cheese. Drake et al. (1994a) had previously shown that cheeses containing SPE were not significantly different in firmness from control full-fat cheeses. It was noted that the average number of fat globule voids remained constant between the control and experimental cheeses, but the average fat globule area decreased for the SPE milk fat cheeses. Crites et al. (1997) suggested that this is due to the greater molecular mass and bulkiness of the SPE compared to milk fat.

Other more esoteric and less commonly investigated fat-based fat replacers include trialkoxytricarballylate (TATCA), trialkoxycitrate (TAC), trialkoxyglyceryl ether (TGE), dialkyl dihexadecylmalonate (DDM), esterified propoxylated glycerols, phenylmethylpolysiloxanes, and poly(dimethylsiloxane) (PDMS). A review by Ognean et al. (2006) provides recent details on these fat-based fat replacers, as does a chapter by Swanson (2006).

22.5 INGREDIENTS THAT DECREASE FAT ABSORPTION IN DEEP FAT FRYING

Deep-fried foods are generally regarded as less than healthful due to the amount of oil absorbed by the food during deep fat frying. This absorption is exacerbated by the presence of so-called "rich" ingredients (sugar, fat, egg, and dairy) in many of the batters and coatings used in deep fat frying, but even when this is not the case, oil absorption can be a problem. It is a problem nutritionally and from a sensory point of view, as foods that absorb excess oil are oily, greasy, and generally unpalatable. One of the strategies for preventing this is the use of ingredients as components of the batters or as coatings that will decrease the amount of absorbed oil. Typically, it is believed that these ingredients function by gelling at the elevated temperatures (thermogelation) involved in deep frying and forming a layer on the outside of the food that is difficult for the hot oil to penetrate. Materials that undergo thermogelation behave as shown in Figure 22.7. During initial stages of heating, the dispersion is flowable and the viscosity decreases as would be expected of a Newtonian fluid behavior. However, at some temperature the viscosity increases dramatically and a gel is formed. As the system is cooled, it returns to its preheated state. Also, gelation at high temperatures decreases batter blow off and pillowing, and it also reduces residual debris levels in the cooking oil.

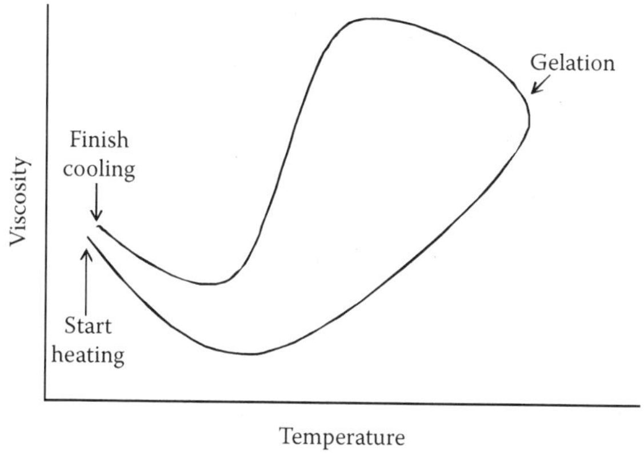

FIGURE 22.7 Idealized viscosity–temperature behavior of a fluid that undergoes thermogelation.

Early reports (Dow Chemical Co. 1991) showed that several methylcelluloses (MCs) and hydroxypropylmethylcelluloses (HPMCs) were effective in reducing absorbed oil. When HPMC was used at levels from 0.25 to 1.0 wt%, there was a 26% reduction in oil absorption in coated chicken nuggets. These nuggets also contained 10%–15% more moisture in the final product, which translates to about a 5% increase in yield. The derivatized celluloses also improved batter coating and pickup in the systems investigated. Sharma et al. (1999) have reported the effect of CbMC in the reduction of fat absorption in paneer, a milk product from India that is obtained by acid precipitation of cow and/or buffalo milk. This product is subsequently deep-fried. The CbMC was either incorporated directly into the milk before acid precipitation or the paneer was dipped in CbMC solutions. The first method was found more useful and decreased the absorption of oil by about 5%. Unfortunately, the sensory ratings for the paneer decreased as the level of CbMC (0.022–0.111 g/L) increased. Texture ratings declined as CbMC increased but taste ratings were only affected at the highest three levels of CbMC.

Soy protein plasticized with gellan gum has also been studied as a means to reduce oil absorption in fried foods. Rayner et al. (2000) examined the use of various soy protein fractions (flour, concentrate, and isolate) modified with gellan gum on fat absorption, penetration force, and sensory attributes of fried doughnut mix, potato disks, and fries. The results showed that only the highest protein fractions (containing the soy isolate) were useful, and that gellan gum was a better plasticizer than glycerin. Fat absorption was reduced from 40% to 55%. The coatings did not seem to have an effect on penetration tests in the case of the potato disks, and the sensory panel could not distinguish the samples with or without the coatings. The recommendation for a coating formulation was 10 wt% soy protein isolate containing 0.05 wt% gellan gum as a plasticizer.

Sahin et al. (2005) examined a number of hydrocolloids (HPMC, guar gum, xanthan gum, and gum arabic) on the quality characteristics of deep fat fried chicken nuggets. They determined that all the gums used increased batter pickup except for gum arabic, and all gums except arabic were able to control the migration of moisture during the frying process. In the initial stages of frying, HPMC was the most effective in controlling water movement. All of the gums except arabic were effective in reducing oil absorption; gum arabic, on the contrary, actually increased oil absorption compared to the control. The poor behavior of gum arabic in this study was suggested to be related to its significant water solubility and low viscosity compared to the other gums used (level of gum was 1 wt% in all cases). Hardness of the chicken nuggets increased with frying time as would be expected, and HPMC and xanthan gum produced softer chicken nuggets compared to the control (no added gum). Overall, HPMC and xanthan gum were the most effective hydrocolloids and gum arabic was the least acceptable.

Recently, gellan was investigated to determine its effect (alone and in combination with other hydrocolloids, e.g., sodium alginate, CbMC, and soy protein isolate) on the characteristics of sev (Bajaj and Singhal 2007). Sev is a traditional Indian snack food made from chickpea flour, which is deep fat fried. Since the oil level in such foods can be quite high, this study was undertaken to determine the effect of gellan (0–0.75 wt%) on oil absorption in this product. Dough texture before frying and sev texture after frying were determined by texture analyzer. At 0.25% gellan, oil absorption was reduced by about 25%. It was suggested that both the hydrophilic nature of gellan and its ability to undergo thermogelation were responsible for this reduction in oil uptake. Combining gellan with alginate, CbMC or soy protein isolate did not have any effect on oil absorption. As gellan concentration in the dough increased, the hardness and stickiness of the dough increased. However, addition of gellan did not affect the hardness, fracturability, or crispness of the fried sev itself.

22.6 CONCLUDING REMARKS

Fat replacers can be carbohydrate, protein, or fat based. The carbohydrate replacers consist primarily of modified starches, starch dextrins (lower molecular weight starch breakdown products), resistant starches, polydextrose, inulin, and food gums. They appear to do their job mainly by influencing the mobility of water in the foods in which they are used. Protein-based replacers are mainly aggregated mixtures of protein from various sources. They have been used in a wide variety of applications but are best suited to uses in dairy products (cheeses, frozen desserts, and yogurts) that do not require much heating in their preparation. Fat-like or fat-based replacers include Olestra, structured lipids (caprenin and Salatrim), and SPE. Certainly Olestra, because of its significant heat stability, is the star of this class of compounds based on its commercial use in fried snack foods. There appears to be much to do in the field of fat replacers as none of the foods containing these replacers (from any class) show properties that are significantly superior to the full-fat foods they are replacing. It may be that using mixtures of these replacers (both within and between classes) may show improvements in functionality due to synergistic interactions, much as is seen in the use of mixtures of alternative sweeteners to replace sucrose. What would be very useful in the development of these replacers would be some fundamental information about what characteristics are important in producing a fat- or oil-like texture. Is viscosity the only important consideration or might suspension/dispersion particle size be important? Little is known about the fundamental sensory detection of fatness/oiliness as it relates to structure–function relationships of the various fat replacers and this is an area where future work may prove invaluable.

REFERENCES

Adapa, S., Dingeldein, H., Schmidt, K.A., and Herald, T.J. 2000. Rheological properties of ice cream mixes and frozen ice creams containing fat and fat replacers. *J. Dairy Sci.* 83:2224–2229.

Akoh, C.C. and Swanson, B.G. 1994. *Carbohydrate Polyesters as Fat Substitutes*. Boca Raton, FL: CRC Press.

Andres, S., Zaritzky, N., and Califano, A. 2006.The effect of whey protein concentrates and hydrocolloids on the texture and color characteristics of chicken sausages. *Int. J. Food Sci. Technol.* 41:954–961.

Aryana, K.J. and Haque, Z.U. 2001. Effect of commercial fat replacers on the microstructure of low-fat Cheddar cheese. *Int. J. Food Sci. Technol.* 36:169–177.

Avellone, G., Di Garbo, V., Abruzzese, G. et al. 2003. Cross-over study on effects of Mediterranean diet in two randomly selected population samples. *Nutr. Res.* 23:1329–1339.

Bajaj, I. and Singhal, R. 2007. Gellan gum for reducing oil uptake in sev, a legume based product during deep-fat frying. *Food Chem.* 104:1472–1477.

Bimal, C. and Guonong, Z. 2006. Olestra: A solution to food fat? *Food Rev. Int.* 22:245–258.

Brandt, L.A. 2000. Novel ingredients for lowfat foods. *Prep. Foods* 169:49, 51–53.

Byars, J. 2002. Effect of a starch-lipid fat replacer on the rheology of soft-serve ice cream. *J. Food Sci.* 67: 2177–2182.

Callaway, C.W. 1998. The role of fat-modified foods in the American diet. *Nutr. Today* 33:156–163.

Camire, M.E., Dougherty, M.P., and Teh, Y.-H. 2006. Frozen wild blueberry-tofu soymilk desserts. *J. Food Sci.* 71:S119–S123.

Crites, S.G., Swanson, B.G., and Drake, M.A. 1997. Microstructure of low-fat Cheddar cheese containing varying concentrations of sucrose polyesters. *Lebensm. Wiss. Technol.* 30:762–766.

Davy, B.M., Davy, K.P., Ho, R.C., Beske, S.D., Davrath, L.R., and Melby, C.L. 2002. High-fiber oat cereal compared with wheat cereal alters LDL-cholesterol subclass and particle numbers in middle-aged and older men. *Am. J. Clin. Nutr.* 76:351–358.

Dow Chemical Co. 1991. Fried foods stabilizer keep moisture in, fat out. *Prep. Foods* 160:61.

Drake, M.A., Nagel, C.W. and Swanson, B.G. 1994a. Milkfat sucrose polyesters as fat substitutes in Cheddar-type cheeses. *J. Food Sci.* 59:326–327, 365.

Drake, M.A., Boutte, T.T., Younce, F.L., Cleary, D.A., and Swanson, B.G. 1994b. Melting characteristics and hardness of milkfat blend sucrose polyesters. *J. Food Sci.* 59:652–654.

Dreher, M., Leveille, G.A., Auerbach, M., Callen, C., Klemann, L., and Jones, K. 1998. Salatrim: A triglyceride-based fat replacer. *Nutr. Today* 33:164–170.

Finley, J.W., Klemann, L.P., Leveille, G.A., Otterburn, M.S., and Walchak, C.G. 1994. Caloric availability of Salatrim in rats and humans. *J. Agric. Food Chem.* 42:495–499.

Gallagher, E., O'Brien, C.M., Scannell, A.G.M., and Arendt, E.K. 2003. Use of response surface methodology to produce functional short dough biscuits. *J. Food Eng.* 56:269–271.

Glicksman, M. 1991. Hydrocolloids and the search for the "oily grail." *Food Technol.* 45:94–103.

Gupta, R., Rathi, P., and Bradoo, S. 2003. Lipase mediated upgradation of dietary fats and oils. *Crit. Rev. Food Sci. Nutr.* 43:635–644.

Hamm, D.J. 1984. Preparation and evaluation of trialkoxytricarballate, trialkoxycitrate, trialkoxyglycerylether, jojoba oil and sucrose polyester as low calorie replacements of edible fats and oils. *J. Food Sci.* 49:419–428.

Inglett, G. E., 2001. New grain products and their beneficial components. *Nutr. Today* 36:66–68.

Jandacek, R.J., Kester, J.J., Papa, A.J., Wehmeier, T.J., and Lin, P.Y.T. 1999. Olestra formulation and the gastrointestinal tract. *Lipids* 34:771–783.

Kavas, G., Oysun, G., Kinik, O., and Uysal, H. 2004. Effect of some fat replacers on chemical, physical and sensory attributes of low-fat white pickled cheese. *Food Chem.* 88:381–388.

Khalil, A. H. 1998. The influence of carbohydrate-based fat replacers with and without emulsifiers on the quality characteristics of lowfat cake. *Plant Foods Human Nutr.* 52:299–313.

Khouryieh, H.A., Aramouni, F.M., and Herald, T.J. 2005. Physical and sensory characteristics of no-sugar-added/low-fat muffin. *J. Food Qual.* 28:439–451.

Kosmark, R. 1996. Salatrim: Properties and applications. *Food Technol.* 50:98–101.

Koutsopoulos, D.A., Koutsimanis, G.E., and Bloukas, J.G. 2008. Effect of carrageenan level and packaging during ripening on processing and quality characteristics of low-fat fermented sausages produced with olive oil. *Meat Sci.* 79:188–197.

Laneuville, S.I., Paquin, P., and Turgeon, S.L. 2005. Formula optimization of a low-fat food system containing whey protein isolate-xanthan gum complexes as fat replacer. *J. Food Sci.* 7:S513–S519.

Lermer, C.M. and Mattes, R.D. 1999. Perception of dietary fat: Ingestive and metabolic implications. *Prog. Lipid Res.* 38:117–128.

Lichenstein, A.H., Kennedy, E., Barrier, P. et al. 1998. Dietary fat consumption and health. *Nutr. Rev.* 56:S3–S28.

Linares, M. 1995. *Methylglucoside Polyesters of Lard Fatty Acids Application in Pork Sausage.* MSc thesis, Washington State University, Pullman, WA.

Liou, B.K. and Grun, I.U. 2007. Effect of fat level on the perception of five flavor chemicals in ice cream with or without fat mimetics by using a descriptive test. *J. Food Sci.* 72:S595-S604.

Mansour, E.H., Khalil, A.H., and El-Soukkary, F.A. 2003. Production of low-fat cookies and their nutritional and metabolic effects in rats. *Plant Foods Human Nutr.* 58:1–14.

Mattes, R.D. 1998. Position of the American Dietetic Association: Fat replacers. *J. Am. Diet. Assoc.* 98:463–468.

Mattes, R.D. 2001. Oral exposure to butter, but not fat replacers elevates postprandial triacylglycerol concentration in humans. *J. Nutr.* 131:1491–1496.

McCoy, S.A., Madison, B.L., Self, P.M., and Weisgerber, D.J. 1989. Sucrose polyesters which behave like cocoa butters. U.S. Patent 4,822,875, filed May 2, 1988, and issued April 18, 1989.

McDowell, M.A., Briefel, R.R., Alaimo, K., Bischof, A.M., Caughman, C.R., Carroll, M.D., Loria, C.M., and Johnson, C.L. 1994. Energy and macronutrient intake of persons age 2 months and over in the United States: *Third National Health and Nutrition Examination Survey-Phase 1988–1991.* Hyattsville, MD: National Center for Health Statistics; Vital and Health Statistics publication 255.

McMahon, D.J., Alleyne, M.C., Fife, R.L., and Oberg, C.J. 1996.Use of fat replacers in low fat mozzarella cheese. *J. Dairy Sci.* 79:1911–1921.

Mela, D.J. 1997. Fat and sugar substitutes: Implications for dietary intakes and energy balance. *Proc. Nutr. Soc.* 56:827–840.

O'Brien, C.M., Mueller, A., Scannell, A.G.M., and Arendt, E.K. 2003. Evaluation of fat replacers on the quality of wheat bread. *J. Food Eng.* 56:265–267.

Ognean, C.F., Darie, N., and Ognean, M. 2006. Fat replacers—Review. *J. Agroaliment. Proc. Technol.* 12:433–442.

Phillips, L.G. and Barbano, D.M. 1997. The influence of fat substitutes based on protein and titanium dioxide on the sensory properties of lowfat milks. *J. Dairy Sci.* 80:2726–2731.

Pivk, U., Godinot, N., Keller, C., Antille, N., Juillerat, M.A., and Raspor, P. 2008. Lipid deposition on the tongue after oral processing of medium-chain tryglycerides and impact on the perception of mouthfeel. *J. Agric. Food Chem.* 56:1058–1064.

Porter, D., Kris-Etherton, P., Borra, S. et al. 1998. Educating consumers regarding choices for fat reduction. *Nutr. Rev.* 56:S75–S100.

Prindiville, E.A., Marshall, R.T., and Heymann, H. 2000. Effect of milk fat, cocoa butter, and whey protein fat replacers on the sensory properties of lowfat and nonfat chocolate ice creams. *J. Dairy Sci.* 83:2216–2223.

Ranhotra, G.S., Gelroth, J.A., and Glaser, B.K. 1994. Usable energy value of a synthetic fat (caprenin) in muffins fed to rats. *Cereal Chem.* 71:159–161.

Rayner, M., Ciolfi, V., Maves, B., Stedman, P., and Mittal, G.S. 2000. Development and application of soy-protein films to reduce fat intake in deep-fried foods. *J. Sci. Food Agric.* 80:777–782.

Roberts, S.B., Pi-Sunyer, F.X., Dreher, M. et al. 1998. Physiology of fat replacement and fat reduction: Effects of dietary fat and fat substitutes on energy regulation. *Nutr. Rev.* 56:S29–S49.

Roller, S. and Jones, S.A. 1996. *Handbook of Fat Replacers.* Boca Raton, FL: CRC Press.

Sahan, N., Yasar, K., Hayaloglu, A.A., Karaca, O.B., and Kaya, A. 2008. Influence of fat replacers on chemical composition, proteolysis, texture profiles, meltability and sensory properties of low-fat Kashar cheese. *J. Dairy Res.* 75:1–7.

Sahin, S., Sumnu, G., and Altunakar, B. 2005. Effects of batters containing different gum types on the quality of deep-fat fried chicken nuggets. *J. Sci. Food Agric.* 85:2375–2379.

Sajilata, M.G. and Singhal, R.S. 2005. Specialty starches for snack foods. *Carb. Polym.* 59:131–151.

Sandrou, D.K. and Arvanitoyannis, I.S. 2000. Low-fat/calorie foods: Current state and perspectives. *Crit. Rev. Food Sci. Nutr.* 40:427–447.

Schmitt, C., Sanchez, C., Desorby-Banon, S., and Hardy, J. 1997. Structure and technofunctional properties of protein-polysaccharide complexes: A review. *Crit. Rev. Food Sci.* 38:689–753.

Sharma, H.K., Singhal, R.S., Kulkarni, P.R., and Gholap, A. 1999. Carboxymethyl cellulose (CbMC) as an additive for oil reduction in deep-fat fried paneer. *Int. J. Dairy Technol.* 52:92–94.

Singh, S.K., Castell-Perez, M.E., and Moreira, R.G. 2000. Viscosity and textural attributes of reduced-fat peanut pastes. *J. Food Sci.* 65:849–853.

Smith, R.E., Finley, J.W., and Leveille, G.A. 1994. Overview of SALATRIM, a family of low-calorie fats. *J. Agric. Food Chem.* 42:432–434.

Softly, B.J., Huang, A.S., Finley, J.W. et al. 1994. Composition of representative SALATRIM fat preparations. *J. Agric. Food Chem.* 42:461–467.

Steenblock, R.L., Sebranek, J.G., Olson, D.G., and Love, J.A. 2001. The effects of oat fiber on the properties of light bologna and fat-free frankfurters. *J. Food Sci.* 66:1409–1415.

Swanson, B.G., 2006. Fat replacers, mimetics and substitutes. In *Nutraceutical and Specialty Lipids and Their Co-Products*, ed. F. Shahidi. Boca Raton, FL: CRC Press, pp. 329–340.

Swanson, R.B. 1998. Acceptability of reduced-fat peanut butter cookies by school nutrition managers. *J. Am. Diet. Assoc.* 98:910–912.

Theophilou, P. and Wilbey, R.A. 2007. Effects of fat on the properties of halloumi cheese. *Int. J. Dairy Technol.* 60:1–4.

Totosaus, A., Alfaro-Rodriguez, R.H., and Perez-Chabela, M.L. 2004. Fat and sodium chloride reduction in sausages using kappa-carrageenan and other salts. *Int. J. Food Sci. Nutr.* 5:371–380.

Turgeon, S.L., Beaulieu, M., Schmitt, C., and Sanchez, C. 2003. Protein-polysaccharide interactions: Phase ordering kinetics, thermodynamic and structural aspects. *Curr. Opin. Colloid Interface Sci.* 8:401–414.

Wei, J.J. 1984. Synthesis and feeding studies of sucrose fatty acid polyesters utilized as simulated milkfat. PhD dissertation, Washington State University, Pullman, WA.

Wekwete, B. and Navder, K.P. 2005. Effect of avocado puree as a fat replacer on the physical, textural and sensory properties of oatmeal cookies. *J. Am. Diet. Assoc.* 105: A-47.

Wekwete, B. and Navder, K.P. 2008. Effects of avocado fruit puree and Oatrim® as fat replacers on the physical, textural and sensory properties of oatmeal cookies. *J. Food Qual.* 31:131–141.

Welty, W.M., Marshall, R.T., Grun, I.U., and Ellersiech, M.R. 2001. Effects of milk fat, cocoa butter, or selected fat replacers on flavor volatiles of chocolate ice cream. *J. Dairy Sci.* 84:21–30.

Xiong, Y.L., Noel, D.C., and Moody, W.G. 1999. Textural and sensory products of low-fat beef sausages with added water and polysaccharides as affected by pH and salt. *J. Food Sci.* 64:550–554.

Xu, X. 2000. Production of specific-structured triacylglycerols by lipase-catalyzed reactions: A review. *Eur. J. Lipid Sci. Technol.* 102:287–303.

Yang, A., Keeton, J.T., Beilken, S.L., and Trout, G.R. 2001. Evaluation of some binders and fat substitutes in low-fat frankfurters. *J. Food Sci.* 66:1039–1046.

Yazici, F. and Akgun, A. 2004. Effect of some protein based fat replacers on physical, chemical, textural, and sensory properties of strained yoghurt. *J. Food Eng.* 62:245–254.

Yildiz-Turp, G. and Serdaroglu, M. 2008. Effect of replacing beef fat with hazelnut oil on quality characteristics of sucuk-A Turkish fermented sausage. *Meat Sci.* 78:447–454.

Zambrano, F., Despinoy, P., Ormenese, R.C.S.C., and Faria, E.V. 2004. The use of guar and xanthan gums in the production of "light" low fat cakes. *Int. J. Food Sci. Technol.* 39:959–966.

Zoulias, E.I., Piknis, S., and Oreopoulou, V. 2000. Effect of sugar replacement by polyols and acesulfame-K on properties of low-fat cookies. *J. Sci. Food Agric.* 80:2049–2056.

Zoulias, E.I., Oreopoulou, V., and Tzia, C. 2002a. Textural properties of low-fat cookies containing carbohydrate- or protein-based fat replacers. *J. Food Eng.* 55:337–342.

Zoulias, E.I., Oreopoulou, V., and Kounalaki, E. 2002b. Effect of fat and sugar replacement on cookie properties. *J. Sci. Food Agric.* 82:1637–1644.

23 Biosurfactants as Emerging Additives in Food Processing

Denise Maria Guimarães Freire, Lívia Vieira de Araújo, Frederico de Araujo Kronemberger, and Márcia Nitschke

CONTENTS

23.1 INTRODUCTION

Surfactant agents are amphipathic molecules constituted by a hydrophilic (it can be nonionic, positively or negatively charged, or amphoteric) and a hydrophobic domain (generally a lipid). The surfactant properties are marked by the interactions of these amphipathic molecules with different polarity degrees and hydrogen bonds at fluid–fluid or fluid–solid interfaces, such as oil/water, air/water, and air/solid, reducing the interfacial tension, the surface tension, and changing the wettability between these coupled phases. Usually, the surfactants are characterized by their physical–chemical properties, such as critical micelle concentration (CMC), and hydrophilic–lipophilic balance (HLB), their chemical structure, and their charge.

Surfactant agents can be divided into synthetic and natural. Synthetic surfactants are, generally, petrochemical-origin products, derived from alkyl or ethylene polymers and polypropylene, whereas natural surfactants are produced by living organisms. Biosurfactants, a class of natural surfactants, are defined as surface-active compounds produced by microorganisms.

When compared to conventional synthetic surfactants, the biosurfactants have some peculiar characteristics (Bognolo 1999):

- One or more functional groups and chiral center
- Large and complex structure
- Higher biodegradability and lower toxicity
- Lower CMC and higher surface activity
- Superior ability to form molecular assembly and liquid crystal
- Biological activity (antimicrobial, antitumor, etc.) (Kitamoto et al. 2002)
- Stability to extremes conditions of pH, salinity, and temperature (Kesting et al. 1996)
- Biosurfactants can be produced from renewable substrates by biotechnological processes (Desai and Banat 1997)

Biosurfactants present distinct interface properties and can be divided into different groups according to their chemical structures: glycolipids, lipopeptides, liposaccharides, phospholipids, neutral lipids, and fatty acids. The great diversity of biosurfactant-producing microorganisms suggests that this biomolecule is an important survival tool for the microbe (Bodour et al. 2003). Thus, biosurfactants are likely to have several roles in the physiology and ecology of these microorganisms.

Microbial physiology is influenced by interfaces, or in other words, by the interaction between two distinct phases (air/liquid, liquid/liquid, solid/liquid, or air/solid), where, generally, the microbial life is plentiful. It can be evidenced by biofilms, superficial films, and the existence of microbial aggregates. Several important phenomena in microbial physiology occur at the interface, such as nutrient mass transfer, excretion and signaling, host–pathogen interaction, and metabolite sequestration. In general, it is through the interfaces that microorganisms interact with the environment, and biosurfactants induce cellular mobility and communication, nutrient access, cellular competition, and pathogenesis in plants and animals. In this context, biosurfactants play a fundamental role in microbial physiology. However, the current understanding of the role played by these molecules in evolutive and physiological processes still presents many gaps (Van Hamme et al. 2006). There are several experimental evidences that point to an increase in bacteria cell growth in the presence of insoluble substrates after the addition of biosurfactants in the medium.

There are clear evidences that emulsification is a cell-density-dependent phenomenon, that is, the higher the cell number, the larger the amount of emulsifier agents produced by them. In spite of these evidences, there are some conceptual difficulties to understand, from an evolutionary point of view, the advantages of the microorganisms able to produce biosurfactants. Firstly, the biomass in open environments, such as a polluted aquifer, will never achieve a cell concentration high enough to effectively emulsify the contaminant oil. On the other hand, it can be questioned what evolutionary advantage should the biosurfactant-producing microorganism have if the production of these biomolecules would also make the emulsified oil bioavailable for the other microorganisms, including its competitors. A formulated hypothesis to reconcile the current evidence with such theoretical considerations is that biosurfactants should have a very important physiological role on oil degradation, not by forming macroscopic emulsions, but in the formation of emulsions close to the cell surface (microemulsions). Thus, each cell would create its own microenvironment and the cell density dependency could be broken through the production of biosurfactants (Ron and Rosenberg 2001).

Mukherjee and Das (2005) observed that *Bacillus subtilis* strains isolated from different habitats (fermented food or petroleum contaminated soil) produced quantitatively and qualitatively different cyclic lipopeptides isoforms under distinct laboratory culture conditions. Their findings reinforce that the type or amount of the hydrophobic substrate(s) available in the parent habitat of the bacteria influences the biosurfactant production and thereby enhances the utilization of a specific substrate group. They concluded that these distinct biosurfactants conferred some kind of competitive advantage to the producing *B. subtilis* strains in their parent ecological niche, playing an important role in the utilization of the hydrophobic substrates available from their natural habitats.

Santos et al. (2002) observed that the proportion of monorhamnolipids (R1) to dirhamnolipids (R2) produced by a *Pseudomonas aeruginosa* strain was related to the carbon source utilized in the culture media. When glycerol (soluble substrate) and vegetable oils (insoluble) were used, the concentration ratios of R2 to R1 were 5.9 and 0.8, respectively. This different proportion of rhamnolipid types can represent an important change in the characteristics of the produced biosurfactant, affecting parameters such as the CMC and interfacial tension, which in turn will affect the substrate availability for the bacteria.

Some aspects of the physiological role of biosurfactants in producing microorganisms and their interaction with the environment include increasing the interfacial area and bioavailability of hydrophobic insoluble substrates; pathogenesis and *quorum sensing*; antimicrobial activity; regulating the attachment–detachment of microorganisms to or from surfaces, heavy-metal binding, and motility (Ron and Rosenberg 2001, Van Hamme et al. 2006).

Apart from their physiological role and nature, the properties exhibited by these biomolecules are the main reason for the increasing interest in their commercial exploitation. The world environmental concern that has been developed in the last few years triggers the attention to biosurfactants, essentially due to their low toxicity and biodegradable nature. This chapter focuses on potential applications of biosurfactant in the food industry, as well as on the new approaches for their production and scale-up.

23.2 POTENTIAL APPLICATIONS OF BIOSURFACTANTS IN FOOD PROCESSING

Biosurfactant properties are of interest to a wide range of industrial fields, from petroleum to pharmaceuticals. In the food industry, these microbial compounds exhibit useful properties as emulsifiers (Banat et al. 2000), antiadhesive, and antimicrobial agents (Singh and Cameotra 2004). Currently, some products based on biosurfactants can be found in the international market, such as PD5® (Pendragon Holdings Ltd—www.pd-5england.co.uk), sold as a fuel additive based on a biosurfactant blend of rhamnolipid and enzymes; EC-601® (EcoChem Organics Co—www.ecochem.com), commercialized as a water insoluble hydrocarbon dispersant, JBR® products (Jeneil Biosurfactant Company), rhamnolipids in aqueous solutions with different purity levels or in a semisolid form, and the Rhamnolipid Inc. products (Rhamnolipid Holdings—www.rhamnolipidholdings.com), among others. Nowadays, the rhamnolipids from *P. aeruginosa* are the most promising class of biosurfactants, since the U.S. Environmental Protection Agency has approved rhamnolipids for the use in food products, cosmetics, and pharmaceuticals (Nitschke and Costa 2007).

23.2.1 Biosurfactants as Food Additives

The basic purpose behind processing food is not only to make it safe to eat at a later date, but also to make it look, taste, and smell as enticing and as close to freshly prepared as possible. Much of the availability, convenience, and variety that people enjoy in their food supply are due to food processing that often uses specific ingredients to maintain the final product quality. Since food that does not taste, look, or smell good will not be eaten, food additives are essential ingredients in our food supply.

Food additives global sales were estimated to be over $25 billion in 2007 and the market has been expected to continue growing at a rate of about 2%–3% per year. In terms of market increase since 2004, the most significant growth rates in food additives were observed in emulsifiers (up 10.5%) and hydrocolloids (up 6.0%) sectors (Letherhead Food International 2008).

The particular combination of characteristics such as emulsifying, antiadhesive, and antimicrobial activities presented by biosurfactants suggests their application as multipurpose ingredients or additives. The development of these biomolecules should represent, in the near future, an increasing significant part of the market for food additives.

23.2.1.1 Emulsifiers

Food makers use emulsifiers to reduce the surface tension between two immiscible phases at their interface allowing them to mix (Shepherd et al. 1995). Due to their functional properties, emulsifiers form an indispensable part of food items such as chilled dairy products and high-priced breads, as well as low-fat spreads that demand organoleptic superiority, stability, improvement in volume, and longer shelf life. And with the food industry changing needs, there has been a growing demand for multipurpose emulsifiers, which includes products that act as low-fat substitutes and stabilizers as well as perform the basic emulsification functions (Nitschke and Costa 2007). Commercially available emulsifiers used in the food and drink sectors comprise two main types, lecithin, derived from soy and egg, and a range of other emulsifiers produced primarily from synthetic sources. The fast growth of functional foods market requiring natural or organic ingredients represents an opportunity for new emulsifiers.

Considering that biosurfactants exhibit surface-active and emulsifier action, there is a great potential market for effective biosurfactants in the food industry not only due to their surface activity, but also for their environmental friendly nature (Mohan et al. 2006), low toxicity (Flasz et al. 1998), their unique structures and properties, and the increasing customer demand for natural or organic over synthetic ingredients.

Biosurfactants may stabilize (emulsifiers) or destabilize (de-emulsifiers) emulsions. High-molecular-mass biosurfactants are, in general, better emulsifiers than the low-molecular-mass ones. Sophorolipids from *Torulopsis bombicola* have been shown to reduce surface and interfacial tension but not to be good emulsifiers (Cooper and Paddock 1984). By contrast, liposan has been shown not only to reduce surface tension, but also to be able to successfully emulsify edible oils (Cirigliano and Carman 1985). Polymeric surfactants offer additional advantages because they coat the oil droplets, thereby forming very stable emulsions that never coalesce. This property is especially useful for making oil/water emulsions for cosmetics and food. In dairy products (soft cheese and ice creams), emulsifier addition improves texture and creaminess. This quality is of special value for low-fat products (Rosenberg and Ron 1999).

A lipopeptide surfactant obtained from *B. subtilis* is able to form stable emulsions with soybean oil and coconut fat (Nitschke and Pastore 2006) and with sunflower, linseed, olive, palm, babassu, and Brazilian nut oils (Costa et al. 2006), suggesting its potential as emulsifying agent in foods. A manoprotein from *Kluyveromyces marxianus* can form emulsions with corn oil that are stable for 3 months; the yeast has been cultivated on whey-based medium suggesting potential application as food bioemulsifier (Lukondeh et al. 2003). The extracellular carbohydrate-rich compound from *Candida utilis* can be successfully used as emulsifying agent in salad dressing formulations (Shepherd et al. 1995). Examples of the chemical structure of some biosurfactants are shown in Figure 23.1.

23.2.1.2 Antimicrobial Activity

A wide range of biosurfactants has shown antimicrobial activity against bacteria, yeast, fungi, algae, and viruses (Nitschke and Costa 2007). Among biosurfactants, lipopeptides form the widely reported class having antimicrobial action. The genus *Bacillus* is responsible for producing the most well-known lipopeptides biosurfactants. *B. subtilis* produces the first biosurfactant presenting this antimicrobial property, that is, surfactin (Das et al. 2008, Fernandes et al. 2007). It also produces other lipopeptides with the same property: fengycin, iturin, bacillomycins, and mycosubtilins (Das et al. 2008). Other *Bacillus* species are known to produce antimicrobial lipopeptides biosurfactants, like *B. licheniformis*, which produces lichenysin, and *B. pumilus*, which produces pumilacidin (Das et al. 2008). Gramicidin S (cyclosymmetric decapeptide) produced by *B. brevis* and polymyxins produced by *B. polymyxa* are other examples of well-known lipopeptide surfactants showing antimicrobial activity. Many lipopeptide syntheses have been observed in early sporulation stages in almost all bacteria from *Bacillus* genus, indicating their important role in the sporulation process (Ron and Rosenberg 2001).

Surfactin has many pharmaceutical applications, such as to inhibit clot formation and to form membrane ionic channels, besides the antibacterial, antiviral, antimicoplasmal, and antitumoral activity; iturin has antifungal activity (Nitschke and Pastore 2002, 2003).

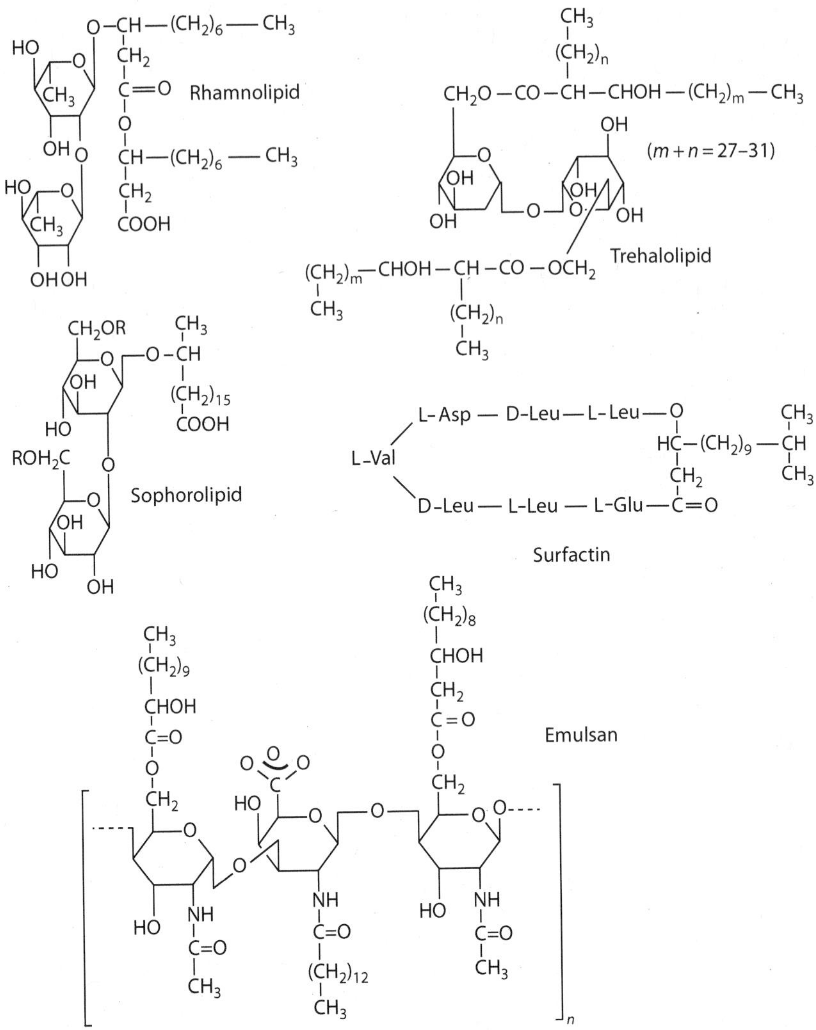

FIGURE 23.1 Chemical structures of some microbial surfactants.

Das et al. (2008) studied the potential use of a lipopeptide biosurfactant produced by a *B. circulans* strain as antimicrobial agent against a wide range of microorganisms. They have found that the biosurfactant has a potent antimicrobial activity against Gram-positive and Gram-negative pathogenic microbial strains.

Fernandes et al. (2007) investigated the antimicrobial activity of biosurfactants obtained by *B. subtilis* R14 cultivation against multidrug-resistant bacteria. A preliminary characterization suggested that two surfactants were produced. The antimicrobial activity evaluation of these compounds was carried out against 29 bacteria. Strains of *Enterococcus faecalis*, *Staphylococcus aureus*, *P. aeruginosa*, and *Escherichia coli* displayed a well-defined drug resistance profile. The tested strains were sensitive to the surfactants, in particular *En. faecalis*. The authors have concluded that lipopeptides have a broad action spectrum, including antimicrobial activity against microorganisms with multidrug-resistant profiles.

Other biosurfactant classes that are reported to have antimicrobial actions are the rhamnolipids produced by *P. aeruginosa* and sophorolipids produced by *C. bombicola*. Mannosylerythritol lipids (MEL-A and MEL-B) produced by a *C. antarctica* strain exhibited antimicrobial action against Gram-positive bacteria (Das et al. 2008).

There are some works developed aiming at the antimicrobial application of a biosurfactant variety. Abalos et al. (2001) studied the antimicrobial activity of a rhamnolipid mixture produced by *P. aeruginosa* AT10 against a wide range of Gram-positive and Gram-negative bacteria, yeast and filamentous fungi. The authors found good inhibitory activity against *E. coli* and *A. faecalis* (32 μg/mL), and also the opportunistic pathogen *Serratia marcescens* (16 μg/mL). No antimicrobial activity was observed on the tested yeast strains. Against the filamentous fungi *Aspergillus niger*, *Gliocadium virens* (16 μg/mL), *C. globosum*, *Penicillium chrysogenum*, and *Aureobasidium pullulans* (32 μg/mL), the results were also promising. The inhibitory activity against phytopathogenic fungi was also assayed and the growth of *Botrytis cinerea*, and *Rhizoctonia solani* was inhibited at 18 μg/mL, that of *Colletotrichum gloesporioides* and *Fusarium solani* at 65 μg/mL, but that of *Pe. funiculosum* at 128 μg/mL. Due to its physicochemical properties and high antimicrobial activity, this mixture of seven homologues rhamnolipids produced by this *P. aeruginosa* strain can be regarded as a useful tool in bioremediation processes, cosmetics, and food industrial applications.

The succinoyl–trehalose lipid, a biosurfactant produced by *Rhodococcus erythropolis*, is able to inhibit herpes simplex virus and parainfluenza virus (Nitschke and Pastore 2002). Despite these activities showing direct application on pharmaceutical products, some of them can also be explored by the food industry, particularly increasing the food shelf life without concern to consumer health, excluding the need of adding synthetic comfits, which, in most of cases, are health harmful.

The combination of nisin with rhamnolipids has extended the shelf life and inhibited thermophilic spores in UHT soymilk. The use of compositions comprising natamycin and rhamnolipids in salad dressing has extended its shelf life and inhibiting mold growth. Compositions comprising natamycin, nisin, and rhamnolipids in cottage cheese have extended shelf life by inhibiting mold and bacterial growth, especially gram-positive and spore-forming bacteria (Gandhi and Skebba 2007).

A 1% emulsion of rhamnolipids can be used in the treatment of leaves of *Nicotiana glutinosa* infected with tobacco mosaic virus, as well as for potato virus-x disease control (Desai and Banat 1997). The Jeneil Biosurfactant Corporation has developed a rhamnolipid biofungicide formulation to prevent crop contamination by pathogenic fungi. This product is considered nonmutagenic and of low acute toxicity to mammals and it is approved by FDA to direct use on vegetables, legumes, and fruits crops (Nitschke and Costa 2007). Kim et al. (2000) have observed that biosurfactant from rhamnolipid B type causes zoospore lysis, zoospore inhibition, spore germination, and hyphal growth inhibition of several fungal species, including *Phytophthora capsici* and *Colletotrichum ribiculare*. Other authors have isolated, from the wheat rhizosphere, several *Pseudomonas* species that are cyclic-lipopeptide-type biosurfactant producers with antifungal activity (de Souza et al. 2003).

Therefore, with all these antimicrobial properties, those compounds can be used to avoid food contamination directly, as food additive (by the proper formulations use), or indirectly, as a detergent formulation to clean surfaces that come in contact with the food.

23.2.1.3 Antiadhesive Properties

It is important to review some concepts to understand why one would use the antiadhesive properties of biosurfactants in the food industry.

Usually, in nature, bacteria do not stand as alone cells, but they organize themselves as communities, which may be more or less structuralized. Previously, it was assumed that microorganisms lived as planktonic cells (free cells). Actually, it is now known that, in most cases, bacteria are found in communities with different complexity degrees and the planktonic existence seems to be only eventual (Kyaw 2006). The communities that grow at the interfaces between phases are called biofilms (Jenkinson and Lappin-Scott 2001). In the food industry, failure in hygienic procedures may lead to the adherence of residues to equipment and surfaces. So, under certain conditions, microorganisms adhere, interact with surfaces, and begin cellular multiplication up to a cellular body development which is able to aggregate nutrients, residues, and other microorganisms (Bagge-Ravn et al. 2003). The biofilm does not include only microorganisms, but also the extracellular material produced all over the surface and any other material that may be inside of its resulting matrix. While bacteria can adhere in a few minutes, the real biofilm takes hours or days to be developed (Hood and Zottola 1995).

As the bulk phase of a fluid containing microorganisms comes into contact with a surface, the first step that occurs is adsorption of molecules to the surface. This step is usually called conditioning. Surface conditioning is generally related to the fluid bulk composition and may have inhibiting or stimulating effects for the adhesion of microorganisms to surface (Hood and Zottola 1995). Biofilms may be formed by populations from only one species or by multiple species, being found in a wide range of surfaces, which may be biotic and/or nonbiotic (Parizzi et al. 2004).

Biofilms protect microorganisms from hostile environments and act facilitating the acquisition of nutrients (Maukonen et al. 2003). On biofilms, microorganisms are more resistant to chemical and physical agents than as planktonic cells, considering the agents used during hygienic procedures (Chae and Schraft 2000, Stepanovic et al. 2004). The resistance of biofilm cells to external influences like antibiotics, host defenses, antiseptic, and shear forces is a constant concern in the food industry and medicine (Jenkinson and Lappin-Scott 2001).

Due to the fact that food processors have zero tolerance levels to pathogens like *Salmonella* spp and *Listeria monocytogenes*, a single cell may be as important as a well-developed biofilm (Hood and Zottola 1995, Nitschke and Costa 2007). The main goal of the food industry is to find a way to produce and furnish a safe, secure, and palatable food, and microorganisms control is essential to reach this goal (Bagge-Ravn et al. 2003, Hood and Zottola 1995). Sanitization of all surfaces in contact with food is a good way to prevent contamination (Somers and Wong 2004). Unfortunately, there are some evidences that current sanitization practices are less effective to adhered microorganisms than to planktonic cells (Hood and Zottola 1995).

New strategies to control biofilms development are constantly under research, such as the control by environmental factors at the processing line and the use of surface active agents, dispersants, biodispersants, enzymes, and new nontoxic or less-toxic biocide compounds. Just like cleaning surfaces, staff training, good manufacturing practices, and structure are important factors to combat hygiene problems in the food industry (Maukonen et al. 2003). Recent studies have shown that biosurfactant preconditioned surfaces can significantly reduce microbial contamination and inhibit or reduce subsequent biofilm development (Meylheuc et al. 2006a).

Davey et al. (2003) have developed studies indicating that purified *P. aeruginosa* rhamnolipids can block initial adherence by fluorescent pseudomonads and *E. coli*, suggesting that surfactants may have a broad and relatively nonspecific ability to interfere with cell-to-cell and cell-to-surface interactions.

Rodrigues et al. (2004) studied the influence of two biosurfactants produced by probiotic bacteria (*Lactococcus lactis* and *Streptococcus thermophilus*) biofilms development by *S. epidermidis*, *St. salivarius*, *S. aureus*, *Rothia dentocariosa*, *C. albicans*, and *C. tropicalis* in voice prostheses. Both biosurfactants showed antimicrobial activity but, depending on the microorganism, they presented different effective concentrations.

Another interesting work was carried out by Rodrigues et al. (2006b). These authors studied the effect of a rhamnolipid biosurfactant on the adhesion, to silicone rubber, of microorganisms (*S. epidermidis*, *St.. salivarius*, *S. aureus*, *R. dentocariosa*, *C. albicans*, and *C. tropicalis*) isolated from explanted voice prostheses. Rhamnolipid showed antiadhesive effect against all microorganisms tested. *R. dentocariosa*, *S. epidermidis*, and *St. salivarius* had inhibitions of 60.9%, 53.1%, and 58.2%, respectively; *S. aureus*, *C. albicans*, and *C. tropicalis* presented only 33.8%, 38.2%, and 35.3% of growth reduction, respectively. These results have confirmed that biosurfactants can have a potential use as a preventive strategy to postpone the initial biofilm formation process.

The preconditioning of stainless steel and PTFE surfaces with a biosurfactant obtained from *P. fluorescens* inhibit the *L. monocytogenes* L028 strain adhesion. A significant reduction (>90%) was attained in microbial adhesion levels on stainless steel, whereas no significant effect was observed in PTFE (Meylheuc et al. 2001). Further work demonstrated that the prior adsorption of *P. fluorescens* surfactant on stainless steel also favored the bactericidal effect of disinfectants (Meylheuc et al. 2006b). The ability of adsorbed biosurfactants, obtained from Gram-negative (*P. fluorescens*) and Gram-positive (*Lactobacillus helveticus*) bacteria isolated from foodstuffs, to inhibit the *L. monocytogenes* adhesion to stainless steel was also investigated. Adhesion tests showed that both biosurfactants were effective, strongly decreasing the surface contamination level. The antiadhesive biological coating reduced both the total adhering flora and the viable/cultivable adherent *L. monocytogenes* on stainless steel surfaces (Meylheuc et al. 2006a). Preliminary studies regarding the corrosion effect of the *P. fluorescens* surfactant on stainless steel have suggested that this surfactant also has a great potential as a corrosion inhibitor (Dagbert et al. 2006).

Kuiper et al. (2004) characterized two lipopeptide biosurfactants produced by *P. putida* PCL1445 and verified that both biosurfactants were able to inhibit biofilm formation from different *Pseudomonas* species and strains (*P. putida* PCL1445, *P. putida* PCL1436, *P. fluorescens* WCS365, and *P. aeruginosa* UCBPP-PA14) on PVC surfaces. Furthermore, these biosurfactants were also able to disrupt pre-existing biofilms.

The use of biosurfactants, which disrupt biofilms and reduce adhesion, in combination with antibiotics could represent a novel antimicrobial strategy, once antibiotics are, in general, less effective against biofilms than planktonic cells; the biofilm disruption by biosurfactant can facilitate the antibiotic access to the cells (Irie et al. 2005).

Surfactin was tested by Mirelles et al. (2001) as an antiadhesive agent to inhibit biofilm development on urethral catheters by *Salmonella enterica*, *E. coli*, *Proteus mirabilis*, and *P. aeruginosa*. The total inhibition by preconditioning the surface was observed for all bacteria, but not for *P. aeruginosa*.

A 24h pretreatment with surfactin reduces *Enterobacter sakazakii* ATCC 29004 cells adhered to stainless steel 304 surfaces by one logarithmic cycle (Pires et al. 2007). Ferreira et al. (2007) compared the surfactin effect on *L. monocytogenes* ATCC 19112 adhesion to stainless steel 304 and polypropylene. Surfactin reduced the adhered cell number from 7.93 to 5.66 log CFU/cm^2 on stainless steel, and from 6.22 to 5.65 log CFU/cm^2 on polypropylene.

Araujo et al. (2007) evaluated crude rhamnolipid broth (7.5g/L) and an aqueous solution of surfactin 0.1% (v/v) as inhibitor agents of *L. monocytogenes* ATCC 19112 biofilm development on polystyrene surfaces. The results showed that the preconditioning with both biosurfactants was able to significantly reduce the adhesion of this pathogen on the polystyrene surface (17% reduction with crude rhamnolipid and 60% with surfactin). These biosurfactants, especially surfactin, have an excellent potential as *L. monocytogenes* antiadhesive agents on polystyrene surfaces.

A biosurfactant produced by a dairy thermophilic *Streptococcus spp* strain can be used for fouling control of heat exchanger plates in pasteurizers, as it delays the *S. thermophilus* colonization, related to fouling formation (Banat et al. 2000).

23.2.1.4 Other Food Applications

Apart from their obvious role as agents that decrease surface and interfacial tension, thus promoting emulsions formation and stabilization, surfactants can have several other functions in food. For example, to control the fat globules agglomeration, stabilize aerated systems, improve texture and shelf life of starch-containing products, modify wheat dough rheological properties, and improve fat-based products consistency and texture (Kachholz and Schlingmann 1987).

In bakery and ice cream formulations, biosurfactants act by controlling consistency, retarding staling, and solubilizing flavor oils. They are also used as fat stabilizer and antispattering agent during oil and fats cooking (Kosaric 2001). Stability, texture, volume, and conservation improvement of bakery products were obtained by rhamnolipid surfactants addition (Van Haesendonck and Vanzeveren 2004). The authors also suggested rhamnolipids use to improve butter cream, croissants, and frozen confectionery products properties. The addition of 0.10% rhamnolipid surfactant to muffins and croissants formulations enhanced the moisture, improved the texture, and maintained freshness for a longer period of time (Gandhi and Skebba 2007). Recently, a bioemulsifier isolated from an *Enterobacter cloacae* marine strain has been described as a potential viscosity enhancement agent. This can be interesting for the food industry especially due to the good viscosity observed at acidic pH allowing its use in products containing citric or ascorbic acid (Iyer et al. 2006).

Biosurfactants can also be used as corrosion inhibitor agents, promoting an interesting application field in the food industry, in which two or more properties can be simultaneously applied. For example, one single biosurfactant can be simultaneously used as antimicrobial agent, as microbial adhesion inhibitor, and to prevent corrosion on stainless steel surfaces. Dagbert et al. (2006) studied the AISI 304 stainless steel corrosion behavior in the presence of the biosurfactant produced by a *P. fluorescens* strain. These authors have concluded that the tested biosurfactant can delay the stainless steel surfaces corrosion, especially if the electrolyte to which the surface will be exposed is not too aggressive.

23.3 CHALLENGES TO PRODUCTION AND WIDESPREAD USE OF BIOSURFACTANT

Food designers may initially formulate their products with additives that provide excellent functionality, only to decide later if the emulsifying system or other ingredient is too expensive. Nevertheless, the ingredient cost should be a consideration from the beginning. Actually, biosurfactants are expensive when compared to other additives, especially the commonly used emulsifiers and preservatives. In spite of their advantages, biosurfactants will only be the preferential additives for industry when they can be supplied at a competitive price.

Economy is often the bottleneck of biotechnological processes and it is currently the main factor that works against the widespread use of biosurfactants (Makkar and Cameotra 2002). An important point that should be considered for developing cheaper bioprocesses is the selection of inexpensive medium components, which account for 10%–30% of overall costs (Cameotra and Makkar 1998). In this regard, good components seem to be agroindustrial by-products or wastes, since these residues generally contain high levels of carbohydrates or lipids to support growth and surfactant synthesis. Moreover, the treatment and disposal costs for these residues are significant to industries that are invariably searching for alternatives to reduce, reuse, recycle, and valorize their wastes (Nitschke and Costa 2007). Another factor that should be focused to reduce biosurfactant production costs is the development of higher yield strains and inexpensive recovery methods (Makkar and Cameotra 2002, Mukherjee et al. 2006).

Also aiming to render the biosurfactants production economically viable, some attempts have been made to increase the biosurfactants yield by manipulation of the medium composition and physiological conditions. Some genes responsible for the biosurfactants production were isolated and characterized. Thus, microbial strains with high yields could be obtained by genetic engineering, which were then used to produce biosurfactants in large amounts using several substrates (Desai and Banat 1997, Mukherjee et al. 2006). As several potential applications considered for biosurfactants depend on their economical production, much effort is still needed for process optimization at biological and engineering levels. It should be pointed out that legal aspects, such as more stringent laws concerning environment pollution and health by industrial activities, also raise biosurfactants chances in substituting their chemical equivalents (Fiechter 1992).

23.3.1 Production of Biosurfactants Using Low-Cost Raw Materials

Recently, with the search of low-cost substrates for biosurfactant production, several works were published about rhamnolipids production from various oils, such as soybean (Cha et al. 2007, Rahman et al. 2002, Raza et al. 2007), olive (Wei et al. 2005), Brazilian nut (Costa et al. 2006), and corn (Tahzibi et al. 2004). In these works, fermentation were carried out in shaken flasks with small volume and with different *P. aeruginosa* strains, except in the case of Raza et al. (2007), who used a *P. putida* strain submitted to mutagenesis. The best result was obtained by Costa et al. (2006) with the Brazil nut oil as carbon source. A rhamnolipid concentration of 9.9 g/L was obtained with a productivity of 83 mg/L/h. Haba et al. (2000) investigated rhamnolipid production from frying oils waste, achieving 2.7 g/L of biosurfactant, expressed in rhamnose units. Thavasi et al. (2007) used peanut oil cake to produce biosurfactants from *Corynebacterium kutscheri*, obtaining 6.4 g/L of biosurfactants in 132 h, resulting in a productivity of 48 mg/L/h. Rodrigues et al. (2006a) produced *L. lactis* and *St. thermophilus* biosurfactants from the cheese whey and molasses, though achieving low product concentrations.

Santos et al. (2002) examined variations related to carbon sources in rhamnolipid synthesis by *P. aeruginosa*. The authors have reported productivities of 23.0, 14.0, 11.2, and 11.0 mg/L/h of biosurfactant expressed in rhamnose units, when using glycerol, ethanol, soy oil, and olive oil, respectively. The work showed the possibility of having biosurfactants for several applications, due to the proportion variation of rhamnolipids obtained for the different carbon sources. Using sodium nitrate as the nitrogen source, 54.1% of monorhamnolipids (R1) and 45.9% of dirhamnolipids (R2) were obtained with soy oil as the sole carbon source, whereas, with glycerol, these values changed to 14.3% (R1) and 85.2% (R2).

23.3.2 Biosurfactant Production Scale-Up

The majority of known biosurfactant producers demands aerobic conditions for their efficient production. These microorganisms have their growth and metabolism affected by different oxygenation conditions (Gomez et al. 2006). The microbial growth is restrained by low oxygen concentrations, being hence necessary to ensure a suitable supply of this nutrient (Alagappan and Cowan 2004). Wei et al. (2005) have also reported that an increase in the dissolved oxygen content has a positive effect on cell growth and rhamnolipid biosurfactant production by a *P. aeruginosa* strain. In a laboratory scale, this production is performed in agitated flasks, with all necessary oxygen to cell metabolism being provided only by the dissolution of atmospheric oxygen at the interface of the fermentation medium and the consequent diffusion of the former through the medium. In this small scale, the oxygen transfer rate is high enough to supply its consumption, since the contact area of the fermentation medium with air, the main variable directly related to the oxygenation, can be considered high in comparison with the total medium volume. However, when a production scale-up is required, this oxygen supply form would only be acceptable if the medium volume/surface area ratio were kept at extremely low values, which is not viable. Thus, one has to resort to the use of bioreactors, with air or pure oxygen being provided through conventional bubbling.

Nevertheless, due to the presence of surfactants, conventional submerged oxygenation can induce very stable foam formation, causing serious operational problems (Benincasa et al. 2002, Gruber and Chmiel 1991, Wu and Ju 1998). The elevated foam production is still increased by the presence of extracellular proteins and by the microbial cells, resulting in high expenditures to its control. Actually, it often renders the production process unfeasible. Mechanical foam breakers are not very efficient and chemical antifoaming agents can alter the product quality and the final effluent pollution potential (Kronemberger et al. 2008).

23.3.2.1 Conventional Process

Several studies have been reported on biosurfactant production in bioreactors, with different techniques for foaming control. Joshi et al. (2008) used a trap installed at the air exit to collect the foam formed in the biosurfactant production from *B. subtilis*. However, this recipient, sterile, should be continuously exchanged, hindering further production scale-up.

Benincasa et al. (2002) studied rhamnolipid production from soapstock by a *P. aeruginosa* strain isolated in oil-contaminated soil. In the production carried out in a batch bioreactor, it was observed that an increase in the oxygen supply enhanced the rhamnolipids final concentration. The experiments were performed in a bioreactor with 1.2 L of useful volume and a foam recycling system.

Another approach is to use foam mechanical separation, as reported by Reiling et al. (1986). The authors developed a pilot-scale process for continuous rhamnolipid production by *P. aeruginosa*. The reaction volume was 23 L, with glucose as a carbon source. This carbon source was maintained in excess, while the nitrogen and iron sources were limiting. A productivity of 147 mg/L/h was achieved, corresponding to a daily biosurfactant production of 80.0 g, and a glucose yield of 77.0 mg/g/L. The bioreactor was also equipped with a foam recycling system, in addition to the foam mechanical separator.

Chen et al. (2007) also studied rhamnolipid biosurfactant production from *P. aeruginosa* S2 in fermentors, but using silicone oil as antifoaming agent. Fermentation were carried out in a bioreactor with 2.0 L of useful volume under a 250-rpm agitation. The best result, 54.7 mg/L/h of rhamnolipids, was achieved with glucose as the carbon source. The dissolved oxygen concentration was not monitored during fermentation.

An alternative possibility is the continuous biosurfactant production in an airlift bioreactor by bacteria immobilized in polyvinyl alcohol beads (Jeong et al. 2004). The authors reported a 0.1 g/h rhamnolipid production by *P. aeruginosa* BYK-2 from fish oil, with the reactor operating with a useful volume of 1.2 L and a dilution rate of 0.018 L/h.

23.3.2.2 Nondispersive Oxygenation

Another option to overcome the difficulties observed with the excessive foam formation in conventional dispersive oxygenation is the use of a nondispersive technology for oxygen transfer. A contactor constituted of polymeric membranes can be used to promote the oxygen transfer from the gas phase to the liquid one, in a bioreactor external recycle, without phase dispersion, as described in the process patented by Petroleo Brasileiro SA—Petrobras (Santa Anna et al. 2004). A similar process has already been described by Gruber et al. (1993) and another one, for alcoholic fermentation, is patented by L'Air Liquide (Cutayar et al. 1989).

Systems integrating biotechnological and membrane processes have already been studied in recent years. The membrane bioreactors (MBRs), used mostly for wastewater treatment, are the most successful integration between membranes and biotechnology. The commercial applications of these systems date back to 1991 and they might still increase in capacity and broaden in application area in future (Yang et al. 2006). Charcosset (2006), Rios et al. (2004), and Wang (2001) presented an overview of the utilization of membrane processes in biotechnology. These processes include the use of microfiltration and ultrafiltration membranes, which can be used to remove any substance that can inhibit the obtention of the product of interest, to retain the cells when purging the bioreactor, in a continuous process, for example, or even to immobilize the cells or any enzyme that catalyses a biochemical reaction. Di Luccio et al. (2002), for instance, investigated ethanol and fructose production in a continuous system integrated to membrane processes. The produced ethanol was continuously removed by pervaporation, while fructose was removed by dialysis through a liquid membrane. Ferraz et al. (2001), in turn, studied an electrodialysis unit coupled to a reactor where gluconic acid and sorbitol were produced by *Zymomonas mobilis* immobilized in a module containing microporous polycarbonate hollow fibers. The electrodialysis unit allowed an efficient gluconic acid removal, maintaining the medium pH constant and thereby avoiding the enzyme inhibition.

In the specific case of biosurfactant production, Gruber et al. (1993) tried to use a continuous stirred tank reactor (CSTR) with cell recycle and membrane contactors for oxygenation in the rhamnolipid production by *P. aeruginosa*. These authors reported that the oxygenation of the culture medium exclusively by membranes was not considered feasible, and therefore conventional bubbling oxygenation was also used, being the contactors utilized only for carbon dioxide removal. However, Kronemberger et al. (2008) demonstrated that biosurfactant production exclusively with nondispersive oxygenation can be achieved, since such technology is in agreement with the process requirements, being able to supply all the necessary oxygen for cell growth and biosurfactant production. With a proper design of the oxygenation system, it is possible to achieve normal process operation, completely avoiding foam formation.

A great advantage of the use of nondispersive oxygenation is the possibility of maintaining the dissolved oxygen concentration at a constant value throughout the process. Great changes in this parameter can affect cellular metabolism, by anoxia or by an excess of oxygen, which can result in the protein production related to oxidative stress, reducing the biosurfactant yield. In most conventional processes for biosurfactant production in bioreactors, the control of this parameter is neglected, being a constant oxygenation condition set throughout the fermentation. With the nondispersive oxygenation, one can use a controller for the constant concentration maintenance, achieved simply by manipulating some process variables. Another great advantage is that, since the oxygenation data are registered, the exact oxygen amount supplied to the culture medium during the process can be determined based on previous characterization of the oxygenation system. Considering negligible oxygen loss and with the cellular concentration data during fermentation, the specific oxygen uptake rate (SOUR) of the microorganism can be determined.

Kronemberger et al. (2008) investigated rhamnolipid production by *P. aeruginosa* PA1 from glycerol in a bioreactor with 3.0 L of useful volume using nondispersive oxygenation. The difference between a fermentation conducted with conventional oxygenation by bubbling and another one with nondispersive oxygenation can be seen in Figure 23.2. The authors have reported rhamnolipid productivity of 30.0 mg/L/h with a substrate to product yield of 20% when the dissolved oxygen concentration was maintained at 4.0 mg/L. In particular, the variation of the specific oxygen consumption rate during the 7 day fermentation period was rather interesting. As it can be seen in Figure 23.3, during the microbial growth exponential phase, this rate rises to values above 80.0 mg of oxygen per gram of cells per hour. Thus, it can be concluded that oxygen is an essential nutrient for microbial growth, since it is eagerly consumed at this stage. Once the stationary phase is reached, the oxygen consumption begins to fall, achieving a constant value equal to 25% of the maximum value, being also important for cell maintaining.

(a) (b)

FIGURE 23.2 Comparison between rhamnolipid production in a bioreactor: (a) with conventional bubbling oxygenation and (b) with nondispersive oxygenation.

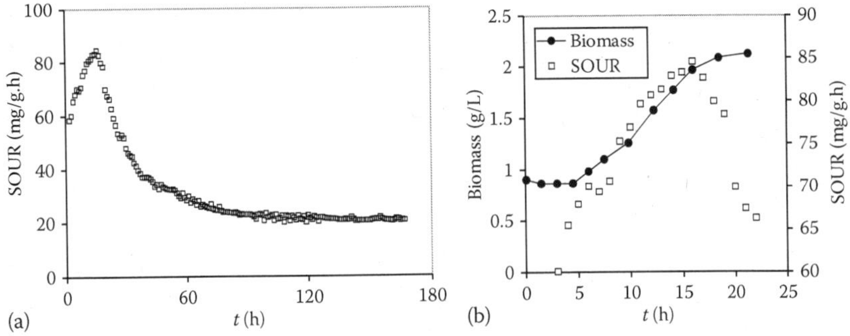

FIGURE 23.3 (a) SOUR during rhamnolipids production by *P. aeruginosa* PA1 with a constant dissolved oxygen concentration (4.0 mg/L). (b) Correlation between microbial growth and oxygen uptake rate.

23.4 CONCLUDING REMARKS

In recent years, many of these biochemically obtained compounds and their production processes have been patented, but only some of them are commercialized. The profitable production of biosurfactants is determined by some factors, such as the raw material costs, the availability of an economic and suitable production procedure, and the product yield of microorganisms. Thus, one can use cheaper or even waste substrates, with engineered strains, and carry out efficient bioprocesses to achieve an economic process for biosurfactant production. The use of engineered strains could be a problem when producing additives for the food industry, due to approval issues, so the problem is reduced to the search for low-cost raw materials and the process scale-up. Raw materials are not a great issue, since biosurfactant can be produced even from nontoxic food industry wastes, which have already been studied at laboratory scale, as cited earlier in this chapter. The great challenge is to develop an economic and viable process for the production of these biomolecules, as the scale-up is difficult due to extreme foam formation during the bioreactions. Mechanical foam breakers and foam recycling systems are not very efficient, and chemical antifoaming agents can alter the product quality and the pollutant potential of the bioreactor final effluent. Thus, the solution that seems to be more reasonable and viable is to avoid the foam production instead of trying to extinguish it after its formation. This could be achieved by using nondispersive oxygenation, which can also lead to a more controlled process, determining the optimum dissolved oxygen concentration for each microorganism and measuring its oxygen uptake rates.

On the other hand, the scant information regarding toxicity of these molecules, combined with the regulations for new food ingredients with approval required by governmental agencies, contributes to the limited use of biosurfactants in the food industry although they have been used in many fields as emulsifiers and as antimicrobial, and antiadhesive agents.

In conclusion, the biosurfactant market tends to increase as the screening programs reveal new microbial strains showing novel molecules with unusual properties, and the search for low-cost raw materials and improved strategies for the processes scale-up provide successful results.

REFERENCES

Abalos, A., Pinazo, A., Infante, M. R., Casals, M., García, F., and Manresa, A. 2001. Physicochemical and antimicrobial properties of new rhamnolipids produced by *Pseudomonas aeruginosa* AT10 from soybean oil refinery wastes. *Langmuir* 17:1367–1371.

Alagappan, G. and Cowan, R. M. 2004. Effect of temperature and dissolved oxygen on the growth kinetics of *Pseudomonas putida* F1 growing on benzene and toluene. *Chemosphere* 54:1255–1265.

Araújo, L. V., Pires, R. C., Siqueira, R. S. de, Freire, D. M. G., and Nitschke, M. 2007. Potential use of preconditioning plastic surfaces with biosurfactants to inhibit *Listeria monocytogenes* 19112 biofilms. Paper presented at the II International Conference on Environmental, Industrial and Applied Microbiology, Sevilla, Spain.

Bagge-Ravn, D., Ng, Y., Hjelm, M., Christiansen, J. N., Johansen., and Gram, L. 2003. The microbial ecology of processing equipment in different fish industries: Analysis of the microflora during processing and following cleaning and disinfection. *Int. J. Food Microbiol.* 87:239–250.

Banat, I. M., Makkar, R. S., and Cameotra, S. S. 2000. Potential commercial applications of microbial surfactants. *Appl. Microbiol. Biotechnol.* 53:495–508.

Benincasa, M., Contiero, J., Manresa, M. A., and Moraes, I. O. 2002. Rhamnolipid production by *Pseudomonas aeruginosa* LBI growing on soapstock as the sole carbon source. *J. Food Eng.* 54:283–288.

Bodour, A. A., Dress, K. P., and Maier, R. M. 2003. Distribution of biosurfactant-producing bacteria in undisturbed and contaminated arid Southwestern soils. *Appl. Environ. Microbiol.* 69:3280–3287.

Bognolo, G. 1999. Biosurfactants as emulsifying agents for hydrocarbons. *Colloid Surf. A.* 152:41–52.

Cameotra, S. S. and Makkar, R. S. 1998. Synthesis of biosurfactants in extreme conditions. *Appl. Microbiol. Biotechnol.* 50:520–529.

Cha, M., Lee, N., Kim, M. Kim, M., and Lee, S. 2007. Heterologous production of *Pseudomonas aeruginosa* EMS1 biosurfactant in *Pseudomonas putida. Bioresource Technol.* 99:2192–2199.

Chae, M. S. and Schraft, H. 2000. Comparative evaluation of adhesion and biofilm formation of different *Listeria monocytogenes* strains. *Int. J. Food Microbiol.* 62:103–111.

Charcosset, C. 2006. Membrane processes in biotechnology: An overview. *Biotechnol. Adv.* 24:482–492.

Chen, S., Lu, W., Wei, Y., Chen, W., and Chang, J. 2007. Improved production of biosurfactant with newly isolated *Pseudomonas aeruginosa* S2. *Biotechnol. Prog.* 23:661–666.

Cirigliano, M. C. and Carman, G. M. 1985. Purification and characterization of liposan, a bioemulsifier from *Candida lipolytica. Appl. Environ. Microbiol.* 50:846–850.

Cooper, D. G. and Paddock, D. A. 1984. Production of a biosurfactant from *Torulopsis bombicola. Appl. Environ. Microbiol.* 47:173–176.

Costa, S. G. V. A. O., Nitschke, M., Haddad, R., Eberlin, M. N., and Contiero, J. 2006. Production of *Pseudomonas aeruginosa* LBI rhamnolipids following growth on Brazilian native oils. *Process Biochem.* 41:483–488.

Cutayar, J., Poillon, D., and Cutayar, S. 1989. Process for the controlled oxygenation of an alcoholic fermentation must or wort. U.S. Patent 4,978,545, filed Mar. 21,1989, and issued Dec. 18, 1990.

Dagbert, C., Meylheuc, T., and Bellon-Fontaine, M. N. 2006. Corrosion behavior of AISI 304 stainless steel in presence of a biosurfactant produced by *Pseudomonas fluorescens. Electrochim. Acta* 51:5221–5227.

Das, P., Mukherjee, S., and Sen, R. 2008. Antimicrobial potential of a lipopeptide biosurfactant derived from a marine *Bacillus circulans. J. Appl. Microbiol.* 104:1675–1684.

Davey, M. E., Caiazza, N. C., and O'Toole, G. A. 2003. Rhamnolipid surfactant production affects biofilm architecture in *Pseudomonas aeruginosa* PAO1. *J. Bacteriol.* 185:1027–1036.

De Souza, J. T., de Boer, M., de Waard, P., van Beek, T. A., and Raaijmakers, J. M. 2003. Biochemical, genetic, and zoosporocidal properties of cyclic lipopeptide surfactants produced by *Pseudomonas fluorescens*. *Appl. Environ. Microbiol.* 69:7161–7172.

Desai, J. D. and Banat, I. M. 1997. Microbial production of surfactants and their commercial potential. *Microbiol. Mol. Biol. Rev.* 61:47–64.

Di Luccio, M., Borges, C. P., and Alves, T. L. M. 2002. Economic analysis of ethanol and fructose production by selective fermentation coupled to pervaporation: Effect of membrane costs on process economics. *Desalination* 147:161–166.

Fernandes, P. A. V., Arruda, I. R. de, Santos, A. F. A. B. dos, Araújo, A. A. de, Maior, A. M. S., and Ximenes, E. A. 2007. Antimicrobial activity of surfactants produced by *Bacillus subtilis* R14 against multidrug-resistant bacteria. *Braz. J. Microbiol.* 38:704–709.

Ferraz, H. C., Alves, T. L. M., and Borges, C. P. 2001. Coupling of an electrodialysis unit to a hollow fiber bioreactor for separation of gluconic acid from sorbitol produced by *Zymomonas mobilis* permeabilized cells. *J. Membr. Sci.* 191:43–51.

Ferreira, F. S., Pires, R. C., Araújo, L. V., Siqueira, R. S. De., and Nitschke, M. 2007. Surfactin inhibits the adhesion of *Listeria monocytogenes* ATCC 19112 to solid surfaces. Paper presented at II Simpósio Latino Americano de Ciências de Alimentos, Campinas, São Paulo (in Portuguese).

Fiechter, A. 1992. Biosurfactants: Moving towards industrial application. *Trends Biotechnol.* 10:208–217.

Flasz, A., Rocha, C. A., Mosquera, B., and Sajo, C. 1998. A comparative study of the toxicity of a synthetic surfactant and one produced by *Pseudomonas aeruginosa* ATCC 55925. *Med. Sci. Res.* 26:181–185.

Gandhi, N. R. and Skebba, V. L. P. 2007. Rhamnolipid compositions and related methods of use. W. O. International Application Patent (PCT) 2007/095258 A3, filed Feb.12, 2007, and issued Aug. 23, 2007.

Gomez, E., Santos, V. E., Alcon, A., Martin, A. B., and Garcia-Ochoa, F. 2006. Oxygen-uptake and mass-transfer rates on the growth of *Pseudomonas putida* CECT5279: Influence on biodesulfurization (BDS) capability. *Energy Fuels* 20:1565–1571.

Gruber, T. and Chmiel, H. 1991. Aerobic production of biosurfactants avoiding foam problems. *Biochem. Eng. Stuttgart* 212–215.

Gruber, T., Chmiel, H., Kappeli, O., Sticher, P., and Fiechter, A. 1993. Integrated process for continuous rhamnolipid biosynthesis. In *Biosurfactants (Surfactants Science Series)*, ed. N. Kosaric. New York: Marcel Dekker, pp. 157–173.

Haba, E., Espuny, M. J., Busquets, M., and Manresa, A. 2000. Screening and production of rhamnolipids by *Pseudomonas aeruginosa* 47T2 NCIB 40044 from waste frying oils. *J. Appl. Microbiol.* 88:379–387.

Hood, S. K. and Zottola, E. A. 1995. Biofilms in food processing. *Food Control* 6:9–18.

Irie, Y., O'Toole, G. A., and Yuk, M. H. 2005. *Pseudomonas aeruginosa* rhamnolipids disperse *Bordetella bronchiseptica* biofilms. *FEMS Microbiol. Lett.* 250:237–243.

Iyer, A., Mody, K., and Jha, B. 2006. Emulsifying properties of a marine bacterial exopolysaccharide. *Enzyme Microb. Technol.* 38:220–222.

Jenkinson, H. F. and Lappin-Scott, H. M. 2001. Biofilms adhere to stay. *Trends Microbiol.* 9:9–10.

Jeong, H., Lim, D., Hwang, S., Ha, S., and Kong, J. 2004. Rhamnolipid production by *Pseudomonas aeruginosa* immobilized in polyvinyl alcohol beads. *Biotechnol. Lett.* 26:35–39.

Joshi, S., Bharucha, C., and Desai, A. J. 2008. Production of biosurfactant and antifungal compound by fermented food isolate *Bacillus subtilis* 20B. *Bioresource Technol.* 99:4603–4608.

Kachholz, T. and Schlingmann, M. 1987. Possible food and agricultural applications of microbial surfactants: An assessment. In *Biosurfactants and Biotechnology*, ed. N. Kosaric, W.L. Carns, and N. C. C Gray. New York: Marcel Dekker, pp. 183–210.

Kesting, W., Tummuscheit, M., Schacht, H., and Schollmeyer, E. 1996. Ecological washing of textiles with microbial surfactants. *Prog. Colloid Polym. Sci.* 101:125–130.

Kim, B. S., Lee, J. Y., and Hwang, B. K. 2000. In vivo control and in vitro antifungal activity of rhamnolipid B, a glycolipid antibiotic, against *Phytophthora capsici* and *Colletotrichum orbiculare. Pest Manag. Sci.* 56:1029–1035.

Kitamoto, D., Isoda, H., and Nakahara, T. 2002. Functions and potential applications of glycolipid biosurfactants—From energy-saving materials to gene delivery carriers. *J. Biosci. Bioeng.* 94:187–201.

Kosaric, N. 2001. Biosurfactants and their application for soil bioremediation. *Food Technol. Biotech.* 39:295–304.

Kronemberger, F. A., Santa Anna, L. M. M., Fernandes, A. C. L. B., Menezes, R. R., Borges, C. P., and Freire, D. M. G. 2008. Oxygen-controlled biosurfactant production in a bench scale bioreactor. *Appl. Biochem. Biotechnol.* 147:33–45.

Kuiper, I., Lagendijk, E. L., Pickford, R., et al. 2004. Characterization of two *Pseudomonas putida* lipopeptide biosurfactants, putisolvin I and II, which inhibit biofilm formation and break down existing biofilms. *Mol. Microbiol.* 51:97–113.

Kyaw, C. M. 2006. Microbial biofilms. Working paper, University of Brasília. http://www.unb.br/ib/cel/microbiologia/index.html (accessed Nov. 10, 2007) (in Portuguese).

Letherhead Food International. 2008. The Food Additives Market. Global Trends and Developments. Market Report. http://www.leatherheadfood.com/lfi/pdf/foodadditives4.pdf (accessed May 28, 2008).

Lukondeh, T., Ashbolh, N. J., and Rogers, P. L. 2003. Evaluation of *Kluyveromyces marxianus* FII 510700 grown on a lactose-based medium as a source of natural bioemulsifier. *J. Ind. Microbiol. Biotechnol.* 30:715–720.

Makkar, R. S. and Cameotra, S. S. 2002. An update to the use of unconventional substrates for biosurfactant production and their new applications. *Appl. Microbiol. Biotechnol.* 58:428–434.

Maukonen, J., Mättö, J., Wirtanen, G., Raaska, T., Mattila-Sandholm, T. and Saarela, M. 2003. Methodologies for the characterization of microbes in industrial environments: A review. *J. Ind. Microbiol. Biotechnol.* 30:327–356.

Meylheuc, T., Van Oss, C. J., and Bellon-Fontaine, M. N. 2001. Adsorption of biosurfactant on solid surfaces and consequences regarding the bioadhesion of *Listeria monocytogenes* LO28. *J. Appl. Microbiol.* 91:822–832.

Meylheuc, T., Methivier, C., Renault, M., Herry, J. M., Pradier, C. M., and Bellon-Fontaine, M. N. 2006a. Adsorption on stainless steel surfaces of biosurfactants produced by gram-negative and gram-positive bacteria: Consequence on the bioadhesive behaviour of *Listeria monocytogenes. Colloid Surf. B* 52:128–137.

Meylheuc, T., Renault, M., and Bellon-Fontaine, M. N. 2006b. Adsorption of a biosurfactant on surfaces to enhance the disinfection of surfaces contaminated with *Listeria monocytogenes. Int. J. Food Microbiol.* 109:71–78.

Mirelles II, J. R., Toguchi, A., and Harshey, R. M. 2001. *Salmonella enterica* serovar Thyphimurium swarming mutants with altered biofilm-forming abilities: Surfactin inhibits biofilm formation. *J. Bacteriol.* 183:5848–5854.

Mohan, P. K., Nakhla, G., and Yanful, E. K. 2006. Biokinetics of biodegradability of surfactants under aerobic, anoxic and anaerobic conditions. *Water Res.* 40:533–540.

Mukherjee, A. K. and Das, K. 2005. Correlation between diverse cyclic lipopeptides production and regulation of growth and substrate utilization by *Bacillus subtilis* strains in a particular habitat. *FEMS Microbiol. Ecol.* 54:479–489.

Mukherjee, S., Das, P., and Sen, R. 2006. Towards commercial production of microbial surfactants. *Trends Biotechnol.* 24:509–515.

Nitschke, M. and Costa, S. G. V. A. O. 2007. Biosurfactants in food industry. *Trends Food Sci. Tech.* 18:252–259.

Nitschke, M. and Pastore, G. M. 2002. Biosurfactants: Properties and applications. *Quim. Nova* 25:772–776 (in Portuguese).

Nitschke, M. and Pastore, G. M. 2003. Evaluation of agroindustrial wastes as substrates for biosurfactant production by *Bacillus. Revista Biotecnologia Ciência & Desenvolvimento* 31:63–67 (in Portuguese).

Nitschke, M. and Pastore, G. M. 2006. Production and properties of a surfactant obtained from *Bacillus subtilis* grown on cassava wastewater. *Bioresource Technol.* 97:336–341.

Parizzi, S. Q. F., Andrade, N. J. de, Silva, C. A. de S., Soares, N. F. F., and Silva, E. A. M. da. 2004. Bacterial adherence to different inert surfaces evaluated by epifluorescence microscopy and plate count method. *Braz. Arch. Biol. Technol.* 47:77–83.

Pires, R. C., Araújo, L. V., Ferreira, F. S., Siqueira, R. S. de, and Nitschke, M. 2007. Potential application of surfactin to inhibit the adhesion of pathogens on stainless steel surfaces. *Revista Higiene Alimentar* 21:530–531 (in Portuguese).

Rahman, K. S. M., Banat, I. M., Thahira, J., Thayumanavan, T., and Lakshmanaperumalsamy, P. 2002. Bioremediation of gasoline contaminated soil by a bacterium consortium amended with poultry litter, coir pith and rhamnolipid biosurfactant. *Bioresource Technol.* 81:25–32.

Raza, Z. A., Khan, M. S., and Khalid, Z. M. 2007. Evaluation of distant carbon sources in biosurfactant production by a gamma ray-induced *Pseudomonas putida* mutant. *Process Biochem.* 42:686–692.

Reiling, H. E., Thanei-Wyss, U., Guerra-Santos, L. H., Hirt, R., Käppeli, O., and Fiechter, A. 1986. Pilot plant production of rhamnolipid biosurfactant by *Pseudomonas aeruginosa. Appl. Environ. Microbiol.* 51:985–989.

Rios, G. M., Belleville, M. P., Paolucci, D., and Sanchez, J. 2004. Progress in enzymatic membrane reactions—A review. *J. Membrane Sci.* 242:189–196.

Rodrigues, L. R., Van der Mei, H. C., Teixeira, J. A., and Oliveira, R. 2004. Influence of biosurfactants from probiotic bacteria on formation of biofilms on voice protheses. *Appl. Environ. Microbiol.* 70:4408–4410.

Rodrigues, L. R., Teixeira, J. A., and Oliveira, R. 2006a. Low-cost fermentative medium for biosurfactant production by probiotic bacteria. *Biochem. Eng. J.* 32:135–142.

Rodrigues, L. R., Banat, I. M., van der Mei, H. C., Teixeira, J. A., and Oliveira, R. 2006b. Interference in adhesion of bacteria and yeasts isolated from explanted voice protheses to silicone rubber by rhamnolipid biosurfactants. *J. Appl. Microbiol.* 100:470–480.

Ron, E. Z. and Rosenberg, E. 2001. Natural roles of Biosurfactants. Minireview. *Environ. Microbiol.* 3:229–236.

Rosenberg, E. and Ron, E. Z. 1999. High-and low-molecular-mass microbial surfactants. *Appl. Microbiol. Biotechnol.* 52:154–162.

Santa Anna, L. M., Freire, D. M. G., Kronemberger, F. A. et al. 2004. Biosurfactant and its uses in bioremediation of oil contaminated sandy soils. PI Patent 0405952-2 A, Petróleo Brasileiro S.A., Brazil, filed Dec. 29, 2004, and issued Jul. 10, 2007.

Santos, A. S., Sampaio, A. P. W., Vasquez, G. S., Santa Anna, L. M., Pereira Jr, N., and Freire, D. M. G. 2002. Evaluation of different carbon and nitrogen sources in the production of rhamnolipids by a strain of *Pseudomonas. Appl. Biochem. Biotechnol.* 98–100:1025–1035.

Shepherd, R., Rockey, J., Sutherland, I. W., and Roller, S. 1995. Novel bioemulsifiers from microorganisms for use in foods. *J. Biotechnol.* 40:207–217.

Singh, P. and Cameotra, S. S. 2004. Potential applications of microbial surfactants in biomedical sciences. *Trends Biotechnol.* 22:142–146.

Somers, E. B. and Wong, A.C. L. 2004. Efficacy of two cleaning and sanitizing combinations of *Listeria monocytogenes* biofilms formed at low temperature on a variety of materials in the presence of ready-to-eat meat residue. *J. Food Protect.* 67:2218–2229.

Stepanovic, S., Cirkovic, I., Ranin, L., and Svabic-Vlahovic, M. 2004. Biofilm formation by *Salmonella* spp. and *Listeria monocytogenes* on plastic surface. *Lett. Appl. Microbiol.* 38:428–432.

Tahzibi, A., Kamal, F., and Assadi, M. M. 2004. Improved production of rhamnolipids by a *Pseudomonas aeruginosa* mutant. *Iran. Biomed. J.* 8:25–31.

Thavasi, R., Jayalakshmi, S., Balasubramanian, T., and Banat, I. M. 2007. Biosurfactant production by *Corynebacterium kutscheri* from waste motor lubricant oil and peanut oil cake. *Lett. Appl. Microbiol.* 45:686–691.

Van Haesendonck, I. P. H. and Vanzeveren, E. C. A. 2004. Rhamnolipids in bakery products. W. O. International Application Patent (PCT) 2004/040984, filed Nov. 4, 2003, and issued May, 21, 2004.

Van Hamme, J. D., Singh, A., and Ward, O. P. 2006. Physiological aspects—Part 1 in a series of papers devoted to surfactants in microbiology and biotechnology. *Biotech. Adv.* 24:604–620.

Wang, W. K. 2001. *Membrane Separations in Biotechnology*. New York: Marcel Dekker, Inc.

Wei, Y., Chou, C., and Chang, J. 2005. Rhamnolipid production by indigenous *Pseudomonas aeruginosa* J4 originating from petrochemical wastewater. *Biochem. Eng. J.* 27:146–154.

Wu, J. and Ju, J. 1998. Extracellular particles of polymeric material formed in n-hexadecane fermentation by *Pseudomonas aeruginosa. J. Biotechnol.* 59:193–202.

Yang, W., Cicek, N., and Ilg, J. 2006. State-of-the-art of membrane bioreactors: Worldwide research and commercial applications in North America. *J. Memb. Sci.* 270:201–211.

Index

C

D

E

F

G

H

I

N

O

P

R

T

U

V

W

X

Y

Z